Fundamental Processes in Radiation Chem

Fundamental Processes in Radiation Chemistry

EDITED BY

P. AUSLOOS

Radiation Chemistry Section
National Bureau of Standards
Washington, D.C.

INTERSCIENCE PUBLISHERS

A DIVISION OF JOHN WILEY & SONS

NEW YORK · LONDON · SYDNEY · TORONTO

QD 601
A95

541.38
A932

Preface

It is only in the past decade that improved instrumentation and new experimental techniques have made it possible to make substantial progress toward unraveling the complexities of the fundamental processes occurring in systems under high energy irradiation. The understanding of such systems has now reached a degree of sophistication such that high energy radiation can be used to obtain accurate information about various well-defined chemical and physical processes which are of general scientific interest. In recent years, as one development has followed another with increasing rapidity, interest and participation in radiation research has grown proportionately. During the time this volume was in preparation, a new international journal and two new series dedicated to reviewing radiation research studies have been announced. At the same time, of course, the books on radiation chemistry already in print have become increasingly out-of-date. Not surprisingly, an "information gap" exists in the wake of such rapid development. Actually, even now it is not uncommon to hear scientists in related disciplines, and even radiation chemists on their darker days, express the opinion that a system under high energy irradiation is essentially chaotic, and can never be a subject for fruitful research. To the degree that the contents of this volume dispel such skepticism, it can be said to fulfill one of its main objectives.

This book represents an attempt to bring together in one volume some critical evaluations of the recent findings on various types of irradiated systems, and to present some of the fundamental information which has been derived from this work. The first chapter presents a general theoretical background for energy deposition processes in irradiated systems. Reviews of specialized topics of intrinsic importance to radiation chemistry in general are presented in the following three chapters. These topics—ionic fragmentation processes, neutral excited molecule decomposition processes, and ion–molecule reactions—were singled out for attention largely because they are areas in which information is available which has not been widely reviewed. The omission of specialized chapters covering certain other fundamental subjects of general interest in radiation chemistry—notably free radical kinetics—is not meant to express a bias in interpretation of radiolytic systems. It was felt that many of these other subjects have been so exhaustively covered elsewhere that the space here could be better

v

devoted to newer findings; free radical reactions, for example, have become an intrinsic part of the physical chemist's vocabulary. The studies discussed in these introductory chapters have not necessarily been carried out in classical radiolytic systems, but rather in experiments designed to "isolate" the phenomenon of interest, e.g., the study of ionic fragmentation in a mass spectrometer (Chapter 2).

In the last six chapters are presented up-to-date views on the understanding of particular systems—both inorganic and organic compounds in the gas, liquid, and solid phases. The emphasis is strongly on the interpretation of experimental results which lead to a better understanding of the basic processes initiated by high energy radiation. Those observations which cannot be understood at the present time, as well as interpretations which are highly speculative, are generally downgraded. In addition, the reader will notice that the discussions in this book are limited to fairly simple systems; macromolecules and biological systems have not been included.

In a field such as radiation chemistry which is advancing so rapidly at the present time, there is probably no single author who could write a book covering all the diverse topics contained herein in an authoritative and comprehensive manner; or, if such an individual could be found, he would be hard-pressed to keep his manuscript up-to-date with the new findings which would appear during the time of its preparation. Thus, in the interest of comprehensive coverage of the various specializations included here, as well as of speeding the appearance of the book in print, the various chapters of this volume have been written by different authors, each an authority in a particular specialized area of the field of radiation research.

Of course, there are certain disadvantages in compiling a book written by ten different authors. Inevitably, there will be a diversity of style, of outlook, even of choices of units in which experimental observations are expressed. In addition, there are areas of overlapping subject matter between two or more chapters. The consequent inconvenience to the reader has been alleviated wherever possible by editing out overlapping discussions, including a comprehensive subject index, and limiting the number of symbols or units used to an absolute minimum. (To trace the meaning of these, the reader is referred to the subject index.) In most cases, the Editor discussed the chapters with the authors during the time the manuscripts were in preparation; in instances of disagreement about the material to be included in a chapter or the interpretation of results for a particular system, the final decision was left to the author. The most recent literature cited in the volume appeared at the end of 1967, or in some cases, at the beginning of 1968.

P. AUSLOOS

Washington, D. C.
August 1968

Contributors

DR. M. ANBAR, *The Weismann Institute of Science, Rehovoth, Israel*

DR. A. R. ANDERSON, *Atomic Energy Research Establishment, Harwell, Didcot, Berks., England*

DR. J. H. FUTRELL, *University of Utah, Salt Lake City, Utah*

DR. R. A. HOLROYD, *Atomics International, Canoga Park, California*

DR. CORNELIUS E. KLOTS, *Health Physics Division, Oak Ridge National Laboratory, Oak Ridge, Tennessee*

PROFESSOR G. G. MEISELS, *Chemistry Department, University of Houston, Houston, Texas*

L. W. SIECK, *Radiation Chemistry Section, National Bureau of Standards, Washington, D.C.*

T. O. TIERNAN, *Office of Aerospace Research, Aerospace Research Laboratories, Wright-Patterson AFB, Ohio*

M. VESTAL, *Scientific Research Instrument Corp., Baltimore, Maryland*

PROFESSOR J. E. WILLARD, *Chemistry Department, University of Wisconsin, Madison, Wisconsin*

DR. T. FFRANCON WILLIAMS, *University of Tennessee, Knoxville, Tennessee*

Contents

Chapter 1

Energy Deposition Mechanisms*

C. E. Klots

Health Physics Division, Oak Ridge National Laboratory
Oak Ridge, Tennessee

* Research sponsored by the U.S. Atomic Energy Commission under contract
with Union Carbide Corporation.

1

I. EXCITATION MECHANISMS

In this section, an abbreviated classification of electronic excitation processes is given as they may occur in isolated atoms or molecules. The intent is to provide a minimum vocabulary for what follows both in this chapter and elsewhere in the volume. The language developed is largely that of photochemistry. This choice seems reasonable in view of the background which the reader is most likely to bring with him. It is further, as will emerge later, of direct relevance to any discussion of charged-particle excitation processes.

The reader should be aware of a number of simplifications. The word *excitation* is used, unless otherwise delineated, in a general sense and may then include *ionization* as a special case. Atoms or molecules undergoing excitation will be assumed to have been in their ground or lowest energy state. This restriction will suffice for all practical purposes. Finally, the discussion will center on electronic transitions, with occasional reference to vibrational structure. No reference is made to rotational structure, the omission being reasonable in the present context.

A. Oscillator Strengths

The Beer-Lambert law offers an empirical description of the decrease in intensity of monochromatic radiation as it traverses a distance dx through a medium containing N identical absorbing centers per unit volume:

$$-dI/I = \sigma N \, dx \tag{1}$$

The quantity σ has the dimensions of area per absorbing unit and may thus be thought of as a cross section; it is, of course, a function of the frequency ν of the incident light.

The absorption of light at discrete wavelengths could be partially understood in classical mechanics in terms of atoms composed of electrons oscillating harmonically at various resonant frequencies. According to this model the number of such presumed oscillators with a given resonant frequency, ν, could be represented in terms of experimental quantities as

$$f_n = \frac{mc}{\pi e^2} \int_n \sigma \, d\nu \tag{2}$$

where m, e are the mass and charge of an electron and c the velocity of light; the integration extends over the range of frequencies comprised by the absorption line. The quantity f_n thus obtained was known as an oscillator strength.

Although the classical model was unable to cope adequately with regions of continuous light absorption, it is nevertheless useful to extend the above terminology, and define a differential oscillator strength as the following

$$df/dv = (mc/\pi e^2)\sigma(v) \qquad (3)$$

First-order perturbation theory of quantum mechanics yields a formula for oscillator strengths:

$$f_n = |X_n|^2 h v / R a_0^2 \qquad (4)$$

in which h is Planck's constant, R the Rydberg of energy, and a_0 the Bohr radius. X_n is the dipole-length matrix element associated with the electronic transition involved. It may be calculated from

$$X_n = \int \Psi_n^* \sum_{j=1}^{Z} x_j \Psi_0 \, d\tau \qquad (5)$$

where Ψ_0 and Ψ_n are the wavefunctions associated with the initial and final states involved in the transition. The summation in the operator extends over all electrons in the absorbing center.

There is a useful theorem, valid in nonrelativistic mechanics, the so-called Kuhn-Thomas sum rule. It states that the sum of all oscillator strengths in discrete transitions plus the integral oscillator strength over all continuous regions of absorption is just equal to the total number of electrons, and thus the atomic number, of the absorbing unit. Thus,

$$\sum f_n + \int (df/dv) \, dv = Z \qquad (6)$$

which correlates nicely with the classical interpretation of oscillator strength.

B. The Franck-Condon Principle

Integrals, such as in Eq. (5), occur frequently not only in conjunction with light absorption and charged-particle collisions, but in many other dynamical situations as well. It is possible, without actually evaluating them, to bring out one of their most important qualitative properties.

Born and Oppenheimer have shown that, with a high degree of precision, one may write for molecules the ground and excited state wavefunctions each as a product of an electronic wavefunction ψ, a function of the coordinates of the electrons and the nuclei, and a nuclear wavefunction χ, a function of only the latter, viz.:

$$\Psi_0 = \psi_0 \chi_0 \qquad (7)$$

$$\Psi_n = \psi_n \chi_n \qquad (8)$$

It is, for example, only with the framework of this approximation that one may attach meaning to the concept of potential energy curves and vibrational levels, as illustrated for hypothetical diatomic molecules in Fig. 1.

Since the operator x in Eq. (5) is a function only of the electronic coordinates, this equation may be written as

$$X_n = \int \chi_n^* \chi_0 \, d\tau' \int \psi_n^* \sum_j x_j \psi_0 \, d\tau \tag{9}$$

Let us denote the integral over the electronic eigenfunctions by X_n. If it is now further assumed that this integral is only a very weak function of the nuclear coordinates, one might with corresponding precision replace it with an average value $\bar{X}_n$ which may then be factored out to obtain

$$X_n = \bar{X}_n \int \chi_n^* \chi_0 \, d\tau' \tag{10}$$

Consider now the absorption of light and resulting excitation of a molecule from, for example, the lowest vibrational level of its ground electronic state to the manifold of vibrational levels of some excited electronic state. The dipole-length matrix element for any such transition and thus the integrated absorption cross section may be seen, in the above approximation, to be proportional to an integral over the vibrational wavefunctions of the levels involved. But this *overlap* integral, as it is commonly labeled, can hardly be large unless the nuclei in the excited state tend to occupy the same spatial positions as in the ground state. That is, the most probable transitions will be those corresponding to the least change in internuclear distances. In the example of Fig. 1a, the transition to the

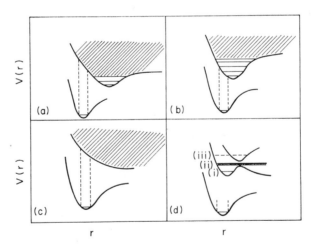

Fig. 1. Representative potential energy curves and illustrations of vertical transitions.

lowest vibrational level of the excited state, the so-called *adiabatic* transition, will scarcely be observable. Instead, the transitions suggested by the vertical lines will probably be the most prominent. This tendency for *vertical* transitions is known as the Franck-Condon principle.

C. Photodissociation and Predissociation

The Franck-Condon principle is of great importance in determining the character of photoabsorption spectra. Thus, two electronic states whose potential energy curves are positioned as in Fig. 1*b* will give rise to a spectrum corresponding, at least in part, to transitions to the stable vibrational levels of the upper electronic state. This situation, which closely approximates the prominent low energy absorption band in iodine, is in sharp contrast with Fig. 1*c*, approximating in turn the analogous band of chlorine. Here, vertical transitions carry the molecule to above the dissociation limit of the excited electronic state. Prompt dissociation and the essentially nonquantized nature of kinetic energies then make the absorption spectrum a true continuum.

These two extremes do not, however, suffice to describe the qualitative character of all absorption spectra. Still another important mechanism must be considered, that of *predissociation*. There is occasionally observed a progression of absorption lines which, on going to shorter wavelengths, suddenly takes on a blurred but not truly continuous character. At still shorter wavelengths the spectrum may again be sharp. This behavior is clearly not to be understood in terms of the previous figures. Its explanation will, in fact, require a role for two excited electronic states in addition to the one from which the transition originates.

With reference to Fig. 1*d*, photoabsorption to level (*1*) will necessarily be sharp in accord with the stability of the upper state. Beginning at approximately level (*2*), however, one might anticipate a continuous spectrum. That this is not necessarily what is observed can then only mean that something prevents dissociation from following promptly upon photoexcitation. The origin of this unusual behavior may be understood by remembering that the very concept of potential energy curves is meaningful only within the framework of the Born-Oppenheimer approximation—itself valid only if the nuclei move with infinitesimal velocity. In real molecules this is, of course, never the case; if one nevertheless wishes to retain the language of potential energy curves, then one must acknowledge the possibility that a system may behave, in the terms of this language, as if it were undergoing *nonadiabatic* skips from one curve to another. The likelihood of such jumps may be large in those regions where two curves nearly approach each other. Dissociation may thus be an improbable and more leisurely process.

In molecules containing three or more atoms an important variation on the predissociation mechanism may occur. While essentially impossible to depict in terms of a two-dimensional diagram, it may be rendered plausible with a simple qualitative description. Photoabsorption may leave a molecule in an excited state with sufficient energy to afford dissociation from that state. This energy will residue initially within the several degrees of freedom of the molecule in accord with the Franck-Condon principle. As such, it may be likely that this excess energy must be redistributed, perhaps concentrated in a particular chemical bond, if dissociation is to occur. This redistribution and the resultant fragmentation may be thought of as a *vibrational predissociation*.

The great majority of molecular absorption bands are rarely of a truly continuous character, even when dissociation is energetically allowed and known to occur. The implication is that dissociation usually does not occur instantly but more leisurely by way of the above predissociation mechanisms. It was, in fact, remarked some time ago (1) that it is extremely difficult to convert photonic into mechanical (kinetic) energy. It is this difficulty then which provides a basis for attempts to treat the rates of unimolecular decompositions following photoexcitation in terms of unimolecular rate theories—as done, for example, in Chapter 2.

D. Photoionization

The absorption by an atom or molecule of a photon of sufficient energy can precipitate the ejection of an electron from the absorbing center. This process of photoionization, known for some time, now underlies a powerful technique for determining the minimum energy for such ejection, the so-called ionization potential.

Much of the discussion of the preceding section may be carried over, without modification, to optical transitions above the first ionization potential. Thus if, upon photon absorption, the electron is ejected sufficiently quickly, before the nuclei can move significantly, then the transition will have been *vertical* and the Franck-Condon principle obeyed. Similarly, if the positive ion so formed has sufficient internal energy, dissociation or predissociation may ensue.

There is, however, one major qualitative difference in the character of absorption spectra above and below the ionization potential (threshold). In the latter region one observes discrete lines, blurred only to the extent that dissociation promptly follows the excitation. Photoabsorption to energy levels above the ionization limit may, however, cause spectra of quite different character to be observed. The energy absorbed in excess of this limit will, immediately subsequent to photoionization, reside either as internal excitation energy of the positive ion or as kinetic energy of the

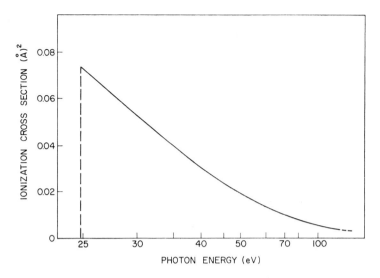

Fig. 2. Photoionization cross section of helium (2).

ejected electron(s). Since the allowed energy levels of the electrons are essentially nonquantized, a truly continuous absorption spectrum can result. This expectation is well illustrated by the photoionization cross section of helium as a function of energy (2) (see Fig. 2). With its discontinuity at onset and continuous variation thereafter this spectrum might be considered a prototype for photoionization, against which much of the following discussion should be measured.

Examination of the photoionization cross section of other atoms and molecules fails, however, to further bear out the above expectations. Figures 3 and 4, for example, illustrate the photoionization cross section of xenon and hydrogen, respectively, at photon energies in the region shortly above the ionization potential (3,4). The banded structure in these spectra is hardly like the continuum of helium but rather more reminiscent of discrete absorption bands below the ionization potential. Indeed this similarity immediately suggests that much of the observed ionization proceeds by way of an indirect mechanism, e.g.,

$$Xe + h\nu \rightarrow Xe^{**} \tag{11}$$

$$Xe^{**} \rightarrow Xe^+ + e \tag{12}$$

Here Xe** would be an atom in an excited state of energy greater than the ionization potential and would then necessarily be metastable with respect

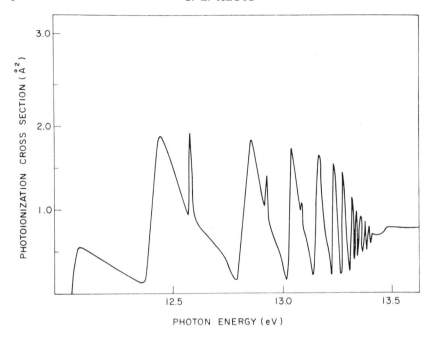

Fig. 3. Photoionization cross section of xenon (3).

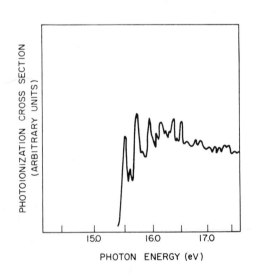

Fig. 4. Photoionization cross section of hydrogen (4).

to electron ejection. This postulated ejection, often referred to as autoionization or preionization, is entirely analogous to the Auger effect.

Such an interpretation of quasi-discrete structure above the ionization threshold would appear intuitively evident. It is further substantiated by the fact that progressions of the indicated type may often be accorded some measure of theoretical understanding (5). Thus one can envisage two possible interpretations of Xe**; such a state might be one in which two electrons have been excited to outlying orbitals such that their total excitation energy would suffice to ionize one of them. Alternatively, such states may arise from excitation of an inner shell electron to some high-lying orbital. In either case their location may often be anticipated in terms of progressions existing below the ionization limit.

The notion that atoms and molecules may absorb energy in excess of their ionization potential, without immediate ejection of an electron, is intriguing and, perhaps, surprising. It is made abundantly evident through consideration that the photoabsorption and photoionization cross sections of molecules are rarely equal. This is illustrated in Fig. 5 for H_2O (6). Clearly, a significant fraction of the absorbed photons do not result in ionization, even at energies considerably in excess of the ionization potential.

It is useful, in order to discuss results of this sort, to introduce the concept of the *photoionization efficiency*, defined as the ratio of the photoionization to the photoabsorption cross section. Thus,

$$\sigma_i = \sigma\eta \tag{13}$$

The photoionization efficiency η thus will be recognized as a quantum yield for photoionization; it will, of course, be a function of wavelength. In the example of Fig. 5, it is seen to be less than unity, as is quite generally true for molecular gases (7,8), whereas for the rare gases it is always unity

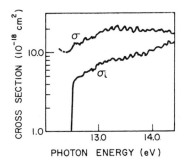

Fig. 5. Photoabsorption and photoionization cross sections of H_2O vapor (6).

(or greater if multiple ionization occurs) (9). This behavior both implies that some other process may intervene in the interval between photo-absorption and electron ejection and identifies decomposition as this intruding mechanism.* Thus, since this mechanism is not available for the rare gases, their ionization efficiencies can hardly be less than unity irrespective of whether their ionization proceeds directly or via an indirect mechanism.

It will be noted that ionization efficiencies of less than unity, while indicating that photodissociation into uncharged fragments occurs even above the ionization threshold, do not directly prove that any of the ionization actually occurs indirectly. Evidence that at least part of the observed ionization is indeed indirect is offered by observations of isotope effects in photoionization efficiencies (10). Thus, it has been found, at any given wavelength, that

$$\eta(C_6H_6) < \eta(C_6D_6) \tag{14}$$

Quite generally, the heavier species has the greater ionization efficiency. This behavior may be rationalized in the following manner. Assume that photoabsorption effects a transition to a neutral state above the ionization potential and that dissociation and autoionization from this state may be described in terms of first order rate constants k_d and k_i, respectively. It is probable (but not imperative) that k_d will be isotopic dependent, being somewhat smaller for the heavier species. It is similarly probable (but again not imperative (11)) that k_i will be roughly isotopic independent. The fraction of such states which autoionize rather than dissociate will then be greater for the heavier species, as will be its photoionization efficiency. Analogous behavior obtains under electron impact (12-14).

It thus seems clear that much of the photoabsorption and at least part of all photoionization processes occur by way of neutral states existing above the ionization potential. Such levels have been called "superexcited states" by Platzman who has vigorously emphasized their importance in photochemistry and radiation chemistry (15). Indeed while the existence of such states had been formally recognized for several decades, it has been largely through Platzman's analysis of quite specialized radiation effects (16,17) that the profusion of superexcited levels became manifest. Absorption spectroscopy in the far vacuum ultraviolet is now a field of vigorous growth, yielding a number of unsuspected features. A useful survey of much of this work has been published (18).

Information from various sources concerning the total oscillator strength distribution below and above the first ionization potential of

---

* The other possibility, radiative deexcitation, is most certainly too slow a process to compete with dissociation or autoionization.

Table I

Total Oscillator Strength and Squared Dipole-Matrix Elements

	Below first ionization potential		Above first ionization potential	
	$\sum f_n$	$\sum M_n{}^2$	$\int (df/dv)\, dv$	$(M_i{}^2)$
H	0.435[a]	0.717[b]	0.565	0.283[b]
H_2	—	0.93[b]	—	0.72[c]
He	0.45[d]	0.27[b]	1.54[f]	0.49[c]
Ne	0.4[e]	—	9.4[g]	1.66[c]
Ar	0.79[h]	0.79[h]	17.2[i]	4.10[c]
Kr	0.91[h]	1.0[h]	35.1[i]	5.51[c]
Xe	2.5[h]	3.1[h]	51.5[i]	7.10[c]

[a] H. A. Bethe and E. E. Salpeter, *Quantum Mechanics of One- and Two-Electron Atoms*, Springer, Berlin, 1957.

[b] R. L. Platzman, *Intern. J. Appl. Radiation Isotopes*, **10**, 116 (1961).

[c] A. E. Kingston, B. L. Schram, and F. J. de Heer, *Proc. Phys. Soc.*, **86**, 1374 (1965).

[d] E. E. Salpeter and M. H. Zaidi, *Phys. Rev.*, **125**, 248 (1962).

[e] D. L. Ederer and D. H. Tomboulian, *Phys. Rev.*, **133**, A1525 (1964).

[f] J. A. R. Samson and F. L. Kelly, GCA Technical Report No. 64-3-N (1964).

[g] J. A. R. Samson, *J. Opt. Soc. Am.*, **55**, 935 (1965).

[h] Estimated from electron scattering data, H. Boersch (unpublished).

[i] Chosen to satisfy the Kuhn-Thomas sum rule.

several atoms has been assembled in Table I. Also included is the available information on the total squared dipole-matrix elements connecting the ground state with all excited states in these two regions. These latter quantities are given in reduced dimensionless form whereby

$$M_n{}^2 = |X_n|^2/a_0{}^2 \tag{15}$$

and thus, within the ionization continuum,

$$M_i{}^2 = \int (df/dv)(R/hv)\, dv \tag{16}$$

The data attest to the preponderance of oscillator strength to be found above the ionization potential of the rare gases. A similar tendency exists, if to a somewhat lesser extent, among molecular species.

E. Threshold Laws

An important area of spectroscopic research, with objectives of considerable interest to the radiation chemist, involves the measurement of the cross sections for various types of electronic transitions, under either electron or photon impact, in the immediate neighborhood of the threshold energies. These studies are concerned not so much with obtaining absolute values of cross sections as they are with the variation of the cross section with energy of the incident particle. Thus, with respect to a transition of energy E_0 in an atom or molecule, to be effected by a particle of energy E (where necessarily $E \geq E_0$), it is the functional dependence of the cross section upon $(E - E_0)$ which is of interest here.

There are some indications of a theoretical nature that certain universal regularities may exist in the behavior of impact cross sections near threshold. The most sweeping statement of this kind has been championed by Morrison (19,20). He suggests the quite simple postulate that a cross section will be, in the threshold region, simply proportional to the excess energy raised to a power one less than the number of electrons leaving the collision complex. This postulate is embodied in the relation

$$Q(E) \sim (E - E_0)^{N-1} \tag{17}$$

and may be illustrated with examples of the excitation processes most commonly encountered. Thus, in direct ionization by electrons

$$e + M \rightarrow M^+ + 2e \tag{18}$$

two electrons leave the collision and the cross-sectional behavior should then be of the form of Fig. 6a. Similarly, both photoionization and electron impact excitation

$$h\nu + M \rightarrow M^+ + e \tag{19}$$

$$e + M \rightarrow M^* + e \tag{20}$$

should follow a step function, as in Fig. 6b.

Note that photoexcitation and electron attachment, being resonance absorption processes,

$$h\nu + M \rightarrow M^* \tag{21}$$

$$e + M \rightarrow M^- \tag{22}$$

will exhibit the functional form of Fig. 6c. This behavior can be included within Morrison's general postulate in the following sense. Figure 6b,

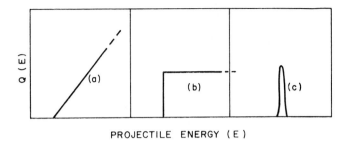

Fig. 6. Hypothetical threshold behavior for frequently encountered excitation processes.

corresponding to processes in which one electron is emitted from the collision complex, is essentially that of the first derivative dQ/dE of Fig. 6a. But Fig. 6c bears in turn a similar relation to Fig. 6b. Provided, then, that Eq. (17) is interpreted in this sense when $N = 0$, it may be said to embrace properly the correct threshold behavior in this case.

Equation (17) is of great simplicity, and, if borne out by experiment, would be of considerable interest. In fact, a fair amount of experimental detail is in agreement with its predictions. Thus the cross sections for single ionization by electron impact are often of the form of Fig. 6a (21,22). The results of Fox for helium provide an unambiguous illustration and are reproduced in Fig. 7a. Similarly, the photoionization cross section of NO (23,24), as illustrated in Fig. 7b, is in accord with expectation; each step correlates quite well with a new vibrational level of the NO^+ ion in its ground electronic state. The photoionization cross section of helium,

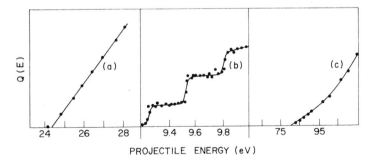

Fig. 7. Experimental cross sections near threshold for (a) electron impact ionization of helium (22), (b) photoionization of nitric oxide (23), and (c) double ionization of helium by electrons (25).

noted earlier (Fig. 2), again displays the proper discontinuity at threshold·
—although the subsequent monotonic decline is not in strict accord.

While the experimental data are not entirely unambiguous, cross
sections for multiple ionization are for the most part consistent with
predictions (25,26). A plot of the cross section for the process

$$e + He \rightarrow He^{2+} + 3e \tag{23}$$

is, for example, shown in Fig. 7c to be consistent with Morrison's
hypothesis.

Thus there is some indication that the threshold laws of Morrison are a
sufficiently close approximation to reality to be of some use. Only in the
case of simple excitation under electron impact to neutral states do
substantial discrepancies occur. While Morrison has found instances
which do seem to conform to his step function behavior (20), other studies
indicate a quite different form (27,28). It is clear that, for this process at
least, no simple threshold law will suffice.

Interest in threshold behavior goes well beyond merely ascertaining the
intrinsic validity of alleged threshold laws. Consider, for example, the
cross section for ionization of neon (29), shown in Fig. 8 as a function of
electron energy. It might be interpreted as a superposition of linear curves
in which each break marks the onset of a new ionization process. Indeed,
in this case the observed breaks correlate quite well with known spectroscopic states of the ion. Electron impact as a spectroscopic tool would thus
seem demonstrated.

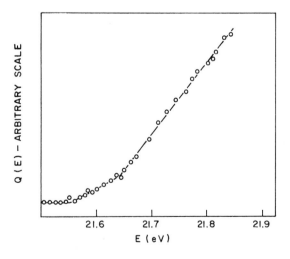

Fig. 8. Electron-induced ionization cross section of neon near threshold (29).

But consider now a similar plot of the ionization cross section of krypton, Fig. 9 (30). While kinks are present at energies corresponding to known energies of the ion, it is evident that a great deal of structure is also present which cannot be accorded so simple an origin. One possible interpretation is that some of the observed ionization raises via an indirect, autoionization mechanism:

$$e + Kr \rightarrow Kr^{**} + e \qquad (24)$$

$$Kr^{**} \rightarrow Kr^+ + e \qquad (25)$$

and indeed much of the extraneous structure can be correlated with known autoionizing states. What makes the present example especially interesting is that not all of the observed structure can be rationalized so easily. Burns has thus suggested (30) that, barring an experimental artifact, the mechanism

$$e + Kr \rightarrow (Kr^*)^- \rightarrow Kr^+ + 2e \qquad (26)$$

is also occurring at selected energies. A similar idea was mentioned some time ago in connection with the electron-induced ionization of hydrogen (31). In an analogous fashion it has been suggested that electron excitation

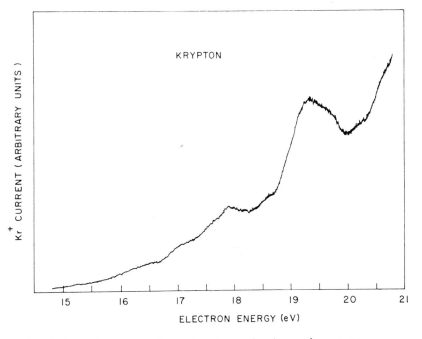

Fig. 9. Ionization cross section of krypton, under electron impact, near threshold (30).

of the metastable state of helium might best be thought of as proceeding by way of $e + \text{He} \rightarrow (\text{He*})^- \rightarrow \text{He*} + e$. The sharp resonance-like behavior of the cross section near threshold is consistent with this picture (32).

By way of further illustration, the photoionization cross section of oxygen near threshold is displayed in Fig. 10 along with the first derivative of the cross section under electron impact (33). The great similarity of the two curves is evident. Neither is of the shape anticipated for direct ionization and again one is led to suspect an important role for an indirect autoionization mechanism.

These conclusions are possible only because one presumes to know in each case what the shapes of ionization curves would be if only direct ionization were occurring. To the extent that this is indeed the case the study of cross section behavior near threshold is clearly of considerable use in identifying the states and mechanisms involved in photo- and electron impact processes. Thus, it will be recognized, we have already made implicit intuitive use of threshold considerations in our earlier discussion of photoionization spectra.

II. EXCITATION CROSS SECTIONS UNDER CHARGED PARTICLE IMPACT

In this section the Born approximation (more correctly, the first Born approximation) is developed as it pertains to the inelastic scattering of charged particles. While restricted in validity to nominally high velocities,

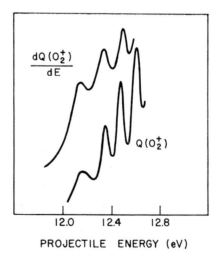

PROJECTILE   ENERGY   (eV)

Fig. 10. Photoionization cross section and first derivative of electron-induced ionization cross section of oxygen near threshold (33).

this approach is apt to be by far the most useful single theoretical construct in so far as the radiation chemist is concerned. In its realm of validity, a remarkable parallel with photochemistry is achieved. This section, then, pursues this parallel, exposing the direct relevance of the preceding discussion.

Again, however, a simplification is to be noted. The formulas developed are nonrelativistic in origin, thereby serving to show more clearly the structure of the theory. Relativistic corrections are often called for, nevertheless, especially with electrons at energies much greater than ~ 1 keV. The appropriate correction terms are given in the references cited.

A. The Born Approximation

We may conveniently begin our discussion of the interaction of charged particles with matter by citing the classic formula, due to Rutherford, relating the cross section for the scattering of a particle of charge z, velocity v, and mass M_1 by an electron of mass m, unbound but initially at rest, with the energy transferred in the collision, W:

$$d\sigma = (2\pi z^2 e^4 / mv^2)/(dW/W^2) \tag{27}$$

Note that the mass of the incident particle does not appear. Although this is a classical equation, it may be retained as a useful landmark for much of the subsequent discussion. It is convenient to make the substitution

$$W = \hbar^2 k^2 / 2m \tag{28}$$

where $k\hbar$ is the magnitude of the vector $\mathbf{k}\hbar = \mathbf{m}\mathbf{v}$ associated with the momentum imparted to the struck electron. Then let $|\epsilon_n(k)|^2$ be the probability that, if an electron receives a momentum k, the struck atom or molecule is sent to an excited state n. Equation (27) then becomes

$$d\sigma_n = \frac{8\pi z^2 e^4}{\hbar^2 v^2} |\epsilon_n(k)|^2 \frac{dk}{k^3} \tag{29}$$

The problem of finding a total cross section σ_n thus involves finding an expression for $|\epsilon_n(k)|^3$ and integrating over all possible values of k.

In the first Born approximation, the form-factor $\epsilon_n(k)$ is given by

$$\epsilon_n(k) = \int \Psi_n^* \Psi_0 \sum_j e^{ik \cdot r_j} \, d\tau \tag{30}$$

where Ψ_0, Ψ_n are ground and excited state wavefunctions and $\mathbf{r}_j$ the position vector of the j electron in the target. In order to expose the realm of applicability of Eq. (30), a brief derivation will be indicated (34).

A first-order perturbation theory formula

$$\omega = (2\pi/\hbar)|H'|^2 \, dn/dE \tag{31}$$

relates the transition probability per unit time ω with the matrix element

$$H' = \langle \Psi_n \chi_n| \, V \, |\Psi_0 \chi_0 \rangle \tag{32}$$

where Ψ_0, Ψ_n and χ_0, χ_n are the initial and final wavefunctions for the target and projectile, respectively. The projectile wavefunctions may be written as plane waves normalized to unit volume,

$$\chi_0 = (v_0)^{-1/2} \exp(ik_0 \cdot r') \tag{33}$$

$$\chi_n = \exp(ik_n \cdot r') \tag{34}$$

with the incident particle further normalized to unit flux so that the transition probability will represent a cross section. The density of states of the scattered particle, dn/dE, may be obtained from particle-in-a-box considerations:

$$dn/dE = (\mu/8\pi^3\hbar^2)k_n \sin\theta \, d\theta \, d\phi \tag{35}$$

which, upon use of the geometric relation,

$$k^2 = k_0{}^2 + k_n{}^2 - 2k_0 k_n \cos\theta \tag{36}$$

yields $dn/dE = k \, dk/4\pi^2\hbar v$ after integration over ϕ.

The perturbation, V, is just the electrostatic interaction of the incident particle with the Z electrons of the target

$$V(r') = \sum_{j=1}^{z} ze^2/|r_j - r'| \tag{37}$$

where $\mathbf{r}_j$, $\mathbf{r}'$ are position vectors associated with a target electron and the projectile, respectively. Upon integration over the latter coordinate, using an identity due to Bethe (35),

$$\int \frac{e^{ik \cdot r'}}{|r - r'|} \, dr' = \frac{4\pi}{k^2} e^{ik \cdot r} \tag{38}$$

Eq. (29) follows.

The Born approximation, and those consequences of it which we shall pursue in this chapter, is thus rooted in the familiar perturbation theory technique of quantum mechanics. As such the requirements that the approximation be valid will inevitably involve the magnitude of the perturbation. In the present instance it is the assumed plane wave description of the incident projectile which must be scrutinized. This description, and thus the Born approximation, will be valid provided (36)

$$ze^2/v\hbar \ll 1 \tag{39}$$

Hence the present treatment need be accurate only at high velocities of the projectile.

It is perhaps also worth remarking at this point on the key role played by $\mathbf{k}$, the momentum transfer vector, in, for example, Eq. (29). In derivations of the Rutherford formula, Eq. (27), a simple and direct correspondence is assumed between the magnitude of the momentum lost by the projectile (and hence acquired by the struck electron) and the energy transfer. When, however, the struck electron is initially bound, as in an atom, no such precise correspondence exists, even classically. This arises because the rest of the atom may recoil in an unspecifiable way, making the net energy transfer *a priori* indeterminate. Hence the total cross section for inducing a given transition will involve an integration over a range of momentum transfer magnitudes. In a sense, then, the first Born approximation is not so much to be differentiated from classical mechanics as it is a supplement to the latter. Indeed in the case of elastic collisions between unbound, elementary but distinguishable charged particles, the first Born approximation reduces to just Rutherford's formula.

It is now convenient to define a generalized oscillator strength $F_n(k)$ by

$$|\epsilon_n(k)|^2 = F_n(k)\hbar^2 k^2 / 2mE_n \tag{40}$$

where E_n is the energy of the n excited state, as measured from the ground state. Equation (29) then becomes

$$d\sigma_n = (16\pi z^2 R^2 a_0^2 / mv^2)/(F_n(k)/E_n)\, d(\ln k) \tag{41}$$

where the substitutions R (the Rydberg, 13.59 eV) $= me^4/2\hbar^2$ and a_0 (the Bohr radius, 0.529 Å) $= \hbar^2/me^2$ have been utilized. Finally, an important sum rule,

$$\sum_n F_n(k) = Z \tag{42}$$

valid for any value of k, has been derived by Bethe.

The inelastic scattering cross section into the small solid angle $d\Omega = 2\pi \sin \theta\, d\theta$ may be obtained from Eqs. (36) and (41) as

$$d\sigma_n = 4R\left(\frac{M_1 z}{m}\right)^2 \frac{k_n}{k_0} \frac{F_n(k)}{E_n} \frac{d\Omega}{k^2} \tag{43}$$

It will be observed that this formula now contains M_1, the projectile mass. Since k_0, k_n are related by energy conservation,

$$k_0^2 - k_n^2 = E_n(2M_1/\hbar^2) \tag{44}$$

the differential cross section for scattering may be obtained at any angle θ in laboratory coordinates. Alternatively, measurements of the inelastic scattering cross section as a function of angle will yield, if the Born approximation is valid, the generalized oscillator strength function $F_n(k)$.

It might be noted that, provided $F_n(k)$ is not a rapidly increasing function of k, the scattering cross section drops off fairly rapidly with k and thus with increasing angle of deflection.

We have, from above,

$$F_n(k) = \left| \int \Psi_n{}^* \left(\sum_j e^{ik\cdot r_j} \right) \Psi_0 \, d\tau \right|^2 E_n/Ra_0{}^2k^2 \tag{45}$$

With the expansion,

$$e^{ik\cdot r} = 1 + ikx + (ikx)^2/2! + \cdots \tag{46}$$

one obtains

$$F_n(k) = \frac{E_n}{Ra_0{}^2} \left[|X_n|^2 + \frac{|X_n{}^2|^2 k^2}{4} + \cdots \right] \tag{47}$$

where, for example,

$$X_n = \int \Psi_n^* \sum_j x_j \Psi_0 \, d\tau \tag{48}$$

is the dipole matrix element connecting the ground and excited states. Thus, the first term in the expansion of $F_n(k)$ is just f_n, the optical oscillator strength. We have then $F_n(k) \to f_n$ as $k \to 0$ and thus as $\theta \to 0$. Extrapolation of scattering cross sections as a function of angle to zero angle can thus yield optical oscillator strengths. Alternatively, if these are known, such an extrapolation may be used as a test of the Born approximation.

B. Total Cross Section

The integration of Eq. (41) to obtain a total cross section formula requires consideration of the limits which k may assume. This may be accomplished most simply through a scrutiny of Eq. (36). In a center-of-mass coordinate system θ may range from 0 to π, yielding

$$k_{\min} = k_0 - k_n \tag{49}$$

$$k_{\max} = k_0 + k_n \tag{50}$$

Further, in this coordinate system, the energy conservation equation becomes

$$k_0{}^2 - k_n{}^2 = \Delta E_n 2\mu/\hbar^2 \tag{51}$$

where μ is the reduced mass of the colliding pair. Approximately, then

$$k_{\min} = \mu E_n/k_0\hbar^2 \tag{52}$$

$$k_{\max} = 2k_0 \tag{53}$$

where $k_c = \mu v/\hbar$. Thus,

$$\sigma_n = \frac{16\pi z^2 R^2 a_0{}^2}{mv^2 E_n} \int_{k_{\min}}^{k_{\max}} F_n(k) \, d(\ln k) \tag{54}$$

In order to perform this integration, one may, following Bethe, utilize the expansion Eq. (47) and retain only its first term. Further, it is convenient to use as an effective upper limit $k_{max} = 1/a_0$. One then obtains

$$\sigma_n = \frac{8\pi z^2 R^2 a_0^2}{mv^2} \frac{f_n}{E_n} \ln \left[\frac{2mv^2 R \gamma_n}{E_n^2} \right] \tag{55}$$

where γ_n is a constant approximating unity correcting for the errors involved in the integration procedure. An elegant graphical interpretation of γ_n has been given by Miller and Platzman (37). It will be seen from their discussion that γ_n, while a function of the state n involved, will tend to be independent of velocity. Equation (55) is often written in the form

$$\sigma_n = \frac{8\pi z^2 R^2 a_0^2}{mv^2} \frac{f_n}{E_n} \ln \left[\frac{mv^2 C_n}{2} \right] \tag{56}$$

where C_n is another constant. Note that Eqs. (55) and (56) refer to the incident projectile only through its charge and velocity. Thus, for example, a medium will, in this approximation, present equal cross sections to protons and electrons of the same velocity.

C. Transitions to the Ionization Continuum

As an example of Eq. (56), one may consider the total cross sections for all transitions to within the ionization continuum of a target species,

$$\sigma_{ion} = \frac{8\pi z^2 R a_0^2}{mv^2} (M_i^2) \ln \left[\frac{mv^2}{2} C_i \right] \tag{57}$$

where $M_i^2 = \int_{IP} (df/dE)(R/E)\, dE$ involves an integration of the oscillator strength above the lowest ionization potential of the medium. Some examples of $(M_i)^2$, typically obtained from an integration of photoabsorption coefficients, have already been presented in Table I. It should be mentioned that Eq. (57) does not give the total ionization cross section of a species unless each such transition produces one and only one charge pair. This will, in general, not be true except for atomic hydrogen. Let us, then, following Platzman (38) define a new quantity

$$(M_i')^2 = \int \frac{df}{dE} \cdot \frac{R}{E} \eta(E)\, dE \tag{58}$$

where $\eta(E)$ is a photoionization efficiency, i.e., the number of charged pairs per photon absorbed at energy E. As discussed earlier, η may be greater than unity where multiple ionization may occur, but, for molecules, may also be less than unity. The total ionization cross section, then, in the Born approximation is

$$Q_{\text{ion}} = \frac{8\pi z^2 R a_0^2}{m v^2} (M_i')^2 \ln\left[\frac{m v^2}{2} C_i'\right]$$ (59)

Representative values of $(M_i')^2$ are given in Table II.

Table II

Values of $(M_i')^2$ for Several Atoms and Molecules as Obtained Primarily from Electron Ionization Cross Sections

H	0.283[a]	c-C_3H_6	10.2
He	0.49[b]	C_3H_6	12.0
Ne	1.87	C_3H_8	13.8
Ar	4.50	1,4-C_4H_6	15.9
Kr	7.51	1-C_4H_8	15.3
Xe	11.75	cis-C_4H_8	16.7
H_2	0.72	$trans$-C_4H_8	15.5
N_2	3.85	n-C_4H_{10}	17.8
O_2	4.75	i-C_4H_{10}	17.4
H_2O	3.14	n-C_5H_{12}	24.4
CH_4	4.28	i-C_5H_{12}	25.0
C_2H_4	7.32	C_6H_6	24.2
C_2H_6	8.63	C_6H_{14}	24.8

[a] R. Platzman, *Intern. J. Appl. Radiation Isotopes*, **10**, 116 (1961).

[b] B. L. Schram, F. J. de Heer, M. J. Van der Wiel, and J. Kistemaker, *Physica*, **31**, 94 (1965); B. L. Schram, M. J. Van der Wiel, F. J. de Heer, and H. R. Moustafa Moussa, *J. Chem. Phys.*, **44**, 49 (1966); J. Schutten, F. J. de Heer, H. R. Moustafa Moussa, A. J. H. Boerboom, and J. Kistemaker, *J. Chem. Phys.*, **44**, 3924 (1966). These references are for all entries except H.

D. Production of Fast Secondary Electrons

It is often desirable to have a cross section formula for the ejection from a target of electrons of especially high energy. Such fast secondary electrons are referred to as delta rays; they are evident in cloud chamber photographs as well-developed branches of the main trajectory. Scrutiny of Eq. (30) when $ka_0 \gg 1$ suggests that $|\epsilon_n(k)|^2$ will be significant only when $k^2 a_0^2 \approx E_n/R$. This and the sum rule reduce Eq. (41) to

$$d\sigma = \frac{16\pi z^2 R a_0^2}{m v^2} Z \frac{dk}{k^3}$$ (60)

which is just the Rutherford formula. Thus, for large momentum transfers the struck electrons can be treated as if they were initially unbound.

Some modification is required when the incident particle is an electron since, after the collision, it is indistinguishable from the ejected electron. In this case the Möller formula should be used:

$$d\sigma = \frac{8\pi z^2 R^2 a_0^2}{mv^2} dW \left[\frac{1}{W^2} + \frac{1}{(T-W)^2} - \frac{1}{W(T-W)}\right] \quad (61)$$

where T is the kinetic energy of the incident electron. It is further understood that the maximum energy transfer is $T/2$ in applications of this formula.

E. Condensed Phase Modifications

Some alteration of the preceding formulas is necessary in matter at high densities. Only a rough indication of how this may be accomplished can be given here. Equation (37) should contain the complex material dielectric constant $|\epsilon|$ in its denominator, which will carry through as $|\epsilon|^2$ in subsequent cross section formulas. The important term f_n/E_n in, for example, Eq. (55) then becomes

$$\frac{df_n}{dE_n} \frac{1}{|\epsilon|^2} \frac{dE_n}{E_n} = \frac{m}{2\pi^2 e^2 \hbar^2} \frac{\text{Im}\left(-\frac{1}{\epsilon}\right)}{N} dE_n$$

$$= \left(\frac{2Z}{\pi}\right)(\hbar\omega_p)^{-2} \text{Im}\left(-\frac{1}{\epsilon}\right) dE_n \quad (62)$$

where ω_p, the so-called *plasma frequency*, is

$$\omega_p = 2\pi \left(\frac{e^2 NZ}{\pi m}\right)^{1/2} \quad (63)$$

The imaginary part of $(-1/\epsilon)$ thus becomes a more accurate index to the relative cross section for various transitions. The very close correlation between the optical properties of matter and its response to charged particles is thus modified but not lost.

F. Stopping Power

The Bethe-Born formalism for charged particle interactions has been most extensively applied in discussions of stopping power, the energy lost per unit path length traversed. Formally, this may be written

$$-dE/dS = N \sum_n \sigma_n E_n \quad (64)$$

where N is the number of atoms or molecules per unit volume and the sum is extended over all accessible final states. This sum is readily evaluated by way of Eq. (65)

$$\int E_n \, d\sigma_n = \frac{16\pi z^2 R^2 a_0{}^2}{mv^2} \int F_n(k) \, d \ln k \tag{65}$$

The most convenient procedure is to divide this summation into two segments: first one integrates Eq. (65) up to some k', using k_{min} from before and the small k approximation

$$F_n(k) \simeq f_n \tag{66}$$

With the definition of a new quantity, I, the average excitation potential:

$$\ln I = \sum f_n \ln E_n \Big/ \sum f_n \tag{67}$$

this first contribution to the stopping power becomes

$$-(dE/dS)_1 = \frac{16\pi z^2 R^2 a_0{}^2}{mv^2} NZ \ln \left(\frac{k'v\hbar}{I} \right) \tag{68}$$

From k' to k_{max}, one uses the Bethe sum rule, Eq. (42), to obtain

$$-(dE/dS)_2 = \frac{16\pi z^2 R^2 a_0{}^2}{mv^2} NZ \ln \left(\frac{k_{max}}{k'} \right) \tag{69}$$

The total stopping power is then, for heavy particles ($M_1 \gg m$)

$$-(dE/dS) = \frac{16\pi z^2 R^2 a_0{}^2}{mv^2} NZ \ln \left(\frac{2mv^2}{I} \right) \tag{70}$$

For fast electrons the analogous formula, after correction for particle indistinguishability, is

$$-(dE/dS) = \frac{16\pi z^2 R^2 a_0{}^2}{mv^2} NZ \ln \left[\frac{mv^2}{2I} \left(\frac{e}{2} \right)^{1/2} \right] \tag{71}$$

where e here is the base of natural logarithms.

The accuracy of these formulas requires not only that the Born approximation be valid but also that there exist a value k' as used in the derivations. That is, k' must be low enough that the expansion Eq. (47) be valid for all σ_n, yet greater than k_{min} for all important transitions. This criterion will not be met at low velocities and the evaluation of Eq. (65) must then be achieved otherwise. Eventually, of course, the Born approximation itself will fail.

The reader is directed to extensive tabulations and discussions of stopping power (39–41). Table III lists a few representative average

Table III

Recommended Values of I for Use in Stopping
Power Calculations[a]

Element	I (eV)
H	18
He	41.8
C	78
N	88
O	101
Ne	128
Ar	190
Kr	430

[a] From Ref. 39.

excitation potentials. Obtained from experimental stopping power measurements, these parameters will permit one to calculate the stopping power of a large variety of media. For molecular media this will entail use of the Bragg additivity law—i.e., that stopping power is merely an additive function of the constituent atoms.

G. Primary Yields

With the preceding formulas it is not difficult to calculate a number of quantities occasionally of interest in discussing the structure of particle tracks. Primary ionizations, that is ionizations caused by the primary incident particle, are occasionally singled out, for example, as if they were of especial importance. The number of these per unit energy deposited and their average spatial separation follow from the stopping power and total ionization cross section formulas. For example, one sees that quite typically ~ 1 ion pair is expected per 100 eV deposited by nominally high energy particles. This is in quite good agreement with visual cloud-chamber observations (42). Similarly, from the formula for delta-ray cross sections, one may estimate yields for fast secondary electrons.

H. Cluster-Size Distributions

Examination of cloud-chamber photographs indicates that the images associated with primary ionizations are often, upon close scrutiny, of a multiple character. These may be attributed to additional ionizations caused by the electron set free in the primary ionization act. A number of studies have been made of the relative frequency of these variously sized groupings

(43). Scrutiny of these studies reveals that the probability that a group or cluster will contain N ionizations is roughly of the form

$$P(N) = e^{-N}(e - 1) \tag{72}$$

whereupon $\overline{N} = 1.58$. There is no obvious theoretical basis for this result nor is there any real basis for focusing on ionizations in discussions of radiation effects. Nevertheless, because of its convenient analytical form, Eq. (72) may find some use in an analysis of track structure.

III. EXPERIMENTAL INVESTIGATIONS OF CHARGED-PARTICLE EXCITATION CROSS SECTIONS

The results of the preceding section provide a natural measure with which experimental investigations of inelastic cross sections might be compared. This course is followed below. It will be evident here, perhaps more than elsewhere in this chapter, that no attempt has been made to give a comprehensive review of the literature. Rather, the works cited have been chosen to indicate the several types of experimental studies which may bear on the overall compass of the chapter. While thus merely representative, the references quoted should nevertheless offer a useful entrée to the pertinent literature.

A. The Franck-Condon Principle

At high velocities a very close correlation exists, as we have seen, between the action of charged particles and of light. As a corollary to this result, one may anticipate that the Franck-Condon principle will be applicable to charged particle-induced excitations. That is, such excitations and ionizations will be "vertical."

This tentative conclusion is important since, as a result of nonadiabatic transition, a molecule will be left in an internally excited state and thus, perhaps, eligible for dissociation. An immediate question then arises concerning the validity of this important result at low velocities where the simple Born-Bethe equations are apt to fail.

Experimental investigations of this point date back a number of years (44). The usual procedure has been to analyze the luminescence following charged particle-induced excitation in terms of the relative population of the emitting vibrational levels. Comparison is then made with expectations derived theoretically or from optical absorption spectroscopy. Recently, a complementary technique has been brought to bear on the problem (45). Here the energy losses of inelastically scattered electrons are examined in the context of Eq. (43) to see if their energy losses conform with the Franck-Condon principle.

Quite generally electron-induced transitions do seem to occur in accord with this principle. Vibrational and rotational temperatures of excited states are consistent with overlap factors and optical selection rules. On the other hand, heavy particle-induced transitions often show, at reduced velocities, considerable deviations from optical expectations. These discrepancies are inevitably in the sense as to indicate an excessive transfer of energy to the internal degrees of freedom of the struck molecule. But this is plausible since the Born-Bethe development has specifically neglected interactions between the nuclei and the projectile.

That electrons should conform to expectations in this respect is rather interesting. This is usually rationalized by saying that a fast electron spends so little time in the vicinity of a target molecule that transitions will inevitably be vertical. But at very low energies, and especially near threshold, this can no longer the case. The applicability of Franck-Condon overlap factors at such energies is thus a question of much current interest (46).

B. Total Ionization Cross Sections

Total ionization cross sections offer a useful entry to a discussion of experimental cross section measurements. They have been extensively studied and are perhaps as precisely known as is any class of excitation parameters. This is fortunate since, as we shall see later, total ionization cross sections can play an especial role in practical formulations of radiation chemistry. More immediately, they present an elegant illustration of the applicability of the theoretical formulations outlined earlier.

The measurement of total ionization cross sections would seem to be tantalizingly simple. By directing a beam of charged particles, of well-defined energy, through a parallel-plate ionization chamber of known dimensions and containing gas at a known density, cross sections should be obtainable merely by collecting the induced ion current. In fact, such measurements are fraught with difficulties; a recent survey of such data suggests that ionization cross sections in simple gases under electron impact can be considered as known with an uncertainty of perhaps 10% (47). Thus, even in this instance of a seemingly most straightforward experiment, the available precision is not earthshaking.

Figure 11 illustrates the total ionization cross section of helium under electron impact. The rise to a broad maximum near 70–100 eV and the subsequent gradual decline at higher energies are both representative features. At energies above 500 eV the data begin to assume the form prescribed by Eq. (59). Again this is characteristic of most atoms and molecules under electron impact. Indeed the parameters given in Table II were obtained (48–52) by fitting just such experimental data to Eq. (59).

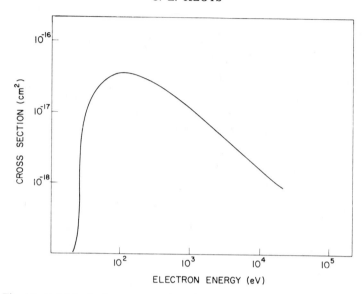

Fig. 11. Total ionization cross section of helium under electron impact (47).

This comparison with the form predicted by Born-Bethe theory is a procedure which ought to be *de rigueur*; in fact, it is scarcely ever made—often at great loss of insight. We might do well then to emphasize the mechanics of the comparison. In Fig. 12 the function $Q(T) \times T$ (where

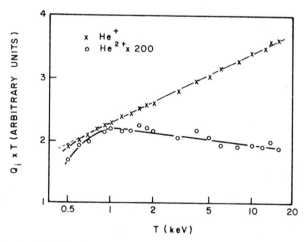

Fig. 12. Total and double ionization cross sections of helium plotted so as to reveal their asymptotic high velocity behavior (51).

$Q(T)$ is the total ionization cross section of helium) is plotted as a function of log T (with T the energy of the projectile electrons). Above 500 eV, the straight line demanded by theory is obtained. Thus, a simple graphical procedure suffices to obtain the parameter $|M_i'|^2$ and C_i' of Eq. (59).

Also included in Fig. 12 is a similar plot of the function $Q_{He^{2+}}(T) \times T$ where $Q_{He^{2+}}(T)$ is the cross section for double ionization of helium. Its shape at high energies is not that of Eq. (59) but more nearly

$$Q \times T \simeq \text{const.} \tag{73}$$

This is most interesting—suggesting in terms of the Born approximation that optically allowed dipole transitions are not responsible for the observed double ionization. Note that if one passes to the third and higher terms in the expansion in Eq. (47) one anticipates a high velocity cross section inversely proportional to energy and thus just of the observed form. It appears then that the multiple ionization of helium occurs predominantly by way of optically forbidden transitions.

But this is, in a simpleminded way, quite plausible. It is easy enough to show that, in terms of an independent electron model of an atom, multiple excitations or ionizations are optically forbidden. Thus, to the extent that electrons in a helium atom are uncorrelated and independent of each other the observed behavior is entirely comprehensible. On the other hand, multiple ionization of heavier atoms does, occasionally, seem to occur predominantly by way of optically allowed transitions. This, too, may be accorded a simple interpretation in the spirit of single-electron wavefunctions; thus, inner-shell ionization followed by an Auger process would account for the observed behavior (52).

Total ionization cross sections under impact by particles other than electrons are the object of much current research. From the Born approximation one expects, at high velocities, that an ionization cross section will depend only on the charge and velocity of the impacting particle. It is indeed found that at sufficiently high velocities observed cross sections do agree with those predictable from the appropriate electron-impact measurements (53–57). At lower velocities real discrepancies occur which are not entirely understood. It should be noted, however, that, in part at least, a divergence between electron and heavy-particle cross sections is to be expected even within the framework of the Born approximation. A careful attention to k_{max} and k_{min} in Eq. (54) shows this and further predicts that heavy particle cross sections will approach those for electrons from above. This is indeed what is observed.

Occasionally one is interested in the cross section for ejection of an electron from some particular shell of an atom. This is especially true if one is concerned with certain specialized phenomena as the Auger effect, X-ray fluorescence and crystalline lattice defects (58), all of which may follow inner shell ionization. Calculations of such cross sections have usually been in the Born approximation, treating the inner electrons with hydrogen-like wavefunctions. The computed results have been reviewed recently (59–61); the agreement with experiment is satisfactory.

C. Excitation Cross Section Measurements

While total ionization cross sections are extremely fundamental parameters, they give no information concerning the excitation of neutral excited states nor even of individual ionic states. Slightly more informative are ionization studies performed in conjunction with a mass spectrometer. Such studies permit one to follow, as a function of energy, the cross section for excitation of that manifold of states which may terminate in an ion of a given mass. An elementary example of this is the measurement for ionization of helium to give He^{2+}, alluded to above. While somewhat more specialized a type of experiment than total ionization measurements, mass analysis still does not give very much in the way of precise information. Nevertheless, it is encouraging that recent workers (62–64) are interpreting their data in the light of Born-Bethe theory, thereby acquiring some insight into the underlying ionization mechanisms.

A seemingly direct technique for determining excitation cross sections of resolved electronically excited states is based upon the fact that many such states will emit radiation of a characteristic wavelength. If the quantum yield of this luminescence is known and a detector sensitive to the appropriate wavelength is available, the rate of emission can be used as a direct measure of the excitation cross section. The technique is then applicable, in principle, to the excitation of any state which readily luminesces.

In practice a number of complications arise. It is necessary to work at quite low pressures so as to minimize the effects of collisional quenching and of radiation imprisonment. Further, there is often present the complication that some of the radiating states of interest may have been excited not only directly but also indirectly as, for example, via a radiative transition from a higher lying level. Corrections for such cascading mechanisms are occasionally possible and are especially well exemplified by a recent study of helium excitation functions (65). More typically, however, the necessity for such corrections introduces sufficient ambiguity into the data analysis as to obviate the technique as one of great applicability. Thus one must carefully distinguish between the excitation cross section for such radiative emission and the more fundamental microscopic cross section for direct excitation of the emitting state. This distinction is not always made clear in the literature.

Two review articles might be consulted for a discussion of the results obtained from the earlier luminescence-excitation studies (66). More recent work is both copious and variegated; excitation mechanisms appropriate to the rare gases and to the upper atmosphere continue, however, to receive much attention (67–69). Of special interest is a recent report that excitation cross sections in helium under electron impact, deduced in the above manner, are in accord with expectations based upon optical oscillator strengths (70). Again, it is regrettable that comparisons of this sort have not always been made.

The techniques outlined above for detecting and measuring inelastic scattering processes are quite inapplicable to an important class of energy-loss mechanisms. Thus when the resultant state is one from which all radiative transitions are strongly forbidden, detection by way of its luminescence is usually impractical. If, further, the species in question is uncharged, its straightforward detection in the usual mass spectrometer is again out of the question. Somewhat more indirect detection techniques are then required.

One such method may be thought of as depending upon the chemical reactivity of the electronically excited state whose population is to be measured. Thus, if a product of some such reaction can itself be measured, it may serve as an index to the excitation process. Consider, for example, excitation of an atom or molecule to a state E_n above the ground state. If E_n is greater than the ionization potential of some second species, the transfer of this excitation energy to the "detector" species can result in its ionization and thus its detection. This scheme has been used most often in conjunction with a metal of low work function as the detector species, whereby the excited atom or molecule sensitizes electron ejection in a fashion analogous to the photoelectric effect. The technique is especially effective in connection with "metastable" states with the requisite radiative lifetime needed to reach the detector. Thus, the well-known metastable states of rare gas atoms as well as some hitherto unknown electronic states of molecules have been studied in this fashion (28,71,72).

These studies reveal, in accord with the metastable character of the induced excitation, cross sections under electron impact which tend to peak shortly above threshold and to drop off rapidly thereafter. It will be recognized, however, that the method is subject to the same complication as in luminescence excitation, viz., the possibility of radiative cascading. Thus it is not clear to what extent the structure observed in excitation cross section functions is fundamental or otherwise due merely to the onset of new excitation mechanisms. When the transition in question is spin-forbidden, its occurrence under electron impact is often said to involve electron exchange and thus thought of as proceeding by way of a negative-ion intermediate. In this context, then, one would indeed expect the excitation cross section to have the form of one or more resonance-like functions.

Detection of excited states need not be accomplished heterogeneously, at the chamber walls. Thus titration of excited rare gas atoms can be accomplished in the gas phase through the sensitized ionization of additive impurities (73). An interesting variation is offered by studies in the pure rare gases. Certain excited states of, for example, argon are known to react according to

$$Ar^* + Ar \rightarrow Ar_2^+ + e \qquad (74)$$

Mass-spectrometric detection of the Ar_2^+ ion then permits one to follow the excitation cross section of the precursor Ar^* species (74). Pulsed electron beams and time-resolved mass spectrometry, in fact, implicate at least three sources of Ar^*; one of these is clearly an optically forbidden transition while at least one other is a strongly allowed transition. Even this is, however, apparently an oversimplification, since recent photo-ionization studies have demonstrated the existence of a multitude of optically accessible precursors to Ar_2^+ (75).

D. Scattering Measurements

We have outlined, above, a number of techniques by which inelastic cross sections under charged particle impact may be measured. In each case one actually detects an interaction through some property of the struck molecule or one of its fragments; in each case some unsatisfactory aspect of the technique has been noted.

A much more powerful technique, and one which bypasses most of the difficulties noted, will now be discussed. It is entirely complementary to those above in the sense that an inelastic event is recorded through a measurement on the scattered *projectile*. Consider a beam of, ideally, monochromatic electrons of energy T_0, directed at a suitable target, with a detector for such electrons located at some angle θ off-axis of the incident beam. If this detector is capable of energy-analyzing the electrons accepted by it, then the current to the detector of electrons with energy $(T_0 - E)$ will be a measure of those scattered into an angle θ with energy loss E. If, further, the density or thickness of the target can be sufficiently reduced, the scattered current will be a consequence of only single interactions and thus of the cross section for an energy loss E (with simultaneous scattering into angle θ). Analysis, in terms of the Born approximation, may then be accomplished with Eq. (43). One will thus have a device for measuring, if the Born approximation is indeed correct, generalized oscillator strengths. Alternatively, an integration of the observed cross section over all angles will give a total cross section for energy loss E.

Programs along these lines have been instigated in recent years in several laboratories (76–84). The work of Lassettre has been especially far-ranging in this respect. Being directed largely at the measurement of scattering cross sections into small angles, this work has offered a number of opportunities to assess the validity of the Born approximation. Lassettre proceeds by defining experimentally a generalized oscillator strength in terms of the observed scattering cross section and Eq. (43). Note, then, that if the Born approximation is indeed correct this quantity, so defined, must, for a given energy loss, possess a number of properties amenable to scrutiny, viz., that $F_n(k)$ must (1) be a function only of k but otherwise not of electron velocity, (2) extrapolate properly to its optical value, and (3) be consistent with the Franck-Condon principle. Thus, Fig. 13 illustrates the oscillator strengths measured by Lassettre for the 12.7 eV transition in hydrogen at several electron energies. The limiting value, obtained by the indicated extrapolation, is in good agreement with the optical oscillator strength.

Electron-impact spectroscopy at small scattering angles is, then, a technique capable of yielding much the same sort of information as optical spectroscopy in the far-vacuum ultraviolet. But because of Eq. (56), it is just such optical data which are of interest in ascertaining the importance of a given transition under high-velocity charged particle impact and thus in radiation chemistry. Figure 14 illustrates the energy loss distribution

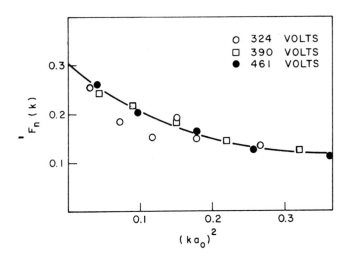

Fig. 13. Generalized oscillator strengths associated with the 12.7-eV transition in hydrogen (76). ○, 324 volts; □, 390 volts; ●, 461 volts.

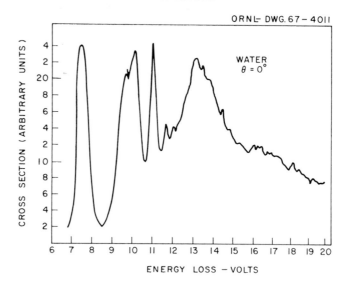

Fig. 14. Energy-loss spectrum, at small scattering angle, for 300-eV electrons in H_2O vapor (45).

from fast electrons, measured at small scattering angle, and hence the optical oscillator strength distribution in water vapor. Insofar as comparison is possible, essential correspondence with the optical absorption spectrum is found.

Data such as those illustrated in Fig. 14 are for the most part of rather recent vintage. That they had not been obtained earlier by the more straightforward light absorption technique is due, largely, to experimental problems which have only recently been solved (and then only in part). Not surprisingly, then, the early literature of radiation chemistry contains numerous references to low-lying energy levels occurring in the easily accessible regions of the ultraviolet. The work of Lassettre has done much to cure this nearsightedness, drawing attention instead to the less familiar but clearly pertinent transitions occurring in the far vacuum ultraviolet.

E. Electron Scattering at Large Angles

If optically accessible transitions predominate in energy-loss spectra at low angles, the reverse is true at wide angles. This is well illustrated in Fig. 15 where the relative magnitudes of the transitions from ground state helium to the forbidden 2^1S and allowed 2^1P state are indicated at two different scattering angles. It thus seems that inelastic scattering measurements at large angle afford a sensitive technique for the detection and

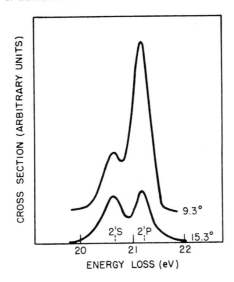

Fig. 15. Relative probabilities of an allowed and forbidden transition in helium at different scattering angles, under impact by 500-eV electrons (76).

analysis of optically forbidden transitions. As such the technique would be an invaluable supplement to ordinary optical techniques and to low-angle scattering measurements. In this context recent descriptions of such measurements are of considerable interest (85–87).

It is by no means clear, however, to what extent such transitions are of relevance to radiation chemistry. One sees from Eq. (56) that, at high velocities, optically allowed transitions will numerically overwhelm those which are forbidden. While, at low velocities, this need not be so, any suggestion that an optically forbidden transition occurs with substantial frequency in an irradiated medium would be injudicious without some more careful analysis. An indication of how such an analysis might be made is given below.

F. Condensed Phases

The measurement of an energy-loss spectrum in a condensed medium is entirely feasible and would proceed in a similar manner. Further, by the use of sufficiently thin targets and, perhaps, a suitable extrapolation one might hope to study energy losses under single collision conditions and thus obtain cross sections. Such studies have, in fact, been undertaken for some time. Of particular interest has been the observation, in metallic media especially, of a number of energy-loss processes not found in the gas phase (88).

That important differences should exist in energy-loss spectra in high and low density media need not be surprising. Density effects and solvent shifts are familiar enough in ordinary optical spectroscopy. Even more fundamentally, and as noted above, the very close correspondence between optical and charged particle excitation spectra is necessarily modified at high densities. One must then inquire to the extent to which such modifications occur. An analysis by Fano (89) has yielded what should be a useful criterion to this end. He defines the parameter μ by

$$\mu = \frac{(\hbar\omega_p)^2}{2RZ} \, [(R/E) \, df/dE] \qquad (75)$$

where ω_p is the "plasma frequency" as before, and shows that when μ is much less than unity the optical loss spectrum will be a reasonable guide to that of fast charged particles. Alternatively, when μ is large, this correspondence will be lost and the more complex description of Eq. (62) must then be considered. Metallic substances, with their loosely bound conducting electrons, are especially prone to such collective effects. The situation is not so clear, however, with respect to nonconductors; characteristic values of μ are somewhat borderline in magnitude here (90). Nevertheless, experimental indications do exist for nominal shifts in energy-loss spectra from purely optical expectations (91–92).

G. Secondary Electrons

Of the several types of experimental information appropriate to an understanding of radiation-induced processes, energy analyses of the electrons ejected in ionizing collisions are the most sparse. This is unfortunate since these "secondary" electrons play an important and even predominant role in effecting radiation action. The range of energies with which they are born, the secondary electron spectrum, is a crucial ingredient in an evaluation of this role.

Presumably, as noted before, the Rutherford and Möller formulas Eqs. (27 and 61) may be used at sufficiently high values of the primary and secondary energies. But this tells us neither what is "sufficiently high" nor what the spectrum will be when these criteria are not met. Experimental measurements are so scant as to provide very little in the way of a guideline. It is encouraging then to find nominally good agreement between these few data and the predictions of the Born approximation Eq. (54) even at quite low velocities (93–95). These predictions had been obtained using hydrogen-like atomic wavefunctions and thus are amenable to scaling to more complex atoms. This is currently under further examination within the context of radiation chemistry (96).

The sort of results one obtains from such a treatment is illustrated in Fig. 16. The ordinate function α can be thought of as a correction factor to the Rutherford or free-electron formula Eq. (27). It is a function of two reduced variables

$$\kappa^2 = \epsilon/I_K \tag{76}$$

and

$$\eta = 2mv^2/I_K \tag{77}$$

where ϵ is the ejected electron energy, I_K is the ionization potential of the target electron, presumed to come from a K-shell, and v the velocity of the incident projectile. These particular results are pertinent to heavy-particle projectiles but similar results are obtained for incident electrons. Note how the form of the free-electron model is approached at large values of ϵ, the ejected electron energy. This tendency is a reflection of the correspondence, noted earlier, between the first Born approximation and classical theory as the binding energy becomes negligible.

It might be mentioned that observations of secondary electron spectra are of interest for quite another reason. Prescriptions such as the Rutherford and Möller formulas imply that the secondary electron spectrum will be a smooth and monotonic function of the electron energy. In fact, this cannot be strictly so. Following an inner shell ionization, for example, an

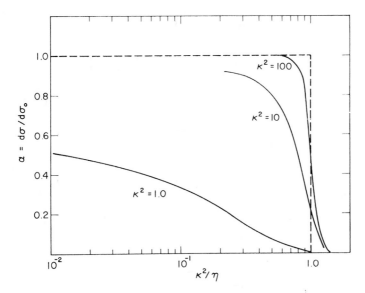

Fig. 16. Correction factors to the Rutherford formula, as obtained with hydrogen-like wavefunctions (96).

Auger cascade will result in the release of one or more groups of roughly monoenergetic electrons. These electrons will then stand out as discontinuities superimposed on the otherwise smoothly varying secondary electron spectrum. Indeed any autoionization process will give rise to such electrons of characteristic and singular energy. We must anticipate, then, that secondary electron spectra will contain much structure arising from such sources. This structure is, in fact, observed (97,98).

IV. AGGREGATE DESCRIPTIONS OF ENERGY DEPOSITION

A complete description of the energy deposition mechanisms occurring in a given irradiated system would include, at the very least, a specification of the nature of the various quantum transitions, via which the radiation is transferred to the medium, as well as some numerical measure of the frequency of each such transition. Previous sections have been concerned with a qualitative description of the detailed energy-loss mechanisms likely to be encountered. It remains then to illustrate in this section various conceptual frameworks within which a more quantitative accounting may be achieved.

A. Units

In order to establish a point which is central to the discussion it will be useful to consider the following observations. Thin layers of sodium salicylate, upon photon absorption at wavelengths between roughly 600–2500 Å, emit fluorescence in the neighborhood of 3500 Å; the number of emitted photons is found to be proportional to and, in fact, roughly equal to the number of absorbed photons (99,100). Thus one may say that the quantum yield of light emission, Φ (fluorescence), is approximately unity between 600 and 2500 Å. When the absorbed photons are of very much greater energy (and thus properly described as X-rays or γ-rays), however, very different results are obtained. The number of emitted photons, per absorption event, is no longer unity but larger and very nearly proportional to the energy of the incident photon. Presumably each fast electron, set free in the primary photoabsorption event, dissipates its energy by inducing a number of electronic transitions in the crystal, each of which, in turn, may terminate with fluorescence. While one might still speak of a quantum yield for fluorescence, it would be strongly dependent upon the energy of the absorbed photon and clearly void of any possible stoichiometric significance. Indeed, whenever the number of induced events per absorbed photon is much greater than unity, the concept of a quantum yield loses its conceptual usefulness as an index for purposes of description.

This example rules out the quantum yield as the natural unit of radiation chemistry and suggests, as an alternate measure, the number of events per unit energy absorbed. In this vein, then, yields in irradiated systems are typically reported in units of G, the number of changes of a specified kind induced in a medium per 100 eV energy absorbed. The unit energy, 100 eV, is an entirely arbitrary magnitude (the Rydberg might have been a happier choice) and, as such, G values can have no intrinsic or stoichiometric significance.

A scheme such as this for enumerating radiation-induced changes requires, if it is to be of use, that the observed yields, the G values, depend only in a minor way on the rate of energy deposition, the total energy absorbed and on the particular kind of radiation involved. Experience bears this out indeed and thus confirms the usefulness of the G value concept. This is not to say that G values are entirely independent of these additional parameters; small effects may indeed occur and, as such, can be of considerable interest. Merely, it has proved fruitful to view them as second-order effects.

The most obvious consequence of the irradiation of matter, in the gas phase at least, is the induced ionization. One may thus speak of a G value for ionization. This and another quantity, W, the average energy (in electron volts) expended per ion pair, are clearly related by

$$G_{\text{ions}} = 100/W \tag{78}$$

Table IV lists W values for a number of gases obtained under alpha particle and fast electron irradiation. The close similarity, in a given medium, of the values associated with these two radiations will be evident.

Yet another parameter often used to characterize radiation effects is the quantity M/N, the ratio of the number of events of some specified type to the number of ion pairs generated by the same dose. This parameter was especially popular at a time when it was believed that virtually all induced chemical changes arose from ionic precursors. If this were the case an M/N would be tantamount to a quantum yield and might thus assume stoichiometric values. But because so simplified an interpretation of radiation action is no longer tenable, the use of M/N can be misleading and hence the practice has been largely abandoned. Its use would seem entirely inappropriate with respect to condensed media where the precise extent of induced ionization is largely unmeasurable and probably meaningless. Note further that, when applicable, M/N and G values are entirely interchangeable, via

$$G = \frac{M}{N} \cdot \frac{100}{W} \tag{79}$$

Table IV

Average Energies per Ion-Pair for Representative Gases

Gas	$W(\alpha)$	W (Fast electron)	Gas	$W(\alpha)$	W (Fast electron)
He	43–46[a,b]	41.9[a,e]	CH_4	29.01[c]	27.0[a,e]
Ne	36.8[a]	36.2[a,e]	C_2H_6	26.47[c]	24.6[a,e]
Ar	26.22[c]	26.20[a,e]	C_3H_8	26.15[c]	23.4[d]
Kr	23.90[c]	23.9[a,e]	$c\text{-}C_3H_6$	25.82[c]	23.7[l]
Xe	21.32[c]	21.8[a,e]	$n\text{-}C_4H_{10}$	25.72[c]	22.9[d]
H_2	36.01[c]	35.9[a,e]	$i\text{-}C_4H_{10}$	26.17[c]	23.0[d]
D_2	35.74[c]	35.6[l]	$i\text{-}C_5H_{12}$	—	23.9[d]
N_2	36.39[g,e]	34.4[a,e]	$c\text{-}C_6H_{12}$	25.05[j]	22.7[h]
O_2	32.23[c]	30.6[a,e]	C_2H_2	27.35[c]	25.8[a,e]
Air	34.96[g]	33.73[e]	C_2H_4	27.87[c]	25.9[a,e]
CO	34.53[c]	32.2[d]	C_3H_6	27.01[c]	24.8[d]
NO	28.86[c]	27.5[d]	$1\text{-}C_4H_8$	26.48[c]	24.4[d]
CO_2	34.16[c]	32.7[a,e]	$2\text{-}C_4H_8$	26.18[c]	23.9[d]
NH_3	—	26.5[d]	$i\text{-}C_4H_8$	26.50[c]	24.4[d]
N_2O	34.4[k]	32.9[d]	CH_3CHO	—	26.4[d]
SO_2	—	30.4[d]	$(C_2H_5)_2O$	—	23.8[d]
BF_3	35.6[e]	—	CH_3OH	—	25.5[h]
H_2O	30.5[f]	29.9[h]	CH_3NH_2	—	23.6[h]
CCl_4	25.79[j]	24.3[h]	$1,3\text{-}C_4H_6$	—	25.0[l]
HCl	—	25.3[l]	C_6H_6	26.93[j]	23.3[l]

[a] W. P. Jesse and J. Sadauskis, *Phys. Rev.*, **97**, 1668 (1955).

[b] T. E. Bortner and G. S. Hurst, *Phys. Rev.*, **93**, 1236 (1954).

[c] C. E. Klots, *J. Chem. Phys.*, **44**, 2715 (1966); **46**, 3468 (1967).

[d] G. G. Meisels, *J. Chem. Phys.*, **41**, 55 (1964).

[e] G. N. Whyte, *Radiation Res.*, **18**, 265 (1963).

[f] R. K. Appleyard, *Nature*, **164**, 838 (1949).

[g] W. P. Jesse, *Radiation Res.*, **13**, 1 (1960).

[h] P. Alder and H. K. Bothe, *Z. Naturforsch.*, **20a**, 1700 (1965).

[i] R. M. LeBlanc and J. A. Herman, *J. Chim. Phys.*, **63**, 1055 (1966).

[j] L. M. Hunter and R. H. Johnsen, *J. Phys. Chem.*, **71**, 3228 (1967).

[k] G. S. Hurst, T. E. Bortner, and R. E. Glick, *J. Chem. Phys.*, **42**, 713 (1965). **46**, 3468 (1967).

[l] W. P. Jesse, *J. Chem. Phys.*, **38**, 2774 (1963).

Whatever advantages accrue from the use of M/N are thus not immediately apparent. Nevertheless, radiation chemists occasionally find the concept useful, not only in the gas phase, but in condensed media as well. This remarkable state of affairs derives from the great refinement in experimental techniques achieved recently and duly exemplified in later chapters of this volume. Further, as will emerge below, the a priori calculation of M/N values may be somewhat more feasible than those of G values. Subsequent work, then, will undoubtedly reflect an increasing role for the M/N unit.

B. The Relative Importance of Neutral and Ionic Precursors in Product Formation

A qualitative but useful first glimpse at the general character of energy deposition processes may be obtained from a simple scrutiny of W values, the average energy expenditure per ion pair. Examination of the data in Table IV indicates these parameters to be, typically, two to three times the ionization potential. One might inquire, then, as to what has happened to the energy in excess of that necessary for ionization. In what follows below we shall closely follow Platzman (101); variations on this sort of analysis have been used by others (102).

A charged particle, of initial energy T_0, in coming to rest will generate a number of ion pairs, N_i, and leave an additional number of atoms or molecules, N_e, in neutral excited states. Each ionization will involve an expenditure of energy equal to at least the ionization potential, the average such expenditure $\bar{E}_i$ exceeding the ionization potential by the average excitation energy of the positive ions. Similarly, generation of the neutral excited states will require some average excitation energy, $\bar{E}_e$. In every ionization one or more electrons are set loose which may, if sufficiently energetic, induce further excitations and ionizations. Eventually, however, they must be reduced to an energy below the lowest electronic excitation energy of the medium. It is convenient to single out and identify electrons having reached this point as *subexcitation* electrons and to define $\bar{\epsilon}_s$ as the average energy which such electrons have at the moment of birth. One has, especially by virtue of this last definition, the energy balance equation:

$$T_0 = N_i(\bar{E}_i + \bar{\epsilon}_s) + N_e\bar{E}_e \qquad (80)$$

Identifying T_0/N_i with the experimental quantity W, one sees that estimates of $\bar{E}_i$, $\bar{E}_e$, and $\bar{\epsilon}_s$ will permit a deduction of the interesting ratio N_e/N_i.

Threshold ionization studies suggest that $\bar{E}_i$ exceeds the ionization potential of most molecules by ~ 3 eV (103,104). No particularly good method for evaluating $\bar{E}_e$ is available. This quantity will necessarily be less than the ionization potential for the rare gas atoms; for molecular gases, because of the very large number of their superexcited states, this will not necessarily be true (105). Thus $\bar{E}_e$ of approximately the ionization potential would seem reasonable. Finally, $\bar{\epsilon}_s$ might be estimated at roughly one-third of the lowest electronic excitation energy and thus perhaps 2 eV (106,107). Taking $W = 25$ eV/ion pair and an ionization potential of 10 eV, N_e/N_i is then ~ 1.

More refined analyses along these lines have been attempted (108) but remain unconvincing, largely because of the uncertainty in estimating $\bar{E}_e$. Nevertheless, the above result is probably correct, within an order of magnitude. Occasional attempts to interpret radiation chemistry solely in terms of ionic precursors or, alternatively, neutral precursors, are surely drastic oversimplifications.

C. The Optical Approximation

In this section we shall undertake the consideration of the general problem of estimating yields with which a given electronic transition is induced. Again a qualitative but useful first approach is available.

Consideration of the high velocity cross section formulae developed earlier (viz., Eqs. (56–59)) indicates an approximate proportionality between a cross section for a given transition σ_n, and the squared dipole matrix element for the transition, $|M_n|^2$. The rate at which such transitions occur, divided by the rate of energy absorption, is then

$$M_n{}^2 \div RZ \tag{81}$$

where the Kuhn-Thomas sum rule has been used in evaluating the denominator. In terms of a yield per 100 eV deposited, this becomes

$$G_n = M_n{}^2\, 100/RZ \tag{82}$$

This simple result, due to Fano (109), represents the optical approximation to radiation chemistry in its most primitive form. In fact, it should prove useful. Thus, upon insertion in it of the quantities $(M_i')^2$ from Table II, along with the appropriate Z, one estimates G values for ionization and thus W values which are always within a factor of two of those observed.

The above procedure might be expected to underestimate yields since it neglects those transitions which are induced by slow secondary electrons. It should be possible, however, to improve upon this formula in a relatively simple fashion. Thus one may write, still in the spirit of the optical approximation,

$$G_n = [M_n{}^2/(M_i')^2]100/W \tag{83}$$

By thus normalizing yields to those of the ions, we have a semiempirical version of the optical approximation, and one which should be a considerable improvement over the Fano version.

Opportunities for a definitive experimental test of Eq. (83) are scant. In Table V we have compared the experimental G values for two sets of excited states in argon with the predictions of Eq. (83). The discrepancies are disappointing.

Table V

Test of Optical Approximation to Yields of Excited
Argon Atoms

ΔE_n(eV)[a]	$M_n{}^2$	G_n	G_n(exptl)[b]
11.6–11.8	0.285	0.23	0.924
14.0–14.3	0.265	0.22	0.456

[a] Each "transition" is actually a cluster of closely spaced transitions.

[b] G. S. Hurst, T. E. Bortner, and R. E. Glick, *J. Chem. Phys.*, **42**, 713 (1965).

D. Degradation Spectra

The optical approximation, in one guise or another, can offer a valuable first insight into the pattern of radiation yields. More powerful techniques for estimating these yields are, however, clearly called for. Thus optically forbidden transitions and nuclear collisions, for example, are not accounted for by this approximation, whereas the cross sections for such processes may in fact be substantial at reduced velocities. Further, it is abundantly evident that low-energy electrons are of the utmost importance; thus, as remarked earlier, the primary particle is responsible for, approximately, one-third of all induced ionization. A comprehensive formalism of radiation chemistry will, then, necessarily involve consideration of a very broad array of energetic particles. One such formalism, likely to be of great conceptual and practical use, will now be considered.

Consider a medium into which particles of energy T_0 are being injected. It will be assumed, for the time being, that the irradiation is uniform and constant. Then every point in the medium is subjected to a flux of charged particles of diverse energies. We may describe this flux by a distribution function $y(T)$, the number of charged particles of energy between T and $T + dT$, crossing an area of 1 cm^2, centered at the point in question, per unit time. The range of energies embraced by $y(T)$ will obviously extend from zero to T_0. The magnitude of $y(T)$ will clearly depend on the rate at which particles of T_0 are being injected. It is useful then to normalize $y(T)$ to an injection rate of one per cubic centimeter per second, whereupon $y(T)$ has the dimensions of cm/eV. The flux spectrum, normalized in this fashion, is also known as a *degradation spectrum* and, synonymously, a *slowing-down spectrum*.

The enormous utility of the degradation spectrum should be evident. Thus the rate at which transitions are induced per cc to state n in a medium containing N molecules per cc is just $\int N y(T) \sigma_n(T)\, dT$. Immediately, then,

$$G_n = \left[N \int y(T) \sigma_n(T)\, dT \right] (100/T_0) \qquad (84)$$

Similarly, one may proceed to write the 100-eV yield of any chemical change as just

$$G_p = \left[N \int dT y(T) \sum_n \sigma_n(T) \Phi_{np} \right] (100/T_0) \qquad (85)$$

where the quantum yield Φ_{np} connects excitation to state n with the possible product p.

The compactness of this result would seem to recommend it. In fact, Eq. (85) serves very nicely to delineate the several factors which enter into radiation chemistry. Thus it is $y(T)$, the degradation spectrum, which describes the radiation flux. Linking it to the medium are the cross section functions $\sigma_n(T)$; through them all electronic transitions arise. Finally, there are the quantum yields, Φ_{np}, within which are contained all subsequent chemistry (and biology).

This simple delineation lends itself to some useful observations. For example, the G values of chemical products are often found, especially in condensed media, to depend somewhat on the type and initial energy of the radiation involved. Such differences, commonly referred to as LET (linear energy transfer (110)) effects, are presumed to arise from spatial correlations of the primary excitations within the track. If this is indeed the case, such differences might be thought of as LET effects on the quantum yields. On the other hand, it is possible that at least some such effects may arise merely from variations in the yields with which various key excitations, the G_n, are induced. If so, such "pseudo LET effects" may be attributed to differences in the degradation spectrum presented to a given medium by different radiations. Thus it appears that some marked "LET effects" in the liquid aromatics are of this latter variety (111).

The central importance of the degradation spectrum concept in any consideration of radiation effects was first emphasized in the classic works of Spencer and Fano (112–113). Virtually all subsequent investigations along such lines are rooted in the principles set forth in their papers. Nevertheless, the concept has been virtually ignored by radiation chemists and biologists—possibly because of its seemingly abstract nature. We might do well, then, to emphasize that degradation spectra are amenable to direct experimental observation. As Spencer and Fano remarked, the degradation spectrum in a given medium is just the array of particles traversing a spherical cavity of unit cross section embedded in that medium. As such, the spectrum is entirely analogous to that of blackbody radiation and so might be similarly measured. Birkhoff and his coworkers have, in fact, succeeded in making such measurements (114–116). Illustrated in Fig. 17 is the measured spectrum arising from ^{64}Cu β-particles in aluminum. The high energy end of this spectrum is necessarily blurred because of the polychromatic nature of beta emission. Nevertheless, the decrease to a minimum followed by a rapid increase in $y(T)$, as one passes to low energies, is entirely characteristic.

These features may be readily understood and some insight gained as to how one might proceed to calculate the degradation spectrum associated with some radiation in a given medium. As noted above, $y(T)$ has the dimension of cm/eV; its magnitude, at any T, is just the total path length

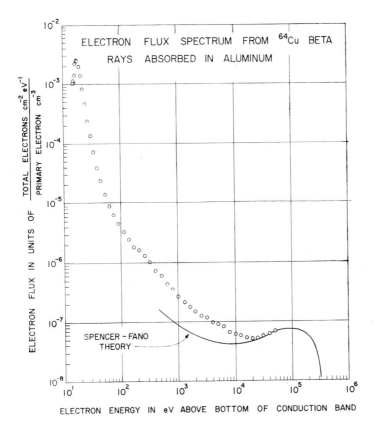

Fig. 17. Experimentally measured degradation spectrum associated with ^{64}Cu β-particles in aluminum (116).

covered by all particles present as they traverse the energy range, dT, at T. In the "continuous slowing-down approximation" this is, for one particle, just the inverse of its stopping power, i.e., $(dE/dx)^{-1}$. Hence, the initial decrease in $y(T)$ of Fig. 17. At lower energies, however, secondary electrons will be present. Let us denote by $N(T)$ the number of such electrons produced, per cc per unit time by the unit source, with energy $> T$. These electrons are eligible to (and, in the continuous slowing-down model, do) pass through dT at T, thus making a contribution to $y(T)$ of $N(T) \div (dE/dx)_T$. The rapid rise in $y(T)$ at low energies, then, just reflects the concurrent rise in $N(T)$.

Construction of the degradation spectrum appropriate to a given radiation and medium requires, thus, at least two ingredients—stopping powers and cross sections for secondary electron ejection. The former may usually be calculated or obtained, with sufficient accuracy, from tabulations and thus will normally not pose any considerable obstacle. The secondary electron ejection cross sections, however, prove to be more difficult to come by. When the incident particle and ejected electron are both of sufficient velocity, one may use the Rutherford or Möller formulas, Eqs. (27) and (61), introduced earlier. Their range of validity, however, effectively confines one to calculation of $y(T)$ at large energies. The low energy region—seen from Fig. 17 to be of great numerical importance— would thus appear inaccessible.

With this limitation, extensive tabulations of slowing-down spectra in a wide variety of media under electron irradiation have been prepared (117). A similar calculation (118) of the spectrum associated with a 5-MeV alpha particle in methane is illustrated in Fig. 18. In this instance, however, an approximate extension to low velocities was possible since the total curve must be such as to reproduce, via Eq. (84), W for the gas. The lower end of the spectrum can thus be adjusted until this requirement is met. Platzman has, in like manner, successfully computed the degradation spectrum associated with 20-keV electrons in helium (101).

Spectra computed in this manner, however, must necessarily be only approximate since Auger and autoionization mechanisms as secondary electron sources are automatically disregarded. Thus Fig. 18 also contains a calculation in which Auger electrons, as well as the K-shell corrections suggested by the results of Fig. 16, have been included. The discontinuity at the Auger electron energy will be noted but the corrected curve remains otherwise remarkably close to that of the free electron model.

Once an approximate degradation spectrum has been mapped, a number of useful enunciations become possible. Thus from a consideration of the function $Q_{ion}(T)y(T)$, where $Q_{ion}(T)$ is the total ionization cross section, one may assess the relative importance of various energy intervals in

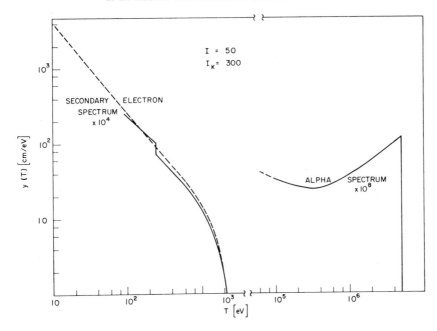

Fig. 18. Calculated degradation spectrum associated with a 5-MeV alpha particle in methane at STP (118).

effecting ionization. We obtain in this manner from Fig. 18, for example, that 44% of the ionization in CH_4 is induced directly by the alpha particle.

Seemingly, one may now also proceed to enter a computed degradation spectrum into Eq. (84) and so obtain G values. In practice, however, this would be quite unrealistic. Equation (84) requires, for reliable results, that the spectrum $y(T)$ be exceedingly accurate; this is not likely to be the case in view of the probable oversimplification introduced in its calculation and the lack of data for a more accurate construction. Fortunately, this difficulty can be largely obviated by the simple device of writing Eq. (84) in a normalized form:

$$G_n = \frac{\int y(T)\sigma_n(T)\,dT}{\int y(T)Q_{\text{ion}}(T)\,dT} \cdot \frac{100}{W} \tag{86}$$

In this formulation, $y(T)$ is used essentially as a device for calculating an average ratio $\overline{\sigma_n/Q_{\text{ion}}}$. To the extent that $\sigma_n(T)/Q_{\text{ion}}(T)$ is independent of T, the computed ratio will be insensitive to the details of $y(T)$ and thus to errors in it. This point is currently under scrutiny (96); we may anticipate, however, that Eq. (86) will provide a most useful formalism within which the yields of specialized electronic transitions may be discussed. Thus the formal origin of two familiar simplifications may be pinpointed. It is, for example, common practice to presume that the excitation processes, produced with 70-eV electrons and manifested in a mass spectrometer, are characteristic of gas phase systems under high energy irradiation. This is equivalent to replacing $y(T)$ in Eq. (86) with a delta function at 70 eV. Note also, that if one replaces the average $\overline{\sigma_n/Q_{ion}}$ in Eq. (86) with its high velocity limit, one recovers the optical approximation. Presumably the discrepancies in Table V are attributable to this extreme simplification.

E. Calculations on an Electron Generation Basis

Two recent and extremely well-informed studies illustrate still another approach to the yields of radiation-induced excitations. These papers, both of which are concerned with atmospheric luminescence observations, proceed on a generation-by-generation basis (119,120). One calculates the number of excitations of a given kind induced by a primary particle as it dissipates its entire energy. One then turns to the secondary electrons, ejected by the primary particle in the course of its slowing down, and calculates the relevant yield from each. Similarly, the tertiary and higher generations of electrons are each considered in turn. The total yield is then just a sum over those of the several generations.

This procedure may be placed, easily enough, on a formal basis. A primary particle, originally of energy T_0, will generate $N\sigma_n(T)(dx/dT)\,dT$ excitations to state n while losing the energy dT. The G value for such excitations, arising solely from this primary as it dissipates its entire energy, is then

$$G_{n,1} = \left[N \int^{T_0} \sigma_n(T)\,(dx/dT)\,dT \right](100/T_0) \qquad (87)$$

Letting $Q(T, \epsilon)$ be the cross section for producing a secondary electron of energy ϵ, the 100-eV yield of such secondaries is, similarly,

$$S(T_0, \epsilon) = \left[N \int^{T_0} Q(T, \epsilon)\,(dx/dT)\,dT \right](100/T_0) \qquad (88)$$

G_n arising from this generation is then just

$$G_{n,2} = \int S(T_0, \epsilon)G_n(\epsilon)\,d\epsilon \qquad (89)$$

An extension to higher generations is straightforward. In this connection an interesting recursion formula has been demonstrated (120).

It will be perceived that this computational procedure requires very much the same sort of input information that a calculation by way of the degradation spectrum would require. Thus in the latter method one utilizes the cross sections for secondary electron emission, $Q(T, \epsilon)$, at the outset to establish the degradation spectrum. Indeed, note that, if there were no such secondary electrons, Eq. (87) would be identical with Eq. (84).

While, therefore, an entirely valid approach to aggregate calculations of radiation-induced transitions, this generation-by-generation form of book-keeping does seem to suffer from certain disadvantages. Thus it would appear not to be easily amenable to a semiempirical formulation, as in Eq. (86), nor to approximate measures analogous to the use of trial degradation spectra. Since it is only rarely that one is apt to be interested in an accounting by generations of radiation yields, calculations on this basis hence seem to offer no advantage over the procedure described earlier.

Other computational procedures have been devised for enumerating the energy deposition processes in irradiated media and have been reviewed by Platzman (101). They, too, tend to suffer from the above disadvantages, in contrast with methods based on the degradation spectrum. We suggest then that a most fruitful area for further investigation will involve the prescription of nominally accurate flux spectra.

F. Quasi-Empirical Cross Section Formulations

Two ingredients, then—cross sections as a function of particle energy and a degradation spectrum—are necessary for an estimation of the yield with which any given transition occurs in an irradiated medium. The latter requirement is likely to cause little difficulty, as discussed above; some general comments might then be made concerning the need for cross section data.

It is more than likely that explicit (much less, reliable) experimental data will not exist for a particular problem at hand. In this context it is thus unfortunate that, as noted above, the optical approximation should seem to be of only qualitative reliability. Its use would require scarcely more than the optical oscillator strength associated with the transition in question. Nevertheless it will undoubtedly find considerable use for purposes of estimation. Further, as additional opportunities arise for a comparison of its predictions with experimental data, it may be possible to work out empirical rules-of-thumb. Tentatively one expects the optical approximation to underestimate low-lying transitions and to overestimate the more energetic transitions as, for example, K-shell ionizations. It will, of course, completely underestimate the yield of optically forbidden excitations.

Alternatively one may wish to consider Eq. (84) and thus the cross section function throughout the entire energy spectrum. Here again we may anticipate the usefulness of semiempirical investigations. Thus a number of efforts are already at hand in which experimental data are being reduced to functional form, often with convenient analytic properties (120–123). Using them and approximate degradation spectra it should be possible to build up quite respectable estimates of yields in radiolyses. Note especially that inelastic scattering studies at a single electron beam energy permit one to deduce, in the framework of the Born approximation, generalized oscillator strength functions and hence the total cross section at all energies. Lassettre has used his data to illustrate this procedure (124).

G. Finite Medium Effects

We have spoken of the flux or degradation spectrum as a property of a given medium, implicitly independent of position within the medium. Further, we have indicated how this spectrum might be calculated solely in terms of the electronic properties of the medium. Both these points may require some modification. The flux at a point in a medium at a distance d from the nearest boundary will contain no contribution from particles which have crossed into the medium from the outside, only if d is greater than the range of any such intruders. The degradation spectrum appropriate to a given irradiated medium will thus, strictly, be independent of surroundings only when the linear dimensions are large compared with any relevant ranges (125).

This is often not the case. Radiation chemistry investigations in gases are quite typically conducted in a variety of glass or metal containers of diverse shapes and materials. It is thus remarkable that much sense at all can be gleaned from such experiments. Indeed, product yields from radiolysis are at least as reproducible, from one laboratory to another, as they are in photochemistry. One might very well expect the prevailing degradation flux and thus the resultant product yields to depend upon the dimensions and nature of the vessel. Only when the "walls" are composed of the same atomic number material as the medium can one be assured that such perturbations will not occur (126). Apparently, then, absolute and relative chemical product yields are nominally insensitive to whatever perturbations do occur.

H. Subexcitation Electrons

We have had occasion above to speak of subexcitation electrons. These are electrons whose energy has dropped to below the level of the lowest electronic transitions from the ground state. It is a status which awaits all components of the degradation spectrum; further it would seem to be an especially uninteresting status since its members are apparently no longer capable of arousing interesting excitations. Nevertheless, as we shall see, there are a number of aspects of radiation chemistry with respect to which such entities can indeed be of interest.

One mode of energy deposition which has received only scant attention in this chapter is that of negative ion formation by electron attachment. A most thorough discussion of this topic has been given elsewhere (127) so that any detailed exposition here would be, at best, superfluous. Nevertheless, certain general aspects of this topic may be discussed usefully at this time.

The water molecule is found to attach electrons according to

$$e + H_2O \rightarrow H^- + OH \tag{90}$$

The process occurs with electrons of between 5.6- and ~8-eV energy, the maximum cross section being ~7×10^{-18} cm^2 at 6.4 eV. One might then ask what fraction of all electrons set loose in a given medium containing some water vapor are actually consumed in this process. While Eq. (85) gives, of course, a formalism from which an answer might be obtained, its use here would be entirely impractical; uncertainties in the prevailing degradation spectrum make such a course much too unreliable. A more suitable approach is required.

Electrons which are born with energy less than 5.6 eV are clearly ineligible for this process. This observation immediately points up an alternative method of estimating the yields of low-energy processes such as the above attachment. Let us consider, analogous to the subexcitation spectrum of before, the spectrum of energies with which electrons are born as they drop below ~8 eV. We may anticipate from the earlier formulas for secondary electrons that this spectrum will be strongly skew toward low energies. Various quasi-theoretical and experimental representations (106,107) of subexcitation spectra have been given, all properly reflecting this anticipated skewness.* Since only that fraction of the spectrum lying above 5.6 eV can undergo attachment to water vapor, an upper limit to the yield of this process may be immediately assigned. A more complete delineation of this yield will require further consideration of the energy-loss mechanisms which may compete with attachment in the 5.6–8 eV range. Nevertheless one may see that an electron attachment process will occur to a significant extent the more nearly the range of energies, over which it may proceed, approaches the thermal energy.

This simple result had been suggested some time ago (128). It is entirely confirmed by experiment (129,130). In a field, such as radiation chemistry,

---

* Platzman (106) has presented an experimental delineation of subexcitation electron spectra. His technique purports to involve the titration of all subexcitation electrons in specified energy intervals. Since this titration cannot be complete, the results cannot be strictly correct. Nevertheless, they remain suggestive.

which scarcely abounds with generalizations, this principle of Magee and Burton has proved both durable and useful. It is, incidentally, the rationalization behind the general neglect of attachment processes in this chapter. Tending to occur at thermal energies, such processes may most usefully be considered as chemical events.

I. Energy Partition in Mixtures

We have, in this chapter, hitherto neglected all discussion of radiation dosimetry. This has been done deliberately because estimates of absorbed dose can usually be obtained, empirically or otherwise, quite apart from any detailed consideration of energy deposition mechanisms. Again, the Bethe stopping power formulas have been pointedly underplayed, being mentioned in an almost parenthetical manner. The course was followed largely by way of reaction against an overemphasis in the past on stopping power formulas; they really provide very little in the way of insight into the detailed energy-absorption processes.

Nevertheless there exists, or has existed until recently, a curious gap in dosimetry theory which can be profitably discussed in terms of the preceding formulations. While one can usually estimate the total energy absorbed in an irradiated medium, it is often necessary to know what fraction of this dose was absorbed by each distinct chemical component of this (presumed homogeneous) medium. Thus energy transfer, sensitization, charge transfer, quenching, scavenging and so forth are often postulated to occur in irradiated mixtures. A quantitative assessment of these mechanisms necessarily requires a knowledge of how the yield of a given electronic transition G_n^0, as generated by the radiation flux in a pure component, is altered upon dilution with other chemical species. In photochemistry one deals with this question, roughly, through a consideration of relative extinction coefficients. We require, then, an analogous method, appropriate to radiation chemistry, for estimating this alteration.

For simplicity the subsequent discussion will be confined to two-component solutions; the extension to multicomponent media is straightforward. Let us write the desired relation in any mixture as

$$G_{ni} = G_{ni}^0[Z_i'N_i/(Z_i'N_i + Z_j'N_j)] \tag{91}$$

where N_i, N_j are the respective concentrations. The ratio Z_i'/Z_j', defined in this manner, is then seen to be a measure of the relative cross section of the components i, j to absorb energy from the prevailing radiation flux. One may then ask if this ratio Z_i'/Z_j' is (1) independent of concentration, (2) independent of the particular state n, and (3) experimentally or theoretically accessible.

An examination of W values, and thus of ion yields, in binary gas mixtures under alpha-particle irradiation, has suggested an affirmative answer to these questions (131). Ion yields from a great number of binary mixtures are in accord with the form of Eq. (91). Through a careful examination of such yields, the weighting coefficients presented in Table VI were obtained (132). With them the initial partitioning of a given dose in any solution containing these several constituents may then presumably be calculated.

Table VI
"Effective Atomic Numbers" for Energy-Partition Calculations[a]

Gas	Z_i'	Gas	Z_i'
He	2.24 ± 0.15	Xe	14.9 ± 0.4
H_2	3.08 ± 0.02	C_2H_4	15.45 ± 0.15
Ne	4.7 ± 0.2	C_2H_6	16.5 ± 0.2
Ar	6.6 ± 0.15	$c\text{-}C_3H_6$	20.5 ± 0.3
N_2	7.02 ± 0.09	C_3H_6	22.75 ± 0.15
O_2	6.24 ± 0.06	C_3H_8	22.8 ± 0.2
NO	7.91 ± 0.07	C_4H_{10}	28.9 ± 0.4
CO	8.39 ± 0.10	$C_4H_8\text{-}1$	29.4 ± 0.1
Kr	9.8 ± 0.15	$i\text{-}C_4H_{10}$	29.5 ± 0.2
CH_4	10.0	$i\text{-}C_4H_8$	29.8 ± 0.2
C_2H_2	14.7 ± 0.3	$cis\text{-}C_4H_8$	30.1 ± 0.2
CO_2	10.9 ± 0.1		

[a] C. E. Klots, *J. Chem. Phys.*, **44**, 2715 (1966); *ibid* **46**, 3468 (1967).

These coefficients, Z_i', might be referred to as "effective atomic numbers," having hitherto been approximated by the atomic number of a given constituent. This approximation was entirely plausible, having been suggested by the near-proportionality between stopping power and atomic number. Interestingly enough, however, it now appears that macroscopic stopping powers are very nearly irrelevant to the present question. A more detailed inquiry suggests instead that a ratio of coefficients Z_i'/Z_j' might better be estimated by

$$Z_i'/Z_j' = (W_i/W_j) \frac{\int y(T)Q_i(T)\,dT}{\int y(T)Q_j(T)\,dT} \tag{92}$$

where $Q_i(T)$, $Q_j(T)$ are total ionization cross sections. The integration is understood to extend over the degradation spectrum actually present in the particular mixture in question. The discussion of a preceding section will suggest, however, that a calculated ratio Z_i'/Z_j' will tend to be insensitive to the details of $y(T)$ and thus might be obtained using only approximate degradation spectra. The apparently real existence of Z_i' parameters, as obtained from the ion-yield measurements, would seem to bear this out. Indeed it has been found possible to reproduce, with reasonable success, many of these coefficients through the use of Eq. (92)—despite the necessary coarseness of the computation.

By way of further illustration we might consider, in closing, a perennial favorite, cyclohexane–benzene mixtures. We estimate, with Eq. (92) and the available total ionization data, a $Z'(C_6H_{12})/Z'(C_6H_6) = 1.0 \pm 0.1$, in accord with experimental ion yields (133) and scarcely distinguishable from the ratio of atomic numbers.

A number of papers, especially pertinent to this chapter, have appeared during the course of its publication. Thus a recent photoionization study of small hydrocarbons (134) suggests that the extent of autoionization in such molecules is less than previously believed. Similarly, an investigation of the electron-induced ionization of helium (135) reemphasizes that simple threshold laws must be regarded with scepticism. Vriens (136) has shown that an expansion technique of Lassettre's (137) for generalized oscillator strengths is of broad applicability. Green (138) has greatly extended his systematic estimates of inelastic cross sections and of transition yields, while Brocklehurst (139) has attempted to unravel the luminescence from irradiated nitrogen to obtain the yields of several primary transitions. Moustafa Moussa has described a careful study of excitation cross sections in helium under electron impact (140). Of especial interest is the accord at high energies with Born-Bethe theory.

REFERENCES

1. J. Franck, *Trans. Faraday Soc.*, **21**, 536 (1926).
2. J. F. Lowry, D. H. Tomboulian, and D. L. Ederer, *Phys. Rev.*, **137**, A1054 (1965).
3. R. E. Huffman, Y. Tanaka, and J. C. Larrabee, *J. Chem. Phys.*, **39**, 902 (1963).
4. V. H. Dibeler, R. M. Reese, and M. Krauss, *J. Chem. Phys.*, **42**, 2045 (1965).
5. R. S. Berry, *J. Chem. Phys.*, **45**, 1228 (1966).
6. K. Watanabe and A. S. Jursa, *J. Chem. Phys.*, **41**, 1650 (1964).
7. R. Schoen, *J. Chem. Phys.*, **37**, 2032 (1962).
8. N. Wainfan, W. Walker, and G. L. Weissler, *J. Appl. Phys.*, **24**, 1318 (1963).
9. J. A. R. Samson, *J. Opt. Soc. Am.*, **54**, 6 (1964).
10. J. Persons, *J. Chem. Phys.*, **43**, 2553 (1965).
11. C. E. Klots, *J. Chem. Phys.*, **46**, 1197 (1967).
12. W. P. Jessee, *J. Chem. Phys.*, **38**, 2774 (1963).

13. S. Meyerson, H. M. Grubb, and R. W. Vander Haar, *J. Chem. Phys.*, **39**, 1445 (1963).
14. F. Fiquet-Fayard and P. M. Guyon, *Mol. Phys.*, **11**, 17 (1966).
15. R. L. Platzman, *Radiation Res.*, **17**, 419 (1962).
16. R. L. Platzman, *J. Phys. Radium*, **21**, 853 (1960).
17. W. P. Jesse and R. L. Platzman, *Nature*, **195**, 790 (1962).
18. U. Fano, *Science*, **153**, 522 (1966).
19. J. D. Morrison, *J. Appl. Phys.*, **28**, 1409 (1957).
20. J. D. Morrison, *Energy Transfer in Gases*, R. Stoop, Ed., Interscience, New York, 1962, pp. 397ff.
21. W. M. Hickam, R. E. Fox, and T. Kjeldaas, *Phys. Rev.*, **96**, 63 (1954).
22. R. E. Fox, *J. Chem. Phys.*, **35**, 1379 (1961).
23. K. Watanabe, *J. Chem. Phys.*, **22**, 1564 (1954).
24. R. M. Reese and H. M. Rosenstock, *J. Chem. Phys.*, **44**, 2007 (1966).
25. R. E. Fox, *J. Chem. Phys.*, **33**, 200 (1960).
26. F. H. Dorman and J. D. Morrison, *J. Chem. Phys.*, **35**, 575 (1961).
27. R. H. McFarland, *Phys. Rev. Letters*, **10**, 397 (1963).
28. G. J. Schultz and R. E. Fox, *Phys. Rev.*, **106**, 1179 (1957).
29. J. D. Morrison, *Advances in Mass Spectrometry*, Vol. 2, A. Elliot, Ed., Pergamon Press, New York, 1963, p. 479.
30. J. F. Burns, *Bull. Am. Phys. Soc.*, **9**, 353 (1964).
31. D. P. Stevenson, *J. Am. Chem. Soc.*, **82**, 5961 (1960).
32. E. Baranger and E. Gerjuoy, *Phys. Rev.*, **106**, 1182 (1957); *Proc. Phys. Soc. (London)*, **A72**, 326 (1958).
33. J. W. McGowan et al., *Phys. Rev. Letters*, **13**, 620 (1964).
34. L. I. Schiff, *Quantum Mechanics*, McGraw-Hill, New York, 1955.
35. H. Bethe, *Ann. Physik*, **5**, 325 (1930).
36. E. J. Williams, *Rev. Mod. Phys.*, **17**, 217 (1945).
37. W. F. Miller and R. L. Platzman, *Proc. Phys. Soc. (London)*, **A70**, 299 (1957).
38. R. L. Platzman, *J. Chem. Phys.*, **38**, 2775 (1963).
39. U. Fano, *Ann. Rev. Nucl. Sci.*, **13**, 1 (1963); *Natl. Acad. Sci.-Natl. Res. Council Publ.*, **1133** (1964).
40. W. Whaling, in *Handbuch der Physik*, Vol. 34, S. Flügge, Ed., Springer, Berlin, 1958, p. 193.
41. A. Nelms, "Energy Loss and Range of Electrons and Protons," NBS Circular 577 (1956).
42. E. C. Pollard, *Advan. Biol. Med. Phys.*, **3**, 153 (1953).
43. A. Ore and A. Larsen, *Radiation Res.*, **21**, 331 (1964).
44. G. O. Langstroth, *Proc. Roy. Soc. (London)*, **A146**, 166 (1934); **A150**, 371 (1935).
45. A. Skebarle and E. N. Lassettre, *J. Chem. Phys.*, **42**, 395 (1965).
46. G. H. Dunn, *J. Chem. Phys.*, **44**, 2592 (1966).
47. L. J. Kieffer and G. H. Dunn, *Rev. Mod. Phys.*, **38**, 1 (1966).
48. B. L. Schram et al., *Physica*, **31**, 94 (1965); **32**, 734 (1966).
49. B. L. Schram et al., *J. Chem. Phys.*, **44**, 49 (1966).
50. J. Schutten et al., *J. Chem. Phys.*, **44**, 3924 (1966).
51. B. L. Schram, A. J. H. Boerboom, and J. Kistemaker, *Physica*, **32**, 185 (1966); B. L. Schram et al., *Physica*, **32**, 197 (1966).
52. F. Fiquet-Fayard, F. Muller, and J. P. Ziesel, *Proc. Intern. Conf. Phys. Electron. At. Collisions, 4th Quebec* (1965), p. 413.
53. F. J. de Heer et al., *Physica*, **32**, 1766, 1793 (1966).

54. D. W. Martin et al., *Phys. Rev.*, **136**, A385 (1964).
55. S. Wexler, *J. Chem. Phys.*, **41**, 1714 (1964).
56. R. H. Schuler and F. A. Stuber, *J. Chem. Phys.*, **40**, 2035 (1964).
57. P. S. Rudolph and C. E. Melton, *J. Chem. Phys.*, **45**, 2227 (1966).
58. J. Durup and R. L. Platzman, *Disc. Faraday Soc.*, **31**, 156 (1961).
59. E. H. S. Burhop, *Proc. Cambridge Phil. Soc.*, **36**, 43 (1940).
60. J. W. Motz and R. C. Placious, *Phys. Rev.*, **136**, A662 (1964).
61. E. Merzbach and H. W. Lewis, in *Handbuch der Physik*, Vol. 34, S. Flügge, Ed., Springer, Berlin, 1958, p. 166.
62. J. E. Monahanan and H. E. Stanton, *J. Chem. Phys.*, **37**, 2654 (1962).
63. P. Kebarle and E. W. Godbole, *J. Chem. Phys.*, **36**, 302 (1962).
64. S. Wexler, *J. Chem. Phys.*, **41**, 2781 (1964).
65. R. M. St. John, F. L. Miller, and C. C. Lin, *Phys. Rev.*, **134**, A888 (1964).
66. H. S. W. Massey, in *Handbuch der Physik*, Vol. 36, S. Flügge, Ed., Springer, Berlin, 1958, p. 307; J. D. Craggs and H. S. W. Massey, *ibid.*, Vol. 37, p. 314.
67. S. Hayakama and H. Nishimura, *J. Geomagnet. Geolec.*, **16**, 72 (1964).
68. W. F. Sheridan, O. Oldenberg, and N. P. Carleton, *Intern. Conf. Phys. Electron. At. Collisions*, Benjamin, New York, 1961.
69. A. H. Gabriel and D. W. O. Heddle, *Proc. Roy. Soc. (London)*, **A258**, 124 (1960).
70. H. Boersch and H. J. Reich, *Optik*, **22**, 289 (1965).
71. J. Olmsted, A. S. Newton, and K. Street, *J. Chem. Phys.*, **42**, 2321 (1965).
72. H. F. Winters, *J. Chem. Phys.*, **43**, 926 (1965).
73. V. Čermák, *J. Chem. Phys.*, **44**, 1318 (1966); **44**, 3774 (1966).
74. P. M. Becker and F. W. Lampe, *J. Chem. Phys.*, **42**, 3857 (1965).
75. R. E. Huffman and D. H. Katayama, *J. Chem. Phys.*, **45**, 138 (1966).
76. E. N. Lassettre, *Radiation Res. Suppl.*, **1**, 530 (1959); E. N. Lassettre et al., *J. Chem. Phys.*, **40**, 1208ff. (1964).
77. H. Boersch, J. Geiger, and H. Hellwig, *Phys. Letters*, **3**, 64 (1962).
78. J. Geiger, *Z. Physik*, **175**, 530 (1963); **177**, 138 (1964); **181**, 413 (1964).
79. J. Geiger and W. Stickel, *J. Chem. Phys.*, **43**, 4535 (1965).
80. C. E. Kuyatt and J. A. Simpson, *Atomic Collision Processes*, North-Holland Publishing Co., Amsterdam, 1964, p. 191.
81. H. G. M. Heideman, C. E. Kuyatt, and G. E. Chamberlin, *J. Chem. Phys.*, **44**, 440 (1966).
82. J. A. Simpson, C. E. Kuyatt, and S. R. Mielzarek, *J. Chem. Phys.*, **44**, 4403 (1966).
83. J. A. Simpson and S. R. Mielzarek, *J. Chem. Phys.*, **39**, 1606 (1963).
84. G. E. Chamberlin and H. G. M. Heideman, *Phys. Rev. Letters*, **15**, 337 (1965).
85. A. Kupperman and L. M. Raff, *J. Chem. Phys.*, **37**, 2497 (1962).
86. J. P. Doering, *J. Chem. Phys.*, **45**, 1065 (1966).
87. G. J. Schulz and J. W. Philbrick, *Phys. Rev. Letters*, **13**, 477 (1964).
88. R. D. Birkhoff, in *Physical Processes in Radiation Biology*, L. Augenstein, R. Mason, and B. Rosenberg, Eds., Academic Press, New York, 1964, Chap. 10.
89. U. Fano, *Phys. Rev.*, **118**, 451 (1960).
90. R. L. Platzman, *Intern. Cong. Radiation Res., 3rd, Cortina* (1966).
91. N. Swanson and C. J. Powell, *J. Chem. Phys.*, **39**, 630 (1963).
92. A. M. Rauth and J. A. Simpson, *Radiation Res.*, **22**, 643 (1964).
93. M. E. Rudd and T. Jorgensen, *Phys. Rev.*, **131**, 666 (1963).
94. C. E. Kuyatt and T. Jorgensen, *Phys. Rev.*, **130**, 1444 (1963).
95. M. E. Rudd, C. A. Sautter, and C. L. Bailey, *Phys. Rev.*, **151**, 20 (1966).

96. C. E. Klots and J. E. Turner, Abstracts, 15th Radiation Research Society Meeting, Puerto Rico, 1967; C. E. Klots, *Advan. Chem. Ser.*, in press.
97. R. B. Barker and H. W. Berry, *Phys. Rev.*, **151**, 14 (1966).
98. M. E. Rudd, *Phys. Rev. Letters*, **15**, 580 (1965).
99. R. Allison, J. Burns, and A. J. Tuzzolino, *J. Opt. Soc. Am.*, **54**, 747 (1964).
100. J. A. R. Samson, *J. Opt. Soc. Am.*, **54**, 6 (1964).
101. R. L. Platzman, *Intern. J. Appl. Radiation Isotopes*, **10**, 116 (1961).
102. G. S. Hurst, T. E. Bortner, and R. E. Glick, *J. Chem. Phys.*, **42**, 713 (1965).
103. W. A. Chupka and M. Kaminsky, *J. Chem. Phys.*, **35**, 1991 (1961).
104. D. P. Stevenson, *Radiation Res.*, **10**, 610 (1959).
105. F. Fiquet-Fayard, *J. Chim. Phys.*, **60**, 651 (1963).
106. R. L. Platzman, *Radiation Res.*, **2**, 1 (1955).
107. S. G. ElKomoss and J. L. Magee, *J. Chem. Phys.*, **36**, 256 (1962).
108. P. Alder and H. K. Bothe, *Z. Naturforsch.*, **20a**, 1700 (1965).
109. U. Fano, *Phys. Rev.*, **70**, 44 (1946).
110. R. E. Zirkle et al., *J. Cellular Comp. Physiol.*, *Suppl. 1*, **39**, 75 (1952).
111. R. H. Schuler, *Trans. Faraday Soc.*, **61**, 100 (1965).
112. U. Fano, *Phys. Rev.*, **92**, 328 (1953).
113. L. V. Spencer and U. Fano, *Phys. Rev.*, **93**, 1172 (1954).
114. R. D. Birkhoff et al., *Health Phys.*, **1**, 27 (1958).
115. W. J. McConnell et al., *Phys. Rev.*, **138**, A1377 (1965).
116. R. D. Birkhoff et al., *Radiation Res.* (in press).
117. R. J. McGinnies, *Natl. Bur. Std. (U.S.)*, Circ. 597 (1959).
118. C. E. Klots, *J. Chem. Phys.*, **46**, 3468 (1967).
119. S. Frankenthal, O. P. Manley, and Y. M. Treve, *J. Chem. Phys.*, **44**, 257 (1966).
120. A. E. S. Green and C. A. Barth, *J. Geophys. Res.*, **70**, 1083 (1965).
121. A. E. S. Green, *AIAA J.*, **4**, 769 (1966).
122. T. R. Carson, *J. Quant. Spectr. Radiative Transfer*, **6**, 563 (1966).
123. L. Vriens, *Physica*, **31**, 385, 1081 (1965).
124. S. M. Silverman and E. N. Lassettre, *J. Chem. Phys.*, **40**, 2927 (1964); **44**, 2219 (1966).
125. L. V. Spencer and F. H. Attix, *Radiation Res.*, **3**, 239 (1955).
126. U. Fano, *Radiation Res.*, **1**, 237 (1954).
127. F. Fiquet-Fayard, in *Actions Chimiques et Biologiques des Radiation*, Vol. 18, H. Haissinsky, Ed., Masson, Paris, 1965, p. 63.
128. J. L. Magee and M. Burton, *J. Am. Chem. Soc.*, **73**, 523 (1951).
129. J. P. Guarino, M. R. Ronayne, and W. H. Hamill, *Radiation Res.*, **17**, 379 (1962).
130. L. A. Rajenbach, *J. Am. Chem. Soc.*, **88**, 4275 (1966).
131. C. E. Klots, *J. Chem. Phys.*, **39**, 1571 (1963).
132. C. E. Klots, *J. Chem. Phys.*, **44**, 2715 (1966).
133. L. M. Hunter and R. H. Johnsen, *J. Phys. Chem.*, **71**, 3228 (1967).
134. W. A. Chupka and J. Berkowitz, *J. Chem. Phys.*, **47**, 2921 (1967).
135. C. E. Brion and G. E. Thomas, *Phys. Rev. Letters*, **20**, 241 (1968).
136. L. Vriens, *Phys. Rev.*, **160**, 100 (1967).
137. E. N. Lassettre, *J. Chem. Phys.*, **43**, 4479 (1965).
138. A. E. S. Green and S. K. Dutta, *J. Geophys. Res.*, **72**, 3933 (1967); and following papers.
139. B. Brocklehurst and F. A. Downing, *J. Chem. Phys.*, **46**, 2976 (1967).
140. H. R. Moustafa Moussa, Thesis, Leiden (1967).

Chapter 2

Ionic Fragmentation Processes

Marvin L. Vestal

Scientific Research Instruments Corporation, Baltimore, Maryland

I. INTRODUCTION

In this chapter we shall be concerned with the fundamental chemistry and physics of the unimolecular reactions of polyatomic ions. The discussion will be limited to singly charged positive ions although the theoretical concepts should also apply to negative ions and neutral molecules.

The subject of negative ions has recently been reviewed by Fiquet-Fayard (1) and by Melton (2).

The subject matter of the present discussion was reviewed in detail by Field and Franklin (3) in their excellent monograph published in 1957. Since that time substantial advances have been made toward a fundamental understanding of ionization and dissociation processes. In more recent years the theory of ionic fragmentation processes has been reviewed by Rosenstock and Krauss (4,5), and by Wahrhaftig (6).

In this chapter we attempt to describe the experimental and theoretical work which has led to recent advances in the theory of ionic fragmentation processes, to present a reasonably coherent picture of the present understanding of these processes, and to suggest fruitful lines for further research.

II. THE QUASI-EQUILIBRIUM THEORY

While other approaches to a theory of the unimolecular fragmentation of polyatomic ions have recently been proposed (7,8), only the quasi-equilibrium theory (QET) developed by Rosenstock, Wallenstein, Wahrhaftig, and Eyring (9) has been extensively applied. Some of the early applications of the theory revealed a number of discrepancies between its predictions and the results of experimental measurements (10–18). More recently those discrepancies have been shown to have been due to the use of an invalid mathematical approximation in the original formulation of the theory (19,20). The reasoning which led to the development of the QET has recently been reviewed in detail by Wahrhaftig (6) and will only be summarized here.

In principle, the interaction of an electron or photon with a molecule and the resulting excitation, ionization, and dissociation processes are understood. The general prescriptions for calculating the pertinent transition probabilities are contained within the framework of quantum mechanics as discussed in Chapter 1.

For a monatomic or diatomic molecule which has a relatively small number of well-separated electronic states, an accurate expression can be written for the transition probability from the ground state of the molecule to each state of the ion. The variation in these transition probabilities as functions of the energy of the ionizing particle can be calculated with sufficient precision in the simplest cases to indicate that the problem is understood. For a diatomic molecule the mass spectrum follows from the shapes of the several potential curves for the states populated by the ionization process (21). In addition, transitions may occur to specific excited or "superexcited" states of the *molecule* (22). A molecule excited to one of these states may dissociate to form a neutral pair or an ion pair, or may autoionize to produce a molecule-ion, again in a specific state.

For even rather small polyatomic molecules the situation is vastly more complicated. The molecule-ion has a great many more low-lying states, in general, than do diatomic molecules and no reasonably precise way is available for calculating the configuration of any of them. Even if the multidimensional potential surfaces were known exactly, the task of calculating a representative set of trajectories on such a surface (for N atoms $3N - 6$ internal coordinates are required) is beyond the capacity of even the most sophisticated of computers.

The quasi-equilibrium theory is the result of an attempt to develop an approximate theory which is consistent with quantum mechanics but which is based on a greatly simplified model. The theory was developed in an attempt to correlate in an approximate way all the different experimental observations on the mass spectra of polyatomic molecules. The idea that a statistical treatment should be valid is supported by the following four quantitative observations:

1. All outer electrons are equally likely to be involved in the ionization process independent of the structure of the molecule.

2. The fragmentation pattern is strongly dependent on structure.

3. Most of the fragment ions are formed with very little kinetic energy.

4. The relative probability for breaking a particular type of bond (e.g., a C—H bond) does not correlate with the relative number of bonds of that type in the molecule.

These observations indicate that a model based on the diatomic molecule is not valid and neither is a model based on the removal of electrons from particular bonds. Rather, these observations imply that dissociation generally does not immediately follow ionization but that the dissociation process is slow enough to permit the transfer of energy from the initial electronic and vibrational excitation caused by "vertical" ionization into the degrees of freedom involved in the observed dissociation. The quasi-equilibrium theory uses a model at the opposite extreme from the diatomic model; namely, that the dissociation is so slow that energy in the reactant ion, irrespective of its initial location, has time to randomize over all degrees of freedom of the ion.

A. The Rate Equation

In the quasi-equilibrium theory the excited ion is treated as an isolated system in a state of internal equilibrium. The quasi-equilibrium rate expression was derived by Rosenstock, Wallenstein, Wahrhaftig, and Eyring (9) by applying the absolute rate theory of Eyring (23) to such a system. The details of the derivation are given in the original papers and have recently been reviewed by Rosenstock and Krauss (4,5) and by Wahrhaftig (6). The model used in this treatment is illustrated very schematically in Fig. 1, where a slice through the potential surface is shown. For an ion of N atoms there are $3N - 7$ additional coordinate axes which are orthogonal to the reaction coordinate illustrated in the figure. In this model the reaction is assumed to occur by passage of reactant systems through an "activated complex" configuration located at a saddle point in the potential surface. The activation energy ϵ_0 for the reaction is the zero point energy of the activated complex measured relative to the zero point energy of the reactant ion. The motion along the reaction coordinate in the vicinity of the activated complex is treated semiclassically; quantum-mechanical reflection and tunneling at the barrier are neglected. The QET rate expression may be written as

$$K(E) = \frac{S}{h} \frac{W\ddagger(E - \epsilon_0)}{\rho(E)} \tag{1}$$

where $K(E)$ is the reaction rate (sec^{-1}) as a function of the internal energy E of the reactant ion, h is Planck's constant, $W\ddagger(E - \epsilon_0)$ is the number of states of the activated complex configuration with energy $\leq E - \epsilon_0$, $\rho(E)\,dE$ is the number of states of the reactant ion with energy between E

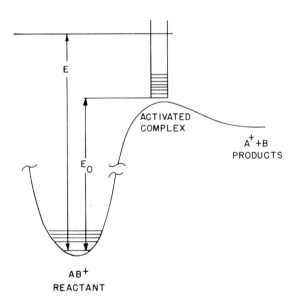

Fig. 1. Schematic diagram of a slice through a potential surface for a polyatomic molecule-ion. The slice is taken along the reaction coordinate axis.

and $E + dE$, and the symmetry factor S is the number of identical reactions (i.e., with the same activation energy and activated complex) yielding the same products. The original derivation of the rate expression used continuous approximations for the expression $W\ddagger$ and ρ; however, as discussed by Rosenstock and Krauss (5), exact enumeration of quantum states can also be used. In the QET the rate is a function only of the internal energy of the reactant ion and is independent of the particular degrees of freedom initially excited by the ionization process.

It has proved exceedingly difficult to devise reasonably direct experimental tests of the validity of the QET rate equation. It should be noted that all the quantities appearing in the rate equation are well defined. If only one precisely defined potential surface were involved in the reaction, then, in principle, the rate as a function of energy could be calculated exactly from the parameters characterizing the potential surface. Unfortunately, many intersecting potential surfaces are generally involved and very little is known in detail about any of them.

It should also be noted that the rate expression as given by Eq. (1) has a discontinuity at the activation energy (24). For $E < \epsilon_0$, $W\ddagger(E - \epsilon_0) = 0$ and the rate is zero. At the activation energy the ground state of the activated complex is the only state available (which may be degenerate) and the minimum rate is given by

$$K_{\min} = S/h\rho(\epsilon_0) \tag{2}$$

This discontinuous behavior results from the semiclassical treatment of the reaction coordinate employed in the derivation of the rate equation. A quantum-mechanical treatment of the reaction coordinate, including the reflection and tunneling effects at the barrier, removes this discontinuity.

The QET was originally thought to be valid only for relatively large molecules (e.g., propane and larger); however, it has recently been applied successfully to rather small molecules such as methane (25) and acetylene (26,27), although it appears that in methane the quantum correction to the reaction coordinate is required to explain the observed "metastable" ions.* Ions decomposing with rates on the order of 10^6 sec^{-1} have been recently observed in some triatomic molecules (28). A particularly severe test of the theory may soon be possible for some of these small molecules. If the potential surface can be determined with sufficient precision to allow accurate determination of $W\ddagger$ and ρ by exact counting techniques, and to allow a correct quantum treatment of the reaction coordinate, then the

---

* In magnetic mass spectrometry the "metastable" ions are those which decompose after acceleration but before magnetic analysis, the apparent mass of the ion $m^* = m_f^2/m_i$, where m_i is the mass of the decomposing ion and m_f is the mass of the fragment ion.

rate may be calculated without introducing additional uncertainties. Comparison of such a calculation with experimental data obtained on the metastable intensity as a function of reaction time might help to resolve the question of the range of validity of the quasi-equilibrium hypothesis.

B. Enumeration of Quantum States

To carry out calculations based on Eq. (1) it is necessary to choose satisfactory approximations to the state density functions. The practice adopted in previous calculations by Wahrhaftig and co-workers (9,10,15) was to neglect all electronic contributions to the state density functions and to represent the nuclear motions by a collection of harmonic oscillators and free rotors. These approximations have been used in most of the applications of the QET to date; however, the classical approximation used previously for calculating the state density functions for a collection of harmonic oscillators has been replaced.

A number of suitable methods are now available for accurately enumerating the states of a collection of harmonic oscillators (20,29–33) all of which are of sufficient complexity to require the use of a computer for detailed calculations. In extensive calculations the method of Vestal, Wahrhaftig, and Johnston (20) has been used in calculating the density functions for a collection of harmonic oscillators. For cases in which free rotors are included in the model, the classical approximation to the density-of-states functions for a free rotor was used in the derivation of the state density functions used in Eq. (1). This method (34) gives the number of states with energy $\leq E$ as

$$W(E) = L\left(\frac{\pi}{\bar{\beta}}\right)^{L/2} \sum_{Q=1} \binom{N - L}{Q} \frac{1}{\Gamma(Q + L/2 + 1)} \left(\frac{h\tilde{\nu}}{\sigma_Q}\right)^{L/2} \left(\frac{E\sigma_Q}{h\tilde{\nu}} - \frac{Q-1}{2}\right)^{Q + L/2}$$

(3)

where h is Planck's constant, $\tilde{\nu}$ is the geometric mean frequency of the harmonic oscillators, $\bar{\beta}$ is the geometric mean energy constant for the classical free rotors, N is the total number of degrees of freedom, L is the number of free rotors, σ_Q are the vibrational coefficients defined in Ref. 20 and Γ is the gamma function. The summation on Q is carried out over all positive terms for $Q \leq (N - L)$. The limit as $L \to 0$ of $W(E)/L$ gives the equation for $W(E)$ for no free rotors used previously (20). Although Eq. (3) appears rather complex, it is quite suitable for computer calculations, and provides a good continuous approximation to the exact quantum-mechanical density of states over the entire energy range of interest.

The validity of this approximate method for enumerating states has been investigated by comparing the results for simple models with the results of exact counting. This comparison is given in Table I. In these

Table I

Calculated Number of States with Energy not Exceeding E for Some
Simple Models Appropriate for Propane

$E/h\nu$	Model A		Model B		Model C	
	Exact	Approx.	Exact	Approx.	Exact	Approx.
$\frac{1}{3}$	1		9	12.6	154	457
$\frac{2}{3}$	1		45	56.0	1.08×10^3	2.43×10^3
1	28	70.9	176	196	4.67×10^3	9.98×10^3
$\frac{4}{3}$			594	636	1.66×10^4	3.41×10^4
$\frac{5}{3}$			1.78×10^3	1.86×10^3	5.21×10^4	1.04×10^5
2	406	982	4.88×10^3	5.04×10^3	1.47×10^5	2.91×10^5
$\frac{7}{3}$			1.25×10^4	1.29×10^4	3.79×10^5	7.61×10^5
$\frac{8}{3}$			3.00×10^4	3.11×10^4	9.21×10^5	1.88×10^6
3	4.06×10^3	9.46×10^3	6.85×10^4	7.20×10^4	2.12×10^6	4.41×10^6
4	3.15×10^4	7.04×10^4	6.41×10^5	7.11×10^5	1.97×10^7	4.49×10^7
5	2.01×10^5	4.32×10^5	4.55×10^6	5.39×10^6	1.38×10^8	3.47×10^8
6	1.11×10^6	2.29×10^6	2.63×10^7	3.34×10^7	7.77×10^8	2.17×10^9
7	5.38×10^6	1.07×10^7	1.30×10^8	1.77×10^8	3.74×10^9	1.15×10^{10}
8	2.35×10^7	4.52×10^7	5.64×10^8	8.20×10^8	1.58×10^{10}	5.34×10^{10}
9	9.41×10^7	1.75×10^8	2.21×10^9	3.41×10^9	6.02×10^{10}	2.21×10^{11}
10	3.48×10^8	6.28×10^8	7.91×10^9	1.29×10^{10}	2.07×10^{11}	8.35×10^{11}
12	3.91×10^9	6.67×10^9	8.19×10^{10}	1.47×10^{11}	—	9.40×10^{12}
15	9.87×10^{10}	1.57×10^{11}	1.80×10^{12}	3.61×10^{12}	—	2.25×10^{14}

calculations model A consists of 27 identical harmonic oscillators; model B consists of 27 harmonic oscillators with 11 oscillators having the geometric mean frequency, 8 having three times the mean frequency, and 8 having one-third the mean frequency; and model C is the same as model B with two of the low frequency oscillators replaced by free rotors. The mean rotational energy constant ($\tilde{\beta}$ in Eq. (3)) was taken equal to the mean oscillator quantum divided by 54. The geometric mean frequency for propane is approximately 1300 cm^{-1}; therefore, 6 mean quanta is approximately equivalent to 1 eV. This rotational energy constant corresponds to a moment of inertia of 4.6×10^{-40} g-cm^2 in agreement with the value given by Pitzer (35). These models represent reasonble first approximations to the oscillator frequencies and moments of inertia for internal rotation in propane.

In all cases the agreement between the exact and the approximate counting is satisfactory. The agreement for model C containing free rotors is not quite as good as for models A and B composed entirely of harmonic oscillators; however, for most applications the error is small compared to that expected due to the uncertainties in the values for the oscillator frequencies.

III. REACTION MECHANISMS

In principle all possible reactions should be included in the reaction scheme used in QET calculations. In practice, however, it is usually possible to eliminate many of these by consideration of the reaction energetics. In general, any reaction can be ignored for which the apparent activation energy (as inferred from differences in electron impact appearance potentials) is more than three times the activation energy for a competitive process.

Essentially two types of dissociation contribute to the primary reaction scheme; namely, simple bond breaks such as

$$CH_3CH_2CH_3{}^+ \rightarrow CH_3CHCH_3{}^+ + H \tag{4}$$

$$CH_3CH_2CH_3{}^+ \rightarrow CH_3CH_2CH_2{}^+ + H \tag{5}$$

$$CH_3CH_2CH_3{}^+ \rightarrow CH_3CH_2{}^+ + CH_3 \tag{6}$$

and four-center reactions such as

$$CH_3CH_2CH_3{}^+ \rightarrow CH_2CH_2{}^+ + CH_4 \tag{7}$$

$$CH_3CH_2CH_3{}^+ \rightarrow CH_3CH^+ + CH_4 \tag{8}$$

$$CH_3CH_2CH_3{}^+ \rightarrow CH_3CHCH_2{}^+ + H_2 \tag{9}$$

Reactions in which the charge ends up on the smaller fragment are also possible; for example,

$$CH_3CH_2CH_3{}^+ \rightarrow CH_3{}^+ + CH_3CH_2 \tag{10}$$

However, with the exception of reaction 10 the difference between the ionization potentials for the two fragments is sufficiently large that the other possible reactions of this type contribute very little. Other reactions such as

$$C_3H_8{}^+ \rightarrow C_2H_6{}^+ + CH_2 \tag{11}$$

should be considered possible; however, the activation energy for this reaction is at least four times the activation energy for reaction (4), for example, and the reaction has not been observed.

The reaction scheme contained in reactions (4)–(10) accounts quite satisfactorily for all the primary reactions of singly charged propane ions which have been experimentally observed. These reactions also account quite satisfactorily for the reactions observed in deuterium-labeled propanes such as $CH_3CD_2CH_3$ or $CD_3CH_2CD_3$ (36). Comparison of calculations with experimental results at low energies indicates that very little, if any, hydrogen rearrangement occurs prior to the *primary* dissociation reactions. However, prior to the *secondary* decompositions, such as

$$C_3H_7{}^+ \rightarrow C_3H_5{}^+ + H_2 \tag{12}$$

$$C_2H_4{}^+ \rightarrow C_2H_2{}^+ + H_2 \tag{13}$$

nearly complete scrambling of the hydrogen atoms occurs. From studies of partially deuterated butanes, McFadden and Wahrhaftig (37) concluded that substantial rearrangement of hydrogen atoms occurred prior to primary fragmentation. Three types of processes have been suggested to account for this apparent scrambling (38). One of these assumes that hydrogen atom interchange occurs which presumably should involve an activated complex similar to that for loss of a hydrogen molecule. Since the loss of hydrogen molecule from n-butane parent ions is a rather unimportant reaction, it seems unlikely that this type of reaction can account for all of the observed rearrangement. Another type of mechanism which has been proposed is an isomerization prior to dissociation. If this is the important mechanism, then hydrogen atom rearrangement should be extensive only for those molecules for which more than one stable isomer is possible. This type of mechanism should produce extensive rearrangement at low energies provided the activation energy for the isomerization reaction is lower than the activation energy for the first dissociation reaction. Evidence is available from appearance potential measurements on n-butane and isobutane which indicates that this may, in fact, be the case for n-butane.

Steiner, Giese, and Inghram (16) found that within an experimental error of ± 0.01 eV the appearance potentials for the production of propyl ion and propylene ion from n-butane are identical at 11.10 for a temperature of 300°K, which when corrected for thermal energy corresponds to 11.19 at 0°K. This has been confirmed by more recent measurements by Chupka and Berkowitz (39). Furthermore, these appearance potentials do not correspond to the ΔH for propyl ion or propylene ion formation as determined from the analysis of photoionization data from propane, isobutane, and propylene (27,36,40).

On the basis of presently available data it is not possible to rule out the possibility that this is mere coincidence; however, it offers strong presumptive evidence for the energy diagram given in Fig. 2. In this diagram the energy zero corresponds to the elements in their standard states at absolute zero. The levels corresponding to the ground states of the molecules are rather uncertain due to the difficulty in measuring the adiabatic ionization potentials. This is discussed in the next section, but the present discussion is not concerned with the precise energy of the ground state. According to this model, the lowest energy reaction for n-butane ions is the isomerization to isobutane ions. This reaction produces isobutane ions with excitation above the dissociation limit for both the production of secondary propyl ion and the propylene ion from isobutane.

The rate-determining step for these reactions is the isomerization reaction and the relative abundance of each product ion is determined by

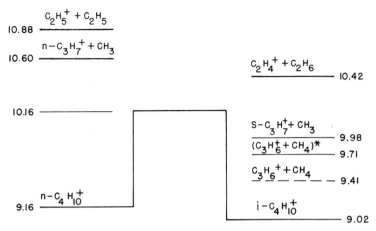

Fig. 2. Energy diagram for butane. The energies given are the standard heats of formation at absolute zero for the indicated reaction products.

the relative rates of the dissociation reactions from the isobutane structure. This model is a plausible hypothesis which appears consistent with presently available data, but is certainly not established. Experiments with labeled butanes at energies near threshold for these reactions should provide the necessary data. In particular photoionization measurements with isobutane labeled with ^{13}C in the tertiary position would be particularly valuable. If the proposed model is correct, then the $C_3H_7^+$ and $C_3H_6^+$ ions should show complete label retention up to a photon energy of 11.5 eV. At higher energies the isomerization to the n-butane structure and back to the isobutane structure should begin to occur to a small extent prior to dissociation and some unlabeled propyl and propylene ions should be produced. Similar experiments with labeled n-butane might help to establish the structure of the activated complex for the isomerization reaction.

Diagrams can also be drawn for other isomeric pairs; for example, a diagram for propyl alcohol is shown in Fig. 3. This diagram is based on rather less substantial evidence than is the one for butane; however, it is in agreement with available data (41,42) and appears to account qualitatively for the substantial differences between the mass spectra for these two molecules. Again the accuracy of the model can be checked by careful measurement of label retention near threshold.

The height of the barrier to isomerization relative to the barrier for dissociation appears to be a very important consideration in the mass spectra of isomers. If activation energy for isomerization is lower than the activation energies for all the dissociation reactions, then the mass spectra

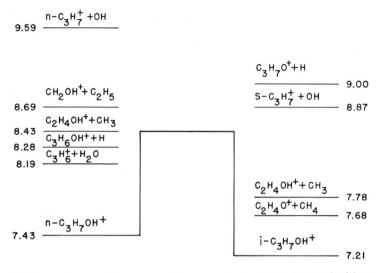

Fig. 3. Energy diagram for propanol. The energies given are the standard heats of formation at absolute zero for the indicated reaction products.

for the isomers will be quite similar, with the principal difference being in the intensity of the parent ion. This is apparently the case for a large number of unsaturated molecules. For these molecules, detailed consideration of the fragmentation must consider reactions from all of the isomeric configurations.

The C_4H_6 isomers, which have been extensively studied, (43–45) provide an excellent example of the latter behavior. An energy level diagram for these molecules is given in Fig. 4. In particular, the lowest energy reaction

$$C_4H_6{}^+ \rightarrow C_3H_3{}^+ + CH_3 \tag{14}$$

has been most extensively studied. Appearance potential measurements (44) show the same heat of formation for the reaction products independent of the initial configuration. Experiments with deuterium labeling in 1-butyne by Dolejsek et al. (43), Fig. 5, show that at low electron energies complete scrambling of the hydrogen atoms occurs, while as the electron energy is increased the relative probability for loss of methyl radical prior to hydrogen rearrangement becomes more probable. This is precisely the behavior expected for the case where the activation energy for isomerization is lower than that for dissociation.

The isomerization reaction is probably best represented as a four-center reaction characterized by a "tight" activated complex configuration, while the dissociation reaction (14) from one of the isomers able to lose

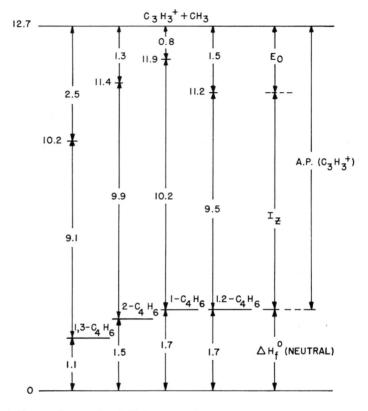

Fig. 4. Energy diagram for C_4H_6 isomers. The energies given are the standard heats of formation at absolute zero for the indicated reaction products.

CH_3 by a simple bond break is characterized by a "loose" activated complex. At parent ion internal energies only slightly above the threshold for dissociation, the rate of the isomerization reaction should be much faster than the rate of dissociation. However, for reactions characterized by a "loose" activated complex, the reaction rate increases more rapidly with energy than the rates for reactions characterized by "tight" complexes. Therefore, at higher internal energies the rate for dissociation may become much faster than the rate for isomerization; hence some of the 1-butyne ions dissociate without hydrogen scrambling.

Reaction mechanisms have been extensively investigated by Meyerson and co-workers (46–55) by employing isotopic labeling techniques. These excellent systematic studies provided the conceptual basis for the models discussed above.

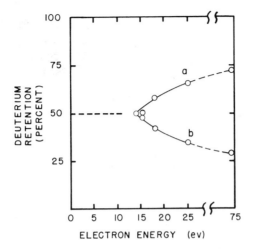

Fig. 5. Dependence on electron energy of deuterium retention (per cent) for the $(M - CH_3)^+$ ions (43). Value for random distribution of deuterium in the molecular ion is marked by the dotted line on the left. (a) 1-butyne-1-d_1. (b) 1-butyne-4-d_1.

IV. ACTIVATED COMPLEX CONFIGURATIONS

If the unimolecular reactions of polyatomic ions involved motion on a single potential surface and if the details of the potential surface were sufficiently well known, then the functions $W\ddagger$ and ρ appearing in the rate equation (Eq. 1) could be evaluated precisely. Unfortunately, it appears that many intersecting potential surfaces are generally involved and very little is known in detail about any of them.

In most calculations using the QET, both the normal configuration and the activated complex configuration have been represented by harmonic oscillators and free rotors, although some work has been done using anharmonic oscillators (33,56). The normal configuration of the ion has been assumed to be very similar to the configuration of the neutral molecule and for the most part the approximate oscillator frequencies of the molecule have been used for the normal configuration of the ion. The torsional motion of methyl groups is usually taken as a free rotation in the ion and in many cases some of the skeletal frequencies are assumed to be somewhat lower in the ion than in the molecule.

The symmetry factor, S in Eq. (1), is the number of identical reactions which contribute to the production of specified products. The total rate is the symmetry factor multiplied by the rate for one such reaction. For simple bond breaks the symmetry factor can usually be derived simply. Considering free rotations frozen and all atoms labeled, the number of different bonds which, when broken, lead to the given products is equal to the symmetry factor for that reaction. For cases in which rearrangement of bonds is involved, a similar procedure is used; however, the number obtained often depends upon some rather arbitrary assumptions about the activated complex configuration and about the structure of the products.

A systematic procedure for deriving the activated complex configurations from the normal configuration has been developed by Wahrhaftig and co-workers (9,10,15). This procedure involves selecting a reaction co-ordinate and then considering what effect distortion of the configuration due to motion along the reaction coordinate may have on the other degrees of freedom. This procedure is quite arbitrary and the activated complexes derived in this manner can hardly be considered unique; however, it does provide a reasonably consistent way for choosing the representation of the activated complex. In recent calculations on a number of molecules (25,27,34,36,40,57,58) the following rules have been found to be of reasonable generality:

1. Four-center reactions (i.e., those producing a molecule-ion and a neutral molecule) are always represented by a "tight" activated complex. The skeletal frequencies are rarely lower than those of the normal con-figuration and some free rotations are usually stopped.

2. C—C bond breaks are always represented by a "loose" activated complex. Some skeletal frequencies are usually lowered and all free rotations are active. In general it appears that the nearer the broken bond is to the center of the molecule, the more the skeletal frequencies are lowered.

3. C—H bond breaks may occur by either a "tight" or a "loose" activated complex. Loss of H from a terminal carbon (or oxygen) usually involves a "loose" complex while loss from an interior carbon involves a rather "tight" complex.

Since the competition between reactions is very sensitive to the repre-sentation used for the activated complex, it is rarely possible to choose the frequencies in a strictly *a priori* fashion. Some fitting to experimental data is nearly always required.

For a "loose" activated complex the density of states increases more rapidly as a function of its internal energy than does the density of states for a tight complex. Thus the rate for a reaction represented by a loose complex, for example, a C—C bond break increases more rapidly with energy than does the rate for a reaction represented by a tight complex, for example a four-center reaction. This may be illustrated by reference to the reaction mechanism for a recent calculation on propane. The set of primary reactions which, together with subsequent decomposition, account for more than 95% of the total ionization, were given in Eqs.

(4)–(10) above. The representation of the normal configuration of propane ion used in a recent calculation (36) is summarized in Table II, where oscillator frequencies are approximately equal to the molecular frequencies determined by Pitzer (35). In the normal configuration of the ion the torsional motion of the methyl groups is taken as a free rotation.

Table II

Representation of the Normal Configuration for the
Propane Molecule-Ion

Designation	Frequency, cm^{-1}	No. of oscillators
C—H stretch	2900	8
C—H bending	1400	8
CH_3 deformation	1200	4
CH_2 deformation	800	2
C—C stretch	900	2
C—C—C bending	400	1
CH_3 torsion	Free rotation[a]	2

[a] The moment of inertia for the free rotor is taken as 4.6×10^{-40} g cm^2 as given by Pitzer (35).

The activated complex configurations for the reactions of propane molecule-ions as derived from the normal configuration are illustrated in Fig. 6.

$$C_3H_8{}^+ \rightarrow sec\text{-}C_3H_7{}^+ + H \tag{4}$$

The activated complex for the formation of secondary propyl ion is taken as a tight ring structure. Both free rotations are assumed stopped and the CH_3 torsional frequency is taken as 300 cm^{-1}. The two CH_3 deformation frequencies are increased by a factor of 1.5 to 1800 cm^{-1} and the C—C—C chain bending frequency to 600 cm^{-1}. The reaction coordinate is taken as the C—H stretch of the secondary hydrogen and the corresponding C—H bending frequency is reduced by a factor of 2 to 700 cm^{-1}. This is the tightest activated complex used in the propane calculation. The symmetry factor is 2.

$$C_3H_8{}^+ \rightarrow n\text{-}C_3H_7{}^+ + H \tag{5}$$

For this reaction the reaction coordinate is taken as the C—H stretch of a primary hydrogen and the corresponding C—H bending frequency is reduced by a factor of 2 to 700 cm^{-1}. The two CH_3 deformation frequencies for the methyl group involved are also reduced by a factor of

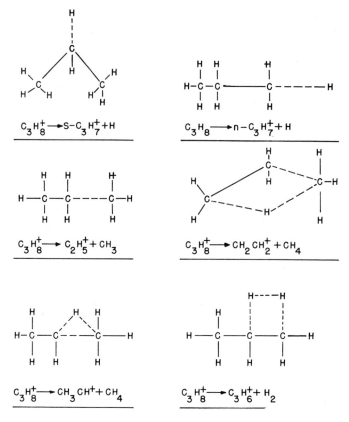

Fig. 6. Schematic diagrams of the activated complex configuration used in the QET calculation on propane (36). The dashed lines indicate the bonds which are broken and the new bonds which are formed.

2 to 600 cm^{-1} and the chain bending frequency is also reduced by a factor of 2 to 200 cm^{-1}. Both free rotations are assumed active. The symmetry factor is 6.

$$C_3H_8{}^+ \rightarrow C_2H_5{}^+ + CH_3 \tag{6}$$

The C—C stretching frequency is taken as the reaction coordinate for this reaction; the CH$_3$ deformation frequencies for the corresponding methyl group are reduced by a factor of 4–300 cm^{-1} and the C—C—C chain bending frequency is also reduced by a factor of 4 to 100 cm^{-1}. Both free rotations are active. The symmetry factor is 2.

$$C_3H_8{}^+ \rightarrow CH_2CH_2{}^+ + CH_4 \tag{7}$$

For the 1,3-elimination of methane the activated complex is taken as a ring structure, similar to that used for the formation of secondary propyl, but slightly looser. The C—C stretch is taken as the reaction coordinate and the C—H stretching and bending frequencies for the hydrogen lost are both reduced by a factor of 2 to 1450 cm^{-1} and 700 cm^{-1}, respectively. Both free rotations are assumed stopped and are taken as torsional vibrations with frequencies of 200 cm^{-1}. The CH_2 deformation frequencies are increased by a factor of 2 to 1600 cm^{-1}. The symmetry factor is 6.

$$C_3H_8{}^+ \rightarrow CH_3CH^+ + CH_4 \tag{8}$$

For the 1,2-elimination of methane the reaction coordinate is again taken as the C—C stretching frequency and the C—H stretching and bending frequencies for the hydrogen lost are reduced by a factor of 2 to 1450 cm^{-1} and 700 cm^{-1}, respectively. One free rotation is stopped and the torsional vibration is taken as 200 cm^{-1}. The symmetry factor is 4.

$$C_3H_8{}^+ \rightarrow C_3H_6{}^+ + H_2 \tag{9}$$

For this reaction the hydrogens lost are assumed to come from adjacent carbon atoms. The reaction coordinate is taken as a C—H stretch and one free rotation is stopped and taken as a torsional vibration with a frequency of 200 cm^{-1}. The symmetry factor is 4.

$$C_3H_8{}^+ \rightarrow CH_3{}^+ + C_2H_5 \tag{10}$$

This reaction is assumed identical to Eq. (6) except that the activation energy is increased by the difference between the ionization potentials of methyl and ethyl radicals.

This set of activated complexes has been found to be satisfactory for ordinary propane, and also, when the changes in these frequencies produced by deuterium substitutions are taken into account, the observed isotope effects (36) are accounted for satisfactorily.

The absolute rates for the unimolecular reactions of propane molecule-ions are plotted as a function of internal energy of the parent ion in Fig. 7. The reactions with low activation energy, Eqs. (4) and (7) correspond in propane to tight activated complexes. Therefore, the reactions with higher activation energies, e.g., (5) and (6), but proceeding by loose activated complexes, overtake the lower energy reactions and are dominant at higher energies.

By contrast, in acetone and biacetyl (59,136) the reactions with lowest activation energy are

$$CH_3(C{=}O)CH_3{}^+ \rightarrow CH_3CO^+ + CH_3 \tag{15}$$

$$CH_3(C{=}O)C{=}OCH_3{}^+ \rightarrow CH_3CO^+ + CH_3CO \tag{16}$$

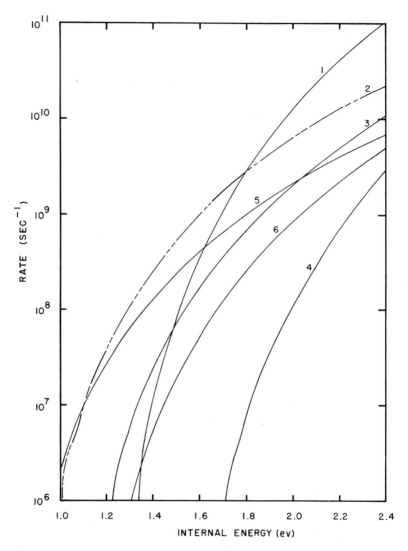

Fig. 7. Calculated rates for the unimolecular reactions of propane molecule-ion as a function of the internal energy of the parent ion (36). (1) $C_3H_8^+ \rightarrow C_2H_5^+ + CH_3$; ($2$) $C_3H_8^+ \rightarrow CH_2CH_2^+ + CH_4$; ($3$) $C_3H_8^+ \rightarrow CH_3CH^+ + CH_4$; ($4$)$C_3H_8^+ \rightarrow$ n-$C_3H_7^+ + H$; (5) $C_3H_8^+ \rightarrow s$-$C_3H_7^+ + H$; (6) $C_3H_8^+ \rightarrow C_3H_6^+ + H_2$.

Being simple C—C bond breaks these reactions are characterized by loose activated complexes. As a result these reactions together with the secondary reaction

$$CH_3CO^+ \rightarrow CH_3^+ + CO \tag{17}$$

account for most of the fragmentation observed for these molecules (60).

V. REACTION ENERGETICS

A considerable amount of the literature of mass spectrometry has been concerned with obtaining thermochemical data from the interpretation of ionization efficiency curves. This important subject has been reviewed by Krauss and Dibeler (61) and more recently by Collin (62). Whether one is concerned with activation energies for ionic fragmentation processes, ionic and radical heats of formation, or bond dissociation energies, the problem is encountered of determining an "appearance potential" from an experimental measurement of ion intensity as a function of electron or photon energy.

One of the difficulties is the fact that several different operational definitions of "appearance potential" occur in the literature, each corresponding to a particular empirical technique for interpreting the ionization efficiency curves (63–68).

None of these methods has a satisfactory theoretical basis, and in general it is impossible to determine the magnitude of the absolute errors which result from their use (69). The QET provides a basis for interpreting the ionization efficiency curves for fragment ions; however, to apply the theory to these problems the heats of formation for the parent ions must be known with some precision.

A. The Adiabatic Ionization Potential

The heat of formation for an ion A^+ is given by

$$\Delta H_f^\circ(A^+) = I_z(A) + \Delta H_f^\circ(A) \tag{18}$$

where $I_z(A)$ is the *adiabatic* ionization potential and $\Delta H_f^\circ(A)$ is the heat of formation for the neutral species. In this discussion all heats of formation are taken for the ideal gas state at absolute zero relative to the standard states of the elements. For the cases in which A is a molecule, the heat of formation for the neutral species is generally known to high accuracy from thermochemical measurements and these values have been extensively tabulated (70,71). Extensive tabulations of molecular ionization potentials also exist (72–75); however, the relationship of the values reported to the adiabatic ionization potential required in Eq. (18) is often not clear. For cases in which the ionization potential has been determined spectroscopically by finding the convergence limit of a Rydberg series, and in which this ionization potential has been confirmed to be the lowest ionization potential by photoionization or electron impact ionization, the adiabatic ionization potential has been determined with quite satisfactory accuracy. For most molecules, however, the experimental ionization potentials in the literature are upper limits on the adiabatic ionization potential.

During the early development of the photoionization techniques for measuring ionization potentials it was noted that the values obtained were quite frequently substantially lower than those measured by electron impact (3). This fact was interpreted to mean that the electron impact values corresponded to the "vertical" ionization potential while the photoionization values were closer to the adiabatic values. For a diatomic molecule the "vertical" ionization potential is a clearly defined and useful concept and corresponds to the transition from the ground state of molecule to the state of the ion for which the Franck-Condon factor is largest (3). For a polyatomic molecule the "vertical" ionization potential may be neither well defined nor useful.

For even a moderate size polyatomic molecule-ion the number of states within an energy increment ΔE, for a single electronic state, increases very rapidly with increasing internal energy. State densities calculated for a few typical cases, using Eq. (3), are given in Table III (143). The total transition probability for ionization of a molecule with an internal energy between E and $E + \Delta E$ is the sum of all of the Franck-Condon factors for transitions

Table III
Calculated State Densities as a Function of
Internal Energy

Internal energy, eV	Density of molecular states[a]		
	CH_4	$1,3\text{-}C_4H_6$	C_3H_8
0.01	1.3×10^1	2.2×10^2	9.4×10^2
0.1	1.2×10^2	1.2×10^3	1.5×10^3
0.2	3.2×10^2	8.8×10^3	5.5×10^3
0.3	7.1×10^2	5.1×10^4	1.8×10^4
0.5	2.6×10^3	9.9×10^5	1.5×10^5
1.0	2.8×10^4	2.6×10^8	1.2×10^7
2.0	6.4×10^5	5.0×10^{11}	8.6×10^9
3.0	5.8×10^6	1.1×10^{14}	1.2×10^{12}

[a] Number of states per electron volt for a single electronic state.

to states within this energy interval. As shown by recent measurements by photoelectron spectroscopy (76–85), for many molecules the energy at which this total transition probability is a maximum is well above the threshold energy determined by either electron impact or photoionization.

For polyatomic molecules the "vertical" ionization potential may reasonably be defined in terms either of the single state of the ion with the largest Franck-Condon factor (see Chapter 1) for transitions from the ground state of the molecule or in terms of the energy at which the total transition probability is a maximum. The first definition is often difficult (if not impossible) to apply, while the second, which is useful in connection with the internal energy distribution for the ion, bears no simple relationship to the minimum ionization energy, and does not, in general, correspond to the vertical transition in the original sense.

The difficulties which may be encountered in attempting to determine the adiabatic ionization potential are illustrated in Table IV. Except for the last three entries these results were selected from the compilation by Kiser (72), and only values with a small quoted error were included. The last three values were obtained with refined photoionization apparatus featuring both high energy resolution and high sensitivity (45,86,87).

Table IV

Experimental Ionization Potentials for Methane

Value	Method[a]	Year	Ref.
13.04 ± 0.02	EICS	1948	72
13.04 ± 0.02	EIVC	1949	72
13.07 ± 0.02	EI	1951	72
13.12 ± 0.02	EIED	1951	72
12.9 ± 0.2	PI	1953	72
13.13 ± 0.02	EIVC	1954	72
12.8 ± 0.2	PI	1955	72
13.07	PI	1956	72
12.99 ± 0.01	PI	1957	72
12.71 ± 0.02	PI	1965	86
12.704 ± 0.008	PI	1965	87
12.55 ± 0.05	PI	1966	45

[a] The methods used for the measurements are as follows: EICS, electron impact critical slope, see Ref. 64; EIVC, electron impact vanishing current, see Ref. 63; EIED, electron impact extrapolated difference, see Ref. 66; PI, photoionization.

Prior to 1954 the ionization potential of methane appeared to be rather well established at 13.08 ± 0.05 eV and good agreement with this value was obtained by careful use of any of the established methods for interpreting ionization efficiency curves. However, the earliest values obtained from photoionization experiments were substantially lower than this seemingly well-established electron impact value, and as the sensitivity of the photoionization technique has been improved, the reported ionization potentials for methane have steadily decreased.

In an attempt to understand these differences, a model calculation was performed in which a uniform distribution of Franck-Condon factors was assumed for the energy region near threshold (88). That is, each transition from the ground state of the molecule to an energetically accessible state of the ion, including the transition to the ground state of the ion, was

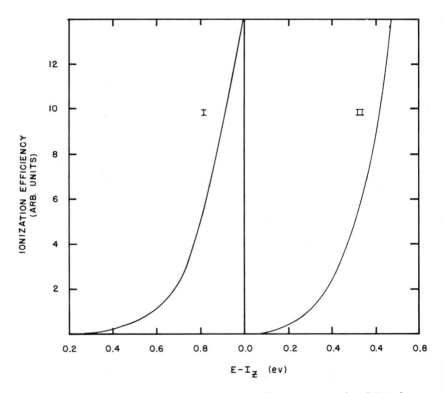

Fig. 8. Linear plots of calculated ionization efficiency curves for CH_4^+ from methane (I) for electron impact ionization and (II) for photoionization. The ionization efficiency is plotted as a function of excess energy ($E - I_z$) of the ionizing particles, where I_z is the adiabatic ionization potential.

assigned equal probability. The total transition probability to ionic states with energy between E and $E + \Delta E$ is, under this model, proportional to the total number of states within the energy interval between E and $E + \Delta E$. Using the linear variation of the cross section with excess energy for electron impact and the step function for photoionization (see Chapter 1), the ionization efficiency* curves given in Fig. 8 were obtained. Any of the conventional methods for interpreting these curves based on such a linear plot would clearly yield values substantially higher than the adiabatic ionization potential (0.6 eV for electron impact and 0.2–0.4 eV for photo-ionization). Brehm's value of 12.55 ± 0.05 was obtained (45) using a semilogarithmic plot of data obtained on apparatus with quite high sensitivity. The calculated photoionization efficiency curve is compared with Brehm's result in Fig. 9 where the ionization potential for the calculated curve was taken as 12.55 eV and the arbitrary ionization efficiency scale was adjusted to agree with the experimental scale in the threshold region. While the good agreement apparent in Fig. 9 hardly constitutes definite proof because of the rather arbitrary assumptions used in the calculations, it provides some basis for a contention that the value of 12.55 ± 0.05 may be within the stated error of the adiabatic ionization potential.

It should be noted that in the calculated curve of Fig. 9 there is a discernible downward break at the ionization potential. Below the ionization potential the curve would drop vertically except for the contributions from "hot bands" in the neutral molecule. Due to the exponential tail on the thermal energy distribution, the thermal energy causes a quasi-linear behavior of the logarithm of ionization efficiency at energies below the adiabatic ionization potential. This calculation was performed for a temperature of 300°K; at lower temperature the break should be more readily apparent.

For molecules in which the electron lost from the molecule is non-bonding, the Franck-Condon factor for the adiabatic transition is apparently very large compared to other transitions to energetically neighboring states and as a result the downward break on the semilog ionization efficiency plot is easily identifiable and can be determined with good precision. Brehm has shown that ionization potentials determined by locating this break are in very good agreement with the spectroscopic values. For molecules in which the threshold process is the removal of a

---

* In this chapter the term "ionization efficiency" is used in the sense familiar to mass spectroscopists, i.e., the ion current measured as a function of the energy of the ionizing particle. This should not be confused with the term "ionization efficiency" used by radiation chemists to denote the fraction of total excitations which produce ions.

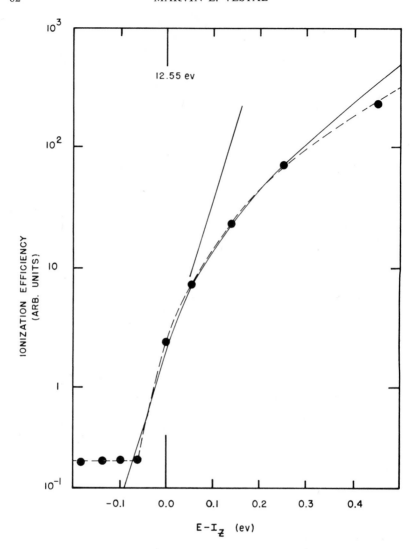

Fig. 9. Comparison of the calculated (solid line) photoionization efficiency curve for $CH_4{}^+$ from methane, on a semilogarithmic plot, with the experimental data (●) of Brehm (45). Only a few of the experimental data points are shown.

bonding electron, the Franck-Condon factor for the adiabatic transition is apparently considerably smaller than when the electron in nonbonding; however, it is not necessarily zero and may be as large or larger than that for other transitions corresponding to somewhat higher energy. The relationship of bonding (or antibonding) character to the Franck-Condon factors has been discussed by Al-Joboury and Turner (79). Because of the very rapid increase in the density of states with energy, the probability of the adiabatic transition may be four orders of magnitude, or more, smaller than the maximum overall transition probability. The presently available data suggest that for the ionization of a bonding electron the adiabatic ionization potential has, in general, not been determined, not necessarily because the adiabatic transition is inaccessible but rather because it is very improbable. The adiabatic ionization potential should correspond to a downward break in the semilog ionization efficiency curve. For large molecules the break may be very difficult to observe because of the smoothing effect of the molecular thermal energy and the very rapid increase in density of states for the ion. However, if the adiabatic ionization potentials can be observed for the smaller members of a structurally similar series of molecules, it may be possible to calculate reasonably accurate values for the larger members by using the method of equivalent orbitals (89–91).

B. The Internal Energy of the Reactant Ion

The rate of an ionic fragmentation reaction can be calculated as a function of the internal energy of the reactant ion using the QET rate equation (Eq. (1)) provided the parameters characterizing the activated complex and normal configuration can be determined. However, in no experiment is the internal energy of the reactant ion determined directly; rather, such related parameters as the energy of the impacting particle and the temperature of the sample gas are usually determined experimentally. In order to compare the results of theoretical calculations with experiment, or to use the theory in the interpretation of experimental results, knowledge is required of the probability distribution functions linking the internal energy of the reactant ion with the experimentally determined energy parameters.

It is well established that the transition probabilities for the ionization of diatomic molecules with a given internal energy can be calculated by invoking the Franck-Condon principle. This principle seems equally valid in the case of polyatomic molecules. However, the necessary information concerning the potential surfaces is not available for similar calculations on polyatomic systems. Furthermore, at energies only slightly above the ground state of the ion the number of vibrational states which must be considered becomes quite large. Nevertheless, a transition probability distribution must exist for a polyatomic molecule; the distribution being composed, for any energy E in the ion, of the totality of all the

Franck-Condon factors for transitions from the ground state of the molecule to states of the ion with energy between E and $E + \Delta E$. That is,

$$Y(E)\,dE = \sum_{i=1}^{n(E)} f_i(E)\,dE \tag{19}$$

where $f_i(E)$ is the Franck-Condon factor for the transition from the ground state of the molecule to the ith state of the ion with energy between E and $E + dE$, and $n(E)$ is the total number of states of the ion with energy between E and $E + dE$.

In many experiments the parameter which can be most readily measured is the energy of the ionizing particle. The variation in the probability of transferring an energy $E + I_z$ in an ionizing process resulting from collision with a particle with energy V can be written as a function of the energy difference $h(V - E - I_z)$. (See Chapter 1.) According to the treatment by Geltman (92) this function is a step function for photo-ionization and a linear function for single ionization by electron impact. For pure charge-exchange ionization (i.e., in which the kinetic energy of the particle does not participate) the delta function $\delta(V - E - I_z)$ (144) is appropriate with V taken as the recombination energy. Thus the internal energy distribution produced as the result of ionization by particles with energy V can be written as:

$$g(V, E) = h(V - E - I_z)Y(E) \tag{20a}$$

where I_z is the adiabatic ionization potential.

For photoionization:

$$h(V - E - I_z) = \begin{cases} 0 & V < E - I_z \\ 1 & V \geqslant E + I_z \end{cases} \tag{21a}$$

For electron impact ionization:

$$h(V - E - I_z) = \begin{cases} 0 & V < E + I_z \\ V - E - I_z & V \geqslant E + I_z \end{cases} \tag{21b}$$

For charge exchange ionization:

$$g(V, E) = \delta(V - E - I_z)Y(E) = Y(V - I_z) \tag{20b}$$

As was shown originally by Morrison (93) and applied more recently by Chupka and Kaminsky (94) and by Inghram and co-workers (59,16), the transition probability distribution is given by the first derivative of the total ionization efficiency for photoionization, and the second derivative of total ionization efficiency by electron impact, provided the Geltman threshold laws are valid. Recent photoionization measurements on simple molecules show that the threshold laws are obeyed only approximately and that substantial autoionization often occurs (95–100). As a result,

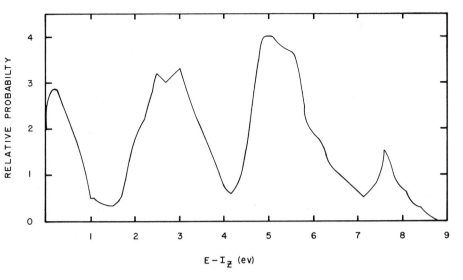

Fig. 10. Transition probability distribution for benzene as determined from the photoelectron spectrum obtained by May and Turner (83).

transition probability distributions obtained by this technique may contain substantial errors, particularly at higher energies.

A more direct method for determining transition probability distributions is the technique of photoelectron spectroscopy developed by Turner and co-workers (76–83) and others (84,85). In this technique both the energy of the vacuum UV photon producing the ionization and the energy of the ejected electron are determined. By observing the ion current as a function of this energy difference the transition probability distribution is obtained directly and with a precision limited only by the energy resolution of the apparatus. As an example the transition probability distribution for benzene determined by Turner and May (83) is given in Fig. 10.

C. Thermal Energy

At best the energy which can be determined experimentally is the energy transferred to the parent ion during the ionizing process. In addition the ion possesses the internal thermal energy of the neutral molecule. The probability that a molecule from an equilibrium ensemble at a temperature T has energy between E and $E + dE$ is given by

$$f(E)\,dE = \frac{\rho(E)\,e^{-E/kt}\,dE}{\displaystyle\int_0^\infty \rho(\epsilon)\,e^{-\epsilon/kt}\,d\epsilon} \tag{22}$$

where k is the Boltzmann constant and $\rho(E)\,dE$ is the number of states of the molecule with energy between E and $E + dE$. The density function ρ can be accurately evaluated using the methods of Eq. (3) with the appropriate molecular frequencies. If an energy E' in excess of the adiabatic ionization potential is transferred to the ions by the ionization process, then the internal energy distribution function for the ions is $f(E - E')$ where E is the internal energy of the ion and f is the thermal energy distribution function given by Eq. (22).

Including thermal energy of the molecules at a temperature T, the distribution in total internal energy distribution for the ion is given by

$$g(E)\,dE = dE \int_0^E f(E - E')\,Y(E')\,dE' \tag{23}$$

where E' is the energy transferred in excess of the adiabatic ionization potential, Y is the transition probability distribution, Eq. (19), and f is the thermal energy distribution, Eq. (22). In the derivation of Eq. (23) it is assumed that the transition probability $Y(E)$ is a function only of the energy transferred and does not depend on the initial state of the molecule. For molecules for which the temperature effects have been extensively investigated this appears to be a very good approximation (101–103,105).

D. Distribution of Internal Energy between the Fragments

In a reaction sequence of the type

$$ABC^+ \rightarrow AB^+ + C$$
$$AB^+ \rightarrow A^+ + B$$

the internal energy of the ion AB^+ is not precisely determined. The kinetic energy of the fragments and the internal energy of the neutral fragment C are not available to the ion AB^+ undergoing further decomposition. The distribution of energy between the fragments has been discussed by Wallenstein and Krauss (104) and by Vestal (105).

In accordance with the quasi-equilibrium hypothesis, the distribution of internal energy between the fragments can be calculated using the microcanonical ensemble. Under this assumption the probability that an energy remains as internal energy of the ion AB^+ is proportional to the number of ways of putting this energy into AB^+ and the remainder of the available energy into the neutral fragment C. Using continuous density functions the probability that the ion AB^+ has an energy between ϵ and $\epsilon + d\epsilon$ is given by

$$\rho(E, \epsilon)\,d\epsilon = \frac{\rho_1(\epsilon)\rho_2(E - \epsilon_0 - \epsilon)\,d\epsilon}{\displaystyle\int_0^{E - \epsilon_0} \rho_1(\epsilon')\rho_2(E - \epsilon_0 - \epsilon')\,d\epsilon'} \tag{24}$$

where E is the internal energy of the reactant ion ABC, ϵ_0 is the activation energy for the primary reaction, $\rho_1(\epsilon)\,d\epsilon$ is the number of states of ion AB^+ with energy between ϵ and $\epsilon + d\epsilon$, and $\rho_2(\epsilon)\,d\epsilon$ is the corresponding number of states for the neutral fragment C.

E. Kinetic Energy in the Reaction Coordinate

The kinetic energy of the decomposing fragments (108) may include, in addition to the "statistical" kinetic energy with which the activated complex passes over the potential barrier, a contribution corresponding to the activation energy for the reverse reaction. In many cases involving decomposition into molecular fragments, some excess energy, E^*, is involved. This energy may appear as internal energy in one or both fragments or as kinetic energy. If excess kinetic energy occurs, this energy as well as the statistical kinetic energy must be subtracted from the energy available to the product ion for further decomposition.

Under the quasi-equilibrium hypothesis the statistical kinetic energy distribution can be readily calculated. The probability that the activated complex has translational energy between ϵ_t and $\epsilon_t + d\epsilon_t$ is

$$p_t(\epsilon_t)\,d\epsilon_t = \frac{p\ddagger(E - \epsilon_0 - \epsilon_t)\,d\epsilon_t}{W\ddagger(E - \epsilon_0)} \tag{25}$$

where $W\ddagger$ is the same function appearing in the rate equation (Eq. (1)) and

$$p\ddagger(E) = \frac{dW\ddagger(E)}{dE} \tag{26}$$

In recent calculations (25,27,34,36,40,57,58,107) it has been found that the energy lost as kinetic energy (provided E^* is small) is usually small compared to the energy lost as internal energy of the neutral fragment. An obvious exception occurs when the neutral fragment is an atom, in which case the kinetic energy of fragmentation is the only energy loss.

The kinetic energy distributions for fragment ions have been studied recently by several investigators (108–114). Examples of the variation of average statistical energy with the internal energy of the reactant ion are given in Fig. 11. The results of calculations (25) on the statistical kinetic energy in the reaction coordinate for the reaction

$$CH_4{}^+ \rightarrow CH_3{}^+ + H \tag{27}$$

are shown in Fig. 12. It should be noted that the statistical or quasi-thermal kinetic energy distributors predicted by the QET are non-Maxwellian, the most probable statistical energy always being zero. As a

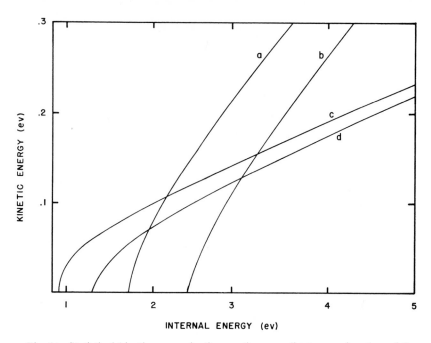

Fig. 11. Statistical kinetic energy in the reaction coordinate as a function of the internal energy of the reactant ion. (a) $CH_4^+ \rightarrow CH_3^+ + H$; (b) $CH_4^+ \rightarrow CH_2^+ + H_2$; (c) $C_3H_8^+ \rightarrow sec\text{-}C_3H_7^+ + H$; (d) $C_3H_8^+ \rightarrow C_2H_5^+ + CH_3$.

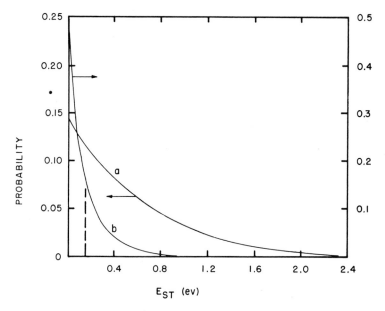

Fig. 12. Statistical kinetic energy distribution for the reaction $CH_4^+ \rightarrow CH_3^+ + H$. (a) 6 eV internal energy in the reactant ion; (b) produced by 70 eV electron impact.

result, the experimental determination of average kinetic energy based on assuming a Maxwellian distribution may result in a slight overestimate of the average energy.

For a kinetic energy E_{st} in the center-of-mass system the energy of the fragment ion in the laboratory system is

$$E_{st}^{+} = E_{st}(\Delta M/M) \tag{28}$$

where ΔM is the mass of the neutral fragment and M is the mass of the reactant ion. Calculated average ion kinetic energies (statistical energy plus thermal kinetic energy of the reactant ion) are compared with the experimental results of Bracher et al. (112) in Table V. This comparison clearly indicates excess kinetic energy in some cases; however, differences smaller than about 0.02 eV may be due to errors either in the calculation or in the determination of the average energy from the experimental results.

Table V
Average Kinetic Energy of Ions in the Electron Impact
Mass Spectra of Ethane and Propane

| m/e | Ethane | | Propane | |
	K.E.[a]	$\bar{E}_{st}$[b]	K.E.[a]	$\bar{E}_{st}$[b]
26	0.119	0.088	0.272	0.197
27	0.104	0.075	0.194	0.178
28	0.069	0.059	0.065	0.083
29	0.069	0.057	0.076	0.093
30	0.052	(0.052)		
39			0.152	0.084
40			0.125	0.077
41			0.091	0.064
42			0.080	0.056
43			0.058	0.054
44			0.052	(0.052)

[a] Average kinetic energy of the indicated ion in eV as reported by Bracher et al. (112).
[b] Calculated average kinetic energy of the indicated ion including the average thermal kinetic energy of the parent molecule in addition to the statistical kinetic energy imparted to the charged fragment (36,58).

Results on a few reactions for which the ionic heats of formation and activation energies indicate some excess energy is required for the reaction are summarized in Table VI. The results of Ottinger (113) were obtained

Table VI

Reactions with Excess Energy

Reaction	$\epsilon_0{}^a$	ΔH, $0°K$	E^{*b}	Experimental (Ref. 113)	(Ref. 111)	E^* Stat.^{c}	Kin.^{d}	.Int^e
$i\text{-}C_4H_{10}{}^+ \rightarrow$								
$C_3H_6{}^+ + CH_4$	0.69	0.39	0.30	—		0.09	0.07	0	0.30
$C_3H_8{}^+ \rightarrow$								
$C_3H_6{}^+ + H_2$	1.27	0.31	0.94	—		0.65	0.10	0.55	0.29
$C_3H_6{}^+ \rightarrow$								
$C_2H_4{}^+ + CH_4$	1.00	0.66	0.34	—		0.09	0.08	0	0.34
$C_2H_6{}^+ \rightarrow$								
$C_2H_4{}^+ + H_2$	1.00	0.65	0.35	—		0.45	0.10	0.35	0
$sec\text{-}C_3H_7{}^+ \rightarrow$								
$C_3H_5{}^+ + H_2$	2.06	1.44	0.62	0.43 ± 0.05		0.03^f	0.4	0.2
$C_2H_5{}^+ \rightarrow$								
$C_2H_3{}^+ + H_2$	2.4	2.1	0.3	0.03 ± 0.01		0.03^f	0	0.3

[a] Activation energies for calculations described in references (36) and (58).
[b] $E^* = \epsilon_0 - \Delta H$.
[c] Calculated for 50 eV electron impact (36,58).
[d] Part of E^* which is in kinetic energy.
[e] Part of E^* remaining as internal energy in the fragments.
[f] Typical for the energy range corresponding to metastable production.

by measuring the kinetic energy of fragmentation for "metastable" reactions in *n*-butane and *n*-heptane. This is a particularly direct and accurate method for determining the excess kinetic energy for particular reactions. These limited data suggest that in the loss of neutral methane, the excess energy remains as internal energy of one or both fragments, while in the loss of hydrogen molecule the excess energy may appear as kinetic energy or as both kinetic and internal energy.

F. Appearance Potentials and Ionization Efficiency Curves for Fragment Ions

The most fundamental quantity which can be calculated using the QET is the so-called "breakdown graph" which gives the relative abundance of the reactant ion and each product ion as a function of the internal energy, E, of the reactant ion. The abundance of the jth product ion at a time t after formation of the reactant ion is given by

$$A_j(E, t) = \frac{K_j(E)}{\sum_i K_i(E)} \left[1 - \exp\left(-\sum_i K_i(E)t\right) \right] \tag{29}$$

where the summation is carried out over the set of competing reactions. The relative abundance of the parent ion is given by

$$A_p(E, t) = \exp\left(-\sum K_j(E)t\right) \tag{30}$$

The "metastable" ions which are observed in the mass spectrum are due to those reactions occurring within a time range determined by the experimental arrangement.

The times appropriate to a particular experiment to be used in Eqs. (29) and (30) and a complete breakdown graph consisting of reactant ion, fragment ions, "metastable" ions, and "missing metastable" (16) ions* may be calculated. Alternatively, the extent of fragmentation may be calculated at a specific characteristic time, e.g., 10^{-5} or 10^{-6} sec.

The QET provides a theoretical basis for interpreting ionization efficiency curves for fragment ions. For primary ions (i.e., fragmentation of the parent ion) the ionization efficiency is given by

$$Q_j(V, T, t) = \int_0^{V - I_z} g(V, T, E)A_j(E, t) \, dE \tag{31}$$

where g is the energy distribution function given by Eq. (20a), A_j is the relative abundance of the jth fragment ion given by Eq. (29) for an internal energy E in the reactant ion at a time t after formation of the neutral ion, T is the temperature of the gas, and V is the energy of the ionizing particle. In the event that the ionizing beam has a spread in energy, the ionization efficiency may be calculated by average Q_j over the energy distribution characterizing the ionizing beam.

Equation (31) includes all the factors which determine an experimental ionization efficiency curve.

In general some of these factors are known only approximately; however, for the lowest energy reaction, the threshold behavior is most strongly influenced by the activation energy for the reaction. Therefore, by choosing the best values available for the other parameters and adjusting the activation energy until good agreement with experimental ionization efficiency curves is obtained, quite precise values for the activation energy

---

* The "missing metastable" ions in a conventional magnetic mass spectrometer are those due to reactions occurring in either the accelerating or deflecting fields. These ions do not fall within any peak in the mass spectrum (see Ref. 16).

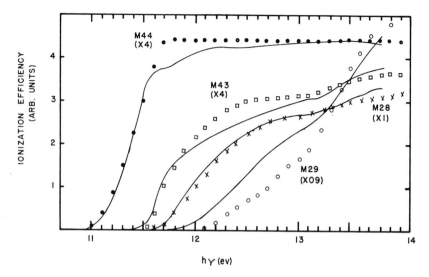

Fig. 13. Calculated (points) photoionization efficiency curves for propane compared
with the experimental results of Chupka and Berkowitz (39).

should result. This method takes into account, with the highest precision
available, the kinetic shifts, thermal shifts, etc., which affect the interpretation. For competing reactions with higher activation energies this method
for interpreting ionization efficiency curves is also applicable; however, the
results may be less precise because of the increased dependence of the
effect of the activated complex configuration on the reaction competition.

Calculated photoionization efficiency curves for several of the major
fragment ions in propane are shown in Fig. 13, where they are compared
with the results of Chupka and Berkowitz (39). The effect of temperature
on ionization efficiency is illustrated in Fig. 14. Calculated curves which
might correspond to those obtained in a typical electron-impact measurement are given in Fig. 15. The additional difficulties inherent in the
interpretation of electron-impact ionization efficiency curves is apparent.

G. Kinetic Shifts

The effect of the unimolecular reaction kinetics on the measured appearance potentials for fragment ions has been discussed previously by
Chupka (106).

In the mass spectrometer the fragment ions which are formed as a result
of reactions requiring on the order of a microsecond or more to occur will
be observed either in a "metastable" peak or not at all. If an appreciable

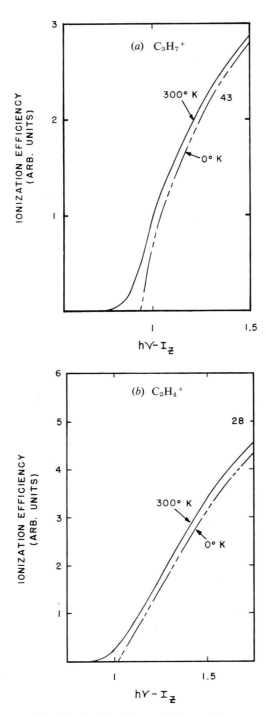

Fig. 14. See Fig. 14*c* on p. 94 for caption.

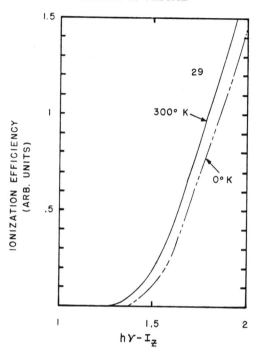

Fig. 14. Calculated photoionization efficiency curves for (a) $C_3H_7^+$, (b) $C_2H_4^+$, and (c) $C_2H_5^+$ from propane at ion source temperatures of 0°K and 300°K. See p. 93 for (a) and (b).

energy increment in excess of the activation energy is required for the rate of the unimolecular reaction, as given by Eq. (1), to exceed a value of ca. 10^6, then the minimum electron or photon energy at which the fragment ion is observed may be somewhat higher than the minimum energy which would be required to produce the fragment ion if the reaction time were not limited.

In the interpretation of experimental data the treatment has varied from neglecting the so-called "kinetic shift" entirely, to gross overestimates based on the incorrect QET rate expression (16) or improper use of the corrected form of the theory (59). The excess energies required to obtain reaction rates in the range appropriate to measurement in the mass spectrometer are summarized for several reactions in Table VII. These reactions are the lowest energy fragmentation occurring in a series of molecules studied in a recent set of calculations using the QET (25,27,34,36, 40,57,58,107). These results indicate that for a great many reactions the kinetic shifts are indeed negligible. However, in other cases—particularly, large unsaturated hydrocarbons for which the activation energy for the

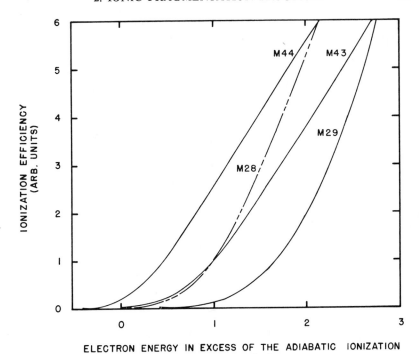

Fig. 15. Calculated electron impact ionization efficiency curves for $C_3H_8^+$(M44), $C_3H_7^+$(M43), $C_2H_4^+$(M28), and $C_2H_5^+$(M29) from propane at an ion source temperature of 250°C. The electron energy distribution was taken as a Maxwellian distribution with temperature of 2500°K.

lowest energy reaction is quite large—the shifts are substantial. The most extreme case of those studied is that of benzene in which the kinetic shift causes an error in excess of 1 eV for the heat of formation for the $C_6H_5^+$ ion. Benzene is discussed in more detail in Section VI.

The results of Table VII were obtained after determining the activation energies by comparing calculated ionization efficiency curves with experimental results as described above. For a reaction of the type

$$AB^+ \rightarrow A^+ + B \tag{32}$$

the activation energy can be written as

$$\epsilon_0 = \Delta H_{\text{reaction}} + E^*$$
$$= \Delta H_f(A^+) + \Delta H_f(B) - \Delta H_f(AB^+) + E^* \tag{33}$$

Table VII

Calculation of Excess Energies Required for Selected Reactions to Attain
Rates in the Range Appropriate to the Mass Spectrometer

Reaction	ϵ_0, eV	N^a	Excess energy ($E - \epsilon_0$) for indicated reaction rate		
			10^5 sec^{-1}	10^6 sec^{-1}	10^7 sec^{-1}
$CH_4^+ \rightarrow CH_3^+ + H$	1.70	9	<0.011	<0.011	<0.011
$C_2H_6^+ \rightarrow C_2H_4^+ + H_2$	1.00	18	<0.01	<0.01	0.06
$C_3H_8^+ \rightarrow C_3H_7^+ + H$	0.94	27	<0.01	0.02	0.17
$i\text{-}C_4H_{10}^+ \rightarrow sec\text{-}C_3H_7^+ + CH_3$	0.69	36	0.03	0.17	0.41
$n\text{-}C_4H_{10}^+ \rightarrow i\text{-}C_4H_{10}^+$	1.0	36	0.02	0.11	0.48
$C_2H_2^+ \rightarrow C_2H^+ + H$	5.8	7	<0.01	0.01	0.04
$C_2H_4^+ \rightarrow C_2H_2^+ + H_2$	2.63	12	<0.01	<0.01	0.01
$C_3H_6^+ \rightarrow C_3H_5^+ + H$	2.07	21	0.05	0.19	0.38
$C_4H_6^+ \rightarrow C_3H_3^+ + CH_3$	2.28	24	0.27	0.45	0.70
$C_6H_6^+ \rightarrow C_6H_5^+ + H$	3.9	30	1.50	2.00	2.60
$C_7H_8^+ \rightarrow C_7H_7^+ + H$	1.95	39	0.60	0.85	1.25
$CH_3OH^+ \rightarrow CH_2OH^+ + H$	1.15	12	<0.01	<0.01	0.01
$C_2H_5OH^+ \rightarrow C_2H_4OH^+ + H$	0.63	21	<0.01	<0.01	0.01
$i\text{-}C_3H_7OH^+ \rightarrow C_2H_4O^+ + CH_4$	0.47	30	<0.01	0.01	0.03
$n\text{-}C_3H_7OH^+ \rightarrow C_3H_6^+ + H_2O$	0.76	30	<0.01	0.08	0.24
$CH_3COCH_3^+ \rightarrow$ $CH_3CO^+ + CH_3$	0.70	30	<0.01	0.01	0.01
$C_2H_5COCH_3^+ \rightarrow$ $C_2H_5CO^+ + CH_3$	0.77	33	<0.01	0.02	0.06

[a] Number of internal degrees of freedom for reactant ion.

where ΔH_f is the heat of formation of the indicated species at absolute zero relative to the standard states of the elements and E^* is the minimum excess energy required for the reaction to occur. E^* may contain contributions due to activation energy for the reverse reaction which appears as excess kinetic energy in the reaction coordinate as well as contributions due to reactions in which the reaction path leads to either of the fragments in a vibrationally or electronically excited state. An energy diagram for the reaction coordinate is given in Fig. 16. In general it has been found that E^* is negligibly small for most reactions involving simple bond breaks. For reactions involving rearrangement of bonds, E^* is often substantial. These reactions are discussed in the section on kinetic energy of the fragment ions. The ionic heats of formation obtained are given in Table VIII. These values were obtained based on the

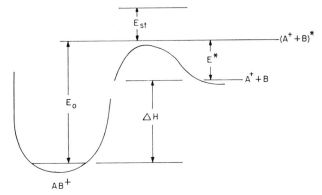

Fig. 16. Energy diagram for the reaction coordinate for a unimolecular reaction, E_0 is the activation energy for the reaction $AB^+ \rightarrow A^+ + B$, ΔH the heat of reaction, E_{st} the statistical kinetic energy, and E^* the excess energy.

Table VIII

Thermochemical Data for Small Organic Radicals obtained by Interpretation of Ionization Efficiency Curves using QET

Radical	$\Delta H_f{}^0$, eV, neutral[a]	$I_z{}^{[b]}$	$\Delta H_f{}^0$, eV, ion[a]
CH_2	3.87	10.4[c]	14.27
CH_3	1.49	9.84[d,e]	11.33
C_2H			17.34
C_2H_3			11.76
C_2H_5	1.23	8.4[e]	9.63
C_3H_3			11.15
C_3H_5			9.93
$sec\text{-}C_3H_7$	0.99	7.5[e]	8.49
$n\text{-}C_3H_7$	1.1	8.1[e]	9.2
C_6H_5			11.9
C_7H_7			9.2
CH_2OH			7.46
CH_3CO	0.18[f]	7.04	6.86
C_2H_5CO			6.51

[a] Standard heat of formation at $0°K$, in electron volts.

[b] Adiabatic ionization potential, in electron volts.

[c] Spectroscopic reference (137).

[d] Spectroscopic reference (137).

[e] Photoionization reference (135).

[f] Derived from the photoionization data of Ref. 59 on acetone and Ref. 136 on biacetyl.

analysis of photoionization efficiencies as described above, using the excellent data obtained recently in a number of different laboratories (16,39,45,36,35). The corresponding bond dissociation energies for the smaller alkanes are given in Table IX.

Table IX
Bond Energies in Small Alkanes corresponding to the Heats of Formation of Table VII

			Bond dissociation energies (eV) at $0°K$		
AB	A	B	$AB \rightarrow A^+ + B$	$AB \rightarrow A + B^+$	$AB \rightarrow A + B$
CH_4	CH_3	H	1.70	5.46	4.42
C_2H_6	C_2H_5	H	1.38	6.58	4.19
C_3H_8	$n\text{-}C_3H_7$	H	1.66	7.16	4.19
C_3H_8	$sec\text{-}C_3H_7$	H	0.94	7.04	4.08
C_2H_6	CH_3	CH_3	2.33	2.33	3.70
C_3H_8	C_2H_5	CH_3	1.33	2.77	3.57
$i\text{-}C_4H_{10}$	$sec\text{-}C_3H_7$	CH_3	0.96	3.30	3.57

VI. APPLICATIONS OF THE QET TO PARTICULAR SYSTEMS

The QET has recently been applied to calculations on a relatively large number of molecules (25–27,34,36,40,57,58,105,107,115,116). For the most part the results of these calculations are in good agreement with experimental results and have clearly shown that the major difficulties observed in the early applications of the theory were due to the incorrect state density functions used in the original formulation of the theory. The present discussion is concerned with recent applications of the theory which have helped to further define the range of molecules for which the QET may be a satisfactory approximation. The principal question, which remains extremely difficult to answer directly, is the extent to which the energy randomization hypothesis is valid.

A. Methane

Recently a detailed calculation on the low energy fragmentation reactions of methane was completed (25). The reactions considered were

$$CH_4^+ \rightarrow CH_3^+ + H \qquad\qquad (27)$$

$$\rightarrow CH_2^+ + H_2 \qquad\qquad (34)$$

$$CH_3^+ \rightarrow CH_2^+ + H \qquad\qquad (35)$$

The photoionization data of Dibeler et al. (86) and of Brehm (45) were used to determine the activation energies for the reaction of the molecule-ion. The activation energy for the secondary reaction was determined from the heats of formation for the methyl and methylene ions inferred from the primary reaction energetics. The calculated breakdown graph for methane is given in Fig. 17, where it is compared with some of the experimental points obtained by von Koch (117) in a tandem mass spectrometer where ionization is effected by charge exchange with ions (such as Xe^+, Hg^+, etc. shown in the figure) thus imparting a specific increment of energy to the molecule-ion so formed. The high energy tail on the CH_3^+ curve is due to the statistical kinetic energy lost in the primary reaction producing CH_3^+ (27).

Calculations were also performed on the deuterated methanes. The calculated portion of the 70 eV mass spectra corresponding to reactions (31, 34, and 35), is compared with the results of Mohler et al (118) in Table X. The calculations on the deuterated methanes were performed using the same models for the activated complex and normal configurations as employed in ordinary methane modified by the appropriate changes in

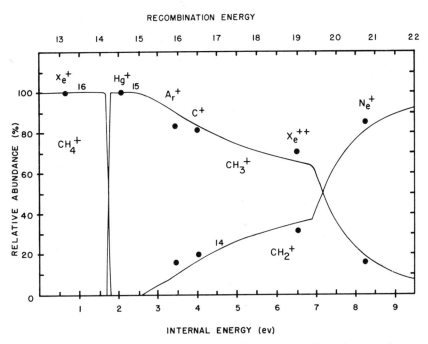

Fig. 17. Calculated breakdown graph for methane compared to charge-exchange results of von Koch (117).

Table X

70 eV Electron Impact Mass Spectra of Methanes

Mass	CH$_4$		CH$_3$D		CH$_2$D$_2$		CHD$_3$		CD$_4$	
	Calc.[a]	Partial[b]	Calc.[a]	Partial[b]	Calc.[a]	Partial[b]	Calc.[a]	Partial[b]	Calc.[a]	Partial[b]
14	8.4	9.3	3.3	5.0[c]	1.1	3.5[c]				
15	43.7	42.8	11.7	11.3	4.9	5.2	3.5	2.7		
16	47.9	47.9	37.1	37.3	17.3	15.3	4.1	5.0	7.9	6.8
17			47.9	46.7	28.8	30.1	26.4	24.4		
18					47.9	45.8	18.0	20.5	44.2	43.2
19							47.9	47.1		
20									47.9	49.9
Parent − H	43.7		37.1		28.8		18.0			
Parent − D			6.9		15.5		16.4		44.2	
Isotope effect[d]	1.79		1.79		1.85		2.04			

[a] Calculated (25).

[b] Partial mass spectrum measured by Mohler et al. (118).

[c] Contains a contribution from CD$^+$ which was not included in the calculation.

[d] (Parent − H)/(Parent − D) × N_D/N_H, where N_D and N_H are the numbers of deuterons and protons, respectively, in the molecule.

oscillator frequencies and activation energies caused by deuterium substitution. These results indicate no detectable failure of the randomization hypothesis for methane. However, some difficulty persists in connection with the so-called "metastable" processes. Dibeler and Rosenstock (119) observed a metastable peak in the mass spectrometer corresponding to the reaction

$$CD_4^+ \rightarrow CD_3^+ + D \tag{36}$$

with an intensity amounting to 0.02% of total ionization. Their data show clearly that this reaction is not collision induced. In the partially deuterated methanes for the corresponding reactions (loss of H or D) the non-collision-induced metastable intensity was at least an order of magnitude smaller. On the other hand, Ottinger (120) observed comparable metastable intensities for loss of H from the partially deuterated methane and loss of D from CD_4. These results are summarized in Table XI. The time scales for these two experiments are apparently not sufficiently different to account for the discrepancy.

Table XI

Metastable Intensities in Methanes

Reaction	Metastable intensity[a]	
	Dibeler and Rosenstock (119)	Ottinger (120)
$CH_4^+ \rightarrow CH_3^+ + H$	—[b]	0.019
$CH_3D^+ \rightarrow CH_2D^+ + H$	—[b]	0.031
$CH_2D_2^+ \rightarrow CHD_2^+ + H$	—[b]	0.033
$CHD_3^+ \rightarrow CD_3^+ + H$	—[b]	0.050
$CD_4^+ \rightarrow CD_3^+ + D$	0.027	0.032

[a] Relative to parent ion intensity of 100.
[b] Only collision-induced "metastable" dissociations were observed.

The theoretical calculations based on the classical treatment of the reaction coordinate predicts that *no* metastables should be observed in the time range appropriate to these experiments. However, an approximate quantum correction to the treatment of the reaction coordinate taking into account reflection at the barrier (25) yields results in good agreement with the measurements of Dibeler and Rosenstock. Further work is required to resolve the apparent discrepancy between the two experimental results.

B. Acetylene

Haarhoff (26) has performed calculations on acetylene and deuterated acetylene and concluded that the QET "provides a satisfactory description of the main processes which determine the decomposition pattern." The study of acetylene also provides an opportunity for determining the extent to which decomposition from separate electronic states in the ion needs to be considered. From photoionization (121), charge exchange (122), and photoelectron spectroscopy (80) measurements it is well established that the threshold for excitation to the first electronically excited state of C_2H_2 occurs at least 3 eV above the ground state. Furthermore, the lowest energy reaction

$$C_2H_2^+ \rightarrow C_2H^+ + H \tag{37}$$

occurs as a result of excitation to this or still higher lying states. Recent calculations (27) show quite unequivocably that if the reaction occurs from the ground electronic states (i.e., with complete randomization) the metastable peak corresponding to the reaction should be observed (calculated as 0.15% of total ionization). On the other hand, if the reaction occurs from the excited state, without randomization to the ground state, the calculation predicts that the metastable should not be observed. The calculated rate curves for these two cases are given in Fig. 18. This metastable is difficult to measure in a single-focusing instrument because of the proximity of the C_2^+ peak; however, it has been reported by King (123).

C. Ethanol

Ethanol has been cited (5) as a case in which the partial failure of the energy randomization hypothesis occurs and dissociation from at least two separated electronic states was required. This conclusion was based on the early charge exchange results of von Koch and Lindholm obtained in the tandem mass spectrometer (124) and on the observation of the metastable ion corresponding to

$$C_2H_5OH^+ \rightarrow C_2H_5^+ + OH \tag{38}$$

by Rosenstock and Melton (125). This reaction is of fairly minor importance in the fragmentation of ethanol and the two major reactions

$$C_2H_5OH^+ \rightarrow C_2H_4OH^+ + H \tag{39}$$

$$C_2H_5OH^+ \rightarrow CH_2OH^+ + CH_3 \tag{40}$$

both occur with lower activation energies. Therefore, the observation of the metastable corresponding to the reaction producing ethyl ion by loss of OH radical is clear evidence of incomplete randomization; however, this metastable has apparently not been observed in other measurements on ethanol (60).

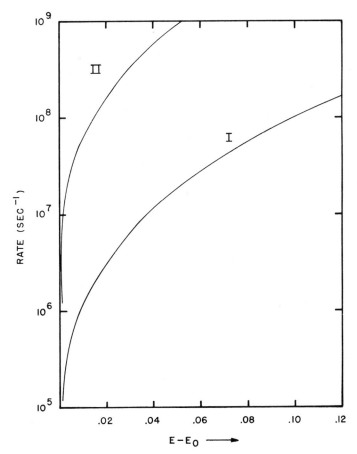

Fig. 18. Calculated rates for the reaction $C_2H_2^+ \rightarrow C_2H^+ + H$ as a function of excess internal energy of the reactant ion. (I) Reaction from the ground state; (II) reaction from the excited state.

A calculated breakdown graph for ethanol (57) is given in Fig. 19 together with some of the experimental points obtained by von Koch and Lindholm (124). These investigators interpreted their data to imply a sharp minimum in the mass 31 curve and a corresponding sharp second maximum in the mass 45 curve. A second maximum in the mass 45 curve is not necessarily inconsistent with the QET since there are three distinct reactions which may produce an ion of this mass, namely

$$C_2H_5OH^+ \rightarrow CH_3CHOH^+ + H \tag{41}$$

$$C_2H_5OH^+ \rightarrow CH_2CH_2OH^+ + H \tag{42}$$

$$C_2H_5OH^+ \rightarrow C_2H_5O^+ + H \tag{43}$$

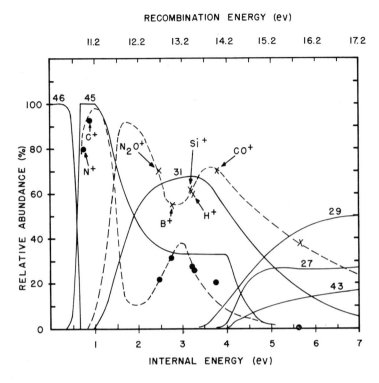

Fig. 19. Calculated breakdown graph for ethanol, compared with some of the experimental points of Lindholm et al. (122). The solid lines are the calculated curves. The dotted lines are those drawn by Lindholm et al.

However, since there can reasonably be only one reaction producing mass 31 (reaction (40)), the sharp dip and subsequent rise in mass 31 is not in accord with the behavior expected from the QET unless decompositions from noninteracting electronic states are allowed. Photoionization measurements on ethanol (41) do not appear to confirm the charge exchange results and are apparently in reasonable agreement with qualitative behavior expected from the QET. It is interesting to note that several of the ions (B^+, H^+, Si^+) which in the charge-exchange data suggest this unusual behavior in ethanol have been shown in more recent work of Lindholm and co-workers (126–129) to react by hydride ion transfer as well as by charge exchange. This possibility, which has been considered in the later charge-exchange measurements, was apparently not taken into consideration in the work on ethanol.

D. Benzene

Benzene has also been cited (5) as a case for which isolated electronic states must be taken into account. This conclusion was drawn from the fact that the range of appearance potentials experimentally determined for the reactions

$$C_6H_6^+ \rightarrow C_6H_5^+ + H \tag{44}$$

$$C_6H_6^+ \rightarrow C_6H_4^+ + H_2 \tag{45}$$

$$C_6H_6^+ \rightarrow C_4H_4^+ + C_2H_2 \tag{46}$$

$$C_6H_6^+ \rightarrow C_3H_3^+ + C_3H_3 \tag{47}$$

$$C_6H_6^+ \rightarrow C_5H_3^+ + CH_3 \tag{48}$$

was much too large to be consistent with the observation of metastable ions corresponding to all five reactions. This conclusion fails to take into account the unusually large kinetic shift predicted by the QET for a very stable ion such as benzene. QET rates calculated for the reactions of benzene are shown in Fig. 20, where the logarithm of the rate is plotted versus internal energy. The very large density of states for the normal configuration at an internal energy equal to the activation energy causes the rate at threshold to be unusually small (3×10^{-2} sec^{-1}) and causes the rate to increase very slowly with energy. As a result the kinetic shifts are quite large.

A summary of the appearance potential data for benzene is given in Table XII. The minimum energy for each reaction ($I_z + \epsilon_0$) is compared

Table XII
Appearance Potentials for Benzene

Mass	$I_z + \epsilon_0$[a]	Calc.[b]	Appearance potential, eV Experimental Ref. 3	Ref. 130
78	9.25	9.25	9.25	9.25
77	13.15	14.4–14.8	14.3–14.6	14.37
76	13.35	14.7–15.3	15.1–18.6	14.59
52	13.85	15.1–15.7	15.6–15.9	15.55
39	13.95	15.3–16.0	16.2–17.0	16?
63	14.05	15.9–17.3	16.2–16.9	—

[a] Ionization potential for the molecule plus the activation energy for the reaction, i.e., the minimum energy for the reaction.
[b] Electron energy at which the fragment attains 0.01–0.1% of total ionization in the mass spectrum.

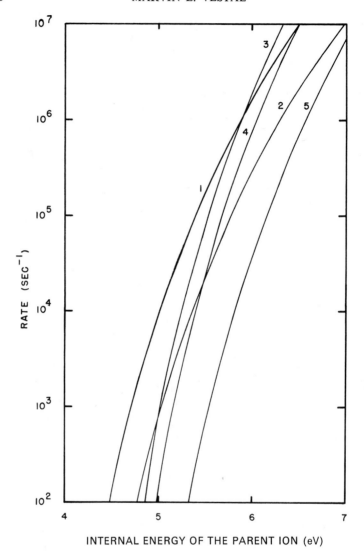

Fig. 20. Calculated rates for the reactions of benzene ions (107). (*1*) $C_6H_6^+ \rightarrow$ $C_6H_5^+ + H$; (*2*) $C_6H_6^+ \rightarrow C_6H_4^+ + H_2$; (*3*) $C_6H_6^+ \rightarrow C_4H_4^+ + C_2H_2$; (*4*) $C_6H_6^+$ $\rightarrow C_3H_3^+ + C_3H_3$; (*5*) $C_6H_6^+ \rightarrow C_5H_3^+ + CH_3$.

to the energy at which the product ion attains an intensity in the range of 0.01–0.1% of the total ionization. The latter values, which are approximately those which would be attained by conventional methods for measuring appearance potentials, are in reasonable agreement with the experimental results quoted in Table XII. A comparison of the calculated metastable intensities with the results of Ottinger (133) is given in Table XIII, and the calculated 70 eV partial mass spectrum is compared with data normalized from the API compilation (60) in Table XIV.

Table XIII
Metastable Ions in the Mass Spectrum of Benzene

	Metastable intensity	
Reaction	Calc.[a]	Exptl.[b]
$78^+ \rightarrow 77^+ + 1$	3.14	1.28
$78^+ \rightarrow 76^+ + 2$	0.46	0.23
$78^+ \rightarrow 52^+ + 26$	0.11	0.36
$78^+ \rightarrow 39^+ + 39$	0.04	0.30
$78^+ \rightarrow 63^+ + 15$	0.004	Not obs.[c]

[a] Calculated for 70 eV electron impact and reaction time between 0.5 and 2.8 μsec (107).

[b] Measured for reaction time between 2 and 6 μsec by Ottinger (131).

[c] This metastable has been observed by Jennings (138), no intensity given.

Table XIV
Benzene Mass Spectrum

Mass	Calc.[a]	Exptl.[b]
78	56.0	66.0
77	16.9	9.1
76	4.2	2.8
63	1.6	2.2
52	13.4	12.9
39	7.9	7.1

[a] Ref. 107.

[b] Normalized to 100 for the ions listed from the API Tables (60).

From these results it is clear that the behavior of benzene ion is in reasonable accord with the predictions of the QET and that isolated electronic states are not required to explain the observed experimental results.

E. Toluene

The ratio of the intensity of the fragment ion to the intensity of the parent ion for the reaction

$$C_7H_8^+ \rightarrow C_7H_7^+ + H \qquad (49)$$

has been observed in toluene as a function of residence time in the ion source by Harrison and Meyer (132) using a pulsed ion source. A QET

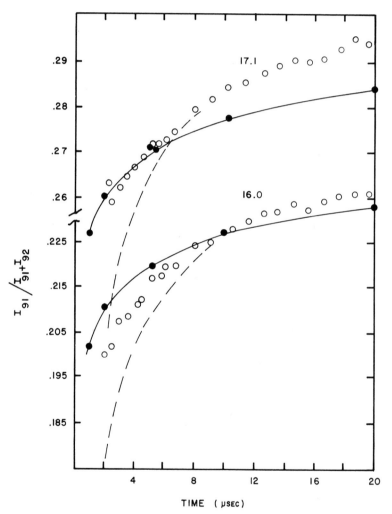

Fig. 21. Calculation (107) of the intensity ratio of mass 91 to the sum of mass 91 and 92 for toluene at electron energies of 16.0 and 17.1 compared with the results of Meyer and Harrison (132).

calculation has recently been performed for this reaction (107). The results of the calculation for two electron energies are given as the ratio of the intensity of mass 91 to the sum of mass 91 and 92 and are plotted as a function of reaction time in Fig. 21, where they are compared with the experimental results. The dotted line in Fig. 21 is the result of a calculation performed by Harrison and Meyer assuming a uniform distribution of rates between 4×10^6 and 5×10^4 sec^{-1}. From the disagreement between this calculated curve and their experimental results these investigators concluded that the QET did not apply. As can be seen from Fig. 21, a calculation based on the QET shows good agreement with the experimental points, and the small discrepancy that does exist is in the opposite direction from that obtained by using the uniform distribution.

VII. APPLICATIONS OF THE THEORY TO RADIATION CHEMISTRY

A complete discussion of the role of ionic fragmentation processes in radiation chemistry is beyond the scope of this chapter. Many aspects of that subject are discussed throughout this volume. This discussion is limited to a summary of the methods by which the experimental results on isolated ions (e.g., mass spectrometry) may be sensibly extrapolated to systems in which very frequent bimolecular collisions occur (e.g., radiation chemistry). In addition, a brief outline is given of the possible application of the quasi-equilibrium hypothesis to theories of ion–molecule reactions and to the stabilization of excited species through collisional deactivation.

A. Fragmentation Pattern as a Function of Reaction Time

The QET rate equation allows a calculation of the absolute rate (sec^{-1}) for each unimolecular reaction of a given ion as a function of the internal energy in the ion.

These calculated rates depend on parameters (e.g., activation energies, oscillator frequencies, etc.) whose values are not known, in general, with sufficient precision to assure the accuracy of the calculated rates. However, measurements of "metastable" intensities in the mass spectrometer provide experimental data in the particular time range defined by the instrument, and comparisons of calculated metastable intensities with experimental results help to determine the accuracy of absolute rates calculated from the QET. Particularly valuable experiments in this regard are the measurements of metastable intensity or of relative reaction probability as a function of reaction time such as those described by Osberghaus and Ottinger (133) and those by Meyer and Harrison (132). Recently metastable intensities have been calculated for a number of

molecules and the results are in good agreement with experimental results
(25,27,34,36,40,57,58,107,115).

The parameters characterizing the reactions can be determined by
comparing results calculated for reaction times corresponding to those for

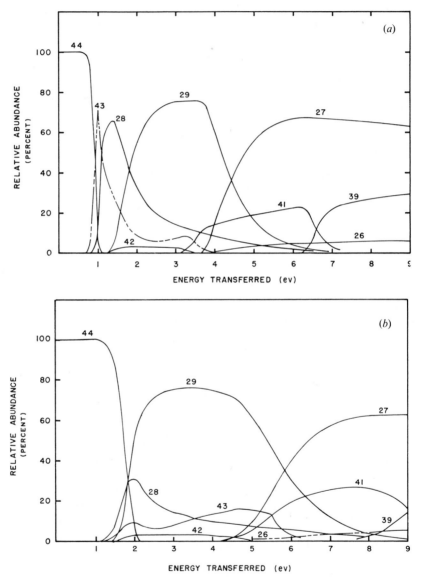

Fig. 22. Calculated breakdown graphs for propane at (a) 10^{-6} sec and (b) 10^{-10} sec
(36). The curves are identified by the mass number for the fragment ion.

which direct mass spectrometric measurements can be made (e.g., 10^{-5} to 10^{-6} sec) with the results of mass spectrometric measurements. The calculated absolute rate curves then provide the basis for calculations at the shorter time scales more appropriate to high pressure systems.

The variation of absolute rates with internal energy for several of the reactions of propane ions were given in Fig. 7. Calculations of breakdown graphs, mass spectra, temperature effects, isotope effects, and metastable intensities based on these rate calculations are in good agreement with experimental results. Calculated breakdown graphs for propane at 10^{-6} and 10^{-10} sec are compared in Fig. 22.

These results are qualitatively typical of those to be expected for most systems. At sufficiently short reaction times the extent of all reactions is somewhat reduced. As a result the parent ion intensity increases with decreasing reaction times; the intensities of the products of the parent ion fragmentation may either decrease or increase; and the intensities of the ions resulting from subsequent fragmentation of the primary ions decrease. These expectations based on the QET are qualitatively in accord with conclusions arrived at by Ausloos and co-workers (134) by studying the effect of pressure on radiation chemical and photochemical yields of products identified with particular ionic precursors.

B. Collisional Deactivation of Excited Ions

In previous attempts to extrapolate mass spectral data (and QET calculations) to fragmentations occurring in high pressure systems, the extent of fragmentation at a time corresponding to the mean collision time has generally been selected as appropriate to the high pressure system (140,141) although it was recognized that the probability for complete deactivation of a highly excited molecule-ion on a single collision might be rather small. Ausloos and Lias (139) have studied the effects of the pressure of several monatomic, diatomic, and polyatomic molecule additives on the fragmentation of n-butane ions. These investigators found that larger molecules were much more effective, at a given pressure, in quenching the unimolecular reactions.

This problem can be treated by methods quite similar to those employed in the discussions on the distribution of energy between the fragments of a unimolecular reaction, provided the quasi-equilibrium hypothesis is a reasonable approximation. The collision of an excited ion with a neutral molecule (in its ground state) can be written as

$$(A^+)' + B \rightarrow (A^+B)' \rightarrow (A^+)'' + B^*$$

where the primes and asterisks are used to indicate internal energy. Since at thermal energies the duration of such a collision is typically an order of magnitude or more larger than the period of the molecular vibrations, it may be reasonable to assume that effective equilibration of the energy occurs among the various internal degrees of freedom. If so, then the energy lost by the ion may be calculated by applying the methods for calculating the statistical energy distributions for unimolecular reaction, discussed in connection with Eqs. (24) and (25), to the unstable, short-lived complex $(A^+B)'$.

C. Ion–Molecule Reactions

The present status of theories of ion–molecule reactions is quite analogous to the status of the theories of mass spectra prior to the development of the quasi-equilibrium theory. Most of the effort up to the present time has concentrated on the collision process just as the early theoretical work on electron impact ionization and dissociation concentrated on the Franck-Condon factors. For polyatomic systems, it is now well established that for the ions observed in the mass spectrometer the unimolecular reactions do not, in general, depend upon the particular state in which the ion is formed, but rather only on the internal energy of the ion. For ion–molecule reactions of polyatomic systems which involve a long-lived intermediate complex, it appears reasonable to suppose that a similar situation exists. Consideration of the collision process is required to determine the probability for complex formation, with a certain energy, just as considertion of vertical ionization is required to determine the probability for parent ion formation, with a certain energy, in electron impact ionization in the mass spectrometer. However, from the viewpoint of the QET, the dissociation of the complex (parent ion) is determined by the reaction energetics and the potential surfaces in the neighborhood of certain activated complex configurations.

The rates for the various possible unimolecular dissociations of the intermediate complex can be calculated by applying the QET to the complex in a manner which is completely analogous to application of the theory to the unimolecular dissociation of parent ions in mass spectra.

There are many experimental studies described in the literature (see Chapter 4 of this volume) of ion–molecule reactions for which a quasi-equilibrium treatment may offer some insight into the processes occurring. As an illustration of the methods which may be employed, the reactions of methylene ions with methane may be considered (142).

In this case the intermediate complex is the ethane ion which has been extensively studied both theoretically (116) and experimentally (39,128).

The reaction scheme which accounts for the major fragments of ethane ion is given by

$$C_2H_6^+ \rightarrow C_2H_4^+ + H_2 \tag{50}$$

$$C_2H_6^+ \rightarrow C_2H_5^+ + H \tag{51}$$

$$C_2H_6^+ \rightarrow CH_3^+ + CH_3 \tag{52}$$

$$C_2H_4^+ \rightarrow C_2H_2^+ + H_2 \tag{53}$$

$$C_2H_5^+ \rightarrow C_2H_3^+ + H_2 \tag{54}$$

The primary reaction

$$CH_2^+ + CH_4 \rightarrow C_2H_6^+ \tag{55}$$

produces ethane ions with an energy relative to the ground state of neutral ethane given by

$$E(C_2H_6^+) = \Delta H_f(CH_2^+) + E_{int}(CH_2^+) + \Delta H_f(CH_4) - \Delta H_f(C_2H_6)$$

$$+ \frac{M(CH_2)}{M(C_2H_6)} \text{ K.E.} \approx 14.3 + E_{int}(CH_2^+) + \tfrac{16}{30} \text{ K.E.} \tag{56}$$

where the energies are expressed in eV. From the charge-exchange studies of methane by von Koch (117) the CH_2^+ ions produced by 50 eV electron impact on methane should have internal energies ranging from zero up to several eV. For the present discussion the internal energy distribution for the methylene ions has been approximated by assuming equally probable excitations of 0 and 2.5 eV for the methylene ions. Under this assumption the ethane ions produced by the ion–molecule reaction at 0.3 eV kinetic energy have energies of approximately 14.5 and 17.0 eV relative to the ground state of ethane. The breakdown graph for ethane has been determined by von Koch (128) using the charge-exchange technique. The arithmetic means of von Koch's data at recombination energies of 14.5 and 17.0 eV were used to obtain the calculated data corresponding to 0.3 eV kinetic energy given in Table XV. For a kinetic energy of 4.2 eV, the corresponding energies in the ethane ion are 16.5 and 19.0 eV, which were used to apply the data of von Koch to the higher kinetic energy case.

The agreement shown in Table XV between the relative abundances of the products of the ion–molecule reaction and the products of dissociation of ethane ions with comparable internal energy appears to support the validity of the quasi-equilibrium approach. For ion–molecule reactions, such as this one, in which the intermediate complex is an ion which can be prepared by direct ionization in the mass spectrometer, the quasi-equilibrium theory provides the basis for applying the mass spectral data to the interpretation of data on ion–molecule reactions. For other reactions, in

Table XV

Comparison of Relative Abundance of Ion–Molecule Reaction Products for $CH_2^+ + CH_4$ with Calculation from Charge-Exchange Data for Ethane

Reaction	K.E.	CH$_3^+$	C$_2$H$_2^+$	C$_2$H$_3^+$	C$_2$H$_4^+$	C$_2$H$_5^+$
			Relative abundance			
$CH_2^+ + CH_4$	0.3	—	11[a]	25[a]	37[a]	27[a]
		8[b]	12[b]	24[b]	36[b]	20[b]
	4.2	—	22[a]	69[a]	5[a]	4[a]
		6[b]	24[b]	48[b]	18[b]	4[b]

[a] Experimental relative abundance (142).
[b] Calculated using the charge-exchange data of von Koch (128) for ethane assuming equally probable CH_2^+ internal energies of 0 and 2.5 eV.

which the intermediate complex is not a species which can be prepared by direct ionization, the QET should also apply provided the intermediate complex is sufficiently long-lived. However, applications to these systems present additional problems, since very little information is available on the properties of these unstable intermediaries.

VIII. PRESENT STATUS

The results described in this chapter support the contention that the QET is a satisfactory approximation for the unimolecular dissociation of many polyatomic molecule-ions. One may also tentatively infer that the quasi-equilibrium hypothesis upon which the theory is based should be equally valid for many other situations involving polyatomic systems—provided that the time scale for the phenomena under investigation is long compared to the vibrational periods for the systems. Application of the quasi-equilibrium hypothesis to other problems involving polyatomic systems (e.g., collisional deactivation and ion–molecule reactions) should, in the next few years, assist greatly in attaining a more satisfactory understanding of the reactions occurring in radiation chemical systems.

REFERENCES

1. F. Fiquet-Fayard, *Actions Chim. Biol. Radiations*, **8**, 31 (1965).
2. C. Melton, in *Mass Spectrometry of Organic Ions*, F. McLafferty, Ed., Academic Press, New York, 1963, Chap. 4.
3. F. H. Field and J. L. Franklin, *Electron Impact Phenomena*, Academic Press, New York, 1957.

4. H. Rosenstock and M. Krauss, in *Mass Spectrometry of Organic Ions*, F. McLafferty, Ed., Academic Press, New York, 1963, Chap. 1.
5. H. Rosenstock and M. Krauss, *Advances in Mass Spectrometry*, Vol. 2, Macmillan, New York, 1963, p. 251.
6. A. Wahrhaftig, *NATO Advanced Study Institute on Mass Spectrometry*, R. I. Reed, Ed., Academic Press, New York, 1964, p. 137.
7. F. H. Mies and M. Krauss, *J. Chem. Phys.*, **45**, 4455 (1966).
8. R. D. Levine, *J. Chem. Phys.*, **44**, 2046 (1966).
9. H. M. Rosenstock, M. B. Wallenstein, A. L. Wahrhaftig, and H. Eyring, *Proc. Natl. Acad. Sci., U.S.*, **38**, 667 (1952).
10. A. Kropf, E. M. Eyring, A. L. Wahrhaftig, and H. Eyring, *J. Chem. Phys.*, **32**, 149 (1960).
11. L. Friedman, F. A. Long, and M. Wolfsberg, *J. Chem. Phys.*, **26**, 714 (1957).
12. L. Friedman, F. A. Long, and M. Wolfsberg, *J. Chem. Phys.*, **30**, 1605 (1959).
13. L. Friedman, F. A. Long, and M. Wolfsberg, *J. Chem. Phys.*, **27**, 613 (1957).
14. J. Collin, *Bull. Roy. Soc. (Liège)*, **7**, 520 (1956).
15. E. M. Eyring and A. L. Wahrhaftig, *J. Chem. Phys.*, **34**, 23 (1961).
16. B. Steiner, C. F. Giese, and M. G. Inghram, *J. Chem. Phys.*, **34**, 189 (1961).
17. W. A. Chupka, *J. Chem. Phys.*, **30**, 191 (1959).
18. W. A. Chupka and J. Berkowitz, *J. Chem. Phys.*, **32**, 1546 (1960).
19. H. M. Rosenstock, *J. Chem. Phys.*, **34**, 2182 (1961).
20. M. Vestal, A. L. Wahrhaftig, and W. H. Johnston, *J. Chem. Phys.*, **37**, 1276 (1962).
21. H. D. Hagstrum, *Rev. Mod. Phys.*, **23**, 185 (1951).
22. R. L. Platzman, *J. Chem. Phys.*, **38**, 2775 (1963).
23. S. Glasstone, K. J. Laidler, and H. Eyring, *Theory of Rate Processes*, McGraw-Hill, New York, 1941.
24. M. Vestal, *J. Chem. Phys.*, **41**, 3997 (1964).
25. M. Vestal, unpublished work, see Ref. 145 for summary of methods.
26. P. C. Haarhoff, *Mol. Phys.*, **7**, 101 (1964).
27. M. Vestal, unpublished work, see Ref. 145 for summary of methods.
28. V. H. Dibeler and H. M. Rosenstock, *J. Chem. Phys.*, **39**, 3106 (1963).
29. E. W. Schlag and R. A. Sondmark, *J. Chem. Phys.*, **37**, 168 (1962).
30. S. H. Lin and H. Eyring, *J. Chem. Phys.*, **39**, 1577 (1963).
31. P. C. Haarhoff, *Mol. Phys.*, **6**, 337 (1963).
32. E. Thiele, *J. Chem. Phys.*, **39**, 3258 (1963).
33. K. A. Wilde, *J. Chem. Phys.*, **41**, 448 (1964).
34. M. Vestal, unpublished work, see Ref. 145 for summary of methods.
35. K. S. Pitzer, *J. Chem. Phys.*, **12**, 310 (1944).
36. M. Vestal, unpublished work, see Ref. 145 for summary of methods.
37. W. H. McFadden and A. L. Wahrhaftig, *J. Am. Chem. Soc.*, **78**, 1572 (1956).
38. A. Langer, *J. Phys. Colloid Chem.*, **54**, 618 (1950).
39. W. A. Chupka and J. Berkowitz, *J. Chem. Phys.*, **47**, 2921 (1967).
40. M. Vestal, unpublished work, see Ref. 145 for summary of methods.
41. W. A. Chupka and J. Berkowitz, ASTM Committee E-14 Conference on Mass Spectrometry and Related Topics, Dallas, 1966.
42. E. Pettersson, *Arkiv Fysik*, **25**, 181 (1963).
43. Z. Dolejsek, V. Hanus, and K. Vokac, in *Advances in Mass Spectrometry*, Vol. 3, W. L. Mead, Ed., Elsevier, Amsterdam, 1966, p. 503.
44. J. Collin and F. P. Lossing, *J. Am. Chem. Soc.*, **79**, 5848 (1957).

45. B. Brehm, *Z. Naturforsch.*, **21a**, 196 (1966).
46. H. M. Grubb and S. Meyerson, in *Mass Spectrometry of Organic Ions*, F. McLafferty, Ed., Academic Press, New York, 1963, Chap. 10.
47. S. Meyerson, *Appl. Spectry.*, **9**, 120 (1955).
48. S. Meyerson and P. N. Rylander, *J. Phys. Chem.*, **62**, 2 (1958).
49. S. Meyerson and P. N. Rylander, *J. Am. Chem. Soc.*, **79**, 1058 (1957).
50. S. Meyerson and J. D. McCollum, in *Advances in Analytical Chemistry and Instrumentation*, Vol. 2, C. Reilley, Ed., Interscience, New York, 1963.
51. P. N. Rylander and S. Meyerson, *J. Chem. Phys.*, **27**, 1116 (1957).
52. P. N. Rylander, S. Meyerson, and H. M. Grubb, *J. Am. Chem. Soc.*, **79**, 842 (1957).
53. S. Meyerson and P. N. Rylander, *J. Chem. Phys.*, **27**, 901 (1957).
54. S. Meyerson, *J. Chem. Phys.*, **34**, 2046 (1961).
55. P. N. Rylander and S. Meyerson, *J. Am. Chem. Soc.*, **78**, 5799 (1956).
56. M. Vestal and H. M. Rosenstock, *J. Chem. Phys.*, **35**, 2008 (1961).
57. M. Vestal, unpublished work, see Ref. 145 for summary of methods.
58. M. Vestal, unpublished work, see Ref. 145 for summary of methods.
59. E. Murad and M. G. Inghram, *J. Chem. Phys.*, **40**, 3263 (1964).
60. Am. Petroleum Inst., Research Project 44, "Catalog of Mass Spectral Data," Texas A and M, College Station, Texas.
61. M. Krauss and V. H. Dibeler, in *Mass Spectrometry of Organic Ions*, F. McLafferty, Ed., Academic Press, New York, 1963, Chap. 3.
62. J. Collin, *NATO Advanced Study Institute on Mass Spectrometry*, R. I. Reed, Ed., Academic Press, New York, 1964, p. 183–222.
63. L. G. Smith, *Phys. Rev.*, **51**, 663 (1937).
64. R. E. Honig, *J. Chem. Phys.*, **16**, 105 (1948).
65. R. H. Vought, *Phys. Rev.*, **71**, 93 (1947).
66. J. W. Warren, *Nature*, **165**, 810 (1950).
67. J. D. Morrison, *J. Chem. Phys.*, **19**, 1305 (1951).
68. J. D. Morrison and A. J. C. Nicholson, *J. Chem. Phys.*, **20**, 1021 (1952).
69. A. Barfield and A. L. Wahrhaftig, *J. Chem. Phys.*, **41**, 2947 (1964).
70. T. Cottrell, *Strengths of Chemical Bonds*, Butterworths, London, 1958.
71. (a) F. D. Rossine, K. S. Pitzer, W. J. Taylor, J. P. Ebert, J. E. Kilpatrick, C. W. Beckett, M. G. Williams, and H. G. Werner, "Selected Values of Properties of Hydrocarbons," American Petroleum Res. Project 44, *Natl. Bur. Std. Circ.* **461**, 1947. (b) F. D. Rossine, D. D. Wagman, W. H. Evans, S. Levine, and I. Jaffe, *Natl. Bur. Std. Circ.* **500**, 1950.
72. R. W. Kiser, "Tables of Ionization Potentials," U.S. Atomic Energy Commission Rept. TID-6142, 1964.
73. K. Watanabe, J. Nakayama, and J. Mottl, "Final Report on Ionization Potentials," University of Hawaii, Honolulu, 1959.
74. J. L. Franklin et al., *Natl. Bur. Std. Publ.*, in preparation.
75. R. R. Bernecker and F. A. Long, *J. Phys. Chem.*, **65**, 1565 (1961).
76. T. N. Radwan and D. W. Turner, *J. Chem. Soc.*, **1966**, 85.
77. D. W. Turner and M. I. Al-Joboury, *J. Chem. Phys.*, **37**, 3007 (1962).
78. M. I. Al-Joboury and D. W. Turner, *J. Chem. Soc.*, **1963**, 5141.
79. M. I. Al-Joboury and D. W. Turner, *J. Chem. Soc.*, **1964**, 4434.
80. M. I. Al-Joboury, D. P. May, and D. W. Turner, *J. Chem. Soc.*, **1965**, 616.
81. M. I. Al-Joboury, D. P. May, and D. W. Turner, *J. Chem. Soc.*, **1965**, 6350.
82. D. W. Turner and D. P. May, *J. Chem. Phys.*, **45**, 471 (1966).

83. D. P. May and D. W. Turner, *Chem. Commun.*, **1966**, 199.
84. R. I. Schoen, *J. Chem. Phys.*, **40**, 1830 (1964).
85. D. C. Frost, C. A. McDowell, J. S. Sandhu, and D. A. Vroom, *J. Chem. Phys.*, **46**, 2008 (1967).
86. V. H. Dibeler, M. Krauss, R. M. Reese, and F. H. Harlee, *J. Chem. Phys.*, **42**, 3791 (1965).
87. A. J. C. Nicholson, *J. Chem. Phys.*, **43**, 1171 (1965).
88. M. Vestal, unpublished work, see Ref. 145 for summary of methods.
89. A. Streitwieser, Jr., "Molecular Orbital Study of Ionization Potential of Organic Compounds," AFOSR Repts. 59-893, 59-912, 59-913, 59-914, 59-915, 1959.
90. A. Streitwieser, Jr. and P. M. Nair, *Tetrahedron*, **5**, 149 (1959).
91. G. G. Hall, *Trans. Faraday Soc.*, **49**, 113 (1953).
92. S. Geltman, *Phys. Rev.*, **102**, 171 (1956).
93. J. D. Morrison, *J. Appl. Phys.*, **28**, 1409 (1957).
94. W. A. Chupka and M. Kaminsky, *J. Chem. Phys.*, **35**, 1991 (1961).
95. R. Botter, V. H. Dibeler, J. A. Walker, and H. M. Rosenstock, *J. Chem. Phys.*, **44**, 1271 (1966).
96. V. H. Dibeler, R. M. Reese, and J. M. Krauss, *J. Chem. Phys.*, **42**, 2046 (1965).
97. V. H. Dibeler and J. A. Walker, *J. Chem. Phys.*, **43**, 1842 (1965).
98. R. M. Reese and H. M. Rosenstock, *J. Chem. Phys.*, **44**, 2007 (1966).
99. R. Botter, V. H. Dibeler, J. A. Walker, and H. M. Rosenstock, *J. Chem. Phys.*, **45**, 1298 (1966).
100. M. E. Akopyan, F. I. Vilesov, and A. N. Terenin, *Bull. Acad. Sci. USSR Phys. Ser.*, **27**, 1054 (1964).
101. A. Cassuto, *Advances in Mass Spectrometry*, Vol. 2, R. M. Elliott, Ed., Macmillan, New York, 1963, p. 296.
102. O. Osberghaus and R. Taubert, *Z. Physik. Chem. (Frankfurt)*, **4**, 264 (1955).
103. H. Erhardt and O. Osberghaus, *Z. Naturforsch.*, **15a**, 575 (1960).
104. M. Wallenstein and M. Krauss, *J. Chem. Phys.*, **34**, 929 (1961).
105. M. Vestal, *J. Chem. Phys.*, **43**, 1356 (1965).
106. W. A. Chupka, *J. Chem. Phys.*, **30**, 1546 (1960).
107. M. Vestal, unpublished work, see Ref. 145 for summary of methods.
108. C. E. Klots, *J. Chem. Phys.*, **41**, 117 (1964).
109. R. Taubert, *Z. Naturforsch.*, **19a**, 484 (1964).
110. R. Fuchs and R. Taubert, *Z. Naturforsch.*, **19a**, 494 (1964).
111. R. Taubert, *Z. Naturforsch.*, **19a**, 911 (1964).
112. J. Bracher, H. Erhardt, R. Fuchs, O. Osberghaus, and R. Taubert, *Advances in Mass Spectrometry*, R. M. Elliott, Ed., Vol. 2, Macmillan, New York, 1963, p. 285.
113. C. Ottinger, *Phys. Letters*, **17**, 269 (1965).
114. J. Appell, J. Durup, and F. Heitz, *Advances in Mass Spectrometry*, Vol. 3, W. L. Mead, Ed., Elsevier, Amsterdam, 1966, p. 457.
115. J. C. Tou, L. P. Hills, and A. L. Wahrhaftig, *J. Chem. Phys.*, **45**, 2129 (1966).
116. Z. Prasil and W. Frost, *J. Phys. Chem.*, **71**, 3166 (1967).
117. H. von Koch, *Arkiv Fysik*, **28**, 529 (1965).
118. F. L. Mohler, V. H. Dibeler, and E. Quinn, *J. Res. Natl. Bur. Std.*, **61**, 171 (1958).
119. V. H. Dibeler and H. M. Rosenstock, *J. Chem. Phys.*, **39**, 1326 (1963).
120. C. Ottinger, *Z. Naturforsch.*, **20a**, 1232 (1965).

121. R. Botter, V. H. Dibeler, J. A. Walker, and H. M. Rosenstock, *J. Chem. Phys.*, **45**, 1298 (1966).
122. E. Lindholm, I. Szabo, and P. Wilmenius, *Arkiv Fysik*, **25**, 417 (1963).
123. J. G. Larson and A. B. King, private communication cited in Ref. 121.
124. H. von Koch and E. Lindholm, *Arkiv Fysik*, **19**, 123 (1961).
125. H. M. Rosenstock and C. E. Melton, unpublished work cited in Ref. 5.
126. E. Pettersson and E. Lindholm, *Arkiv Fysik*, **24**, 49 (1963).
127. E. Pettersson, *Arkiv Fysik*, **25**, 181 (1963).
128. H. von Koch, *Arkiv Fysik*, **28**, 559 (1965).
129. W. A. Chupka and E. Lindholm, *Arkiv Fysik*, **25**, 349 (1963).
130. J. Momigny, L. Brakier, and L. D'or, *Bull. Acad. Roy. Soc. Belg.*, **48**, 1002 (1962).
131. C. Ottinger, *Z. Naturforsch.*, **20a**, 1229 (1965).
132. F. Meyer and A. G. Harrison, *J. Chem. Phys.*, **43**, 1778 (1965).
133. O. Osberghaus and C. Ottinger, *Phys. Letters*, **16**, 121 (1965).
134. P. Ausloos, S. G. Lias, and I. B. Sandoval, *Discussions Faraday Soc.*, **36**, 66 (1963).
135. F. A. Elder, C. Giese, B. Steiner, and M. Inghram, *J. Chem. Phys.*, **36**, 3292 (1962).
136. E. Murad and M. G. Inghram, *J. Chem. Phys.*, **41**, 404 (1964).
137. G. Herzberg, *Can. J. Phys.*, **39**, 1511 (1961).
138. K. R. Jennings, *J. Chem. Phys.*, **43**, 4176 (1965).
139. P. Ausloos and S. G. Lias, *J. Chem. Phys.*, **45**, 524 (1966).
140. J. H. Futrell, *J. Chem. Phys.*, **35**, 353 (1961).
141. M. Vestal, A. L. Wahrhaftig, and W. H. Johnston, "Theoretical Studies in Basic Radiation Chemistry," Aeronautical Res. Lab. Rept. 62-426, 1962.
142. J. P. Abramson and J. H. Futrell, *J. Chem. Phys.*, **45**, 1925 (1966).
143. M. L. Vestal, unpublished work, see Ref. 145 for summary of methods.
144. B. Friedman, *Principles and Techniques of Applied Mathematics*, Wiley, New York, 1956, p. 136.
145. M. Vestal and G. Lerner, "Fundamental Studies Relating to the Radiation Chemistry of Small Organic Molecules," Aerospace Res. Lab. Rept. 67-0114 (1967).

Chapter 3

The Fragmentation of Highly Excited Neutral Molecules

L. Wayne Sieck

National Bureau of Standards, Washington, D.C.

I. INTRODUCTION

To the reader not familiar with the field of radiation chemistry, the inclusion of a chapter describing the fragmentations of highly excited neutral molecules would perhaps appear to be irrelevant in a volume dedicated to a discussion of the physical and chemical effects of what is commonly referred to as "ionizing radiation." Chapter 1, of course, clearly points out the importance of neutral excitation in systems subjected to radiolysis. Indeed, one needs only to compare W values (the average energies necessary to cause an ionization act, which are mostly in the range of 20–35 eV) with molecular ionization potentials (8–15 eV) in order to verify that much of the energy deposited in a system will be found in nonionized levels of neutral molecules. As pointed out in Chapter 1, it is very difficult to establish what fraction of the absorbed energy results in neutral excitation. This problem has been considered in some detail in another volume (1) and is discussed in other chapters of this book. Although no attempt will be made in this particular chapter to assess the relative importance of neutral excited states in radiation chemistry, the information contained here does provide a basis for interpreting those aspects of the overall decomposition which cannot be associated with ion formation.

In recent years, the literature of chemical physics has been characterized by a widespread interest in the unimolecular fragmentations of highly excited neutral molecules. Prior to the 1960's, most of the attention was focused on diatomic and triatomic systems, particularly by physicists, because the states of these molecules are subject to analysis by absorption and emission spectroscopy and the data were sorely needed by theoreticians to test the advancing art of applied quantum mechanics (molecular orbital theory). Although many of the intensive properties of these molecules and the associated fragments (force constants, bond energies and angles, position and symmetries of energy levels, etc.) (2) were established during this period, we had no accurate insight into the decomposition patterns of highly excited molecules until the first detailed far ultraviolet photolysis studies were reported (3). Mechanisms involving neutral molecules had been invoked for years to account for the appearance of various products in radiolysis and discharge tube experiments, but the polychromatic nature of the incident radiation always prevented an accurate estimate of the excitation energies involved in the parent molecules. The behavior of molecules in lower states (0–5 eV) has now very nearly been defined by photolysis in the visible and near ultraviolet, pyrolysis and shock heating, Hg sensitization, and other methods. However, the dissociative events occurring at these excitation energies have been adequately reviewed elsewhere (4,5) and will not be examined in detail in this chapter. The emphasis here will be placed on the fragmentations of relatively complex systems (mainly organic) excited to nonionized levels situated well above the dissociation limit of the ground state, especially where the excitation is monochromatic and the primary processes are reasonably well understood.

Very few theoretical descriptions of dissociating organic systems at high excitation energies have appeared in the literature, which may be ascribed to the fact that the necessary chemical information has only been accumulated recently. A related and somewhat more discouraging observation is that little, if any, structure can be found in the vacuum ultraviolet absorption spectra of saturated hydrocarbons, which have received the greatest attention in chemical studies. The absence of structure indicates transitions to repulsive levels, and the properties of these potential surfaces are not easily predicted since extensive intercombination and mixing is expected in these regions of high oscillator strength. It is no surprise, then, that only the fragmentations of the simplest hydrocarbons, methane (6,7) and ethane (7) have been treated in any detail by molecular orbital theory in its various forms. Hopefully, the recent explosion of chemical data will provide the proper justification and incentive for more theoretical scrutiny in the near future.

The molecules under consideration may be excited to regions of high state densities and many channels are available for dissociation. Consecutive fragmentations of vibrationally excited transients are favored at these energies, and in many cases this condition tends to obscure both the identities and relative probabilities of competing primary processes. Since the unraveling of parent–daughter relationships at high energies presents such a formidable kinetic problem, an effort has been made to review those studies which most clearly develop the relationship between a particular mode of fragmentation and the energy content of the parent molecule.

A. Methods For Producing Excited Molecules

The great majority of studies have involved exposure of a sample either to energetic photons or electrons. Sensitization experiments, which involve energy transfer to the sample from atoms or molecules in metastable states, have also been reported for a few isolated systems (8). As discussed in Chapter 1, the secondary electrons released in a sample by the interactions of high energy radiation are capable of inducing molecular excitation. Highly excited molecules are also generated during the radiolysis when ionic intermediates are neutralized, since an amount of energy equivalent to the ionization energy of the ionic species may be available if a free electron is involved in the neutralization step. The unimolecular chemistry which follows ion–electron combination is very difficult to pinpoint, however, and the unraveling of the ensuing fragmentation schemes remains one of the challenging problems in radiation chemistry. What is known of the chemical consequences of neutralization will be discussed in other chapters. Some of the other common methods which have been used in an attempt to specifically investigate dissociative channels in highly excited neutral molecules will be described in the following passages.

1. **Electron Impact.** *a. Mass Spectrometry.* Whenever a chemical medium is exposed to a flux of energetic electrons a fraction of the electron–molecule encounters will result in electronic excitation. Consequently, excited neutral molecules are necessarily generated even as one is determining the mass spectrum of a gaseous sample in a conventional analytical mass spectrometer. The use of a mass spectrometer for the specific purpose of determining fragmentation channels in highly excited neutral molecules requires extensive instrumental design changes, since under normal operating conditions only the ionic products of electron (or photon) impact are detected. Melton (9), for example, has recently achieved conditions which are essentially equivalent to low pressure gas phase radiolysis in such an apparatus. This instrument is discussed in some detail in Chapter 4, Section III.

b. Applied Fields during the Radiolysis. The application of an electric field to a sample under irradiation, a technique originally introduced by Essex and his co-workers (10), has been used extensively as an interpretive tool in gas phase studies. In the region of field strengths below the threshold for electron multiplication, electrons accelerating under the influence of the applied field only have sufficient energy to excite molecules to neutral electronic levels, and the yields of those products which can be traced to fast ion–molecule reactions and the fragmentation of excited ions are essentially unaffected. As the field strength is increased in the saturation current region the average electron energy increases and upper electronic levels may be populated. Detailed explanations of those experimental conditions necessary to carry out meaningful experiments are reported elsewhere (11,12).

A representative set of data derived from the radiolysis of a C_2H_6–C_2D_6–NO mixture (1:1:0.05) is shown in Fig. 1. The resulting yields of some labeled methanes and ethylenes are plotted as M/N (molecules/ion pair) at various field strengths in the saturation current region. It can be seen that $M(C_2D_4)/N$ and $M(CD_4)/N$ increase markedly at higher voltages while $M(C_2D_3H)/N$ and $M(CD_3H)/N$ are essentially constant.

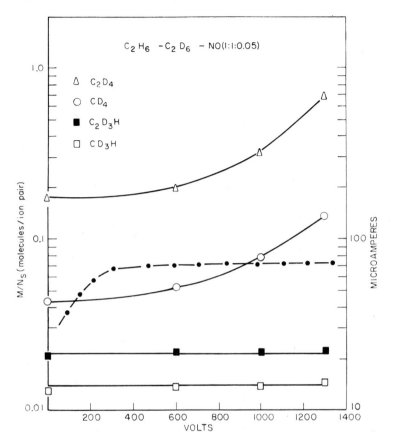

Fig. 1. Effect of an applied electric field during the radiolysis of ethane. (———) Ion pair yield of a particular product; (–––) ion current in microamperes. (Data taken from Ref. 13.)

On the basis of such data Carmichael et al. (13) ascribed the enhancement in the yields of C_2D_4 and CD_4 to the dissociations

$$C_2D_6^* \rightarrow C_2D_4 + D_2 \quad (\text{or } 2D) \tag{1}$$

$$C_2D_6^* \rightarrow CD_4 + CD_2 \tag{2}$$

The constancy of $M(C_2D_3H)/N$ and $M(CD_3H)/N$ over the same range may be taken as evidence that these products are formed either in fast ion–molecule processes or other bimolecular reactions involving energetic radicals produced during ionic fragmentation.

For several reasons, the exact amount of energy imparted to a decomposing molecule in an applied field experiment cannot be defined. In the first place, the secondary electron spectrum generated by radiolysis exhibits a distribution of energies which is not necessarily characteristic of the steady-state degradation spectrum. Superimposed on the primary distribution is a Boltzmann distribution of velocities reflecting the mean free paths of the accelerating electrons before they lose energy through inelastic collision. Since the greater portion of the oscillator strength found in subionization states is situated in the region of the ionization threshold (3) most of the additional neutral decomposition induced by applied fields can often be related to the high energy "tail" of the electron swarm.

The suggestion may be made that the excitation of neutral molecules in applied field experiments proceeds mainly through spin or symmetry-forbidden transitions. This argument appears to be well founded in view of the observed contours of excitation efficiency curves for forbidden and allowed transitions. Optically forbidden excitations exhibit sharp maxima at electron energies corresponding to the transition energy, while cross sections for allowed transitions increase essentially linearly from the threshold value and exhibit broad maxima in the region 25–50 eV. Therefore one might suspect that an applied electric field, which produces additional excitation by low energy electrons, would result in a greater relative population of optically forbidden states than is representative of the unperturbed radiolytic system.

For saturated hydrocarbons, however, which have received the greatest attention in applied field experiments, there is no evidence that the mechanism of neutral decomposition is influenced by the multiplicity of the initial state in either the spin or Δ function. Furthermore, a far greater portion of the subionization oscillator strength lies in a region of high state densities where intersystem crossing is quite likely, and the physical characteristics of the initial excitation act are probably of little consequence. For olefins, carbonyl compounds, and some inorganic systems exhibiting low energy transitions in optical absorption (4,14), however, an applied field may induce preferential excitation of low-lying levels. For example, this condition has been clearly demonstrated in pentanones (15).

In Fig. 2 are displayed the ion pair yields of the major isotopic ethylenes obtained during the radiolysis of $CD_3COCD_2CH_2CH_3$ conducted in the presence of electric fields. Preferential excitation of low-lying levels is clearly indicated in this system since ethylene-C_2H_4, which is formed in a low energy elimination process (4), increases quite rapidly in the saturation current region while the ion-pair yield of CD_2CH_2 (the major isotopic ethylene obtained in the vacuum ultraviolet photolysis at 8.4 and 10.0 eV) only increases at high field strengths where electron multiplication becomes noticeable.

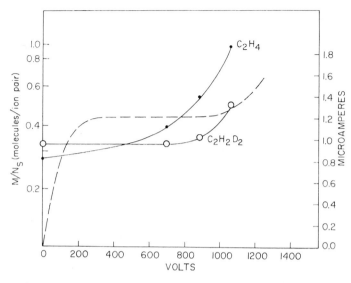

Fig. 2. Effect of an applied electric field on the yields of $C_2H_2D_2$ and C_2H_4 obtained during the radiolysis of 2-pentanone- $-d_5$. (———) Ion pair yields; (‐‐‐) ion current in microamperes. (Data taken from Ref. 15.)

c. Acceleration of Photoelectrons.   This potentially useful gas phase technique for producing electrons with energies insufficient to cause ionization has not been exploited on a widespread basis following the initial experiments of Williams (16). In such experiments photoelectrons, generated at a cathode by the action of ultraviolet light, are accelerated through a reaction vessel under the influence of an applied electric field. The average electron energies may then be varied over a wide range by altering the field strength. In many respects the technique is analogous to the application of external fields during the radiolysis. However, in the Williams technique the role of ions is unimportant at low field strengths and the interpretation is greatly simplified.

d. Controlled Discharges.   The use of glow and spark discharges for the specific purpose of investigating primary modes of decomposition of neutral excited molecules is subject to several criticisms, though a great wealth of related information has been provided by such studies. The major emphasis has been placed, and not incorrectly, in evaluating the spectral features of transient species in such systems. The extensive work in this regard has resulted in the accumulation of a multitude of free radical absorption and emission spectra as well as the cataloging of luminescence features of excited molecular species. Characterization of the spectra has in turn led to a detailed understanding of electronic potential energy configurations for a variety of compounds of interest to radiation and photochemists.

Unfortunately, high concentrations of intermediates are often generated during discharge-induced decompositions and the resultant secondary reaction mechanisms are quite difficult to interpret. These secondary effects may be illustrated by examining the decomposition of methane, which provides an attractive system for discussion in the present context because a detailed knowledge of the reaction mechanism is not required. One needs only to consider the relative rate of formation of a single product of the vapor phase decomposition, acetylene. The facts are these: (a) acetylene is formed in trace amounts in the low intensity vacuum ultraviolet photolysis (17) (10.0 and 11.7 eV), and in the radiolysis (18), where a broad spectrum of excited levels is populated; (b) at very high dose rates in the radiolysis the yield is slightly enhanced (19), and in the high intensity flash photolysis (20) at 10.0 eV, acetylene is a major product of the decomposition; (c) when methane is subjected to a microwave or dc discharge in either the positive or negative glow region (21), as well as to spark discharge at higher pressures (22), acetylene and acetylenic polymers appear in very high yields even when attempts are made to quickly remove stable products from the reaction zone. The reason for this varied behavior is easily explained.

At high rates of energy input two situations are favored: (a) additional decomposition of otherwise stable intermediates through a two-photon or two-electron process sequence; $CH_3 + h\nu$ or $e^- \rightarrow CH + H_2 (+e^-)$ or $CH_2 + H (+e^-)$, $CH_2 + h\nu$ or $e^- \rightarrow CH + H (+e^-)$ or $C + H_2 (+e^-)$, etc., and (b) radical–radical interactions. In the absence of these competing reactions, CH_2, CH, and C disappear rapidly by insertion into methane (20). The observation that acetylene is a major product may be related to either of the two mechanisms favored at high intensities; degradation of radical intermediates to C atoms which, following insertion, yield acetylene $(C + CH_4 \rightarrow C_2H_4{}^* \rightarrow C_2H_2 + H_2)$ and/or radical intercombinations involving CH_2, CH, and C atoms. Unfortunately, one cannot choose between the two possibilities even in this relatively simple system. Another general feature of discharge-induced decompositions is the surface heating which unavoidably occurs in the electrode regions from ion and electron bombardment (except in the microwave discharge which has no electrodes), favoring pyrolysis of stable end products unless they are rapidly removed. Invariably, the decomposition products of an organic system subjected to discharge will contain a greater proportion of olefins than is characteristic of other modes of energy deposition (23,24). In many cases the excess yields can be traced to incomplete removal of stable chemical products from the reaction zone in addition to the other mechanisms peculiar to high intensity conditions. Although such investigations are interesting in their own right it is doubtful that the chemical information (based on product analysis) derived from such studies can be applied other than in a qualitative manner to the problem at hand.

2. Photolysis. Certainly the most useful technique for producing highly excited molecules is photolysis in the far ultraviolet. Such investigations have received considerable impetus in the last few years following the development of relatively stable rare gas resonance radiation sources of high intensity and spectral purity. The resonance levels may be easily excited by a low pressure electrodeless discharge sustained with a microwave generator, and the resultant photons transmitted into the reaction vessel through lithium fluoride, sapphire, or calcium fluoride windows. The reader is referred to the literature for construction details (25,26). The exciting wavelengths provided by such sources include xenon, 1295 Å (9.6 eV) and 1470 Å (8.4 eV); krypton, 1165 Å (10.6 eV) and 1236 Å (10.0 eV); argon, 1048 Å (11.8 eV) and 1067 Å (11.6 eV) and the Lyman alpha radiation, 1216 Å (10.2 eV). The helium resonance line, 584 Å (21.2 eV) has also been used in a windowless apparatus (27). Less energetic photons are provided by Hg, 1849 Å (6.7 eV) and argon + Br_2 mixtures, 1633 Å (7.6 eV) and 1528 Å (8.1 eV) (28).

a. Advantages. There are several inherent advantages to high energy photolysis when compared with other methods of excitation:

1. In view of the quantized nature of photoabsorption, the energy initially contained in the parent molecule may be defined exactly. In electron impact (radiolysis) a wide distribution of excited levels is populated and the situation is more or less analogous to photolysis with white light of all wavelengths. When high intensities of incident light are not required, a monochromator in conjunction with a continuum source is often used and essentially any wavelength in the far ultraviolet may be chosen at will. The rare gas sources provide higher fluxes and are used mainly in those experiments where chemical analysis of end products is required. Since a number of exciting wavelengths are available from these lamps, dissociative channels may be investigated over a wide range of energies.

2. Experiments may be carried out in the absence of ions when the photon energy is less than the ionization potential. This feature of photolysis is particularly attractive since ionic fragmentation, ion–molecule reactions, and neutral molecule dissociations often generate chemically equivalent intermediates and end products.

3. When the photon energy exceeds the ionization potential, it is possible in certain cases to examine the fragmentation patterns of superexcited molecules. In view of the predicted importance of superexcitation in the radiolysis (Chapter 1, Section I), this area of investigation provides particularly useful information for the radiation chemist.

The interpretation of the neutral chemistry occurring in photoionization experiments often depends to a large extent on a prior knowledge of the products associated with the ionic species formed. Parent molecule–ions are formed exclusively when the photon energy is only slightly above the ionization potential and the excess energy is insufficient to cause fragmentation of the molecule–ion. As the photon energy is increased, however, ionic fragmentation is more important and the quantum yield for ionization is usually enhanced (see Section I-A-2-b).

The argon resonance lamp, which provides a high energy photon source (11.6–11.8 eV), has been used successfully in photoionization studies involving a number of molecules exhibiting ionization potentials lower than 11.6 eV. A few investigations of olefinic systems and cycloalkanes have also been reported for the krypton photolysis (10.0 and 10.6 eV).§

4. Collisional deactivation of internally excited transients may be attempted by the introduction of large excesses of additives which are transparent to the incident light. In the radiolysis, of course, secondary electrons will lose energy via inelastic interaction with all components of a mixture; hence the resulting excitation will not be as selective.

b. Actinometry. In spite of the advanced state of the art it is still quite difficult to evaluate quantum yields in the short wavelength photolysis. This situation is simplified in photoionization studies because the products may be expressed as ion–pair yields and quantum yields may then be determined if the ionization efficiency of the parent molecule is known at that particular wavelength (41). Consider the photolysis of the molecule depicted in Fig. 3 having an ionization energy E_I and characterized by an optical absorption spectrum (A). At wavelength λ_1, corresponding to an energy E_1 which is less than E_I, the molecular system decomposes exclusively by a neutral mechanism(s). However, since ionization efficiencies (ions formed/photons absorbed) are considerably less than unity in the region near threshold, the resulting photochemistry at λ_2 also involves excited neutral molecules for the most part even though E_2 is in excess of E_I. For the hypothetical case, the ionization efficiency approaches unity only when the photon energy is in the range of E_3 (λ_3). The absolute value of the ionization efficiency at λ_2 or λ_3 can be obtained by the application of an external electric field and

§ As of this writing photoionization studies have been reported for the following compounds; ethane (29), propane (30,31), *n*-butane (32), isopentane (33), cyclopropane (34), cyclobutane (35), cyclopentane (36), cyclohexane (37), ethylene (38), and propylene (39,40).

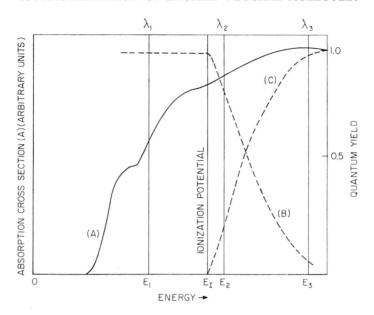

Fig. 3. Quantum yields for various processes as a function of wavelength (energy) for a hypothetical system. (A), optical absorption curve; (B), quantum yield for formation of excited neutral molecules which do not ionize; (C), quantum yield for ionization.

the collection of all charges provided that the number of incident quanta are known. In the event that the photon flux is not known, the saturation current can be compared with that obtained (at the same lamp intensity) for another molecule exhibiting a well-established ionization efficiency at that wavelength. When the comparison is made in the same cell and the values of the saturation currents corresponding to complete absorption of all photons in each system are used, the ionization efficiency of the unknown may be derived from the relationship

$$\eta_{unknown} = \frac{I_{unknown}\,\eta_{standard}}{I_{standard}} \qquad (3)$$

where $I_{unknown}$ and $I_{standard}$ are the respective saturation currents corresponding to absorption of all incident photons and $\eta_{unknown}$ and $\eta_{standard}$ are the respective ionization efficiencies. Nitric oxide has been used as a standard for measuring incident fluxes because the ionization efficiency is reasonably well established at 1236 Å ($\eta \approx 0.77$) and at 1067–1048 Å ($\eta \approx 0.7$) (42). The sum of the quantum yield for ionization and the quantum yield for population of superexcited states which do not subsequently autoionize is, of course, unity.

The yield of CO obtained from the photolysis of CO_2 has also been used in a number of laboratories as a chemical actinometer. The quantum yields so derived for other systems at the various wavelengths are based on the assumption that CO is formed with a quantum yield of unity in the far ultraviolet region. However, this value (unity) is controversial (see Section III-B).

B. Methods for Determining Unimolecular Decay Patterns in Excited Molecules

Details of ionic fragmentation schemes may be quickly established by introduction of gaseous samples into an analytical mass spectrometer. Successive dissociations are represented in the mass spectrum, and by varying the energy of the ionizing electron beam one can determine the appearance potentials reflecting the threshold energy requirements for primary, secondary, and tertiary decompositions. Although the observed mass spectrum is sensitive to changes in pressure, temperature, and electron energy (see Chapter 2), the radiation chemist at least has a feeling for which ions to expect in a given system based on the parent–daughter relationships which may be derived from appearance potential measurements and thermochemical heats of formation.

Unfortunately, no conventional instrumentation has been developed which provides the analogous information for excited neutral molecules. The various experimental approaches which are available will be discussed in some detail in the following passages. These approaches fall into two categories; direct methods, in which the detection of intermediates and stable products is carried out while the experiment is in progress, and indirect methods, which require chemical analysis of end products.

1. Direct Methods. *a. Mass Spectrometry.* Beck and Osberghaus (43–45) have used a modified mass spectrometer in an attempt to directly determine neutral dissociative paths in some hydrocarbons. Initial ionization and excitation of a gaseous sample is induced by a chopped electron beam (200 eV) at low pressures in a collision chamber. Unaffected parent molecules, as well as neutral molecular products and radicals resulting from the decomposition of excited ions and neutral molecules, are allowed to diffuse into a second chamber which serves as a source for a conventional mass spectrometer. Ions formed in the collision chamber are rejected during diffusion by means of an interposed grid. The neutral species are then ionized in the second chamber and mass analysis is performed. The detected mass spectrum has two components; a dc signal and a superimposed ac signal. The dc component, reflecting ionization and fragmentation of parent molecules which were not decomposed in the collision chamber, appears as a constant background. The ac component can be related to ionization of the diffusing neutral molecular and free radical products formed by the chopped electron beam in the collision chamber. This ac component, of course, contains the features of interest.

However, it is very difficult to investigate the primary decomposition patterns of neutral molecules by this particular method. A major interpretive dilemma exists since neutral products are also formed during the fragmentation of excited *ions* in the first chamber. In order to subtract these contributions it is necessary to find at least one neutral product which is associated *only* with the formation of ions. Unfortunately there are few, if any, neutral products which can be uniquely correlated with either fragmentation of the parent ion or fragmentation of neutral molecules following impact with nominally 200 eV electrons. The energy content of the parent molecule prior to neutral dissociation also cannot be defined since a broad distribution of excited levels is generated by high energy electron bombardment.

More recently, highly sophisticated mass spectrometric instrumentation capable of low energy operation has been developed by Melton (9) and preliminary studies of the electron-impact induced decomposition of ammonia (9a) and methane (9b) have been reported over a wide range of energies (variable from zero to several hundred eV) and pressures (up to 10 torr.). Although the construction details will not be considered here (see Chapter 4), this instrument provides a powerful tool for establishing the energy requirements pertinent to neutral dissociations because one can actually derive appearance potentials for the formation of free radical intermediates and molecular products. Hopefully, more complex molecules will soon be subjected to investigation.

Due to instrumental limitations, mass spectrometric investigations of neutral decompositions *in situ* are necessarily restricted to low gas densities where consecutive fragmentations of excited transient species are favored. Consequently, the distribution of neutral products reflects all processes occurring from 10^{-14} to 10^{-6} sec after excitation, and a detailed knowledge of parent–daughter relationships is required to define the primary processes. Even when a unique mass spectrum of neutral decomposition products is obtained, it is very difficult to establish the relative concentrations of free radical intermediates and stable molecular products actually formed in the reaction chamber because ionization cross sections for free radicals must be assumed. In addition, corrections must be made for the fact that lighter fragments diffuse more rapidly in the mass spectrometer, especially at reduced pressures. Free radical loss at the walls of the reaction chamber will be particularly important for hydrogen atoms, and the calculated neutral spectrum is very sensitive to the choice of diffusion coefficients.

The mass spectrometer has also been used to identify the dissociation products of photochemical decompositions. The products, both free radical and molecular, are allowed to diffuse into the ion source of a conventional mass spectrometer where the flow components are ionized with low energy electrons. A comparison is then made between the mass spectrum obtained with and without illumination. As pointed out earlier, ionization by electron impact suffers from the disadvantage that parent molecules will also be ionized, and the subsequent dissociations of the parent ion often lead to the production of fragment ions which have the same mass as radical and molecular species formed during the photolysis. When mass overlap occurs very high concentrations of products are often required. To minimize this difficulty Beckey, and co-workers (46,47) have used the field ionization technique (ionization occurring on or near an emissive surface placed in a very high electric field), which induces very little ionic fragmentation of the diffusing sample. Consequently, mass overlap due to fragmentation of parent ions is greatly reduced and much lower concentrations of photochemical products are required to carry out the analysis. Unfortunately, little is known about relative cross sections for field emission ionization, so that at present this method only establishes the existence of a given radical or molecular product and does not provide a measure of relative concentrations.

b. Evaluation of Transient Spectral Features. Highly energized decomposition products often dissipate excess energy via radiative deexcitation processes which may be monitored under certain conditions. Luminescence is quite common for diatomic and triatomic radical species formed in exothermic dissociations of inorganic gases in the vapor phase, and emission has also been observed from several small organic radicals. Although the observation of characteristic spectral features in either absorption or emission defines the transient formation of a given species, it is difficult to determine absolute yields by this method. In condensed media, however, concentrations of relatively large organic radicals have been measured with some success through a combination of absorption techniques and chemical analysis in systems subjected to radiolysis (see Chapter 7). Spectroscopic identification of absorbing or emitting species *in situ* is most helpful in elucidating mechanisms when the excitation energy contained in the parent molecule is known. A useful method for defining the threshold energy requirements for formation of photoactive intermediates is to follow the emission spectrum of the system as a function of exciting wavelength. In this way the "appearance wavelength" may be determined. This approach has been used with some success for inorganic molecules in the vapor phase (see Section IV).

2. Analysis of Stable Chemical Products.   Our insight into the complex mechanisms associated with the fragmentation of highly excited molecules has been based, for the most part, on a thorough examination of the stable chemical products of the decomposition obtained under various experimental conditions. The observed end products may be classified into two distinct categories: (*1*) molecular products and (*2*) bimolecular products. In the context of this chapter a *molecular* product will be defined as a *molecule* formed as result of a unimolecular dissociation of the highly excited parent molecule or an excited fragment thereof. *Bimolecular* products, on the other hand, include all of those others which may be ascribed to free radical reactions involving the substrate or other radical species.

Molecular and *bimolecular* products may be distinguished from one another in various ways, including the introduction of radical scavengers and interceptors (see Section I-B-2-a) and by the use of isotopically labeled analogs of the parent molecule (see Section I-B-2-b). Additional details of the fragmentation may also be established by examination of the isotopic composition of the *molecular* products which are recovered following the decomposition of selectively labeled parent molecules (see Section I-B-2-b). Consequently, it is often necessary to perform isotopic analysis in addition to merely achieving separation of the various products from each other and determining their relative or absolute yields.

Although the most useful data is usually obtained when one deliberately introduces an impurity into the decomposing system (in the form of radical scavengers, interceptors, etc.) it is also advantageous in certain cases to examine the products obtained in the pure§ system. When a pure system which may decompose by a free radical mechanism is investigated, the necessity for reporting yields obtained at very low conversions cannot be overemphasized. This criterion demands special attention when the parent molecule exhibits a low sensitivity to free radical attack since accumulated products of the decomposition act as competing internal scavengers for reactive intermediates in these situations and it is exceedingly difficult to derive the primary processes of interest from the observed product distribution. Failures to obtain "initial" distributions characterized the literature until the early 1960's and in many cases precluded identification of important transient species. The analytical difficulties

§ In this chapter, starting material subjected to a purification procedure and into which no additives have been deliberately introduced will be referred to as a "pure" system.

encountered in low conversion experiments (10^{-2} to $10^{-3}\%$) have largely been resolved by recent advances in flame ionization gas chromatography and mass spectroscopy.

Occasionally, conversion effects may be traced to secondary decomposition of otherwise stable reaction products by the incident radiation. For example, the vapor phase flash photolysis of propane was recently reported by Griffiths and Back (48) using an argon continuum source (with quartz optics) which transmits all photons with $\lambda \geq 1500$ Å. The product distribution was determined at various conversions, and it was found that the yield of propylene, which is the major hydrocarbon product at these wavelengths, was depleted very rapidly following repetitive flashing of the sample. This study provides an excellent illustration of those complicating effects which are favored when a continuum is used as the excitation source. In general, the rapid depletion of primary olefinic products in the flash photolysis can be ascribed to the fact that the absorption curves for olefins extend to much longer wavelengths than those found for saturated hydrocarbons. Consequently, the longer wavelength components of the continuum may be absorbed quantitatively by accumulated olefinic products of the decomposition. The monochromatic photolysis at shorter wavelengths conducted at equivalent conversions will not exhibit such drastic secondary effects since the parent molecule is not transparent to any of the photons incident upon the system.

Conversion effects were particularly severe in the experiments of Griffiths and Back since the extinction coefficient of propylene is at least an order of magnitude higher than that of propane in the region where propane did absorb (wavelengths less than 1650 Å, but limited by the transmission of quartz). This example is but one of many which may be extracted from the literature to illustrate the fact that one must be extremely careful when basing the interpretation of a particular experiment on the yields of those products which survive exposure of the sample.

   a. Removal and Detection of Free Radicals.   The addition of scavengers is a powerful tool for assessing the role of thermal free radicals in kinetic investigations. In all organic systems and those inorganic systems which may generate hydrogen atoms, two general types are recognized which differ in the mechanism for interception:

$$R\cdot + X \to RX\cdot \qquad \text{Class I} \qquad\qquad (4)$$

$$R\cdot + YH \to RH + Y\cdot \quad \text{Class II} \qquad\qquad (5)$$

Additives falling into Class I include O_2, NO, and the olefins. Although the eventual chemical fate of adducts such as $RO_2\cdot$ and $RNO\cdot$ is not usually defined there is no conclusive evidence that the addition of any scavenger affects the distribution of *molecular* products. The effectiveness of olefins is largely restricted to removal of hydrogen atoms and small alkyl radicals at low concentrations.

Measurement of individual free radical yields is often possible in those mixtures containing Class II additives, which include H_2S, HI, and the heavier halogens. The radical interception products obtained in the presence of H_2S or HI may be distinguished from *molecular* products by using the perdeuterated analog of the molecule under investigation, since isotopic analysis will indicate the fraction of H-containing components resulting from abstraction. Consider a system in which methane is formed by two distinct mechanisms: (*1*) *molecular* elimination and (*2*) abstraction or disproportionation reactions involving methyl radicals and other radical species. When H_2S or HI is added to the completely deuterated analog of this particular molecule, *molecular* methane appears as CD_4 and methyl radicals are titrated as CD_3H [$CD_3 + H_2S$ (HI) $\rightarrow CD_3H + HS\cdot(I\cdot)$]. Among all Class II interceptors, H_2S shows the most promise because of its relative ease of manipulation in the laboratory and demonstrated high reactivity toward alkyl radicals containing up to four carbon atoms (1).

Although the use of such additives is fairly straightforward when the decomposition involves excited neutral molecules, complications may arise in systems containing ions (as in photoionization, radiolysis, etc.) due to interactions of charged intermediates with the scavenger. Charge exchange reactions of ions with scavengers exhibiting lower ionization potentials (such as NO, IP = 9.24 eV) have been found to be particularly important and in many cases lead to the formation of stable chemical products which would otherwise be ascribed to decomposition of neutral molecules. Hydrogen sulfide is also very reactive toward ions, and although proton transfer and charge exchange reactions have been observed (49) it has been used with great success as a radical interceptor in the vapor phase radiolysis at relatively low concentrations ($<20\%$) (50).

b. Isotopic Labeling.   Perhaps the greatest single contributing factor to our understanding of the complex fragmentation mechanisms exhibited by excited neutral molecules in photochemical and radiolytic systems has been the availability of labeled compounds of high isotopic purity. Deuterium labeling is most helpful in organic compounds, and the investigation of equimolar mixtures of perdeuterated-perprotonated analogs is particularly useful in separating molecular and bimolecular products from one another. In addition, examination of the isotopic composition of the various molecular products obtained from a selectively labeled parent molecule often reveals details of the unimolecular fragmentation pattern. Various other isotopes including ^{18}O, ^{15}N, and ^{13}C have also been used to great advantage in several inorganic systems.

II. THE FRAGMENTATION PATTERNS OF
HYDROCARBONS

Before proceeding with the discussion it is important to point out that secondary decomposition of internally excited intermediates is commonly observed in the vapor phase due to the extra energy which is available for partitioning among primary dissociation products. This situation often creates a major interpretive problem since these secondary processes are found to be particularly important for free radical intermediates:

$$RCH_2CH_2R' \rightarrow RCH_2 + R'CH_2 \tag{6}$$

followed by

$$RCH_2, R'CH_2 \rightarrow \text{decomposition products} \tag{7}$$

In order to determine which chemical end products are formed in these degradations, the density may be increased to the point where the excess energy of RCH_2 and $R'CH_2$ is dissipated by collision with neighboring molecules. In the vapor phase, however, it is very difficult to achieve the pressures required for complete deactivation. Extension of the experiments into condensed media, where stabilization is quite effective, is often of little diagnostic value because a new interpretive problem is created. Although collisions may prevent the further dissociation of RCH_2 and $R'CH_2$ in the solid, the dense environment does not permit the radicals to freely diffuse away from one another. As a general rule, larger radical pairs may be considered effectively trapped at the site of production and a mutual radical–radical interaction will follow unless one of the reactants is removed by reaction with the substrate. Two such interactions are known; disproportionation and combination. A recombination event generates the parent molecule and the disproportionation yields an alkane of lower molecular weight and an olefin.

$$RCH_2 + R'CH_2 \rightarrow RCH_3 + \text{olefin} \tag{8}$$

The interpretive problems are compounded by the fact that disproportionation/combination ratios depend both on the structure of the radical pair and the temperature (51,52). Consequently, in condensed phases the simultaneous production of an olefin and an alkane may be due to either a geminate radical disproportionation or an elimination from the excited parent molecule ($RCH_2CH_2R'^* \rightarrow RCH_3 + \text{olefin}$), or both. Selective deuterium labeling is often useful as a diagnostic tool in these ambiguous situations since the products of *molecular* elimination may exhibit an isotopic composition which differs from the composition of the disproportionation products.

The following discussion will be based for the most part on the results of those photolysis studies which have incorporated the rare gas resonance lamps. Two methods of presentation were considered; (*1*) a review of the current level of understanding on a molecule-by-molecule basis and (*2*) a less specific treatment in which the emphasis would be placed on the dissociative modes which are characteristic of a family series. The latter alternative was considered appropriate for the alkanes because the modes of fragmentation which have been observed appear to be characteristic of all members (although the relative probabilities vary, of course). Cyclic and olefinic hydrocarbons, on the other hand, will be examined individually since these molecules often exhibit diverse fragmentation patterns when compared with their neighbors in the homologous series.

As pointed out earlier, quantum yields have not been measured accurately except in photoionization experiments where all charges have been collected. Fortunately, one can derive valuable information merely from the *alterations* in the product distribution which are induced by changes in excitation energy and density. Much of the data presented in the following sections are normalized to the major decomposition product of a particular system. This method only provides the most expedient basis for discussion and does not infer that the absolute yields of these products are necessarily independent of variations in the usual experimental parameters.

A. Alkanes

1. Elimination of Hydrogen. The elimination of hydrogen represents a major mode of energy dissipation in electronically excited alkanes, particularly at lower excitation energies.§ Considerable caution must be exercised in determining absolute *molecular* yields in highly energized systems as hydrogen atoms with translational energies in excess of the potential barrier for abstraction react with a collision efficiency close to unity. In practice, bimolecular contributions to the total hydrogen yield obtained in the presence of scavengers are evaluated through examination of the isotopic composition of the hydrogen fraction obtained from equimolar mixtures of perdeutero-perprotonated analogs. For example, the relative importance of bimolecular contributions to the total yield of unscavenged hydrogen in propane (53) at higher excitation energies is reflected by the distribution of isotopic hydrogens obtained from the

§ Numerous references will be made to "lower" and "higher" excitation energies. Since most of the data examined here has been derived from photolysis experiments using the rare gas resonance sources, the term "lower" excitation energies refers to absorption of 8.4 eV photons (xenon photolysis), and is to be compared with the "higher" excitation energies (10.0 and 11.7 eV) available with the krypton and argon lamps.

Table I

Isotopic Hydrogens and Methanes Obtained in the Vapor Phase
Photolysis of Variously Labeled Propanes[a,b]

	$\lambda(\text{Å})$	D_2	HD	H_2	CD_4	CD_3H	CH_3D	CH_4
I. $C_3H_8 + C_3D_8 + NO$	1470	31.9	7.6	60.5	40.0	2.7	1.2	56.1
	1236	30.9	21.8	47.3	52.4	0.8	0.8	45.9
	1067–48[c]	Not reported			42.0	5.8	0.2	52.0
II. $CH_3CD_2CH_3 + NO$	1470	45.9	23.9	30.2			74.8	25.2
	1236	26.0	29.9	44.1			84.6	15.4
	1067–48[c]	Not reported					93.7	6.3
III. $CD_3CH_2CD_3 + NO$	1470	8.9	23.6	67.4	9.8	90.2		
	1236	19.2	57.2	28.6	7.7	92.3		

[a] Total pressure ~ 10 torr. Unless otherwise noted, all data taken from Ref. 53.
[b] 2% NO added in all experiments.
[c] Taken from Ref. 31.

photolysis of C_3H_8–C_3D_8–NO mixtures (Table I). Assuming 100%
isotopic purity of C_3D_8 and quantitative removal of thermal species,
hyperthermal atoms contribute 15.2% (2 × HD, or 2 × 7.6%) and
43.6% (2 × 21.8%) to the total hydrogen yield at 1470 and 1236 Å. The
factor of 2 is applicable irrespective of the relative H and D atom
concentrations, but is based on the assumption that abstraction of H
from C_3H_8 and D from C_3D_8 occur with equal probabilities. The residual
yield provides a measure of unimolecular detachment, which may be represented as

$$C_nH^*_{2n+2} \rightarrow C_nH_{2n} + H_2 \tag{9}$$

In certain cases, selective labeling has revealed the stereospecificity of
hydrogen detachment. The most intriguing mechanism uncovered is the
elimination originating at a single carbon atom, which has been found
in a number of hydrocarbons (3,33). For example, 2,2 detachment is
clearly defined by the formation of D_2 and H_2 in the respective vapor
phase photolyses of $CH_3CD_2CH_3$–NO and $CD_3CH_2CD_3$–NO mixtures
(Table I).

Elimination of *molecular* hydrogen from a single carbon atom has also
been observed in the branched alkanes. Following an analysis (54) of the
isotopic composition of the hydrogen fraction obtained during the
photolysis of $(CH_3)_3CD$ at two wavelengths in the far ultraviolet it was
suggested that the probability for loss from a single methyl group may be
favored over the 1,2 mechanism involving the tertiary site as the excitation
energy is increased. However, simultaneous detachment from a single
methyl group is of minor importance in the 1470 and 1236 Å photolysis of
neopentane (55).

Process (9) also contributes to the formation of pentene found in the photolysis of isopentane at lower excitation energies (33). The fact that HD is obtained in high yields in the 1470 and 1236 Å decomposition of $(CH_3)_2CDCH_2CH_3$ and the observation of 2-methyl-2-butene,

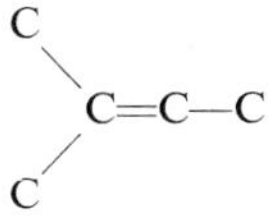

$$
\begin{array}{c}
C \\
\diagdown \\
C{=}C{-}C \\
\diagup \\
C
\end{array}
$$

as the major component of the pentene fraction suggest simultaneous involvement of both secondary and tertiary skeletal sites in the detachment step. Simultaneous elimination at a single carbon atom and other mechanisms also contribute to the *molecular* yield, but a detailed assessment of the various sites involved has not been performed for this molecule (as well as others) because of the excessive number of selectively labeled analogs required. The situation is further complicated by the fact that *molecular* hydrogen may also be eliminated from internally excited primary dissociation products.

The isomerization and dissociations of alkylidene (carbene) transients formed in primary processes have been investigated in varying detail for a number of systems. For example, methylcyclopropane has been recovered as a minor photolysis product in the 1470 Å photolysis of isobutane (56), indicating that cyclization of the isobutylidene diradical (formed when

Table IIA

Products from the Decomposition of Neutral Excited Ethanes[a,b]

	Wavelength	
	1470 Å	1236 Å
CH_3CD_3–NO(1 : 0.02)		
Hydrogen	2.32	2.10
Methane	0.05	0.655
Ethylene	1.00	1.00
Acetylene	0.74	0.545
Propane	0.013	0.123
Composition of ethylene fraction		
CD_2CDH	51.0	26.6
CD_2CH_2	27.0	66.8
$CDHCH_2$	22.0	6.6

[a] Pressure, 30 torr.
[b] Taken from Ref. 13.

Table IIB

Photolysis of CH_3CD_3 at 1236 Å: The Effect of Pressure[b,c]

P (cm)	Hydrogen	Methane	Ethylene	Acetylene	Propane	Ethylene distribution		
						C_2H_3D	$C_2H_2D_2$	C_2HD_3
0.23	3.11	1.00	1.21	0.77	0.044	5.8	70.5	23.7
3.04	3.29	1.00	1.53	0.82	0.188	6.6	66.8	26.6
58.4	2.56	1.00	1.75	0.67	0.390	8.8	59.4	31.8

[c] 2% NO added in all experiments.

molecular hydrogen is detached from a single methyl group) can compete with the isomerization to butene. Perhaps the most intensive effort has been directed toward an understanding of the nature of the ethylidene diradical formed in ethane; $CH_3CH_3{}^* \rightarrow CH_3CH + H_2$. In a series of articles by McNesby and co-workers, (29,57–60) as well as an article by Carmichael et al. (13), which describe the photodecomposition of CH_3CD_3 at 1470, 1236, and 1067–48 Å, several observations have been recorded which will be examined in light of their general applicability (see Table IIA, B).

At these excitation energies, a large fraction of the absorbed energy is available for partitioning among the primary decomposition products. For the molecule under consideration, there is evidence that the excess internal energy can be dissipated by the elimination of a second molecule of hydrogen, forming acetylene. Hampson and McNesby (58) have also suggested that rearrangement of the ethylidene radical into a potentially repulsive configuration of excited ethylene precedes this elimination because they found no evidence for chemistry reflecting a long-lived biradical structure. The overall kinetic progression may be represented as §

$$CH_3CH_3{}^* \rightarrow CH_3CH^\ddagger + H_2 \tag{10}$$

$$CH_3CH^\ddagger \rightarrow CH_2CH_2{}^\ddagger \tag{11}$$

$$CH_2CH_2{}^\ddagger \rightarrow CHCH + H_2 \tag{12}$$

An additional step, the collisional deactivation of $C_2H_4{}^\ddagger$, must also be included for discussion:

$$CH_2CH_2{}^\ddagger + M \rightarrow CH_2CH_2 + M \tag{13}$$

EFFECT OF ENERGY. A decrease in wavelength from 1470 Å (8.6 eV) to 1236 Å (10.0 eV) *at constant pressure* increases the observed ratio of

§ In this chapter the notation * will be used to indicate excited parent molecules and the notation ‡ will be used to denote excess energy content in dissociation products.

dissociation (12) to stabilization (13), henceforth referred to as R_d/R_s. This parameter characterizes the kinetic sequence and is readily evaluated by photolyzing CH_3CD_3 since the stabilized transient of interest is recovered as CD_2CHD and CH_2CDH in the ethylene fraction (formation of ethylene by successive loss of atomic hydrogen most likely generates CH_2CD_2). The total yield of acetylene, on the other hand, provides a measure of R_d. On this basis, and utilizing the data contained in Table II, $R_d/R_s = 1.0$ at 1470 Å and 1.6 at 1236 Å for a pressure of 30 torr. The effect of wavelength is not surprising since more energy is available as internal energy of $CH_2CH_2^\ddagger$ at 1236 Å, favoring the dissociation.

EFFECT OF COLLISIONAL DEACTIVATION.   An increase in density at any particular wavelength decreases the value of R_d/R_s. In the 1470 Å photolysis Hampson and McNesby (60) have investigated the competition between decomposition and collisional deactivation at various pressures both at room temperature and 310°C in the absence of scavengers, and some experiments were also reported incorporating N_2 and Ar as quenchers at a fixed pressure of ethane. Assuming a collision efficiency of unity for deactivation, a dissociative lifetime of 6×10^{-10} sec was derived for $CH_2CH_2^\ddagger$. It was also tentatively concluded that two distinct states of $CH_2CH_2^\ddagger$ are eventually populated following the "carbene" elimination of hydrogen in reaction (1). One state, incapable of crossing to a repulsive internuclear configuration, was ascribed to singlet state molecules excited to high vibronic levels and the other level was suggested to be a triplet. In a related investigation Pirog and McNesby (59) found that collisional deactivation is completely effective ($R_d/R_s = 0$) in the 1470 Å photolysis of dilute solutions of ethane in liquid nitrogen. At 1236 Å, an increase in density also favors stabilization as R_d/R_s falls from 2.1 at 2.3 torr to 0.94 at 584 torr in the photolysis of CH_3CD_3–NO mixtures (Table II).

These trends observed in the elimination of hydrogen from ethane are characteristic of those found for other n-alkanes. Ethane, of course, represents a rather simple case in so much as further decomposition occurs mainly via loss of another molecule of hydrogen. Complex olefinic structures formed as transients in higher homologs may be degraded by alternative paths involving C—C cleavage. However, as the molecular complexity increases, additional degrees of freedom are available for internal flow of energy, a condition which tends to increase the dissociative lifetime.

The emphasis placed on the formation of carbenes following elimination of hydrogen from a single carbon atom is not intended to imply that other mechanisms are not contributing. Actually, examination of the data obtained from the variously labeled propanes and other alkanes (3,32,33) verifies that 1,2 detachments may play a role in certain cases.

2. Elimination of Alkanes. The unimolecular elimination of lower molecular weight alkanes from highly excited parent molecules has been observed for every member of the homologous series (except methane, of course).

$$CnH^*_{2n+2} \rightarrow C_{n'}H_{2n'+2} + C_{n''}H_{2n''} \tag{14}$$
$$(n' + n'' = n)$$

Three examples will be cited here.

In propane, the lack of significant isotopic mixing in the methanes formed in the photolysis of C_3H_8–C_3D_8–NO mixtures (Table I) clearly indicates that *bimolecular* contributions are negligible in the presence of 2% NO at 1470 and 1236 Å. Selective labeling has again demonstrated that the mechanism which forms a carbene diradical is favored. For example, CH_3D is the major methane obtained in the photolysis of $CH_3CD_2CH_3$; $(CH_3CD_2CH_3^* \rightarrow CH_3CD + CH_3D)$. The competing "1,3" elimination, which forms CH_4 and CH_2CD_2 as end products, is less probable and decreases in relative probability as the photon energy is varied from 8.4 to 11.7 eV. An analogous trend is indicated by the isotopic methanes obtained from the decomposition of $CD_3CH_2CD_3$. The ethylidene intermediate initially formed in the "1,2" detachment of methane is stabilized in the solid and rearranges quantitatively to ethylene.

Methane is also eliminated from the branched paraffins, although the quantum yields vary from molecule to molecule. Tertiary methyl groups are almost exclusively involved in isopentane (33) since at least 90% of the methanes obtained from $(CH_3)_2CHCH_2CD_3$ consist of CH_4. However, two separate mechanisms are operating since both CH_3D and CH_4 are observed in the photolysis of $(CH_3)_2CDCH_2CH_3$;

$$(CH_3)_2CDCH_2CH_3^* \rightarrow CDH_3 + CH_3CCH_2CH_3 \text{ (1,2 carbene elimination)} \tag{15}$$
$$\rightarrow CH_4 + CH_3CD{=}CHCH_3 \text{ (1,3 olefin elimination)} \tag{16}$$

A decrease in wavelength from 1470 to 1236 Å favors the 1,2 elimination, and a similar trend with energy is also found for the elimination of ethane (see Ref. 33). It is pertinent to note that the unsaturate yields obtained in the photolysis of the glassy solid at 77°K agree very well with the corresponding alkane yields which can be ascribed to reaction (14) (33b).

The photolysis of $CD_3CH_2CH_2CD_3$ (32) and $CH_3CD_2CD_2CH_3$ (61) has clarified the mechanism for the *molecular* detachment of ethane in the neutral decomposition of *n*-butane. Two possibilities exist:

$$CD_3CH_2CH_2CD_3^* \rightarrow CD_3CH_3 + CHCD_3 \text{ (a 2,3 "carbene" mechanism)} \tag{17}$$
$$CD_3CH_2CH_2CD_3^* \rightarrow CD_3CH_2D + CD_2CH_2 \text{ (a 2,4 olefin elimination)} \tag{18}$$

In the vapor phase, at least 85% of the *molecular* ethane is CD_3CH_3 (process (17)). At very high pressures, the yield of acetylene is considerably reduced and C_2D_3H is equivalent to CD_3CH_3, indicating stabilization of the ethylidene intermediate ($CHCD_3$) formed via the 2,3 "carbene" mechanism. The remainder of the ethane yield consists entirely of $C_2D_4H_2$.

Consistent patterns are observed for the elimination of lower molecular weight alkanes as the excitation energy is increased. As a general rule, the disproportionations which form a carbene diradical and an alkane (processes such as (15) and (17)) are favored relative to the mechanisms which yield the olefin directly (processes such as (16) and (18)) at higher excitation energies.

The *molecular* formation of an alkane containing one less carbon atom than the decomposing parent molecule indicates that methylene (CH_2) elimination must be considered:

$$C_nH^*_{2n+2} \rightarrow C_{n-1}H_{2n} + CH_2 \tag{19}$$

For example, the propanes formed during the photolysis of C_2H_6–C_2D_6–NO mixtures (13) exhibit an isotopic composition (propane-d_0, -d_2, -d_6, and -d_8) which can only be explained in terms of $C_2H_6^* \rightarrow CH_4 + CH_2$ followed by $CH_2 + C_2H_6 \rightarrow C_3H_8^{\ddagger}$. The exothermicity of the insertion reactions renders $C_3H_8^{\ddagger}$ highly unstable; e.g., methane $\gg$ propane in the 1236 Å photolysis of CH_3CD_3, at pressures below 1 atm (Table II). The insertion product is assumed to decompose into radical fragments (CH_3 and C_2H_5), which is stoichiometrically equivalent to abstraction of H by CH_2. An increase in pressure does stabilize $C_3H_8^{\ddagger}$, and at high densities in the vapor (> 30 mg/cc) the propane yield is approximately equal to the molecular methane (13). Concrete evidence for collisional stabilization of methylene insertion products has also been obtained in the radiolysis and photolysis of methane (62,63) and propane (30). The pressures at which stabilization occurs are much higher than those found by Kistiakowsky and others (64), who studied the reactions of methylene radicals generated from other sources. Apparently, more energy is contained in the methylene fragments formed in the far ultraviolet photolysis of hydrocarbons. The systematic investigation of the C_1 through C_6 alkanes has revealed that methylene formation is an important primary process only for the lower molecular weight members of the series (mainly C_1 through C_3).

3. Production of Hydrogen Atoms and Other Free Radicals. In view of the HD yields observed in C_nD_{2n+2}–H_2S mixtures (1) in far ultraviolet photolysis experiments, an additional dissociative mode of highly excited alkanes is indicated;

$$C_nH^*_{2n+2} \rightarrow C_nH^{\ddagger}_{2n+1} + H \tag{20}$$

Perhaps the most important aspect of H atom evolution is the competition which is observed between this mode of fragmentation and *molecular* detachment of hydrogen (process (9)). Radical counting, as accomplished both by the H_2S interception technique and by the introduction of other scavengers, has clearly established that loss of H atoms increases in relative importance as the excitation energy is increased (see, for example, Table III). This trend is of general occurrence and has been well documented in the systematic vacuum ultraviolet photolysis of many hydrocarbons.

The successive fragmentations of the energy-rich intermediates $C_nH_{2n+1}^{\ddagger}$ formed in reaction (20) have prevented identification of those primary processes which generate free radicals by C—C cleavage:

$$C_nH_{2n+2}^* \rightarrow C_{n'}H_{2n'+1} + C_{n''}H_{2n''+1} \qquad (21)$$

where $n' + n'' = n$. The H_2S titration data clearly indicates the magnitude of the interpretive problem, and in most cases it is impossible to decide which of the radical species was generated in the primary step. For example, the relative yield of CD_3 is considerably in excess of the C_2D_5 yield in the 1236 Å photolysis of C_3D_8–H_2S mixtures (see Table III). Methyl radicals may be formed by at least two mechanisms; dissociation of the parent molecule ($C_3D_8^* \rightarrow CD_3 + C_2D_5$) and secondary elimination from an internally excited propyl radical; $C_3D_8^* \rightarrow C_3D_7^{\ddagger} + D$ followed by $C_3D_7^{\ddagger} \rightarrow CD_3 + C_2D_4$. However, the yield of C_2D_5 does not provide a measure of the primary process because the ethyl radicals so formed may also decompose further by loss of a deuterium atom. The overall stoichiometry is equivalent in both schemes ($C_3D_8^* \rightarrow C_2D_4 + CD_3 + D$) and even selective labeling is of little value in establishing either the basis for the inequality $CD_3 > C_2D_5$ or the actual parent–daughter relationships involved.

Comparison of the H_2S interception results for the various noncyclic hydrocarbons (1) verifies that an increase in methyl substitution increases the probability for methyl radical production, as one would have perhaps

Table III

Photolysis of C_3D_8 in the Presence of $H_2S^{(a,b)}$

										Total radicals
										total molecular
Wavelength	D_2	$D\cdot$	CD_4	$CD_3\cdot$	C_2D_4	$C_2D_5\cdot$	$C_2D_3\cdot$	C_3D_6	C_2D_6	products
1470 Å (8.4 eV)	5.1	3.8	0.47	1.2	*1.00*	0.13	0.28	2.6	0.19	0.56
1236 Å (10.0 eV)	1.0	1.5	0.58	1.0	*1.00*	0.15	0.24	0.80	0.27	0.78
1067–48										
(11.6–11.8 eV)	1.5	2.6	0.64	2.1	*1.00*	0.22	0.32	0.68	0.16	1.3

[a] 10% H_2S added.

[b] Taken from Ref. 31.

expected on the basis of first principles. Unfortunately, it is again not possible to determine whether or not primary dissociations of the parent molecule are responsible for the observed yields.

The alterations in the fragmentation patterns induced by changes in excitation energy are exemplified by the pattern found for propane, which exhibits many of the same trends found for other alkanes. The relative free radical and *molecular* products obtained in the vapor phase photolysis of C_3D_8–H_2S mixtures at 1470, 1236, and 1067–48 Å are compiled in Table III for reference. Consideration of the $D_2/D\cdot$ and $CD_4/CD_3\cdot$ ratios reported there at the various wavelengths leaves no doubt that free radical production is favored at higher excitation energies. The same trend is indicated by the yields of $C_2D_3\cdot$ and $C_2D_5\cdot$. The relative yields will depend, of course, on the collision frequency, since many of the radical species are formed by secondary and tertiary decompositions. A rigorous examination of the competition which exists among the various *primary* modes of decomposition is a very tedious bookkeeping task, and for many systems even an approximate calculation of such patterns at various excitation energies cannot be carried out because of the complexity of the product distribution. For this particular molecule (propane), the accumulated data has conclusively established that the probability for *molecular* detachment of hydrogen in a *primary* process is drastically reduced at shorter wavelengths, and is compensated for, to a large extent, by an increase in the probability for the dissociation which gives a hydrogen atom and a highly energized propyl radical. More data is necessary before one can uniquely define the *primary* patterns for higher homologs and branched alkanes, especially at higher excitation energies.

In any event, these data demonstrate that the fragmentation of neutral excited molecules in radiolysis systems will strongly depend on the amount of energy extracted from the secondary electrons, and point out the difficulty which will be encountered in using low energy photolysis data to predict the behavior of systems which may be excited to upper levels.

The data reported in Table III also indicate that superexcited propane behaves chemically much in the same way as molecules excited to sub-ionization levels, since a continuous trend in the fragmentation pattern is observed as the excitation exceeds the ionization potential (IP propane = 11.2 eV). Superexcitation also occurs in *n*-butane during the 1048–1066 Å photolysis, while the photochemistry observed at 1470 and 1236 Å reflects transitions to excited levels below the ionization potential (IP = 10.6 eV). The product distributions obtained in the vapor phase photolysis of this molecule are compiled in Table IV.

Although the various arguments will not be discussed here, Ausloos and Lias (32) have demonstrated that neither decomposition of parent molecular

Table IV

Molecular Product Yields Obtained by Various Methods
in the Gas Phase

Normal Butane

Method	Meth-ane	Ethyl-ene	Ethane	Pro-pane	Acetyl-ene	Propyl-ene	Total butene
Photolysis							
1479 Å[a]	0.06	*1.00*	0.46	0.03	0.22	0.36	1.06
1236 Å[a]	0.15	*1.00*	0.46	0.03	0.28	0.40	0.31
1067–48 Å[a]	n.d.[b]	*1.00*	0.27	—[c]	0.32	0.21	—[c]
Radiolysis[a]	n.d.	*1.00*	0.25	—[c]	0.22	0.23	n.d.
Mass spectrometer[d]	0.24	*1.00*	0.03	n.d.	0.44	0.15	0.09

Normal Pentane[e]

Method	Ethylene	Propane	Propylene	Ethane	Butene	Pentene
Photolysis						
1470 Å	*1.00*	0.62	1.59	0.52	0.35	11.1
1236 Å	*1.00*	0.27	0.57	0.22	0.14	2.3
Radiolysis						
Applied field	*1.00*	0.13	0.54	0.15	0.21	0.9
direct	*1.00*	0.19	0.53	—[c]	0.16	—[c]

[a] Taken from Ref. 32.

[b] n.d. = not determined.

[c] Yields resulting from decomposition of neutral molecules cannot be uniquely defined.

[d] Taken from Ref. 44.

[e] Taken from Ref. 65.

ions nor ion–molecule reactions contribute to the yields of acetylene, ethylene, ethane, or propylene in the photoionization of *n*-butane at 1066–48 Å at pressures above 1 torr. Regardless of the nature of the neutral mechanisms responsible for the formation of these products, the data verify that no abrupt changes occur in the product distribution when superexcited states are populated. Moreover, the distribution reflects a continuous trend with increasing energy and the variations are qualitatively what one would expect; namely that secondary fragmentations are favored when more energy is available for partitioning among the products of primary dissociations (the yields of C_2 unsaturates are higher). In many respects the continuous trend is surprising since there are indications that some superexcited states are Rydberg levels associated with higher ionization potentials of the molecule (see Chapter 1) and one might expect to observe a change in the neutral photochemistry when more tightly bound electrons are excited. Based on the chemical analysis of many molecules subjected to photoionization in the threshold region, however, no abrupt changes are evident. In fact, the product distribution resulting from the dissociations of superexcited molecules may be predicted fairly accurately simply by a consideration of the photochemistry observed at other wavelengths in the far ultraviolet.

4. Comparison of Photon and Electron Impact.   In several of the alkanes it is possible to compare data obtained in the photolysis, direct radiolysis, and the radiolysis conducted in the presence of an applied electric field. Included in Table IV is a tabulation of those yields which can be uniquely ascribed to neutral mechanisms in the gas phase radiolysis of *n*-butane. Matching this data with the photochemical distribution indicates that the average excitation energy is the direct radiolysis in the region of the ionization potential, verifying the predicted importance of superexcited states in radiation chemistry. Beck and Niehaus (44) have also obtained a neutral fragmentation pattern for this molecule (see Table IV) using the modification of mass spectrometry discussed in Section I-B-1-a. In many respects the agreement with other methods is quite good. These studies were carried out at much lower pressures, however, where secondary fragmentations are favored. This effect is clearly indicated by the relatively large yield of acetylene obtained by this method, since the formation of this product can be uniquely associated with secondary and tertiary dissociations.

Correlation of the radiolysis and photolysis data for *n*-pentane (IP = 10.55 eV) also reported in Table IV confirms the importance of super-excited levels in the radiolysis. In this system, as well as many others, differential ion-pair yields resulting from the application of electric fields in the radiolysis have also been reported. These incremental values, reflecting the additional decomposition induced by accelerated electrons, compare very favorably with the 1236 Å photolysis and also indicate an average excitation in the region of the ionization potential.

B. Cycloalkanes

The cycloalkanes comprise a relatively simple homologous series for examination because the products of the decomposition are few in number and the primary dissociative modes may be rather easily derived by considering the effect of a variation in density. The vacuum ultraviolet photolyses of cyclopropane, cyclobutane, cyclopentane, and cyclohexane have been reported for $\lambda = 1470$, 1236, and 1066–1048 Å under a variety of conditions in both the vapor (34–37,66–68) and condensed phases

Table V
Molecular Products[a] Obtained in the Vacuum Ultraviolet
Photolysis of Cycloalkanes

Cyclopropane[b]

$\lambda(\text{Å})$	Phase[c]	H_2	C_2H_2	C_2H_4	C_3H_6	Allene	Methylcyclopropane
1470	Solid	T[d]	4.6	58	*100*	2.3	30
1236	Solid	T	0.4	33	*100*	7.1	20
1236	Vapor	T	30	*100*	3.7	58	T

Cyclobutane[a]

$\lambda(\text{Å})$	Phase[c]	C_2H_2	C_2H_4	C_4 unsaturates	C_3 unsaturates
1470	Solid[e]	4.5	*100*	9.5	0.0
1236	Solid[e]	0.4	*100*	9.7	0.0
1067–48	Solid[e]	0.5	*100*	11.5	0.0
1236	Vapor[f]	28	*100*	4.5	4.0

Cyclopentane

$\lambda(\text{Å})$	Phase[c]	H_2	C_2H_4	C_2H_2	C_3H_6	Cyclopentene	Allene	1-Pentene	Pentadiene
1470	Vapor[g]	206	*100*	10	24	175	8.1	T	13.0
1236	Vapor[g]	80	*100*	9	20	37	16	T	7.8
1236	Solid[g]	n.d.[h]	3	T	2	*100*	T	17	T

Cyclohexane

$\lambda(\text{Å})$	Phase[c]	H_2	C_2H_4	$1,3\text{-}C_4H_6$	Cyclohexene
1470	Vapor[i]	200	*100*	54	150
1236	Vapor[i]	80	*100*	40	32
1067–48	Vapor[i]	n.d.	*100*	37	23
1236	Solid[j]	n.d.	*100*	400	6,140

[a] Major products only.
[b] Taken from Ref. 34.
[c] NO or O_2 added in all vapor phase experiments.
[d] T = trace amount.
[e] Taken from Ref. 67.
[f] Taken from Ref. 35.
[g] Taken from Ref. 36.
[h] n.d. = not determined.
[i] Taken from Ref. 37.
[j] Taken from Ref. 67.

(34,36,67,68). Major products which can be uniquely ascribed to decompositions of neutral molecules are listed in Table V.

In order to provide a common basis for comparison, three general processes involving the highly excited ring system must be considered:

1. Molecular detachment of hydrogen, retaining the cyclic structure:

$$\text{cyclo } (C_nH_{2n})^* \rightarrow \text{cyclo } (C_nH_{2n-2})^{\ddagger} + H_2 \tag{22}$$

2. Initial C—H cleavage, retaining the cyclic structure:

$$\text{cyclo } (C_nH_{2n})^* \rightarrow \text{cyclo } (C_nH_{2n-1})^{\ddagger} + H \tag{23}$$

3. Initial C—C cleavage, resulting in ring opening:

$$\text{cyclo } (C_nH_{2n})^* \rightarrow (CH_2)_n^{\ddagger} \tag{24}$$

where $n = 3, 4, 5,$ or 6 and $\ddagger$ denotes possible internal energy content.

1. Cyclopropane.   At low densities (vapor phase) ethylene is the major carbon-containing product obtained from the decomposition of cyclopropane (34,66). The overall mechanism

$$\Delta^* \rightarrow CH_2CH_2 + CH_2 \tag{25}$$

has been confirmed experimentally by the recovery of the stabilized methylene insertion product, methylcyclopropane, in the solid and the observation that CD_2H_2 is formed in large yields in the vapor phase photolysis of $(CD_2)_3$–H_2S mixtures (34). Methylcyclopropane is also recovered at higher pressures in the vapor phase.

The photolysis of $(CD_2)_3$–H_2S mixtures has verified that molecular detachment of hydrogen in a primary process is of negligible importance at all wavelengths in the far ultraviolet. However, radical titration does indicate a high yield of hydrogen atoms, particularly at shorter wavelengths.

In the solid (34), the extensive formation of the structural isomer, propylene, at all wavelengths provides evidence for C—C cleavage in the parent molecule. Propylene is not obtained in the vapor phase photolysis, (see Table V). These data suggest that allene, which is a major product in the vapor phase, is most probably formed during the subsequent dissociation of a trimethylene diradical which is stabilized by collision in the solid and rearranges into propylene. Since molecular detachment of hydrogen in a primary process is of minor importance, the overall reaction

$$\Delta^* \rightarrow C_3H_4 + H + H \tag{26}$$

most likely represents the major source of those hydrogen atoms indicated by radical counting techniques.

2. Cyclobutane. Cyclobutane is also rather unique among the lower molecular weight hydrocarbons because the decomposition of the excited molecule leads almost exclusively to the formation of a single product, ethylene. The channel of lowest energy is again that described by general process 24 ring opening, followed by symmetric cleavage at the central C—C bond in the tetramethylene diradical.

$$\square^* \rightarrow \cdot CH_2—CH_2—CH_2—CH_2 \cdot^{\ddagger} \rightarrow 2CH_2=CH_2^{\ddagger} \tag{27}$$

The relatively minor yields of acetylene and hydrogen are sensitive to both wavelength (in the vapor) and density, and reflect the further decomposition of internally excited ethylenic structures. As in cyclopropane, *molecular* detachment of hydrogen is of negligible importance. This is evidenced by the very small molecular yield obtained in $(CD_2)_4$: $(CH_2)_4$:NO mixtures (35) and the fact that both cyclobutene and the expected rearrangement products of vibrationally excited cyclobutene (mainly 1,3-butadiene) are only recovered in small quantities, even in the solid (67).

3. Cyclopentane. In contrast to cyclopropane and cyclobutane, Doepker et al. (37) have determined that the major mode of fragmentation is the *molecular* detachment of hydrogen and the formation of internally excited cyclic structures (process (22)). This mechanism, followed by subsequent unimolecular reactions of cyclo-$(C_5H_8)^{\ddagger}$, is clearly reflected by the large relative yields of *molecular* hydrogen and cyclopentene obtained at 1470 Å in the vapor phase (see Table V). At higher excitation energies (1236 Å) the relative yields of cyclopentene and hydrogen are reduced somewhat because other dissociative modes (see next paragraph) compete more favorably. The excess energy available at this wavelength also favors secondary fragmentation of cyclo-$(C_5H_8)^{\ddagger}$. Not surprisingly, the *molecular* products obtained from the photolysis of cyclopent*ene*–NO mixtures (ethylene and smaller quantities of acetylene and C_3 unsaturates) are among those which are reduced by a substantial increase in density during the photolysis of cyclopent*ane*.

At 10.5 eV (1236 Å) and 11.7 eV (argon resonance photolysis, not shown in Table V), comparison of the free radical yields obtained by H_2S titration with the yields of lower molecular weight products indicates that the subsequent dissociations of cyclo-$(C_5H_8)^{\ddagger}$ cannot account quantitatively for the photochemistry observed in cyclopentane. In order to account for the observed product distribution, Doepker et al. invoked the general process (24) followed by dissociation of the pentamethylene intermediate into free radicals and various molecular products including ethylene and cyclopropane (or propylene). This sequence was corroborated by the recovery of the stabilized pentamethylene rearrangement product, 1-pentene, in the solid phase photolysis and the drastic reduction in C_2–C_3 olefinic products which followed the increase in density.

4. Cyclohexane. Cyclohexane, which has received considerable attention among radiation chemists, especially in condensed phases, has recently been the subject of several photolysis studies. Reference data obtained at 1470 and 1236 Å is recorded in Table V.

Clearly the most important dissociative mode of the highly excited molecule is *molecular* detachment of hydrogen (process (22)). Doepker and Ausloos (37), who photolyzed the selectively labeled analog cyclohexane-1,1,2,2,3,3-d_6 have suggested that the elimination of hydrogen results mainly in carbene formation. This conclusion was based on a comparison of the experimentally observed isotopic hydrogens with the distribution assumed for simultaneous detachment from adjacent carbon atoms (4:2:1 for H_2, HD and D_2, assuming an intramolecular isotope effect favoring H_2 and HD detachment). Although the majority of the resulting carbene diradicals rearrange into cyclohexene at both wavelengths, secondary degradation into C_2H_4 and 1,3-C_4H_6 is observed in the vapor phase at pressures as high as 50 torr. An increase in pressure reduces the yields of these products and they are predictably negligible in the liquid phase (68).

The vapor phase data indicate a stoichiometric excess of ethylene which cannot be correlated with further decomposition of internally excited cyclo-$(C_6H_{10})^{\ddagger}$. Selective labeling has revealed that ring rupture in the cyclohexane parent molecule is responsible (process (24)). This mode is particularly important in the 1236 Å photolysis, and is reminiscent of the increasing role of C—C cleavage observed at higher excitation energies in the noncyclic alkanes. The recovery of the stabilized hexamethylene rearrangement product (probably 1-hexene, by analogy with cyclopentane) in condensed phases has not been reported, however, in spite of the fact that ring cleavage may account for 20% of the primary processes in the vapor phase photolysis at 1236 Å. Although most of the observed decomposition of cyclohexane at 8.4 and 10 eV can probably be ascribed to *molecular* detachment of hydrogen (reaction (22)), the gas phase photolysis of C_6H_{12}–C_6D_{12} mixtures (37) and the photolysis of liquid C_6H_{12} in the presence of scavengers (68) indicate a small but not negligible yield of H atoms (process (23)).

5. General Considerations. The relative probabilities observed for reactions 22–24 within the homologous series may be rationalized in very simple terms. For example, ring cleavage represents the major dissociative mode only for those members characterized by higher strain within the cyclic structure (cyclopropane and cyclobutane). *Molecular* detachment of hydrogen is of little importance for these systems but represents the major mode of energy dissipation in the more stable five- and six-membered rings (cyclopentane and cyclohexane). However, for the latter pair the probability for C—C cleavage does exhibit a rather direct functional dependence on the degree of excitation. It is very difficult to assess directly the import of reaction (23) within the series since H atoms may be formed in secondary processes.

C. Unsaturated Hydrocarbons

1. Monoolefins. Relatively few monoolefins (only ethylene, propylene, and butene) have been investigated in any detail at high excitation energies. Vacuum ultraviolet experiments incorporating the rare gas resonance lamps are also somewhat more difficult to interpret since photoionization occurs in all monoolefins (except ethylene) at energies below 10eV.

a. Ethylene. The vapor phase photolysis of ethylene has been reported at 1849 Å (69), 1470 Å (38,69), 1236 Å (38,69), and 1066–1048 Å (38). The major chemical products in the presence of radical scavengers are hydrogen and acetylene at all wavelengths. Simultaneous detachment from a single carbon atom as well as 1,2 elimination contributes to the *molecular* hydrogen yield. For example, H_2, HD, and D_2 are obtained in the photolysis of CH_2CD_2 at all of the above mentioned wavelengths (see Table VI).

$$CH_2CD_2{}^* \rightarrow H_2 + CCD_2 \tag{28}$$

$$CH_2CD_2{}^* \rightarrow D_2 + CCH_2 \tag{29}$$

$$CH_2CD_2{}^* \rightarrow HD + CDCH \tag{30}$$

Details of the 1,2 mechanism have been derived from the photolysis of *trans*-CHDCHD by Okabe and McNesby (69). Since 25–30% of the

Table VI
Vapor-Phase Photolysis of CH_2CD_2: Formation of
Hydrogen and Acetylene

$\lambda(\text{Å})$	H_2	HD	D_2
1849[a]	40.0	40.0	20.0
1470[a]	40.3	40.0	19.7
1236[a]	41.7	40.7	17.6
1067–48[b]	41.8	41.5	16.7

[a] Ref. 69.
[b] Ref. 38.

hydrogen from this analog consists of H_2 and D_2, ethylene molecules excited to Rydberg levels apparently rotate internally about the C—C bond prior to the 1,2-elimination. More recently, Tschuikow-Roux et al. (70) found that isomeric butenes (especially 1-butene) are formed in high yields when ethylene is photolyzed in the solid phase and suggested the formation of an ethylidene-like structure of the excited molecule ($CH_3CH^{\ddagger}$) which yields butene upon insertion into a neighboring molecule. An additional source of acetylene (other than rearrangement of vinylidene diradicals and 1,2 molecular elimination of hydrogen) is indicated by the inequality $C_2H_2 > H_2$, and has been ascribed to the overall process

$$C_2H_4^* \rightarrow C_2H_2 + 2H \tag{31}$$

Ethylene also exhibits some variations in the fragmentation pattern which are reminiscent of alkanes as the excitation energy is increased: (1) A decrease in wavelength favors loss of hydrogen atoms relative to loss of a hydrogen molecule; (2) The distribution of isotopic hydrogens obtained from CH_2CD_2 (Table VI) is insensitive to changes in wavelength above and below the ionization potential, and it is apparent that no discontinuities occur in the fragmentation pattern when superexcited states are populated (argon photolysis).

b. *Propylene and Butene.*   The vapor phase photolysis of propylene at 1470 and 1236 Å (ionization efficiency ≈ 0.20 at 1236 Å) has also been reported in some detail (39). In the presence of radical scavengers, the major decomposition products are lower molecular weight unsaturates (ethylene and acetylene). Preliminary data is also available for the isomeric butenes, (47,71) but the excitation energy was only 6.7 eV (1849 Å) in these studies. Perhaps the most important characteristic of polycarbon monoolefins is that they behave much in the same way as saturated hydrocarbons at higher excitation energies. Many unimolecular channels are available for decomposition, including *molecular* elimination of hydrogen, methylene radicals, and lower molecular weight alkanes. Extensive C—C and C—H cleavage generating free radical intermediates also occurs at higher excitation energies, and the terminal methyl group appears to be a particularly labile site.

2. Acetylenic Hydrocarbons.   A rather detailed investigation of the 1470 and 1236 Å vapor-phase photolysis of acetylene has been reported by Stief et al. (72). Major chemical products of the decomposition are various polyolefins, including some solid material, and small quantities of hydrogen.

In addition to a primary process which yields hydrogen atoms via the rapid predissociation $C_2H_2^* \rightarrow C_2H + H$, there is considerable evidence for a mechanism which involves long-lived excited electronic states of the parent molecule since the addition of quenchers (N_2, CO_2) at fixed pressures of acetylene *reduces* the quantum yield for conversion to polymeric products. Tanaka (73), who observed the sensitized ionization of NO in a mass spectrometer, has provided additional support for the long-lived excited state hypothesis:

$$C_2H_2^* + NO \rightarrow NO^+ + C_2H_2 + e^- \qquad (32)$$

Stief et al. were also able to show that C_2^* formed by *molecular* loss of hydrogen from a long-lived state of acetylene was excited to the $A^3\Pi_g$ level in the 1236 Å photolysis and the intensity of the luminescence $(C_2(^3\Pi_g) \rightarrow C_2(^3\Pi_u) + h\nu)$ and the associated hydrogen yield were also very sensitive to pressure and added quenchers. It is interesting to note that the overall reaction

$$C_2H_2^* \rightarrow H_2(^1\Sigma_g) + C_2^*(A^3\Pi_g) \qquad (33)$$

is spin-forbidden assuming that $C_2H_2^*$ is in a singlet state produced in an optically allowed transition from the ground state. Since triplet C_2^* has been detected, the rate-determining step for hydrogen detachment may well be the rate at which intersystem crossing occurs in the highly excited acetylene molecule.

Stief and DeCarlo (74) have also derived quantum yields (based on CO_2 actinometry) for several of the primary processes occurring in the photo-decomposition of methylacetylene at 1470 and 1236 Å. At pressures below 1 torr, the major products of the decomposition are hydrogen and acetylene.

Both D_2 and HD are obtained in the scavenged photolysis of $CD_3C\equiv CH$ at 1470 Å, indicating that both 1,1 detachments (yielding D_2) and 1,3 mechanisms are contributing.

Acetylene, the other major product at both wavelengths, may be formed by either two mechanisms;

$$CH_3C\equiv CH^* \rightarrow CH_2 + C_2H_2 \qquad (34)$$

$$CH_3C\equiv CH^* \rightarrow CH_3 + C_2H \qquad (35)$$

followed by

$$C_2H + CH_3C\equiv CH \rightarrow C_2H_2 + CH_3C\equiv C\cdot \qquad (36)$$

Since the isotopic acetylenes obtained for $CD_3C\equiv CH$ at 1470 Å are 90% C_2H_2, the dissociation into CD_3 and C_2H is apparently favored over methylene elimination. The methyl-methyl combination product, ethane, is also recovered in high yields in the absence of scavengers at this wavelength. At 1236 Å, ethane is reduced to trace amounts and the yields of

C_2H_2 and C_2HD are nearly equivalent, suggesting that the methylene detachment process which forms CD_2 and C_2HD is an important primary process only at higher excitation energies.

Unfortunately, quantitative investigation of acetylenic homologs in the far ultraviolet is complicated by the relative ease with which such compounds may be polymerized at higher pressures by free radical attack. This condition has prevented a determination of the possible role of excited state mechanisms in chain propagation, which, by analogy with acetylene, may be quite important.

III. OTHER ORGANIC COMPOUNDS

The investigation of organic compounds other than hydrocarbons has been largely restricted to photolysis in the near ultraviolet. Recently, Calvert and Pitts (4) have published an extensive and thorough review of these researches, including detailed mechanistic discussions of the photophysical and photochemical processes occurring in a variety of electronically excited molecules. Repetition of these discussions, however brief, was not deemed appropriate because of the intended emphasis here on more highly excited systems. Unfortunately, little attention has been given to the fragmentations of aldehydes, ketones, acids, amines, and other hetero-derivatives at shorter wavelengths, although some representative molecules including several aldehydes (75), methylamine (76), and methyl cyanide (77) have been investigated in the 1800–2000 Å region.

An interesting analytical approach for monitoring the photodecomposition of oxygenated compounds has been reported by Harrison and co-workers (78). The exciting lamp in the experiments was a continuum source (hydrogen discharge lamp), and formation of photochemical products was detected by following the alterations in the absorption spectrum (parent molecule plus products) resulting from prolonged exposure. High conversions were sometimes required before the characteristic absorption features of photochemical products could be detected, however, and saturated hydrocarbons are not easily identified by this method since the absorption curves of these molecules are continuous. Nevertheless, these experiments represented the first attempts in the far ultraviolet photolyses of complex organic systems. Only three systems have been investigated at $\lambda < 1800$ Å: acetone, pentanone, and methanol. What is known of the fragmentation patterns of these molecules will be discussed briefly.

A. Acetone

Although the near ultraviolet photolysis of acetone has long served as one of the classic systems for kinetic study (4), Leiga and Taylor (79) have conducted the only reported investigation of this compound at $\lambda < 1500$ Å. As expected, at both 1470 and 1236 Å the major dissociative path is

$$(CH_3)_2C=O^* \rightarrow CH_3C=O^\ddagger + CH_3 \qquad (37)$$

followed by

$$CH_3C=O^\ddagger \rightarrow CO + CH_3 \qquad (38)$$

In addition to process (37) the dissociation paths shorter wavelengths include C—H cleavage [$(CH_3)_2C=O^* \rightarrow H_2$ or $H + CH_3COCH$ or CH_3COCH_2] and the *molecular* elimination of methane. The overall quantum yield derived for the evolution of CO was higher at 1236 Å (~ 1.0) than at 1470 Å (~ 0.67). Evidence was also found for the formation of methyl radicals and hydrogen atoms with excess translational energy at both wavelengths. However, in view of the high conversions and the possible role of ion–molecule interactions in the 1236 Å photolysis (10.0 eV, IP acetone $= 9.7$ eV), the interpretation invoked by Leiga and Taylor should be considered only as tentative.

B. Pentanone

A detailed investigation of the 1470–1236 Å photolysis of the pentanones in the vapor phase has been completed by Scala and Ausloos (15). Major *molecular* products include CO, H_2, C_2H_4 and lesser amounts of C_1–C_3 alkanes and olefins.

The important dissociative channel in the far ultraviolet photolysis of 3-pentanone is that found at longer wavelengths, alpha C—C cleavage;

$$(C_2H_5)_2C=O^* \rightarrow C_2H_5CO^\ddagger + C_2H_5 \qquad (39)$$

followed by

$$C_2H_5CO^\ddagger \rightarrow C_2H_5 + CO \qquad (40)$$

Hydrogen atom loss from the 1 and 2 positions, as well as methyl loss, were also detected. At these high excitation energies secondary decomposition of carbonyl and alkyl transients is quite extensive in the vapor phase. Ethylene, for example, which is a major product, may be eliminated either from internally excited β-pentanonyl radicals ($CH_3CH_2COCH_2CH_2^\ddagger \rightarrow C_2H_4 + C_2H_5 + CO$) or ethyl radicals ($C_2H_5^\ddagger \rightarrow C_2H_4 + H$). Intramolecular elimination of H_2 and CH_4 from the parent molecule also contribute to the overall decomposition but play a minor role compared to alpha carbon cleavage.

Selective labeling, in the form of $CD_3COCD_2CH_2CH_3$, has provided some very useful information in the 2-pentanone system.

In the near ultraviolet (3130 and 2537 Å) *molecular* ethylene is formed exclusively via a primary photoelimination (4). For pentanone-α-d_5, this detachment would yield ethylene-C_2H_4 and acetone-d_5 after rearrangement of the enol product:

$$CD_3\overset{\overset{\textstyle O}{\|}}{C}CD_2CH_2CH_3^* \rightarrow CD_3(OH)C=CD_2 + C_2H_4 \tag{41}$$

Although some *molecular* C_2H_4 and acetone was detected following photolysis at 1470 and 1236 Å, the major component of the ethylene fraction was CH_2CD_2. The formation of this analog has been ascribed to secondary fragmentation of internally excited carbonyl and alkyl (propyl) radical intermediates formed in highly exothermic primary C—H and C—C cleavages in the parent molecule.

Propylene, which consists entirely of CD_3CHCH_2, was also assigned to secondary fragmentations. Contributions from a primary process such as

$$CD_3\overset{\overset{\textstyle O}{\|}}{C}CD_2CH_2CH_3^{*'} \rightarrow CD_2HCHO + CD_3CH=CH_2 \tag{42}$$

were excluded since acetaldehyde was not recovered.

C. Methanol

The photochemistry of methanol vapor has been investigated in the 1850–2000 Å region (80,81) and at 1236 Å (81) in various admixtures (1850 Å; C_2H_4, CO_2, O_2, and I_2, and 1236 Å; C_2H_4). Major products of the decomposition in both wavelength regions are G_2, CH_2O, and $(CH_2OH)_2$. Three primary processes have been proposed to describe the fragmentation:

$$CH_3OH^* \rightarrow CH_2O + H_2 \tag{43}$$
$$\rightarrow CH_3O + H \tag{44}$$
$$\rightarrow CH_2OH + H \tag{44a}$$
$$\rightarrow CH_3 + OH \tag{45}$$

1850–2000 Å: The hydrogen formed in the photolysis of CD_3OH–scavenger mixtures is mainly HD ($\sim 80\%$), indicating that a "1,2" mechanism is favored in reaction (43). It is interesting to note that carbene detachments of the general type characteristic of hydrocarbons are not important in methanol since no molecular D_2 is obtained from CD_3OH. On the basis of the decrease in hydrogen when H atom scavengers are added it has been determined that reaction (43) accounts for approximately 20–30% of the fragmentation in the 1850–2000 Å region. Experimental distinction between reactions (44) and (44a) is quite difficult since methoxy radicals are quickly converted to CH_2OH by abstraction of hydrogen atoms ($CH_2O + CH_3OH \rightarrow CH_3OH + CH_2OH$). However, little D_2 is formed from CD_3OH in the absence of scavengers. This observation provides evidence for reaction (44) assuming that $H(D) + CD_3OH \rightarrow HD(D_2) + CD_2OH$ represents the mechanism for removal of hydrogen atoms. Very little water and methane are obtained in the pure system in this wavelength region, which has been taken as evidence that reactions (43) and (44) are the only important processes when methanol is photolyzed in the first absorption band (lowest singlet state).

Although the decomposition at 1236 Å (81) would seem to be more difficult to unravel since more energy is available for partitioning among dissociation products, Hagege et al. found that the product distribution obtained at this wavelength was very nearly the same as that obtained at 1849 Å at equivalent pressures. On this basis, they concluded that a radiationless internal conversion occurred to the lower singlet state and that collisions were required to achieve the configuration necessary for dissociation since an increase in pressure enhanced the quantum yield for decomposition (81,81a).

IV. INORGANIC COMPOUNDS

In many respects, the primary fragmentation patterns of inorganic molecules are more easily characterized than those patterns found for hydrocarbons and other organic compounds since the extra energy available for partitioning during primary fragmentations may be dissipated via radiative transitions occurring from bound states in electronically excited dissociation products. This luminescence may be monitored spectroscopically, and in many cases end product analysis is not required to establish details of the unimolecular decay sequence. This behavior is also of considerable diagnostic value to the photochemist, since the "appearance wavelength" for the formation of an emitting species may be determined by varying the exciting wavelength absorbed by the parent molecule. In those instances where the radiative transitions are spin- or symmetry-forbidden, the sensitized luminescence and/or chemiluminescence resulting from interactions of metastable dissociation products with selected additives has provided very useful information. Certain features of the fragmentation patterns associated with those inorganic compounds containing substituent hydrogen atoms have also been established by indirect techniques, including the introduction of radical scavengers (usually olefins) and other titrants for radical intermediates. Through a combination of these two methods, chemical analysis and kinetic spectroscopy, a great deal of information has been accumulated, especially for the lower molecular weight compounds.

A detailed examination of inorganic molecules was considered to be beyond the scope of this chapter, since much of the earlier work has been reviewed elsewhere (3). More recently, certain aspects of the vacuum ultraviolet photolysis of HN_3 (82,83), N_2H_4 (74,84), H_2O_2 (74), NO_2 (85,86), NOCl (86), and CS_2 (87) have also been reported, but these compounds will not be discussed here. The fragmentation patterns derived for four typical systems of general interest will be treated in some detail; these include water, carbon dioxide, nitrous oxide, and ammonia.

A. Water

The calculated threshold wavelength requirements for the formation of dissociation products in various excited states are compiled in Table VII.

Table VII
Threshold Wavelength Requirements for Formation for
the Various Possible Dissociation Products of Water

Process	E (eV)	λ(Å)
(46) $H_2O^* \rightarrow H_2 + O(^3P)$	5.0	2468
(47) $H_2O^* \rightarrow H(^2S) + OH(X^2\Pi)$	5.1	2420
(48) $H_2O^* \rightarrow H_2 + O(^1D)$	7.0	1763
(49) $H_2O^* \rightarrow H(^2S) + OH(A^2\Sigma^+)$	9.1	1356
(50) $H_2O^* \rightarrow H_2 + O(^1S)$	9.4	1320
(51) $H_2O^* \rightarrow 2H(^2S) + O(^3P)$	9.5	1299
(52) $H_2O^* \rightarrow 2H(^2S) + O(^1D)$	11.5	1073
(53) $H_2O^* \rightarrow H(^2P) + OH(X^2\Pi)$	15.3	807

In the vacuum ultraviolet, water exhibits three regions of optical absorption, two of which are essentially continuous (88). The lowest energy continuum is situated between 1430 and 1860 Å, with a maximum at approximately 1655 Å, and the second diffuse system extends from 1430 to approximately 1250 Å. Below 1250 Å a number of bands are observed and the absorption spectrum is highly structured. Most experiments have been performed with 1470, 1236, and 1216 Å radiation, corresponding to photolysis in the first continuum and the 1240 Å band system, respectively.

Only two empirically distinguishable processes are possible at any wavelength:

$$H_2O^* \rightarrow H + OH \tag{54}$$

$$H_2O^* \rightarrow H_2 + O \tag{55}$$

The competition between these modes and the nature of the dissociation products has been investigated spectroscopically (OH radical) and by the introduction of additives which provide a measure of the hydrogen atom yield from reaction (54).

There is general agreement that process 55 is unimportant in the first continuum ($\lambda > 1430$ Å). Stief (89), who photolyzed D_2O at 1470 Å in the presence of varying amounts of C_2H_4, concluded that reaction (54) accounts for approximately 94% of the primary photochemical processes at this wavelength. Vermeil et al. (90,91), on the other hand, were unable to find any evidence for reaction (55) in the first continuum when H_2O was photolyzed in the presence of D_2. In H_2O-D_2 mixtures, H atoms are titrated as HD ($H + D_2 \rightarrow HD + D$) and the *molecular* yield, including unintercepted H atoms, appears as H_2. The H_2/HD ratio, when extrapolated to infinite dilution ($H_2O/D_2 = 0$) was found to pass through the origin ($H_2/HD = 0$) in the 1470 Å photolysis. A finite intercept for the H_2/HD ratio would, of course, have provided evidence for molecular elimination of hydrogen.

A finite intercept for the H_2/HD ratio at infinite dilution in H_2O-D_2 mixtures was obtained at 1236 Å (91). Independent evidence for a significant molecular yield at this wavelength ($\sim 25\%$ of primary processes) was derived by McNesby et al. (92) during the photolysis of $H_2O-C_2D_4$ mixtures. Based on the variation in the value of the infinite dilution intercept plotted for H_2O-D_2 mixtures at various pressures of H_2O, Cottin et al. (91) concluded that the relative probabilities for processes (54) and (55) were pressure dependent. This hypothesis was based on an extrapolation, however, and the values of the various intercepts depend to a large extent on how one constructs curves through the experimental points. If the reported intercepts are considered valid, water represents one of the few systems where the relative probabilities for competing *primary* processes are effected by collision processes in the low pressure region (5–15 torr). Evidence has also been obtained for the formation of hydrogen atoms with kinetic energies considerably in excess of thermal values at 1236 Å (91).

At $\lambda > 1356$ Å it is possible to produce hydroxyl radicals excited to the $A^2\Sigma^+$ level (see Table VII). The formation of OH* was first noted by Neurimin and Terinin (93), who photolyzed water vapor with the hydrogen continuum and detected the characteristic fluorescence ($OH(A^2\Sigma^+) \rightarrow OH(X^2\Pi) + h\nu$) in the near ultraviolet (0–0 band ~ 3062 Å). High resolution studies have been reported by Carrington (94) at several exciting wavelengths ($\lambda \leq 1300$ Å). At 1216 Å (Lyman α), the population of rotational levels associated with $OH(A^2\Sigma^+)$ were found to peak in the region of $K = 18$–22, with no significant population in higher states ($K > 22$) (the equilibrium values correspond to $K \leq 8$ with a maximum intensity at $K = 3$). These nonequilibrium rotational distributions, which are also observed at 1236 and 1302 Å, indicate that most of the excess energy above that required to form $OH(A^2\Sigma^+)$ in process (49) is found in rotational levels of OH* and relatively little is given to the H atom in the form of translational energy. The efficiency with which $OH(A^2\Sigma^+)$ is formed at these wavelengths is relatively small, however, and there is every reason to believe that H atoms may be quite hot when the primary dissociation forms hydroxyl radicals in the ground electronic state at these exciting wavelengths. Recently, Welge and Stuhl (95) found that $OH(X^2\Pi)$ was formed in very low rotational levels when water vapor was photolyzed at $\lambda > 1400$ Å, indicating that the hydrogen atoms would have average kinetic energies on the order of 72 kcal/mole in the 1470 Å photolysis. This observation is not compatible with the conclusions of Cottin et al. (91), who found no chemical evidence for hot atoms at this wavelength.

B. Carbon Dioxide

Carbon dioxide has been extensively used as a chemical actinometer in the far ultraviolet using a quantum yield of unity for evolution of CO in the vapor phase. At $\lambda > 1070$ Å, three primary processes are possible, all of which form CO directly:

$$CO_2^* \rightarrow CO(^1\Sigma) + O(^3P) \quad (\lambda < 2240 \text{ Å})§ \tag{56}$$

$$CO_2^* \rightarrow CO(^1\Sigma) + O(^1D) \quad (\lambda < 1650 \text{ Å})§ \tag{57}$$

$$CO_2^* \rightarrow CO(^1\Sigma) + O(^1S) \quad (\lambda < 1273 \text{ Å})§ \tag{58}$$

Mahan (96), was unable to detect the characteristic green chemiluminescence $O(^3P) + NO \rightarrow NO_2 + h\nu$ in both the 1270 and 1236 Å photolysis of CO_2–NO mixtures. This negative evidence for process 56 is compatible with the Wigner rules, which require $CO(^1\Sigma)$ and $O(^1D)$ (or $O(^1S)$ at $\lambda < 1273$ Å) as the dissociation products of the upper $^1\Delta$ levels of CO_2^* excited at these wavelengths. Although there is general agreement that process (57) is the only important dissociative mode in the xenon and krypton resonance photolyses, various values have been reported for the quantum yield of CO. In fact, the most recent detailed investigation (97) of the 1470 Å photolysis has yielded a value of 0.25, which is far below the "accepted" value of approximately 1.0 found earlier in several laboratories (96,98). Since one molecule of CO is expected to result from every primary dissociation, a value lower than unity is either suggestive of back reactions which destroy accumulated CO in the system or collisional deactivation of CO_2^*.

There is considerable evidence that metastable $O(^1D)$ reacts rapidly with CO_2 to form the relatively long-lived species CO_3^*. The transient formation of this adduct is suggested by the isotopic exchange observed in the

§ Wavelengths correspond to calculated thresholds.

photolysis (99) of $^{18}CO^{18}O$–$^{16}CO^{16}O$ mixtures (the exchange reflecting dissociation of the isotopically mixed CO_3*) and the recent detection of the infrared absorption features of CO_3 after photolysis of a solid CO_2 matrix at 1470 Å (100). The spectrum of this species was also found after condensation of the products of a gas phase electric discharge in CO_2. Furthermore, in the vapor phase photolysis the ratio of CO/O_2 is usually greater than 2.0, which is the stoichiometric value assuming that O atoms eventually combine to form O_2 molecules. The O_2 deficiency has most recently been ascribed to chemical adsorption of CO_3 on the walls of the photolysis vessel (97), presumably via carbonate formation. Other mechanisms involving ozone formation $(O + O_2 \rightarrow O_3)$ have also been proposed (96,101). Regardless of the secondary kinetics which may be involved and the nature of the back reaction (if any), the quantum yield for formation of CO in the vacuum ultraviolet photolysis of CO_2 does not seem to be established with any degree of certainty. The recent article by Ung and Schiff (97) includes a detailed summary of our present level of understanding in this very complex system.

C. Nitrous Oxide

Although only two general processes are possible

$$N_2O^* \rightarrow N_2 + O \tag{59}$$

$$N_2O^* \rightarrow NO + N \tag{60}$$

the atomic fragments (N and O atoms) and molecular products (N_2 and NO) may be formed in a variety of excited state reflecting the excess energy contained in the parent molecule prior to dissociation. The calculated threshold wavelengths requirements for the possible dissociations are listed in Table VIII. Also included is a tabulation of the energy levels associated with excited atoms and molecules which may be formed as primary dissociation products.

Little is known concerning the kinetics of excited oxygen and nitrogen atom reactions, as well as N_2*, with N_2O. As early as 1934 (102), excited NO molecules had been detected spectroscopically in the vacuum ultraviolet photolysis of N_2O conducted at high conversions and the characteristic emissions

$$NO^*(A^2\Sigma^+) \rightarrow NO(X^2\Pi) + h\nu \quad (\gamma \text{ bands}) \tag{72}$$

$$NO^*(B^2\Pi_2) \rightarrow NO(X^2\Pi) + h\nu \quad (\beta \text{ bands}) \tag{73}$$

were found to appear at wavelengths *longer* than those required to populate

Table VIII[a,b]

Calculated Threshold Energy and Wavelength Requirements for the
Neutral Dissociation of N_2O

Processes	Threshold wavelengths below which reaction is possible	
	eV	Å
(61) $N_2O \rightarrow N_2(^1\Sigma^+) + O(^1S)$	5.86	2115
(62) $N_2O \rightarrow NO(X^2\Pi) + N(^2D)$	7.31	1696
(63) $N_2O \rightarrow N_2(A^3\Sigma_u^+) + O(^3P)$	7.89	1570
(64) $N_2O \rightarrow NO(X^2\Pi) + N(^2P)$	8.50	1458
(65) $N_2O \rightarrow N_2(B^3\Pi_g) + O(^3P)$	9.06	1368
(66) $N_2O \rightarrow N_2(B'^3\Sigma_u^-) + O(^3P)$	9.89	1254
(67) $N_2O \rightarrow N_2(a^1\Pi_g) + O(^3P)$	10.26	1208
(68) $N_2O \rightarrow NO(A^2\Sigma^+) + N(^4S)^c$	10.37	1195
(69) $N_2O \rightarrow NO(B^2\Pi_r) + N(^4S)^c$	10.62	1168
(70) $N_2O \rightarrow N_2(B^3\Pi_g) + O(^1D)$	11.03	1124
(71) $N_2O \rightarrow NO(A^2\Sigma^+) + N(^2D)$	12.76	972

[a] Energy levels of excited atoms in eV: $N(^2D) = 2.38$, $N(^2P) = 3.57$, $O(^1D) = 1.97$, $O(^1S) = 4.19$.

[b] Energy levels of excited molecules in eV: $NO(A^2\Sigma^+) = 5.45$, $NO(B^2\Pi_r) = 5.69$, $N_2(A^3\Sigma_u^+) = 6.22$, $N_2(B^3\Pi_g) = 7.39$, $N_2(B'^3\Sigma_u^-) = 8.22$, $N_2(a^1\Pi_g) = 8.59$.

[c] Signifies spin-forbidden processes.

these states via primary photodissociation. It was realized that secondary reactions of excited N, O, or N_2 with accumulated NO were responsible for this effect, and efforts were made to determine the origin of the luminescence by photolyzing mixtures of N_2O and NO (103–105). Okabe (105) has reported a definitive study in which a monochromator was used and the β, γ fluorescence (2000–4000 Å) from N_2O and N_2O/NO mixtures was monitored as a function of exciting wavelength from 2000 to 1000 Å. Other investigations in the far ultraviolet have been restricted to the xenon and krypton resonance lines (1470 and 1236 Å). A typical set of data is displayed in Fig. 4. In Fig. 4a the percent absorption of N_2O (defined as the number of photons absorbed per number of incident photons) is plotted as a function of wavelength from 1600–1100 Å. Figure 4b shows the luminescence intensity of the gamma system from a N_2O/NO mixture per

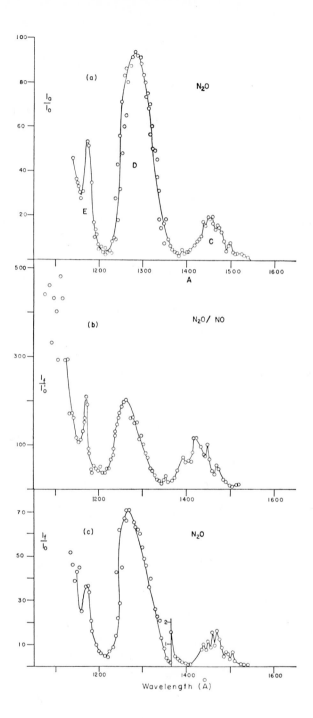

incident photon (arbitrary scale on ordinate) and Fig. 4c displays the analogous data (β system of NO*) obtained in the photolysis of pure N_2O. When pure N_2O is excited, only the β system of NO* is obtained and the fluorescence intensity is very much lower than that obtained in N_2O/NO mixtures, which is essentially pure gamma. Okabe interprets these observations as follows:

N_2O/NO mixtures: The gamma bands appear at a threshold wavelength of approximately 1520 Å (Fig. 4b). Although processes (61)–(63) are energetically allowed, $O(^1S)$, $N(^2D)$, and $O(^3P)$ cannot form $NO^*(A^2\Sigma^+)$ and the fluorescence in this region can only be ascribed to $N_2(A^3\Sigma_u{}^+)$ + NO $\rightarrow$ $NO^*(A^2\Sigma^+)$ + N_2. As the photon energy is increased, another source of NO* is indicated at approximately 1350 Å. This feature was correlated with process (65) followed by $N_2(B^3\Pi_g)$ + NO $\rightarrow$ NO (gamma). At much shorter wavelengths (< 1250 Å) the increase corresponds to the calculated threshold for process (66); $N_2O^* \rightarrow N_2(B'^3\Sigma_u{}^-)$ + $O(^3P)$ and the gamma bands are excited after energy transfer from $N_2(B'^3\Sigma_u{}^-)$.

Pure N_2O: Only the β bands are excited in pure N_2O at low conversion. The emission is first obtained (Fig. 4c) at 1540 Å and was ascribed to the dissociation into ground state NO* and $N(^2P)$ (process (64) followed by $N(^2P)$ + $N_2O \rightarrow N_2$ + $NO(B^2\Pi_r)$). Other reactions involving $N(^2P)$ and $N_2(B^3\Pi_g)$ were invoked at shorter wavelengths to explain the contour of the chemiluminescence curve, but Okabe found no evidence at any wavelength for the production of NO* in a *primary* process, indicating that the spin-forbidden photodissociations (68) and (69) are not important.

Although the relative probabilities of the general processes may not be established by chemiluminescence measurements alone, process (60) is apparently the major mode of fragmentation at 1470 Å while the dissociation to give molecular N_2 directly is favored at shorter wavelengths (3). There are some experimental indications that N_2O^* assumes a bent (triangular) configuration at longer wavelengths. This structure may account for the expulsion of the labeled internal nitrogen atom in the photolysis of ^{15}NNO and $N^{15}NO$ (106). Whether or not $O(^1D)$ or $O(^1S)$ is formed in the lowest energy transition in N_2O ($\lambda \leq 2140$ Å) is still a matter of controversy although Okabe's work has confirmed the formation of $O(^3P)$ at higher excitation energies.

Fig. 4. Absorption and fluorescence of N_2O and N_2O/NO mixtures. (a) absorption of N_2O at a pressure of 0.2 torr as a function of wavelength; I_a/I_0 = number of photons absorbed/number of incident photons in arbitrary units. (b) luminescence intensity of NO (gamma system) in a mixture of 0.2 torr of N_2O and 0.2 torr of NO as a function of the wavelength absorbed by N_2O; I_f/I_0 = number of photons emitted/number of incident photons in arbitrary units; (c) luminescence of NO (beta system) in pure N_2O at a pressure of 0.2 torr; I_f/I_0 = number of photons emitted/number of incident photons in arbitrary units. (All data taken from Ref. 105.)

Table IX

Threshold Wavelength and Energy Requirements
Calculated for the Neutral Decomposition of NH_3[a]

Reaction	Threshold (eV)	Wavelength (Å)
(77) $NH_3^* \rightarrow NH + H_2$[b]	3.91	3170
(78) $NH_3^* \rightarrow NH_2(X^2B_1) + H$	4.43	2798
(79) $NH_3^* \rightarrow NH(a'\Delta) + H_2$	5.54	2237
(80) $NH_3^* \rightarrow NH_2^2(A_1\Pi_u) + H$	5.70	2175
(81) $NH_3^* \rightarrow NH(b'\Sigma^+) + H_2$	6.61	1875
(82) $NH_3^* \rightarrow NH(A^3\Pi) + H_2$[b]	7.62	1627
(83) $NH_3^* \rightarrow NH(X^3\Sigma^-) + H + H$	8.44	1468
(84) $NH_3^* \rightarrow NH(c'\pi) + H_2$	9.35	1325
(85) $NH_3^* \rightarrow NH(a'\Delta) + H + H$	10.07	1231

[a] Based on the value of 1.6 eV for the energy of $NH(a'\Delta)$ with respect to $NH(X^3\Sigma^-)$ (H. Okabe, private communication).

[b] Spin-forbidden.

D. Ammonia

Three chemically distinguishable processes are likely to occur in the neutral decomposition of ammonia at energies below the ionization potential

$$NH_3^* \rightarrow NH + H_2 \tag{74}$$

$$NH_3^* \rightarrow NH_2 + H \tag{75}$$

$$NH_3^* \rightarrow NH + H + H \quad \text{(overall process)} \tag{76}$$

In Table IX are listed the calculated wavelength and energy requirements for formation of primary products in ground and excited states. The photolysis of NH_3–C_2D_4 mixtures has provided a measure of the relative contributions from process (74) at 1849, 1470, and 1236 Å since in such mixtures atomic hydrogen is rapidly removed by addition to ethylene and molecular hydrogen appears as H_2. At 1849 Å (92) only trace amounts of molecular hydrogen are observed ($\sim 4\%$ of the unscavenged yield), which rules out significant contributions from reaction (74) at low excitation energies. These results correlate very well with a vapor phase flash photolysis investigation by Bayes, Becker, and Welge (107) and the photolysis of NH_3 deposited in an argon matrix at 4.2°K by Schnepp and Dressler (108). In both laboratories only the absorption spectrum of NH_2 radicals was detected at wavelengths above 1550–1600 Å.

Although molecular detachment of hydrogen is negligible at $\lambda > 1500$ Å, scavenging studies indicate that this primary mode can account for 12 and 14% of the photodecomposition at 1470 (109) and 1236 Å (92), respectively. These results reflect upper limits since the possible role of hot hydrogen atoms has not been examined. At $\lambda < 1500$ Å, the formation of the NH radical has been independently confirmed by absorption spectroscopy in several laboratories (107,108), and Becker and Welge (110) have detected the luminescence $NH^*(c'\pi) \rightarrow NH(a'\Delta) + h\nu$ in the 1165, 1236, and 1295 Å photolysis (verifying process 84). Since the emission transition $NH(A^3\Pi) \rightarrow NH(X^3\Sigma) + h\nu$ was not detected, process (82) was considered to be relatively unimportant. However, the threshold for formation of molecular hydrogen and NH radicals is approximately 1500 Å, indicating that either reaction (79) or (81) must be contributing to the overall decomposition in this wavelength region.

Very recently, Okabe and Lenzi (111) have determined the "appearance wavelengths" for formation of $NH^*(c'\pi)$, (process (84)) and $NH_2(A^2A_1)$ (process (80)). By comparing the fluorescence intensity observed from NH_2^* with the contour of the absorption spectrum of NH_3 it was determined that formation of $NH_2(A^2A_1)$ is associated with several structured band systems in the region 1655–1286 Å, as well as the underlying continuum (1250–1600 Å). Production of $NH^*(c'\pi)$, on the other hand, was found to be associated with the continuous absorption of NH_3 at $\lambda \leq 1325$ Å. No evidence was found for violation of the spin conservation rule in a primary dissociative process.

Melton (9) has also investigated the electron impact-induced decomposition of ammonia in a modified mass spectrometer, and an appearance potential of 4.2 eV (corresponding to approximately 3000 Å) was obtained for general process (reaction (75)). The neutral dissociations are very difficult to characterize by this method, however, due to the very high cross section observed for dissociative electron attachment in ammonia at low impacting energies. These dissociative attachment processes, which also generate neutral fragments, effectively mask the formation of those dissociation products which can be ascribed to fragmentation of excited neutral molecules.

Lindholm (112) has also recently completed a molecular orbital treatment of the photodissociations occurring in the vacuum ultraviolet. The theoretical findings were in agreement with the observed chemical data; namely that loss of H atoms is favored over molecular detachment of hydrogen. A detailed examination of the physical nature of the various excited states of ammonia is included in this paper as well as in the article by Okabe and Lenzi.

REFERENCES

1. P. Ausloos and S. G. Lias, *Actions Chim. Biol. Radiations*, **11**, 1 (1967).
2. G. Herzberg, *Spectra of Diatomic Molecules*, Van Nostrand, Princeton, N. J. (1955).
3. J. R. McNesby and H. Okabe, *Advances in Photochemistry*, Vol. III, b, W. A. Noyes, G. S. Hammond, and J. N. Pitts, Eds., Interscience, New York, 1964.
4. J. G. Calvert and J. N. Pitts, *Photochemistry*, Wiley, New York, 1966.
5. W. A. Noyes, Jr., G. S. Hammond, and J. N. Pitts, Eds., *Advances in Photochemistry*, Interscience, New York, 1962 to present.
6. D. Peters, *J. Chem. Phys.*, **41**, 1046 (1964).
7. E. Lindholm, *Arkiv Fysik*, to be published.
8. G. von Bunau and R. N. Schindler, *J. Chem. Phys.*, **44**, 420 (1966).
9. (a) C. E. Melton, *J. Chem. Phys.*, **45**, 4414 (1966). (b) C. E. Melton and P. S. Rudolph, *J. Chem. Phys.*, **47**, 1771 (1967).
10. N. T. Williams and H. Essex, *J. Chem. Phys.*, **17**, 995 (1949).
11. P. Ausloos and R. Gorden, Jr., *J. Chem. Phys.*, **41**, 1278 (1964).
12. T. W. Woodward and R. A. Back, *Can. J. Chem.*, **41**, 1463 (1963).
13. H. H. Carmichael, R. Gorden, Jr., and P. Ausloos, *J. Chem. Phys.*, **42**, 343 (1965)
14. P. G. Wilkinson, *J. Mol. Spectry.*, **6**, 1 (1961).
15. A. Scala and P. Ausloos, *J. Phys. Chem.*, **70**, 260 (1966).
16. R. R. Williams, *J. Phys. Chem.*, **63**, 776 (1959).
17. R. Gorden, Jr. and P. Ausloos, *J. Chem. Phys.*, **46**, 4823 (1967).
18. P. Ausloos and S. G. Lias, *J. Chem. Phys.*, **38**, 2207 (1963).
19. R. W. Hummel, *J. Phys. Chem.*, **70**, 2685 (1966).
20. W. Braun, K. H. Welge, and J. R. McNesby, *J. Chem. Phys.*, **45**, 2650 (1966).
21. A. W. Tickner, *Can. J. Chem.*, **39**, 87 (1961); C. Ponnamperuma and F. Woeller, *Nature*, **203**, 272 (1964).
22. M. Burton and J. L. Magee, *J. Chem. Phys.*, **23**, 2194 (1955).
23. J. Sturm, Thesis, University of Notre Dame, Notre Dame, Indiana, 1956.
24. A. Kupperman and M. Burton, *Radiation Res.*, **10**, 636 (1959).
25. P. Ausloos and S. G. Lias, in *The Chemistry of Ionization and Excitation*, G. R. A. Johnson and G. A. Scholes, Eds., Taylor and Francis, Ltd., London, 1967.
26. H. Okabe, *J. Opt. Soc. Am.*, **54**, 478 (1964).
27. R. A. Back and D. C. Walker, *J. Chem. Phys.*, **37**, 2348 (1962).
28. B. A. Thompson, R. R. Reeves, and P. Harteck, *J. Phys. Chem.*, **69**, 3964 (1965).
29. A. H. Laufer and J. R. McNesby, *J. Chem. Phys.*, **42**, 3329 (1965).
30. R. E. Rebbert and P. Ausloos, *J. Chem. Phys.*, **46**, 4333 (1967).
31. P. Ausloos and S. G. Lias, paper presented at the International Conference on Photochemistry, Munich, Germany, Sept., 1967.
32. P. Ausloos and S. G. Lias, *J. Chem. Phys.*, **45**, 524 (1966).
33. (a) A. A. Scala and P. Ausloos, *J. Chem. Phys.*, **45**, 847 (1966). (b) A. A. Scala and P. Ausloos, *J. Chem. Phys.* **47**, 5129 (1967).
34. A. A. Scala and P. Ausloos, *J. Chem. Phys.*, to be published.
35. R. D. Doepker and P. Ausloos, *J. Chem. Phys.*, **43**, 3814 (1965).
36. R. D. Doepker, S. G. Lias, and P. Ausloos, *J. Chem. Phys.*, **46**, 4340 (1967).
37. R. D. Doepker and P. Ausloos, *J. Chem. Phys.*, **42**, 3746 (1965).

38. R. Gorden, Jr. and P. Ausloos, *J. Chem. Phys.*, **47**, 1799 (1967).
39. D. A. Becker, H. Okabe, and J. R. McNesby, *J. Chem. Phys.*, **69**, 538 (1965).
40. R. Gorden, Jr., R. D. Doepker, and P. Ausloos, *J. Chem. Phys.*, **44**, 3733 (1966).
41. P. Ausloos and S. G. Lias, *Radiation Res. Rev.*, **1**, 75 (1968).
42. K. Watanabe, F. M. Matsunage, and H. Sakai, *Appl. Opt.*, **6**, 391 (1967).
43. D. Beck and O. Osberghaus, *Z. Physik*, **160**, 406 (1960).
44. D. Beck and A. Niehaus, *J. Chem. Phys.*, **37**, 2705 (1962).
45. D. Beck, *Discussions Faraday Soc.*, **36**, 602 (1963).
46. H. D. Beckey and W. Groth, *Z. Physik. Chem. (Frankfurt)*, **20**, 307 (1959).
47. H. Okabe, H. D. Beckey, and W. Groth, *Z. Naturforsch.*, **21**, 135 (1966).
48. D. W. L. Griffiths and R. A. Back, *J. Chem. Phys.*, **46**, 3913 (1967).
49. L. I. Bone and J. H. Futrell, *J. Chem. Phys.*, **47**, 4366 (1967).
50. P. Ausloos and S. G. Lias, *J. Chem. Phys.*, **44**, 521 (1966).
51. J. A. Kerr and A. F. Trotman-Dickenson, *Progr. Reaction Kinetics*, **1**, 105 (1961).
52. P. S. Dixon, A. P. Stefani, and M. Szwarc, *J. Am. Chem. Soc.*, **85**, 3344 (1963).
53. P. Ausloos, S. G. Lias, and I. B. Sandoval, *Discussions Faraday Soc.*, **36**, 66 (1963).
54. H. Okabe and D. A. Becker, *J. Am. Chem. Soc.*, **84**, 4004 (1962).
55. S. G. Lias and P. Ausloos, *J. Chem. Phys.*, **43**, 2748 (1965).
56. E. Tschuikow-Roux and J. R. McNesby, *Trans. Faraday Soc.*, **62**, 2158 (1966).
57. R. F. Hampson, J. R. McNesby, H. Akimoto, and I. Tanaka, *J. Chem. Phys.*, **40**, 1099 (1964).
58. R. F. Hampson, Jr., and J. R. McNesby, *J. Chem. Phys.*, **42**, 2200 (1965).
59. J. A. Pirog and J. R. McNesby, *J. Chem. Phys.*, **42**, 2490 (1965).
60. R. F. Hampson, Jr., and J. R. McNesby, *J. Chem. Phys.*, **43**, 3592 (1965).
61. H. Okabe and D. A. Becker, *J. Chem. Phys.*, **34**, 2549 (1963).
62. P. Ausloos, R. Gorden, Jr., and S. G. Lias, *J. Chem. Phys.*, **40**, 1854 (1964).
63. P. Ausloos, R. E. Rebbert, and S. G. Lias, *J. Chem. Phys.*, **42**, 540 (1965).
64. W. B. DeMore and S. W. Benson, *Advances in Photochemistry*, Vol. 2, W. A. Noyes, G. S. Hammond, and J. N. Pitts, Eds., Interscience, New York, 1964, p. 219.
65. P. Ausloos and S. G. Lias, *J. Chem. Phys.*, **41**, 3962 (1964).
66. C. L. Currie, H. Okabe, and J. R. McNesby, *J. Phys. Chem.*, **67**, 1494 (1963).
67. A. A. Scala and P. Ausloos, private communication.
68. R. A. Holroyd, J. Y. Yang, and F. M. Servedio, *J. Chem. Phys.*, **46**, 4540 (1967).
69. H. Okabe and J. R. McNesby, *J. Chem. Phys.*, **36**, 601 (1962).
70. E. Tschuikow-Roux, J. R. McNesby, and J. L. Faris, *J. Phys. Chem.*, **71**, 1531 (1967).
71. P. Borrell and F. C. James, *Trans. Faraday Soc.*, 62, 1 (1966).
72. L. J. Stief and V. J. DeCarlo, *J. Chem. Phys.*, **42**, 3113 (1965).
73. I. Tanaka, private communication.
74. L. J. Stief and V. J. DeCarlo, Final Report, NASA Contract No. NASW-1417, 18 May, 1967, Melpar, Inc., Falls Church, Virginia.
75. F. E. Blacet and R. A. Crane, *J. Am. Chem. Soc.*, **76**, 5337 (1954).
76. J. V. Michael and W. A. Noyes, Jr., *J. Am. Chem. Soc.*, **85**, 1228 (1963).
77. D. E. McElcheran, M. H. J. Wijnen, and E. W. R. Steacie, *Can. J. Chem.*, **36**, 321 (1958).
78. A. J. Harrison and J. S. Lake, *J. Phys. Chem.*, **63**, 1489 (1959).
79. A. A. Leiga and H. A. Taylor, *J. Chem. Phys.*, **41**, 1247 (1964).

80. R. P. Porter and W. A. Noyes, Jr., *J. Am. Chem. Soc.*, **81**, 2307 (1959).
81. (a) J. Hagege, S. Leach, and C. Vermeil, *J. Chim. Phys.*, **62**, 736 (1965). (b) J. Hagege, P. C. Roberge, and C. Vermeil, paper presented at the International Conference on Photochemistry, Munich, Sept. 1967.
82. K. H. Welge, *J. Chem. Phys.*, **45**, 166 (1966).
83. K. H. Welge, *J. Chem. Phys.*, **45**, 4373 (1966).
84. L. J. Stief, V. J. DeCarlo, and R. J. Mataloni, *J. Chem. Phys.*, **46**, 592 (1967).
85. R. J. Cvetanovic, *J. Chem. Phys.*, **43**, 1850 (1965).
86. K. H. Welge, *J. Chem. Phys.*, **45**, 1113 (1966).
87. M. de Sorgo, A. J. Yarwood, O. P. Strausz, and H. E. Gunning, *Can. J. Chem.*, **43**, 1886 (1965).
88. K. Watanabe and M. Zelikoff, *J. Opt. Soc. Am.*, **43**, 753 (1953).
89. L. J. Stief, *J. Chem. Phys.*, **44**, 277 (1966).
90. M. Cottin, C. Vermeil, and J. Masanet, *Compt. Rend.*. **263**, 753 (1966).
91. M. Cottin, J. Masanet, and C. Vermeil, *J. Chim. Phys.*, **63**, 959 (1966).
92. J. R. McNesby, I. Tanaka, and H. Okabe, *J. Chem. Phys.*, **36**, 605 (1962).
93. H. Neuimin and A. N. Terenin, *Acta Physicochim, USSR.*, 465 (1936).
94. T. Carrington, *J. Chem. Phys.*, **41**, 2012 (1964).
95. K. H. Welge and F. Stuhl, *J. Chem. Phys.*, **46**, 2440 (1967).
96. B. H. Mahan, *J. Chem. Phys.*, **33**, 959 (1960).
97. A. Y. Ung and H. I. Schiff, *Can. J. Chem.*, **44**, 1981 (1966).
98. P. Warneck, *J. Chem. Phys.*, **43**, 1849 (1965).
99. D. L. Baulch and W. H. Breckenridge, *Trans. Faraday Soc.*, **62**, 2768 (1966).
100. N. G. Moll, D. R. Clutter, and W. E. Thompson, *J. Chem. Phys.*, **45**, 4469 (1966).
101. P. Warneck, *Discussions Faraday Soc.*, **37**, 57 (1964).
102. P. K. Sen Gupta, *Proc. Roy. Soc. (London)*, **A146**, 824 (1934).
103. K. H. Becker and K. H. Welge, *Z. Naturforsch.*, **20a**, 442 (1965).
104. K. H. Welge, *J. Chem. Phys.*, **45**, 166 (1966).
105. H. Okabe, *J. Chem. Phys.*, **47**, 101 (1967).
106. J. P. Doering and B. H. Mahan, *J. Chem. Phys.*, **36**, 1682 (1962).
107. K. D. Bayes, K. H. Becker, and K. H. Welge, *Z. Naturforsch.*, **17a**, 676 (1962).
108. O. Schnepp and K. Dressler, *J. Chem. Phys.*, **32**, 1683 (1960).
109. W. Groth, H. Okabe, and H. J. Rommel, *Z. Naturforsch.*, **19a**, 507 (1964).
110. K. H. Becker and K. H. Welge, *Z. Naturforsch.*, **18**, 600 (1963).
111. H. Okabe and M. Lenzi, *J. Chem. Phys.*, **47**, 5241 (1967).
112. E. Lindholm, *Arkiv Fysik*, to be published.

Chapter 4

Ion–Molecule Reactions

Jean H. Futrell

Department of Chemistry, University of Utah, Salt Lake City, Utah

and

Thomas O. Tiernan

Aerospace Research Laboratories, Chemistry Research Laboratory
Wright-Patterson Air Force Base, Ohio

I. INTRODUCTION

The investigation of ion–molecule reactions by various means currently constitutes one of the most active areas of chemical physics. It is therefore impossible that the present chapter be all-inclusive, but it is hoped that it will supplement past reviews on this topic (1–10,289), serve as a bridge to new discoveries, and provide a useful framework for discussing some aspects of radiation chemistry.

As is often the case of "new" fields of research, closer inspection reveals fairly extensive historical roots. The origins of ion–molecule reaction studies may be found in the pioneering researches in mass spectroscopy (11). The archetype ion–molecule product ion, H_3^+, was identified by Dempster in 1916 (12) and the mechanism for its production

$$H_2^+ + H_2 \rightarrow H_3^+ + H \tag{1}$$

was correctly established by 1925 (13,14). Hogness and Harkness (15) studied the formation of I_3^- and I_3^+ in iodine vapor subjected to electron impact in 1928. By varying the ion path length they showed that primary ion intensities were only slightly affected but that secondary processes were substantially increased by increasing the ion path through neutral I_2 vapor. It is interesting to note that these workers also observed tertiary ions in their experiments.

The obvious parameter for characterizing ion–molecule reactions in mass spectrometry is the pressure dependence of ion intensity. Direct ionization is a unimolecular process and exhibits a first-order pressure dependence, while ion–molecule reactions are bimolecular, second-order processes. The ratio of intensities of secondary/primary ions is therefore directly proportional to pressure. Shortly after Urey's discovery of deuterium, Bleakney (16) showed that such a plot of $(m/e\ 3)/(m/e\ 2)$ vs. pressure for hydrogen gave a finite intercept indicating the presence of HD^+ to the extent of about 1 part per 2500 parts H_2^+. It is a source of chagrin to mass spectroscopists that this most important isotope of hydrogen remained undiscovered for many years because the origin of $m/e\ 3$ ions in the hydrogen mass spectrum via reaction (1) was so well understood!

Another criterion for identifying ion–molecule reaction products was introduced in 1951 by Washburn, Berry, and Hall (17), who demonstrated the technique of varying the electric field strength within the ion source. Their objective was analytical in scope, and they showed that the ion extraction field could readily be increased to suppress the formation of H_3O^+, permitting a direct determination of HDO^+ in water samples. The principal effect of increasing the extraction field is, of course, the reduction of ion residence time, and hence the reduction of reaction probability. This technique forms the basis of the most frequently applied method for the quantitative estimation of ion–molecule reaction rates.

For the two decades from 1930–1950 the main interest in mass spectroscopy was in the basic physics of ionization and dissociation, in the precise determination of isotopic masses and abundances and in the development of analytical mass spectrometry. With improvement in instrumentation and technique, particularly in vacuum technology, the nuisance of these secondary processes was largely eliminated. Great strides were made in understanding and measuring the properties of isolated gaseous ions (18). One is inclined to believe, however, that this progress was made partly at the expense of studies of bimolecular reactions of gaseous ions with molecules.

It is interesting to note that one of the earliest theoretical papers (19) concerning radiation chemical problems was a treatment of reaction (1). Taking the activation energy as the difference between the polarization and rotational energies of the system, this treatment calculated a rate constant for the reaction of $2.07 \times 10^{-9}\kappa$ cm^3-molecule^{-1} sec^{-1}, where κ is the transmission coefficient for the reaction. This rate is indeed large and suggests immediately that ion–molecule reactions will play a central role in radiation chemistry. This fact was largely unrecognized by contemporary investigators, however, and it remained for a new generation of chemists, beginning in 1952, to rediscover this important class of reactions (20–21). These workers independently discovered the reaction

$$CH_4^+ + CH_4 \rightarrow CH_5^+ + CH_3 \tag{2}$$

which not only reawakened the interests of radiation chemists in such reactions but also caused some reorientation of our concepts of valency and bonding.

With this much of a historical introduction we shall now proceed in this chapter to outline our current understanding of ion–molecule reactions. We shall describe in some detail a theoretical framework for the kinematics of ion–molecule reactions and the various experimental approaches now used for such investigations. We shall discuss at length some of the systems which are relatively well understood, and then proceed to discuss systems of more general chemical interest. Because of their relevance to radiation chemistry we also include a brief discussion of dissociative charge transfer reactions.

II. THEORETICAL CONSIDERATIONS

A. The Microscopic Cross Section

If we consider the generalized ion–molecule reaction

$$P^+ + M \rightarrow [PM]^+ \rightarrow S^+ + N \tag{3}$$

we may conveniently divide the problem of describing it theoretically into two parts—the formation of the ion–molecule complex $[PM]^+$ and its subsequent dissociation into secondary ion products. We shall defer for the moment attaching any particular significance to the complex and shall discuss the microscopic cross section for the collision of slow ions with molecules. We shall present the impact parameter treatment of the problem, although the activated complex approach assuming loose complexes yields equivalent results (23).

The classical impact parameter treatment for structureless point particles was first solved by Langevin (24) and later elaborated by Gioumousis and Stevenson (25) for the charge—induced dipole long-range interaction of ions with molecules. Consider two particles m_1 and m_2 which interact with one another through a potential $V(r)$ which is a function of the vector $\mathbf{r}$ separating them. Let the position vectors of the two particles be $\mathbf{r}_1$ and $\mathbf{r}_2$ and take the origin of coordinates at the center-of-mass,

$$r_1' = -\frac{m_2}{m_1 + m_2}\,\mathbf{r} \tag{4a}$$

and

$$r_2' = \frac{m_1}{m_1 + m_2}\,\mathbf{r} \tag{4b}$$

The kinetic energy of the particles* in this coordinate system is

$$T = \tfrac{1}{2}(m_1\dot{r}_1'^2 + m_2\dot{r}_2'^2) = \tfrac{1}{2}\mu\dot{r}^2 \tag{5}$$

where μ is the reduced mass and the dotted variables denote time derivatives. Similarly the angular momentum about the center of mass is

$$L = m_1 r_1'^2\dot{\theta} + m_2 r_2'^2\dot{\theta} \tag{6}$$

where $\dot{\theta}$ is the angle between the line of centers and the velocity vector of the center of mass. This reduces to

$$L = \mu r^2\dot{\theta} \tag{7}$$

In polar coordinates the equations of motion are

$$\frac{d}{dt}(\mu r^2\dot{\theta}) = 0 \tag{8}$$

$$\frac{d}{dt}(\mu\dot{r}) - \mu r\dot{\theta}^2 + \frac{\partial V}{\partial r} = 0 \tag{9}$$

Substituting Eq. (7) into Eq. (9) and differentiating the first term,

$$\mu\ddot{r} - \frac{L^2}{\mu r^3} = -\frac{\partial V}{\partial r} = f(r) \tag{10}$$

---

* We simplify the equations somewhat by considering motion with respect to the center of mass. For the complete problem one should add the kinetic energy of the center of mass, $\tfrac{1}{2}\mu v^2$, where v is the relative velocity.

where $f(r)$ is the central force between the particles. It is convenient to write this equation in the form

$$\mu\ddot{r} = \phi(r) \tag{11}$$

which reduces the problem still further to one dimensional terms. Here $\phi(r)$ is the sum of the central force $f(r)$ and a fictitious outwardly directed centrifugal force $L^2/\mu r^3$. This force may be written as an effective potential

$$V_{\text{eff}}(r) = V(r) + \frac{L^2}{2\mu r^2} \tag{12}$$

which may be considered in the same manner as usual potential energy diagrams.

Such potential curves are presented in Figs. 1–3 for inverse first power, third power, and fourth power attractive potentials, respectively. Dashed lines show the centrifugal and attractive potentials which are added to obtain $V_{\text{eff}}(r)$. Obviously a second power potential is a special case because of the r^{-2} dependence of the centrifugal potential; it will not be presented here. Figure 1 shows that the effective potential is repulsive at sufficiently small r and stable orbits are possible for values of E lying within the well. This is generally the case for central potentials of the form $V(r) = -kr^{-n}$ with $n < 2$.

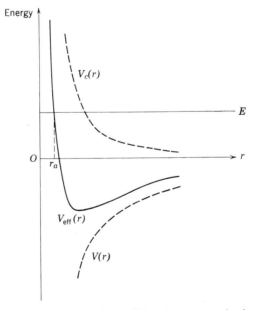

Fig. 1. Potential function for analyzing collisions in an attractive inverse-first-power potential field. (E. W. McDaniel, ref. 27, p. 69.)

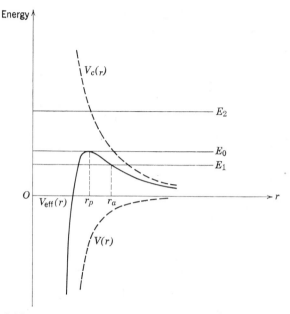

Fig. 2. Potential function for analyzing collisions in an attractive inverse-third-power potential field. (E. W. McDaniel, ref. 27, p. 69.)

In Fig. 2 the case for $n > 2$ is shown to be quite different. For distances of closest approach greater than r_a (energy $< E_1$), for example, the particle is reflected. For greater energy than E_0, however, the mass point senses a repulsion only for $r > r_p$ and senses an attraction at shorter r. It will therefore pass through the scattering center, which is here assumed to be a point center. For $E' = E_0 + \delta E$ an interesting case of an unstable condition ensues, as the particle will orbit the scattering center.

The case of a polarizable molecule interacting with a point charge is obviously a case of special interest to us here. Here the long-range potential is $V(r) = -(e^2\alpha/2r^4)$ where e is the electronic charge and α the polarizability of the neutral. This example is illustrated in Fig. 3. The effective potential is

$$V_{\text{eff}}(r) = -(e^2\alpha/2r^4) + (L^2/2\mu r^2) \qquad (13)$$

At the critical distance of approach r_k (corresponding to r_p of Fig. 2) we note that $V_{\text{eff}}(r)$ is a minimum. Solving for r at this point,

$$\frac{\partial V_{\text{eff}}(r)}{\partial r} = 0 = \frac{2e^2\alpha}{r^5} - \frac{L^2}{\mu r^3} \qquad (14)$$

$$r_k = \frac{e}{L}\sqrt{2\alpha\mu} \qquad (15)$$

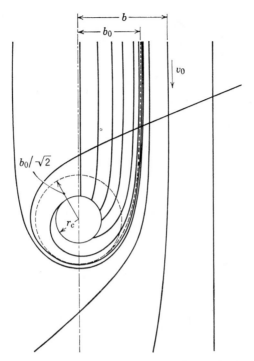

Fig. 3. Potential function for analyzing collisions in the point
charge-induced dipole potential field.

and the critical energy E_c, for surmounting the barrier is obtained by
evaluating $V_{\text{eff}}(r)$ at r_k.

$$V_{\text{eff}}(r = r_k) = E_c = L^4/8e^2\mu^2\alpha \tag{16}$$

Since in center of mass coordinates the angular momentum about the
center of mass is (27)

$$L = \mu b v_0 \tag{17}$$

where b is the impact parameter and v_0 is the relative velocity, and the
kinetic energy is

$$T = \mu v_0^2/2 = E_c \tag{18}$$

we may combine Eqs. (16–18) and solve for the critical impact parameter
$r_k = b_0$

$$b_0 = \left[\frac{4e^2\alpha}{\mu v_0^2}\right]^{1/4} \tag{19}$$

We may identify this with a cross section

$$\sigma(v) = \pi b_0^2 = \frac{\pi}{v_0}\left(\frac{4e^2\alpha}{\mu}\right)^{1/2} \tag{20}$$

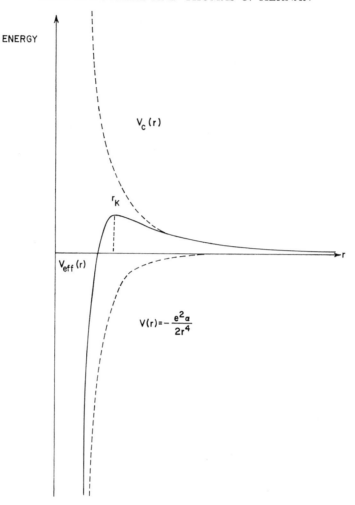

Fig. 4. Various trajectories for the polarization potential as a function of impact parameter b for given initial velocity v_0. Only the incoming branch of each spiraling trajectory is shown. (E. W. McDaniel, ref. 27, p. 73.)

Hence for the inverse fourth power attraction potential the microscopic cross section varies as $E^{-1/2}$. In fact, for a potential $V(r) = -kr^{-n}$ the cross section may be shown to vary as $E^{-2/n}$ for $n > 2$.

In Fig. 4 are plotted trajectories for the fourth power polarization potential. For greater clarity the outgoing branches of spiraling trajectories are not shown. For impact parameters $b > b_0$ the particles approach no closer than $b_0/\sqrt{2}$, while those with $b < b_0$ pass through the origin if no repulsive core potential pertains.

We now suppose that there exists some critical separation r_c between the reactants at which chemical forces come into play and drive the reaction to completion and that reaction is impossible for greater separation. If r_c lies between 0 and $b_0/\sqrt{2}$, all collisions for which $b < b_0$ must lead to reaction while no collisions with $b > b_0$ do. These are the conditions under which Eq. (20) is properly identified with a reaction cross section. Thus the model is unrealistic for higher velocities when $r_c > b_0/\sqrt{2}$ in general. The high energy limit of Eq. (20) is zero, while the limit should be πr_c^2. These parameters and the various possible ion paths for a hypothetical case are illustrated in Fig. 4.

An alternative expression for the cross section at high energies (28), assuming that approach within a critical radius is both the necessary and sufficient condition for reaction is, for potentials of arbitrary form,

$$\sigma' = \pi r_c^2\left[1 - \frac{2V(r)}{\mu v_0^2}\right] \tag{21}$$

Equation (21) is applicable at all energies for attractive potentials which do not lead to orbiting. It applies only at high energies for orbiting-type potentials and for repulsive potentials to impact energies above a threshold energy corresponding to the potential energy evaluated at r_c. For the case of the inverse fourth power polarization potential, therefore, the microscopic cross section has the form

$$\sigma = \pi\left(\frac{2e^2\alpha}{E}\right)^{1/2} \quad \text{for} \quad E < \frac{e^2\alpha}{2r_c^4} \tag{20a}$$

and the form

$$\sigma' = \pi r_c^2\left(1 + \frac{e^2\alpha}{2r_c^4 E}\right) \quad \text{for} \quad E > \frac{e^2\alpha}{2r_c^4} \tag{21a}$$

Such a model predicts a monotonic decrease from a high value at low velocity to πr_c^2 at high velocity. A conceptually similar treatment has been used by Hamill et al. (29–31) and by Stevenson (5) to deduce phenomenological cross sections which agree with experiment over a considerable range of ion energy.

Rosenstock pointed out some years ago that the simple theory outlined above does not consider an elementary constraint imposed by conservation of angular momentum (6). The ion–molecule reaction complex has angular momentum about the center of mass. As we have shown, the kinetic energy of the system must be sufficient to overcome the barrier created by the centrifugal potential (exemplified by r_p in Fig. 2 and E_c in Eq. (16), q.v.). It may readily be seen that the products of reaction must surmount an equivalent barrier in order to separate, as the product ion and molecule interact by an equivalent potential function. Since in general the reduced mass, polarizability, and kinetic energy change in the course of the reaction, it may happen that not all complexes can decompose into products. In some cases, particularly involving isotopic hydrogen, quite sizable effects are predicted from these considerations (32,33). In point of fact the body of evidence on ion–molecular reactions suggests that such an effect on reaction rates is rarely observed. This in turn implies that the angular momentum of the ion–molecule reaction pair must be converted efficiently into rotational angular momentum of the products.

Now that we have developed the "slow-ion" theory in some detail we shall mention briefly additional considerations which may modify certain of the deductions from this form of the microscopic cross section. It is obvious, for example, that the assumption of a central potential for the interaction which is independent of position is never strictly true. For molecules which have a permanent dipole moment this assumption may be grossly in error. Theard and Hamill (31) and Moran and Hamill (34) treated polar molecules using the simplifying assumption that the dipole always aligns itself with the ion field. In this way an essentially central potential of the form

$$V(r) = -(e^2\alpha/2r^4) - (e\mu_D/r^2) \tag{22}$$

where μ_D is the dipole moment of the molecule, is deduced. This results in an additional term in the cross section

$$\sigma = \pi\left[\left(\frac{2e^2\alpha}{E}\right)^{1/2} + \frac{e\mu_D}{E}\right] \tag{23}$$

which accounts for the substantial increase in cross section for polar molecules often noted at low energies. A more elaborate treatment has been developed along similar lines by Dugan and Magee (35). These authors include quantum mechanical effects, treat both linear and symmetric-top polar molecules, and discuss the regime of ion energies where orientation of the dipolar species is probable. At higher energy orientation becomes quite improbable, and the cross sections reduce at the limit to gas kinetic values. It may be noted that analytic solutions do not exist for models of this degree of complexity and numerical solutions are required.

An interesting related problem is consideration of the anisotropy of the polarizability tensor. It is most frequently the case that polarizabilities along different axes of the molecule are different, 40% differences not being uncommon. Gioumousis has considered this problem as well as the effect of the quadrupole moment of the neutral reactant in a critical discussion of ion–molecule reaction rates (36). The calculation is quite straightforward, although numerical methods are again required. For the case considered

$$Ar^+ + H_2 \rightarrow ArH^+ + H \tag{24}$$

the result is probably not distinguishable experimentally from the simpler calculation assuming an isotropic polarization potential.

The approaches discussed thus far refer only to strong-coupling collisions in which it is presumed that thorough mixing occurs and all information about initial configuration is lost. For exothermic ion–molecule reactions at low kinetic energy, the long range strongly attractive forces make such strong-coupling collisions plausible. At high impacting ion energy, however, the requirements of the strong coupling model are not fulfilled and a different reaction mechanism becomes operative. Binding energies of atoms become negligible in this approximation and a hard-core collision model is appropriate. Light and Horrocks (37) have advanced a model for the generalized stripping reaction

$$\{1\} + \{2, 3\} \rightarrow \{1, 3\} + \{2\} \tag{25}$$

It is anticipated from the molecular dynamics involved that such a reaction is most important when $m_1 \simeq m_2$. An alternative scheme, which may best be described as a "pickup" mechanism, was proposed by Henglein, Lacmann, and Jacobs (38) for certain high energy ion–molecule reactions. It may be formulated

$$\{1\} + \{2, 3\} \rightarrow \{1, 3\} + \{2\} \tag{26}$$

in which target atom 3 is considered quasi-free and atom 2 does not participate in the reaction. Some data for N_2^+–H_2 and Ar^+–H_2 reactions have been obtained which support this hypothesis (39,40). Both stripping models lead to very large kinetic isotope effects, approaching ∞ as a limit.

B. Dissociation of the Reaction Complex

We now turn to the question of dissociation of the intermediate complex (PM^+) of Eq. (3). This question has been approached for triatomic systems using the phase space theory advanced by J. C. Light (41–43) and by other related statistical approaches (44–46). Wolf has given a particularly enlightening discussion of the statistical approach (46).

The basic postulate of such an approach is that in the initial formation of the complex it loses all information on its initial states and that its decomposition is therefore governed by the phase space available to various output channels. For three particles the phase space element is

$$d\Gamma = \prod_{i=1}^{3} d^3r_i d^3p_i \tag{27}$$

where r_i and p_i are the position and momentum vectors of particle i. The various individual output reaction channels are computed as a fraction of the total phase space available.

By separating out the integrals which yield constants of motion independent of channel and vibrational energy, the relevant reduced phase space available which makes possible a choice of reaction channel is

$$\Gamma = \int\int \left[1 + \left(\frac{J_{\text{rot}}}{J_t}\right)^2 - 2\left(\frac{J_{z\text{rot}}}{J_t}\right)\right]^{-1/2} dJ_{\text{rot}} \, dJ_{z\text{rot}} \tag{28}$$

For convenience the total angular momentum $J_t = J_{\text{orb}} + J_{\text{rot}}$ is taken along the z axis. The constraints of conservation of energy are then applied for each individual channel. The requirement for stability of the product molecule is introduced by requiring that the rotational energy of the molecule in the ith channel be,

$$(J_{\text{rot}}^2/2I) < D_v^i \tag{29}$$

where D_v^i is the dissociation energy of the vth vibrational level for channel i and I is the moment of inertia of the diatomic molecule. Finally it is required that the products must separate from the complex, i.e., they must overcome the centrifugal potential introduced into the one-dimensional central-field potential, of Eq. (12) and Figs. 1–3. The cross section for dissociation is found by difference,

$$\sigma_{\text{diss}} = \pi b_0^2 - \sum_i \sum_v (E_{\text{tr}}^0, i, v) \tag{30}$$

where E_{tr}^0 is the center of mass translational energy. The total cross section, πb_0^2 has been computed for the polarization potential but could, of course, be computed for other potentials. For convenience the angular momentum and internal energy of the reactant molecule were taken (46) to be negligible in comparison with E_{tr}^0 and total angular momentum; corrections may of course be made for particular cases for which this approximation is invalid.

A somewhat simplified statistical approach was developed by Tannenwald (45) to explain the fact that certain highly exothermic three particle ion–molecule reactions have very small cross sections. It is concluded that, for such highly exothermic reactions, three-body breakup is quite generally more likely than any two-body reactions because the density of states for three free particles can become much greater than for two. As the energy increases the relative density of states becomes increasingly greater.

Unfortunately, there is no straightforward procedure for extending these statistical treatments to molecules of more than two atoms. For more complex molecules of chemical interest an alternative approach is therefore required. The situation is quite analogous to the problem of describing the dissociation of complex ions or molecules treated by Vestal elsewhere in this volume (Chapter 3) and there seem to be no fundamental reasons why such a treatment should not apply to the dissociation of the ion–molecule reaction complex.

Several arguments may be advanced to support the plausibility of a randomized complex of significant lifetime (57–59). For example, from the reaction of parent ion in ethylene,

$$C_2H_4^+ + C_2H_4 \rightarrow [C_4H_8^+] \rightarrow Products \tag{31}$$

the $[C_4H_8^+]$ intermediate has actually been observed by several investigators (22,26,63,99,324). More recently isotopic experiments have established that complete randomization of hydrogens in the complex occurs prior to dissociation (49). Similar conclusions have been drawn from isotopic labelling experiments for the $[C_2H_7^+]$ complex resulting from the reaction (47)

$$CH_3^+ + CH_4 \rightarrow [C_2H_7^+] \rightarrow C_2H_5^+ + H_2 \tag{32}$$

In still other cases it has been shown that collisional stabilization of the intermediate complex is possible, and lifetimes of the order of 10^{-7} sec have been estimated for such complexes (48,50,101). "Metastable" ion spectra have also been observed (51,52) indicating lifetimes for unimolecular dissociation of fractions of a microsecond. Finally, several examples are known where the ionic dissociation fragments from the postulated intermediate resemble the electron impact mass spectrum of the equivalent molecule.

It should be noted that there is also some evidence which suggests that in certain types of reactions well-defined intermediates are involved. Such evidence is provided by radiolysis studies of H_2^- transfer (331), and H_2 transfer reactions (332,333), using isotopic techniques, which indicate that a strongly stereospecific transfer is involved. These results would appear to rule out a randomized complex for these cases. On the basis of the data already noted, however, we may presume that, in some cases at least, the collision complex has a lifetime which is quite long compared with vibrational and rotational periods. In these cases, structural information related to the mode of formation of the complex is lost because of extensive internal rearrangement, and the available energy is distributed among the various degrees of freedom of the complex. Within the framework of such a model, the properties of the complex with regard to various decomposition channels are characterized solely by its energy content.

The application of such a quasi-equilibrium statistical theory to the microcanonical ensemble of complexes is relatively straightforward, at least for complexes corresponding to normal ion structures (60). In general it is necessary to apply the methods of Chapter 3 twice to solve the ion–molecule reaction problem. The fragment ions from the primary dissociation of complex molecules contain excess energy above the minimum energy for dissociation and this energy will carry over into internal energy of the ion–molecule reaction complex. This distribution of energies is determined by the ion, the neutral fragment or fragments and by the frequently factor and activation energy describing the primary dissociation process. If the same ion is produced by more than one primary reaction an even more complex distribution of reactant ion internal energies is obtained. The ion–molecule reaction complex is then considered as the entity resulting from the combination of primary ions with this distribution of internal energy with neutral molecules at the temperature of the experiment. Thermochemical considerations define the minimum energies for various dissociation paths of the resulting complex, and the frequency factors are approximated in the usual fashion. A "mass spectrum" is then calculated for a time scale appropriate to the method of measurement.

A partial application of similar considerations to the computation of kinetic energies of ion–molecule reaction products has been carried out by Stanton and Wexler (53). They used two approximations—the semiclassical method of Klots (54) and the "temperature" method of Chupka (55)—to compute the kinetic energy release in fragmentation for complete energy equilibration in the complex. They considered three ion–molecule reactions, two in the ethylene system and one for methane, and compared the calculated results with experimental measurements for those reactions.

In all three cases the experimental values were quite small, but slightly higher than those calculated. In two cases the differences were less than experimental error, but it was concluded that a discrepancy existed for the third reaction. It should be noted, however, that these authors assumed the reactant ions were not vibrationally or electronically excited, i.e., they did not allow for the probable distribution of excitation energy in reactant ions alluded to above. Inclusion of this factor would probably bring the results into quite satisfactory agreement.

It may be noted that Tal'roze and Frankevich (56) also measured kinetic energies of product ions from several ion–molecule reactions, including

$$NH_3^+ + NH_3 \rightarrow NH_4^+ + NH_2 \tag{33}$$

$$H_2O^+ + H_2O \rightarrow H_3O^+ + OH \tag{34}$$

In all cases the energy distribution indicated that part of the exothermicity was apparently converted into translational energy. With increasing internal degrees of freedom for the complex less energy appeared in translation, suggesting approximate equipartition of energy over all degrees of freedom. This is consistent with the concept of a quasi-equilibrium complex in which exchange of the available energy occurs.

C. Macroscopic Cross Sections and the Rate of Reaction

The rate of reaction, w, of two structureless particles P and M of Eq. (3) at a given point may be expressed in terms of the integrals,

$$w = \int_{-\infty}^{+\infty} \int_{-\infty}^{+\infty} f_p(v_p) f_m(v_m) v\sigma(v) \, dv_p \, dv_m \tag{35}$$

where $f_p(v_p)$ and $f_m(v_m)$ are the velocity distribution functions of the reactants, v_i are the particle velocities, v is the relative velocity $|v_p - v_m|$ of the reacting pair, and $\sigma(v)$ is the velocity dependent microscopic cross section.

The velocity distribution of the neutral reactants will generally be a Maxwell-Boltzman distribution for the temperature of the experiment but the ion velocity distribution will not generally be such in a mass spectrometry experiment. Gioumousis and Stevenson (25) have treated the velocity distribution problem for the usual ion source experiment. Taking the z coordinate along the path, the Maxwellian distribution is assumed for the x and y directions. Along z the distribution is

$$f_p(v_z, z) = \frac{\eta(v_z, z) \dfrac{P^+ m_p}{2kT} \exp\left[-(1/kT)(\tfrac{1}{2}mv_z^2 - eE_r z)\right]}{[(\pi/kT)(\tfrac{1}{2}mv_z^2 - eE_r z)]^{1/2}} \tag{36}$$

where P^+ is the number of ionizations per unit area in the $z = 0$ plane and E_r is the ion extraction field, k is the Boltzmann constant, and T is the absolute temperature. The coefficient, η, assumes the following values

$$\eta = 1 \quad \text{for} \quad z < 0$$

$$\eta = 2 \quad \text{for} \quad \begin{cases} z > 0 \\ v_z > 0 \\ (\tfrac{1}{2}mv_z^2 - eE_r z) > 0 \end{cases}$$

$$\eta = 0 \quad \text{for} \quad \begin{cases} v_z < 0 \\ (\tfrac{1}{2}mv_z^2 - eE_r z) < 0 \end{cases}$$

The rate of production of secondary ions is

$$i_s = W = \int w \, d\tau \tag{37}$$

where $d\tau$ is an element of volume in the ion source, and the integration is over the entire ion source volume.

For the special case of the inverse fourth power polarization force the microscopic cross section and velocity may be integrated and removed from the integral. The expression for the point rate of reaction then simplifies to

$$w = 2\pi e(\alpha/\mu)^{1/2} \int_{-\infty}^{+\infty} \int_{-\infty}^{+\infty} f_p(v_p) f_m(v_m) \, dv_p \, dv_m \tag{38}$$

In general for an r^{-n} attractive potential the cross section is a function of $v^{-4/n}$ and the relative velocity must be retained in the integral.

For the simplified case the appropriate distributions are inserted and the expression is integrated to give

$$i_s = W = 2n_m i_p \pi e(\alpha/\mu)^{1/2}(2m_p/eE_r l)^{1/2} \tag{39}$$

where l is the ion path length. In this integration the additional usually quite good approximation is made that the thermal energy of the ions is negligible in comparison with $eE_r l$.

The phenomenological cross section for production of secondary ions as usually measured experimentally is

$$Q = (l n_m)^{-1} \ln (1 + i_s/i_p) \tag{40}$$

which reduces in the limit $i_s \rightarrow 0$ to

$$Q = i_s/l n_m i_p \tag{40a}$$

where i_s and i_p are the measured secondary and primary ion currents. For the classical interaction potential chosen, and with the indicated approximation in the integration, this may be identified with the macroscopic cross section

$$Q = 2\pi e(\alpha/\mu)^{1/2}(2m_p/eE_r l)^{1/2} \tag{41}$$

if we equate the rate of production of secondary ions with the measured ion current, i_s.

For the zero-field case, assuming a Maxwellian distribution of velocities for both ions and molecules the rate of production of secondary ions is

$$w = n_p n_m v\sigma(v) \tag{42}$$

$$w = n_p n_m [2\pi e(\alpha/\mu)^{1/2}] \tag{42a}$$

from which it is readily apparent that we may define the bimolecular rate constant for the reaction

$$k = 2\pi e(\alpha/\mu)^{1/2} \tag{43}$$

In terms of the phenomenological cross section, Q, the rate constant becomes

$$k = (eE_r l/2m_p)^{1/2} Q \tag{44}$$

which justifies a common treatment of experimental data.

Since the distinction between phenomenological, macroscopic, and microscopic cross sections is not always apparent in the literature it is perhaps worthwhile to reiterate the distinction from an experimental viewpoint (64). The phenomenological cross section defined by Eq. (40) is also the average over the ion path of the microscopic cross section

$$Q = l^{-1} \int_0^l \sigma[v_p(x)] \, dx \tag{45}$$

where dx is the increment of path length and $v_p(x)$ is the ion velocity at point x of the ion path. Using the relationship

$$v_p = (2E_r ex/m_p)^{1/2} \tag{46}$$

the equation may be transformed into

$$Q = m_p/2E_r el \int_0^{v_{p,\max}} v_p \sigma(v_p) \, dv_p \tag{47}$$

and $v_{p,\max}$ is evaluated from Eq. (46) at $x = l$. This equation may be inverted to yield the microscopic cross section as a function of $Q(E)$

$$\sigma_p(v_{p,\max}) = Q(E) + E \frac{dQ}{dE} \tag{48}$$

In the case we have discussed most fully, where the macroscopic cross section depends on $E^{-1/2}$ we have the result

$$\bar{\sigma}_p(v_{p,\max}) = \tfrac{1}{2} Q(E) \tag{49}$$

Therefore experimental Q's are strictly comparable only for experiments where the terminal ion energy is the same.

In general for arbitrary $\sigma(v)$ the thermal rate constant defined above is

$$k(T) = \langle v\sigma(v) \rangle \tag{50}$$

averaged over the Boltzmann distribution of relative velocity. This rate constant is temperature dependent and is equal to the product $Q\bar{v}$ only in the case of v^{-1} dependence of σ. In other words, the temperature independence of ion–molecule reaction rates normally assumed is a direct consequence of the functional form of the microscopic cross section, and deviations are expected for other forms of the interaction potential.

We note in passing that Eliason, Stogryn, and Hirschfelder (65) have derived collision cross sections including approximate quantum corrections for generalized pair potentials $\phi(r) = -ar^{-\delta}$. The microscopic cross section is

$$\sigma_{ESH} = 4\pi e/v(\alpha/\mu)^{1/2}[A^{(1)}(4)] \tag{51}$$

where $[A^{(1)}(4)]$ is a molecular collision integral. Substituting the numerical value for the collision integral of 0.55259 corresponding to the $\delta = 4$ polarization potential (66) gives a result in quite good agreement with the classical value. For exact agreement, of course, the collision integral would equal 0.5.

The classical treatment appears to be adequate for most purposes considering the precision currently possible in measurements of total cross sections for ion–molecule reactions. Some time ago Durup (2) and, more recently, Giese (67) discussed possible resonance effects qualitatively. These authors conclude that there may be an enhancement in reaction rate as the recombination energy of the ion approaches the vertical ionization potential of the neutral. Although there is some evidence to support this viewpoint (2,67,68) the significance of such correlations cannot now be considered as well established.

D. Charge Transfer Processes

The theory of non-thermal charge transfer between gaseous ions and atoms has been discussed by Rapp and Francis (69). The simplest case is that of symmetric resonant charge transfer

$$A^+ + A \rightarrow A + A^+ \tag{52}$$

which is usually treated as a one electron problem even in cases where A is a complex polyelectronic system. The complex A_2^+ formed from A^+ and $(A^+ + e^-)$ considers both A^+ species as point centers of charge. The nonstationary state A_2^+ is expressed in terms of the symmetric and antisymmetric stationary states of energy E_s and E_a, respectively. The cross section is deduced to be

$$\sigma(v) = 2\pi \int_0^\infty P(b, v)b\ db \tag{53}$$

as a function of the impact parameter, b. The probability of charge transfer is (70)

$$P(b, v) = \sin^2 \left[\int_{-\infty}^{+\infty} \frac{E_a - E_s}{2hv}\ dx \right] \tag{54}$$

along the collision orbit x. The energies E_a and E_s depend only on internuclear separation. They may be evaluated exactly for H^+ and approximately for other systems (71). The end result of an approximate solution leads to the form of the cross section (69)

$$\sigma^{1/2} = -k_1 \ln(v) + k_2 \tag{55}$$

often used for extrapolating data over velocity ranges. These authors point out that a two state approximation is used in this theory (i.e., truly symmetric resonant charge transfer) while experimental studies often include an unknown mixture of excited states of A^+ which may react by a nonresonant process.

The asymmetric charge transfer process

$$A^+(i) + B(j) \to A(k) + B^+(l) \tag{56}$$

is discussed by Rapp and Francis in terms of the asymptotic states $(A^+ + B)$ and $(A + B^+)$ to which the electronic states of the collision intermediate AB^+ must extrapolate. Only those states with identical symmetry need be considered. A zeroth-order approximation is obtained via the time-dependent Schroedinger equation using atomic orbitals and energies for the simplified problem

$$A^+ + (B^+ + e^-) \to (A^+ + e^-) + B^+ \tag{57}$$

The results are shown to be consistent with the Massey-Burhop criterion (72), which was deduced from more qualitative considerations of the "near-adiabatic" hypothesis. This criterion states that the maximum cross section occurs at

$$a|\Delta E|/hv \simeq 1 \tag{58}$$

where a is a constant of the order of 8 Å and ΔE is the energy difference between initial and final states.

It is interesting to note that the theoretical cross sections for asymmetric charge transfer in Ref. 69 should be multiplied by a statistical factor, f, determined by the requirement of identical symmetry in reactant and product states. This was tested in comparing the rates of the reactions

$$O^+(^4S) + H(^2S) \to O(^3P) + H^+ \tag{59}$$

$$H^+ + H(^2S) \to H(^2S) + H^+ \tag{60}$$

where reaction (59) is an interesting case of "accidental resonance" (ΔE is very small, so the maximum in cross section from Eq. (58) occurs at near-thermal velocity). The statistical weighting factor f for reaction (59) is 3/8, so the ratio of rates of reactions σ_{68}/σ_{69} is predicted to be 3/8 at energies above the maximum. Experimentally a value 0.35 is found (73).

The Massey-Burhop criterion appears to provide a satisfactory description of the maximum observed for endothermic charge transfer. The threshold behavior is less well understood. Maier (74) has developed arguments for a threshold law of the approximate form

$$\sigma = B(E - \Delta H)^n / E^{1/2} \qquad \text{for} \quad E \geq -\Delta H, \quad 1 \leq n \leq 2 \qquad (61)$$

where B is a constant, E is the kinetic energy in the center-of-mass, and ΔH is the heat of reaction. This equation is derived assuming that the phenomenon is governed primarily by the density of states in the momentum space of the separating particles, analogous to the threshold law developed by Wannier for ionization by electron impact (75). A more detailed statistical treatment of rearrangement collisions has been given by Pechukas and Light (42), which can also be applied to endothermic charge transfer, yielding similar results.

The above considerations describe charge transfer in the "intermediate velocity range between approximately 10^5–10^8 cm/sec. At low velocity where the polarizability of the target molecule significantly perturbs the ion trajectory a different approach is indicated. Here we may expect the same arguments applied to ion–molecule reactions to be applicable. In addition to the long range, grazing incidence sort of impact in which electron transfer is possible we expect orbiting collisions in which electron transfer is a probable process. For the point charge-induced dipole model developed earlier the orbiting charge-transfer contribution should be

$$\sigma_{\text{oct}} = P_x (2\pi e/v)(\alpha/\mu)^{1/2} \qquad (62)$$

where P_x is the probability for electron exchange. For the resonant case and with no possibility for ion–atom exchange processes, P_x would assume a value of the order of 0.5. For the nonresonant case, P_x would assume a value of the order of $f/2$, and endothermic processes would not occur, in general, in this velocity regime. In all cases low velocity charge exchange from orbiting collisions may best be considered simply as one possible output channel of the ion–molecule reaction complex.

III. EXPERIMENTAL TECHNIQUES FOR STUDYING ION–MOLECULE REACTIONS

A. Single Source Mass Spectrometric Methods

Most investigations of ion–molecule reactions involve some type of mass spectrometric method. The earliest method utilized conventional single-stage mass spectrometers, often of commercial design, with only

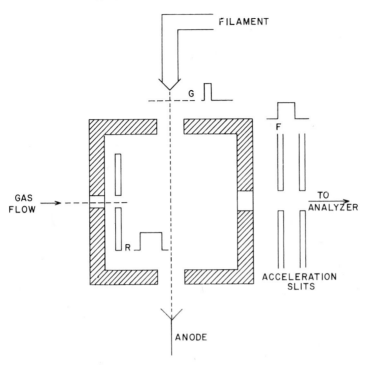

Fig. 5. Schematic representation of ion source. For pulsed operation sequential voltage pulses are applied to the control grid, G, repeller, R, and focus slits, F.

modest modifications at most. The ion source for such an instrument is shown schematically in Fig. 5. The simplest approach using an instrument of this degree of sophistication is the measurement of the relative abundances of primary and secondary ions as a function of pressure and electric field gradient within the source. By varying the voltage on the repeller, R, the average ion energy may be altered. Friedman and coworkers (76) and Hamill and co-workers (29–31) have utilized this method extensively. Because the ions are formed in an electric field and react throughout the distance between the electron beam and the exit slit of the source, however, the energy at which the ions react cannot be directly determined. In fact, such an experiment allows ions to react at any energy between thermal and the calculated exit energy, and the results should be considered as representing some average value between these experimental limits. Because the reaction cross section normally decreases with energy, these experiments tend to emphasize low energy reactions.

In order to minimize this problem, Ryan et al. (77,78) and Harrison (79) combined the pulse techniques of Tal'roze (80) with a small continuous repeller field. In this mode of operation, ions are formed by a very short ionizing pulse burst and are allowed to react under the influence of a small dc field for a selected delay time. The reaction is then quenched by application of a large (80 V/cm, for example), repeller pulse superimposed on the dc repeller field. In this fashion, the energy spread of the ions is due only to the finite duration of the ionizing pulse and the magnitude of the repeller field. This nearly monoenergetic ion "bunch" is accelerated by the impressed repeller field and achieves collectively an amount of kinetic energy depending upon the distance they have travelled before the reaction is quenched. By studying the extent of reaction as a function of delay time, one may observe the effects of translational energy. The derivative of a plot of ratio of product to reactant ion intensities versus time is the product of the concentration of the neutral reactant and the rate constant at delay time t and may be used to deduce the rate constant k as a function of t. The mean energy of the ion bunch is simply related to t; hence k as $f(E)$ is easily derived from these data.

In Fig. 6 are presented the results of such a pulsed electron beam–dc repeller-pulsed extraction experiment for the reaction

$$CH_3^+ + CH_4 \rightarrow C_2H_5^+ + H_2 \tag{32}$$

The slope of the line is the product of the rate constant and the con-

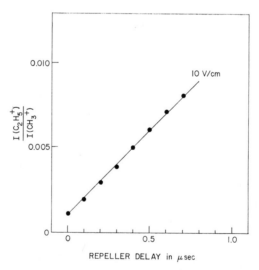

Fig. 6. Pulsed source study of the reaction $CH_3^+ + CH_4 \rightarrow C_2H_5^+ + H_2$. Ratio $C_2H_5^+/CH_3^+$ as a function of delay time for 10 V/cm extraction field. (K. R. Ryan and J. H. Futrell, ref. 77.)

centration of the neutral reactant, which can be measured independently. Since the data fall on a straight line, which is also independent of the repeller field, the rate constant of this reaction is shown to be energy independent over the range of the experiment. As we have shown in Section II this implies a $1/v$ energy dependence of the cross section. In general the plot of Fig. 6 will be a curved line and the tangent at any point will give the rate constant for that particular ion energy.

In another variation of the impulse technique Tal'roze (81,82) applied a deflection potential to the focus electrode (F) of Fig. 5 to reject ions at all periods of the cycle except during the extraction of ions by the high amplitude pulse. In this way any possibility that ions created adventitiously (as by surface reactions, for example) and diffusing out of the source out of phase with the experiment cycle might contribute to the amplitude of collected ions during the extraction pulse was eliminated. Harrison (83) has also used the focus pulse technique to measure the ion residence time for primary ions directly by varying the time delay for the pulse for optimum transmission of the primary ion "bunch." By such measurements one is able to measure the residence time directly and is not forced to rely on a calculated value for this parameter. Similar procedures have been followed in this laboratory (84).

Another "triple-pulse" technique which follows fairly directly from these considerations may be used to do constant energy experiments in a more elegant manner. This involves turning the repeller pulse on, holding it constant for a time period t_2, then turning it off sharply. It may easily be shown that this is a constant momentum experiment so long as the time t_2 is short compared with source residence time. For a given mass primary ion the kinetic energy is

$$T_m = \frac{e}{2m_p} (E_2 t_2)^2 \tag{63}$$

where E_2 is the amplitude of the (assumed perfectly square) extraction pulse and t_2 is its time duration. A variation of this method devised by Henchman (85) has been used to measure product and reactant ion residence times and hence, kinetic energies, to deduce some features of the kinematics of reaction.

Tal'roze has also used the mass dependence of ion residence times in connection with a pulse technique for studying charge-transfer processes (82). Provided that a heavy mass primary ion is used as a charge exchange reactant ion, one simply delays the extraction pulse until the primary ions have diffused away or have been swept from the sampling region. Ions formed at a later time may therefore be attributed to charge exchange products from the heavy ions which remain. It is naturally very desirable

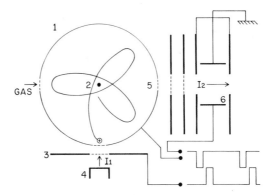

Fig. 7. Long residence time "Ion Trap" source for studying ion–molecule reactions at low pressure. Ions are produced by electron impact by ionizing pulse admitted by control grid, *3*. Ions formed in cylindrical source chamber, *1*, are trapped by a small attractive potential applied to central wire, *2*. After time delay, extraction potential, between *1* and *2* ejects a known fraction of ions through slit *5*. Unwanted ions diffusing through *5* are rejected at other times by deflection pulse on electrode *6*. (V. L. Tal'roze and G. V. Karachevtsev, ref. *7b*.)

to use a rejection pulse to eliminate spurious signals in such experiments. Since quite generally there is a phase shift between primary and secondary ions it would seem that phase sensitive methods might be used to advantage to ensure that the appropriate ion signal is uncontaminated in a variety of experiments.

Tal'roze has also developed an unconventional ion source for pulse studies which permits very long ion residence times (7b). This is shown schematically in Fig. 7. The axial symmetry of the source and the small diameter of the central electrode permit the use of a small dc field gradient to achieve a trapping potential for ions injected into the source or created *in situ* by an electron pulse. The low surface area of the central wire results in spiralling orbits of great length before the ions are eventually captured. Thus the decay of primary ions is quite slow for this configuration. Therefore ions are formed and allowed to react for variable time periods quite long in comparison with more conventional pulse experiments, then pulsed out by a large amplitude attractive pulse on the outer electrode. The cylindrical geometry also has the advantage that its extraction efficiency is readily calculable.

Appearance potential measurements and the shapes of ionization efficiency curves have been used to determine parent–daughter relationships in single source ("low-pressure") experiments. A situation which commonly arises, however, is that a product ion has more than one primary ion precursor (192). A technique for handling such a case termed the "ratio-plot method" has been devised by Harrison (86). Consider the reactions

$$P_1^+ + M \rightarrow S^+ + N_1 \tag{3a}$$

$$P_2^+ + M \rightarrow S^+ + N_2 \tag{3b}$$

for which the relationship

$$I_s = f_1 Q_1 l M I_{p_1} + f_2 Q_2 l M I_{p_2} \tag{64}$$

applies. Here I_i is ion intensity, f_i the collision efficiency, Q_i the phenomenological cross section, l the path length, and M the concentration of neutral reactant molecules. This relationship may be rearranged to give

$$I_s/I_{p_1} = f_1 Q_1 l M + f_2 Q_2 l M (I_{p_2}/I_{p_1}) \tag{65}$$

and

$$I_s/I_{p_2} = f_1 Q_1 l M (I_{p_1}/I_{p_2}) + f_2 Q_2 l M \tag{66}$$

Thus from plots of the respective intensity, ratios the slopes and intercepts give the $(f_i Q_i)$ product for the individual reactions. For acceptable self-consistency it is required that the $(f_i Q_i)$ value from the intercept of one ratio plot agree with the $(f_i Q_i)$ value deduced from the slope of the companion plot and vice-versa. A convenient way of changing the relative intensity of the primary ions is to vary the ionizing voltage, and favorable cases yield an unambiguous determination of the individual $(f_i Q_i)$ values. The technique may also be used in combination with pulse techniques (87) and be used to demonstrate that a particular ion does not participate in a given reaction. It should be pointed out, however, that the tandem spectrometer method is the authoritative technique for determining parent–daughter relationships, for appearance potential-ionization efficiency curve comparisons can lead to quite erroneous conclusions (88).

A logical further exploitation of the techniques of single-source mass spectrometry of special interest to radiation chemistry has been the development of multistage differential pumping to permit high pressure operation. These developments by Field (89,90), Melton (91,92), Wexler (62,93), Volpi (94), Henchman (95), Green (96), and others (84), have extended the range of operation using conventional electron-impact ionization to 3 torr source pressure. Melton (91,97), Kebarle (61,98,99) and Volpi (100,101) have used radioactive isotopes for ionization media, increasing the possibilities for a high pressure differential between the source and analyzer. Kebarle (61,99) has operated at source pressures as high as one atmosphere. Wexler (62) has used a similarly tightly enclosed source and a Van de Graaff accelerator to produce a well-defined 2-MeV proton beam to produce primary ions. Among the advantages of such a technique, in addition to the gas-tightness of the source, are the fact that attenuation and scattering by the source gas are negligible, the ionization source is well defined, and source electron beam collimating magnets are eliminated. With these refinements many experimental objections to the high pressure–internal ionization methods were removed. The good correspondence of the data thus obtained with earlier work validates, in large measure, the less sophisticated high-pressure techniques which have been used.

An unfortunate consequence of experiments at the higher pressure limits is that reaction times and ion kinetic energies are ill-defined. Consequently meaningful rate constant measurements cannot be made except through deconvolution methods which, in the authors' opinion, demand more sophisticated knowledge of the reaction dynamics and mechanism than is now available. It is of interest to note, however, that interesting thermodynamic data on ionic solvation may be obtained in high pressure experiments (99).

A hybrid instrument which does not naturally fall into the category of either a single source or tandem instrument has been designed by Melton and Rudolph, (92) and is of particular interest for radiation chemistry investigations. The ion source of this instrument is shown schematically in Fig. 8. This is a differentially pumped system and the pressures drop successively from one chamber to another, being as high as 10 torr in the reaction region, while the analyzer tube pressure is typically of the order of 10^{-6} torr. Electron guns are mounted on each of the chambers and the apparatus can be used to study primary processes for production of both ions and free radicals and for the study of the secondary reactions of these species as well. The third chamber was designed as a high sensitivity ion source which is used for identifying reaction products produced in the preceding compartments. It may also be used to measure ionization cross sections. The irradiation of the sample in the high pressure region in chamber one produces ions, excited molecules, and free radicals which undergo further reactions yielding products which may be ionized by the electron gun in chamber three. The electron beam in chamber one is quite close to the repeller electrode, 1, which can be used to collect positive ions shortly after formation and to quench any positive ion-molecule reactions which would ordinarily occur in this reaction volume. Alternatively it is biased positive and operated as an ordinary high pressure ion source. Chamber two is operated at an intermediate pressure and can be used to bridge the gap between low pressure (that is, low pressure mass spectrometric investigations of the early stages of ion molecule reactions) and the high pressure region of reaction chamber one.

Free radicals are also produced by electron bombardment and the high sensitivity ion source of chamber three may be used to ionize products

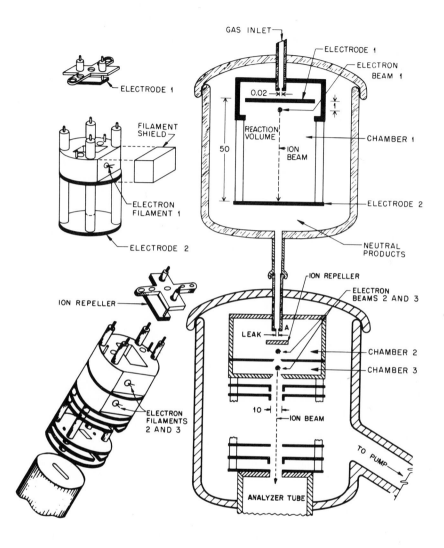

Fig. 8. Three-chambered ion source for studying neutral and ionized products produced by electron impact. Dimensions are in millimeters, and the pressure decreases in stages from chamber 1, which may be as high as 10 torr, to chamber 3, typically 10^{-6} torr. Electron guns on each chamber are used for direct ionization and excitation or for detecting products effusing from preceding chambers. (C. E. Melton, ref. 92.)

diffusing from chamber two. The energy of the electrons in chamber two may be varied to obtain pseudo-appearance potential curves for these unstable species. It is convenient in this type of experiment to apply a negative voltage to the ion repeller of chamber two to prevent positive ions formed in this chamber from entering chamber three. Under these conditions only neutral species will effuse through the slit where they may subsequently be ionized and mass analyzed.

In actual experiments using such a three chamber source Melton found it necessary to consider the following effects of electric field on the radiolysis of gas. If a negative field is applied to the repeller of chamber one, those reactions which can be attributed to positive ion–molecule reactions decrease, while negative ion–molecule reactions are enhanced. The number of free radicals which are produced as neutral products of positive ion–molecule reactions are correspondingly decreased and, similarly, free radicals produced by negative ion–molecule reactions are increased. Finally the number of free radicals, negative ions, and excited states produced by secondary electrons is somewhat increased. Reversing the field to make the repeller repulsive to positive ions produces a corresponding inversion of these effects. It is also necessary to make some approximations in evaluating the concentration of species in the reaction chambers which is to be attributed to a signal of a given level at the final detector. For ammonia, the only system for which an investigation using this instrument has been reported thus far, the results are quite consistent with what is known about ammonia radiolysis with regard to the relative importance of positive and negative ion–molecule reactions, free radical and other neutral contributions to the overall decomposition.

A modification of a single stage mass spectrometer closely akin to tandem experiments was devised by Čermák and Herman (102). In this method, the potential between the filament and the source chamber (see Fig. 5) is kept below the ionization potential of the gas. The potential between the chamber and anode is then adjusted so that the energy of the electrons is sufficiently high to cause ionization in the anode region and to accelerate those ions back into the source region. By varying the anode potential, the average kinetic energy of the primary ions may be adjusted. Because of their unfavorable transverse velocity component, at low ion repeller fields the observed spectrum will consist entirely of secondary ions. This method is well suited to the study of charge transfer reactions and certain types of ion–molecule reactions. If ion–molecule reactions involve a significant amount of momentum transfer, however, severe discrimination against that product is expected. Hence, reactions which proceed via complex formation or stripping reactions involving transfer of a relatively massive moiety either are not observed, or are registered at somewhat distorted intensities. An additional complication is that elastic or nonreactive scattering collisions may allow a primary ion to be detected as a secondary. Simple charge transfer reactions and proton transfer reactions are therefore the classes of ion–neutral interaction most appropriately studied by this technique.

A very recent addition to the arsenal of tools available for the study of ion–molecule reactions is the use of photoionization for producing reactant ions (103–106). Perhaps the principal advantage of this technique is the good energy resolution of which photoionization is capable as compared to electron impact. In addition, since the threshold law for photoionization is a step function, while direct ionization by electron impact is linear with excess energy, photon impact may actually be capable of producing a higher intensity of primary ions in the neighborhood of threshold (106). The spatial definition of the ionizing beam is usually superior, and it is not affected by adventitious electric or magnetic fields. Finally there is no primary space charge effect associated with the ionizing medium.

An interesting refinement of the photoionization method due to Warneck (104) is the use of a pulsed repetitive arc light source and an oscilloscope technique for measuring ion source residence times directly at various applied repeller fields. By analysis of the ion pulse shapes, drift velocities, diffusion coefficients, and equivalent ion temperatures may be derived in favorable cases.

B. Tandem Mass Spectrometers

A considerably more complex method for studying the effects of translational energy involves the use of tandem mass spectrometers. The advantages of such beam machines may be illustrated by reference to Fig. 9, which is a schematic description of an idealized beam apparatus. In such an instrument, the primary ions are produced by thermal means or by electron or photon impact, accelerated to some appropriate energy, mass and energy selected, retarded to final energy and formed into a well-focused and collimated beam at the collision region. A state and velocity-selected molecular beam may be used to react ions with unstable molecules, free radicals, or stable species. Alternatively a collision chamber may be used with gain in intensity of products but loss of information on details of the reactive scattering mechanism. The postcollision velocity analyzer, mass analyzer, and ion detector are indicated as rotatable so that differential cross sections can be measured by studying the angular distribution of mass and energy analyzed product ions.

It should be noted that no such apparatus as that described above now exists, although a number of tandem instruments incorporating several

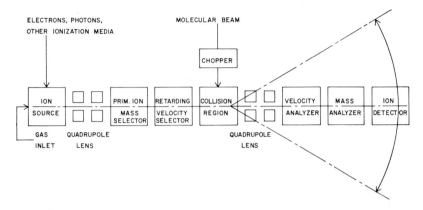

Fig. 9. Schematic diagram of an idealized tandem mass spectrometer and molecular beam apparatus for measuring differential cross sections and ion energetics.

of these features have been developed. The earliest tandem mass spectrometer was built by Lindholm (107), and since that time several of these instruments have been constructed for investigating translational energy effects (108–114). A tandem spectrometer recently constructed at our laboratories (108), is unique in the method employed to control the energy spread of the primary ion beam and in its capabilities for operation at very low primary beam energy with moderate mass resolution. Also noteworthy are the double-beam apparatus of Turner, Fineman, and Stebbings (112) which can be used to study the angular dependence of reaction cross sections, and the single-beam apparatus of Bailey (113) which provides mass and energy selection for both reactant and product ion beams. With such apparatus a great deal of progress toward understanding the kinematic details of ion–molecule reactions for some simple systems has been made.

As already mentioned, in a tandem mass spectrometer the primary beam is extracted from some ionization region, mass selected and then decelerated or accelerated before passage into the reaction chamber. On most of the smaller instruments (109,114), very low (below 10 V) primary acceleration potentials are used so that the beam may enter the reaction chamber with little, if any, change in kinetic energy. Such instruments often suffer from inadequate primary ion mass resolution, so that it may be necessary to use an isotopically enriched gas for gases which have several abundant isotopes. In larger machines, up to a thousand-fold decrease in primary ion kinetic energy is obtained by using a very strong focusing deceleration lens system. The details of these lens systems have been described in other articles (108,115). The low energy ions leaving the reaction chamber are reaccelerated for conventional mass analysis. Many of these instruments use a pair of quadrupole lenses (116) following reacceleration to increase the intensity of secondary ions. Such a lens system is particularly well adapted to this application because of its large physical size and strong focusing properties.

There are two basic types of fixed-angle tandem machines: transverse and in-line. The transverse instruments, such as are used in Stockholm, (107,117), Moscow (111), Baltimore (114), and Birmingham (110) are designed to discriminate very strongly against mass transfer processes and therefore favor collection of the products of charge transfer reactions. Even with this geometry, however, mass transfer products may be observed at low translational energies. It is also possible, as Koski does (114,118), to employ higher than normal repeller potentials in order to extract the primary beam from the reaction chamber along with those secondaries which result from mass transfer reactions. By employing the in-line configuration, as in the ARL (108), and Giese (109), designs, the products of mass or momentum transfer processes may be observed more efficiently.

Henglein (38) has constructed a machine for studying stripping reactions which does not fall into any of the above categories. It consists of an ion gun followed by a flight tube which also serves as a reaction chamber. A velocity selector scans the ions which have suffered little or no change in direction, and energy analysis of the secondary ion beam is employed to deduce cross sections and reaction mechanisms in chosen simple cases.

C. Other Methods

A method for the measurement of thermal velocity ion–molecule reaction rates which approximate the flow methods of classical kinetics has been developed by Fehsenfeld, Ferguson, Schmeltekopf, and colleagues (119–121). This procedure makes use of a large amount of excess carrier gas, typically helium, which both thermalizes ions and transports them to the detector mass spectrometer. Diffusive mixing produces a uniform concentration of molecules which is calculated from the flow rate. The velocity of gas flow is determined by pump parameters and the reaction time is equal to the distance from the gas injection port to the spectrometer entrance port divided by flow velocity. The decrease in primary ion signal and increase in secondary ion signal can be interpreted in terms of a reaction rate.

This flowing afterglow apparatus operates in the following manner. A plasma is created by either continuous or pulsed microwave discharge which produces He^+ and He^*. The first reactant gas is injected by means of a nozzle and reacts with the ions and excited species of the weakly ionized plasma. The ions thus formed are thermalized by collision with the He carrier gas before arriving at the nozzle through which the second reactant gas is injected. Reactions of the thermalized ions in the reaction zone between this nozzle and the mass spectrometer sampling port are monitored as a function of flow parameters. A quadrupole mass spectrometer is a particularly appropriate detector.

Another plasma technique is the pulsed discharge-afterglow method of Dickenson and Sayers (122) which has been used by a number of workers to estimate rate constants for ion–molecule reactions (123–125) particularly for systems of aeronomic interest. This method consists of ionizing the gases under investigation with a high power radiofrequency discharge pulse and observing the ion density decay/formation rate with time.

A related technique suggested by Hasted (126) consists of injecting a mass resolved ion beam into a mobility apparatus which, by the development into two or more peak profiles, can be resolved into the rate constants for postulated ion–molecule reactions. A further development of this idea would add a second mass analyzer for positive identification of the mobility ion peaks. Variations of this theme have been applied by a number of workers (127–131). One of the most versatile devices of this type has recently been placed in operation by McDaniels and colleagues (132,133) and has been used to measure the three body rate constant for the reaction

$$H^+ + 2H_2 \rightarrow H_3^+ + H_2 \tag{67}$$

Perhaps the greatest utility for such techniques lies in the measurement of similar high kinetic order processes for simple systems.

A new instrumental technique for the study of ion-neutral reactions which appears highly promising is the technique of ion cyclotron resonance. The principle of ion cyclotron resonance may be described by considering the motion of an ion placed in a strong magnetic field. It is constrained by the field to move in a helical path. This motion is the superposition of translation along a field line and of circular motion in the plane perpendicular to that field line. Motion along the field line is independent of the presence of the field, and circular motion in the perpendicular plane can be considered independent of its translation along the field lines. The angular frequency of rotation of a given ion is related to its charge, mass and magnetic field by the relation $\omega_c = eB/m$ where ω_c is the characteristic cyclotron frequency.

Consider now that a linearly polarized rf electric field is applied in a direction perpendicular to the magnetic field. Such a linear field may be treated as the superposition of two circularly polarized rf fields with opposed senses of rotation. If the applied field is at the cyclotron frequency, the circular component with the same sense of rotation gives rise to a continual acceleration. The ion moves in an Archimedes spiral whose radius increases linearly with time, $r = Et/B$, where E is electric field strength. This process continues until the ion collides with a neutral molecule or with the wall, extracting energy from the applied electric field and increasing the translational energy of the ion.

Since little energy conversion occurs unless the applied field is at or near the cyclotron frequency, this process provides a basis for mass analysis. This is accomplished by measuring the B/ω_c ratio, since $M/e = B/\omega_c$. This is the operating principle of the omegatron mass spectrometer (134). This device uses ion collection, but it is also possible to monitor absorption of rf energy or induced potential. The intensity of absorption or induced potential is directly proportional to the number of resonant ions in the cavity.

This particular mass spectrometric method is of interest for studying ion–molecule reactions mainly because of the very long, spiral ion path and the related long residence time of ions in the cavity. Consequently the probability of ion-neutral collision is quite high even at every low pressure in the chamber. Moreover, the possibility of simultaneously irradiating the ions with two or more oscillators provides a rather sensitive probe for coupled chemical reactions (135,136). In this double resonance technique one rf oscillator is set at the cyclotron frequency of a given ion and a second, variable frequency oscillator is swept as an analyzer. In this way the effect on the remainder of the mass spectrum of exciting a particular ion may be directly determined. If two ions are coupled, for example, as reactant and product of an ion–molecule reaction, the only requirement for an effect to be observed by double resonance is that the rate constant be a function of ion velocity. Since many reactions are thought to exhibit the Langevin $1/v$ velocity dependence of cross section leading to an energy-independent bimolecular rate constant, however, it is well to keep in mind this ambiguity of a null result in cyclotron double resonance. Nevertheless, one can anticipate many valuable contributions from the application of this very promising technique.

In addition to the foregoing techniques, it should be mentioned that mass spectrometric sampling of flames and discharges have yielded information on ion–molecule reactions. Estimates for rates of reaction of a number of unstable species are available from such investigations, and it has been shown that ion–molecule reactions may substantially alter the course of such reactions. In particular, relatively slow reactions may be studied by such techniques since fast ion–molecule reactions either take place too rapidly for meaningful sampling by mass-spectrometric probes or the products are substantially altered in extraction through the static sheath which surrounds the probe. The reaction schemes are generally quite complex and ion–molecule reactions in such systems are in competition with a number of simultaneous processes which render interpretation difficult. Consequently information provided by these experiments is at present somewhat less quantitative than that obtained from the other techniques we have discussed. These topics have been reviewed recently elsewhere (139,140) and will not be considered further here.

All the approaches described thus far have been used to deduce the various reaction schemes by identifying and characterizing the ionic reactant and ionic product of ion–molecule reactions. Obviously equivalent information can be obtained from the identification and characterization of the neutral species as well, if possible complicating ambiguities can be eliminated. In addition, investigation of the neutral product may lead to structural information which is unavailable from mass spectrometric investigations. Such studies are in a sense complimentary to the methods which have been discussed and may lead to the discovery of new types of reactions and changes in the mode of reaction under radiolytic conditions. This leads in turn to information concerning the reaction complex for thermal velocity ions, to accurate values for relative reaction rates, and to information on pressure effects.

The technique has been developed particularly by Ausloos and co-workers (137) and generally requires keen insight into the probable reaction mechanisms in a given system. Deuterium labelled compounds are generally used to distinguish bimolecular from unimolecular reactions and to distinguish structural specificity whenever it exists. More recently photoionization has been used for selective ionization of components in mixtures and for studying the kinetics of various competing processes for ions formed with specific amounts of electronic excitation (138). The utilization of these techniques will be discussed further in the next section.

IV. EXAMPLES OF ION–MOLECULE REACTIONS

Several hundred ion–molecule reactions have by now been observed and many of these have been extensively studied. A detailed discussion of all these reactions is not possible within the confines of this review. Instead we have chosen to discuss in some detail a few systems which have received the attention of many investigators, and to briefly mention and list some of the more important of the remaining ion–molecule reactions.

A. Reactions of Hydrogen

Because it is the simplest molecular system, the ion–molecule reactions of hydrogen have occupied a central role in our understanding of general features of reactions involving molecules. Although studies of hydrogen and hydrogen isotopes were among the earliest and the most thoroughly investigated of ion–molecule reaction systems, this extensive literature will not be reviewed here. Rather the results of some very recent pertinent experiments will be cited to illustrate some features which are of general significance. Suffice it to say that the hydrogen system was examined theoretically by Eyring, Hirschfelder, and Taylor in 1936 (19) who calculated for thermal velocity reactants a specific rate constant for reaction (1) of 2.07×10^{-9} cm^3 molecule^{-1} sec^{-1} which is an excellent agreement with Stevenson and Schissler's (21) and Klein and Friedman's (141) experimental value of 2.02×10^{-9} cm^3 molecule^{-1} sec^{-1}. At low velocity, therefore, the orbiting complex theory appears to provide an adequate interpretation of this reaction.

As the relative kinetic energy of the reactants is increased, it is clear that at some limit the collision complex description can no longer apply, if indeed it applies even at low energy. Beam experiments have been quite enlightening regarding the higher energy reactions of isotopic hydrogen. The reaction of HD$^+$ with HD was investigated by Giese (142), and by Futrell and Abramson (143). At low velocity the overall isotope effects in the reaction

$$HD^+ + HD \rightarrow H_2D^+ + D \qquad (1a)$$
$$\rightarrow D_2H^+ + H \qquad (1b)$$

are such that the ratio H_2D^+/HD_2^+ is about 0.8. It increases monotonically with increasing energy to a maximum value of about 5 at 3.5 eV center of mass kinetic energy and decreases to approximately unity at higher energy. The initial increase at lower energies probably is a manifestation of a kinetic isotope effect reflecting a changeover from complex formation to atom pickup at higher energies. The decrease is H_2D^+/D_2H^+ above 4 eV may be attributed to the conversion of translational energy of the initial HD$^+$ ion into vibrational excitation of the products which, through normal vibrational isotope effects, causes H_2D^{+*} to decompose more readily than D_2H^{+*}. The observation of D$^+$ and H$^+$ as products above 4 eV provides support for this hypothesis.

The reaction of D$_2^+$ with HD also illustrates several features related to kinetic energy (143). The percentage of secondary reactions as a function of kinetic energy represented by each of several reaction paths is given in Fig. 10. The reactions which describe this system are

$$D_2^+ + HD \rightarrow HD_2^+ + D \qquad (1c)$$
$$\rightarrow D_3^+ + H \qquad (1d)$$
$$\rightarrow HD^+ + D_2 \qquad (1e)$$
$$\rightarrow D^+ + D + HD \qquad (1f)$$

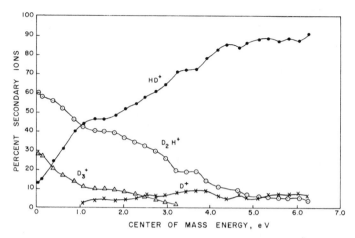

Fig. 10. Products of the reaction of D_2^+ with HD as a function of kinetic energy. (F. P. Abramson and J. H. Futrell, ref. 143.)

There are actually two types of reactions which are included in (1c). One is a deuteron transfer, but an abstraction reaction analogous to (1d) would also produce HD_2^+. The type of behavior shown in Fig. 10 is typical of many ion-molecule reactions inasmuch as the probability of ion–molecule reaction decreases relative to charge transfer as the kinetic energy of the incident ion increases. At low velocity, exothermic charge transfer and ion–molecule reaction are competitive processes, and one or the other may predominate depending on the properties of the system.

In a recent elegant experiment the kinematics of the reactions of D_2^+ with D_2 and with H_2 were investigated in considerable detail by Dover-spike and Champion (144). These workers measured both the angular distribution and velocity distribution of reactant and product ions from the reaction. By such an analysis of the reaction dynamics they were able to establish several reaction channels and to study their energy dependence.

The results of these experiments may be summarized as follows: The most probable reaction channels for D_3^+ and D_2H^+ formation in the two experiments obey the kinematic predictions of a simple atom pickup model from a laboratory collision energy of 2 eV up to collision energies where the product ions are unable to store any more internal energy. Above about 4 eV center of mass collision energy the reactions deviate from a simple pickup model, as is anticipated since there is an upper limit to the amount of excitation energy which can be stored in the product ions without dissociation. A deuteron transfer product, H_2D^+, and a low velocity D_3^+ peak followed identical kinematics which could not be explained by any simple collision model. In addition a velocity group of product ions were observed with intensity maxima corresponding to a completely inelastic process. The most probable interpretation of these peaks is that they correspond to reaction channels of the excited complex ions $(D_4^+)^*$ and $(D_2H_2^+)^*$ which then dissociate into the observed products D_3^+, D_2H^+, H_2D^+ with low relative velocity of separation of the product ion and neutral.

Typical data from this experiment are shown in Fig. 11 for the reaction D_2^+ with H_2 at zero scatter angle. The solid lines of Fig. 11a give the energy profile of the nominally 3.95 eV D_2^+ primary ion, while the open circles give the energy profile of the D_2H^+ product ions. The data are treated by a deconvolution technique to remove the influence of such factors as finite energy resolution of the velocity selectors, etc., to give the more highly structured presentation of Fig. 11b. Here peak A is a back-scattered ion peak corresponding to the forward scattered maximum at about 3.2 eV. Both peaks correspond to an atom pickup model. Peak B is the inelastic process which is suggested to proceed via a $(D_2H_2^+)$ complex. Peak C, if it is not an artifact of the experiment, results from the exothermicity of the reaction in this channel being converted into translational energy of the products.

Champion and Doverspike also observe D^+ product ions which, from their velocity spectra, are attributed to collision induced dissociation and to dissociation of excited product ions. Above about 4 eV center of mass collision energy the velocity maxima of the D_3^+ and D_2H^+ ions deviate from the kinematics of the simple pickup model. If it is assumed that the dissociation limit of D_2H^+ is about 3 eV and that it can dissociate isotropically into $H^+ + D_2$ and $D^+ + HD$ one can calculate velocity distributions for H^+ and D^+ which agree very well with the experimentally determined values. Another group of D^+ ions, well removed in energy from both these D^+ ions and the H_2^+ ions produced by charge exchange, is attributed to collision-induced dissociation. It is therefore concluded that post-reaction dissociation is a prominent feature of the experiment at moderate collision energies.

These results are in qualitative agreement with conclusions of Durup and Durup (145) using a quite different experimental approach. These workers measured the kinetic energy of reactant and product ions after the method of Berry (146) but used a two-chambered source to separate the ion formation and collision regions. The possible mechanisms, complex formation, atom pickup, and deuteron transfer could be distinguished by the energy of the product ions. Using the velocity of the reactant D_2^+ ions as a standard with energy, eV, the D_3^+ product ions should have $\frac{5}{8}$ eV for complex formation $\frac{1}{3}$ eV for D-pickup, and $\frac{5}{6}$ eV for deuteron transfer. The results are shown in Fig. 12 from which we deduce that the

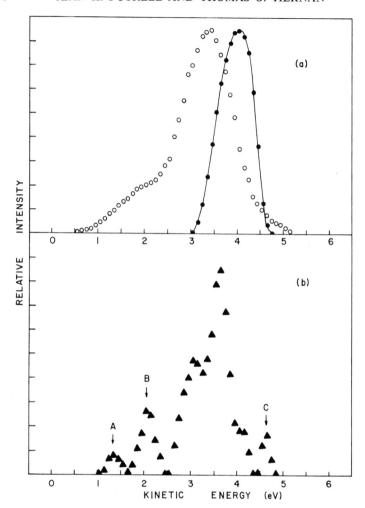

Fig. 11. Kinetic energy of D_2^+ and D_2H^+ ions from the zero-angle reaction of 3.95 eV D_2^+ with H_2 measured with apparatus of Fig. 13. (a) Experimental data for D_2^+ ○ and D_2H^+ ●. (b) These data deconvoluted to remove apparatus effects. See text for discussion of designated peaks. (L. D. Doverspike and R. L. Champion, ref. 144.)

complex model applies at low energy (below 1–2 eV) but that deuteron transfer best describes the reaction over the range 4–10 eV. These workers also studied the formation of D^+ via a second order process and conclude that two mechanisms apply. At high energies (10 eV) collision induced dissociation is the most important, but at lower energy dissociations of $(D_4^+)^*$ and $(D_3^+)^*$ complexes appear to contribute significantly.

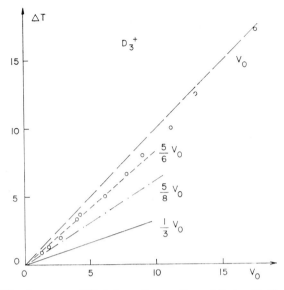

Fig. 12. Kinetic energy of D_3^+ from the reaction of D_2^+ with D_2 referenced to the kinetic energy of D_2^+. Lines at $1/3 \ V_0$, $5/8 \ V_0$, and $5/6 \ V_0$ refer to kinematic models of atom pickup, complex formation, and deuteron transfer, respectively. (J. Durup and M. Durup, ref. 145.)

These two investigations are unfortunately not in good agreement concerning the relative importance of atom pickup and deuteron transfer as the high energy reaction mechanism. The fact that Durup and Durup examine averaged cross sections while Bailey's apparatus was used to examine mainly the zero-angle mechanism may be partly responsible for the discrepancy. It is also possible that the internal excitation of the D_2^+ may be different in the two experiments. Regardless of these details, however, the picture emerges that formation of orbiting complexes precedes the proton transfer reaction at low velocity. With increasing relative kinetic energy the mechanism goes smoothly over to a stripping mechanism which may be an atom pickup, a proton transfer or a mixture of these reaction modes.

The effect of internal energy on the cross section for proton transfer in hydrogen has been studied by Weingartshofter and Clarke (147) who used an electron monochromator to excite selectively the different vibrational levels of the hydrogen molecule ion. In this manner a characteristic ionization efficiency curve with breaks corresponding to the excitation of discrete levels of H_2^+ was constructed. On increasing the pressure in the source a correspondingly structured ionization efficiency curve for H_3^+ was

Table I

Relative Cross Sections for Reaction

$$H_2^+(v^1) + H_2 \rightarrow H_3^+ + H$$

Vibrational level (v^1)	0	1	2	3	4	5
Relative cross section	1.0	0.5	0.4	0.5	0	0

observed. From the slopes of the corresponding segments relative cross sections were obtained and are reported in Table I.

As discussed previously, the theoretical cross section for forming H_3^+ agrees with the experimentally determined cross section. Since this treatment assumes unit probability of reaction for all H_2^+ ions which form orbiting complexes the results of Weingartshofer and Clarke are somewhat puzzling. It would appear that either an interaction potential stronger than the charge-induced isotropic dipole is required (so that the integrated transmission coefficient can be less than unity) or that there is an error in either the measured total cross section or in the relative cross sections of the individual vibrational levels. It is interesting to speculate on yet another possibility that there is a translational energy threshold for some vibrational levels. Two measurements of the H_3^+ and D_3^+ cross sections using pulse methods (148,149) have indicated that thermal velocity ions react more slowly than is indicated by more conventional methods. Both laboratories involved in these measurements point out that there are possibly severe mass discrimination effects in the pulsed source measurement for isotopic hydrogen. Also, it should be noted that the velocity selector method of Weingartshofer and Clarke can only be used for low velocity ions. Furthermore, Friedman and co-workers (141,150,161) have observed translational energy thresholds for certain rare gas–hydrogen reactions, while the reaction $O^+ + N_2 \rightarrow NO^+ + N$ has a well-established ion kinetic energy threshold (152).

B. Rare Gas–Hydrogen Reactions

Ion–molecule reactions of this category are of particular interest because many properties of the reactants and products are either well known or may be calculated. The intermediate three body complexes formed in these systems are sufficiently simple to permit their treatment by tractable theoretical methods. It is apparent that for these reactions, the same product can be considered to result from either hydrogen ion or rare gas ion precursors. Thus two reaction paths are possible,

$$X^+ + H_2 \rightarrow XH^+ + H \tag{68}$$

$$H_2^+ + X \rightarrow XH^+ + H \tag{69}$$

where X represents the rare gas atom, and it is only with fairly recent experiments that the relative contributions of these two processes have been assessed.

Isotope effects and energy dependences of the cross sections for these reactions have been examined by several investigators. The hydrogen–helium and hydrogen–neon systems were among the first to be subjected to detailed scrutiny. Friedman and co-workers, (150,151) found that the He^+–H_2 and Ne^+–H_2 reactions exhibited negligible cross sections even though these reactions are exothermic. The major portion of the product results from the endothermic reactions involving H_2^+, and strong kinetic energy dependence of the cross sections of these reactions was observed with thresholds of 0.35 eV and 0.18 eV for HeH^+ and NeH^+ formation, respectively. This energy threshold was explained (150) by assuming that a competing radiative process for H_2^+ ions exists, since the indications are that these ions are vibrationally excited. Light's phase space theory can also account for the observed thresholds and isotope effects for these reactions on the basis of efficient conversion of translational energy into internal energy of the system. It remains for these energy effects to be confirmed in a beam experiment, something which has not yet been done, possibly because of the difficulty of obtaining low energy H_2^+ ions in abundance.

In contrast to the reactions just discussed, in the argon–hydrogen and krypton–hydrogen systems both reactions (68) and (69) occur with comparable cross sections. Isotope effects for the abstraction reaction with argon,

$$Ar^+ + H_2 \rightarrow ArH^+ + H \tag{24}$$

have been studied as a function of translational energy in several laboratories. For comparison purposes, the data from several sources are all plotted in Fig. 13. The solid circles represent the data of Klein and Friedman (141), triangles represent the data of Henglein and co-workers (167), and the data points of Futrell and Abramson (143) are shown as open circles. Considering the different experimental techniques employed in obtaining this data, the agreement is reasonably good. Klein and Friedman used a single stage mass spectrometer and apparently neglected to allow for the complementary reaction,

$$H_2^+ + Ar \rightarrow ArH^+ + H \tag{69a}$$

which would result in a slight shift of their data. Henglein and co-workers used a beam instrument in which the secondary ions are energy analyzed but not mass analyzed. The data of Futrell and Abramson were obtained with a tandem mass spectrometer, as previously described (143). It should be noted that the latter results are also in agreement with the data of Berta,

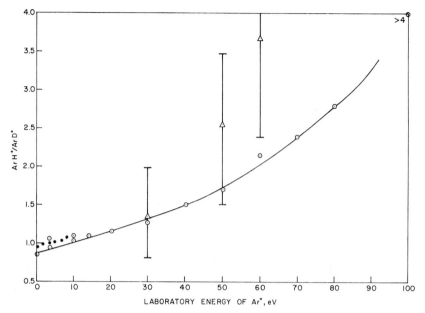

Fig. 13. ArH$^+$/ArD$^+$ as a function of Ar$^+$ energy. (J. H. Futrell and F. P. Abramson, ref. 143.)

Ellis, and Koski (108) not shown in this figure (114). Figure 13 indicates a large isotope effect at high ion energies and this has been explained by Henglein (39,167) as a "spectator stripping" reaction. The pronounced isotope effect results from the dissociation of product ions when the internal energy deposited in the product exceeds its dissociation energy. According to the "spectator stripping" or atom pickup model, the energy deposited in the argon hydride product is given by,

$$E_i = E_{Ar}^+ \frac{M_i}{M_{Ar^+} + M_i} - \Delta H \tag{70}$$

where E_{Ar}^+ represents the impacting ion energy, M_i the mass of the target atom and ΔH is the heat of the reaction. A stable product will be formed only if $E_i <$ D(ArH$^+$), the dissociation energy of the argon hydride ion. Using appropriate values of the quantities in equation (70), one calculates that ArD$^+$ should not be observed above a maximum energy of 38 eV while the limit for observance of ArH$^+$ should be 75 eV. Thus above 38 eV an infinite kinetic isotope effect is predicted by the stripping model. Figure 13 indicates that while the data are not in strict agreement with the stripping theory, the trend observed support, at least qualitatively, the predictions of this theory.

The moderate isotope effects observed at low energies are substantially in agreement with the observations of Klein and Friedman (141) who were able to explain their results satisfactorily as a combination of isotope effects. The separation of center of mass from center of polarizability in the HD molecule produces a configuration isotope effect because of the resulting difference in activation energies for the XDH^+ and XHD^+ complexes. Also the decomposition of these complexes involves a vibrational isotope effect, owing to the differences in zero-point energies of the probable transition states. A third isotope effect in the dissociation of product ions becomes increasingly important with increasing reactant ion kinetic energy.

Attempts to understand the argon–hydrogen system are further complicated by the fact that the complementary reactions (68) and (69) also exhibit isotope effects. Although not in quantitative agreement, the trends observed for these reactions with changing kinetic energy are similar. This is shown by Fig. 14 in which isotope effects for both reactions are plotted as a function of center of mass energy. However, the reactions exhibit distinctive differences, as Giese and Maier have noted (109). The differences would appear to reflect the consequences of formation at low center-of-mass energies of reaction complexes with differing amounts of internal energy in the two cases. This likely results from the fact that the reacting HD^+ species is vibrationally excited, and the results for the HD^+–Ar reaction therefore represent the superposition of reactions of a distribution of vibrationally excited HD^+.

Bailey and co-workers (169–171) have recently employed a beam scattering instrument to study reaction (69a) as well as the corresponding charge transfer and collisional dissociation reactions.

$$H_2^+ + Ar \rightarrow H_2 + Ar^+ \tag{71}$$

$$H_2^+ + Ar \rightarrow H^+ + H + Ar \tag{72}$$

It is significant that the charge transfer cross-section plotted as a function of the impacting ion kinetic energy exhibited three successive maxima. In terms of the Massey adiabatic hypothesis, the maxima are interpreted as indicating vibrationally excited levels of the H_2^+ ions, and the energy defects observed suggest a distribution of vibrational states. Hertel and Koski (168) have observed similar resonances in the HD^+–Ar reaction.

In a more elaborate beam experiment in which both angular and kinetic energy distributions of product ions were determined, Bailey and co-workers have obtained detailed results for the reaction,

$$Ar^+ + D_2 \rightarrow ArD^+ + D \tag{24a}$$

From this experiment the total internal energy change in the reaction is

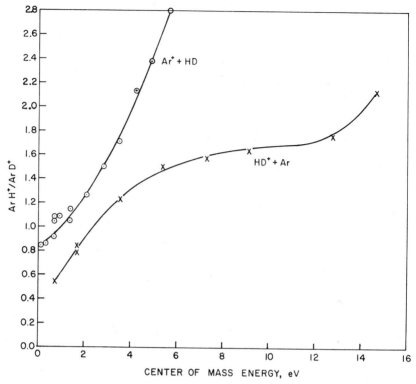

Fig. 14. ArH$^+$/ArD$^+$ from the reactions of Ar$^+$ with HD and HD$^+$ with Ar as a function of energy in the center of mass system. (J. H. Futrell and F. P. Abramson, ref. 143.)

determined. This value is found to be strongly dependent upon primary ion–kinetic energy, exhibiting a maximum which corresponds to the formation of internally excited ArD$^+$ product with an excitation energy approximately equal to the dissociation energy for ArD$^+$ into Ar + D$^+$. The results obtained in this study at moderate ion energies (up to about 50 eV) are consistent with the simple atom pickup stripping mechanism of Henglein, but again quantitative agreement is not obtained. At low collision energies, a comparatively large amount of back scattering of product ions was observed, suggesting that at these low velocities a collision complex is formed in this reaction.

C. Reactions of Methane

1. Reaction with CH$_4^+$. One of the most interesting ion–molecule reactions is the protonation of methane, first reported by Tal'roze and Lyubimova in 1952 (20) and subsequently investigated by many workers

(47,52,80,83,90,93,148,153–163). The methane parent ion and methyl ion together account for some 85% of the primary mass spectrum of methane. There is quite general agreement that the principal reactions of these ions are

$$CH_4^+ + CH_4 \rightarrow CH_5^+ + CH_3 \tag{2}$$

$$CH_3^+ + CH_4 \rightarrow C_2H_5^+ + H_2 \tag{32}$$

In addition, a small quantity of $C_2H_3^+$ (ca. 4% of the total reaction) is formed by CH_3^+ while charge transfer is the only competing reaction for low velocity CH_4^+ (47). The product ions (CH_5^+ and $C_2H_5^+$) are unreactive toward methane and the concentrations of these ions at elevated pressure asymptotically approach those of the primary precursor ions.

The cross sections for CH_4^+ disappearance and CH_5^+ formation have been redetermined recently by Giardini-Guidoni and Friedman (162) and compared quantitatively with theoretical estimates. Figure 15 presents their data as a function of the reciprocal maximum ion velocity $(eEl)^{-1/2}$. The phenomenological cross section for CH_4^+ loss includes both collision-induced dissociation and charge exchange losses in the analyzer as well as

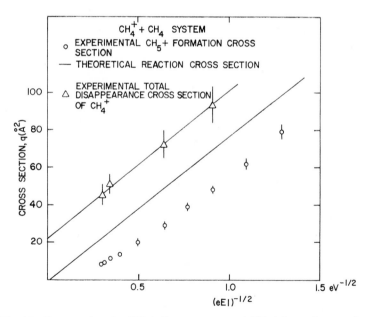

Fig. 15. Cross sections for CH_4^+ disappearance and CH_5^+ formation as a function of reciprocal average kinetic energy. The orbiting collision polarization potential cross section for methane is shown for comparison. (A. Giardini-Guidoni and L. Friedman, ref. 162.)

disappearance via reaction in the source. These losses may be evaluated from Fig. 15.

The intercept is interpreted as the limiting case of all losses occurring in the analyzer tube. If this intercept is subtracted, the experimental points for the CH_4^+ disappearance cross section fall on the solid line computed for the Langevin cross section. The CH_5^+ formation cross section, however, falls consistently below the theoretical line and does not exhibit the anticipated $E^{-1/2}$ velocity dependence. As was pointed out earlier by Field, Franklin, and Lampe (154) and by Kubose and Hamill (164) a better fit over most of the range of the experiment is obtained by an E^{-1} dependence of cross section on ion kinetic energy. Because of the discrepancy noted between the CH_4^+ disappearance and CH_5^+ formation cross sections, Giardini-Guidoni and Friedman suggested that alternative reaction channels might account for the unexpected energy dependence.

The tandem mass spectrometer is well suited to investigating the possible multiplicity of reaction channels and their dependence on ion kinetic energy. Using such an instrument, Abramson and Futrell (163) investigated reactions of CH_4^+ with CD_4 as a function of kinetic energy. Unlike the reaction of CH_4^+ with CH_4, this system permits the resolution of collision-induced dissociation from the formation of methyl ions via dissociation of hydronated methane secondary ions. The results are shown in Fig. 16, for center-of-mass energies ranging from 0.17 to 6.8 eV.

As previously reported (47) there are a number of products from this reaction other than the simple proton transfer at m/e 21. The product at m/e 20 certainly represents charge transfer to CD_4 but also might represent $CD_3H_2^+$ formed by an exchange reaction in the intermediate complex (155). Similarly the peak at m/e 19 is also an exchange product, either $CD_2H_3^+$ or CD_3H^+. The very rapid decay of the m/e 19 channel with increasing kinetic energy combined with the maximum in the CD_4H^+ channel suggests that at higher kinetic energies an intermediate complex does not live long enough for exchange reactions to occur. At higher kinetic energy, therefore, these channels appear as simple proton transfer, which explains the initial increase in the m/e product. Such exchange reactions and their diminution at higher energies have also been observed for other reactions which supposedly go through a "loose" intermediate complex (50).

There are several reactions which may contribute to the product at m/e 18. At lower kinetic energies the predominant reaction is D abstraction by the impacting ion producing CH_4D^+. This abstraction reaction has been shown to require a longer-lived complex than does proton transfer. The decrease in the m/e 18 product with kinetic energy up to 2.5 eV$_{c.m.}$ is repeated evidence for this. At higher energies two other reactions are

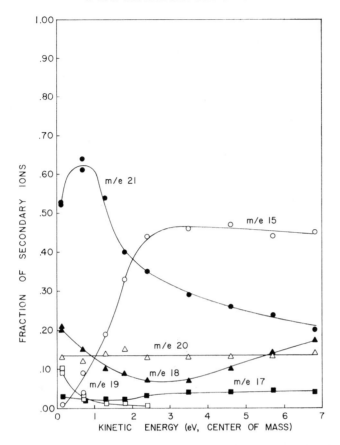

Fig. 16. Secondary ions from the reaction of CH_4^+ with CD_4 as a function of ion kinetic energy. (F. P. Abramson and J. H. Futrell, ref. 163.)

possible but no unambiguous measure of their relative importance is available. These are the loss of HD from an excited CD_4H^+ and dissociative charge transfer, each of which produces CD_3^+. The combination of these two processes is undoubtedly responsible for the increase in m/e 18 above 2.5 eV.

The product at m/e 17, however, can only result from loss of D_2 from an excited CD_4H^+ and should be a reliable measure of that type of dissociation which leads to methyl ions. Similarly the m/e 15 peak can only be CH_3^+ and surely must result from collisionally induced dissociation of the impacting methane ion. Despite the ambiguity in interpreting the CD_3^+ product, the results clearly show that over virtually the entire range of the experiment dissociation of the impacting ion is the most important source of methyl ions formed by collision of accelerated CH_4^+ ions with methane molecules.

The production of CH_3^+ ions in this system is endothermic by 1.37 eV for unexcited methane ions. Since about half of these dissociation reactions occur below 1.37 $eV_{c.m.}$ it is apparent that excited methane ions possessing internal energy up to and perhaps over 1 eV are reacting to produce the various products. Futrell and Abramson (163) further conclude that the combination of collision-induced dissociation and dissociation following reaction mainly account for the discrepancy between the CH_4^+ disappearance and CH_5^+ formation cross sections noted by Giardini-Guidoni and Friedman (162). For the principal ion–molecule reaction in methane, therefore, it appears that the Langevin theory quite adequately describes the low velocity reaction once competing modes of decomposition are considered.

To gain some further insight into energy conversion effects the dissociation of protonated methane was studied briefly. For this purpose CD_4 and CH_4 were reacted with CHO^+ ions produced by ionizing methanol and acetaldehyde in the source of the first stage spectrometer (47,166). The relative amounts of protonated deuteromethane as well as decomposition products were measured at 3 energies and are reported in Table IIA. The reaction sequence is

$$CHO^+ + CD_4 \rightarrow CD_4H^+ + CO \tag{73}$$

$$CD_4H^+ \rightarrow CD_3^+ + HD \tag{73a}$$

$$\rightarrow CD_2H^+ + D_2 \tag{73b}$$

With CH_4 target molecules, of course, reactions (73a) and (73b) are indistinguishable.

The ratio CHD_2^+/CD_3^+ in column 7 of Table IIA is of interest, as it represents the relative probability of D_2/HD elimination from CD_4H^+. For equivalent hydrogens and no isotope effect in CD_4H^+ we would expect 1.5 for this ratio. The slowly changing isotope effect indicated by the results may represent the superposition of isotope effects resulting from the differences in zero point energies of C—H and C—D bonds and a kinetic isotope effect from the different vibrational frequencies of the two bonds. The latter effect becomes more important as the internal energy, and consequently the probability of dissociation, increases.

As discussed by von Koch (161) both charge exchange and dissociative proton transfer of CHO^+ with methane are endothermic for ground state ions. At our lowest impacting ion energy the total energy of the system is insufficient to drive these reactions even if the total center of mass translational energy of the complex is available for the reaction. Hence the ion

Table II

Effect of Ion Translational Energy on the Reaction of CHO^+ with Methane

A. Reactions of CHO^+ (CH_3OH) with CD_4

Ion energy	Relative intensity, counts/sec				$\dfrac{CD_4H^+}{CHD_2^+ + CD_3^+}$	$\dfrac{CHD_2^+}{CD_3^+}$
eV	CHD_2^+	CD_3^+	CD_4^+	CD_4H^+		
0.3	201	244	57	1337	3.00	0.82
1.9	219	228	61	704	1.57	0.96
4.2	242	223	79	453	0.97	1.09

B. Reactions of CHO^+ (CH_3OH) with CH_4

Ion energy	Ratios of secondary ions CH_5^+/CH_3^+
0.2 eV	3.9
0.4	3.3
0.6	3.0

C. Reactions of CHO^+ (CH_3CHO) with CH_4:

Ion energy	Ratio of secondary ions CH_5^+/CH_3^+
0.3 eV	19
0.6	17
1.6	10
2.5	7.4
5.6	5.3
8.6	2.7

beam in both experiments must contain excited CHO^{+*} ions. The increasing extent of dissociation with energy reported in column 6 of Table IIA indicates conversion of translational energy into internal energy. This effect was further investigated using CHO^+ ions from both methanol and acetaldehyde (166,167). The data are reported in Tables IIB and IIC and demonstrate a relatively efficient conversion of translational energy into excitation of the product CH_5^+ ions with subsequent dissociation into CH_3^+. In addition, the results show that the average internal energy of CHO^+ derived from acetaldehyde is lower than the average internal energy of CHO^+ derived from methanol.

2. Reactions with CH$_3$$^+$. As stated earlier there is general agreement that CH$_3$$^+$ reacts mainly to form C$_2$H$_5$$^+$. It can be postulated that the reaction proceeds via formation of a C$_2$H$_7$$^+$ intermediate complex, although this species has never been observed in pure methane at pressures to 160 torr (184). That it does have a finite lifetime is indicated by the isotopic distribution of ethyl ions from the reaction of isotopically labeled reactants. Randomization of an intermediate C$_2$H$_n$D$_{7-n}$ complex has been reported previously (169,185) from studies of CH$_4$–CD$_4$ mixtures, and was studied extensively in our laboratory (175). Identical distributions of isotopic ethyl ions from the dissociation of [C$_2$H$_3$D$_4$$^+$] intermediates prepared from several isotopic reaction pairs were observed in this study. Similar results for [C$_2$H$_4$D$_3$$^+$] were also obtained.

The concept of an intermediate complex of the formula (C$_2$H$_7$$^+$) with a significant lifetime prior to dissociation into the observed products seems fully substantiated. The existence of stable C$_2$H$_7$$^+$ is indicated by studies of ion molecule reactions of ethane (158,170,171,167,189,188) and it can be produced in quite high yield by proton transfer from CHO$^+$ to ethane (166). Therefore its stability is probably determined by its energy content. If we assume that single collision stabilization is possible for C$_2$H$_7$$^+$ and a collision frequency dictated by charge-induced dipole forces one can estimate that the lifetime of the complex must be less than 10^{-10} sec for it not to be observed. This, however, does allow ample time for isotopic rearrangement prior to dissociation.

The effect of translational energy of the methyl ion on product distribution was also investigated briefly. For quasi-thermal CH$_3$$^+$ the ratio C$_2$H$_3$$^+$/C$_2H_5$$^+$ is 0.04 and increases to 2.4 and 4.2 eV ion kinetic energy (47). These results clearly demonstrate that kinetic energy may be converted rather efficiently into internal energy. The sequential breakdown of the complex

$$C_2H_7^{+*} \rightarrow C_2H_5^{+*} + H_2 \tag{32a}$$

$$C_2H_5^+ \rightarrow C_2H_3^+ + H_2 \tag{32b}$$

is further indicated by the absolute intensities of the product ions. At 4.2 eV the absolute intensity of C$_2$H$_3$$^+$ is more than twice the intensity at 0.3 eV, while the C$_2$H$_5$$^+$ has decreased drastically.

D. Other Hydrocarbon Systems

The majority of early mass spectrometric studies of ion–molecule reactions were concerned with hydrocarbons. In addition, it is for certain hydrocarbon systems that the radiation chemical techniques for investigating these reactions (which were outlined in the Experimental Section) have proved to be particularly informative. Data derived from this combination of sources has resulted in the characterization of several different categories of ion–molecule reactions which are important in hydrocarbons.

The earliest major ion–molecule reaction type to be identified was the hydride transfer reaction,

$$X^+ + C_nH_{2n+2} \rightarrow XH + C_nH_{2n+1}^+ \tag{74}$$

(where X^+ represents an alkyl ion). Such reactions were established by the mass spectrometric studies of Field and Lampe (335). More recently, significant studies of the hydride transfer reaction have been reported by Ausloos and co-workers who employed the experimental techniques of radiation chemistry. Lias and Ausloos (336), for example, devised a simple competition method to determine relative rates of these reactions. The method is based on the competition between the reactions,

$$C_2D_5^+ + C_3D_8 \xrightarrow{k_A} C_2D_6 + C_3D_7^+ \tag{75}$$

$$C_2D_5^+ + XH_2 \xrightarrow{k_B} C_2D_5H + XH^+ \tag{75a}$$

where XH_2 represents the perprotonated hydrocarbon added to C_3D_8, the latter also serving as the source of the $C_2D_5^+$ ions. From such a reaction scheme, there follows the kinetic expression,

$$\frac{k_B}{k_A} = \left(\frac{C_2D_5H}{C_2D_{6corr}}\right)\left(\frac{C_3D_8}{XH_2}\right) \tag{76}$$

The C_2D_6 yield here must be corrected for the contribution from the unimolecular decomposition of excited C_3D_8 molecules. Similar studies of hydride transfer reactions for $sec\text{-}C_3D_7^+$ ions and $sec\text{-}C_4D_9^+$ ions with various hydrocarbons including several cyclic alkanes have also been accomplished (137,337). These relative rate data may be converted to an absolute basis using the rate constant value of 8×10^{-10} cc molecule^{-1} sec^{-1} obtained for the specific hydride transfer reaction,

$$C_2H_5^+ + (CH_3)_4C \rightarrow C_2H_6 + C_5H_{11}^+ \tag{77}$$

Kinetic data obtained from these studies indicates that the rate of the H^- transfer reaction for a given ion generally increases with an increase of the collision cross section (which is proportional to $(\alpha/\mu)^{1/2}$, see Section II A). However, it was observed in all cases that the reaction rate increases faster with increase in molecular weight of XH_2 than does the collision cross section. This may be taken as an indication that reaction does not occur at every collision, even though all the cases of H^- transfer reactions considered in these studies are exothermic. It would appear, however, that the degree of exothermicity determines to some extent the probability of reaction. The reactions of $C_2D_5^+$ ions, to which H^-

transfer is more exothermic than to either $sec\text{-}C_3D_7{}^+$ or $sec\text{-}C_4D_9{}^+$ ions, show a much closer correspondence between relative reaction rates and collision cross sections than is observed for the latter ions.

It should be noted that mass spectrometric studies of hydride ion transfer reactions do not support the trends in reactivity which are exhibited by the radiolytic data. In fact, Field and Lampe (335) reported an appreciable decrease in "effective cross sections" for these reactions with increasing molecular weight, a trend also observed in another recent investigation (338). The study of Munson and co-workers (185) suggests a less pronounced dependence for these reactions. This discrepancy between radiolytic and mass spectrometric observations possibly results from the fact that kinetic data derived in the latter studies are based on the detection of the $C_nH_{2n+1}^+$ species which could also be formed by reactions other than hydride transfer.

Radiolytic studies of ion-molecule reactions are particularly useful for establishing the structure of the reacting ion and possibly of the reaction complex, topics about which mass spectrometric studies can provide less information. Thus Ausloos and co-workers (137,337) have shown that for ethyl, propyl, and butyl ions, the relative cross sections of the hydride transfer reactions are independent of the origin of the carbonium ion. These observations suggest that the reacting ions have the same structure regardless of their origin. Still more convincing is the isotopic evidence reported by these authors. Structure analysis of the C_3D_7H and C_4D_9H formed respectively in the reactions,

$$C_3D_7{}^+ + XH_2 \rightarrow C_3D_7H + XH^+ \qquad (78)$$

$$C_4D_9{}^+ + XH_2 \rightarrow C_4D_9H + XH^+ \qquad (79)$$

showed that these products consisted almost entirely of CD_3CDHCD_3 and $CD_3CDHCD_2CD_3$. It follows from these observations that mainly $sec\text{-}C_3D_7{}^+$ and $sec\text{-}C_4D_9{}^+$ ions are involved in the H^- transfer reactions. Therefore, H or D atom rearrangements to achieve the most thermodynamically stable configurations can occur readily in these carbonium ions. No carbon skeleton rearrangements appear to take place, however.

Information concerning the reaction complex formed in hydride transfer reactions is provided by a radiolysis study of butane. It was observed (339) that some 90% of the propane formed in the radiolysis of $(CH_3)_3CD$ consists of propane-d_2. Since propane is formed mainly by a hydride transfer reaction to the propyl ion in this system, this observation requires the reaction sequence,

$$(CH_3)_3CD^+ \rightarrow C_3H_6D^+ + CH_3 \qquad (80)$$

$$C_3H_6D^+ + (CH_3)_3CD \rightarrow C_3H_6D_2 + C_4H_9{}^+ \qquad (81)$$

However, if the complex formed in reaction (81) were subject to extensive rearrangement, the statistically favored product would be C_3H_7D. Thus, apparently, no randomization of hydrogens in the intermediate complex takes place in this case. Similar conclusions may be drawn from observation of the propane product produced in the radiolysis of $CD_3CH_2CH_2$-CD_3 which is almost entirely propane-d_3 (340).

From the observation that the ion pair yields of products whose precursors are the reactant ions (for example $C_4D_{10} + C_4D_9H$ for the $C_4D_9{}^+$ ion) are effectively constant for all C_nD_{2n+2}–XH_2 mixtures, it has been concluded that hydride ion transfer is usually the only mode of reaction between an alkyl ion and a neutral alkane or cyclo-alkane (137). An exception to this is the case for which XH_2 is cyclopropane. In this instance, the expected hydride transfer product is negligible even though transfer of an H^- ion from cyclopropane is exothermic. However, additional evidence indicates that here there is an alternative mode of reaction which competes with hydride transfer. A similar competition with hydride transfer is noted in the case where alkyl ions are reacted with unsaturated hydrocarbons, where proton transfer and condensation reactions are possible alternatives.

There is also evidence to indicate that H^- transfer occurs from alkanes to smaller olefinic ions (331,341),

$$C_nH_m{}^+ + XH \rightarrow C_nH_{m+1} + X^+ \qquad (82)$$

and to carbonium ions having the formula $C_nH_{2n-1}^+$ (308),

$$C_nH_{2n-1} + XH \rightarrow C_nH_{2n} + X^+ \qquad (83)$$

Here again, hydride transfer is in competition with other modes of reaction. Examination of H^- transfer to olefinic ions by radiolytic techniques is difficult because the neutral product in this case is a free radical. Some success has been realized in such studies by using H_2S as a radical scavenger (331). From such experiments it has been found that the probability of H^- transfer from various alkanes to $C_3D_6{}^+$ is strongly dependent upon the alkane structure. Hydride transfer appears to be favored when the H^- donor is a branched alkane containing one or more tertiary H atoms.

Another type of reaction which has importance for some hydrocarbon systems involves transfer of a proton from certain donor ions to hydrocarbon molecules. As already mentioned protonated methane was one of the first ion–molecule reaction products identified mass spectrometrically (20), and many workers have subsequently studied this species (see Section IV-C). Proton transfer from ions such as $H_3{}^+$ and CHO^+ to alkanes, followed by decomposition of the protonated alkane was first demonstrated to occur by Lindholm and co-workers (219,220) in experiments conducted with a tandem mass spectrometer. The occurrence of proton transfer reactions from CH_5^+ and H_3^+ to higher alkanes was established at about the same time by analysis of the products formed in the radiolysis of CH_4 or H_2 to which trace amounts of deuterated higher alkanes had been added (342–344). Thus upon addition of $n\text{-}C_5D_{12}$ to CH_4, evidence was obtained for the production of $C_5D_{12}H^+$, followed by the decomposition reactions,

$$C_5D_{12}H^+ \rightarrow CD_3H + C_4D_9^+ \tag{84}$$

$$\rightarrow C_2D_5H + C_3D_7^+ \tag{84a}$$

$$\rightarrow C_3D_7H + C_2D_5^+ \tag{84b}$$

The alkyl ions were assumed to undergo deuteride transfer reactions with the pentane to yield the respective perdeutero-alkanes. The observed equivalence of the neutral products which would result from such a reaction sequence was considered to corroborate the proposed scheme. Similar data were obtained with hydrogen mixtures. It is significant that in the radiolysis of a $H_2\text{-}n\text{-}C_5D_{12}$ mixture, more than 95% of the methane product was CD_3H, which suggests that the protonated precursor ion must have had the structure $(CD_3CD_2CD_2CD_2CD_3H^+)$. It appears, therefore, that very little rearrangement of hydrogen atoms occurs in the protonated alkane. Also of interest is the observation that proton transfer to cyclopropane results mainly in the formation of a stable $C_3H_7^+$ ion (345). On the basis of ion interceptor studies Ausloos and Lias (345), assert that this ion acquires the secondary structure $(CH_3CHCH_3^+)$ prior to reaction with the interceptor molecule. Thus these authors conclude that if a protonated cyclopropane ring structure does exist, as Rylander and Meyerson suggested (346), its lifetime must be relatively short. In the same study it was also shown that proton transfer to ethylene, propylene, and 2-butene leads mainly to $C_2H_5^+$, $sec\text{-}C_3H_7^+$ and $sec\text{-}C_4H_9^+$, respectively.

Another interesting application of radiolysis techniques to the investigation of ion–molecule reactions was accomplished by Lawrence and Firestone who established the occurrence of a proton transfer chain reaction. Thus the reaction,

$$CH_{5-i}D_i^+ + CD_4 \rightarrow CH_{4-i}D_i + CD_4H^+ \tag{85}$$

was shown to take place in the radiolysis of $D_2\text{-}CH_4$ mixtures. These authors suggested that in all cases the protonated alkane decomposes shortly after its formation. It was also demonstrated that over the temperature range -78 to $25°C$, there is no thermal activation energy for reaction (85). Lawrence and Firestone (347,348) also observed chain character for the ammonium ion–ammonia reaction sequence. No evidence was found, however, for proton transfer from $C_2H_7^+$ to C_2H_6 by these investigators.

On the other hand, in a high pressure mass spectrometric investigation of proton transfer reactions, Aquilanti and Volpi (349), have presented evidence for a reaction between $C_2H_7^+$ and C_2H_6. In addition, the reaction sequence suggested by their data,

$$H_3^+ + C_2H_6 \rightarrow (C_2H_7^+)^* + H_2 \tag{86}$$

$$(C_2H_7^+)^* \rightarrow C_2H_5^+ + H_2 \tag{86a}$$

$$(C_2H_7^+)^* + M \rightarrow C_2H_7^+ + M \tag{86b}$$

indicates that the excited protonated ion can be deactivated by collision. These workers obtained a value of 3×10^{16} molecule/cm³ for the ratio k_{86a}/k_{86b}, and from this arrived at a lifetime of 10^{-7} sec for $C_2H_7^+$. The ions $C_3H_9^+$ and $C_4H_{11}^+$ were not observed from appropriate proton transfer reactions at pressures below 0.3 torr, indicating shorter lifetimes for these species.

As already mentioned in previous sections of this review, the concept of collisional deactivation of ion–molecule reaction products seems quite well established. Such a notion is further supported in the case of proton transfer reactions by the fact that, at pressures above 100 torr, there are virtually no differences between the decomposition modes of the protonated alkane produced by H_3^+ and by CH_5^+ even though the H_3^+ reaction produces a protonated ion which has about 2 eV more energy.

Other investigations of proton transfer reactions from carbonium ions to larger olefins or other organic compounds have been mainly devoted to reactions of ethyl and *sec*-propyl ions. Thus, it was observed that ethyl ions undergo proton transfer to polar compounds such as CH_3OH, CH_3NO_2, CH_3OCH_3, and $(CH_3)_2N_2$ (350). The rate of the proton transfer reaction for these compounds is 10–100 times that of the competing hydride transfer reaction. In another study by Ausloos et al. (337), evidence was presented for the reaction,

$$C_3H_7^+ + X \rightarrow C_3H_6 + XH^+ \tag{87}$$

where X represents CH_3NO_2, CH_3OH; CH_3OCH_3, 1-pentene or 4-methyl-*cis*-pentene. In cases where X is an olefin, hydride transfer and condensation reactions with carbonium ions compete effectively with proton transfer. Proton transfer has also been observed in a number of other organic systems including alcohols, ketones, and amines. Many of these reactions are tabulated in the next section of this review.

In the above discussions of hydride and proton transfer reactions, it has been noted that for the case where an unsaturated hydrocarbon ion reacts with a saturated hydrocarbon molecule a competion exists between several possible reactions. One such reaction which is observed in many of these systems is the H_2^- transfer reaction, which is of the general type,

$$C_nH_m^+ + XH_2 \rightarrow C_nH_{m+2} + X^+ \tag{88}$$

and which is the class of hydrocarbon reactions which we will consider at this point. The first reaction of this type which was reported was observed in propane by several investigators (185,308,351).

$$C_2D_4^+ + C_3D_8 \rightarrow C_2D_6 + C_3D_6^+ \tag{88a}$$

In more recent studies, Ausloos and co-workers observed that the fragmentation of cycloalkane ions usually yields large amounts of olefinic ions. They subsequently utilized the radiolysis of cycloalkanes to produce such ions for the measurement of relative rates of H_2^- transfer reactions (137,337). Their most extensive study involved the reactions of $C_3D_6^+$ ions with various hydrocarbons for which the rate of H_2^- transfer reaction,

$$C_3D_6^+ + XH_2 \rightarrow C_3D_6H_2 + X^+ \tag{89}$$

was obtained relative to the reaction,

$$C_3D_6^+ + C_5D_{10} \rightarrow C_3D_8 + C_5D_8^+ \tag{89a}$$

using the kinetic relationship,

$$\frac{k_{89}}{k_{89a}} = \left(\frac{C_3D_6H_2}{C_3D_8}\right)\left(\frac{C_5D_{10}}{XH_2}\right) \tag{90}$$

The relative rate constants derived in this manner indicate that there is a gradual increase in the rate of the H_2^- transfer with increasing molecular weights of the neutral molecule when this is a normal alkane. However, a strong dependence of the rate of this reaction upon structure was observed for different isomers of a particular saturated hydrocarbon, the rate appearing to decrease with increasing degree of branching of the hydrocarbon. But this decrease in the H_2^- transfer rate is generally balanced by an increase in the rate of the competing H^- transfer reaction so that the total reactivity for different isomers is effectively constant. These observations were confirmed by a recent mass spectrometric study of these reactions by Sieck and Futrell (341). Information concerning H_2^- transfer reactions has also been derived from photoionization of C_3D_6–AH_2 mixtures (352), and from radiolytic studies of cyclo-C_5H_{10}–cyclo-C_5D_{10} mixtures to which free radical scavenger was added (331).

It has been generally concluded from all of the investigations mentioned above that the H_2^- transfer reaction involves only a loosely bound complex and that no rearrangement or exchange of H atoms occurs during the lifetime of the reaction complex. In addition, the H_2^- transfer reaction is highly stereospecific in nature, as demonstrated by the fact that the terminal hydrogen atom of n-butane or isobutane is transferred almost exclusively to the center atom of $C_3D_6^+$. It was also established that the reaction involves exclusively the transfer of hydrogen atoms located on two adjacent carbon atoms of the XH_2 molecule.

A new type of ion–molecule which may be designated as an H_2 transfer reaction has recently been observed for certain hydrocarbon systems. It may be written generally as,

$$C_nH_m + XH_2^+ \rightarrow C_nH_{m+2} + X^+ \tag{91a}$$

where C_nH_m is cyclopropane or an unsaturated hydrocarbon with fewer carbon atoms than the parent alkane or cycloalkane ion, XH_2^+. A reaction which may be expected to compete with (92) is the H atom transfer reaction,

$$C_nH_m + XH_2^+ \rightarrow C_nH_{m+1} + XH^+ \tag{91b}$$

Ausloos and co-workers have observed H_2 transfer reactions in both the vapor phase (333,353) and in the condensed phase (354), using radiolytic techniques. Quite recently, a mass spectrometric investigation of H and H_2 transfer reactions has been reported by Abramson and Futrell (355) for a large number of hydrocarbon ions. Relative rates for H_2 transfer from cyclohexane ions which were determined in both the radiolytic and the mass spectrometric experiments were in close agreement, indicating that there is a clear correlation between ion–molecule studies carried out by these two different techniques. As in most of the other transfer reactions just discussed, the collision complex formed in the H and H_2 transfer reactions would appear to permit no randomization of hydrogens. These reactions demonstrate a marked stereospecificity. It was shown in the mass spectrometric study that for butane ions, the hydrogens are removed from adjacent carbon atoms and are lost preferentially from the 2,3 positions in the H_2 transfer. Also radiolytic studies showed that n-butane formed in cyclohexane containing 1-butene-d_8 consisted exclusively of $CD_2HCDHCD_2CD_3$.

It is important to note that the relative probabilities for H_2 transfer in the gas phase differ appreciably from those found in the liquid phase studies. It must therefore be assumed that the probability of H_2 transfer relative to that of H transfer varies with the density of the medium. This in turn suggests that collisional deactivation of the reaction complex may be an important factor for these reactions.

The last general type of ion–molecule reactions for hydrocarbons which will be considered includes those reactions which may be loosely termed condensation reactions. This category comprises those reactions in which substantial rearrangement occurs and which therefore presumably involve the formation of a strongly bound complex. As we have mentioned previously in considering the reaction complex (Section II-B), this entity may decompose to form fragments or may be collisionally stabilized, depending on the conditions of the experiment. A collisionally stabilized complex may react further and sequential condensation products have been observed for several hydrocarbons, particularly in unsaturated systems (22,26,49,99,324). Radiolytic techniques are of little value in the study of condensation reactions since the origin of high molecular weight ionic species cannot readily be traced from neutral product analysis. Mass spectrometric studies, especially at elevated pressures, can, however, provide much useful information about these processes and about the nature of the reaction complexes involved. Much of the data pertinent to this topic has already been discussed in this review and an extensive tabulation of observed condensation reactions appears in the last section.

V. ENERGETICS OF ION–MOLECULE REACTIONS

Energetic considerations for ion–molecule interactions have been a subject of great interest since the earliest investigations of these reactions. This section will cover some of the more important topics of this nature which have been considered. No thermochemical data of significance for ion–molecule reactions are listed in this review since extensive compilations of such data are provided in other sources (18,171–176).

A. Energy Effects for Exothermic Reactions

Energy effects in exothermic reactions have already been discussed in connection with the hydrogen, rare gas–hydrogen, and methane systems. In addition, translational energy effects have been rather extensively discussed in several recent articles (76,143,117,201). Accordingly the coverage here will be limited to certain additional points of interest.

The temperature dependence of several ion–molecule reactions observed in the mass spectrometer was examined in early studies by Stevenson and Schissler (21,153,177) who found negative temperature coefficients for the rate constants. It was therefore concluded that there is effectively no activation energy for ion–molecule reactions. Unfortunately, since temperature effect studies have been accomplished for comparatively few ion–molecule reactions, it is clear that the absence of an activation energy for all such reactions is by no means established. In fact, in one instance (but not involving a mass spectrometric observation), an activation energy of 7.4 kcal/mole has been reported for the reaction (178)

$$N_2^+ + O_2 \rightarrow NO^+ + NO \tag{92}$$

Moreover, several authors (1,2,5,179,180) have noted that reactions having activation energies greater than about 5 kcal/mole would be too slow to be observable in a mass spectrometer. This conclusion was based on consideration of the pressures commonly employed in ion sources and on the geometry and sensitivity of typical mass spectrometers. Consequently, it was generally conceded that only exothermic (or at least, athermic) ion–molecule reactions could be observed in a mass spectrometer (20,21,56, 179,180,181). If this criterion that $\Delta H \leq 0$ for all reactions observed is accepted, then it is possible to calculate upper limits to the heats of formation of product ions by the use of Hess's Law of constant heat summation. That is,

$$\Delta H_f \text{ (product ion)} \leq \sum \Delta H_f \text{ (reactants)} - \sum \Delta H_f \text{ (neutral products)} \quad (93)$$

Some workers, carrying this line of reasoning still further, postulated that the requirement of exothermicity was not only a "necessary" but also a "sufficient" condition for the observation of a given reaction (20,182,183). Several objections may be raised to such a postulate (180), the most significant being that in many cases plausible exothermic reactions are simply not observed in mass spectrometric investigations (2,90,184, 185), while in other instances exothermic reactions are observed to exhibit comparatively small cross sections (68,90,202). If these reactions are viewed as taking place via an intermediate collision complex, this presumably indicates either that the complex is dissociating into the original reactants or that there are alternate reaction channels available.

As a consequence of recent theoretical developments (41,44) it is no longer assumed that all exothermic processes will necessarily occur. Accordingly several investigators have adopted the notion of competing alternate channels of reaction to explain the nonobservance or observance with small cross section of certain exothermic ion–molecule reactions. Where energetically possible charge transfer is a likely competing process. For example, Tal'roze (7) has observed charge transfer with thermal Ar^+ ions,

$$Ar^+ + H_2 \rightarrow Ar + H_2^+ \quad (94)$$

in competition with

$$Ar^+ + H_2 \rightarrow ArH^+ + H \quad (24)$$

Wilmenius and Lindholm (203) assert, on the basis of their studies of the reactions of a large number of ions with methanol, that the energetically allowed hydride ion transfer is actually of importance only when charge transfer is not possible. Whenever the recombination energy of the incident ion is greater than the ionization potential of the molecule charge transfer is the dominant reaction.

These observations of Wilmenius and Lindholm (203) are taken by Giese (67) as being consistent with the idea of a resonance potential, as previously mentioned. The importance of resonance forces in ion–molecule reactions was initially suggested by Durup (2) and was recently reconsidered by Giese (67). This approach considers the classical inter-action potential to be inaccurate at the ranges appropriate to observed collisions and suggests that a better approximation to the true potential is obtained by adding to the classical potential the quantum mechanical resonance potential. Such a treatment results in the prediction of enhanced reaction rates as the recombination energy of the ion approaches the ionization potential of the molecule. For values of ion recombination energy exceeding the ionization potential of the molecule, the reaction will be impeded. These considerations also offer a possible explanation for trends observed in the reaction series,

$$HD^+ + X \rightarrow XH^+ + D \tag{68a}$$

$$\rightarrow XD^+ + H \tag{68b}$$

where X is He, Ne, Ar, Kr, and Xe (142). Here the total cross sections are in order of magnitude, $Xe \ll He < Ne < Ar, Kr$. The small cross sections for He and Ne are understandable since these reactions are endothermic, but the falloff for Xe is almost certainly not explained in this manner. It is suggested, therefore, that the explanation perhaps lies in the fact that $RE(HD^+) < I(Ar)$ or $I(Kr)$ but $RE(HD^+) > I(Xe)$. Shannon and Harrison (202) have obtained qualitative agreement with the predictions of the resonance hypothesis in their study of concurrent D atom and D^+ ion transfer reactions occurring in several systems. Their results suggest that the interaction potential for these reactions is other than the $1/r^4$ potential. The resonance approach, of course, is based on highly speculative arguments and its application must be regarded as limited to the realm of qualitative predictions.

Another explanation for the failure to observe certain exothermic reactions is that the expected product is formed with internal energy exceeding its dissociation energy and decomposes prior to detection. The work of several investigators has suggested this possibility (162,182,184, 185,202–204). Thus, observation of the deuteron transfer reaction (162)

$$D_2^+ + CH_3CN \rightarrow CH_3CND^+ + D \tag{95}$$

with a very low cross section may be explained on this basis since the calculated exothermicity exceeds the energy required for decomposition of CH_3CND^+ by either of two paths. The fact that CH_3CND^+ is observed at all indicates that this ion can contain energy in excess of the minimum required for dissociation or that the neutral D atom carries off a part of the exothermicity as kinetic energy. Volpi and co-workers (101,205–208) have recently reported the results of an extensive group of experiments dealing with proton and deuteron transfer reactions to several hydrocarbons and alcohols. They concluded that protonated ions were formed in excited states with lifetimes of the order of 10^{-5} to 10^{-9} sec, depending on the species, which then decompose. Several exothermic dissociative channels appear to be available and evidence was obtained that the excited intermediates could be collisionally deactivated at higher source pressures. It was noted that the distribution of reaction exothermicities is important in determining which channels are populated. In one of these investigations (208), the effect of changing the ion energy was determined. The results suggested that kinetic-to-internal energy transfer can occur, thereby increasing the total energy available to the system and favoring dissociative reaction channels.

The partition of the energy of reaction between the internal and translational modes of the products is one of the most important problems in the study of reactive collisions. Until quite recently, very few experimental measurements bearing on this point were available. A number of techniques have been developed for the measurement of ion kinetic energies and are now being used to answer this problem. One such method which was devised by Reese and Hipple (209) and Berry (146) has been applied to ion–molecule reaction product ions by Tal'roze (56) and by Durup (210). In this method, the mass spectrometer is modified by placing a pair of deflecting electrodes behind the second slit with the plane of the electrodes perpendicular to the long dimension of the ion slit system. It is apparent that the components of initial velocity of ions directed along the length of the slit will produce a spread in the ion beam in this direction in the analyzing region. If the length of the collector slit is decreased appreciably, this slit will transmit only that portion of the ions entering the analyzer which have a small velocity component along the length of the slit system. However, by applying suitable potentials to the deflecting electrodes, a certain fraction of the ions having larger velocity components in this direction will be collected. From a measurement of the collected ion current as a function of deflecting voltage the distribution function for one component of the initial kinetic energy is obtained. This method is expected to give quantitatively reliable data for initial kinetic energies less than about six times thermal energy. This method was used by Tal'roze (56) to study several reactions, including

$$H_2O^+ + H_2O \rightarrow H_3O^+ + OH \tag{34}$$

$$NH_3^+ + NH_3 \rightarrow NH_4^+ + NH_2 \tag{33}$$

Figure 17 shows the observed velocity distribution curve for H_3O^+ product ions and thermal energy H_2O^+ reactant ions. The dotted line in the figure represents the distribution of ions which would result if all of the reaction energy is converted into kinetic energy of the products. These results therefore indicate that the reaction exothermicity appears mainly as internal energy. In general, Durup's observations (210) are also in agreement with this conclusion. This evidence led Tal'roze (7,56) to postulate, for such cases, a comparatively long-lived collision complex with equipartition of the exothermicity into all degrees of freedom.

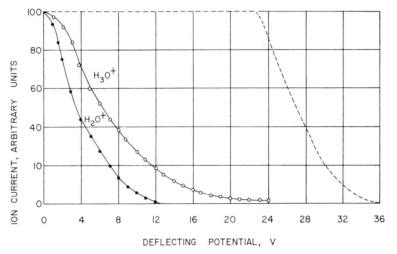

Fig. 17. Distribution of H_2O^+ and H_3O^+ ions with respect to velocity component
(V. L. Tal'roze and E. L. Frankevich, ref. 56.)

Another method which can be applied to the determination of ion kinetic energies is the retarding potential technique (211). A variation of this method, in which the retarding potential is applied to the ion repeller in the mass spectrometer, has been used by Friedman and co-workers (212,213) to estimate the kinetic energies of ion–molecule reaction products. This technique depends upon the fact that for each applied retarding potential only ions with sufficient kinetic energy to overcome the repeller field can leave the source and be focused at the collector. Relative velocity distributions for ions can thus be obtained. The errors inherent in this method may be considerable, owing to accelerating field penetration into the source region and possible surface charge effects. Reuben and Friedman (212) applied this method to isotopic hydrogen ion-molecule reactions and observed identical kinetic energy distributions for reactant and product ions. From this they concluded that most of the reaction exother-

micity goes into internal energy of the product, assuming that the theoretically calculated heats of reaction were not in error. The inverted repeller
technique has also been applied to the investigation of the helium–
hydrogen system (213).

Another example for which enlightening studies of product ion kinetic
energies were made is the He^+/O_2 system, for which Moran and Friedman
have measured the kinetic energy distribution of the O^+ product (214).
In this case experiments with both ^4He and ^3He isotopes were conducted and the repeller-retarding voltage curves obtained are shown in
Fig. 18. Contributions from O^+ produced by electron impact on O_2 have
been subtracted from these curves. The data obtained from these studies
indicate a smaller average value of kinetic energy deposited in O^+ in
reactions of ^3He with O_2. If the reaction mechanism were one of resonant
charge transfer followed by dissociation, the $^3He^+$ and $^4He^+$ isotopes

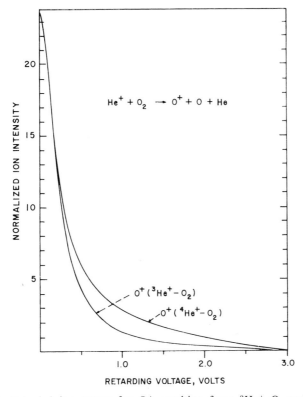

Fig. 18. Retarded ion curves for O^+ resulting from $^3He^+–O_2$ and $^4He^+–O_2$
interactions. Normalized ion intensities are plotted as a function of retarding voltage.
(T. F. Moran and L. Friedman, ref. 214.)

would deposit almost identical amounts of energy in O_2. The O_2^+ transient species could dissociate into O^+ and O in their respective ground states with the liberation of 2.93 eV kinetic energy. The difference between the O^+ kinetic energy distribution obtained with $^3He^+$ and $^4He^+$ is taken as strong evidence for a mechanism which proceeds via $^4HeO^+$ with $^3HeO^+$ decomposing and leaving behind a lower velocity O^+.

With the development of more specialized apparatus, it has become possible to obtain much more reliable measurements of product ion kinetic energies for ion-neutral interactions. Some laboratories have now utilized electrostatic velocity selectors for energy analysis of the product ion beam. Bailey and co-workers (144,170,190) have used a 127° electrostatic velocity selector in their apparatus to measure kinetic energies of product ions. A brief description of this apparatus and the significance of some of the experimental results obtained have been presented in other sections. It seems clear that studies such as these, which involve determination of both kinetic energy and angular distributions of product ions, are of the greatest significance for the determination of the mechanisms of reactive collisions and the ultimate solution of the problem of energy partition.

Dissociative charge transfer reactions comprise yet another class of exothermic reactions for which energy effects have been extensively investigated. While the role of primary ion kinetic energy in these reactions is not fully understood, it appears that the cross-sections are relatively independent of translational energy. The cross-section dependence typically exhibited is illustrated by the $C_2H_2^+$ curve shown in Fig. 19. With increasing energy endothermic processes appear, as indicated by this figure. It is commonly observed for these exothermic charge transfer reactions that the cross sections rise in the low energy region (< 10 eV). This may be attributed to a change in mechanism. At lower energies, charge transfer occurs to some extent through an intermediate complex and is governed by the principles applicable to complex formation and subsequent decomposition. At higher ion energies (> 10 eV), the "slow-ion" theory is inadequate to describe the process. In this case, long-range electron transfer predominates and intimate collision complexes are unimportant.

In a lengthy series of investigations, (107,194,203,215–225), Lindholm and co-workers have investigated dissociative charge transfer processes for a large number of molecules. It is generally assumed in these studies that ion kinetic energy is not a significant factor and that the ion transfers to the target molecule only its recombination energy (that is, the difference between the total energy of the initial ion, which may be excited, and the neutralized ion, which may also be excited). Because of uncertainties with

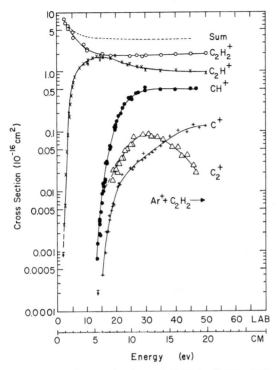

Fig. 19. Cross sections for production of $C_2H_2^+$, C_2H^+, CH^+, C^+, and C_2^+ from the reaction of Ar^+ with C_2H_2 as a function of Ar^+ kinetic energy. (W. B. Maier II, ref. 199.)

regard to excited states, the recombination energy can assume several values in many cases. These workers were, nevertheless, able to extract rather detailed fragmentation patterns from the cross section measurements for a number of compounds. Comparison of results obtained by this technique with data from electron and photon impact studies has been made in several cases (203,217,219,220), and reasonably quantitative agreement is obtained.

Giese and Maier, (74,198,199) have criticized the Lindholm experiments for failing to consider the role of kinetic energy, noting that this is negligible only at very low velocities of the incident ion. However, as already discussed, the perpendicular geometry of Lindholm's apparatus effectively discriminates against momentum transfer processes, while the in-line configuration transmits the products of these reactions. For the Xe^+/C_2H_4 system, for example, the reaction,

$$Xe^+ + C_2H_4 \rightarrow Xe + CH_2^+ + CH_2 \tag{96}$$

is endothermic by 5.0 eV. While both Maier (199) and Tal'roze (195) observed fairly large intensities of CH_2^+ from this reaction, Szabo (223) using the Stockholm apparatus, observed only a very small intensity which decreases smoothly toward zero with decreasing velocity. Lindholm therefore concluded that Szabo's spectra can readily be extrapolated to zero velocity and that kinetic energy need not be considered in his experiments.

A number of other investigators have also examined kinetic energy effects for exothermic charge transfer reactions, (102,158,171,227–231) and these reactions have recently been reviewed by Stebbings (197).

B. Endothermic Reactions

The overwhelming majority of ion–molecule reactions which have been detected mass spectrometrically have proved to be exothermic, in agreement with the considerations just presented. Recently, however, several reactions which are endothermic if one considers only the participation of ground state species have been reported. For these cases it appears that the participation of either reactant ion translational energy or excitation energy, or both, are effective in promoting the reaction. Herzberg and co-workers (186), for example, observed the endothermic process,

$$H_2^+ + He \rightarrow HeH^+ + H \qquad \Delta H = +1.1\ eV \qquad (69b)$$

and suggested that it may proceed at the expense of the translational energy. A more detailed study of this reaction, as well as the corresponding reaction with neon,

$$H_2^+ + Ne \rightarrow NeH^+ + H \qquad \Delta H = +0.6\ eV \qquad (69c)$$

has been made by Friedman and co-workers (150,151). They conclude that while there is a pronounced dependence of reaction cross section upon kinetic energy at low ion velocities, the behavior at higher velocities indicates that the heat of reaction is supplied largely by internal energy of H_2^+. An analysis of the vibrational energy levels of H_2^+ and consideration of the energy requirements of the reaction indicates that, for the He reaction, only H_2^+ ions in the fifth vibrational state and above react. Similarly, only those in the second vibrational state and above for the Ne reaction are sufficiently energetic to react. These results support the Polanyi assertion that internal energy is much more effective than translational energy (187) in promoting endothermic reactions.

In another instance, Refaey and Chupka (188) have observed the endothermic reactions,

$$O_2^+ + H_2 \rightarrow O^+ + H_2O \qquad \Delta H = +1.6\ eV \qquad (97)$$

$$O_2^+ + H_2 \rightarrow OH^+ + OH \qquad \Delta H = +2.1\ eV \qquad (98)$$

at higher ion velocities, (40–360 eV). It was indicated for these reactions that the O_2^+ ion was mainly in the electronic ground state. The cross sections observed, however, were considerably lower than the geometric cross sections, being of the order of 0.2 $Å^2$ or less, again suggesting that the utilization of translational energy for overcoming the reaction potential barrier is relatively inefficient for simple systems.

It is interesting to note that the reactions just discussed exhibit certain aspects common to those observed for endothermic charge transfer. Thus the O_2^+/H_2 reactions cited show a cross section dependence on kinetic energy which passes through a maximum, while the H_2^+–He and H_2^+–Ne systems exhibit kinetic energy thresholds. It was suggested (15), that the thresholds observed for the latter may occur because of a competing reaction, that of enforced dipole radiation, which could occur with a sufficiently long-lived collision complex. That is, induced-radiation-deactivation of the vibrational levels of H_2^+ could result because of the perturbation created by formation of the complex. The significant effect of kinetic energy in this case, then, is to determine the time scale for a reactive collision. It should also be noted that the phase space theory of Light and co-workers (42,43), can successfully rationalize the observed energy threshold for these reactions. This theory, however, accounts for the kinetic energy dependence of the reactions on the basis of the conversion of the energy of relative motion into internal energy to permit reaction of H_2^+ ions in low vibrational states.

Some very recent investigations made possible by the use of sophisticated beam techniques shed additional light on the role of translational energy in endothermic ion-molecule reactions involving the transfer of heavy particles. In one such study, Maier (189) has investigated the energy dependence of the cross section for the endothermic reaction,

$$C^+ + D_2 \rightarrow CD^+ + D \tag{99}$$

which involves transfer of a deuterium atom. The observed behavior is shown in Fig. 20 which indicates a rather sharp threshold for the reaction. The sensitivity of the reaction to kinetic energy is evident from the fact that the cross-section attains 80% of its maximum value at approximately 0.6 eV (on the E_{CM} scale) above the apparent threshold. Moreover, the cross-section for this endothermic process is fairly large. Again, the use of the Light theory (42) enables one to fit the experimental data quite well, as the curve drawn through the experimental points in Fig. 20 shows.

In addition to the endothermic rearrangement and atom transfer reactions mentioned above, a number of endothermic charge transfer and dissociative charge transfer reactions have now been observed in the mass spectrometer. The energy dependence for some of these has been examined

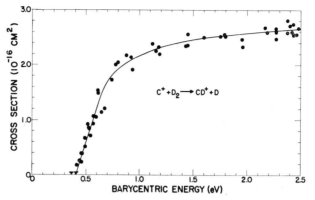

Fig. 20. Cross section for $C^+ + D_2 \rightarrow CD^+ + D$ as a function of the barycentric energy of the reactants. (W. B. Maier II, ref. 189.) Solid circles represent experimental data points; curve represents that obtained by fitting the data using the theory of ref. 42.

in detail. Tal'roze (195), measured the cross section for the reaction,

$$Xe^+ + C_2H_4 \rightarrow Xe + CH_2^+ + CH_2 \tag{96}$$

as a function of kinetic energy. He observed an abrupt onset of the reaction at a threshold energy of about 40 eV (E_{lab}). The cross section rises sharply with increasing energy to a plateau region. He also observed comparatively large cross sections for a variety of other endothermic charge transfer reactions. These results suggested that kinetic energy is transferred into internal energy in these processes.

A consideration of the magnitude of the parameters of the Massey adiabatic hypothesis (see Section II-D) for some observed endothermic charge transfer processes indicates that for the low velocity threshold region, very small cross sections are expected. This is obviously not in agreement with Tal'roze's results. He postulated that the reason for disagreement with the adiabatic hypothesis may be the crossing of the potential levels in the system (195). It is expected that such crossings would be common for complex polyatomic systems because of the large number of energy levels. Therefore, resonance restrictions should be of little consequence for charge exchange in complex systems. Similar failures of the Massey criterion had been observed for comparatively simple systems (196), even prior to the observations of Tal'roze. Stebbings (197) has discussed some of the problems encountered in attempting to apply the Massey theory. A most serious complication arises in the assignment of the energy defect, since one or more of the interacting particles may be in an unknown excited state.

Experimental evidence suggests that non-resonance charge transfer processes occurring at low energies proceed by a more intimate encounter than that to which the "long-range" theory is applicable. Support for this view was provided by the comprehensive studies of Giese and Maier (109,198) and of Maier (74,199) who studied dissociative ionization processes of N_2, CO, and N_2O, as well as some hydrocarbons, induced by impact of rare gas ions. All of these reactions exhibited sharp thresholds in the kinetic energy dependence of the reaction cross sections. The energy dependence is shown in Fig. 20 for several dissociative processes occurring in the Ar^+/C_2H_2 system. The cross sections for these reactions are in the range of geometrical cross sections, suggesting that the interconversion of translational and electronic energy proceeds quite efficiently.

It is possible, of course, that the kinetic energy absorbed in these dissociative endothermic processes could be used in breaking a bond in the target particle. However, Maier (199) has also studied the endothermic pure charge transfer reaction,

$$Kr^+ + D_2 \rightarrow D_2^+ + Kr \qquad (68c)$$

and observes both the same threshold behavior and a cross section of comparable magnitude. Obviously, this reaction cannot proceed unless the kinetic energy of the colliding particles is by some means converted into electronic energy. The cross section in the vicinity of threshold is described by the equation

$$\sigma_D = B(E + \Delta E)^n E^{-1/2} \qquad (100)$$

from which the energy defect, ΔE, may readily be deduced.

In a recently reported study (113), of the endothermic dissociative charge transfer reaction,

$$N_2^+ + H_2 \rightarrow N_2 + H + H^+ \qquad (101)$$

Vance and Bailey found that Maier's relationship described their data quite accurately. This is illustrated by the solid line in the lower curve of Fig. 21, which is obviously a good fit to the experimental points. The upper curve in Fig. 21 shows the energy dependence of the cross section for the endothermic charge transfer reaction,

$$N_2^+ + H_2 \rightarrow N_2 + H_2^+ \qquad (102)$$

exhibiting the characteristic fall off toward a threshold at lower kinetic energies. It was noted by these investigators that the theory of Rapp and Francis (69) for nonsymmetric, nonresonant charge-transfer processes of

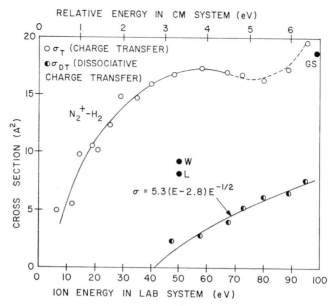

Fig. 21. Charge transfer (σ_T) and dissociative charge transfer (σ_{DT}) cross sections for the $N_2{}^+$–H_2 system. (D. W. Vance and T. L. Bailey, ref. 113.)

this type predicts a cross section of resonant form at high energies, rising to a maximum at the point where (see Section II-D)

$$v \approx \frac{P_0 \, \Delta E}{h} \tag{103}$$

This theory would predict a peak in the cross section at about 70 eV for this $N_2{}^+/H_2$ reaction, and the data in Fig. 21 suggest the existence of a peak in this neighborhood.

Another endothermic charge transfer process which has been observed (200) is

$$N_2{}^+ + CO \rightarrow N_2 + CO^+(A^2\Pi) \tag{104}$$

Although this excitation reaction is endothermic by 1 eV, the cross section was found to vary by less than a factor of two when the kinetic energy in the barycentric system was decreased from 100 to 8 eV. This suggests an efficient conversion of translational energy to excitation energy just above threshold.

It appears to be well established that the cross section of these endothermic processes is very much dependent upon kinetic energy. For endothermic charge transfer there will be some threshold kinetic energy below which the reaction is not possible. This critical ion kinetic energy is that energy, expressed in barycentric units, which equals the energy defect of the reaction. Although many problems still exist with regard to the assignment of energy defects and with the application of existing theories to these endothermic reactions, these studies offer considerable promise for yielding much information on ion and neutral energetics.

C. Participation of Excited Species in Ion–Molecule Reactions

A considerable amount of data has now accumulated which provides evidence for the participation of excited species in ion–molecule reactions. Since the majority of such reactions have been investigated by mass spectrometric techniques, in which the primary ions are produced by impact of comparatively high energy, it is plausible that many primary ions react in excited ionic states. In addition, excited atoms or molecules may be produced and undergo inelastic collisions with other species, resulting in the production of ions. Two processes of this type are of particular interest: the so-called "Hornbeck-Molnar" process of associative ionization

$$X^* + Y \rightarrow XY^+ + e^- \tag{105}$$

and its nonassociative counterpart, Penning ionization,

$$X^* + Y \rightarrow X + Y^+ + e^- \tag{106}$$

The excitation energy may be electronic, vibrational or rotational. Electronically excited ions may be metastable, those in excited states from which a radiative transition to the ground state is forbidden, or simply long-lived, those in states from which radiation is allowed, but which nevertheless have comparatively long lifetimes. The latter situation may occur when the energy separation between ground and excited state is small or whenever, because of the Franck-Condon principle, a vertical transition cannot readily occur. Metastable states have mean lifetimes which range from a millisecond to many seconds, while the range of mean lifetimes for long-lived states is of the order of a microsecond to hundreds of microseconds. Mass spectrometric techniques typically exhibit an instrumental lifetime for removal of a reactant species—either by a reactive collision or by extraction from the chamber—of the order of a few microseconds. Thus excited ionic states with a lifetime of a microsecond or more, including radiative states, may be important for ion-molecule reactions studied by these methods.

Investigations accomplished to date on excited ion–molecule reactions point to the participation of many states of the reactant ion in the collision process. In some cases the contribution of excited ions is estimated from the known energetics for a given reaction by taking into account the estimated relative populations of energy levels of the reactant ion. As we have already noted, Friedman and co-workers (150,151) were able to explain the observed cross sections for the H_2^+/He and H_2^+/Ne reactions by invoking the participation of vibrationally excited H_2^+ ions.

In the case of charge transfer reactions, experimental data which are not in agreement with theory have sometimes been interpreted as suggesting the involvement of excited species. By considering the energy defect for individual excited states, Hasted and co-workers (232) determined cross sections for several charge transfer reactions in the energy region 0.06–10 eV. They found that the observed cross sections for the processes,

$$Ne^+ + Ar \rightarrow Ne + Ar^+ \tag{107}$$

and

$$Ar^+ + Ne \rightarrow Ar + Ne^+ \tag{108}$$

were in good agreement with the Rapp and Francis theory if appropriate combinations of excited states were assumed. Further, Hertel and Koski (168) found it necessary to postulate the formation of excited product ions in the reaction of He^+ ions with NH_3 in order to account for discrepancies between the observed results and those predicted by the Massey criterion. The role played by vibrational excitation in charge transfer is also indicated by another study by Berta and Koski (32) of the kinetic energy dependence of the reaction

$$HD^+ + Ar \rightarrow Ar^+ + HD \tag{69d}$$

They found the dependence to exhibit several maxima which approximated the spacing of vibrational levels in H_2^+. This behavior is interpreted in terms of the Massey criterion.

Several techniques have been developed to provide more direct experimental evidence for excited species in ion–neutral reactions. Lindholm's techniques have already been discussed. It will be recalled that in his experiments the reactant ion is varied, thereby effectively changing the energy transferred to the molecule. Participation of excited reactants and products is often evident from structure of these curves.

Another method involves the observation of the reaction cross-section dependence upon the properties of a given reactant ion beam. The properties of the beam are varied by changing the ion velocity at fixed electron energy or by varying electron energy while ion velocity is held constant. For both types of experiments marked changes in the values of cross sections are often observed. It therefore appears that observed cross sections are total cross sections summed over a number of states of the reactant ion. Thus for the generalized case of a charge transfer reaction,

$$A^{+*} + B \rightarrow A + B^+ \pm E \qquad (109)$$

the total cross section associated with all states of the reactant ion beam can be expressed as,

$$Q_t(A^+) = \sum_i Q_i(P_i / \sum_i P_i) \qquad (110)$$

where Q_i represents the cross section for a particular state of A^+ and the term in brackets is the relative population of this state in the ion beam.

In a strict sense the Q_i's are themselves composite cross sections, since the states of the product ion and neutral particle must also be considered. For the moment, however, we will consider the simpler case where the

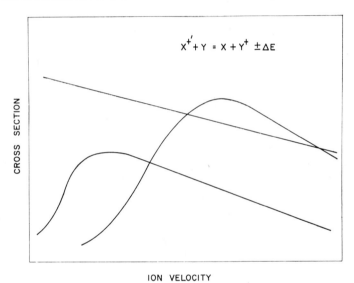

Fig. 22. Charge exchange cross sections associated with three different states of the reaction ion. (J. W. McGowan, ref. 233.)

products are formed only in the ground state. McGowan (233) has illustrated the cross-section dependence which might be observed for each of three states of the reactant ion in such a hypothetical situation. This is illustrated in Fig. 22. Here the ground state ion in the charge transfer has a resonance form (linear plot), while the first excited state shows a maximum at low velocities and the second excited state has a maximum at a somewhat larger velocity. To indicate the influence of the excited states on the total cross section, we show in Fig. 23 several of the latter curves which can be observed when different proportions of excited ions are present in the reactant beam (233). The resonant curve (linear), is obtained when the reactant ions are all in the ground electronic state. This would be the case for electron energies below the appearance potential of the first excited state. The other curves shown are for several different relative proportions of ground, first and second excited states. With increasing amounts of excited ions, structure associated with these states becomes increasingly evident. From such structure in the total cross section curves, several authors have argued for the existence of different excited states in their reactant ion beams (194,196).

It is clear then that the population of states in the ion beam is a definite function of the energy of the ionizing electrons. This has been established in several studies which show the influence of vibrational states of $H_2{}^+$

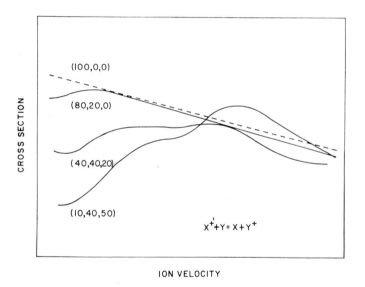

Fig. 23. Total cross section curves when the partial curves of Fig. 22 are combined in different proportions. Proportions of the ground, first and second excited states, respectively, are indicated in brackets. (J. W. McGowan, ref. 233.)

in certain reactions of this ion (234,235). The technique of obtaining cross sections for reactions as a function of electron energy at constant ion velocity has also been recently utilized by Turner, Stebbings, and Rutherford to study several charge transfer reactions between nitrogen and oxygen ions and between these same ions and nitric oxide (229,230). From the observed cross-section dependences, and from the known composition of the beam as determined in other experiments, partial cross sections for reactions of specific ionic states may be determined.

A direct experimental method for determining the concentration of excited ions in an ion beam has been reported by Turner (236). The apparatus permits a mass analyzed beam of ions to be directed into a gas-filled attenuation chamber. The loss of ions from the beam is measured as the beam passes through the gas at low pressure. A beam composed entirely of ground-state ions passing through a fixed path-length of gas at pressure, P, will be attenuated due to various processes. The fraction of ions reaching a collector at the end of the path will decrease exponentially with increasing gas pressure. For a single state of the ions, a semilogarithmic plot of the fraction of ions reaching the collector versus gas pressure should then be a straight line. Ions in an excited state will have a different, typically larger, reaction cross section and would thus give a line with a different slope. An ion beam containing both ground and excited state ions would then give a composite curve and by extrapolation of the plot to zero pressure, the fraction of ions in the excited state can be determined. Plots for such an experiment are shown for O_2^+ ions in N_2 in Fig. 24.

In a study of long-lived excited ions of oxygen and nitrogen, McGowan and Kerwin have also observed a marked dependence of cross sections upon the population of these excited species in the ion beam. The $a^4\Pi_u$

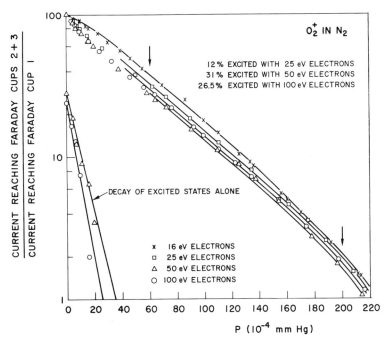

Fig. 24. Attenuation of an O_2^+ ion beam in N_2. (B. R. Turner, ref. 236.)

and $b^4\Sigma_g$ excited states of O_2^+ were identified in the reaction,

$$O_2^+(^4\Pi_u, {}^4\Sigma_g) + O_2 \rightarrow O^+ + O + O_2 \tag{111}$$

and thresholds for several higher states of O_2^+ were obtained.

Yet another technique for the investigation of excited ions in ion–neutral reactions, and the one which has been most widely used, is that of examining the structure of ionization efficiency curves. Space limitations do not permit a discussion here of the techniques for measurement of these curves or the problems associated with such measurements. These topics are covered at some length in other sources (237,238). It is sufficient here to note that the major difficulty for electron impact studies is one of obtaining a monoenergetic electron beam. It is only for such a beam that the structure exhibited in ionization efficiency curves is sufficiently detailed to permit association of this structure with specific excited states (239). Even without special efforts to eliminate the spread of electron energies, some inferences about excited species can be drawn from appearance potential measurements.

The manner in which appearance potential measurements can be utilized for such purposes is typified by the investigation of N_3^+ produced in the reaction,

$$N_2^{+*} + N_2 \rightarrow N_3^+ + N \tag{112}$$

by Kaul and Fuchs (240). They found an appearance potential of N_3^+ of about 20 eV. Since this is well above the ionization potential of N_2^+ but below the appearance potential of N^+ and since the N_3^+ ion exhibited a definite velocity dependence in the source, it was concluded that excited N_2^+ ions must be involved. Čermák and Herman (241) later noted that the N_2^+ excited state $^4\Pi_u$ crosses the Franck-Condon region close to the energy corresponding to the appearance potential of N_3^+ (21.1 $\pm$ 0.1 eV). They therefore associated N_3^+ production with this long-lived state of the nitrogen molecule ion. This observation was confirmed in other recent experiments (242,243).

In a similar fashion, Čermák and Herman (244) measured appearance potentials for a large number of ion–molecule product ions and concluded that specific excited states of reactant ions were involved in these reactions. Dong and Cottin made similar studies of some of these systems (245). Henglein has used appearance potential measurements to obtain information about the reactions of excited ions in carbon disulfide and in various aromatic compounds (246). Munson and co-workers (247) employed similar measurements to investigate the formation of N_3^+ and other ions resulting from the reaction of electronically excited nitrogen ions with various rare gases. Munson (248) also found that appearance potentials suggested the participation of excited $C_2H_2^+$ in the reactions of acetylene. Melton and Hamill have applied the retarding potential difference technique to determine appearance potentials for excited ion–neutral reactions in the same system (249). The reactions of $C_2H_2^+$ and $C_2H_2^{+*}$ are discussed in more detail elsewhere (48).

More elaborate experimental methods have been utilized in some laboratories to investigate threshold behavior of electron impact ionization. Thus McGowan and Fineman (250) have utilized an electron spectrometer with energy resolution of 0.05 eV to investigate vibrational and rotational structure for H_2^+ and H^+. Other similar studies have also been made by McGowan and co-workers (251,252). In another illustration of the use of monoenergetic electron beams, the cross section for the reaction,

$$H_2^{+*} + H_2 \rightarrow H_3^+ + H \tag{113}$$

was shown to have a maximum for ions in the ground vibrational state, (147). Curiously, the cross section was found to decrease with increasing excitation of H_2^+ and the reaction was not observed for H_2^+ in states above the third excited state. This trend may be due to increased probability of a back-reaction of the complex as its excitation energy is increased.

A novel electron capture "indicator" method reported by Henglein (253) is probably generally applicable to the study of excitation and ionization processes in which slow electrons are produced. In this technique, compounds with high electron affinities—CH_3I and SF_6 in this example—were used to detect excited levels in SO_2, CS_2, and Xe. Thus the reaction sequence,

$$SO_2 + e^-(\text{fast}) \rightarrow SO_2^* + e_0^-(\text{slow}) \tag{114}$$

$$e_0^-(\text{slow}) + SF_6, SO_2 \rightarrow SF_6^-, SO_2^- \tag{115}$$

for example, was used to identify 3 levels of SO_2^* between 2 and 8 eV. For Xe, the onsets of Xe*, Xe^+, Xe^{2+}, and Xe^{3+} formation were reflected in the SF_6^- signal.

Čermák (254) has recently applied the technique of Penning ionization to detect long-lived excited states of neutral molecules. The Penning ionization reaction is of the type,

$$X^* + YZ \rightarrow X + YZ^+ + e^- \tag{116}$$

In general, ionization occurs when the energy of excited particle X* is greater than or equals the ionization potential of molecule YZ. If a series of molecules is chosen with increasing ionization potentials, then YZ serves as a "detector" for excited states of X*. The energy of a particular excited state is then given by IP $(M_1) \leq E < $ IP (M_2), where M_2 is a detector molecule which is not ionized and M_1 is a molecule with a lower IP which can be ionized. Long-lived excited states in several diatomic and polyatomic molecules have been identified in this manner (255,256). In later experiments, the kinetic energy of electrons released in the Penning ionization reaction was also measured to yield additional information.

For a system of considerable aeronomic interest, the reaction of ions with vibrationally excited molecules has been studied (257). Vibrational excitation of N_2 in the O^+/N_2 system resulted in an enhancement of the rate of the reaction

$$O^+ + N_2 \rightarrow NO^+ + N \tag{117}$$

Ion kinetic energy also has a pronounced effect on the cross section of this reaction. This behavior is not fully understood, but several factors which may be significant have been discussed in a review by Giese (152).

In Table III a list is given of some of the excited ion–molecule reactions which have been observed. A more extensive compilation of reactions of

Table III
Excited Ion–Molecule Reactions Investigated

Ion	Probable excited state	Reaction	References
Ar^+	$^2P_{1/2}$	$Ar^{+*} + N_2 \rightarrow Ar + N_2^+$	268
Ar^+	Near Ar^{2+} threshold	$Ar^{+*} + Ar \rightarrow Ar^{2+} + Ar + e$	274
Xe^+	$^2P_{1/2}$	$Xe^{+*} + CH_4 \rightarrow Xe + CH_4^+$	275, 228, 161, 195
Kr^+	$^2P_{1/2}$	$Kr^{+*} + Kr \rightarrow Kr + Kr^+$	271
Kr^+	$^2P_{1/2}$	$Kr^{+*} + CH_4 \rightarrow Kr + CH_4^+$	107, 276
Kr^+	$^2P_{1/2}$	$Kr^{+*} + CH_4 \rightarrow Kr + CH_3^+ + H$	275
CS_2^+	$^2\Pi_{3/2u}$	$CS_2^{+*} + CS_2 \rightarrow CS_3^+ + CS$	246, 266, 277
CS_2^+		$CS_2^{+*} + CS_2 \rightarrow C_2S_2^+ + S_2$	246
CS_2^+		$CS_2^{+*} + CS_2 \rightarrow C_2S_3^+ + S$	246
CS_2^+		$CS_2^{+*} + H_2O \rightarrow H_2OS^+ + CS$	246
CS_2^+		$CS_2^{+*} + I_2 \rightarrow CSI^+ + SI$	246
CS_2^+		$CS_2^{+*} + I_2 \rightarrow SI^+ + CSI$	246
H_2^+	$^1\Sigma_g^+$ (range of v)	Several reactions	246
H_2^+	$^1\Sigma_g, v_0$	$H_2^{+*} + He \rightarrow HeH^+ + H$	151
H_2^+	$^1\Sigma_g$, all v	$H_2^{+*} + H_2 \rightarrow H_3^+ + H$	147
H_2^+	$^1\Sigma_g$, all v	$H_2^{+*} + H_2 \rightarrow H^+ + H + H_2$	264, 234
N^+	^3P	$N^{+*} + O_2 \rightarrow N + O_2^+$	229
N^+	^1D	$N^{+*} + N_2 \rightarrow N + N_2^+$	233
N^+	^1D	$N^{+*} + Kr \rightarrow N + Kr^+$	196
N^+	^1S	From dissociative charge transfer	196, 219

(continued)

Table III (*continued*)

Ion	Probable excited state	Reaction	References
N^+	^3P	$N^{+*} + NO \rightarrow N + NO^+$	230
N_2^+	$^4\Sigma_u^+$	$N_2^{+*} + Ar \rightarrow ArN^+ + N$	240
N_2^+	$^4\Sigma_u^+$	$N_2^{+*} + Kr \rightarrow KrN^+ + N$	240
N_2^+	$A^2\Pi_u$	Several reactions	233, 229
N_2^+	$^4\Sigma_u^+$	$N_2^{+*} + N_2 \rightarrow N_3^+ + N$	240, 266, 267, 247
N_2^+	$X, v = 0, 1, 2$	$N_2^{+*} + Ar \rightarrow N_2 + Ar^+$	268
I_2^+		$I_2^{+*} + H_2O \rightarrow H_2OI^+ + I$	269
O_2^+	$A^2\Pi_u; b^4\Sigma_g$	$O_2^{+*} + O_2 \rightarrow O_3^+ + O$	246, 270
O_2^+	$A^2\Pi_u$	$O_2^{+*} + NO \rightarrow O_2 + NO^+$	230
O_2^+	$a^4\Pi_u$	$O_2^{+*} + D_2 \rightarrow O_2D^+ + D$	271
O_2^+	$b^4\Sigma_g; a^4\Pi_u$	$O_2^{+*} + N_2 \rightarrow O_2 + N_2^+$	233
O_2^+	$a^4\Pi_u; b^4\Sigma_g$	$O_2^{+*} + O_2 \rightarrow O^+ + O + O_2$	233
O_2^+	Higher states	$O_2^{+*} + O_2 \rightarrow O^+ + O^- + O_2^+$	233
O^+	$^2P, ^2D$	From dissociative charge transfer	194, 272
O^+	^2S	$O^{+*} + Xe \rightarrow O + Xe^+$	219
O^+	^2D	$O^{+*} + O_2 \rightarrow O + O_2^+$	233, 229
O^+	^2D	$O^{+*} + N_2 \rightarrow O + N_2^+$	229
C^+	^4P	Several reactions	194, 217
CO^+		$CO^{+*} + CO \rightarrow C^+ + O + CO$	273
CO^+		$CO^{+*} + CO \rightarrow C_2O^+ + O$	247
$C_6H_5CN^+$		$C_6H_5CN^{+*} + C_6H_5CN \rightarrow C_{12}H_9CN^+ + HCN$	246
$C_6H_5Cl^+$		$C_6H_5Cl^{+*} + C_6H_5Cl \rightarrow C_{12}H_9^+ + HCl + Cl$	246
$C_6H_5Br^+$		$C_6H_5Br^{+*} + C_6H_5Br \rightarrow C_{12}H_{10}^+ + Br_2$	246

excited species present in the upper atmosphere has been given by McGowan (258).

D. Thermochemical Quantities of Significance to Ion–Molecule Reaction Studies

In single source mass spectrometric experiments, one of the principal means of establishing the reaction mechanism involves a comparison of appearance potentials for reactant and product ions. Moreover, appearance potentials and the structure exhibited by ionization efficiency curves often provide indications of the participation of excited ions in ion–neutral reactions. If the reaction producing a given ion does not form excited products, the appearance potential is effectively the heat of the reaction. With this assumption heats of formation of ions can be obtained from such data by the usual thermochemical methods. Thus for a given reaction, induced by electron impact,

$$R_1R_2 + e^- \rightarrow R_1^+ + R_2 + 2e^- \tag{118}$$

one can write the relation,

$$A(R_1^+) = \Delta H \text{ reaction} = \Delta H_f(R_1^+) + \Delta H_f(R_2) - \Delta H_f(R_1R_2) \tag{119}$$

Appearance potentials have also been used by many investigators to obtain bond strengths and radical ionization potentials by means of the relationship,

$$A(R_1^+) = I(R_1) + D(R_1 - R_2) \tag{120}$$

which is applicable to such processes. These quantities are of signal significance for assessing the energetics of ion–molecule reactions and their often determinative influence on reaction pathway. A comprehensive and authoritative compilation of ionization potentials and heats of formation for selected ions can be found in references (171,172). Such data have been extensively reported in the literature over the past several years (278–287,295–303).

Obviously with ionic heats of formation available, it is possible to determine the heat of reaction for a given ion–neutral interaction. To a very good approximation, reactions observed in conventional mass spectrometers for low velocity ions may be considered either exothermic or athermic, that is, $\Delta H \leq 0$. Therefore, if the ion-neutral reaction producing the secondary ion can be established and if the heats of formation of reactants and neutral products are known, upper limits to the heats of formation of product ions may be calculated by using Hess's Law of constant heat summation. As we have pointed out, the use of ground state ion energetics in such calculations is but a first approximation, and the possibility that both reactants and products may contain excitation energy should not be excluded.

Another useful thermochemical quantity which can be evaluated from ion–molecule reactions is the proton affinity. In some cases this can be determined directly by observation of the reaction,

$$B + H^+ \rightarrow BH^+ \tag{121}$$

The proton affinity is, by definition, the negative heat of this reaction,

$$P(B) = -\Delta H \text{ reaction} = D(B - H^+) \tag{122}$$

that is,

$$P(B) = -[\Delta H_f(BH^+) - \Delta H_f(H^+) - \Delta H_f(B)] \tag{123}$$

Recently, Munson (259) obtained relative proton affinities indirectly for some polar molecules by observing proton exchange reactions between molecules. The order of such proton affinities can be readily determined from such experiments. If the proton transfer reaction,

$$BH^+ + C \rightarrow CH^+ + B \tag{124}$$

occurs, then $P(C) > P(B)$. In this manner, he was able to establish the order, $P(H_2O) < P(CH_3OH) < P(CH_3OCH_3); P(HCHO) < P(CH_3CHO) < P(CH_3COCH_3); P(CHOOH) < P(CH_3COOH); P(NH_3) < P(CH_3NH_2) < P(CH_3)_2NH < P(CH_3)_3N$. Several authors have applied the relation given above to calculate proton affinities from heat of formation data (290,293). Such calculations are subject to the same limitations which apply to heats of formation determined in analogous experiments. Proton affinities have also been calculated from quantum mechanical considerations for simple hydrides (288,291). Some proton affinities have been estimated from thermodynamic cycle calculations (171,292,294).

Giese and Maier (74,198,199) have pioneered the study of endothermic dissociative charge transfer reactions for determining bond energies. A sharp threshold is observed for processes in which the translational energy of the impacting ion is used to drive a reaction. The amount of energy E_r which is available for reaction with a neutral particle of mass M_n is related to the laboratory energy of the impacting ion of mass M_i by the expression,

$$E_r = E_i\left(\frac{M_n}{M_n + M_i}\right) \tag{125}$$

At threshold then this energy should correspond to just that energy necessary to overcome this endothermicity of the dissociative charge transfer reaction. This is equated to the appropriate bond energy of the dissociating species. The threshold energies observed by Giese and Maier for several molecules, including CO, N_2, and H_2O, produced excellent agreement with accepted bond energy values for these systems. This suggests that this technique may be as valuable as electron or photon impact for the determination of molecular properties. Moreover, the precision of the determination, because of the mass effect shown in Eq. (125), is much greater for a given energy spread in the ionizing medium.

Another interesting application of ion–molecule reactions for the evaluation of thermodynamic quantities is that reported by Kebarle and co-workers (260–262). These workers have utilized a high pressure mass spectrometer to observe ion clusters such as $NH_4^+ \cdot mNH_3$ ($m = 1$–20) and $H_3O^+ \cdot nH_2O$ ($n = 1$–8). Equilibrium constants were calculated on the assumption that these clusters represent the equilibrium concentration. From the variation of cluster distribution with temperature, enthalpies and entropies of solvation for individual solvation steps were evaluated. This technique offers considerable promise for investigating ion–solvent molecule interactions. A similar method has been applied to $O_4{}^+$ ion clusters by Yang and Conway (263) and to hydrates of ions formed in mobility experiments by Saporoshenko (129,131).

VI. TABLES OF ION–MOLECULE REACTIONS AND KINETIC DATA

In the following tables, we have considered six of the most important categories of ion–molecule reactions. Examples of each type are listed along with rate constants or cross section measurements for these reactions for which such data was reported. We have endeavored to include most reactions which have been investigated subsequent to earlier reviews. We have emphasized the most recent kinetic data since much of the information obtained in the very early stages of ion–molecule reaction investigations is now suspect. The type of experiment which was employed in studying each reaction is also cited according to the following code: MS, single source mass spectrometer; MT, tandem mass spectrometer; MP, mass spectrometer pulse technique; MCRS, multichambered radiolysis source; B, beam technique; CB, cross beam technique; CH, Čermák-Herman experiment; F, flame studies; A, flowing afterglow system; D, discharge studies; R, radiation chemistry techniques. In cases where data are reported from several sources for a given reaction, all experiments are listed on a single line, the individual entries being separated by semicolons and listed in the same order under each heading. A dash appearing under a specific heading for a given reaction in the Tables signifies no available information. Additional comments are appended as footnotes following Table IX.

Comments on Tables

1. Relative rate reported.
2. Structure of C_4D_9H product found to be $CD_3CD_2CDHCD_3$.
3. Structure of C_3D_7H product found to be CD_3CDHCD_3.
4. Structure of $C_3D_6H_2$ found to be CD_3CDHCD_2H.
5. Reaction cross section measured as function of ion translational energy.
6. FS = 10 V/cm.
7. Reaction of $CH_2{}^+$ also postulated.
8. FS = 6.5 V/cm.
9. FS = 4 V/cm.
10. FS = 12.5 V/cm.
11. Rate constant reported at FS = 12.6 V/cm; k also measured over FS range 5–126 V/cm.
12. Products were also observed which result from dissociation of the proton transfer product.

13. No proton transfer adduct was observed, only dissociation products were detected.
14. Relative cross section reported.
15. The first rate constant listed is for thermal energy ions; the second rate constant is for ions reacting at FS = 10.5 V/cm.
16. Harrison and co-workers have recently improved the accuracy of the pressure measurements in their experiments with a resulting revision in their rate data for several reactions. The value of the rate constant cited in this instance has not been corrected in this manner and can be considered to be approximately one-half the true value.
17. Rate constant reported for thermal ions.
18. Reactant ion considered to be in excited state.
19. Considered to be dissociative proton transfer by investigators.
20. Rate constant given here for repeller potential of 6 V. Rate constant data reported over repeller potential range of 4–12 V.
21. Both ion–atom interchange and charge transfer observed for this reaction by use of isotopic measurements.
22. These data obtained for ions of 3.4 eV ion exit energy.
23. Cross-section and rate constant reported for FS = 10.5 V/cm.
24. Products mass and energy analyzed.
25. Effect of vibrationally excited molecules evaluated.

Table IV

Positive Atomic Ion and Atom Transfer Reactions

Reaction	X	Type of experiment	Rate constant ($\times 10^{10}$ cc molecule^{-1} sec^{-1})	Cross section, Q ($\times 10^{16}$ cm^2 molecule^{-1})	Ref.	Comments[a]
$H_2^+ + X \rightarrow XH^+ + H$	H_2	MS; MP; B; MT	19.4;(519,13.3); —;—	—;—;—;—	212; 149; 113; 109	11;(15, 16); (5, 24); 5
	He	MS; MT	—;—	—;—	151; 109	5; 5
	Ne	MS	—	—	150	5
	Ar	B	—	—	169	5
	NH_3	MCRS	11.0	34.0	92	—
$D_2^+ + X \rightarrow XD^+ + D$	D_2	MS; MP	13.9; (5.8, 15.0)	—;—	212; 149	11; (15, 16)
	Ar	MS; MT	—;—	68.5; —	322; 109	—; 5
	HCN	MP	30.5	47.7	322	23
	CH_3CN	MS	—	9.9	322	—
	N_2	MT	—	—	109	5
	HCl	MP	75.2	117.4	322	23
	CO_2	MS	—	46.0 ± 9	322	—
	CO	MS	—	74.6	322	—
$D_2^+ + XH \rightarrow D_2H^+ + X$	NH_2	MS	0.3	—	305	—
	PH_2	MS	0.3	—	305	—
	AsH_2	MS	0.3	—	305	—
	Cl	MP	25.5	39.9	322	23
$HD^+ + X \rightarrow XH^+ + D$	HD	MP	3.8, 8.4	—	149	15, 16
	Ar	MT	—	—	32	5
$HD^+ + X \rightarrow XD^+ + H$	HD	MP	2.9, 6.9	—	149	15, 16
	Ar	MT	—	—	32	5

Reaction	X					
$H_3^+ + X \rightarrow XH^+ + H_2$	Xe	MS	10.0	—	100	9
	Kr	MS	<0.03	—	100	9
	Ne	MS	<0.03	—	100	9
	He	MS	<0.03	—	100	9
	N_2	MS	10.0	—	100	9
	CH_4	MS	7.5	—	101	12
	C_2H_2	MS	10.8	—	205	—
	C_2H_4	MS	9.9	—	205	12
	C_2H_6	MS	8.3	—	101	12
	C_3H_4	MS	12.1	—	205	12
	C_3H_6	MS	8.7	—	205	12
	$c\text{-}C_3H_6$	MS	9.7	—	205	12
	CH_3OH	MS	15.6	—	206	9, 12
	C_2H_5OH	MS	15.6	—	206	9, 12
	C_3H_7OH	MS	15.6	—	206	9, 12
	$i\text{-}C_3H_7OH$	MS	15.6	—	206	9, 12
$H_3^+ + X \rightarrow$ Products	C_3H_8	MS	9.7	—	101	13
	C_4H_{10}	MS	10.5	—	101	13
	$i\text{-}C_4H_{10}$	MS	10.1	—	101	13
$D_3^+ + X \rightarrow XD^+ + D_2$	CH_4	MS	6.9	—	101	12
	C_2H_2	MS	7.8	—	205	—
	C_2H_4	MS	6.8	—	205	12
	C_2H_6	MS	7.5	—	101	12
	C_3H_4	MS	9.1	—	205	12
	C_3H_6	MS	6.5	—	205	12
	$c\text{-}C_3H_6$	MS	6.3	—	205	12
	CH_3OH	MS	11.0	—	206	9, 12
	C_2H_5OH	MS	11.0	—	206	9, 12
	C_3H_7OH	MS	11.0	—	206	9, 12

(continued)

Table IV (*continued*)

Reaction	X	Type of experiment	Rate constant ($\times 10^{10}$ cc molecule^{-1} sec^{-1})	Cross section, Q ($\times 10^{16}$ cm^2 molecule^{-1})	Ref.	Comments[a]
$D_3^+ + X \rightarrow$ Products	C_3H_8	MS	8.6	—	101	13
	C_4H_{10}	MS	10.0	—	101	13
	$i\text{-}C_4H_{10}$	MS	9.5	—	101	13
$C^+ + X-D \rightarrow CD^+ + X$	D	MT	—	—	189	5
$CH^+ + X \rightarrow XH^+ + H$	H_2	MS	5.8	—	317	10
$CH_2^+ + X-H \rightarrow CH_3^+ + X$	H	MS	2.3	—	317	10
$CH_4^+ + X \rightarrow XH^+ + CH_3$	CH_3	MS; MT	—; —	15.0; —	159, 304	6; 7
	CH_4	MS; CH; MT; MP	—; —; —; (12.2, 11.7); (9.6, 10.7)	32.0; —; —; —; —	159; 158; 304; 312; 78	6; —; —; 15; 15
	H_2O	CH	—	—	158	—
$CH_4^+ + X-D \rightarrow CH_3D^+ + X$	D	MS	0.03	—	317	10
$CH_4^+ + X-H \rightarrow CH_5^+ + X$	H	MS	0–0.07	—	317	10
$CH_3D^+ + X \rightarrow XD^+ + CH_3$	CH_3D^+	MP	2.3, 3.0	—	312	15
$CH_3D^+ + X-H \rightarrow CH_4D^+ + X$	CH_2D	MP	7.8, 10.2	—	312	15
$CH_2D_2^+ + X \rightarrow XH^+ + CHD_2$	CH_2D_2	MP	4.1, 7.3	—	312	15
$CH_2D_2^+ + X \rightarrow XD^+ + CH_2D$	CH_2D_2	MP	3.8, 5.9	—	312	15
$CHD_3^+ + X \rightarrow XH^+ + CD_3$	CHD_3	MP	2.9, 3.9	—	312	15
$CHD_3^+ + X \rightarrow XD^+ + CHD_2$	CHD_3	MP	5.9, 8.6	—	312	15

Reaction	X	Method				Ref.
$CD_4^+ + X \to XD^+ + CD_3$	CD_4	MP	8.2, 11.0	—	312	15
	CH_3CHO	MP	4.7, 9.0	—	313	15, 16
	CO	MP	2.0, 5.2	—	312	15
	O_2	MS	0.15	—	320	—
	HCN	MP	≤0.4	≤2.0	322	23
$C_2D_2^+ + X \to XD^+ + C_2D$	C_2D_5OH	MT	—	—	204	12
$C_2H_3^+ + X \to XH^+ + C_2H_2$	C_2H_4	MP; MS	(3.6, 3.6); 3.6	—; —	87; 324	(15, 16); 10
$C_2D_3^+ + X \to XD^+ + C_2D_2$	C_2D_5OH	MT	—	—	204	12
$C_2D_5^+ + X \to XD^+ + C_2D_4$	C_2D_5OH	MT	—	—	204	12
$C_2H_6^+ + X \to XH^+ + C_2H_5$	C_2H_6	MS	—	0.3	159	6
$C_3H_7^+ + X \to XH^+ + C_3H_6$	C_3H_7OH	MT	—	—	221	—
$CH_2OH^+ + X \to XH^+ + CH_2O$	CH_3OH	MS; MP	—; (21.0, 16.0)	—; —	206; 312	9; 15
	C_2H_5OH	MS	—	—	206	—
	C_3H_7OH	MT; MS	—; —	—; —	221; 206	—; 9
$CH_2OD^+ + X \to XD^+ + CH_2O$	CH_3OD	MT	—	—	218	—
$CD_2OH^+ + X \to XH^+ + CD_2O$	CH_3OH	MT; MT	14.4, 12.4	—; —	204; 218	—; 12
$CD_2OH^+ + X \to XD^+ + CDOH$	C_2D_5OH	MT	<0.3, 0.5	—	204	1, 12
$CH_3OH^+ + X \to XH^+ + CH_3O$	CD_3OH	MP	25.0, 25.0	—	312	15
$CH_3OH^+ + XH \to CH_3OH_2^+ + X$	CD_3OH	MP	—	—	87	15, 16
$CD_3OH^+ + X \to XD^+ + CD_2OH$	CH_3OH	MT	—	—	312	15
	CD_3O	MT; MT	—; —	—; —	204	—
$CD_3OH^+ + X \to XH^+ + CD_3O$	CH_3OH	MT; MT; MP	—; —; (12.0, 16.5)	—; —; —	204; 218	—; 15
	CD_3OH	MT; MT	—; —; (21.0, 16.0)	—; —	204; 218; 312	—; —; 15
$CH_3OD^+ + X \to XH^+ + CH_2OD$	CH_3OH	MT; MT; MP	—	—	204; 218	—; 15
	CD_3OH	MT	—	—	204; 218; 312	—; —; 15
	CH_3OD	MT	—	—	218	—

(continued)

Table IV (*continued*)

Reaction	X	Type of experiment	Rate constant ($\times 10^{10}$ cc molecule^{-1} sec^{-1})	Cross section, Q ($\times 10^{16}$ cm^2 molecule^{-1})	Ref.	Comments[a]
$CH_3OD^+ + X \rightarrow XD^+ + CH_3O$	CH_3OD	MT	—	—	218	—
$C_2H_4OH^+ + X \rightarrow XH^+ + CH_3CHO$	C_2H_5OH	MS	—	—	206	9
	$i\text{-}C_3H_7OH$	MS	—	—	206	9
$C_2D_4OH^+ + X \rightarrow XH^+ + C_2D_4O$	C_2D_5OH	MT	—	—	204	12, 14
$CH_3OCD_3^+ + X$ $\rightarrow XD^+ + CH_3OCD_2$	CH_3OCD_3	MP	9.4, 8.9	—	312	15
$CH_3CHO^+ + X \rightarrow XH^+ + CH_3CO$	CH_3CHO	MP	12.3, 11.2	—	313	15, 16
$CHO^+ + X \rightarrow XH^+ + CO$	H_2O	F; F	100.0–250.0; —	—; —	139; 311	—; —
$CHO^+ + X \rightarrow XH^+ + CO$	CH_3CHO	MP	8.3, 13.6	—	313	15, 16
$CH_3CN^+ + X \rightarrow XH^+ + CH_2CN$	CH_3CN	MP	19.6, 22.7	—	312	15
$CH_3CN^+ + XD \rightarrow CH_3CND^+ + X$	D	MS	—	4.7	322	—
$CH_3Cl^+ + X \rightarrow XH^+ + CH_2Cl$	CH_3Cl	MP	17.0, 12.1	—	312	15
$CH_2Cl^+ + X \rightarrow XH^+ + CHCl$	CH_3Cl	MP	4.6, 3.2	—	312	15
$(CH_3)_2NH^+ + X$ $\rightarrow XH^+ + (CH_3)_2N$	$(CH_3)_2NH$	MS	6.0 ± 1.0	—	325	8
$C_2D_5OH^+ + X \rightarrow XH^+ + C_2D_5O$	C_2D_5OH	MT	—	—	204	14
$C_2D_5OH^+ + XD$ $\rightarrow C_2D_5OHD^+ + X$	C_2D_4OH	MT	—	—	204	14

Reaction	X	Method				Ref
$C_2H_5OHD^+ + X \to XH^+ + C_2H_5OD$	C_2H_5OH	MT	—	—	204	—
$C_3H_6OH^+ + X \to XH^+ + C_2H_5CHO$	C_3H_7OH	MS, MT	—; —	—; —	206, 221	9; —
$C_3H_6OH^+ + X \to XH^+ + CH_3COCH_3$	$i\text{-}C_3H_7OH$	MS	—	—	206	9
$C_3H_7OH^+ + X \to XH^+ + C_3H_7O$	C_3H_7OH	MT	19.3, 19.1	—	221	—
$(CH_3)_2O^+ + X \to XH^+ + CH_3OCH_2$	$(CH_3)_2CO$	MP	11.5, 10.2	—	312	15
$CH_3OCD_3^+ + X \to XH^+ + CH_2OCD_3$	CH_3OCD_3	MP		—	312	15
$CH_3NHCH_2^+ + X \to XH^+ + C_2H_5N$	$(CH_3)_2NH$	MS	4.0 ± 1.0	—	325	8
$CH_2N^+ + X \to XH^+ + HCN$	$(CH_3)_2NH$	MS	5.0	—	325	8
$CH_3NH_2^+ + X \to XH^+ + CH_2NH_2$	CH_3NH_2	MP	21.0	—	321	22
$CH_2NH_2^+ + X \to XH^+ + CH_3N$	CH_3NH_2	MP; MS	$24.0; 7.0 \pm 1.0$	—; —	321; 325	22; 8
$CH_3N^+ + X \to XH^+ + CH_2N$	CH_3NH_2	MP; MS	$26.0; 9.0 \pm 1.0$	—; —	321; 325	22; 8
$CH_2N^+ + X \to XH^+ + HCN$	CH_3NH_2	MP; MS	$21.0; 9.0 \pm 1.0$	—; —	321; 325	22; 8
$NH_4^+ + X \to XH^+ + NH_3$	$(CH_3)_2NH$	MS	7.0	—	325	8
$NH_3^+ + X \to XH^+ + NH_2$	NH_3	MS; MS; MCRS; MP	—; —; 15.0; (10.0, 14.8)	58.0; —; 51.0; —	94; 305; 92; 312	6; —; —; 15
$NH_3^+ + XD \to NH_3D^+ + X$	D	MS	< 0.1	—	305	—
$NH_2^+ + X \to XH^+ + NH$	NH_3	MP	6.5, 9.7	—	312	15
$N_2H_3^+ + X \to XH^+ + N_2H_2$	N_2H_4	MP	1.0, 3.8	—	87	15, 16
$N_2H_4^+ + X \to XH^+ + N_2H_3$	N_2H_4	MP	1.4, 5.5	—	87	15, 16
$H_3O^+ + X \to XH^+ + H_2O$	C_2H_5OH	MS	—	—	206	9
$H_2O^+ + X \to XH^+ + OH$	H_2O	CH; MP	—; (16.0, 19.6)	—; —	312; 158	—; 15
$H_2O^+ + X \to XH^+ + OH$	CH_4	CH	—	—	158	—
$OH^+ + X \to XH^+ + O$	H_2O	MP	15.3, 19.2	—	312	15
$OD^+ + X \to XD^+ + O$	D_2O	MP	7.3, 8.6	—	312	15

(continued)

Table IV (*continued*)

Reaction	X	Type of experiment	Rate constant ($\times 10^{10}$ cc molecule^{-1} sec^{-1})	Cross section, Q ($\times 10^{16}$ cm^2 molecule^{-1})	Ref.	Comments[a]
$D_2O^+ + X \rightarrow XD^+ + OD$	D_2O	MP	11.9, 11.5	—	312	15
$HS^+ + X \rightarrow XH^+ + S$	H_2S	MP	9.9, 7.2	—	312	15
$H_2S^+ + X \rightarrow XH^+ + HS$	H_2S	MP	7.7, 5.8	—	312	15
$HCN^+ + X \rightarrow XH^+ + CN$	HCN	MP	6.0, 5.6	22.8	322	15
$HCN^+ + XD \rightarrow HCND^+ + X$	D	MP	4.5	18.3	322	23
$HCN^+ + XD \rightarrow HCND^+ + X$	CD_3	MP	11.4, 15.0	61.0	322	15
$HCN^+ + X \rightarrow XH^+ + CN$	CD_4	MP	0.2	0.8	322	23
$HCl^+ + X \rightarrow XH^+ + Cl$	HCl	MP	0.7, 1.5	7.0	322	15
$HCl^+ + XD \rightarrow HClD^+ + X$	D	MP	15.1	72.7	322	23
$HCl^+ + X \rightarrow XH^+ + Cl$	CD_4	MP	4.8	23.3	322	23
$HCl^+ + XD \rightarrow HClD^+ + X$	CD_3	MP	23.9	115.2	322	23
$O_2^+ + XH \rightarrow O_2H^+ + X$	H	MS	0.4	—	100	9
$O_2^+ + X \rightarrow XO^+ + O$	C_2H_2	MS	~0.2	—	320	18
$O^+ + XO \rightarrow O_2^+ + X$	N_2	MS	0.2	1.0	306	8
$O^+ + XN \rightarrow NO^+ + X$	N	MS; MT; A	0.05; —; —	1.0; —; —	309; 315; 257	9; 5; 25
$N_2^+ + XD \rightarrow N_2D^+ + X$	D	(B, MT); MS; (B, CB)	—; 17.0; —	—; —; —	(40, 109); 25; (190, 193)	5; —; (5, 24)
$N_2^+ + XH \rightarrow N_2H^+ + X$	H	B; MS	—; 21.0	—; —	40; 100	5; 8
$N_2^+ + XN \rightarrow N_3^+ + X$	N	MS	—	—	328	18

Reaction						
$N_2^+ + XH \rightarrow N_2H^+ + X$	OH	D	—	115.0	140	—
$N_2H^+ + X \rightarrow XH^+ + N_2$	H_2O	D	—	440.0	140	—
$N^+ + XO \rightarrow NO^+ + X$	N_2	MS	5.5	27.0	306	9
$N^+ + XO \rightarrow NO^+ + X$	O	MS	3.0 ± 1.0	—	309	9
$He^+ + XH \rightarrow HeH^+ + X$	H	MS; MS	<0.04; —	—; —	100; 191	9; —
$Xe^+ + XH \rightarrow XeH^+ + X$	H	MS	<0.2	—	100	9
$Xe^+ + XH \rightarrow XeH^+ + X$	CH_3	MS	<0.38	—	319	10
$Ne^+ + XH \rightarrow NeH^+ + X$	H	MS	<0.08	—	100	9
$Ar^+ + XH \rightarrow ArH^+ + X$	NH_2	MS	<0.2	—	305	—
$Ar^+ + XH \rightarrow ArH^+ + X$	DH_2	MS	<0.2	—	305	—
$Ar^+ + XH \rightarrow ArH^+ + X$	AsH_2	MS	<0.2	—	305	—
$Kr^+ + XH \rightarrow KrH^+ + X$	H	MS; MS	6.0; 5.0	—; —	100; 25	9; —
$Kr^+ + XH \rightarrow KrH^+ + X$	CH_3	MS	0.12	—	318	10
$Ar^+ + XH \rightarrow ArH^+ + X$	H	MS; (B, MT); MS; MS	15.6; —; 16.0; 17.0	—; —; —	305; (39, 143); 100; 25	—; 5; 9; —
$Ar^+ + XD \rightarrow ArD^+ + X$	D	MT; B	—; —	—;	109; (39, 190)	5; (5, 24)
$Ar^+ + XH \rightarrow ArH^+ + X$	CH_3	MS	0.20	—	318	10
$CO^+ + XH \rightarrow COH^+ + X$	H	B	—	—	40	5
$CO^+ + XC \rightarrow C_2O^+ + X$	O	MS	—	—	328	18
$CO^+ + XD \rightarrow COD^+ + X$	D	B; MP; MS	—; 7.2; —	—; 32.7	40; 160; 322	5; 17; —
$CO^+ + XD \rightarrow CDO^+ + X$	CD_3	MP	2.7, 5.6	13.3	149	15
$CO_2^+ + XD \rightarrow CO_2D^+ + X$	D	MS	—	—	322	—
$C_2F_3^+ + XF_2 \rightarrow C_2F_5^+ + X$	C_2F_2	MS	—	0.01	307	10
$PH_3^+ + X \rightarrow XH^+ + PH_2$	PH_3	MS	9.3	—	305	—
$PH_3^+ + XD \rightarrow PH_3D^+ + X$	D	MS	<0.1	—	305	—
$AsH_3^+ + X \rightarrow XH^+ + AsH_2$	AsH_3	MS	7.2	—	305	—
$AsH_3^+ + XD \rightarrow AsH_3D^+ + X$	D	MS	<0.1	—	305	—

[a] See Comments on Tables, p. 252.

Table V

Hydride Ion Transfer Reactions

Reaction	Type of experiment	Rate constant ($\times 10^{10}$ cc molecule^{-1} sec^{-1})	Cross section, Q ($\times 10^{16}$ cm^2 molecule^{-1})	Ref.	Comments[a]
$CH_3^+ + C_2H_6 \rightarrow C_2H_5^+ + CH_4$	MS	17	—	185, 310	10
$CH_3^+ + C_3H_8 \rightarrow C_3H_7^+ + CH_4$	MS	—	29	308	6
$CH_3^+ + CH_3OH \rightarrow CH_2OH^+ + CH_4$	MT	—	—	218	
$CH_3^+ + C_3H_7OH \rightarrow C_3H_6OH^+ + CH_4$	MT	—	—	221	
$C_2H_2^+ + C_2H_6 \rightarrow C_2H_5^+ + C_2H_3$	MS, MT	—	—	185, 310	10
$C_2H_3^+ + C_2H_6 \rightarrow C_2H_5^+ + C_2H_4$	MS, MT	—	—	185, 310	10
$C_2H_3^+ + C_3H_8 \rightarrow C_3H_7^+ + C_2H_4$	MS	—	17	308	8
$C_2H_4^+ + C_2H_6 \rightarrow C_2H_5^+ + C_2H_5$	MS	—	—	185	10, 18
$C_2H_4^+ + C_3H_8 \rightarrow C_3H_7^+ + C_2H_5$	MS	—	18	308	6
$C_2H_5^+ + C_nH_m \rightarrow C_nH_{m-1}^+ + C_2H_6$	R	—	—	10	1, 3
$C_2H_5^+ + C_3H_8 \rightarrow C_3H_7^+ + C_2H_6$	MS	—	20	308	6
$C_3H_3^+ + C_3H_8 \rightarrow C_3H_7^+ + C_3H_4$	MS	—	8	308	6
$C_3H_5^+ + C_3H_8 \rightarrow C_3H_7^+ + C_3H_6$	MS	—	15	308	6
$C_3H_7^+ + C_nH_m \rightarrow C_nH_{m-1}^+ + C_3H_8$	R	—	—	10	1, 5
$C_4H_9^+ + C_nH_m \rightarrow C_nH_{m-1}^+ + C_4H_{10}$	R	—	—	10	1, 2
$CH_2OH^+ + C_3H_8 \rightarrow C_3H_7^+ + CH_3OH$	MT	—	—	219	—
$CH_2OH^+ + C_3H_7OH \rightarrow C_3H_6OH^+ + CH_3OH$	MT	—	—	221	—
$CH_2NH_2^+ + CH_3NH_2 \rightarrow CH_2NH_2^+ + CH_3NH_2$	MP	6	—	321	22
$CHO^+ + CD_3OH \rightarrow CD_2OH^+ + CDHO$	MT	—	—	218	—
$CDO^+ + C_3H_8 \rightarrow C_3H_7^+ + CDHO$	MT	—	—	219	—
$OH^+ + CH_4 \rightarrow CH_3^+ + H_2O$	CH	—	—	158	—
$S^+ + C_2D_6 \rightarrow C_2D_5^+ + SD$	MT	—	—	310	—
$SH^+ + C_2D_6 \rightarrow C_2D_5^+ + HDS$	MT	—	—	310	—

H_2 and H_2^- Transfer Reactions

Reaction	X	Type of experiment	Rate constant ($\times 10^{10}$ cc molecule^{-1} sec^{-1})	Cross section, Q ($\times 10^{16}$ cm^2 molecule^{-1})	Ref.	Comments[a]
$C_2H_2^+ + XH_2 \rightarrow C_2H_4^+ + X$	C_2H_4	MS	—	—	185	10
$C_2H_3^+ + XH_2 \rightarrow C_2H_5^+ + X$	C_2H_4	MS	—	—	185	10
$C_2H_4^+ + XH_2 \rightarrow X^+ + C_2H_6$	C_3H_6	MS; MS	—; —	—; 10	185; 308	10; 6
$C_5H_{10}^+ + X \rightarrow C_5H_8^+ + XH_2$	C_3H_6	R	—	—	10	1
	C_2H_4	R	—	—	10	1
	C_2H_2	R	—	—	10	1
	$c\text{-}C_3H_6$	R	—	—	10	1
$C_5H_{12}^+ + X \rightarrow C_5H_{10}^+ + XH_2$	C_3H_6	R	—	—	10	1
	C_2H_4	R	—	—	10	1
	C_2H_2	R	—	—	10	1
	$c\text{-}C_3H_6$	R	—	—	10	1
$C_6H_{12}^+ + c\text{-}C_3D_6$ $\rightarrow C_6H_{10}^+ + CD_2HCD_2CD_2H$	—	R	—	—	10	1
$C_3D_6^+ + XH_2 \rightarrow C_3D_6H_2 + X^+$	$c\text{-}C_4H_6$	R	—	—	10	1, 4
	$c\text{-}C_5H_8$	R	—	—	10	1, 4
	$c\text{-}C_6H_{10}$	R	—	—	10	1, 4
	C_4H_8	R; MS	—; 3.94	—; —	10; 204	(1, 4); —
	$i\text{-}C_4H_8$	R; MS	—; 1.41	—; —	10; 204	1, 4
	$i\text{-}C_5H_{10}$	R	—	—	10	1, 4
	C_5H_{10}	R	—	—	10	1, 4
$C_3D_6^+ + (CH_3)_3CD$ $\rightarrow C_3D_7H + C_4H_8^+$	—	R	—	—	10	1, 4
$C_3D_6^+ + CD_3CH_2CH_2CD_3$ $\rightarrow C_3D_7H + CD_2CHCH_2CD_3^+$	—	R	—	—	10	1, 4

[a] See Comments on Tables, p. 252.

Table VII

Condensation Reactions

Reaction	Type of experiment	Rate constant ($\times 10^{10}$ cc molecule^{-1} sec^{-1})	Cross section, Q ($\times 10^{16}$ cm^2 molecule^{-1})	Ref.	Comments[a]
$CH^+ + CH_4 \rightarrow C_2H_2^+ + H_2 + H$	MS	—	11	159	6
$CH_2^+ + CH_4 \rightarrow C_2H_2^+ + 2H_2$	MT	—	—	304	—
$CH_2^+ + CH_4 \rightarrow C_2H_3^+ + H_2 + H$	MT	—	—	304	—
	MS	—	25	159	6
$CH_2^+ + CH_4 \rightarrow C_4H_4^+ + H_2$	MT	—	—	304	—
$CH_3^+ + CH_4 \rightarrow C_2H_3^+ + 2H_2$	MT	—	—	304	—
$CH_3^+ + CH_4 \rightarrow C_2H_5^+ + H_2$	MT	—	—	304	—
	MS	—	12	159	6
	MP	8.6; 6.0	—;—	87	15, 16
$CH_4^+ + D_2 \rightarrow CH_3D_2^+ + H$	MS	0.001	—	317	10
$CH_4^+ + D_2 \rightarrow CH_3D^+ + HD$	MS	0.1	—	317	10
$CH_3^+ + D_2 \rightarrow CH_2D^+ + HD$ $CHD_2^+ + H_2$	MS	1.7	—	317	10
$CD_3^+ + H_2 \rightarrow CD_2H^+ + HD$ $CH_2D^+ + D_2$	MS	1.6	—	317	10
$CD_3^+ + CD_4 \rightarrow C_2D_5^+ + D_2$	MP	6.1; 4.6	—;—	87	15, 16
$CH_4^+ + CH_4 \rightarrow C_2H_6^+ + H_2$	MT	—	—	304	—
$CH_4^+ + O_2 \rightarrow CH_3O^+ + OH$	MS	0.26	—	320	—
$C_2^+ + C_2H_4 \rightarrow C_4H_2^+ + H_2$	MS	—	54.8	323	6
$C_2^+ + C_2H_4 \rightarrow C_4H_3^+ + H$	MS	—	1	323	6
$C_2H^+ + C_2H_6 \rightarrow C_4H_4^+ + H_2 + H$	MS	—	1	159	6

Reaction	Method				
$C_2H^+ + C_2H_4 \rightarrow C_4H_2^+ + H_2 + H$	MS	—	21	323	6
$C_2H^+ + C_2H_4 \rightarrow C_4H_3^+ + H_2$	MS	—	34.9	323	6
$C_2H^+ + C_2H_6 \rightarrow C_3H_3^+ + CH_4$	MS	—	30	159	6
$C_2H^+ + C_2H_6 \rightarrow C_4H_5^+ + H_2$	MS	—	2	159	6
$C_2H_2^+ + CH_4 \rightarrow C_3H_5^+ + H$	MP	1.6; 0.92	—; —	87	15; 16
$C_2H_2^+ + C_2H_2 \rightarrow C_4H_3^+ + H$	MP	5.3; 5.0	—; —	87	15; 16
$C_2H_2^+ + C_2H_2 \rightarrow C_4H_2^+ + H_2$	MP	2.3; 1.9	—; —	87	15; 16
$C_2H_2^+ + C_2H_4 \rightarrow C_3H_3^+ + CH_3$	MS	—	17.5	323	6
	MS	2.3	—	324	10
$C_2H_2^+ + C_2H_4 \rightarrow C_4H_5^+ + H$	MS	—	10.2	323	6
	MS	1.4	—	324	10
$C_2H_2^+ + C_3H_6 \rightarrow C_5H_5^+ + H_2 + H$	MS	—	34.1	323	6
$C_2H_3^+ + C_2H_6 \rightarrow C_3H_5^+ + CH_4$	MS	—	3.7	159	6
$C_2H_3^+ + C_2H_4 \rightarrow C_3H_3^+ + CH_4$	MS	—	2	323	6
$C_2H_3^+ + C_2H_4 \rightarrow C_4H_5^+ + H_2$	MS	—	1.7	323	6
$C_2H_3^+ + C_2H_6 \rightarrow C_4H_6^+ + H_2 + H$	MS	—	0.03	159	6
$C_2H_3^+ + C_2H_6 \rightarrow C_4H_7^+ + H_2$	MS	—	0.3	159	6
$C_2H_3^+ + C_2H_6 \rightarrow C_3H_4^+ + CH_4 + H$	MS	—	0.1	159	6
$C_2H_3^+ + C_3H_8 \rightarrow C_3H_5^+ + C_2H_6$	MS	—	6	308	6
$C_2H_3^+ + C_2H_6 \rightarrow C_3H_7^+ + CH_3$	MS	—	0.3	159	6
$C_2H_4^+ + C_2H_4 \rightarrow C_3H_5^+ + CH_3$	MS	—	42.3	323	6
$C_2H_4^+ + C_2H_4 \rightarrow C_4H_5^+ + H_2 + H$	MS	2.6	—	324	6
$C_2H_4^+ + C_2H_4 \rightarrow C_4H_8^+$	MS	—	3.5	323	6
$C_2H_4^+ + C_2H_4 \rightarrow C_4H_7^+ + H$	MS	2.0	—	324	10
$C_2H_5^+ + C_2H_6 \rightarrow C_3H_6^+ + CH_4 + H$	MS	0.21	—	324	10
$C_2H_5^+ + C_2H_6 \rightarrow C_4H_8^+ + H_2 + H$	MS	—	0.08	159	6
$C_5H_5^+ + C_2H_6 \rightarrow C_4H_9^+ + H_2$	MS	—	0.07	159	6

(continued)

Table VII (continued)

Reaction	Type of experiment	Rate constant ($\times 10^{10}$ cc molecule^{-1} sec^{-1})	Cross section, Q ($\times 10^{16}$ cm^2 molecule^{-1})	Ref.	Comments[a]
$C_3H^+ + C_3H_6 \rightarrow C_4H_3^+ + C_2H_4$	MS	—	46.4	323	6
$C_3H^+ + C_3H_6 \rightarrow C_4H_4^+ + C_2H_3$	MS	—	6.4	323	6
$C_3H_2^+ + C_3H_6 \rightarrow C_4H_3^+ + C_2H_5$	MS	—	2.6	323	6
$C_3H_2^+ + C_3H_6 \rightarrow C_4H_4^+ + C_2H_4$	MS	—	18.2	323	6
$C_3H_2^+ + C_3H_6 \rightarrow C_5H_5^+ + CH_3$	MS	—	30.6	323	6
$C_3H_2^+ + C_3H_4 \rightarrow C_4H_3^+ + C_2H_3$	MP	5.5; 3.2	—; —	87	15; 16
$C_3H_3^+ + C_3H_6 \rightarrow C_4H_5^+ + C_2H_4$	MS	—	18.2	323	6
$C_3H_4^+ + C_3H_4 \rightarrow C_6H_7^+ + H$	MP	3.7; 1.5	—; —	87	15; 16
$C_3H_4^+ + C_3H_4 \rightarrow C_3H_5^+ + C_3H_3$	MP	2.0; 2.2	—; —	87	15; 16
$C_3H_4^+ + C_3H_6 \rightarrow C_4H_6^+ + C_2H_4$	MS	—	16.3	323	6
$C_3D_4^+ + C_3D_4 \rightarrow C_6D_7^+ + D$	MP	4.7; 3.3	—; —	87	15; 16
$C_3H_5^+ + C_3H_6 \rightarrow C_4H_7^+ + C_2H_4$	MP	—	32.2	323	6
$C_3H_5^+ + C_3H_6 \rightarrow C_5H_7^+ + CH_4$	MP	—	6.6	323	6
$C_3H_6^+ + C_3H_6 \rightarrow C_4H_7^+ + C_2H_5$	MS	—	9.9	323	6
$C_3H_6^+ + C_3H_6 \rightarrow C_4H_8^+ + C_2H_4$	MS	—	13.8	323	6
$C_3H_6^+ + C_3H_6 \rightarrow C_5H_9^+ + CH_3$	MS	—	11	323	6
$CH_2OH^+ + C_3H_8 \rightarrow C_2H_5^+ + CH_4 + CH_2O$	MT	—	—	219	19
$CD_3OHD^+ + CD_3OH \rightarrow C_2D_6OH^+ + HDO$	MT	—	—	204	—
$CD_3OH_2^+ + CD_3OH \rightarrow C_2D_6OH^+ + H_2O$	MT	—	—	204	—
$C_2H_5OH_2^+ + C_2H_5OH \rightarrow (C_2H_5OH)_2H$	MT	—	—	204	19
$CHO^+ + C_2D_6 \rightarrow CO + C_2D_4H^+ + D_2$	MT	—	—	310	19
$CHO^+ + C_2D_6 \rightarrow C_2D_5^+ + CO + HD$	MT	—	—	310	19
$CHO^+ + C_2D_6 \rightarrow C_2D_4H^+ + CO + D_2$	MT	—	—	310	19
$CHO^+ + C_3D_8 \rightarrow C_3D_7^+ + CO + HD$	MT	—	—	310	19
$CDO^+ + C_3H_8 \rightarrow C_2H_5^+ + CH_3D + CO$	MT	—	—	310	19

Reaction	Method			Ref.[a]	
$N_2^+ + N_2 \rightarrow N_4^+$	D		0.01	140	—
$H_3^+ + 2H_2 \rightarrow H_5^+ + H_2$	MS			243	—
$NH_2^+ + NH_3 \rightarrow N_2H_4^+ + H$	MS			130	6
$NH_2^+ + NH_3 \rightarrow N_2H_4^+ + H$	MCRS	0.01	0.07	94	—
$NH_3^+ + NH_3 \rightarrow N_2H_5^+ + H$	MCRS		0.1	92	6
$NH_3^+ + NH_3 \rightarrow N_2H_5^+ + H$	MS		0.3	92	6
$N^+ + NH_3 \rightarrow N_2H^+ + H_2$	MCRS	0.6	0.02	94	6
$N^+ + NH_3 \rightarrow N_2H^+ + H_2$	MS		1.8	92	6
$NH^+ + NH_3 \rightarrow N_2H^+ + H + H_2$	MS		1.1	94	6
$NH^+ + NH_3 \rightarrow N_2H^+ + H + H_2$	MCRS		0.89	94	6
$NH^+ + NH_3 \rightarrow N_2H_2^+ + H_2$	MS		0.78	94	6
$NH^+ + NH_3 \rightarrow N_2H_2^+ + H_2$	MS		0.9	92	6
$NH^+ + NH_3 \rightarrow N_2H_3^+ + H$	MCRS	0.03	0.35	94	6
$NH^+ + NH_3 \rightarrow N_2H_3^+ + H$	MS		0.05	94	6
$NH_2^+ + NH_3 \rightarrow N_2H_3^+ + H_2$	MS		0.3	92	6
$O_2^+ + C_2H_2 \rightarrow CHO^+ + CHO$	MS	0.05		320	18
$O_2^+ + CH_4 \rightarrow CH_3O_2^+ + H$	MS	12.6		320	18
$O_2^+ + CH_4 \rightarrow CH_2O^+ + H_2O$	MS	0.2		320	18
$Ar^+ + CH_4 \rightarrow ArCH_2^+ + H_2$	MS	0.008		318	10
$Ar^+ + CH_4 \rightarrow ArCH_3^+ + H$	MS	0.003		318	10
$Kr^+ + CH_4 \rightarrow KrCH_2^+ + H_2$	MS	0.01		318	10
$Kr^+ + CH_4 \rightarrow KrCH_3^+ + H$	MS	0.03		318	10
$Xe^+ + CH_4 \rightarrow XeCH_2^+ + H_2$	MS	0.005		318, 319	10
$Xe^+ + CH_4 \rightarrow XeCH_3^+ + H$	MS	0.022		318, 319	10
$Xe^+ + C_2H_2 \rightarrow XeC_2H^+ + H$	MS			326	18
$XeC_2H_2^{+*} + M \rightarrow XeC_2H_2^+$	MS			326	10
$C_2F_2^+ + C_2F_4 \rightarrow C_3F_3^+ + CF_3$	MS		16	307	10
$C_2F_3^+ + C_2F_4 \rightarrow C_3F_4^+ + CF_3$	MS		0.05	307	10
$C_2F_3^+ + C_2F_4 \rightarrow C_4F_6^+ + F$	MS		0.2	307	10
$C_2F_4^+ + C_2F_4 \rightarrow C_3F_5^+ + CF_3$	MS		2.4	307	10

[a] See Comments on Tables, p. 252.

Table VIII

Negative Ion Reactions

Reaction	Type of experiment	Rate constant ($\times 10^{10}$ cc molecule^{-1} sec^{-1})	Cross section, Q ($\times 10^{16}$ cm^2 molecule^{-1})	Ref.	Comments[a]
$O^- + CH_3I \rightarrow OI^- + CH_3$	MS	—	~2	157	—
$O^- + I_2 \rightarrow OI^- + I$	MS	—	~0.8	157	—
$O^- + IBr \rightarrow OI^- + Br$	MS	—	~0.05	157	—
$O^- + CH_3NO_2 \rightarrow CH_2NO_2^- + OH$	MS	—	~3	157	—
$O^- + CH_3NO_2 \rightarrow CHNO_2^- + H_2O$	MS	—	~2	157	—
$O^- + NO_2 \rightarrow NO_2^- + O$	MS	13	—	314	20
$O^- + NO_2 \rightarrow O_2^- + NO$	MS	0.2	—	314	20
$O^- + N_2O \rightarrow NO^- + NO$	MS; MS	—; 0.37	—	329; 314	; 20
$O^- + N_2O \rightarrow O_2^- + N_2$	MS	—	—	329	—
$O^- + N_2O \rightarrow NO_2^- + N$	MS	0.02	—	329	—
$O^- + N_2O \rightarrow N_2O^- + O$	MS	0.3	—	314	20
$O^- + O_2 \rightarrow O_2^- + O$	MS	—	~3	314	20; 21
$SO^- + SO_2 \rightarrow SO_2^- + SO$	MS	—	—	157	—
$I_2^- + I_2^- \rightarrow I_3^- + I$	MS	—	—	15	—
$Cl_2^- + Cl_2 \rightarrow Cl_3^- + Cl$	MS	—	—	131	—
$Br_2^- + Br_2 \rightarrow Br_3^- + Br$	MS	—	—	131	—
$NH^- + NH_2 \rightarrow NH + NH_2^-$	MCRS	—	—	92	—

Reaction	MCRS				
$NH_3^- + NH_3 \rightarrow N_2H_6^-$	MS	—	—	92	—
$CN^- + (CN)_2 \rightarrow (CN)_2^- + CN$	MS	—	—	334	—
$H^- + H_2O \rightarrow OH^- + H_2$	MS	46.3	0.3	316	—
$D^- + D_2O \rightarrow OD^- + D_2$	MS	—	~20	314	20
$C_6H_5NO_2^- + SO_2 \rightarrow SO_2^- + C_6H_5NO_2$	MS	—	—	157	—
$CH_2S^- + CH_3SSCH_3 \rightarrow CH_3SSCH_2^- + CH_3S$	MS	—	—	330	—
$CH_3S^- + CH_3SSCH_3 \rightarrow CH_2SSCH_2^- + CH_3SH$	MS	—	—	330	—
$HCO_2^- + N_2 \rightarrow CN^- + NO_2H$	MS	~0.08	—	204	—
$HCO_2^- + HCOOH \rightarrow HCOOHCOOH^-$	MS	~0.03	—	204	—
$HCO_2^- + DCOOH \rightarrow HCOODCOOH^-$	MS	~0.03	—	204	—
$DCO_2^- + HCOOH \rightarrow DCOOHCOOH^-$	MS	~0.02	—	204	—
$DCO_2^- + DCOOH \rightarrow DCOODCOOH^-$	MS	~0.02	—	204	—
$DCO_2^- + DCOOD \rightarrow DCOODCOOD^-$	MS	~0.02	—	204	—

[a] See Comments on Tables, p. 252.

Table IX

Miscellaneous Positive Ion Reactions

Reaction	Type of experiment	Rate constant ($\times 10^{10}$ cc molecule^{-1} sec^{-1})	Cross section, Q ($\times 10^{16}$ cm^2 molecule^{-1})	Ref.	Comments[a]
$H_3O^+ + Pb \rightarrow Pb^+ + H_2O + H$	F	10	—	139	—
$H_3O^+ + Mn \rightarrow Mn^+ + H_2O + H$	F	8.5	—	139	—
$H_3O^+ + Cr \rightarrow Cr^+ + H_2O + H$	F	8.0	—	139	—
$H_3O^+ + Li \rightarrow Li^+ + H_2O + H$	F	7.5	—	139	—
$H_3O^+ + Zn \rightarrow Zn^+ + H_2O + H$	F	1.0	—	139	—
$CO^+ + K \rightarrow CK^+ + O$	MS	—	8	327	—
$CO^+ + K \rightarrow OK^+ + C$	MS	—	8	327	—
$CO^+ + Na \rightarrow CNa^+ + O$	MS	—	4	327	—
$(HCOOC_2H_5 + Hg)^+ \rightarrow C_3H_6O_2Hg^+$	MS	—	—	327	—
$C_2H_5OH^+ + Hg \rightarrow C_2H_5OHg^+ + H$	MS	—	—	327	—
$C_2H_5OH^+ + Hg \rightarrow C_2H_4OHg^+ + H_2$	MS	—	620	327	—
$C_2H_5OH^+ + Hg \rightarrow CH_3OHg^+ + CH_3$	MS	—	12	327	—
$C_2H_3O^+ + Hg \rightarrow CH_2OHg^+ + CH$	MS	—	210	327	—

[a] See Comments on Tables, p. 252.

REFERENCES

1. M. Pahl, *Ergeb. Exact. Naturw.*, **34**, 182 (1962).
2. J. Durup, *Les reactions entre ions positifs et molecules en phase gaseuse*, Gauthier-Villars, Paris, 1960.
3. F. W. Lampe, J. L. Franklin, and F. H. Field, in *Progress in Reaction Kinetics*, G. Porter, Ed., Pergamon Press, New York, 1961, p. 69.
4. C. E. Melton, in *Mass Spectrometry of Organic Ions*, F. W. McLafferty, Ed., Academic Press, New York, 1963, p. 65.
5. D. P. Stevenson, in *Mass Spectrometry*, C. A. McDowell, Ed., McGraw-Hill, New York, 1963, p. 589.
6. H. M. Rosenstock, U. S. Atomic Energy Commission Report JLI-650-3-T, TID-4500 (1959).
7. (*a*) V. L. Tal'roze, *Pure Appl. Chem.*, **5**, 455 (1962); (*b*) V. L. Tal'roze and G. V. Karachevtsev, in *Advances in Mass Spectrometry*, Vol. III, W. L. Mead, Ed., The Institute of Petroleum, London, 1966, p. 211.
8. A. F. Trotman-Dickenson, *Ann. Rep. Progr. Chem. (Chem. Soc. London)*, **55**, 36 (1958).
9. M. J. Henchman, *Ann. Rev.*, **62**, 39 (1965).
10. *Advan. Chem. Ser.*, **58** (1966).
11. (*a*) J. J. Thomson, *Rays of Positive Electricity*, Longmans Green, London, 1933; (*b*) H. D. Smyth, *Rev. Mod. Phys.*, **3**, 347 (1931).
12. A. J. Dempster, *Phil. Mag.*, **31**, 438 (1916).
13. T. R. Hogness and E. G. Lunn, *Phys. Rev.*, **26**, 44 (1925).
14. H. D. Smyth, *Phys. Rev.*, **25**, 452 (1925).
15. T. R. Hogness and R. W. Harkness, *Phys. Rev.*, **32**, 784 (1928).
16. W. Bleakney, *Phys. Rev.*, **41**, 32 (1932).
17. H. W. Washburn, C. E. Berry, and L. G. Hall, "Mass Spectrometry in Physics Research," No. 20, *Natl. Bur. Std. Circ.*, **522** (1953).
18. F. H. Field and J. L. Franklin, *Electron Impact Phenomena*, Academic Press, New York, 1957.
19. H. Eyring, J. O. Hirschfelder, and H. S. Taylor, *J. Chem. Phys.*, **4**, 479 (1936).
20. V. L. Tal'roze and A. K. Lyubimova, *Dokl. Akad. Nauk, SSSR*, **86**, 909 (1952).
21. D. P. Stevenson and D. O. Schissler, *J. Chem. Phys.*, **23**, 1353 (1955).
22. S. Wexler and R. Marshall, *J. Am. Chem. Soc.*, **86**, 781 (1964).
23. K. Yang and T. Ree, *J. Chem. Phys.*, **35**, 588 (1961).
24. M. P. Langevin, *Ann. Chim. Phys.*, **5**, 245 (1905). An English translation of this classic paper is given in appendix 1, reference 27.
25. G. Gioumousis and D. P. Stevenson, *J. Chem. Phys.*, **29**, 294 (1958).
26. I. Szabo, *Arkiv Fyzik*, **33**, 57 (1967).
27. E. W. McDaniel, *Collision Phenomena in Ionized Gases*, Wiley, New York, 1964.
28. Reference 27, Section 3-5B, Chapter 3.
29. R. P. Pottie, A. J. Lorquet, and W. H. Hamill, *J. Am. Chem. Soc.*, **84**, 529 (1962).
30. N. Boelrijk and W. H. Hamill, *J. Am. Chem. Soc.*, **84**, 730 (1962).
31. L. P. Theard and W. H. Hamill, *J. Am. Chem. Soc.*, **84**, 1134 (1962).
32. M. Berta and W. S. Koski, *J. Am. Chem. Soc.*, **86**, 5098 (1964).
33. M. Berta, Ph.D. Thesis, Johns Hopkins University (1964).
34. T. F. Moran and W. H. Hamill, *J. Chem. Phys.*, **39**, 1413 (1963).

35. J. V. Dugan and J. L. Magee, NASA TN-D-3229 (Feb. 1966).
36. G. Gioumousis, "Cross Sections and Rate Constants for Ion-Molecule Reactions," Lockheed Missiles and Space Company Technical Report 2-12-66-4, AD 644622 (October 1966).
37. J. C. Light and J. Horrocks, *Proc. Phys. Soc.*, **84**, 527 (1964).
38. A. Henglein, K. Lacmann, and G. Jacobs, *Ber. Bunsenges. Phys. Chem.*, **69**, 279 (1965).
39. K. Lacmann and A. Henglein, *Ber. Bunsenges. Phys. Chem.*, **69**, 286 (1965).
40. K. Lacmann and A. Henglein, *Ber. Bunsenges. Phys. Chem.*, **69**, 292 (1965).
41. J. C. Light, *J. Chem. Phys.*, **40**, 3221 (1964).
42. P. Pechukas and J. C. Light, *J. Chem. Phys.*, **42**, 3281 (1965).
43. J. C. Light and J. Lin, *J. Chem. Phys.*, **43**, 3209 (1965).
44. O. B. Firsov, *Soviet Phys.*, *JETP*, **15**, 906 (1962), Eng. Trans. of *Zh. Eksperim. i Teor. Fiz.*, **42**, 1307 (1962).
45. A. Tannenwald, *Proc. Phys. Soc.*, **87**, 109 (1966).
46. F. A. Wolf, *J. Chem. Phys.*, **44**, 1619 (1966).
47. F. P. Abramson and J. H. Futrell, *J. Chem. Phys.*, **45**, 1925 (1966).
48. J. H. Futrell and T. O. Tiernan, *J. Phys. Chem.*, **72**, 158 (1968).
49. T. O. Tiernan and J. H. Futrell, *J. Phys. Chem.*, (1968) in press.
50. L. I. Bone and J. H. Futrell, *J. Chem. Phys.*, **46**, 4084 (1967).
51. G. A. W. Derwish, A. Galli, A. Giardini-Guidoni, and G. G. Volpi, *J. Am. Chem. Soc.*, **87**, 1159 (1965).
52. F. H. Field, J. L. Franklin, and M. S. B. Munson, *J. Am. Chem. Soc.*, **85**, 3575 (1963).
53. H. E. Stanton and S. Wexler, *J. Chem. Phys.*, **44**, 2959 (1966).
54. C. E. Klots, *J. Chem. Phys.*, **41**, 117 (1964).
55. W. A. Chupka, ASTM Committee E-14, 11th Annual Meeting, San Francisco, May, 1963, cited in reference 53.
56. V. L. Tal'roze and E. L. Frankevich, *Proc. 1st All-Union Sci. Tech. Conf. Appl. Radioact. Isotopes, Moscow, 1957*, p. 13.
57. R. F. Pottie and W. H. Hamill, *J. Phys. Chem.*, **63**, 877 (1959).
58. A. Henglein, *Z. Naturforsch.*, **77a**, 44 (1962).
59. A. Henglein, G. Jacobs, and G. A. Muccini, *Z. Naturforsch.*, **18a**, 98 (1963).
60. H. M. Rosenstock, M. B. Wallenstein, A. L. Wahrhaftig, and H. Eyring, *Proc. Natl. Acad. Sci.*, **38**, 667 (1952).
61. P. Kebarle, R. M. Haynes, and S. Searles, *Advan. Chem. Ser.*, **58**, 210 (1966).
62. S. Wexler, A. Lifshitz, and A. Quattrochi, *Advan. Chem. Ser.*, **58**, 193 (1966).
63. G. G. Meisels, *Advan. Chem. Ser.*, **58**, 243 (1966).
64. J. C. Light, *J. Chem. Phys.*, **41**, 586 (1964).
65. M. A. Eliason, D. E. Stogryn, and J. O. Hirschfelder, *Proc. Natl. Acad. Sci.*, **42**, 546 (1956).
66. J. O. Hirschfelder, private communication, cited in reference 3, p. 78.
67. C. F. Giese, in *Advances in Mass Spectrometry*, Vol. III, W. L. Mead, Ed., The Institute of Petroleum, London, 1966, p. 321.
68. T. W. Shannon and A. G. Harrison, *J. Chem. Phys.*, **43**, 4201 (1965).
69. D. Rapp and W. E. Francis, *J. Chem. Phys.*, **37**, 2631 (1962).
70. E. F. Gurnee and J. L. Magee, *J. Chem. Phys.*, **26**, 1237 (1957).
71. A. Dalgarno, *Phil. Trans. Roy. Soc.*, **A250**, 426 (1958).
72. H. S. W. Massey and E. H. S. Burhop, *Electronic and Ionic Impact Phenomena*, Oxford University Press, London, 1952, p. 514.

73. W. L. Fite, R. F. Stebbings, D. G. Hummer, and R. T. Brackmann, *Phys. Rev.*, **119**, 663 (1960).
74. W. Maier II, *J. Chem. Phys.*, **41**, 2174 (1964).
75. G. H. Wannier, *Phys. Rev.*, **100**, 1180 (1955).
76. L. Friedman, *Advan. Chem. Ser.*, **58**, 87 (1966), and references cited therein.
77. K. R. Ryan and J. H. Futrell, *J. Chem. Phys.*, **43**, 3009 (1965).
78. K. R. Ryan, J. H. Futrell, and C. D. Miller, *Rev. Sci. Instr.*, **37**, 107 (1966).
79. A. G. Harrison, T. W. Shannon, and F. Meyer, in *Advances in Mass Spectrometry*, Vol. III, C. W. Mead, Ed., The Institute of Petroleum, London, 1966, p. 377.
80. V. L. Tal'roze and E. L. Frankevich, *Zh. Fiz. Khim.*, **34**, 2709 (1960).
81. G. V. Karachevtsev, M. I. Markin, and V. L. Tal'roze, *Izv. Akad. Nauk, SSSR, Otd. Khim. Nauk.*, **1961**, 1528.
82. G. V. Karachevtsev, M. I. Markin, and V. L. Tal'roze, *Kinetica i Kataliz*, **5**, 377 (1964).
83. T. W. Shannon, F. Meyer, and A. G. Harrison, *Can. J. Chem.*, **43**, 149 (1965).
84. J. H. Futrell, T. O. Tiernan, C. D. Miller, and F. P. Abramson, *Rev. Sci. Instr.*, (1968) in press.
85. L. Matus, D. J. Hyatt, and M. J. Henchman, *J. Chem. Phys.*, **46**, 2439 (1967).
86. A. G. Harrison and J. M. S. Tait, *Can. J. Chem.*, **40**, 1986 (1962).
87. A. G. Harrison, J. J. Myher, and J. C. J. Thynne, *Adv. Chem. Series*, **58**, 150 (1966).
88. L. W. Sieck, F. P. Abramson, and J. H. Futrell, *J. Chem. Phys.*, **45**, 2859 (1966).
89. J. L. Franklin and F. H. Field, *J. Am. Chem. Soc.*, **83**, 3555 (1961).
90. F. H. Field and M. S. B. Munson, *J. Am. Chem. Soc.*, **87**, 3289 (1965).
91. P. S. Rudolph and C. E. Melton, *J. Phys. Chem.*, **63**, 916 (1959).
92. C. E. Melton, *J. Chem. Phys.*, **45**, 4414 (1966).
93. S. Wexler and N. Jesse, *J. Am. Chem. Soc.*, **84**, 3425 (1962).
94. G. A. W. Derwish, A. Galli, A. Giardini-Guidoni, and G. G. Volpi, *J. Chem. Phys.*, **39**, 1599 (1963).
95. M. J. Henchman, H. T. Otwinowska, and F. H. Field, in *Advances in Mass Spectrometry*, M. L. Mead, Ed., The Institute of Petroleum, London, 1966, p. 359.
96. B. C. de Souza and J. H. Green, private communication.
97. P. S. Rudolph and C. E. Melton, *J. Chem. Phys.*, **32**, 1128 (1960).
98. P. Kebarle and E. W. Godbole, *J. Chem. Phys.*, **36**, 302 (1962).
99. P. Kebarle and A. M. Hogg, *J. Chem. Phys.*, **42**, 668 (1965).
100. V. Aquilanti, A. Galli, A. Giardini-Guidoni, and G. G. Volpi, *J. Chem. Phys.*, **43**, 1969 (1965).
101. V. Aquilanti, A. Galli, A. Giardini-Guidoni, and G. G. Volpi, *J. Chem. Phys.*, **44**, 2307 (1966).
102. V. Čermák and Z. Herman, *Nucleonics*, **19** (9), 106 (1961).
103. W. Poschenrieder and P. Warneck, *J. Appl. Phys.*, **37**, 2812 (1966).
104. P. Warneck, *J. Chem. Phys.*, **46**, 502 (1967).
105. P. Warneck, *J. Chem. Phys.*, **46**, 513 (1967).
106. W. A. Chupka, paper presented at the 15th Annual Conference on Mass Spectrometry, ASTM Committee E-14, Denver, Colorado, May 14–19, 1967.
107. E. Lindholm, *Z. Naturforsch.*, **9a**, 535 (1954).
108. J. H. Futrell and C. D. Miller, *Rev. Sci. Instr.*, **37**, 1521 (1966).
109. C. F. Giese and W. B. Maier II, *J. Chem. Phys.*, **39**, 739 (1963).

110. J. B. Homer, R. S. Lehrle, J. C. Robb, M. Takahashi, and D. W. Thomas, in *Advances in Mass Spectrometry*, Vol. II, R. M. Elliot, Ed., Pergamon Press, New York, 1963, p. 503.
111. G. K. Lavrovskaya, M. I. Markin, and V. L. Tal'roze, *Kinetika i Kataliz*, **2** (1), 21 (1961).
112. B. R. Turner, M. A. Fineman, and R. F. Stebbings, *J. Chem. Phys.*, **42**, 4088 (1965).
113. D. W. Vance and T. L. Bailey, *J. Chem. Phys.*, **44**, 486 (1966).
114. E. R. Weiner, G. R. Hertel, and W. S. Koski, *J. Am. Chem. Soc.*, **86**, 788 (1964).
115. E. Gustafsson and E. Lindholm, private communication (1958) cited in *Atomic and Molecular Processes*, D. R. Bates, Ed., Academic Press, New York, 1962, p. 705.
116. C. F. Giese, *Rev. Sci. Instr.*, **30**, 260 (1959).
117. E. Lindholm, *Advan. Chem. Ser.*, **58**, 1 (1966) and references cited therein.
118. G. R. Hertel and W. S. Koski, *J. Am. Chem. Soc.*, **87**, 1686 (1965).
119. F. C. Fehsenfeld, A. L. Schmeltekopf, P. D. Goldan, H. I. Schiff, and E. E. Ferguson, *J. Chem. Phys.*, **44**, 4087 (1966).
120. P. D. Goldan, A. L. Schmeltekopf, F. C. Fehsenfeld, H. I. Schiff, and E. E. Ferguson, *J. Chem. Phys.*, **44**, 4095 (1966).
121. F. C. Fehsenfeld, E. E. Ferguson, and A. L. Schmeltekopf, *J. Chem. Phys.*, **44**, 3022 (1966).
122. P. H. G. Dickenson and J. Sayers, *Proc. Phys. Soc.*, **76**, 137 (1960).
123. W. L. Fite, J. A. Rutherford, W. R. Snow, and W. A. J. van Lint, *Discussions Faraday Soc.*, **33**, 264 (1962).
124. W. L. Fite and J. A. Rutherford, *Discussions Faraday Soc.*, **37**, 192 (1964).
125. J. Sayers and D. Smith, *Discussions Faraday Soc.*, **37**, 167 (1964).
126. C. H. Bloomfield and J. B. Hasted, *Discussions Faraday Soc.*, **37**, 176 (1966).
127. J. A. Morrison and D. Edelson, *J. Appl. Phys.*, **33**, 1714 (1962).
128. D. Edelson, J. A. Morrison, and K. B. McAfee, *J. Appl. Phys.*, **35**, 1682 (1964).
129. M. Saporoschenko, *Phys. Rev.*, **139**, 349 (1965).
130. M. Saporoschenko, *J. Chem. Phys.*, **42**, 2760 (1965).
131. C. E. Melton, G. A. Rapp, and P. S. Rudolph, *J. Chem. Phys.*, **29**, 968 (1958).
132. G. E. Keller, D. W. Martin, and E. W. McDaniel, *Phys. Rev.*, **140**, A1535 (1965).
133. T. M. Miller, J. T. Moseley, D. W. Martin, and E. W. McDaniel, paper presented at the 5th International Conference on the Physics of Electronic and Atomic Collisions, Leningrad, July, 1967.
134. H. Sommer, H. A. Thomas, and J. A. Hipple, *Phys. Rev.*, **82**, 697 (1951).
135. J. D. Baldeschwieler, paper presented at the 152nd Annual Meeting of the American Chemical Society, New York, September, 1966.
136. L. R. Anders, J. L. Beauchamp, R. C. Dunbar, and J. D. Baldeschwieler, *J. Chem. Phys.*, **45**, 1062 (1966).
137. P. Ausloos, G. G. Lias, and A. A. Scala, *Advan. Chem. Ser.*, **58**, 264 (1966), and references cited therein.
138. P. Ausloos and S. G. Lias, *J. Chem. Phys.*, **45**, 524 (1966).
139. H. F. Calcote and D. E. Jensen, *Advan. Chem. Ser.*, **58**, 291 (1966).
140. M. M. Shahin, *Advan. Chem. Ser.*, **58**, 315 (1966).
141. F. S. Klein and L. J. Friedman, *J. Chem. Phys.*, **41**, 1789 (1964).
142. C. F. Giese, *Bull. Am. Phys. Soc.*, **9**, 189 (1964).
143. J. H. Futrell and F. P. Abramson, *Advan. Chem. Ser.*, **58**, 107 (1966).

144. L. D. Doverspike and R. L. Champion, *J. Chem. Phys.*, **46**, 4718 (1967).
145. J. Durup and M. Durup, *J. Chim. Phys.*, **64**, 386 (1967).
146. C. E. Berry, *Phys. Rev.*, **78**, 597 (1950).
147. A. Weingartshofer and E. M. Clarke, *Phys. Rev. Letters*, **12**, 591 (1964).
148. K. R. Ryan and J. H. Futrell, *J. Chem. Phys.*, **42**, 824 (1965).
149. A. G. Harrison, A. Ivko and T. W. Shannon, *Can. J. Chem.*, **44**, 1351 (1966).
150. T. F. Moran and L. Friedman, *J. Chem. Phys.*, **39**, 2491 (1963).
151. H. von Koch and L. Friedman, *J. Chem. Phys.*, **38**, 1115 (1963).
152. C. F. Giese, *Advan. Chem. Ser.*, **58**, 20 (1966).
153. D. P. Stevenson and D. O. Schissler, *J. Chem. Phys.*, **24**, 926 (1956).
154. F. H. Field, J. L. Franklin, and F. W. Lampe, *J. Am. Chem. Soc.*, **79**, 2419 (1957).
155. C. D. Wagner, P. A. Wadsworth, and D. P. Stevenson, *J. Chem. Phys.*, **28**, 517 (1958).
156. R. Fuchs, *Z. Naturforsch.*, **16a**, 1026 (1961).
157. A. Henglein and G. A. Muccini, *J. Chem. Phys.*, **31**, 1426 (1959).
158. A. Henglein and G. A. Muccini, *Z. Naturforsch.*, **17a**, 452 (1962); *ibid.*, **18a**, 753 (1963).
159. G. A. W. Derwish, A. Galli, A. Giardini-Guidoni, and G. G. Volpi, *J. Chem. Phys.*, **40**, 5 (1964).
160. C. W. Hand and H. von Weyssenhoff, *Can. J. Chem.*, **42**, 195 (1964).
161. H. von Koch, *Arkiv Fysik*, **28**, 529 (1965).
162. A. Giardini-Guidoni and L. Friedman, *J. Chem. Phys.*, **45**, 937 (1966).
163. F. P. Abramson and J. H. Futrell, *J. Chem. Phys.*, **46**, 3264 (1967).
164. D. A. Kubose and W. H. Hamill, *J. Am. Chem. Soc.*, **85**, 125 (1963).
165. L. Melander, *Isotope Effects on Reaction Rates*, Ronald Press, New York, 1960.
166. F. P. Abramson, private communication.
167. A. Henglein, K. Lacmann, and B. Knoll, *J. Chem. Phys.*, **43**, 1048 (1965).
168. G. R. Hertel and W. S. Koski, *J. Am. Chem. Soc.*, **86**, 1683 (1964).
169. M. G. Menendez, B. S. Thomas, and T. L. Bailey, *J. Chem. Phys.*, **42**, 802 (1965).
170. R. L. Champion, L. D. Doverspike, and T. L. Bailey, *J. Chem. Phys.*, **45**, 4377 (1966).
171. V. I. Vedeneyev, L. V. Gurvich, V. N. Kondrat'yev, V. A. Medvedev, and E. L. Frankevich, *Bond Energies, Ionization Potentials and Electron Affinities*, Edward Arnold, London, 1966. It should be noted that this volume also tabulates heats of formation of atoms and radicals and proton affinities as well as the information suggested by the title.
172. A revised edition of *Electron Impact Phenomena* (ref. 18) is in preparation which will provide an up to date tabulation and critical evaluation of data available in the literature.
173. R. R. Bernecker and F. A. Long, *J. Phys. Chem.*, **65**, 1565 (1961).
174. T. L. Cottrell, *The Strengths of Chemical Bonds*, Butterworths, London, 1958.
175. A. G. Gaydon, *Dissociation Energies and Spectra of Diatomic Molecules*, Chapman and Hall, London, 1953.
176. "JANAF Thermochemical Tables," Thermal Research Laboratory, The Dow Chemical Co., Michigan, 1964.
177. D. P. Stevenson and D. O. Schissler, *J. Chem. Phys.*, **29**, 282 (1958).
178. M. T. Dmitriev, *Zh. Fiz. Khim.*, **32**, 2418 (1958).
179. F. W. Lampe and F. H. Field, *J. Am. Chem. Soc.*, **79**, 4244 (1957).
180. F. W. Lampe and F. H. Field, *J. Am. Chem. Soc.*, **81**, 3242 (1959).

181. V. L. Tal'roze and E. L. Frankevich, *Dokl. Akad Nauk.*, **111**, 376 (1956).
182. V. L. Tal'roze and E. L. Frankevich, *J. Am. Chem. Soc.*, **80**, 2344 (1958).
183. V. L. Tal'roze and E. L. Frankevich, *Zh. Fiz. Khim.*, **33**, 955 (1959).
184. M. S. B. Munson and F. H. Field, *J. Am. Chem. Soc.*, **87**, 3294 (1965).
185. M. S. B. Munson, J. L. Franklin, and F. H. Field, *J. Phys. Chem.*, **68**, 3098 (1964).
186. M. Herzberg, D. Rapp, J. B. Ortenburger, and D. D. Briglia, *J. Chem. Phys.*, **34**, 343 (1961).
187. J. C. Polanyi, *J. Chem. Phys.*, **31**, 1338 (1959).
188. K. M. Refaey and W. A. Chupka, *J. Chem. Phys.*, **43**, 2544 (1965).
189. W. B. Maier II, Los Alamos Scientific Laboratory, Preprint LA-DC-8545 (1967).
190. L. D. Doverspike, R. L. Champion, and T. L. Bailey, *J. Chem. Phys.*, **45**, 4385 (1966).
191. H. B. Gutbier, *Z. Naturforsch.*, **12A**, 499 (1957).
192. D. Hutchison, A. Kuppermann and L. Pobo, 9th Annual Mass Spectrometry Conference, ASTM Committee E-14, 1961, p. 75.
193. Z. Herman, J. D. Kerstetter, T. L. Rose, and R. Wolfgang, *J. Chem. Phys.*, **46**, 2844 (1967).
194. E. Gustafsson and E. Lindholm, *Arkiv Fysik*, **18**, 219 (1960).
195. V. L. Tal'roze, *Proc. Symp. Chem. Effects Nucl. Transformations (Engl. transl.)*, *Prague, 1960* (Publ. 1961); G. K. Lavrovskaza, M. I. Markin, and V. L. Tal'roze, *Kinetics and Catalysis (USSR) (Eng. transl.)*, **2**, 18 (1961).
196. H. B. Gilbody and J. B. Hasted, *Proc. Roy. Soc. (London)*, **A238**, 334 (1957).
197. R. F. Stebbings, in *Advances in Chemical Physics*, Vol. X, J. Ross, Ed., Interscience, New York, 1966.
198. C. F. Giese and W. B. Maier II, *J. Chem. Phys.*, **39**, 197 (1963).
199. W. B. Maier II, *J. Chem. Phys.*, **42**, 1790 (1965).
200. N. G. Utterback and H. P. Broida, *Phys. Rev. Letters*, **15**, 608 (1965).
201. D. J. Hyatt, E. A. Dodman, and M. J. Henchman, *Advan. Chem. Ser.*, **58**, 131 (1966).
202. T. W. Shannon and A. G. Harrison, *J. Chem. Phys.*, **43**, 4206 (1965).
203. P. Wilmenius and E. Lindholm, *Arkiv Fysik*, **21**, 97 (1962).
204. C. E. Melton, G. A. Rapp, and T. W. Martin, *J. Phys. Chem.*, **64**, 1577 (1960).
205. V. Aquilanti and G. G. Volpi, *J. Chem. Phys.*, **44**, 3574 (1966).
206. V. Aquilanti, A. Galli, and G. G. Volpi, in *Simp. Dinamica Reazioni Chim.*, *Consiglio Nazl Ric., Roma, 1966*.
207. V. Aquilanti, A. Galli, and G. G. Volpi, private communication.
208. V. Aquilanti, A. Galli, and G. G. Volpi, private communication.
209. R. M. Reese and J. A. Hipple, *Phys. Rev.*, **75**, 1332 (1949).
210. J. Durup and F. Heitz, *J. Chim. Phys.*, **61**, 470 (1964).
211. H. D. Hagstrum, *Rev. Mod. Phys.*, **23**, 185 (1951).
212. B. G. Reuben and L. Friedman, *J. Chem. Phys.*, **37**, 1636 (1962).
213. L. Friedman and T. F. Moran, *J. Chem. Phys.*, **42**, 2624 (1965).
214. T. F. Moran and L. Friedman, *J. Geophys. Res.*, **70**, 4992 (1965).
215. E. Lindholm, *Arkiv Fysik*, **8**, 257 (1954).
216. E. Lindholm, *Arkiv Fysik*, **8**, 433 (1954).
217. H. von Koch and E. Lindholm, *Arkiv Fysik*, **19**, 123 (1961).
218. E. Lindholm and P. Wilmenius, *Arkiv Kemi*, **20**, 255 (1963).
219. E. Pettersson and E. Lindholm, *Arkiv Fysik*, **24**, 49 (1963).

220. W. A. Chupka and E. Lindholm, *Arkiv Fysik*, **25**, 349 (1963).
221. E. Pettersson, *Arkiv Fysik*, **25**, 181 (1963).
222. H. Sjogren, *Arkiv Fysik*, **29**, 565 (1965).
223. I. Szabo, *Arkiv Fysik*, **31**, 287 (1966).
224. H. Sjogren, *Arkiv Fysik*, **31**, 159 (1966).
225. H. Sjogren, *Arkiv Fysik*, **32**, 539 (1966).
226. C. F. Giese, in *Advances in Chemical Physics*, Vol. X, J. Ross, Ed., Interscience, New York, 1966.
227. J. H. Futrell and T. O. Tiernan, *J. Chem. Phys.*, **39**, 2539 (1963).
228. A. Galli, A. Giardini-Guidoni, and G. G. Volpi, *Nuovo Cimento*, **31**, 1145 (1964).
229. R. F. Stebbings, B. R. Turner, and J. A. Rutherford, *J. Geophys. Res.*, **71**, 771 (1966).
230. B. R. Turner, J. A. Rutherford, and R. F. Stebbings, *J. Geophys. Res.*, **71**, 4521 (1966).
231. W. L. Fite, A. C. H. Smith, and R. F. Stebbings, *Proc. Roy. Soc.*, *(London)*, **A268**, 527 (1962).
232. Y. Kaneh, L. R. Magill, and J. B. Hasted, private communication.
233. J. W. McGowan, *Long Lived Excited Ions in Collision Mass Spectrometry*, to be published.
234. J. W. McGowan and L. Kerwin, *Can. J. Phys.*, **41**, 316 (1963).
235. J. W. McGowan and L. Kerwin, *Can. J. Phys.*, **42**, 972 (1964).
236. B. R. Turner, Rept. GA-7225, General Atomic, July 18, 1966.
237. M. Krauss and V. H. Dibeler, in *Mass Spectrometry of Organic Ions*, F. W. McLafferty, Ed., Academic Press, New York, 1963, p. 119.
238. C. A. McDowell, in *Mass Spectrometry*, C. A. McDowell, Ed., McGraw-Hill, New York, 1963, p. 506.
239. P. Marmet and L. Kerwin, *Can. J. Phys.*, **38**, 787 (1960).
240. W. Kaul and A. Fuchs, *Z. Naturforsch.*, **15a**, 326 (1960).
241. V. Čermák and Z. Herman, *Collection Czech. Chem. Commun.*, **30**, 1343 (1965).
242. M. C. Cress, P. M. Becker, and F. W. Lampe, *J. Chem. Phys.*, **44**, 2212 (1966).
243. R. K. Asundi, G. J. Schulz, and P. J. Chantry, Westinghouse Research Laboratories Scientific Paper 67-9E2-113-P1, January 4, 1967.
244. V. Čermák and Z. Herman, *J. Chim. Phys.*, **57**, 717 (1960).
245. P. Dong and M. Cottin, *J. Chim. Phys.*, **57**, 557 (1960).
246. A. Henglein, *Z. Naturforsch.*, **11a**, 37 (1962).
247. M. S. B. Munson, F. H. Field, and J. L. Franklin, *J. Chem. Phys.*, **37**, 1790 (1962).
248. M. S. B. Munson, *J. Phys. Chem.*, **69**, 572 (1965).
249. C. E. Melton and W. H. Hamill, *J. Chem. Phys.*, **41**, 1469 (1964).
250. J. W. McGowan and M. A. Fineman, IVth International Conference on Phys. of Electron and Atomic Collisions, Université Laval, Canada, 1965.
251. M A. Fineman, E. M. Clarke, H. P. Hanson, and J. W. McGowan, IVth International Conference on Phys. of Electron and Atomic Collisions, Université Laval, Canada, 1965.
252. J. W. McGowan, M. A. Fineman, E. M. Clarke, and H. P. Hanson, General Atomic Rept. GA-7387, Part I and Part II.
253. V. G. Jacobs and A. Henglein, *Z. Naturforsch.*, **19a**, 906 (1964).
254. V. Čermák, *J. Chem. Phys.*, **44**, 1318 (1966).
255. V. Čermák, *J. Chem. Phys.*, **44**, 3774 (1966).
256. V. Čermák, *J. Chem. Phys.*, **44**, 3781 (1966).

257. A. L. Schmeltekopf, F. C. Fehsenfeld, G. I. Gilman, and E. E. Ferguson, *Planetary Space Sci.*, **15**, 401 (1967).
258. J. W. McGowan, General Atomic Report No. GA-7590, January 3, 1967.
259. M. S. B. Munson, *J. Am. Chem. Soc.*, **87**, 2332 (1965).
260. P. Kebarle and A. M. Hogg, *J. Chem. Phys.*, **42**, 798 (1965).
261. A. M. Hogg and P. Kebarle, *J. Chem. Phys.*, **43**, 449 (1965).
262. A. M. Hogg, R. M. Haynes, and P. Kebarle, *J. Am. Chem. Soc.*, **88**, 28 (1966).
263. Jae-Hyun Yang and D. C. Conway, *J. Chem. Phys.*, **40**, 1729 (1964).
264. N. N. Tunitskii, *Dokl. Akad. Nauk SSSR*, **101**, 1083 (1955).
265. J. B. Hasted, *Proc. Roy. Soc. (London)*, **A212**, 235 (1952).
266. V. Čermák and Z. Herman, *J. Chim. Phys.*, **55**, 51 (1959).
267. M. Saporoschenko, *Phys. Rev.*, **111**, 1550 (1958).
268. P. Marmet and J. D. Morrison, *J. Chem. Phys.*, **36**, 1238 (1962).
269. A. Henglein and G. A. Muccini, *Z. Naturforsch.*, **15a**, 584 (1960).
270. J. T. Herron and H. I. Schaff, *Can. J. Chem.*, **36**, 1159 (1958).
271. S. B. Karmohapatrd, *J. Chem. Phys.*, **35**, 1524 (1961).
272. Ya. M. Fogel, *Soviet Phys., Usp.*, **3**, 390 (1960).
273. S. E. Kuprianov, *Soviet Phys., JETP* **3**, 390 (1960).
274. J. W. McGowan and L. Kerwin, *Proc. 2nd Conf. Phys. Electrons Atomic Collisions, Boulder, 1961*, p. 36.
275. G. G. Meisels, *J. Chem. Phys.*, **31**, 284 (1959).
276. C. E. Melton, *J. Chem. Phys.*, **33**, 647 (1960).
277. J. Collin, *J. Chem. Phys.*, **57**, 424 (1960).
278. K. Watanabe, *J. Chem. Phys.*, **26**, 542 (1957).
279. R. E. Honig, *J. Chem. Phys.*, **16**, 105 (1948).
280. J. Collin and F. P. Lossing, *J. Am. Chem. Soc.*, **81**, 2064 (1959).
281. A. G. Harrison, R. F. Pottie, and F. P. Lossing, *J. Am. Chem. Soc.*, **83**, 3204 (1961).
282. F. L. Mohler, V. H. Diebeler, L. Williamson, and H. M. Dean, *J. Res. Natl. Bur. Std.*, **48**, 188 (1952).
283. P. G. Wilkinson, *Can. J. Phys.*, **34**, 596 (1956).
284. V. J. Hammond, W. C. Price, J. P. Teegan, and A. D. Walsh, *Discussion Faraday Soc.*, **9**, 52 (1950).
285. F. D. Vilesov, *Dokl. Akad. Nauk SSSR*, **132**, 632 (1960).
286. E. W. C. Clarke and C. A. McDowell, *Proc. Chem. Soc.*, **1960**, 69.
287. S. N. Foner and R. J. Hudson, *J. Chem. Phys.*, **25**, 602 (1956).
288. J. R. Hoyland and F. W. Lampe, *J. Chem. Phys.*, **37**, 1066 (1962).
289. F. W. Lampe and F. H. Field, *Tetrahedron*, **7**, 189 (1959).
290. E. L. Frankevich and V. L. Tal'roze, *Zh. Fiz. Khim.*, **33**, 1093 (1959).
291. F. W. Lampe and J. H. Futrell, *Trans. Faraday Soc.*, **59**, 1957 (1963).
292. J. Sherman, *Chem. Rev.*, **11**, 164 (1932).
293. V. N. Kondratyev and N. D. Soklov, *Zh. Fiz. Khim.*, **29**, 1265 (1955).
294. A. P. Altshuller, *J. Am. Chem. Soc.*, **77**, 3480 (1955).
295. G. Briegleb, *Z. Elektrochem.*, **53**, 350 (1949).
296. A. S. Russell, C. M. Fontana, and J. H. Simons, *J. Chem. Phys.*, **9**, 381 (1941).
297. J. Higuchi, *J. Chem. Phys.*, **31**, 563 (1959).
298. J. L. Franklin, F. W. Lampe, and H. E. Lumpkin, *J. Am. Chem. Soc.*, **81**, 3152 (1959).
299. M. S. B. Munson and J. L. Franklin, *J. Phys. Chem.*, **68**, 3191 (1964).
300. E. W. Godbole and P. Kebarle, *Trans. Faraday Soc.*, **58**, 1897 (1962).

301. W. Kaul, U. Lauterback, and R. Taubert, *Z. Naturforsch.*, **16a**, 624 (1961).
302. F. H. Field and F. W. Franklin, *J. Am. Chem. Soc.*, **80**, 5583 (1958).
303. M. A. Berta, B. Y. Ellis, and W. S. Koski, *Advan. Chem. Ser.*, **58**, 80 (1966).
304. I. Szabo, *Arkiv Fysik*, private communication.
305. A. Giardini-Guidoni and G. G. Volpi, *Nuovo Cimento*, **17**, 919 (1960).
306. G. A. W. Derwish, A. Galli, A. Giardini-Guidoni, and G. G. Volpi, *J. Chem. Phys.*, **40**, 3450 (1964).
307. G. A. W. Derwish, A. Galli, A. Giardini-Guidoni, and G. G. Volpi, *J. Am. Chem. Soc.*, **86**, 4563 (1964).
308. G. A. W. Derwish, A. Galli, A. Giardini-Guidoni, and G. G. Volpi, *J. Chem. Phys.*, **41**, 2998 (1964).
309. V. Aquilanti and G. G. Volpi, *Ric. Sci.*, **36**, 359 (1966).
310. H. von Koch, *Arkiv Fysik*, **28**, 559 (1965).
311. J. A. Green and T. M. Sugden, *Ninth International Symposium on Combustion*, Academic Press, New York, 1963.
312. S. K. Gupta, E. G. Jones, A. G. Harrison, and J. J. Myher, *Can. J. Chem.*, **45**, 3107 (1967).
313. F. K. Amenu-Kpodo, Master's Thesis, Univ. of Toronto, 1965.
314. J. F. Paulson, *Advan. Chem. Ser.*, **58**, 28 (1966).
315. B. R. Turner, J. A. Rutherford, and R. F. Stebbings, 18th Annual Gaseous Electronics Conference, Minneapolis, Minn., October, 1965.
316. E. E. Muschlitz, Jr., *J. Appl. Phys.*, **28**, 1414 (1957).
317. M. S. B. Munson, F. H. Field, and J. L. Franklin, *J. Am. Chem. Soc.*, **85**, 3584 (1963).
318. F. H. Field, H. N. Head, and J. L. Franklin, *J. Am. Chem. Soc.*, **84**, 1118 (1962).
319. F. H. Field and J. L. Franklin, *J. Am. Chem. Soc.*, **83**, 4509 (1961).
320. J. L. Franklin and M. S. B. Munson, Tenth International Symposium on Combustion, The Combustion Institute, 1965, p. 561.
321. E. G. Jones and A. G. Harrison, *Can. J. Chem.*, **45**, 3128 (1967).
322. A. G. Harrison and J. C. J. Thynne, private communication.
323. A. G. Harrison, *Can. J. Chem.*, **41**, 236 (1963).
324. F. H. Field, *J. Am. Chem. Soc.*, **83**, 1523 (1961).
325. M. S. B. Munson, *J. Phys. Chem.*, **70**, 2034 (1966).
326. P. S. Rudolph, S. C. Lind, and C. E. Melton, *J. Chem. Phys.*, **36**, 1031 (1962).
327. Z. Herman and V. Čermák, *Collection Czech. Chem. Commun.*, **30**, 2114 (1965).
328. V. Čermák and Z. Herman, *Collection Czech. Chem. Commun.*, **30**, 1343 (1965).
329. B. J. Burtt and Jay Henis, *J. Chem. Phys.*, **41**, 1510 (1964).
330. K. Jager and A. Henglein, *Z. Naturforsch.*, **21**, 1251 (1966).
331. R. D. Doepker and P. Ausloos, *J. Chem. Phys.*, **44**, 1951 (1966).
332. P. Ausloos and S. G. Lias, *J. Chem. Phys.*, **43**, 127 (1965).
333. R. D. Doepker and P. Ausloos, *J. Chem. Phys.*, **42**, 3746 (1965).
334. C. E. Melton and P. S. Rudolph, *J. Chem. Phys.*, **33**, 1594 (1960).
335. F. H. Field and F. W. Lampe, *J. Am. Chem. Soc.*, **80**, 5587 (1958).
336. S. G. Lias and P. Ausloos, *J. Chem. Phys.*, **37**, 877 (1962).
337. R. P. Borkowski and P. Ausloos, *J. Chem. Phys.*, **40**, 1128 (1964).
338. L. M. Draper and J. H. Green, *J. Phys. Chem.*, **68**, 1439 (1964).
339. R. P. Borkowski and P. Ausloos, *J. Chem. Phys.*, **38**, 36 (1963).
340. P. Ausloos and S. G. Lias, *J. Chem. Phys.*, **45**, 524 (1966).
341. L. W. Sieck and J. H. Futrell, *J. Chem. Phys.*, **45**, 560 (1966).
342. P. Ausloos, S. G. Lias, and R. Gorden, Jr., *J. Chem. Phys.*, **39**, 818 (1962).

343. P. Ausloos, S. G. Lias, and R. Gorden, Jr., *J. Chem. Phys.*, **40**, 1854 (1964).
344. P. Ausloos and S. G. Lias, *J. Chem. Phys.*, **40**, 3599 (1964).
345. P. Ausloos and S. G. Lias, *Discussions Faraday Soc.*, **39**, 36 (1965).
346. P. N. Rylander and S. Meyerson, *J. Am. Chem. Soc.*, **78**, 5799 (1956).
347. R. H. Lawrence, Jr. and R. F. Firestone, *J. Am. Chem. Soc.*, **87**, 2288 (1965).
348. R. H. Lawrence, Jr., and R. F. Firestone, *Advan. Chem. Ser.*, **58**, 278 (1966).
349. V. Aquilanti and G. G. Volpi, *J. Chem. Phys.*, **44**, 2307 (1966).
350. I. B. Sandoval and P. Ausloos, *J. Chem. Phys.*, **38**, 2454 (1963).
351. P. Ausloos, S. G. Lias, and I. B. Sandoval, *Discussions Faraday Soc.*, **36**, 66 (1963).
352. R. Gorden, Jr., R. D. Doepker, and P. Ausloos, *J. Chem. Phys.*, **44**, 3733 (1966).
353. P. Ausloos and S. G. Lias, *J. Chem. Phys.*, **43**, 127 (1965).
354. P. Ausloos, A. A. Scala, and S. G. Lias, *J. Am. Chem. Soc.*, **88**, 1583 (1966).
355. F. P. Abramson and J. H. Futrell, *J. Phys. Chem.*. **71**, 1233 (1967).

Chapter 5

Inorganic Gases

A. R. Anderson

Chemistry Division, A.E.R.E., Harwell, England

1. HISTORICAL SURVEY

With the contemporary emphasis on the radiolysis of solids and liquids, it is salutary to recall that the true origins of radiation chemistry can be traced to the observations of chemical changes induced by electrical discharges in gases. These were first observed in the 18th century and studied quite extensively during the 19th century. In many of these early studies it was recognized, either at the time or subsequently, that some of these reactions offered the possibility of converting electrical energy into chemical free energy and a review in 1909 (1) discussed many reactions which are still widely studied at present. Indeed the detailed understanding of some of these reactions and their quantitative description still remain elusive, e.g., the radiolytic decomposition of carbon dioxide, the oxidation of nitrogen, and the formation of ozone from irradiated oxygen.

While these early studies revealed a new and interesting branch of chemistry, the establishment of the radiation chemistry of gases as an exact quantitative science is largely due to the pioneering work of Lind in the U.S.A. and of Mund in Belgium. By using carefully prepared sources of radon, Lind and his co-workers studied a wide series of α-particle radiation induced reactions in the gas phase, involving both decomposition and synthesis. Lind's experiments were designed to show more clearly the relationship between the number of ion pairs formed in a gas and the number of molecules undergoing chemical change. While much of the earlier interpretation of the data is no longer valid, or is subject to fresh insight, many of his quantitative data survive rigorous reexamination.

The work developed so rapidly that Lind published his first monograph on the subject in 1921 (2), which was later revised in 1928 and brought up to date in a new publication in 1961 (3).

Mund's work (4,5) was also directed principally to studies of the α-particle radiolysis of gases and included investigations of many classical systems, viz., H_2/D_2 exchange, ozone formation, ammonia synthesis, together with the radiation induced decomposition of many organic molecules.

Lind and Mund interpreted their observations in terms of the "ion cluster" hypothesis, an approach which dominated the interpretation of gas phase radiation chemistry until the mid-1930's. According to this thesis, the radiolytically produced ions were surrounded by "clusters" of neutral molecules. It was argued that when charge recombination between the electrons (or negative ions) and the positive ion cluster occurred, the energy released was shared among all the molecules in the cluster, so that several of them might decompose or participate in subsequent chemical reactions.

A drastic shift in emphasis to neutral excited molecule decompositions and the subsequent free radical mechanisms was occasioned by the publication of two papers in 1936 by Eyring, Hirschfelder, and Taylor (6). These authors pointed out that the average energy required to form an ion pair (" W value") is inevitably greater than the lowest ionization energy of an atom or molecule, and that the excess energy must be taken up in the formation of electronically excited molecules. They also suggested that ion formation would be followed by charge neutralization to form electronically excited molecules, which would undergo dissociation to form free radicals whose reactions would account for the formation of all observed products.

Support to these views was apparently given by the results obtained by Essex and his co-workers (7,8) who demonstrated that radiolysis was still observed in the presence of an applied electric field. Since it was believed that under these conditions, all ions would be collected at the electrodes and the dissipation of the recombination energy at the electrode surface would reduce or eliminate the resultant formation of free radicals, it was argued that any products formed under these conditions must have precursors other than ions. As a result of the interpretation by Eyring, Hirschfelder, and Taylor, ions were relegated to a relatively minor role as precursors of free radicals, a bias which dominated the interpretation of radiation chemical systems for nearly twenty years.

A renewed recognition of the important part played by ions in determining the chemistry of irradiated systems was brought about by the elucidation of the details of ionic fragmentation processes (Chapter 2) and of ion–molecule reactions (Chapter 4) in mass spectrometric investigations. Moreover, recent mass spectrometric investigations of ions formed in irradiated gases at modest pressures (≤ 200 torr) (9–13) have indicated that ion clusters may be formed in polar gases. It is now clearly recognized that the radiation chemistry of gases can be described adequately only by considering the detailed fate of both ions and electronically excited molecules, formed by the absorption of high energy radiation.

Interest in the radiation chemistry of certain inorganic gases has been stimulated by various practical problems. For example, questions about the radiolysis of carbon dioxide were raised by its use as a coolant in nuclear reactors. Other examples include the irradiation of ammonia to produce hydrazine as a rocket fuel, or radiation studies directed at obtaining a commercially viable method for the fixation of atmospheric nitrogen. In addition, interest in studies involving oxygen, ozone, oxides of nitrogen, water vapor, and carbon dioxide, has been aroused by the close correlation between the radiation chemistry of these systems and the chemistry of the upper atmosphere.

II. THE ROLE OF IONS AND ELECTRONICALLY EXCITED MOLECULES

The initial overall effect of the absorption of high energy radiation is the formation of ions and electronically excited species along the track of the ionizing particle. The details of the individual processes and their mathematical descriptions have been given in Chapter 1.

A. Positive Ions

It has been shown in Chapters 1 and 2 that positive ions formed by the absorption of high energy radiation will often be in excited states. The excited ions may dissociate (fragment) or undergo internal rearrangement before collisional deactivation or charge recombination occur.

It should be emphasized that the distribution of fragment ions observed in a mass spectrometer is not necessarily reproduced in other radiation chemical systems, since most radiation chemical studies are carried out at pressures several orders of magnitude greater than the pressures used in the typical mass spectrometer. At these higher pressures, the probability of reaction or collisional deactivation of the excited ions is greatly increased.

Reactions of ions with neutral parent molecules are now known to be of paramount importance in radiolytic product formation. These reactions are discussed in detail in Chapter 4, and specific ion–molecule reactions of special importance in inorganic systems are presented in later sections of this chapter dealing with the radiolysis of specific gases. However, another general class of reaction involving ions which merits particular attention in the consideration of inorganic systems is the formation of stable ion–molecule clusters, for example, ions such as $H_3O^+(H_2O)_7$ and $NH_4^+(NH_3)_{20}$ have been observed in the mass spectrometer (12,13).

B. Reactions of Negatively Charged Species

The moderation of fast secondary electrons produced in ionization, the calculation of electron degradation spectra, and the role of subexcitation electrons are discussed in Chapter 1. It should be emphasized that several inorganic molecules (e.g., O_2, halogens) have strong electron affinities and are able to form negative ions in the vapor phase. Attachment may be either dissociative or nondissociative:

$$X + e^- \rightarrow X^- \tag{1}$$

$$X + e^- \rightarrow A^- + B \tag{2}$$

Several ion–molecule reactions involving negative inorganic ions such as O^-, Cl^-, or OH^- have been identified in the mass spectrometer (15,16) and in afterglow studies (17) but the main importance of negative ions in influencing the chemistry of an irradiated system probably lies in their role in charge recombination reactions. Thus, in a system where electron capture occurs, there is a possibility that the charge recombination step may involve a negative ion rather than an electron.

Recombination of a positive ion and electron (or negative ion) gives an excited molecule as the initial product, which may then dissociate:

$$X^+ + e^- \rightarrow X^* \qquad (3a)$$
$$\downarrow$$
$$A + B \qquad (3b)$$

One or more of the subsequent dissociation products, which may be stable molecules or free radicals, will probably be excited, and translationally or vibrationally excited radicals formed in this way may be very reactive chemically. The high chemical reactivity of the excited radicals means that they are not susceptible to the normal scavenging techniques used for removing thermal radicals by reaction with low concentrations of chemically reactive substances. The term "hot" is often applied to these excited radicals, but they should be distinguished clearly from the "hot" atoms formed in nuclear reactions (e.g., $^3He(n,p)^3H$) which may possess several kilo-electron volts of kinetic energy.

The recombination processes can be modified by secondary reactions occurring before, and in competition with, neutralization by combination with an electron. Capture of electrons by molecules with high electron affinities are exothermic reactions so that the energy released on subsequent recombination is lowered by an amount equal to the electron affinity. If this leads to an energy release less than the lowest dissociation levels of species in the charge neutralization complex, then no bond scission can occur, although the formation of excited molecules may lead to further reaction on subsequent collision.

Association of the positive ion can also modify the charge neutralization reaction if the solvation of the positive ion is strongly exothermic as for water vapor (12) and ammonia (13). Detailed knowledge of the products resulting from charge neutralization is rather sparse, but there appears to be an increasing amount of evidence in the radiolysis of inorganic gases that recombination reactions are not necessarily dissociative and therefore not of the type shown in reaction (3b). This evidence is mainly deductive and stems from observations in a number of gases where the interpretation of product yields and reaction mechanisms suggests that the recombination processes are nondissociative, e.g., in the radiolysis of O_2 (18), N_2O (19), CO (20,21), CO_2 (22), and N_2-O_2 mixtures (23).

C. Electronically Excited Molecules

Electrically neutral excited molecules are formed in two ways under the action of ionizing radiation, viz., by the direct excitation to a specific electronic level above or below the ionization potential, and as products in charge recombination reactions. It is important to emphasize that there is an appreciable probability of a molecule receiving energy in excess of its lowest ionization potential without immediate ejection of an electron, thus forming electrically neutral excited molecules possessing energy greater than the ionization energy ("superexcited molecules"). Platzman (24) has described these superexcited states and has pointed out the implications for radiation chemistry. A superexcited molecule, rather than undergo ionization, may, like molecules excited to states below the ionization potential, undergo dissociation to form smaller molecules or free radicals, one or both of which may be electronically excited. It is known (24) that typical molecular excitation levels produced from the absorption of ionizing radiation lie very high.

As pointed out in Chapter 1, the cross section for excitation to various electronic states is proportional to the ratio of oscillator strength/energy transition, and it is a corollary of this relationship that only optically allowed transitions can occur. However, at low energies, (e.g., electrons with kinetic energies < 100 eV) optically forbidden transitions can be induced by electron exchange. In addition, the absorption of energy from ionizing radiation is highly nonspecific, and can excite any part of the molecule. These facts, together with the high quantum energies of ionizing radiation and the large proportion of highly excited molecules which are produced mean that, while data from photochemical experiments can be useful in understanding radiation chemical processes, such data must be extrapolated with care both quantitatively and qualitatively.

An electronically excited molecule is thermodynamically unstable, and can lose energy rapidly by several competitive pathways. The actual lifetime of an excited molecule depends on its nature, on the complexity of the molecule, and the possible alternative degradation processes. The magnitude of such lifetimes are generally in the very wide range from 10^{-13} to 10^{-3} sec. The processes which are of prime interest to the chemist are those which lead to chemical reaction, the most obvious being unimolecular dissociation (or isomerization), and bimolecular collisions involving electron transfer, abstraction, or addition reactions. Unimolecular dissociation of excited molecules usually leads to the formation of smaller molecules and/or chemically reactive free radicals which can participate in reactions involving abstraction, addition, disproportionation, rearrangement, and combination, details of which are discussed more fully in Chapter 3. In contrast the most likely processes leading to energy degradation without reaction are radiation conversion (fluorescence), or nonradiative conversion (internal conversion) to the ground state. The latter is generally less probable than internal conversion to the lowest excited state followed by fluorescence to the ground state. Internal conversion is a rapid process ($\sim 10^{-13}$ sec) so that while higher electronic states are formed initially and can lead to rapid unimolecular dissociation, bimolecular reactions of excited states are generally those of the lowest excited states. Other energy degradation processes include intersystem crossing which involves internal conversion with change of multiplicity, i.e., transition from a low-lying singlet state to a lower-lying triplet excited state. Triplet states are potentially very important in radiation chemistry, since light emission with change of multiplicity (phosphorescence) is a slow process ($> 10^{-4}$ sec) and the electronic energy is available for comparative long times to promote chemical reaction. Triplet states may also be formed by direct excitation by slow electrons and in the recombination reaction of a positive ion and an electron.

It is clear that fluorescent energy emitted by one molecule could be absorbed by another but energy transfer can also occur from excited molecules by a nonradiative resonance process; this is formally equivalent to the emission of a photon by the excited molecule and its absorption by another molecule whose absorption spectrum overlaps the emission spectrum of the emissive molecule. This process is not restricted to situations involving collisions between molecules, but can occur when the distance separating the molecules is less than the wavelength of the emitted photon and can take place efficiently over distances of 50–100 Å (25).

All these energy transfer processes afford ways to concentrate the effect of energy absorbed by a mixture into chemical reactions of a minor component, and emphasize the important role which can be played by traces of foreign material (present as impurities or added deliberately) in determining the overall radiation chemistry which is observed.

III. GAS PHASE DOSIMETRY

A. Radiation Chemical Yields

Radiation chemical yields in the gas phase may be expressed either in terms of the ion pair yield, (M/N), or of the G value. The ion pair yield has the units of molecules or atoms (M) formed or converted per ion formed (N); the G value is the number of molecules or atoms formed or destroyed by the absorption of 100 eV of energy (see Chapter 1). Interconversion between G values and ion pair yields is simple when the W value is known: $M/N \times 100/W = G$. W values of most inorganic molecules are in the region of 25–35 eV.

The ion pair yield is the more fundamental unit, and was used exclusively in early radiation chemistry studies but increased emphasis on condensed phase radiation chemistry led to extensive use of the G value. This practice at least has the advantage of providing continuity in the comparison of quantitative data from the three phases, and will be used throughout this chapter. Back et al. have, however, suggested (26) that gas phase yields should again be expressed in terms of M/N (based on direct measurements of ion currents) in view of the difficulties of measuring absorbed dose in gases. This suggestion has considerable merit and highlights the difficulties of gas dosimetry.

B. Primary Dosimetry

The basic requirement of any primary dosimetric method is an accurate measurement of the energy input to the system, which can be achieved by calorimetry in condensed media. In gases at modest pressures, however, the use of calorimetry becomes impossible since the energy absorbed by the container walls is generally much greater than that absorbed by the gas and corrections become unreasonably large. In general, all the problems of gas phase dosimetry stem from the disparity in energy absorption between the gas and the container walls since the energy absorbed by the gas is due mainly to secondary electrons ejected from the walls (27a) (see Chapter 1).

There are three methods which can be used for primary dosimetry in gases, viz.,

1. measurement of ion currents
2. calculation of energy input from a calibrated internal source
3. the complete absorption of an ion beam (of precisely defined energy) from an accelerator and measurement of the ion current

None of these methods is completely satisfactory, however, since the latter two are applicable only in special cases and the first measures only ion currents and not absorbed dose. Ionization methods require a great deal of care in their use to ensure that the requisite physical conditions are fulfilled, and care in interpretation of the data to ensure that the necessary corrections are rigorously and accurately applied (27b). The suggestion of Back et al. (26) to measure ion currents in a simple vessel which can also be used for conventional chemical measurements is the most attractive to radiation chemists but it has some potential drawbacks. In order to exploit the method fully it is necessary *inter alia* to ensure that the chemistry is independent of the nature of the electrode surfaces, the geometry of the vessel (i.e., electrode separation distances, surface/volume ratio) and gas pressure.

When a beam of ions is absorbed completely by a gas under irradiation e.g., with low energy H^+, D^+, or He^{2+} particles in gases at ca. 1 atm pressure the range is only a few cm, the whole cell is a Faraday cup if electrically insulated, and the indirect measurement of energy absorption can be made with great precision. The incident energy of the ions from the accelerator is known with high precision (better than $\pm 1\%$), the integrated ion current can be measured very accurately, and window loss corrections (which are generally small with the thin windows used) can be measured or calculated precisely. There are further corrections which have to be applied due to the forward ejection of electrons from windows made of insulating materials but again these corrections are small for thin windows and the over-all errors in the method can be readily reduced to $\pm 2\%$. The drawbacks in the method lie in its specificity of application, the nonuniform dose rate in the irradiated zone and the possible perturbing effect on the chemistry of the small internal electrode for current measurements.

The dosimetry of internal radiation sources, e.g., Rn, ^{3}H, ^{32}P, ^{35}S relies on a knowledge of the absolute value of the amount of radioactivity present, the half-life of the isotope, the energy of the particle(s) emitted, and the short range of the α or β particle compared with the size of the irradiated volume. If the latter condition is fulfilled it is assumed that the total energy liberated by the disintegrating nucleus constitutes the absorbed dose. Corrections can be made for absorption by the walls, the absorption or loss of energetic γ-rays, and for energy transferred to the recoil nucleus. In general, however, these corrections are (or can be made) small, and high precision can be claimed for this method, which suffers mainly from its specificity. It can be used on a comparative basis for dosimetry in the absence of LET effects, e.g., ^{3}H beta particle radiation has been used as a comparative standard for studies with ^{60}Co γ radiation.

The complex problems of accurate dosimetry in nuclear reactors have not been discussed since they are outside the scope of this book and are the subject of an authoritative review (28).

C. Secondary Dosimetry

While the careful establishment of primary dosimetry methods is of paramount importance, the general requirement of radiation chemists is for a secondary dosimeter which is accurate and simple to use. Ideally the gas used as a dosimeter should be easily obtainable, stable under laboratory conditions, and its chemistry should be independent of traces of impurities, or the effects of impurities should be readily suppressable. Moreover the product yield(s) should be reproducible, independent of dose rate and total dose (over a range dictated by the nature of the work), independent of pressure and temperature, and either independent of LET or well characterized for radiation of differing quality. Many of these requirements have been met by the Fricke dosimeter for condensed media but, as yet, any comparable success in developing a gas phase dosimeter has been rather limited. There is little general agreement on the dosimeter to use and, where a choice has been made, there are often wide and inexplicable variations in the reported yields.

The principal gas phase chemical dosimeters which have been used are the production of nitrogen from nitrous oxide, the polymerization of acetylene, and the production of hydrogen from ethylene. Nitrous oxide has been used most frequently but there is no accepted agreement on yields, on the effect of pressure, or on the effect of temperature (Section V-H). The use of ethylene is becoming more widespread and there appears to be a general measure of agreement on the yield of hydrogen in the range $G(H_2) = 1.2$–1.3 for γ radiation and for electrons. Moreover it has been shown recently that with H^+ and He^+ radiation from accelerators (29a) the yield of hydrogen ($G(H_2) = 1.31 \pm 0.02$) appears to be independent of dose rate up to a value of at least 10^{18} eV cm^{-3} sec^{-1}, and independent of percentage decomposition up to an overall conversion of 10%. This latter result is probably unique to the irradiation conditions used, however, since the polymer which is formed in high yield ($G > 11$) settles on to the walls of the irradiation vessel and thus its concentration in the irradiated zone, defined by the beam geometry, is largely independent of dose.

Janssen (30) et al. have suggested the use of CS_2 as a gas phase dosimeter, and report that the decrease in pressure is independent of dose, dose ratio and pressure of CS_2.

The detailed radiation chemistry of nitrous oxide is described in Section V-H and that of ethylene and acetylene, in Chapter 6.

IV. EXPERIMENTAL TECHNIQUES

Although the classical work on the radiation chemistry of gases used alpha particle radiation from internal sources, most of the contemporary work is carried out with external sources of radiation, which include ^{60}Co γ radiation, ions and electrons from accelerators, and mixed gamma and fast neutron radiation in nuclear reactors. While the particular method of dosimetry employed may depend to some extent on the radiation source being used, many of the experimental techniques of gas phase radiolysis are independent of the source. Some of the general techniques are described first, followed by a brief description of some more specific techniques.

A. Scavenging Techniques

Trace impurities may effect major changes in the radiolytic reaction mechanism of a compound. For instance, the absorbed energy can effectively be concentrated in a compound present only in trace amounts if its ionization energy or electronic excitation energy levels are lower than those of the major component of the mixture. A trace component with high electron affinity can alter the neutralization mechanism; some specific trace impurities may undergo ion–molecule reactions with intermediates resulting from irradiation of the major compound, or may be very reactive toward free radicals. These properties can be utilized as important diagnostic tools. Deliberate addition of small concentrations of foreign compounds ("scavengers") which will affect the reaction mechanism in a specific, known way is a valuable method for deducing the details of radiolytic processes.

Electron scavengers such as N_2O, SF_6, and CCl_4, which capture electrons efficiently, are used to determine the extent of the reaction due to processes involving electron capture or electron–ion recombination, e.g., in the radiolysis of water vapor (31,32). Radical scavengers, which react efficiently with free radicals, are used to inhibit the normal secondary reactions of these species and to change the product distribution in an understandable way. Ethylene and propylene are commonly added to remove H atoms without the resultant formation of molecular hydrogen (32), and O atom scavengers, SO_2 and NO_2, have been used to determine primary yields in the radiolysis of CO_2 (33,34) by inhibiting the oxidation of the product CO. More detailed applications of these techniques are described in the sections dealing with the radiolysis of specific gases.

Positive ion or charge scavenging can also be important. Water is a common impurity and since its ionization potential is lower (12.59 eV) than that of many inorganic gases, e.g., H_2, N_2, CO, CO_2 it could comprise a significant contaminant in these systems unless the gases are thoroughly dried before irradiation.

B. Isotopic Studies

Isotopic labeling used in conjunction with product analysis is a powerful tool for the elucidation of mechanisms of radiation induced processes and indeed in gases where no net chemical change takes place it is the only method which can be employed to examine the primary processes, e.g., the radiolysis of H_2 (35,36), and N_2 (23,37). Other investigations in which isotopic labeling has been used include the radiolysis of water vapor (38–40), of NH_3 (41), CO_2 (42–44), and N_2O (45). By irradiating mixtures of D_2O vapor with various light hydrogen radical scavengers, Baxendale and Gilbert (46) showed the existence of a residual yield of D_2, and in a similar fashion Anderson et al. (47) obtained a residual yield of H_2 in the radiolysis of H_2O/C_6D_6 mixtures. In both cases the "residual" hydrogen could not be scavenged and the use of the isotopic technique showed that it originated from the water and not from the added scavenger. In the radiolysis of CO_2, Dominey (42,43) and others (22) have used the radioactive isotope ^{14}C to elucidate details of the radiation-induced exchange reaction between CO and CO_2 and to gain an insight into the principal processes which account for the radiation stability of CO_2. Anbar and Perlstein (44) have used a combination of various stable isotopes of carbon and oxygen in mixtures of $CO_2/CO/O_2$ for a similar purpose. Details of all these applications are given in the relevant sections.

C. The Effect of Applied Electric Fields

The principles and complexities involved in the application of electrostatic fields to irradiated gases are described fully in Chapter 6. As pointed out, complications due to negative ion formation may occur in the irradiation of inorganic gases. The radiolysis of NH_3 affords a good example of the effects which can be studied with applied electrical fields in gases. In this system, charge neutralization occurs predominantly with the ammonium ion NH_4^+ formed in the ion–molecule reaction (4).

$$NH_3^+ + NH_3 \rightarrow NH_4^+ + NH_2 \tag{4}$$

It is clear that charge neutralization of this secondary ion leads to a radical which is intrinsically unstable,

$$NH_4^+ + e^- \xrightarrow[\text{surface}]{\text{electrode}} NH_4^* \tag{5a}$$

$$\downarrow$$

$$NH_3 + H \tag{5b}$$

The predominance of the NH_4^+ ion in NH_3 radiolysis was not known to Smith and Essex when they assumed that charge neutralization at the electrode does not lead to decomposition (48), so that their deductions about the contribution of ionic processes to the radiolysis are incorrect. It is possible that reactions such as (5) at a surface could lead to diminution in product yields due to the recombination of H atoms on the surface or to their reactions with surface defects, at the expense of the homogeneous reaction (6)

$$H + NH_3 \rightarrow NH_2 + H_2 \tag{6}$$

The heterogeneous processes would, however, depend on the nature of the surface and could vary with the state and composition of the electrode material. Examples of additional work on the effect of electrostatic fields on NH_3 radiolysis (41,49) and on N_2O radiolysis (19,50) are given in Sections V-I and V-H, respectively.

D. Pulse Radiolysis

The technique of pulse radiolysis has been used increasingly since 1960 to identify and to study the kinetic behavior of chemically reactive intermediates in liquids, but the method has not been readily applied to gases due to the much smaller energy absorption in gases at modest pressures (ca. 1 atm). This restriction was first overcome by Sauer and Dorfman (51) who used a high pressure cell to investigate the pulse radiolysis of oxygen in the presence of inert gases at a total pressure of 50 atm. Sauer (52) has used this technique to extend studies of the kinetics of ozone formation from oxygen, and has measured the relative efficiencies of various gases acting as third bodies in the reaction (Section V-E). Meaburn and Perner (53) observed transient species in CO_2, CO, and CH_4 at lower pressures (~ 400 torr) by using the very high dose rates available from a Febetron generator (Field Emission Corporation) producing 250 keV electrons with a pulse length of 2×10^{-7} sec and a dose in the pulse of 6×10^{18} eV. They identified one of the transient species produced in these systems as a carbon atom from the coincidence with a line in the carbon arc.

E. Luminescence Studies

Another potentially useful technique, which has been little exploited, is the observation of luminescence from gases under ionizing radiation. Work of this type has been reported by several authors (54–57) but its usefulness has been restricted, mainly due to the low light intensities produced by nonpulsed sources and the correspondingly long exposure times required to obtain satisfactory spectra. With the development of image intensifying techniques, however, this method may be used more extensively in the future since it offers the opportunity to study competition between light emission from specific excited states and collisional deactivation, under a variety of conditions.

V. RADIOLYSIS OF SPECIFIC GASES

In describing the radiation chemistry of specific gases particular attention has been given to work which has been reported since the publication of previous monographs or texts (3,5,58), although for completeness relevant earlier work is included to give full descriptions of the systems. This treatment means that greater emphasis is given to certain gases than in previous texts, for since 1964 (58) a considerable volume of new work has been published on studies of water vapor, carbon dioxide, nitrous oxide, and ammonia. In addition new contributions have been made to the interpretation of classical systems, e.g., the radiation-induced exchange reaction between hydrogen isotopes, and the radiolysis of hydrogen halides, and rather surprisingly the first measurements using the isotope exchange technique on pure nitrogen have been reported.

It is difficult to give equal treatment to all published work so that, as far as possible, data have been selected to describe the major features of the particular reactions and to provide a basis for understanding the suggested interpretations. Where conflict in data or interpretation exists, the temptation to make ex cathedra statements has been largely resisted and the various views are expressed as objectively as possible; if any subjective emphasis in interpretation is detected it is unintentional. Many of the controversies in interpretation which exist at present are due to lack of adequate data to give a complete description of the system and, moreover, mechanisms are being revised continually in the light of new experimental evidence. These changes in interpretation are often due to rapid developments in techniques leading to greater precision in conventional measurements and to new methods of detecting and measuring intermediates, developments which help to ensure that the radiation chemistry of gases remains a live and vital part of chemical kinetics.

A. Hydrogen

The radiolysis of hydrogen produces no chemical change but the primary radiolytic processes have been elucidated by studies of the radiation-induced exchange reactions in the systems H_2/D_2 and H_2/T_2, and of the o-p H_2 conversion. These reactions all proceed by a chain mechanism which was considered to be exclusively a free radical process until the work of Schaeffer and Thompson (35,59) showed that the predominant chain carrier in the H_2/D_2 radiation-induced exchange is the H_3^+ ion. In the absence of additives they found exchange yields for the alpha radiation induced reaction as high as $M/N = 1.8 \times 10^4$ at 100 mm pressure but these yields were markedly reduced by the presence of low concentrations ($< 1\%$) of Kr and Xe. With low concentrations of the rare gases He, Ar, and Ne (ionization potentials greater than that for H_2) there is a small enhancement of the exchange rate but at higher concentrations ($> 10\%$) the reaction is again inhibited. The thermal or photolytically induced exchange reactions are not affected by the presence of any of these rare gases and these observations led Schaeffer and Thompson to propose an ionic chain mechanism for the radiation-induced process. Initially (59) they ascribed the inhibition by Xe and Kr to a charge transfer process but later (35) proposed a more energetically favorable scheme involving proton transfer. The details of the mechanism, based on some known ion–molecule reactions and calculated bond energies are as follows:

Initiation	$H_2^+ + H_2 \rightarrow H_3^+ + H$	(7)
Ionic chain	$H_3^+ + D_2 \rightarrow HD_2^+ + H_2$	(8)
	$HD_2^+ + H_2 \rightarrow H_2D^+ + HD$	(9)
	$H_2D^+ + D_2 \rightarrow HD_2^+ + HD$	(10)

Initiation inhibition R is any rare gas	$H_2^+ + R \rightarrow RH^+ + H$	(11)
Chain inhibition R is Kr or Xe	$H_3^+ + R \rightarrow RH^+ + H_2$	(12)
Enhancement	$R^+ + H_2 \rightarrow RH^+ + H$	(13)
	$RH^+ + H_2 \rightarrow R + H_3^+$	(14)
R is He, Ar, Xe only		
Termination	$H_3^+ + M^- \rightarrow H_2 + M + H$	(15)
	$H_3^+M + e^- \rightarrow H_2 + H + M$	(16)

where M^- is H^- or an impurity ion and H_3^+M is most probably $H_3^+H_2$.

This mechanism explains all the main features of the reaction for inhibition of the formation of chain carriers, i.e., competition between reactions (7) and (11) will only be effective when the concentration of rare gas is comparable to that of hydrogen. On the other hand, inhibition of the chain carrier (reaction (12)) would be expected at low concentrations of additive. It is not clear whether the termination reaction is homogeneous or heterogeneous, since the measured chain length is very susceptible to traces of impurities. On the assumption that termination is due to H_3^+ diffusing to the walls, Schaeffer and Thompson (35) calculated a theoretical yield for the exchange reaction of $M/N = 1.8 \times 10^7$ which is about 10^3 times the values measured under very stringent conditions of purity. Calculations of the chain length with H atoms as carriers gave $M/N = 40$ which is very close to the value obtained in the presence of traces of Xe.

The ionic chain involving H_3^+ must also be important in the conversion of *para* to *ortho* hydrogen, which reaction proceeds with a chain length of 10^3 and previous descriptions of the mechanism based on H atom reactions alone are incomplete.

In a mass spectrometric study of the rare gas/hydrogen system at high pressure (~ 0.3 torr), Aquilanti et al. (36) measured changes in ion intensities which were largely in accord with the mechanism of Schaeffer and Thompson. In the presence of small amounts of Xe and N_2 the hydrides XeH^+ and N_2H^+ were formed at the expense of H_3^+ which clearly indicates occurrence of the reactions (17) and (18).

$$H_3^+ + Xe \rightarrow XeH^+ + H_2 \tag{17}$$
$$H_3^+ + N_2 \rightarrow N_2H^+ + H_2 \tag{18}$$

In view of the inhibiting effects of Kr on the exchange reaction, surprisingly they found that added Kr ($\leq 10\%$) had no effect on the intensity of H_3^+. However they argue that the observed inhibition by Kr may be explained with a value for k_{12} of 10^{-14} cm^3 molecule^{-1} sec^{-1} if $k_8 = 8 \times 10^{-13}$ cm^3 molecule^{-1} sec^{-1}, as they calculate. They also estimate that with this value for k_8 the chain length observed by Schaeffer and Thompson

$(M/N = 1.8 \times 10^4)$ could be accounted for by a partial pressure of nitrogen impurity as low as 5×10^{-6} torr. In the presence of argon they observed an increase in $H_3{}^+$ intensity due to the reactions,

$$Ar^+ + H_2 \rightarrow ArH^+ + H \qquad (19)$$

$$ArH^+ + H_2 \rightarrow H_3{}^+ + Ar \qquad (20)$$

With Ne and He added however, they found no change in the intensities of the ions $H_3{}^+$, Ne^+, and He^+ at partial pressures up to 0.03 torr, at a total pressure of 0.3 torr. These observations agree with previous studies (60,61) which have shown that neon or helium hydride ions are formed mainly by the reactions of vibrationally excited $H_2{}^+$ ions with little contribution from reaction (13).

The apparently anomalous role of krypton in its inability to react with the chain carrier $H_3{}^+$ to form KrH^+ was further emphasized in studies of the H_2–D_2 exchange reaction at low gas pressures (~0.3 torr) in the negative glow of the dc discharge, by Dawson and Tickner (62). With the exception of the behavior of Kr, which was found to enhance the exchange reaction, these authors found a good qualitative agreement between the discharge work and the radiolysis experiments. In the case of added Kr they observed the ion $KrH_3{}^+$ and suggested that in the radiolysis experiments of Schaeffer and Thompson (35,59) this ion could be neutralized in the gas phase thus leading to chain termination and inhibition at the higher gas pressures. Dawson and Tickner proposed this alternative inhibition mechanism for Kr in the radiolysis experiments since they suggest that $k_8 \sim 3 \times 10^{-10}$, about 400 times the value deduced by Aquilanti et al., so that inhibition by Kr cannot be attributed to reaction (21),

$$Kr + H_3{}^+ \rightarrow KrH^+ + H_2 \qquad (21)$$

for which these authors give $k_{21} = 3 \times 10^{-12}\ cm^3$ molecule^{-1} sec^{-1}. Dawson and Tickner also show that the formation of the ion $H_5{}^+$ in the uninhibited radiolysis experiments could account for chain termination and that it is unnecessary to postulate the presence of impurities.

B. Hydrogen Halides

The radiation induced reaction between hydrogen and chlorine is also a chain process which results in very high yields of hydrogen chloride ($G = 10^4$–10^5). In this case there is no evidence for ionic chain carriers; the chain is propagated by hydrogen and chlorine atoms, in a manner similar to the classic photochemical synthesis. The atoms are produced as a result of primary ionization and excitation of the elements, and as a result of electron capture by Cl_2 molecules,

$$Cl_2 + e^- \rightarrow Cl + Cl^- \qquad (22)$$

$$Cl_2^+ + Cl \rightarrow Cl_3^+ \tag{23}$$

$$Cl_3^+ + Cl^- \rightarrow 4Cl \tag{24}$$

$$H + Cl_2 \rightarrow HCl + Cl \tag{25}$$

$$Cl + H_2 \rightarrow HCl + H \tag{26}$$

Termination is principally by atom–atom reactions:

$$Cl + Cl + M \rightarrow Cl_2 + M \tag{27}$$

$$H + H + M \rightarrow H_2 + M \tag{28}$$

$$H + Cl + M \rightarrow HCl + M \tag{29}$$

In contrast, however, the radiation-induced reaction between hydrogen and bromine at room temperature is not a chain process and results in a maximum yield, $G(HBr) \sim 9$ (3). The difference is due to the fact that while reaction (26) between Cl atoms and H_2 is approximately thermoneutral, the corresponding reaction with Br atoms is ~ 17 kcal/mole endothermic (with the compounds and elements in their standard states at 18°C and 1 atm pressure). Thus the principal fate of the bromine atoms will be recombination (Eq. (30)) which is equivalent to one of the possible chain termination steps in the H_2/Cl_2 reaction.

$$Br + Br + M = Br_2 + M \tag{30}$$

Gaseous HBr is decomposed (63) with $G(-HBr) \sim 12$ both in the absence and presence of inert gases. The similarity of the yield in the presence of inert gases (based on total energy absorption in the mixtures) is explained by complete charge transfer to the HBr followed by the relevant part of the reaction scheme proposed by Eyring Hirschfelder and Taylor (6b) (Eqs. (30–36)).

$$HBr \rightarrow HBr^+ + e^- \tag{31}$$

$$HBr \rightarrow HBr^* \tag{32}$$

$$HBr^* \rightarrow H + Br \tag{33}$$

$$e^- + HBr \rightarrow H + Br^- \tag{34}$$

$$HBr^+ + Br^- \rightarrow H + 2Br \tag{35}$$

$$H + HBr \rightarrow Br + H_2 \tag{36}$$

However in recent studies concerned with reactions of electrons, Armstrong et al. (64a) have found evidence for three-body capture reactions of the type:

$$e^- + 2HX = HX^- + HX \tag{37a}$$

$$e^- + 2HX \rightarrow H + HX_2^- \tag{37b}$$

and suggest that the dissociative two body reaction (34) is not tenable. These authors have also measured the yield of H_2 from gamma irradiated

(high temperature and low pressure) the charge neutralization reaction will change when a negative ion replaces the electron. With nitrous oxide as an electron scavenger, the negative ion which will participate is O^- or OH^-;

$$N_2O + e^- \rightarrow N_2 + O^- \tag{50}$$

$$O^- + H_2O \rightarrow OH + OH^- \tag{51}$$

In the absence of solvation effects the neutralization reaction (52a) is > 200 kcal exothermic and since the dissociation energy of the OH bond in water is 119 kcal/mole there is sufficient energy available to dissociate a water molecule, reaction (52b),

$$H_3O^+ + OH^- \rightarrow 2H_2O^* + 211\text{--}215 \text{ kcal/mole} \tag{52a}$$
$$\downarrow$$
$$H + OH + H_2O + 92\text{--}96 \text{ kcal/mole} \tag{52b}$$

If the negative ion is O^- in reaction (53), similar arguments apply

$$H_3O^+ + O^- \rightarrow H_2O^* + OH + 213 \text{ kcal/mole} \tag{53a}$$
$$\downarrow$$
$$H + OH + 94 \text{ kcal/mole} \tag{53b}$$

In both cases, however, when the positive and negative ions are solvated there is insufficient energy released in the charge recombination to dissociate a water molecule. Since Melton has recently observed the H_3O^- ion (68a), its participation in the charge neutralization reaction will have to be considered when more data are available on its abundance, stability, etc. The electrically neutral species, H_3O, has also been identified but its mode of formation and its possible role in the radiolysis of water vapor are not known (68b).

Another complicating factor is introduced by the presence of scavengers with higher proton affinities than that for water, e.g., CH_3OH, NH_3, so that proton transfer reactions (54) and (55) can occur:

$$H_3O^+ + CH_3OH \rightarrow H_2O + CH_3OH_2^+ \tag{54}$$

$$H_3O^+ + NH_3 \rightarrow H_2O + NH_4^+ \tag{55}$$

As a consequence of proton transfer the central ion in the ion–molecule cluster is no longer the hydronium ion; in addition the number distribution of molecules in the cluster is not necessarily directly proportional to their concentration in the mixture (13). Thus the products of the charge neutralization process could be changed by the presence of a small amount of material with a high proton affinity.

1. Yields. In contrast to the vast amount of work on the radiolysis of liquid water, published data on water vapor are rather sparse and it is only since 1963 that a significant amount of new work has been reported in the literature. Attempts to obtain meaningful data from the radiolysis of pure water vapor in the absence of deliberately added scavengers have, however, met with little success since reported values of yields show a 500-fold variation between different workers, viz. $G(H_2) = 0.06$ (71), 3 (72), 0.015–1.4 (38), 0.11–1 (73) and 5.9 (74). The following experimental difficulties probably explain these wide variations in reported yields: (1) lack of careful and consistent cleaning of the irradiation vessels, (2) the difficulties in avoiding trace contaminations on transferring water before irradiation, (3) the effects caused by variations in the nature and condition of vessel walls, and (4) the determination of integral G values without carrying out a complete study of yield as a function of dose. Anderson et al. (75) report such a yield vs dose study using ^{60}Co γ radiation, and although their data are somewhat scattered they appear to point clearly to two main conclusions: (1) that once a low steady-state concentration of hydrogen is built up no further net radiolysis occurs, in agreement with predictions from measurements with liquid water, and (2) the previous treatment and history of the cell can have a profound influence on the steady-state concentrations of hydrogen. Although the above integral yields calculated for different doses have no significance if steady-state conditions obtain, it is of interest to note that the lowest value reported by Firestone (38), $G(H_2) = 0.015$ for a measurement at a single dose is very close to the value which can be calculated from Anderson's data at approximately the same total dose (10^{21} eV g^{-1}).

While the previous data are rather unsatisfactory, the essential stability of water vapor to high energy radiation has been demonstrated unequivocally by Anderson and Best (29b) using 1.5 MeV protons from a van de Graaff accelerator. In this work the same silica cell was used repeatedly for consecutive irradiation and sampled through an array of break seals thus minimizing any effect of changes in surface conditions. At the high dose rates used, $7 \times 10^{19} - 10^{22}$ eV g^{-1} sec^{-1}, consistent steady-state concentrations of hydrogen and oxygen were obtained (Fig. 1) showing a linear dependence on (dose rate)$^{1/2}$. At the highest dose rate used the ratio of $[H_2]/[O_2]$ was close to 2 but this ratio fell progressively as the dose rate was lowered. This observation was consistent with the loss of an almost constant amount of oxygen to the walls of the vessel since only traces of hydrogen peroxide were detected at 125°C. Extrapolation of these data to the lower dose rates used in earlier work (75) with γ radiation (10^{17} eV g^{-1} sec^{-1}) lead to a calculated steady state concentration of hydrogen of 3×10^{-4} mole % in comparison with the measured value of 10^{-3} mole %, showing a satisfactory measure of agreement in view of the difficulties in obtaining reproducible data at low dose rates.

While it is of interest to demonstrate the radiation stability of pure water vapor, data of fundamental significance to the elucidation of the

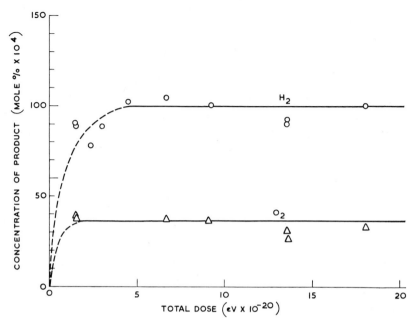

Fig. 1. Steady state concentrations of H_2 and O_2 in the proton radiolysis of pure water vapor. Temperature: 125°C; pressure, 1 atm; dose rate: 5×10^{20} eV g^{-1} sec^{-1}.

role of ions and of excited states in the radiolysis are only obtained by the addition of chemically reactive scavengers. Much progress has been made along these lines since 1963, and the details will be discussed under the following headings rather than in a strictly chronological order:

>   nonscavengeable yield of hydrogen
>   radical and electron yields
>   isotope effects
>   effects of temperature and pressure

2. Nonscavengeable Hydrogen. It has now been shown by various workers (32,40,46,47) that a nonscavengeable yield of hydrogen is obtained in the presence of both electron and H atom scavengers in the radiolysis of water vapor with a G value of ~ 0.5; these data are summarized together with radical and electron yields in Table I. Anderson et al. found no change in the yield at temperatures up to 200°C, while Johnson and Simic found a constant yield up to 400°C (76) and observed a slight increase in the yield (from 0.64 to 0.9) in the presence of high concentrations of electron scavengers ($>20\%$ N_2O and SF_6) at 140°C (32). This latter result might suggest that the nonscavengeable hydrogen originates from excited water molecules as in the photolysis where molecular detachment of hydrogen (reaction 40) becomes important with light of wavelength 1236 Å (77).

With ionizing radiation, however, other contributions could arise from resonance capture of electrons by water molecules (46) (reactions 56 and 57) or charge neutralization reactions with a specific orientation of water

Table I

Yields from the γ Radiolysis of Water Vapor

System	Temp., °C	$G(R)^a$	$G(H)$	$G(e^-)$	$G'(H_2)^b$	$G(OH)$	Ref.
H_2O/D_2	84–140	11.7					38
H_2O/C_6H_{12}	116	6.8					73
$H_2O/C_2H_5OC_2H_5$		6.8					
H_2O/CH_3OH		7.3					
H_2O/C_2H_5OH		8.1					
H_2O/C_6D_6	120–200				0.47		47
H_2O/NH_3	120	5.7^c				6.7^c	75
		$(5.9)^d$				$(7.0)^d$	
H_2O/NH_3	180–220	6.4^c					31
		(6.6)					
$H_2O/NH_3/O_2$	120	5.6			0.6	6.5	
		(5.8)			(0.62)	(6.8)	
H_2O/C_6H_{12}	120–240	6.5	3.5	3.3			31
H_2O/C_2H_5OH	120–140	(6.7)	(3.6)	(3.4)			
$H_2O/CH_3OH/N_2O$	115			3.0			80
H_2O/CH_3OH		7.4					79
H_2O/C_3H_7OH		8.1					
$H_2O/cyclo-C_6H_{12}$		6.5					
H_2O/N_2O				3.1			
		$G(R)$	$G(D)$	$G(e^-)$	$G'(D_2)$		
D_2O/H_2	116	10.5			0.56		46
D_2O/C_3H_8		7.0			0.48		
D_2O/CH_3OH		7.0			0.80		
D_2O/C_3H_8	100–185	7.6	4.2	2.7	0.64		32
		(7.0)	(3.9)	(2.5)	(0.59)		

[a] $G(R)$ is total yield of reducing species and where necessary is derived from the total yield of hydrogen assuming $G'(H_2) = 0.5$.

[b] $G'(H_2)$ is the yield of nonscavengeable H_2.

[c] Calculated from stoichiometry.

[d] Yields in brackets are adjusted to the arbitrary dosimetry standard, $G(N_2) = 11$ from N_2O, used in references 46 and 73.

molecules around the hydronium ion.

$$e^- + H_2O \rightarrow H^- + OH \tag{56}$$

$$H^- + H_2O \rightarrow H_2 + OH^- \tag{57}$$

Yang and Marcus' suggestion (40) that the nonscavengeable hydrogen arises simply from residual recombination of H atoms does not seem to be supported by the bulk of the evidence.

3. Radical and Ion Yields. The first significant measurement of radical yields was made by Firestone (38), who studied the tritium β radiation induced exchange reaction (Eq. (58)) between D_2 and H_2O over a temperature range from 84 to 278°C.

$$D_2 + H_2O \rightarrow HD + HDO \tag{58}$$

At temperatures above 150°C the exchange takes place via a chain reaction but at lower temperatures the yield of HD approached a constant value $G(HD) = 11.7 \pm 0.6$ which was equated with $G(-H_2O)$. This value is much higher than the value deduced for liquid H_2O ($G(-H_2O) = 4$–6) but a detailed calculation by Fiquet-Fayard (78) considering the fate of various ions and excited states showed that if charge neutralization produced 2 H atoms (reaction 49b) the calculated $G(H) = 11.1$ was in agreement with Firestone's results.

Since 1963, however, a number of independent measurements (31,32,46, 73,75,79) made in the presence of various H atom scavengers have led to values for $G(-H_2O)$ close to 7 under conditions of temperature and pressure similar to those used by Firestone. These data which are summarized in Table I show a large measure of agreement on the total yield of reducing species, for irradiations carried out at temperatures < 200°C and at pressures ~ 1 atm especially when the reported yields are adjusted to the same dosimetry standard. The only reported value for $G(OH)$ from measurement with $H_2O/NH_3/O_2$ (75) is compatible with the total yield of reducing species, but the suggested correlation between $G(-NH_3)$ and $G(OH)$ in this work is not necessarily unambiguous. The assumed stoichiometry implies that reactions such as,

$$NH_4^+ + O_2^- \rightarrow NH_2O + H_2O \tag{59}$$

which would lead to $G(-NH_3) > G(OH)$ do not occur to any significant extent.

The measurement of electron yields by the suppression of hydrogen production in the presence of electron scavengers (31,32,79,80) e.g., N_2O, SF_6, CCl_4, and by deductions based on the effect of temperature on total yields (31,81) have led to values of $G(e^-) = 3 \pm 0.3$. Values for the yields of H atoms from neutral excited molecules have also been deduced from these studies giving $G(H) \sim 3.5$, which is much higher than the maximum possible value suggested for liquid water, i.e., $G(H) = 0.6$. In the liquid phase it is possible that the lower yield is due to collisional deactivation of excited molecules or to rapid recombination of radicals within the solvent cage, but neither of these processes appears to be important in the photolysis of liquid water (at 1470 Å) since the quantum yield $\phi(-H_2O)$ is close to unity (82).

While there is general agreement on the yields measured in the presence of chemical scavengers at temperatures $< 150°C$ and ca 1 atm pressure (Table I), two separate measurements of the radiation-induced exchange reaction between hydrogen and water vapor give significantly higher yields. Baxendale's value (46) for the yield of the H_2/D_2O exchange $G(HD) = 10.5$, measured under the same conditions as his results with chemical scavengers seems to preclude any explanation based on dosimetry artifacts and suggests that there is a fundamental difference between the "exchange" and "chemical" yields. Various explanations have been advanced to explain this difference (31,73) but no unequivocal interpretation has been clearly demonstrated.

4. Isotope Effects. Yang and Marcus (40) have measured isotope effects in the γ radiolysis of tritiated water vapor and in H_2O/D_2O mixtures, each containing cyclopentane. Isotopic analysis of the hydrogen produced led to calculated values of $k_H/k_T = 1.7 \pm 0.1$ and $k_H/k_D = 1.2 \pm 0.1$ as the relative rates for H, T, and D atom formation. These values are much lower than the values of 2–6 reported for the deuterium effect in liquid water, but Johnson and Simic (32) have queried the validity of these conclusions. These authors have examined the isotope effect in the presence of electron scavengers in the vapor phase and conclude that there is a significant isotope effect for atoms derived from the dissociation of the excited species formed on charge neutralization of the hydronium ion i.e. $k_H/k_D = 2.3$; for hydrogen atoms formed in processes other than neutralization of the hydronium ion they find $k_H/k_D = 1.1$. They suggest that the ratios obtained by Yang and Marcus, together with the low values these authors obtained for the yield of hydrogen in the presence of cyclopentane ($G(H_2) = 5.0$) could be explained by the presence of low concentrations of an efficient electron scavenger.

5. Effect of Temperature and Pressure. Measurements on the effects of temperature and pressure are rather limited but Anderson and Winter (31) report a series of measurements with three different scavengers, viz., NH_3, cyclo-C_6H_{12}, and C_2H_5OH in which they observe a stepwise increase in $G(H_2)$ with increasing temperature. They attribute this increase to a change in the ion–molecule equilibrium (Eq. 46) leading to increased production of H atoms on neutralization of the ion $H_3O^+(H_2O)_n$ when $0 \leq n \leq 2$. The increased hydrogen yield $G(H_2) \sim 3.5 \sim G(e^-)$, is consistent with this interpretation and by calculating relative equilibrium constants for the ion–molecule equilibrium they deduce a value of 78–100 kcal/mole for the solvation energy of the hydronium ion in the vapor phase, which is close to the solvation value in liquid water (~ 100 kcal/mole). However, while these data and interpretation are consistent with the mass spectrometric measurements of Kebarle and Hogg (12) a preliminary investigation by Johnson and Simic (76) of the radiolysis of D_2O/C_6H_{12} shows no evidence for a stepwise increase in the hydrogen yield. They suggest that the increase in yields with temperature is entirely due to the radiation-induced thermal pyrolysis of the cyclohexane. There may be a difference in the behavior of D_2O and H_2O on charge recombination, as indicated by the isotope effect and more work is clearly needed to resolve what appears to be a genuine conflict in both experimental data and in interpretation.

Anderson's hypothesis on the chemical effects resulting from a change in the degree of solvation of the H_3O^+ ion is not entirely consistent with his measurements of the effects of electron scavengers (31) since $G(H_2)$ is reduced by the same amount (~ 3.5) in both the high and low temperature regions. Other anomalous results, however, have been observed in the presence of electron scavengers; Johnson and Simic (32) found that with D_2O/C_3H_8 the increase in $G(N_2)$ from added N_2O was always less than the required stoichiometry $\Delta G(N_2) = 2G(e^-)$, while with H_2O/CH_3OH Dixon (79) found $\Delta G(N_2) = 2 \Delta G(H_2)$ but that $-\Delta G(H_2)$ was significantly less than $G(e^-)$. These apparently anomalous data tend to emphasize the present uncertain state of knowledge concerning details of charge recombination reactions.

There is little change in observed yields in the pressure range 1–3 atm but at lower pressures Anderson and Winter (31) have reported increased yields of hydrogen. Further work has shown, however, that part, at least, of this change is connected with a wall effect, since the yields change with changes in surface area/volume ratio at different pressures. Thus the simple tentative postulate of an excited water molecule readily deactivated by collision is not tenable. These measurements further illustrate the great importance which attaches to the experimental requirement to vary all the adjustable parameters in gas phase radiolysis before it is possible to make an unambiguous interpretation of the data.

D. Recombination of Hydrogen and Oxygen

Measurements of the radiation-induced recombination of oxygen and hydrogen give initial yields for $G(-H_2) = G(H_2O) = 12$. In the T_2/O_2

reaction at 25°C Dorfman and Hemmer (83) found that the initial rate of reaction is directly proportional to radiation intensity and independent of changes in the T_2/O_2 ratio from 0.16–1.6. They found an isotope effect and quote initial values of $(M/N)_o^T = 3.25$ and $(M/N)_o^H = 4.2$ (corresponding to $G_0(-T_2) = 9.8$ and $G_0(-H_2) = 12.7$ with their assumed W value of 33 eV); the latter yield is very close to the value reported by Lind (3) for α-particle radiolysis, $M/N = 3.92$. No measurable hydrogen peroxide is found in static systems but it must be formed and disappear in subsequent steps. There is no evidence for a chain reaction, since only a 3-fold increase in exchange rate is reported over the temperature range from 25–500°C.

In the presence of water vapor, yields for the recombination of hydrogen and oxygen are in the range $G(H_2O) = G(-H_2) = 3–6$; they are independent of temperature up to 138°C but appear to increase at higher temperatures (84). At a fixed dose (1.1×10^{20} eV g^{-1}) reported yields for the disappearance of hydrogen are apparently linearly dependent on (dose rate)$^{1/2}$, which is consistent with a mechanism involving bimolecular recombination of species whose rates of formation depend linearly on dose rate.

E. Oxygen

The important processes induced by the absorption of ionizing radiation by oxygen are as follows:

$$O_2 \rightarrow O_2^+ + e^- \tag{60}$$
$$O_2 \rightarrow O^+ + O + e^- \tag{61}$$
$$O_2 \rightarrow O_2^* \tag{62}$$
$$O_2^* \rightarrow O + O \tag{63}$$
$$O_2 + e^- \rightarrow O_2^- \tag{64}$$
$$O_2 + e^- \rightarrow O^- + O \tag{65}$$

The most probable charge recombination reactions will involve the ions O_2^- and O^-, e.g.,

$$O_2^+ + O_2^- + M \rightarrow 2O_2^* + M \tag{66a}$$
$$\downarrow 2O + O_2 \tag{66b}$$
$$O_2^+ + O^- + M \rightarrow O_3^* + M \tag{67a}$$
$$\downarrow O + O_2 \tag{67b}$$

Fueki and Magee (85) have considered these reactions together with 22 reversible secondary processes involving ions, atoms and excited molecules in making an a priori calculation of the yield of ozone from oxygen radiolysis. They obtain a value of $G(O_3) = 13$ for irradiations at 0.1–1 atm pressure on the assumption that only one excited molecule is formed per ion pair. At the time it was difficult to compare this calculated value with previous experimental measurements since the reported yields vary over a very wide range ($G(O_3) = 2 - 124$) (3), but in a subsequent careful determination of the initial yield for γ radiolysis, Johnson and Warman (18) obtained a value of $G(O_3) = 12.8 \pm 0.6$ (based on $G(N_2) = 12.5$ for N_2O) at oxygen pressures from 100 to 760 mm. From a study of the effect of various rare gases on the yield of ozone, they postulated the following mechanism:

Reactions (60–67) followed by:

$$O_2^{*+} + O_2 \rightarrow O_3^+ + O \tag{68}$$

$$O^+ + O_2 \rightarrow O + O_2^+ \tag{69}$$

$$O_3^+ + O_2 \rightarrow O_3 + O_2^+ \tag{70}$$

$$O_2^* + O_2 \rightarrow O + O_3 \tag{71}$$

$$O + O_2 + M \rightarrow O_3 + M \tag{72}$$

The primary yields which they deduce are $G(O_2^+) = 0.8$; $G(O^+) = 0.5$ $G(O_2^{*+}) = 1.9$; $G(O_2^*) = 4.1$. They assume that neutralization of ground state O_2^+ does not lead to dissociation or to ozone formation. The reasons for suggesting that charge neutralization of O_2^+ is nondissociative are based principally on measurements in the presence of added Ar and Kr where the charge transfer reactions (73) and (74) can occur

$$Ar^+ + O_2 \rightarrow O_2^+ + Ar \tag{73}$$

$$Kr^+ + O_2 \rightarrow O_2^+ + Kr \tag{74}$$

Experimental values of ozone yields calculated on the basis of contributions from energy absorption in the rare gases are much lower than calculated values based on charge transfer followed by dissociative neutralization of O_2^+, plus energy transfer to oxygen from neutral excited rare gas atoms. Better agreement is obtained when it is assumed that ozone results principally from energy transfer to oxygen from excited Ar and Kr atoms. While the agreement between Fueki and Magee's calculations and Johnson and Warman's experimental yields is apparently excellent it is somewhat fortuitous, since the former authors include contributions from the charge neutralization of ground state O_2^+ in their calculations but Johnson and Warman suggest that this reaction is nondissociative. The calculations of Fueki and Magee also predict a decrease in yields at pressure > 1 atm and it is of interest to note that the measurements of Kircher et al. (86) give $G(O_3) = 9$ at pressures of 26–136 atm, in excellent qualitative agreement with the calculations.

Sauer (52) has used the technique of pulse radiolysis to extend studies of the kinetics of the reaction (72). He has used oxygen, carbon dioxide, and nitrous oxide as source molecules for the O atoms in the presence of various third bodies, viz., He, Ar, and finds that the rate of reaction (72) is independent of the source of O atoms. The efficiencies of the third bodies He, CO_2, N_2O relative to Ar = 1, are 0.8, 5, 5, respectively; the specific rate constant in the presence of Ar is $k_{72} = 5 \times 10^8$ M^{-2} sec^{-1}, which is similar to that reported for reactions of $O(^3P)$ atoms. The lack of dependence of the rate constants on the source of O atoms suggests that while different excited states may be formed initially, at the high gas pressures used (ca. 50 atm) they are rapidly deactivated before reacting to form ozone.

In order to determine initial yields Johnson and Warman (18) carried out irradiations at 90 and 200°K where the yield of ozone was linearly dependent on dose up to ozone concentrations of $\sim 10^{-2}$ mole %. At room temperature, however, the ozone yield decreases with increasing dose at ozone concentrations corresponding to $\sim 10^{-4}$ mole %. Kircher et al. (86) found an apparent activation energy of ~ 2 kcal/mole for the ozone decomposition reaction over the temperature range from -78 to 70°C. The marked dependence of ozone yields on dose is a common feature of the factors contributing to the wide discrepancies in measured yields. Studies in gas flow systems have shown that the yields of ozone increase with increasing gas flow rate and yields as high as $G(O_3) = 3.9$–7.8 have been reported for alpha particle radiolysis (3). Although thermal effects and the presence of impurities can contribute to the decomposition of ozone, the dose effect is attributed principally to a radiation induced chain process; chain lengths of 1.5×10^4 have been reported for its radiolytic decomposition. Reactions of both O atoms and of ions and the interpretation of data by Kircher et al. (86) suggest that more than one chain decomposition process occurs. Possible reactions are as follows:

$$O + O_3 \rightarrow 2O + O_2 \tag{75}$$

$$O_2^- + O_3 \rightarrow O_3^- + O_2 \tag{76}$$

$$O_3^- + O_3 \rightarrow O_2^- + 2O_2 \tag{77}$$

Lampe et al. (87) have advanced some qualitative arguments for the existence of an LET effect on the production of ozone in the radiolysis of gaseous oxygen; they suggest that at high LET, charge neutralization of O_2^+ ions by free electrons will lower the yield relative to that at low LET where negative oxygen ions will be the dominant species (Eqs. (66) and (67)). These ideas, however, do not coincide with Johnson and Warman's conclusions; moreover there are no adequate experimental data to test the effect of LET since it is unlikely that results reported for α-particle radiolysis are true initial yields.

Ozone is produced in liquid oxygen with a yield of $G(O_3) = 6$ (86); this observation is consistent with results from other irradiated systems where product yields are decreased in the liquid due to the cage effect.

F. Nitrogen

No chemical change occurs on the irradiation of pure nitrogen but deductions concerning the primary radiolytic processes have been made recently from measurements of the radiation induced isotopic exchange reaction between $^{14,14}N_2$ and $^{15,15}N_2$ (23,37,88). Unfortunately there is a quantitative difference of 20% between reported values for G(exch), which difference is crucial in trying to deduce the mechanism of the exchange reaction. Dawes and Back (23) find G(exch) $= 7.3 \pm 0.5$ while Anbar and Perlstein (37,88) find G(exch) $= 9.5 \pm 0.5$ for pure nitrogen. The discrepancy is not due to dosimetry, since in the exchange reaction between $^{15,15}N_2$ and ^{14}NO Anbar and Perlstein (37) find $G(^{14,15}N_2) = 2.56 \pm 0.13$ in agreement with Dawes and Back's value of 2.57, and they suggest that the lower value of 7.3 for the exchange in pure nitrogen may be due to the presence of a trace impurity.

It is clear from a consideration of the dissociation energy of the nitrogen molecule that simple atom recombination is inadequate to account for the magnitude of the radiolytic exchange. Dawes and Back proposed that there is an additional non-atomic process which may take place via the N_4^+ ion (Eqs. (78) and (79)).

$$^{14,14}N_2 + {}^{15,15}N_2^+ \rightarrow N_4^+ \tag{78}$$

$$N_4^+ + e^- \rightarrow {}^{14,15}N_2 + {}^{14}N + {}^{15}N \tag{79}$$

However, Anbar and Perlstein (37) show that the quantitative estimate of exchange based on reaction of the N_4^+ ion is in error. They also point out (37) that their earlier conclusion (88) that excited nitrogen molecules do not exchange with ground state nitrogen molecules is probably erroneous. This conclusion is only applicable to the two lowest triplet states of N_2^*, since reaction (80) is energetically feasible for other excited states of N_2^* down to 6 eV above the ground state

$$^{14,14}N_2^* + {}^{15,15}N_2 \rightarrow 2^{14,15}N_2 \tag{80}$$

The inclusion of this reaction leads to calculated values for the exchange yield much closer to the experimental values than in the case where only dissociative excitation of the nitrogen molecule (9.67 eV) is considered in addition to ionic processes, but further work is needed to resolve the experimental discrepancies.

Additional evidence that species other than N atoms must participate in the exchange reaction is obtained from studies in the presence of oxygen and nitric oxide. In the presence of NO the reaction

$$N + NO \rightarrow N_2 + O \qquad (81)$$

which is very efficient, would eliminate N atoms from the system, but exchange still occurs between nitrogen molecules to the extent of $G(\text{exch}) = 3.12$ (23). Both with added NO and O_2 it is suggested that charge neutralization of the ion N_4^+ by the negative ions does not lead to the formation of N atoms (Eq. (79)) but simply to dissociation into nitrogen molecules (Eqs. (82) and (83))

$$N_4^+ + O_2^- \rightarrow 2N_2 + O_2 \qquad (82)$$

$$N_4^+ + NO^- \rightarrow 2N_2 + NO \qquad (83)$$

G. Nitrogen and Oxygen

Irradiation of air or synthetic mixtures of nitrogen and oxygen produces principally nitrogen dioxide, together with small amounts of nitrous oxide, nitric oxide and ozone (89–97). These reactions are of great practical interest because of their possible application in industrial methods for nitrogen fixation and the potential health hazard associated with the formation of these products in the proximity of large radiation sources.

The initial yields of products depend on the ratio $N_2:O_2$ and on pressure, with $G(NO_2) = 2.16$ for an equimolar mixture at 1 atm pressure; the yield increases slightly at lower pressures and rises to a limiting value of 5–6 at pressures > 100 atm (91). Harteck and Dondes (94–97) also found an effect of varying $N_2:O_2$ ratio, and showed further (97) that on prolonged irradiation of dry air the oxygen is completely consumed, but with an equimolar mixture of N_2 and O_2 a small amount of O_2 (6.4%) remained in the equilibrium mixture. The reaction schemes have been described by a large and complex series of reactions, but Dmitriev and his colleagues differ from Harteck and Dondes in their interpretation of the role of ions in the reaction. In general, Harteck and Dondes write the charge neutralization reactions as dissociative, while Dmitriev argues that the nondissociative reaction (84) is predominant; this latter interpretation is in accord with the suggested role of O_2^- in the radiation-induced N_2 exchange reaction (23).

$$N_2^+ + O_2^- \rightarrow N_2 + O_2 \qquad (84)$$

From the results of experiments with electrons at energies < 16 eV where they found no reaction, Dmitriev and Pshezhetskii (90) conclude that reactions involving N_2^+ ions are important since the ionization potentials of O_2 and N_2 are, respectively, 12.08 eV and 15.58 eV. In more recent work Dmitriev (93b) has shown further that the main reactions in the radiation induced oxidation of nitrogen are ion–molecule reactions involving the ions N_2^+, N^+, O^+, and O_2^+. Lampe et al. (87) have reviewed some of the studies concerned with nitrogen fixation and suggest further speculative reactions involving N_3^+ and N_4^+ ions.

Dmitriev has also presented data recently (93a) to show a correlation between the yield of NO_2, the ion recombination coefficient, and the concentration of positive ions as a function of pressure. From electron bombardment experiments at controlled voltages he shows that the ratio of concentrations of O_2^-: free electrons is very high at atmospheric pressure and increases further as the pressure is raised at 150 atm, so charge neutralization must occur exclusively with the O_2^- ion. The ion recombination coefficient, however, decreases with increasing pressure (from 1.6 cm^3 sec^{-1} at 1 atm to 0.06 cm^3 sec^{-1} at 150 atm), and since the yield of NO_2 increases with pressure he argues that this presents evidence in favour of the nondissociative charge recombination reaction (84). Presumably as the pressure is increased reactions such as (85) compete with the charge recombination reaction, leading to an enhanced yield of NO_2 in subsequent steps (Eqs. (86) and (87))

$$N_2^+ + O_2 \rightarrow NO^+ + NO \tag{85}$$

$$NO^+ + O_2^- \rightarrow NO + O_2 \tag{86}$$

$$2NO + O_2 \rightarrow 2NO_2 \tag{87}$$

When N_2—O_2 mixtures are irradiated in the presence of water vapor the major product is nitric acid. (Even in the presence of liquid water the reaction is controlled by gas phase reactions (98).) Nitrous oxide is also formed initially, and nitric acid production continues until all the water is consumed, after which the nitric acid is destroyed, with the formation of a stoichiometric amount of nitrogen dioxide (99). The reactions show a marked dependence on the concentration of oxygen and maximum yields are found with about 15% of oxygen by volume (99): maximum yields reported are $G(HNO_3) = 2.25$ and $G(N_2O) \sim 0.7$; after water exhaustion, $G(-HNO_3) = G(NO_2) = 10.0$. Wright et al. (98) reported that nitric acid production occurs in the absence of added oxygen, but Jones (99) found no acid production at concentrations of oxygen $<4\%$ and suggests that in the previous work enough water may have been decomposed to provide sufficient oxygen for reaction to occur.

H. Nitrous Oxide

Nitrous oxide has been studied quite extensively because of its suggested use as a gas phase dosimeter, but there are wide variations in reported yields (Table II) and in the reported effects of temperature and pressure. The initial radiolytic products appear to be nitrogen, oxygen, and nitric oxide, with nitrogen dioxide being formed as a secondary product,

Table II

Yields in the Radiolysis of Nitrous Oxide

Radiation	$G(N_2)$	$G(O_2)$	$G(NO)$	$G(NO_2)$	Notes	Ref.
			Yields			
γ	11.0 ± 0.4	3.0 ± 0.3		$G(NO + NO_2) =$ 5.75 ± 0.5	Also x-ray and ^3H β's	104
	9.67	1.35		4.64	Also reactor radiation and fission recoils	101
	12.8 ± 0.4	2.4		$G(NO_2{}^-) = 2.3 \pm 0.2$	Measured yields	103
		5.1	4.6 ± 0.4		Dose $< 1.2 \times 10^{20}$ eV g^{-1} Derived yields	103
	$11.9 \pm 0.7 = G(N_2) + G(O_2)$				Also reactor radiation	102
	11.5 ± 0.5				Dose $<6.1 \times 10^{19}$ eV g^{-1} Also reactor radiation	100
	12.1				Dose: 2×10^{19} eV g^{-1}	(a)
	12.7 ± 0.4				Dose $<5.1 \times 10^{19}$ eV g^{-1}	(b)
	$G(N_2 + O_2) = 13.6 - 12.2$				$P = 1$–4 atm	(c)
1 MeV e^-	10.0 ± 0.2	4.0 ± 0.4	3.9 ± 0.3	$G(NO + NO_2) = 3.9 \pm 0.3$	Initial yields	50
		1.0			Measured	
X-ray	11.3	3.95	6.82		$P = 200$ mm	19
	14.4	3.95	13.1		$P = 50$ mm	
Reactor radiation and fission fragments	9.7	1.2		5.0		(d)

(a) M. Steinberg, Brookhaven National Laboratory Research Report, BNL 612 (1960).
(b) D. A. Kubose, *Trans. Am. Nucl. Soc.*, **7**, 318 (1964).
(c) K. Furukawa and S. Shida, *J. Nucl. Sci. Technol.*, **3**, 41 (1966).
(d) F. Mosely and A. E. Truswell, Harwell Research Report AERE-R 3078 (1960).

particularly by postirradiation reactions between NO and O_2 when the gaseous mixture is trapped at $-196°C$. Variations in reported yields of oxygen and oxides of nitrogen can be attributed, at least in part, to the analytical procedures adopted, but even when the yield of nitrogen oxides has been measured by dissolution of the irradiated gas in deaerated alkaline solution and determination of the nitrite ion produced, the agreement between yields reported by various workers is still poor. For dosimetry purposes, there is general agreement that it is preferable to measure the yield of nitrogen alone, since it is unaffected by the analysis method used. Even so there is no agreement on the initial yield of nitrogen, and the best mean value which can be deduced from all the available data is $G(N_2) = 11 \pm 1$ for nitrous oxide at ~ 1 atm pressure, $20°C$, and for absorbed doses $< 10^{20}$ eV g^{-1}. Some of the reported discrepancies in the experimental data can be resolved to some extent by considering separately the effect of the variables, total dose, temperature, and pressure on the measured products.

1. Total Dose. It is clear that the yields of decomposition products decrease with increasing dose (or percentage conversion), but there is disagreement on the dose at which departure from the true initial yields occurs. In part this may reflect the precision of individual measurements but in itself it is insufficient to explain the discrepancies in yields. Several authors agree that yields decrease at doses $< 2 \times 10^{20}$ eV g^{-1}; within this range the data of Jones and Sworski (50) show deviations at doses $\leq 10^{20}$ eV g^{-1}, and Linacre (100) finds deviations at doses $\leq 6 \times 10^{19}$ eV g^{-1}. On the other hand, Harteck and Dondes (101) claim linearity of nitrogen yields with dose to 6×10^{20} eV g^{-1} and "near linearity" to a dose of 6×10^{21} eV g^{-1}, while Flory (102) reports yields of $N_2 + O_2$ linear with dose to 10^{21} eV g^{-1}.

Jones and Sworski argue that the higher yields reported by Johnson (103), and by Burtt and Kircher (19) may be due to inherent difficulties associated with establishment of true saturation currents in the cylindrical geometry of their ionization chambers and the possibility of low collection efficiency leading to low apparent doses. However, these criticisms are not applicable to the measurements of Hearne and Hummel (104), which gave $G(N_2) = 11.0 \pm 0.4$ for γ and β radiation, nor to the data of Linacre giving $G(N_2) = 11.5 \pm 0.5$ for both γ and reactor radiation (100).

The suggestion that the yields reported by Hearne and Hummel may be affected by wall thickness and cell geometry appears to be invalidated by the fact that Linacre finds no effect of varying surface area/volume ratio from 3 to 16 at 3 different pressures in the range from 0.1–1 atm, nor at 30 atm.

2. Effect of Temperature. There seem to be only small increases in the decomposition yield at temperatures $< 120°C$ but Harteck and Dondes' claim (101) of temperature-independent yields up to 200°C is not substantiated by other work. Flory (102) finds a maximum increase in $G(N_2 + O_2)$ of 20% over the temperature range from -90 to 120°C but a rapid increase at temperatures $> 150°C$; Jones and Sworski (50) find an increase of $\sim 50\%$ in $G(N_2 + O_2)$ at temperatures $< 150°C$ which is due to a decrease in the oxygen yield accompanying an increase in the nitrogen yield (80%). Above 150°C the yield of noncondensable gases increases sharply, but this is shown to be due to an increase in N_2 and NO rather than an increase in N_2 and O_2 as suggested by Flory. ($G(O_2)$ is essentially zero at 150°C.) The overall effect of temperature is, however, close to that reported by Flory but less than that found by Gorden and Ausloos (45).

As with the effect of dose there are clear discrepancies in the reported effects of temperature, but Linacre (100) has recently observed a greater effect of temperature at doses $> 6 \times 10^{19} \text{ eV g}^{-1}$ than on the initial yields. There may be an interrelationship between temperature and total dose which could explain some of the discrepancies in the literature.

3. Effect of Pressure. Changes in pressure appear to have less effect on yields than changes in either temperature or dose, but some variations have been reported in experimental data. Harteck and Dondes (101) report no change in decomposition yields from 1 atm pressure to almost the critical pressure and other authors report no effect of changes in pressure over more limited pressure ranges 220–555 mm (19), 250–1000 mm (102), 200–760 mm (103), 21–660 mm (104).

Burtt and Kircher (19), however, report an increase in the yields of nitrogen and nitric oxide below 200 mm, while Jones and Sworski (50) find that decreasing the pressure from 600 to 50 mm has only a small effect on the N_2 yield or the primary decomposition yield of N_2O, but results in an increase in the primary yield of NO and a decrease in the primary yield of oxygen. Linacre finds no change in the yield of N_2 at pressures between 30 atm and 1 atm but the yield of oxygen increases from $G(O_2) = 1.2$ at 10 atm to $G(O_2) = 3.5$ at 0.1 atm.

4. Effect of Additives and Applied Electric Fields. The presence of hydrogen or methane, added to test the effect of probable impurities (104), has little effect on the yield of nitrogen but results in a decrease in the yields of both oxygen and the oxides of nitrogen. Addition of Xe and Kr enhances the decomposition; energy is transferred efficiently from Xe to N_2O, although the ionization potential of N_2O (12.63 eV) is above that for Xe (12.13 eV). The product distribution is different in the rare gas sensitized radiolysis, giving an increase yield of oxygen and a lower value of the ratio $^{30}N_2/^{29}N_2$ in the radiolysis of $^{14}N^{15}NO$ (45).

In the presence of applied electric fields there is a decrease in the yields of N_2 and NO at low field strengths (1 V cm^{-1} mm^{-1}) (19) but at higher field strengths and voltages insufficient to cause secondary ionization (50) product yields increase.

5. Mechanism. In view of the conflicting experimental data no author has attempted to give a complete interpretation of N_2O radiolysis, but several features of the mechanism seem to have been established by various work. In the radiolysis of $^{14}N^{14}NO$–^{15}NO mixtures, Gorden and Ausloos (45) established that $G(^{14}N) = 1.52$ (based on $W_{N_2O} = 32.9$ eV), but also showed that molecular nitrogen can be formed by processes other than N atom reactions with NO (Eq. (81)) or by molecular detachment.

In the radiolysis of $^{14}N^{15}NO$ the presence of 26 mole % ^{14}NO did not surpress the formation of $^{30}N_2$, which suggests that one or more as yet unidentified processes involving N_2O molecules can lead to its formation. They also found that the yield of $^{30}N_2$ from the radiolysis of $^{14}N^{15}NO$ is not affected by changes in temperature and that the enhanced yields at higher temperatures can be attributed almost entirely to processes which produce molecular nitrogen directly. They suggest that the effect of temperature is best explained by an ionic chain mechanism

$$^{14}N^{15}NO + e^- \rightarrow {}^{29}N_2 + O^- \tag{88}$$

$$O^- + {}^{14}N^{15}NO \rightarrow {}^{14}NO + {}^{15}NO^- \tag{89}$$

$$O^- + {}^{14}N^{15}NO \rightarrow {}^{15}NO + {}^{14}NO^- \tag{90}$$

$$O^- + {}^{14}N^{15}NO \rightarrow {}^{29}N_2 + O_2^- \tag{91}$$

$$NO^- + N_2O \rightarrow NO + N_2 + O^- \tag{92}$$

$$O_2^- + N_2O \rightarrow O_2 + N_2 + O^- \tag{93}$$

It is clear from this work and from that of Jones and Sworski (50), who find a sharp decrease in $G(O_2)$ and an increase in $G(NO)$ with increasing temperature, that reaction (92) must be more important than reaction (93).

From the cracking pattern of $^{14}N^{15}NO$ at 70 eV Gorden and Ausloos have deduced that only about 35% of the parent ion N_2O^+ decomposes, with the major fragment ion being NO^+ ($>50\%$). In the radiolysis at higher pressures the following ion molecule reactions will predominate:

$$N_2O^+ + N_2O \rightarrow 2N_2 + O_2^+ \tag{94}$$

$$N_2O^+ + N_2O \rightarrow N_2 + NO + NO^+ \tag{95}$$

Direct electron attachment reactions can also be expected;

$$N_2O + e^- \rightarrow N_2 + O^- \tag{96}$$

Dissociation of excited molecules can give O atoms or N atoms:

$$N_2O^* \rightarrow N_2 + O \tag{97}$$

$$N_2O^* \rightarrow NO + N \tag{98}$$

Reactions of ground state O atoms and N atoms with N_2O (Eqs. (99–101)) are endothermic but N atoms react efficiently with NO to give nitrogen (Eq. (81)).

$$N + N_2O \rightarrow N_2 + NO - 83 \text{ kcal/mole} \tag{99}$$

$$O + N_2O \rightarrow N_2 + O_2 - 79 \text{ kcal/mole} \tag{100}$$

$$O + N_2O \rightarrow 2NO - 35 \text{ kcal/mole} \tag{101}$$

Hearne and Hummel (104) have used the results of their measurements in the presence of hydrogen to account for the possible mode of formation of some of the products. The absence of any effect of hydrogen on measured yields of nitrogen is used to suggest that reaction (100) cannot be important since this would be suppressed by reaction (102).

$$H_2 + O \rightarrow OH + H \tag{102}$$

However, as the authors point out, this deduction is not unequivocal since reaction (103) could participate in the presence of hydrogen.

$$H + N_2O \rightarrow N_2 + OH \tag{103}$$

However, they argue that the decrease in $G(NO)$ in the presence of hydrogen produces good evidence in favor of reaction (101) as one source of NO.

The decreased yields at low gas pressures under the influence of low strength applied electric fields are attributed to a decrease in ion re-combination, and the absence of an effect of low field strengths at higher pressures suggest that recombination does not lead to decomposition, probably due to the formation of ion–molecule clusters (19). From their isotopic studies with $^{14}N^{15}NO$ Gorden and Ausloos (45) concluded that the increased yields of nitrogen in the presence of applied electric fields are due to electronic excitation of N_2O by accelerated electrons. Jones and Sworski (50) show by kinetic analysis that the pressure dependence of $G(N_2)$ and $G(NO)$ in an applied electric field (11 V cm^{-1} mm^{-1}) is consistent with a process involving collisional deactivation of N_2O^* in competition with its dissociation. However they cannot discriminate between two possibilities involving dissociative attachment of all electrons and bimolecular deactivation of N_2O^*, or no significant electron capture and termolecular deactivation of N_2O^*.

I. Ammonia

There are many similarities between the primary processes which occur following the absorption of ionizing radiation by ammonia and by water vapor, but the subsequent radiation chemistry of the two molecules shows marked differences in the absence of scavengers. In contrast to pure water vapor which is essentially stable to high energy radiation (Section 5-V-C), equilibrium between ammonia and its radiolysis products, principally nitrogen and hydrogen, is not attained until about 85–97% of the ammonia is decomposed.

The predominant ions produced in the mass spectrometric cracking pattern of NH_3 (105) are NH_3^+ and NH_2^+, with NH^+ and N^+ present in trace amounts. Small yields of other hydrides, viz. N_2H^+, $N_2H_2^+$, $N_2H_3^+$, $N_2H_4^+$ and $N_2H_5^+$ have also been observed (106). Through mass spectrometric measurements at higher pressures however, Hogg and Kebarle (13) have shown that the major, and probably the only, ion present in the radiolysis of ammonia is the solvated ammonium ion $NH_4^+(NH_3)_n$. The primary ions are converted to the ammonium ion by the ion–molecule reactions (104) and (105) followed by electrostatic attraction of NH_3 molecules (reaction (106)), to form the ammoniated complex with values of $n \leq 20$ at pressures < 200 torr.

$$NH_3^+ + NH_3 \rightarrow NH_4^+ + NH_2 \tag{104}$$

$$NH_2^+ + NH_3 \rightarrow NH_4^+ + NH \tag{105}$$

$$nNH_3 + NH_4^+ \rightleftharpoons NH_4^+(NH_3)_n \tag{106}$$

Charge neutralization of the simple ammonium ion by electrons can occur by the exothermic reactions (107–109),

$$NH_4^+ + e^- \rightarrow NH_3 + H + 111 \text{ kcal/mole} \tag{107}$$

$$NH_4^+ + e^- \rightarrow NH_2 + 2H + 7 \text{ kcal/mole} \tag{108}$$

$$NH_4^+ + e^- \rightarrow NH_2 + H_2 + 111 \text{ kcal/mole} \tag{109}$$

If the full solvation energy of the ammonium ion is realized in the vapor phase (~ 84 kcal/mole) then the reaction equivalent to reaction (108) is endothermic to the extent of 77 kcal/mole.

Evidence from photochemistry shows that excited ammonia molecules can dissociate to give H, NH_2, NH, and H_2; at wavelengths > 1500 Å the predominant decomposition is by reaction (110) but at shorter wavelengths a significant contribution from reactions (111) and (112) is observed (77)

$$NH_3^* \rightarrow H(^2S_{1/2}) + NH_2 \tag{110}$$

$$NH_3^* \rightarrow H_2 + NH(^1\pi) \tag{111}$$

$$NH_3^* \rightarrow H + H + NH(^3\Sigma) \tag{112}$$

While recognizing that the photolysis data can only give a guide to some of the processes which can occur with ionizing radiation, it is of interest to note that Jones and Sworski (41) have observed a yield of nonscavengeable hydrogen ($G = 0.75$) and that Gordon (107) has observed the NH radical, which reacts with second-order kinetics at the high dose rates used in pulse radiolysis (10^{23} eV cm^{-3} sec^{-1}).

Melton (108) has reported a near-ideal radiation chemistry investigation by using a 3-compartment source so that reactions can be studied over a range of pressure from 10^{-9}–10 torr. Each compartment has a separate electron source, and with this equipment he is able to measure inter alia ionization cross sections, ion–molecule reactions, cross sections for the production of free radicals, and G values. He shows that the most important transient species are NH_2, H, and $NH_4{}^+$ and that eventually all the primary ions formed react with NH_3 to produce $NH_4{}^+$ and H or NH_2; in addition the following secondary radicals were observed, NH_4, N_2H_2, N_2H_3, and N_2H_5. The primary products of the irradiation are 58.8% positive ions (0.4% negative ions) and 40.8% free radicals, while positive ion reactions account for 54% of the hydrogen formed and for 65% of the nitrogen.

1. Mechanism. Although the detailed steps in the radiolysis of ammonia have not been established unequivocally, the mechanism is described adequately by the primary processes (Eqs. (104–112)) and the following sequence of reactions

$$H + H + M \rightarrow H_2 + M \tag{113}$$

$$H + NH_2 \rightarrow NH_3 \tag{114}$$

$$NH_2 + NH_2 \rightarrow N_2H_4 \tag{115}$$

$$NH_2 + NH_2 \rightarrow NH_3 + NH \tag{116}$$

$$H + N_2H_4 \rightarrow NH_3 + NH_2 \tag{117}$$

$$H + N_2H_4 \rightarrow H_2 + N_2H_3 \tag{118}$$

$$NH_2 + N_2H_4 \rightarrow NH_3 + N_2H_3 \tag{119}$$

$$NH + NH_3 \rightarrow 2NH_2 \tag{120}$$

$$NH + NH \rightarrow N_2 + H_2 \tag{121}$$

$$2N_2H_3 \rightarrow N_2H_4 + N_2 + H_2 \tag{122}$$

$$H + NH_3 \rightarrow H_2 + NH_2 \tag{123}$$

Reaction (122) is not written as a disproportionation reaction but simply as a stoichiometric representation showing the final products from reactions of the N_2H_3 radical; the detailed mechanistic steps involve either a number of speculative radical–radical reactions or disproportionation of the N_2H_3 radical and of other possible intermediates, e.g., N_2H, N_2H_5. The fact that hydrazine is not detected as a product from the static irradiation of gaseous ammonia in contrast to liquid ammonia is attributed to its rapid destruction by H and NH_2 radicals in reactions (117–119),

and Jones and Sworski (41) have furnished kinetic evidence that reaction
(118) predominates over reaction (117). In a gas flow system, however,
where hydrazine is removed rapidly from the irradiation zone, it is observed
as a product with the yield depending *inter alia* on gas flowrate (41).

Gordon (107) found no evidence for reaction (120) at very high dose
rates, but it may be important at lower dose rates where the steady state
concentration of NH and other radicals is much lower than under pulse
radiation conditions.

The radiolysis of liquid NH_3 (109,110) differs in two main respects
from radiolysis in the gas phase; the overall decomposition yield is lower
and hydrazine is produced in small yields in static systems. These differences are probably due to collisional deactivation of excited molecules
and to the high probability of the return of the secondary electrons to the
parent spur in the polar liquid (110).

2. Yields. The overall radiolytic decomposition yield of ammonia
has been measured by several workers, and there is general agreement
that the only decomposition products in the absence of additives are
hydrogen and nitrogen in the stoichiometric ratio 3:1. There is reasonable
agreement on the value of the product yields at room temperature
(Table III) with the following average values, within the limits of $\pm 10\%$:
$G(H_2) = 6$; $G(N_2) = 2$; $G(-NH_3) = 4$. The observed decomposition
yields increase with increasing temperature, and there is again agreement
on the limiting yield at high temperature, viz, $G(H_2) = 14.5$–15. There is
more spread in the values of yields at intermediate temperatures, but this
may be explained, at least partially, by a dose-rate effect depending on
competition between reactions (123) and radical–radical reactions
(113–116).

The only data which appear to be inconsistent with other work are the
results of Toi et al. (111) at the lowest pressure which they used at 137°C,
but their lower results may be associated with wall effects due to the use
of a stainless steel autoclave for measurements at high pressure. With
increasing ammonia density in the range from 0.05 to 0.15 g cm^{-3}, well
below the critical density of 0.235 g cm^{-3}, they observed a marked decrease in the yields of hydrogen and nitrogen. Small yields of hydrazine
were observed in the high density region and from the most probable
limit of these yields they estimate the decrease in $G(-NH_3)$ to be between
57% and 76%. They attributed the decrease in yields principally to the
formation of ion–molecule clusters and the consequent reduced probability of dissociative charge neutralization due to the distribution of the
recombination energy over the molecules in the cluster. However, there
is no conclusive evidence for this postulate, and their alternative suggestion

of collisional deactivation of excited species is more consistent with Kebarle's recent mass spectrometric measurements (13).

Decomposition yields have been shown to decrease with increasing dose rate (41,112) although there is one report of the opposite effect at pressures < 500 mm over very limited range of intensity (113). Jones and Sworski (41) find that the dose rate effect depends upon pressure and define two pressure regions; < 400 torr where the decomposition yield is independent of pressure and > 400 torr where decomposition increases with decreasing pressure. (There is a rapid and irreproducible decrease in yields at pressures < 100 torr, which is probably due at least in part to wall effects.) At 200 torr they observed negligible effects with an 80-fold increase in the dose rate but at 700 torr the yield $G(-NH_3)$ fell from 3.6 to 2.6 with an 100-fold increase in dose rate. They attribute the effect to possible competition between reactions (113,117,118) or to the quenching of excited NH_3^* molecules formed by $H + NH_2$ recombination.

Horscroft (112) found a decrease in $G(-NH_3)$ from 4.0 to 1.6 over a surprisingly wide intensity range (10^{14} eV ml^{-1} sec^{-1} to 4×10^{19} eV ml^{-1} sec^{-1}) in the proton irradiation at a pressure of 300 torr, and offered an explanation similar to that proposed by Luyckx (114) based on collisional deactivation of excited molecules. The extensive intensity range over which Horscroft observed the effect might be partially due to the effect of temperature, for at the high beam currents used the steady-state temperature in the irradiated gas could be considerably in excess of ambient (21). Since increasing temperature increases the decomposition yields an increase of temperature with beam current would have the effect of increasing the dose rate dependent region, since the primary decomposition yield would increase with increasing beam current, i.e. temperature.

The effect of temperature has been studied most extensively by Jones and Sworski (41), who also made some important measurements of the primary H atom yields as distinct from the overall decomposition yields measured by other workers. At 23°C they found $G(H) = 10.0 \pm 1.5$ at 200 torr and $G(H) = 11.7 \pm 0.5$ at 600 torr from measurements with added deuterium and with added hydrazine, and a maximum primary decomposition yield $G(-NH_3) = 13.2$. On increasing the temperature the measured hydrogen yield increased to a value of $G(H_2) = 15.1 \pm 0.5$ in a static system and 15.9 ± 1.0 in a flow system. These values are higher than the calculated value of 12.5 based on the primary yield of H atoms ($G = 11.7$) and molecular hydrogen ($G = 0.7$), and on the reasonable assumption that at high temperatures the reaction $H + NH_3$ predominates over H atom combination. (If at high temperatures, H atoms are produced in place of molecular hydrogen, then the calculated upper limit for the measured $G(H_2)$ is 13.2). Thus the measured G value for hydrogen

Table III
Yields in the Radiolysis of Ammonia

Radiation	Temp. °C	Yields				Notes	Ref.
		$G(-NH_3)$	$G(H_2)$	$G(N_2)$			
1 MeV e^-	Room temp.	3–4	4.5–6	1.5–2		$P = 400$ torr	41
	100	4.8	7.2	2.6		Data derived from	
	150	7.0	10.2	3.6		figure	
	200	8.4	12.8	4.4			
	250	9.6	14.5	4.9			
	300	10	15	5			
		$G(-NH_3) = 13.2$		$G(NH_2) = 13.2$		Primary yields	
		$G(H) = 11.1 \pm 0.5$		$G(H_2) = 0.75 \pm 0.5$			

From measurements in a flow system at temperature $\leqslant 300°C$ the limiting yield of hydrazine is given as $G(N_2H_4) = 15.9\ e^{-1930/RT}$

Radiation	Temp. °C	$G(-NH_3)$	$G(H_2)$	$G(N_2)$	Notes	Ref.
$e^- \leqslant 100$ eV	Room temp.	6.1	8.8	2.9	$G(N_2H_4) = 0.03$	108
^{60}Co γ	Room temp.	(4.2)	6.3	2		115
	$> 200°C$	9.6	14.4	4.6		
X	Room temp.	3.8				(a)

Radiation	Conditions					Ref
β	137°C		6.2 1.13	2.1 0.32 $G(N_2H_4) = 0.2$	5.5×10^{-4} g ml^{-1} 0.312 g ml^{-1}	111
H$^+$	Room temp.	4	6	2	Low dose rate limiting values	112
D$^+$	−78°C to room temp.			$G(N_2H_4) = 0.5$		(b)
α	18°C	3.0	4.5			(c)
	108	6.0	9.0			
	220	8.8	13.2			
	315	9.6	14.4			48
	30	5.2	7.8			
	100	9.1	13.6			
	Room temp.	3.7–4.0 3.15–3.8			$P = 20$ cm $P = 50$ cm	49
Reactor radiation	~70	3.5	5.2			(d)
Fission fragments		6.48	7.11	2.26 $G(N_2H_4) = 1.11$		(e)

(a) P. Gunther and L. Holzapfel, Z. Phys. Chem., **38B**, 211 (1937).

(b) F. W. Lampe, E. R. Weiner, and W. H. Johnston, Intern. J. Appl. Radiation Isotopes, **14**, 231 (1963).

(c) E. E. Wourtzel, Radium, **11**, 289, 332 (1914).

(d) P. C. Davidge, Harwell Research Report, AERE—C/R 1569 (1959).

(e) D. A. Landsman and C. M. Noble, Harwell Research Report, AERE-M921 (1961).

at high temperatures is most probably about 2.6–3.4 units higher than the calculated value, and the discrepancy could be as much as 5.1, taking the limits of the experimental data. They suggest that the increase may be due either to more efficient decomposition of NH_3^* molecules at the higher temperature or to a decrease in the W value. On the basis of similar but less detailed observations Anderson and Winter (115) have suggested, however, that the temperature effect may be due to changes in the ion molecule equilibrium (Eq. (106)), so that the higher temperatures charge neutralization of the bare ammonium ion predominates to give 2 H atoms. This mechanism leads to a theoretical increase in $G(H_2)$ of 3.8 based on $W = 26$ eV, which is in reasonable agreement with the experimental increase in view of some of the uncertainties involved.

Horscroft (112) found that the addition of xenon to NH_3 increased the decomposition yield, N_2 has little effect, and H_2 appears to inhibit the decomposition, although these latter data were rather inconsistent. It appears unlikely that the enhancement caused by Xe is due simply to the charge transfer reactions (124) and (125) since the ionization potentials of both N_2 (15.6 eV) and H_2 (15.4 eV) are higher than that for NH_3 (10.15 eV), and reactions of electrically neutral excited states of Xe must be important.

$$Xe^+ + NH_3 \rightarrow NH_3^+ + Xe \tag{124}$$

$$Xe^+ + NH_3 \rightarrow NH_2^+ + H + Xe \tag{125}$$

The roles of added N_2 and H_2 are not clear, although it is known that ammonia can be radiolytically synthesized from its elements and any possible enhancement of ammonia decomposition by charge or energy transfer must be compensated by its reformation. Hydrogen is thought to inhibit the decomposition through the reaction (126)

$$NH_2 + H_2 \rightarrow NH_3 + H \tag{126}$$

Smith and Essex (48) found a decrease of 30% in the decomposition yield on the application of an electric field at a gas pressure of 62 cm, but found no effect at a pressure of 20 cm; they attributed the decrease to ion neutralization without decomposition at the electrodes. However, Burtt and Baurer (49) found no decrease at pressures of both 20 cm and 62 cm, while Jones and Sworski (41) found no decrease at 20 cm. At high field strengths enhanced decomposition is observed, presumably due to acceleration of secondary electrons.

The decrease in yield observed by Smith and Essex is somewhat difficult to explain in the light of the knowledge that the predominant ion in the radiolysis is undoubtedly the solvated ammonium ion $(NH_4^+(NH_3)_n)$, which must decompose on charge neutralization whether it be a homogeneous or heterogeneous process. If, of course, the predominant ion in the vapor phase is the bare ammonium ion, which could give two H atoms on neutralization, then neutralization at a surface would lead to the production of only one H atom. However, in the light of Kebarle's data this seems unlikely, and it would appear that any reduction in yield due to heterogeneous recombination must be attributed to reactions with impurities or defects at the surface.

Jones and Sworski (41) have demonstrated the qualitative similarities between the photolysis and radiolysis of ammonia in a study of the effects of gas flowrate, pressure and dose rate in the electron radiation induced decomposition. They have measured yields of hydrazine as high as $G(N_2H_4) = 4$ in a flow system at 300°C and deduce that the limiting yield of hydrazine is given by the expression $G(N_2H_4) = 15.9\,e^{-1930/RT}$.

The radiolysis of liquid NH_3 is of special interest in its comparison with the radiation chemistry of water (Chapter 10). In the $^{60}Co\,\gamma$ radiolysis of liquid NH_3 at 20°C, Dainton et al. (110) measure the initial yields; $G(H_2) = 0.84 \pm 0.02$, $G(N_2) = 0.23 \pm 0.01$, $G(N_2H_4) = 0.18$ and deduce the following primary yields from this and other work; $G(H_2) = 0.84 \pm 0.25$, $G(N_2H_4) = 0.9 \pm 0.5$, $G(-NH_3) = 2.1$, $G(e^-) \simeq G(NH_2) \simeq 0.35$. Sutherland and Kramer (116) find no N_2 at doses $< 30 \times 10^3$ rads but with further increases in dose, $\leq 6 \times 10^5$ rads, they find yields of $G(H_2) = 0.96$, $G(N_2) = 0.21$, $G(N_2H_4) = 0.18$, close to those reported by Dainton et al. (110). From their considerations of material balance Sutherland and Kramer (116) deduce that additional products are formed in each of the dose ranges, $< 30 \times 10^3$ rads and $> 40 \times 10^3$ rads. In the high dose region the additional product has been identified as HN_3 by ultraviolet and infrared spectroscopic studies; in the low dose region the unknown species contains an $-N{=}N-$ group and was identified as one of a mixture of the products N_2H_2 (diazene), N_3H_3 (triazene), or N_4H_4 (tetrazene).

3. Ammonia Synthesis. Radiolysis of mixtures of nitrogen and hydrogen lead to the formation of ammonia, but reported yields are low, viz., $G(NH_3) = 0.7$–1.1 with α radiation (3), 0.76 with protons at low dose rate (112), and 0.7 with $^{60}Co\,\gamma$ radiation (117). Cheek and Linnenbom (117) studied the effect of pressure, of varying H_2/N_2 ratios, and the presence of rare gases. They found that radiation energy absorbed primarily by hydrogen did not seem to contribute significantly to ammonia production, and that the rare gases He, Ar, Xe, and Kr all enhanced the radiolytic synthesis. Sensitization by rare gases does not appear to involve charge transfer, since the greatest enhancement was caused by Xe and Kr with ionization potentials lower than those for hydrogen and nitrogen. The data suggest that the principal processes leading to the formation of ammonia involve neutral excited nitrogen species, and nitrogen ions are involved only as precursors of the neutral species.

J. Carbon Dioxide

The radiolysis of carbon dioxide has been studied extensively but there still remain surprising discrepancies in the reported yields, and consequently there is little agreement on the details of the radiolytic mechanism. Indeed, at the time of this writing there does not appear to be one single mechanism which explains adequately all the experimental data, in spite of what appears to be a simple overall process represented stoichiometrically by Eq. (127):

$$CO_2 \underset{b}{\overset{a}{\rightleftharpoons}} CO + O \tag{127}$$

This simple equation apparently obscures a series of complex reactions which have been widely studied but which are inadequately understood.

It has been shown that the predominant ions formed from CO_2 in the mass spectrometer are CO_2^+ and CO^+ although O^+, C^+ (118,119) and CO_3^+ (120) have also been identified. Dissociation of electronically excited molecules leads predominantly to carbon monoxide and oxygen atoms (65), although the formation of carbon atoms is also possible. The energy requirements for the dissociation reactions leading to the various excited states are as follows:

$$CO_2(^1\Sigma g^+) \rightarrow CO(^1\Sigma^+) + O(^3P) - 5.5 \text{ eV} \tag{128}$$
$$\rightarrow CO(^1\Sigma^+) + O(^1D) - 7.5 \text{ eV} \tag{129}$$
$$\rightarrow CO(^1\Sigma^+) + O(^1S) - 9.7 \text{ eV} \tag{130}$$
$$\rightarrow CO(\alpha^3\pi) + O(^3P) - 11.5 \text{ eV} \tag{131}$$
$$\rightarrow C + O + O - 16.5 \text{ eV} \tag{132}$$

Ion-molecule reactions which could be important in determining the radiation chemistry include the following (14,67,121,122):

$$CO^+ + CO_2 \rightarrow CO_2^+ + CO \tag{133}$$
$$CO_2^+ + O_2 \rightarrow CO_2 + O_2^+ \tag{134}$$
$$CO_2^+ + CO_2 \rightarrow C_2O_4^+ \tag{135}$$
$$CO + O^- \rightarrow CO_2 + e^- \tag{136}$$
$$CO + O_2^- \rightarrow CO_2 + O^- \tag{137}$$
$$CO_2 + O_2^- \rightleftharpoons CO_4^- \tag{138}$$
$$CO_2 + O^- \rightleftharpoons CO_3^- \tag{139}$$

There is no direct evidence for electron attachment to CO_2 in the gas phase although the CO_2^- ion is formed in irradiated aqueous solutions of CO_2

(123). If the CO_2^- ion is formed in the gas phase it would have only a transitory existence in the presence of O_2 since the electron transfer reaction (140) would be efficient,

$$CO_2^- + O_2 \rightarrow CO_2 + O_2^- \tag{140}$$

In spite of the variations in some of the reported yields for the primary processes, the radiation stability of CO_2 is well established (3), since measurements in static gas systems with γ radiation (34,124), with fission fragments (33), with α particle radiation (33), with reactor radiation (33,124), lead to little overall decomposition at dose rates in the range 10^{13}–5×10^{14} eV cm^{-3} sec^{-1}. The radiation stability is only apparent, however, since studies in the presence of chemical scavengers and measurements of the radiation-induced exchange process show that carbon dioxide is decomposed. This points to the conclusion that in the irradiation of the pure gas, the initial decomposition products react to reform CO_2. It is the nature of this reformation step which is ill understood.

The initial decomposition of CO_2 in the absence of additives has been measured at high dose rates (3×10^{16}–1.3×10^{18} eV cm^{-3} sec^{-1}) using low energy protons from a van de Graaff accelerator (22). The steady state concentrations of the products CO and O_2 are essentially proportional to (dose rate)$^{1/2}$ and if this relationship holds at lower dose rates it would predict a steady state concentration of carbon monoxide of ~ 20 ppm at the low dose rate used by Dominey and Palmer (42,43). Thus while radiolytic decomposition can be measured at high dose rates the data are entirely consistent with experimental observations of apparent stability at lower dose rates. The dependence of steady state values on (dose rate)$^{1/2}$ suggests a process where the decomposition reaction is directly proportional to dose rate and in which the subsequent reformation reactions are kinetically of second order. However, these data give no direct information about the nature of the reacting species involved.

In contrast to the situation in the gas phase, liquid carbon dioxide decomposes at relatively low dose rates (33b,125) to yield carbon monoxide, oxygen, and ozone, with a decomposition yield close to the values reported from scavenger and exchange reaction studies in the gas.

While gaseous CO_2 is essentially stable under ionizing radiation its decomposition by vacuum ultraviolet radiation is well established (70,126–133) but there are many details of the reaction which are not understood (see Chapter 3).

1. Yields. Attempts to measure primary yields in the gas phase radiolysis of CO_2 have been made in three general ways as follows:

1. measurements with the pure gas at high dose rates

2. measurements in the presence of known oxygen atom scavengers

3. by measuring radiation-induced exchange reactions in mixture of carbon dioxide, carbon monoxide, and oxygen with isotopic labeling of the carbon or oxygen atoms

The latter two methods have been most extensively used, but there is still a disappointing degree of disagreement in the data as shown in Table IV, which is attributable to differences in dosimetry and to differences in interpretation of the fundamental processes involved.

The only decomposition yields measured in the absence of added

Table IV

Yields in the Radiolysis of Carbon Dioxide

Radiation	Method	Yields, $G(CO)$[a]	Ref.
$^{60}Co\ \gamma$	NO_2 scavenger	4.5 ± 0.4	34
	NO_2/SO_2 scavenger	3.5 ± 0.23	124
	^{14}CO exchange	4.2 ± 0.5	42
	^{14}CO exchange	4.4 ± 0.4	43
	^{14}CO exchange	4.5 ± 0.5	(a)
	$^{14}CO + N_2O$	4.4	(a)
	$^{14}CO_2$ exchange	4.15 ± 0.29	(b)
	Liquid CO_2	$5 \rightarrow 3.5; G(O_2) = 0.2\text{–}0.6$ $G(O_3) < 0.7$	125
X-ray	$C^{18}O_2$ exchange	~ 11	44
H^+	Fast gas flow	$4.25 \pm 0.25; G(O_2) = 2.24 \pm 0.1$	22
	^{14}CO exchange	$4.2\text{–}4.6$	
Radon α	NO_2 scavenger	10	33
	Solid CO_2	$9\text{–}10$	33
Reactor radiation	NO_2/SO_2 scavengers	$7\text{–}8.5$	33
	Liquid CO_2	$4\text{–}5$	33
	Liquid $CO_2 + NO_2$	$4\text{–}5$	33
	^{14}CO exchange	3.0 ± 0.3	42
		4.10 ± 0.19	(b)
	$^{14}CO_2$ exchange	4.19 ± 0.12	(b)
	NO_2 scavenger	2.49 ± 0.18	124
		2.8 ± 0.3	42

[a] Assuming $G(CO)_{initial} = G(-CO_2)_{initial}$.

(a) G. R. A. Johnson, private communication.

(b) D. A. Dominey, private communication.

scavengers or of isotopically labeled molecules at relatively high concentrations are those reported by Anderson and Best (22), who used a fast gas-flow system and found limiting yields of $G(CO) = 4.25 \pm 0.25$ and $G(O_2) = 2.24 \pm 0.1$ at high gas flowrates and at two different dose rates. The ratio $G(CO)/G(O_2) = 1.90 \pm 0.14$ is close to the theoretical ratio and although small amounts of ozone were found they had no significant effect on the stoichiometry. These measured yields were obtained under conditions where the instantaneous concentrations of carbon monoxide and oxygen in the irradiation volume were $<4\%$ of the relevant steady state concentrations and the authors concluded that they were measuring the initial yields in the absence of any recombination reaction.

Interpretation of measurements in the presence of oxygen atom scavengers is based on the assumption that the back reaction is suppressed and that the measured yields of carbon monoxide represent the primary radiolytic decomposition yield of carbon dioxide. Results from these measurements seem to fall into two general regions giving in one case an average value of $G(CO) = 4.0 \pm 0.5$ (34,42,124) and in the other a yield of carbon monoxide in the range $G(CO) = 7$–10 (33). These variations have been discussed (124) on the basis of discrepancies in dosimetry but the differences have not been resolved satisfactorily. There appears to be a significant difference between the effects of γ radiation and reactor radiation (42,43) but in view of the difficulties of precise gas phase dosimetry in mixed radiation fields it is not certain that this difference has been established unequivocally. Since a large fraction of the energy absorption from reactor radiation is due to recoil ions produced by fast neutron scattering, this could lead to a difference in the initial distribution of absorbed energy to form ions and excited molecules, and hence a difference in the subsequent chemistry.

Measurement of the radiation-induced exchange reaction of ^{14}C between ^{14}CO (concentration $\sim 1\%$) and CO_2 of normal isotopic composition have been used by Dominey and others (22,42,43,134) in a further attempt to elucidate the primary yields. The philosophy underlying these (and other) measurements of radiolytically induced exchange reactions is to study the nature of the recombination reaction in the presence of additives (CO and/or O_2) and to relate these measurements to the primary processes in the radiolytic decomposition reaction. It should be noted, however, that in general the concentrations of additive are always in excess of the steady state concentrations for radiolysis of pure carbon dioxide and so may influence the deductions about the primary processes.

In measurements of the ^{14}C exchange reaction there is no detectable chemical change in the system and the yield of the exchange reaction

$(G(^{14}CO_2))$ is taken as a measurement of the primary radiolytic decomposition of carbon dioxide. The rate of exchange of ^{14}C between ^{14}CO and CO_2 and the rate of formation of carbon monoxide from mixtures of carbon dioxide and nitrogen dioxide depend on the rate of energy absorption in the carbon dioxide, and it seems likely that the rate-controlling step in both reactions is the initial breakdown of the carbon dioxide molecule (42,43). The value for the yield of the exchange reaction agrees well with the value of the yield of carbon monoxide obtained from CO_2/NO_2 mixtures in separate experiments. Anderson and Best (22) have extended these measurements to higher dose rates and find $G(ex) = 4.2–4.6$ in the absence of added oxygen. However, their results indicate that the full yield of the exchange reaction is not obtained until the appropriate steady state level of molecular oxygen has accumulated in the gas.

Dominey et al. (42,43) find an enhancement of the exchange in the presence of added oxygen for both γ radiation and reactor radiation, and this result has been confirmed at higher dose rates (22). In addition, however, the measurements at high dose rates show that on further increasing the oxygen concentration the yield of the exchange reaction falls, with the maximum values for $G(ex)$ depending on the concentration of oxygen, carbon monoxide and on the dose rate (29b) (Fig. 2).

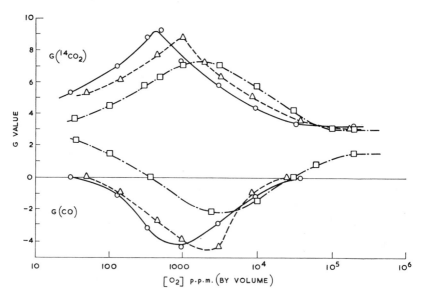

Fig. 2. Effect of O_2 on yields in the $^{14}CO/CO_2$ exchange reaction. Radiation 1.5 MeV H^+; temperature, 20°C; dose rate, ○, 3×10^{16} eV cm^{-3} sec^{-1}; △, 10^{17} eV cm^{-3} sec^{-1}; □, 10^{18} eV cm^{-3} sec^{-1}.

Measurements on the exchange of isotopically labeled oxygen atoms between CO and CO_2 have produced results apparently at variance with the ^{14}C exchange data. Hirota et al. (135) have made measurements of the exchange reaction (141),

$$C^{18}O_2 + C^{16}O_2 \rightarrow 2C^{16}O^{18}O \tag{141}$$

and report a value of $G(ex)_{total} = 51$, which they argue should represent the primary decomposition yield of carbon dioxide. However, this conclusion presupposes the absence of a chain reaction, and later work (44) which gives values of $G(ex) > 10^3$ provides clear evidence for a chain mechanism. Anbar and Perlstein (44) also give a value of $G(CO) \sim 11$ for the primary yield of carbon monoxide from carbon dioxide decomposition, deduced from measurements on the x-ray induced exchange reaction between $C^{18}O_2$ and $C^{16}O$. The reported yields are $G(C^{18}O) = 11.8 \pm 0.25$ and 8.2 ± 0.5 at CO concentrations of 22 and 42 mole %, respectively, but the applicability of these yields, calculated on the basis of total energy absorption, to primary processes in carbon dioxide radiolysis must be questionable. When the yields are calculated on the basis of the energy absorbed by the CO_2 alone they are comparable ($G(C^{18}O) \sim 14$) but, as the authors point out, this agreement seems fortuitous when based on the assumption that there is no energy transfer from carbon monoxide to carbon dioxide in a mixture containing 42 mole % CO. Moreover, they report a large drop in the yield of $C^{18}O$ (from 11 to 8) on increasing the total pressure from 200 to 500 mm at the same concentration of carbon monoxide, but the significance of this observation is not discussed.

2. Mechanism. The apparent radiation stability of carbon dioxide is almost certainly due to a very rapid recombination of the primary products, but the precise nature of the back reaction is not understood. Hirschfelder and Taylor (136) ascribed the ineffectual decomposition of carbon dioxide to the oxidation of carbon monoxide by ozone (Eq. 142)

$$CO + O_3 \rightarrow CO_2 + O_2 \tag{142}$$

$$O_2 + O + M \rightarrow O_3 + M \tag{143}$$

This mechanism, however, is invalidated by the experiments of Harteck and Dondes (137), who found that the reaction has an activation energy of > 28 kcal/mole and does not proceed fast enough at room temperature to account for the observed stability. These authors proposed an alternative scheme in which carbon suboxides, produced in reactions involving carbon atoms, react rapidly with oxygen atoms or ozone molecules (in the absence of O atom scavengers) to reform carbon dioxide. This complex mechanism, however, is not consistent with the following observations (138,139):

1. that CO_2 is reduced to CO by C_2O

2. C_2O was found to produce CO rather than CO_2 + C in reactions with O atoms

3. in the reaction C_3O_2 + O, the formation of CO is 10 times faster than the formation of CO_2.

These measurements with oxygen atoms generated by the absorption of energy from other than ionizing radiation are not conclusive, since different energy states may be involved, but they suggest that reactions of carbon suboxides do not lead to the reformation of carbon dioxide in radiolysis.

The mechanism proposed by Dominey and Palmer (42) on the basis of the ^{14}CO exchange work appears to explain many of the features of the reaction. From measurements of competitive reactions between NO_2 and CO for the oxidizing species they reject the simple reformation reaction (127b) based on known rate constants for ground state O atoms, although they do not exclude the possibility of a mechanism based on excited O atoms. They suggest that the recombination reaction could proceed through a step involving either the O_2^+ ion or an oxidizing species produced on its subsequent charge neutralization in the sequence of reactions (144–148)

$$CO_2 \overset{\sim\!\sim\!\sim}{\longrightarrow} CO_2^* \tag{144a}$$

$$\downarrow$$

$$CO + O \tag{144b}$$

$$CO_2^+ + e \rightarrow (CO_2^*) \tag{145a}$$

$$\downarrow$$

$$CO + O \tag{145b}$$

$$O + O + M \rightarrow O_2 + M \tag{146}$$

$$CO_2^+ + O_2 \rightarrow O_2^+ + CO_2 \tag{134}$$

$$(O_2^+) + CO \rightarrow CO_2 + O^+ \tag{147}$$

$$O_2^+ + e^- \rightarrow O_2^* \tag{148}$$

(The representation (O_2^+) implies that it is not necessarily this species which participates in the reaction; Dainton (140) suggests that O_2^- is more feasible.)

Dominey and Palmer calculate that complete charge transfer (Eq. (134)) will occur in competition with charge neutralization (Eqs. (145,148)) at very low concentrations of oxygen at the dose rate used. Thus once a low steady state concentration of molecular oxygen has accumulated in the system participation of the efficient oxidation reaction (147) will lead to no further net radiolysis. This mechanism appeared to be supported by measurements at high dose rates (22) on the exchange yield with ^{14}CO, where it was shown that the full yield of the exchange reaction was only obtained when the concentration of molecular oxygen had attained steady state. In addition, these authors suggested tentatively that the effect of oxygen in enhancing the exchange reaction could be explained if the charge neutralization reaction (145a) is nondissociative due to the formation of ion–molecule clusters (and possible participation of O^- and O_2^- ions). The retarding effect of oxygen at higher concentrations (Fig. 2) has been ascribed to competition between oxygen and carbon monoxide for an electrically neutral oxidizing species which behaves kinetically as an O atom (29b). It is further deduced that the recombination reaction between carbon monoxide and oxygen can only be explained adequately on the basis of at least two participating oxidizing species (possibly one electrically charged and the other electrically neutral), and this view is further supported by the measurements of Dominey et al. (141) of the inhibitory effect of methane on the $^{14}CO/CO_2$ reaction.

These ideas based on an oxidizing species derived from molecular oxygen appear to be invalidated however, by some of the observations of Anbar and Perlstein (44). They suggest that the principal oxidizing species involved in the efficient oxidation of CO is the CO_3 radical and, moreover, that the presence of molecular oxygen may suppress the overall production of excited CO_2^* molecules formed in the charge recombination reaction. They propose a mechanism based on the following reactions

$$CO_2^* \rightarrow CO + O \tag{144b}$$

$$CO_2^* + CO_2 \rightarrow CO_3 + CO \tag{149}$$

$$CO_2^* + O \rightarrow CO_3 \tag{150}$$

$$CO_3 + CO_2 \rightarrow CO_2 + CO_3 \text{ (exchange reaction)} \tag{151}$$

$$CO_3 + CO \rightarrow 2CO_2 \tag{152}$$

There seems little doubt that possible reactions of the CO_3 species must be considered in understanding the radiolysis of CO_2, since there is an increasing amount of evidence demonstrating the existence of both CO_3 ions and radicals. The ion CO_3^+ has been observed by Čermák and Herman (120) and the negative ion CO_3^- has been observed in flames (142), in afterglow studies (143), and in electron drift experiments (67,144). The CO_3 radical was postulated as an intermediate in the gas phase photolysis exchange reaction between $O(^1D)$ and CO_2 (145), and in the oxidation of CO by electronically excited O_2 (146). Moreover from an analysis of the photolysis of mixtures of CO and CO_2, $CO + O_2$, and $CO + O_2 + CO_2$, Young and Ung (147) conclude that $O(^1D)$ reacts with CO_2 to form CO_3 at least five times more rapidly than it is deactivated to $O(^3P)$, but Preston and Cvetanovic (148) suggest that further experimental evidence is required before this mode of formation of CO_3 can be considered proved. However, Moll et al. (149) have identified the species CO_3 by infrared

spectroscopic analysis of the products formed in three reactions, viz.,
(1) Xe resonance UV photolysis of solid CO_2 at $77°K$, (2) photolysis of
O_3 in a CO_2 matrix at $50–60°K$ using light of wavelength 2537 Å, and
(3) rf discharge in gaseous CO_2 followed by trapping the products at
$50–70°K$.

Anbar and Perlstein (44) find no significant enhancement of the radiation
induced exchange reaction in the presence of molecular oxygen
$(CO_2:{}^{18}O_2:CO = 98:1:1)$ but this in itself does not invalidate Dominey's
"molecular oxygen" hypothesis since the concentration of oxygen used is
possibly high enough to give yields on the retardation side of the maxima
for the curves of $G(ex)$ vs. oxygen concentration (Fig. 2). In a very significant experiment, however, they irradiated a mixture of CO_2, ^{13}CO, CO,
and $^{18,18}O_2$ (98:0.6:0.4:1) and found only small yields $(G \leq 0.15)$ of
$^{13}C^{16}O^{18}O$. This result suggests that the reaction (153),

$$^{13}CO + ({}^{18,18}O_2{}^+) \rightarrow {}^{13}C^{16}O^{18}O + (O^+) \tag{153}$$

cannot be dominant, and that molecular oxygen ions per se cannot account
for the efficient back reaction. However, this conclusion does not exclude
the role of a molecular oxygen species as an intermediate, if it can lead
to the formation of CO_3. It has been proposed (146) that CO_3 radicals
can be formed in reaction (154), the excited $O_2{}^*$ molecule being formed by
energy transfer from CO_2 or in charge neutralization reactions involving
the $O_2{}^+$ or $O_2{}^-$ ions, under ionizing radiation.

$$^{18}O_2(^3\Sigma_u{}^+) + {}^{13}CO \rightarrow CO_3{}^* \tag{154}$$

The $CO_3{}^*$ radical $(^{13}C^{18}O^{18}O^{16}O)$ will then, according to Anbar's measurements on the $C^{18}O_2/C^{16}O_2$ exchange reaction, undergo a rapid exchange
reaction with CO_2 so that the predominant species which subsequently
participates in the oxidation of ^{13}CO is a $C^{16}O_3$ radical. Johnson has also
proposed the participation of the $CO_3{}^+$ ion in the radiolytic oxidation of
carbon monoxide (section 5-V-K), and since this ion could be formed
from reactions involving $O_2{}^+$ and could exchange rapidly with CO_2,
isotopic scrambling would again result in a predominance of ^{16}O atoms
in the oxidizing species. Alternatively, the ions $CO_4{}^-$ or $CO_3{}^-$, which
have been observed in electron drift experiments (67) in mixtures of CO_2
and O_2, may undergo exchange with CO_2 molecules and participate in
the recombination reaction. The very significant associative detachment
reaction (136) has been observed (17,67), and if $CO_3{}^-$ and $CO_4{}^-$ ions can
also oxidize CO these reactions provide additional ways of reforming CO_2.

$$CO + O^- \rightarrow CO_2 + e^- \tag{136}$$

While these ideas are tentative, they do suggest that the mechanisms
proposed by Anbar and by Dominey are not mutually exclusive but could

be complementary. An objective summary of the present situation lies in the recognition that no single mechanistic proposal appears to explain all the experimental data. However the measurements on isotopic exchange reactions are providing a valuable insight into possible alternative pathways, and it may be confidently anticipated that further studies under a variety of conditions including changes in $CO_2/CO/O_2$ ratios, changes in dose rate, and measurements in the presence of electron scavengers will lead to an unambiguous understanding of the processes which account for the radiation stability of carbon dioxide.

3. Liquid Carbon Dioxide. The decomposition of liquid CO_2 is of great theoretical interest, for it provides the somewhat unusual situation of a material which is less stable in the liquid phase than in the gas phase. In addition to measurements of the decomposition yields, Baulch et al. (125) have also shown the absence of a recombination reaction in the liquid phase by measurement of the $C^{16}O_2/C^{18}O_2$ exchange reaction which gives little isotopic mixing, in marked contrast to the gas phase measurements (132,145).

Baulch et al. (125) attribute the major differences in the observed behavior in the two phases to the much larger spatial separation and lifetime before recombination of the ionic species in the gas phase. As discussed above the ions O_2^- and O_2^+ could participate in gas phase reactions but in liquid CO_2 (dielectric constant 1.6) it is unlikely that the electrons escape the coulombic field of the parent ion, so that geminate recombination predominates. Thus in the liquid, much higher concentrations of molecular oxygen are required for effective competition between electron capture and geminate recombination to occur; if therefore the ion O_2^- is responsible (either directly or as a precursor) for oxidation of CO this affords a rational explanation of the different behavior in the two phases. The measurements of the $^{14}CO/CO_2$ exchange reaction in the presence of molecular oxygen at high dose rates in the gas phase are consistent with this general hypothesis. Baulch et al. also suggest that several mechanisms can be derived by which a short chain reaction could occur, e.g.,

$$CO + O_2^- \rightarrow CO_2 + O^- \tag{137}$$

$$O^- + O_3 \rightarrow O_2 + O_2^- \tag{155}$$

Mechanisms of this type might also explain the high yields ($G(^{14}CO_2) = 11$) which have been observed in the $^{14}CO/CO_2$ exchange reaction in the presence of O_2 in the gas phase (29b).

4. The Reaction between CO_2 and Graphite. The radiolytically induced oxidation of graphite by CO_2 is extremely important in the development of CO_2-cooled, graphite-moderated nuclear reactors. The introduction of heterogeneous reactions at the graphite surface further complicates the understanding of an already complex reaction. A recent survey by Lind and Wright (150) discusses the main parameters affecting the reaction and includes data on the effect of additives, viz. CO, H_2, H_2O, and CH_4, on the role of surface oxide, and on the effect of graphite open pore volume.

K. Carbon Monoxide

Less effort has been expended on the radiolysis of carbon monoxide than on that of carbon dioxide, but the main features of its decomposition are clearly established. The major radiolysis products are carbon dioxide and a carbonaceous solid with an O:C ratio < 1 and with an empirical composition which varies with the conditions of formation.

Evidence from spectroscopy and photochemistry (65) shows that CO absorbs light in a relatively weak band in the region 1150–1500 Å which produces the transition,

$$CO + h\nu \rightarrow CO(A^1\pi) \tag{156}$$

The electronically excited CO molecules can react in one of two ways, by dissociation (reaction (157)) or by reaction on collision (reactions (158) and (159))

$$CO^* \rightarrow C + O \tag{157}$$

$$CO^* + CO \rightarrow C_2O + O \tag{158}$$

$$CO^* + CO \rightarrow CO_2 + C \tag{159}$$

The relative abundance of the ions CO^+, O^+, $C^+ = 94:2:4$ has been measured in the mass spectrometer (118). The energy requirements for various primary dissociation processes are as follows:

$$CO \rightarrow CO^+ + e^- \qquad\qquad -14.1\ eV \tag{160}$$

$$CO \rightarrow C^+(^2P) + O(^3P) + e^- \qquad -22.4\ eV \tag{161}$$

$$CO \rightarrow C(^3P) + O^+(^4S) + e^- \qquad -24.7\ eV \tag{162}$$

$$CO \rightarrow C^+(^3P) + O^- \qquad\qquad -9.6\ eV \tag{163}$$

$$CO \rightarrow C^+(^2P) + O^-(^2P) \qquad -20.9\ eV \tag{164}$$

1. Yields. The reported yields for carbon monoxide decomposition with different types of ionizing radiation are shown in Table V. It is seen that there is reasonable agreement for the initial values of $G(CO_2)$ and that the largest discrepancies lie in the values for $G(-CO)$. This is due to two principal reasons (1) direct measurement of $G(-CO)$ involves studying the reaction to large percentage decompositions, and (2) derivation of

Table V

Yields in the Radiolysis of Carbon Monoxide

Radiation	Initial yields			Ref.
	$G(-CO)$	$G(CO_2)$	$G(C)_{solid}$	
^{60}Co γ	7.84[a]	1.96 ± 0.2	—	21
	9.2[a]	2.3 ± 0.3	—	158
	(7.8)[b]	(1.45 ± 0.26)[b]	—	
	9.3	1.9	7.4	152
Radon α	11.6–19.1			3
	5.4			151
	8.4–9.6			14
H^+	7.45 ± 0.4[c]	2.05 ± 0.15	5.4 ± 0.3	21
Reactor radiation	5–6			(a)

[a] Assuming stoichiometry: $4nCO = nCO_2 + (C_3O_2)_n$.
[b] Adjusted to same dosimetry standard as used in Reference 21.
[c] Limiting yields at low dose rate.
(a) W. R. Marsh and J. Wright, Harwell Research Report, AERE-R 4198 (1964).

$G(-CO)$ by total product analysis is difficult, since it involves quantitative collection and analysis of the ill-defined solid. The high value for $G(-CO) = 11.6$–19.1 found by Lind (3) may have been influenced by traces of O_2, which has a marked effect on the reaction, for it seems in later work that the mean value for the decomposition yield can be adequately expressed as $G(-CO) = 8 \pm 1$. The reasonable agreement between results from different radiation sources suggests that there is no significant effect of LET at low gas pressures (ca. 1 atm) over the range of ionizing densities studied, although measured yields in nuclear reactors are subject to some doubts about the dosimetry. Since $G(CO_2) \sim 2$ this leads to the overall stoichiometry represented by Eq. (165),

$$4nCO \rightarrow nCO_2 + (C_3O_2)_n \qquad (165)$$

Lind and Bardwell (151) showed that the solid could be oxidized by irradiated CO_2; although this reaction is slow this is due, at least in part, to the heterogeneous nature of the process. These authors also suggested that the solid is a mixture of C_3O_2 and carbon, but it seems unlikely that elemental carbon and a carbon–oxygen complex would be produced simultaneously and would coexist. Recent work by Meaburn (53) indicates that the likely fate of carbon atoms produced in the radiolysis is reaction with CO molecules (Eq. 166).

Dondes et al. (152) derived their yields of solid by postirradiation oxidation in the irradiation cell without prior exposure to air, as did Anderson et al. (21). These latter workers also determined the empirical composition of the solid by vacuum pyrolysis followed by oxidation of the residual carbon. In this work they demonstrated a mass balance by calculating the total amounts of carbon and oxygen in both the gaseous and solid products and found a maximum deviation of $\leq 10\%$ from the theoretical C:O ratio of unity. They found an empirical composition $(C_{1.45}O)_n$ for the solid formed at room temperature with a marked increase in the C:O ratio with increasing temperature of irradiation ($C_{14}O$ at 700°C); these changes in composition were accompanied by increases in $G(CO_2)$ and decreases in $G(C)_{solid}$. Similar changes were observed on postirradiation heating of the gas irradiated at room temperature and the authors concluded that the major effect of temperature lies in the subsequent thermal decomposition of the solid $(C_3O_2)_n$ initially formed during radiolysis.

Lind's suggestion (3) that the primary reactions resulting in solid formation occur in the gas phase is strongly supported by this work. However, these observations cannot lead to a unique assignment of the origin of the solid, since heterogeneous nucleation processes could compete with homogeneous processes, particularly at low dose rates and at low gas pressures. In addition, the subsequent deposition of solids formed in the gas phase will be a complex function of surface condition and material, temperature, and flow conditions.

2. Mechanism. The detailed mechanism of CO radiolysis has not been elucidated, since among other uncertainties there appears to be little unequivocal evidence on the relative importance of ions and excited molecules in the radiolysis. Rudolph and Lind (14) conclude that in the presence of a small amount of product CO_2 the CO^+ ion is deactivated by the charge-transfer reaction and only excited CO* molecules react. This suggests, however, that the decomposition would be inhibited by inert gases with ionization potentials lower than that for CO, but they found no effect on the addition of Xe (IP = 12.13 eV).

Stewart and Bowlden (153) conclude that ionization is the predominant process in their interpretation of the effect of added inert gases, but in a later paper Dondes et al. (152) present new data and conclude that the effect of inert gases is best explained by considering energy transfer from specific excited states.

In addition to the retardation of CO radiolysis attributed to the charge transfer reaction (133) (14) or to subsequent radiolytic oxidation of the carbonaceous solid by CO_2 (151), Anderson et al. (21) found a dose-rate inhibition of its decomposition at high dose rates ($> 10^{15}$ eV cm^{-3} sec^{-1}). They attribute this effect to competition between reactions e.g., leading to its decomposition (166–168) and reactions such as (169) and (170) leading to its reformation, e.g., Eqs. (169) and (170).

These competitive processes are not significant at low dose rates.

$$C + CO \rightarrow C_2O \tag{166}$$

$$O + CO \rightarrow CO_2 \tag{167}$$

$$C_2O + CO \rightarrow C_3O_2 \tag{168}$$

$$C_2O + O \rightarrow 2CO \tag{169}$$

$$O + C + M \rightarrow CO + M \tag{170}$$

It seems clear that in pure carbon monoxide the simple charge neutralization reaction (171) cannot predominate since the ion molecule reactions (172–175) have been observed in the mass spectrometer (154,155)

$$CO^+ + e^- \rightarrow CO^* \tag{171a}$$
$$\downarrow$$
$$C + O \tag{171b}$$

$$CO^+ + CO \rightarrow C_2O_2^+ \tag{172}$$

$$CO^+ + CO \rightarrow C_2O^+ + O \tag{173}$$

$$CO^+ + CO \rightarrow C^+ + O + CO \tag{174}$$

$$CO^+ + CO \rightarrow C + O^+ + CO \tag{175}$$

The participation of these reactions can lead to several alternative pathways to the observed products but it is not possible to assign the relative importance of any of the steps. Reaction (172) could be extremely important if the $C_2O_2^+$ ion can attach further CO molecules or intermediates, e.g., C_2O, C_3O_2, C atoms at higher pressures than those used in the mass spectrometer. Such ion association reactions followed by charge neutralization could lead to the production of large molecular aggregates (carbonaceous solid) or to a mechanism whereby the role of ionization processes is minimized through distribution of the recombination energy over a large molecular complex $(CO)_n^+$. Since the energy released on recombination with an electron is 14.1 eV and the dissociation energy of the C—O bond is 11.1 eV, neutralization of even the bimolecular complex $(C_2O_2)^+$ will not necessarily lead to immediate dissociation if the recombination energy is shared equally between the two molecules. Moreover, equal distribution of the recombination energy between three molecules will not produce electronic excitation to the lowest level ($^3_a\pi$ at 6.03 eV).

Reactions such as those represented schematically by Eq. (176).

$$C + CO \longrightarrow C_2O \xrightarrow{+CO} C_3O_2 \qquad (176)$$

with nCO and nCO branches leading to Solid

must contribute to the overall process and further evidence for the precise nature of some of these steps will undoubtedly arise from pulse radiolysis studies. It seems clear, however, that the carbonaceous solid is not formed by simple polymerization of C_3O_2 molecules since the infrared spectra of thermally and photolytically polymerized C_3O_2 (156,157) are quite different from that of the solid formed in CO radiolysis (21).

3. Influence of Oxygen. The pronounced effect of oxygen on the radiolytic decomposition of carbon monoxide has been further investigated by studies of the CO/O_2 radiation-induced reaction. With a stoichiometric mixture of $2CO:O_2$, Lind and Bardwell (3,151) found that carbon dioxide was the only product, with a much higher yield than for carbon monoxide alone.

Clay, Johnson, and Warman (158) have studied the $^{60}Co\ \gamma$ radiation induced oxidation in much greater detail with CO/O_2 mixtures ranging from 1–98 mole % of O_2 and have investigated the effect of dose rate, pressure, and the presence of various additives. Yields of CO_2 are at a maximum with O_2 concentrations of ~ 50 mole % and the highest reported value is $G(CO_2) = 8,000$ at a total gas pressure of 60 mm. The initial yield of CO_2 also depends on gas pressure for a fixed $CO:O_2$ ratio but is independent of dose rate over a 30-fold range; it is markedly decreased by the presence of low concentrations of Hg vapor, CO_2, Xe, and CH_4 but is unaffected by small concentrations of Kr, Ne, Ar, or N_2. These inhibitory effects plus the low rate of ozone formation lead the authors to suggest that O atoms are not involved as chain carriers and that the chain process most probably involves ionic intermediates. A reaction scheme involving the following reactions is postulated:

Initiation
$$CO^+ + O_2 \rightarrow CO_3^+ + O \qquad (177)$$
$$O_2^+ \ (^4\pi_u \text{ or higher states}) + CO \rightarrow CO_3^+ + O \qquad (178)$$

Propagation
$$CO_3^+ + CO \rightarrow CO_2^+(^2\pi_u?) + CO_2 \qquad (179)$$
$$CO_2^+(^2\pi_u) + O_2 \rightarrow O_2^+(^4\pi_u) + CO_2 \qquad (180)$$

Termination
$$CO_3^+ + CO + M \rightarrow CO_2^+ + CO_2 + M \qquad (181)$$

Inhibition by Hg, CO_2, Xe, and CH_4 probably involves reactions with one or more of the ionic species involved in the chain initiation or propagation, e.g.,

$$CO^+ + Xe \rightarrow Xe^+ + CO \tag{182}$$

$$CO^+ + CO_2 \rightarrow CO + CO_2^+ \tag{133}$$

The postulated reaction scheme explains many of the experimental measurements but the role of inert gases is somewhat anomalous: Ar^+ and Ne^+ ions cannot initiate chains although their ionization potentials are higher than that of CO, and there is no indication that at sufficiently high concentrations of Ar the postulated intermediates O_2^+ ($^4\pi_u$) and CO_2^+($^2\pi_u$) are removed by charge exchange.

At high dose rates (21) the effect of O_2 is dose rate dependent, which the authors attribute to competition between charge recombination and chain initiation by charge transfer (Eqs. (177 and 178)); this postulated competition will be unimportant at the lower dose rates used by Clay, Johnson, and Warman (158).

REFERENCES

1. E. Warburg, *Jahrb. Radioakt. Elektronik.*, **6**, 181 (1909).
2. S. C. Lind, *The Chemical Effects of Alpha Particles and Electrons*, The Chemical Catalogue Co., New York, 1st ed. 1921, 2nd ed. 1938.
3. S. C. Lind, *Radiation Chemistry of Gases*, Reinhold, New York, 1961.
4. W. Mund, *L'Action Chimique des Rayons Alpha en Phaze Gazeuze*, Hermann et Cie., Paris, 1935.
5. W. Mund, "Les Effects Chimiques Produits Par Les Rayons Ionisants en Phaze Gazeuze," in *Actions Chimiques et Biologiques des Radiations*, Vol. II, M. Haissinsky, Ed., Masson et Cie., Paris, 1956.
6. (a) H. Eyring, J. O. Hirschfelder, and H. S. Taylor, *J. Chem. Phys.*, **4**, 479 (1936); (b) *ibid.*, **4**, 570 (1936).
7. H. Essex and D. Fitzgerald, *J. Am. Chem. Soc.*, **56**, 65 (1934).
8. H. Essex, *J. Phys. Chem.*, **58**, 42 (1954).
9. A. W. Tickner and P. F. Knewstubb, *J. Chem. Phys.*, **38**, 464 (1963).
10. A. W. Tickner, P. H. Dawson, and P. F. Knewstubb, *J. Chem. Phys.*, **38**, 1031 (1963).
11. P. H. Dawson and A. W. Tickner, *J. Chem. Phys.*, **40**, 3745 (1964).
12. P. Kebarle and A. M. Hogg, *J. Chem. Phys.*, **42**, 798 (1965).
13. A. M. Hogg and P. Kebarle, *J. Chem. Phys.*, **43**, 449 (1965).
14. P. S. Rudolph and S. C. Lind, *J. Chem. Phys.*, **33**, 705 (1960).
15. C. E. Melton, G. A. Rapp, and P. S. Rudolph, *J. Chem. Phys.*, **29**, 968 (1958).
16. A. Henglein and G. A. Muccini, *J. Chem. Phys.*, **31**, 1426 (1959).
17. F. Fehsenfield, E. E. Ferguson, and A. L. Schmeltekopf, *J. Chem. Phys.*, **45**, 1844 (1966).
18. G. R. A. Johnson and J. M. Warman, *Discussions Faraday Soc.*, **37**, 87 (1964).
19. B. J. Burtt and J. F. Kircher, *Radiation Res.*, **9**, 1 (1958).
20. S. Dondes, P. Harteck and H. von Weyssenhoff, *Z. Naturforsch.*, **19a**, 13 (1964).
21. A. R. Anderson, J. V. F. Best, and M. J. Willett, *Trans. Faraday Soc.*, **62**, 595 (1966).

22. A. R. Anderson and J. V. F. Best, *Trans. Faraday Soc.*, **62**, 610 (1966).
23. D. H. Dawes and R. A. Back, *J. Phys. Chem.*, **69**, 2385 (1965).
24. (a) R. L. Platzman, *Vortex*, **23**, 372 (1962); (b) R. L. Platzman, *Radiation Res.*, **17**, 419 (1962).
25. Th. Förster, *Discussions Faraday Soc.*, **27**, 7 (1959).
26. R. A. Back, T. W. Woodward, and K. A. McLauchlan, *Can. J. Chem.*, **40**, 1380 (1962).
27. (a) *Radiation Dosimetry*, G. J. Hine and G. I. Brownell, Eds. Academic Press, New York, 1956; (b) J. W. Boag, "Ionization Chambers" in *Radiation Dosimetry*.
28. J. K. Linacre, Ed., *The Determination of Absorbed Dose in Reactors*, to be published by I.A.E.A. (1968).
29. A. R. Anderson and J. V. F. Best, (a) *Nature*, **216**, 576 (1967); (b) unpublished work.
30. V. O. Janssen, A. Henglein, and D. Perner, *Z. Naturforsch.*, **19b**, 1005 (1964).
31. A. R. Anderson and J. A. Winter, "Radiolysis of Water Vapor" in *Radiation Research*, G. Silini, Ed., North Holland, 1967.
32. G. R. A. Johnson and M. Simic, *J. Phys. Chem.*, **71**, 1118 (1967).
33. (a) P. Harteck and S. Dondes, *J. Chem. Phys.*, **23**, 902 (1955); (b) *ibid.*, **26**, 1727 (1957).
34. M. Steinberg, Brookhaven National Laboratory Research Report No. B.N.L. 665 (1961).
35. O. A. Schaeffer and S. O. Thompson, *Radiation Res.*, **10**, 671 (1959).
36. V. Aquilanti, A. Galli, A. Giardini-Guidoni, and G. G. Volpi, *J. Chem. Phys.*, **43**, 1969 (1965).
37. M. Anbar and P. Perlstein, *J. Phys. Chem.*, **70**, 2052 (1966).
38. R. F. Firestone, *J. Am. Chem. Soc.*, **79**, 5593 (1957).
39. J. Y. Yang and L. H. Gevantman, *J. Phys. Chem.*, **68**, 3115 (1964).
40. J. Y. Yang and I. Marcus, *J. Am. Chem. Soc.*, **88**, 1625 (1966).
41. F. T. Jones and T. J. Sworski, (a) *Trans. Faraday Soc.*, **63**, 2411 (1967); (b) F. T. Jones, T. J. Sworski, and J. M. Williams, *ibid.*, **63**, 2426 (1967).
42. D. A. Dominey and T. F. Palmer, *Discussions Faraday Soc.*, **36**, 35 (1963).
43. D. A. Dominey, T. F. Palmer, and J. C. Robertson, A.E.R.E. Harwell Research Report No. AERE-R 4481 (1964).
44. M. Anbar and P. Perlstein, *Trans. Faraday Soc.*, **62**, 1803 (1966).
45. R. Gorden, Jr., and P. Ausloos, *J. Res. Natl. Bur. Std.*, **69A**, 79 (1965).
46. J. H. Baxendale and G. P. Gilbert, *J. Am. Chem. Soc.*, **86**, 516 (1964).
47. A. R. Anderson, B. Knight, and J. A. Winter, *Nature*, **201**, 1026 (1964).
48. C. Smith and H. Essex, *J. Chem. Phys.*, **6**, 188 (1938).
49. B. P. Burtt and T. Baurer, *J. Chem. Phys.*, **23**, 466 (1955).
50. F. T. Jones and T. J. Sworski, *J. Phys. Chem.*, **70**, 1546 (1966).
51. M. C. Sauer, Jr., and L. M. Dorfman, *J. Am. Chem. Soc.*, **86**, 4218 (1964).
52. M. C. Sauer, Jr., *J. Phys. Chem.*, **71**, 3311 (1967).
53. G. M. Meaburn and D. Perner, *Nature*, **212**, 1042 (1966).
54. R. W. Nicholls, E. M. Reeves and D. A. Bromley, *Proc. Phys. Soc.*, **74**, 87 (1959).
55. (a) B. Brocklehurst, AERE Harwell Research Report No. AERE-C/R 2669 (1963); (b) *Trans. Faraday Soc.*, **60**, 2151 (1964); (c) *Trans. Faraday Soc.*, **63**, 274 (1967).
56. (a) S. Dondes, P. Harteck, and C. Kunz, *Radiation Res.*, **27** 174 (1966); (b) F. Morse, P. Harteck, and S. Dondes, *Radiation Res.*, **29**, 317 (1966); (c) C. Kunz,

S. Dondes, and P. Harteck, *Abstracts of 153rd Am. Chem. Soc. Mtg.*, *April*, *1967*.

57. G. von Bunau, *Fortschr. Chem. Forsch.*, **5**, 347 (1965).
58. J. W. T. Spinks and R. J. Woods, *An Introduction to Radiation Chemistry*, Wiley, New York, 1964.
59. S. O. Thompson and O. A. Schaeffer, *J. Am. Chem. Soc.*, **80**, 553 (1958).
60. H. von Koc and L. Friedman, *J. Chem. Phys.*, **38**, 1115 (1963).
61. T. F. Moran and L. Friedman, *J. Chem. Phys.*, **39**, 2491 (1963).
62. P. H. Dawson and A. W. Tickner, *J. Chem. Phys.*, **45**, 4330 (1966).
63. E. G. Zubler, W. H. Hamill, and R. R. Williams, Jr., *J. Chem. Phys.*, **23**, 1263 (1955).
64. (a) D. A. Armstrong, private communication; (b) R. A. Lee, R. S. Davidow, and D. A. Armstrong, *Can. J. Chem.*, **42**, 2906 (1964); (c) R. S. Davidow, R. A. Lee, and D. A. Armstrong, *J. Chem. Phys*, **45**, 3364 (1966).
65. J. G. Calvert and J. N. Pitts, Jr., *Photochemistry*, Wiley, New York, 1966.
66. (a) M. Cottin, *J. Chim. Phys.*, **56**, 1024 (1959); (b) M. M. Mann, A. Hustrulid, and J. T. Tate, *Phys. Rev.*, **58**, 340 (1940).
67. J. L. Moruzzi and A. V. Phelps, *J. Chem. Phys.*, **45**, 4617 (1966).
68. (a) C. E. Melton, unpublished work reported by T. J. Sworski in Proceedings of the Fifth Informal Conference on the Radiation Chemistry of Water (Notre Dame), A.E.C. Document No. COO-38-519, p. 41 (1966); (b) C. E. Melton and H. W. Joy, *J. Chem. Phys.*, **46**, 4275 (1967).
69. J. C. J. Thynne and A. G. Harrison, *Trans. Faraday Soc.*, **62**, 2468 (1966).
70. R. P. Bell, *The Proton in Chemistry*, Methuen, London, 1959.
71. W. Duane and O. Scheuer, *Radium*, **10**, 33 (1913).
72. P. Gunther and L. Holzapfel, *Z. Phyzik. Chem.*, **42B**, 346 (1939).
73. J. H. Baxendale and G. P. Gilbert, *Discussions Faraday Soc.*, **36**, 186 (1963).
74. Von S. Hofmann, N. Riehl, W. Rupp, and R. Sizman, *Radiochem. Acta*, **1**, 203 (1963).
75. A. R. Anderson, B. Knight, and J. A. Winter, *Trans. Faraday Soc.*, **62**, 359 (1966).
76. G. R. A. Johnson and M. Simic, *Nature*, **212**, 1570 (1966).
77. J. R. McNesby, I. Tanaka, and H. Okabe, *J. Chem. Phys.*, **36**, 605 (1962).
78. F. Fiquet-Fayard, *J. Chim. Phys.*, **57**, 453 (1960).
79. R. S. Dixon, Proceedings of the Fifth Informal Conference on the Radiation Chemistry of Water (Notre Dame); A.E.C. Document No. COO-519, p. 115 (1966).
80. J. H. Baxendale and G. P. Gilbert, *Science*, **147**, 1571 (1965).
81. A. R. Anderson, B. Knight, and J. A. Winter, *Nature*, **209**, 199 (1966).
82. A. Sokolov and G. Stein, *J. Chem. Phys.*, **44**, 2189 (1966).
83. L. M. Dorfman and B. A. Hemmer, *J. Chem. Phys.*, **22**, 1555 (1954).
84. B. M. Benjamin and H. S. Ibsin, *Energia Nucl. Milan*, **13**, 165 (1966).
85. K. Fueki and J. L. Magee, *Discussions Faraday Soc.*, **36**, 19 (1963).
86. J. F. Kircher, J. S. McNulty, J. L. McFarling, and A. Levy, *Radiation Res.*, **13**, 452 (1960).
87. F. W. Lampe, W. S. Koski, E. R. Weiner, and W. H. Johnston, *Industrial Uses of Large Radiation Sources*, Vol. 1, 41, I.A.E.A. Salzburg 1963.
88. M. Anbar and P. Perlstein, *J. Phys. Chem.*, **68**, 1234 (1964).
89. M. T. Dmitriev, *Zh. Fiz. Khim.*, **32**, 2418 (1958).
90. M. T. Dmitriev and S. Ya. Pshezhetskii, *Intern. J. Appl. Rad. Isotopes*, **5**, 67 (1959).

91. M. T. Dmitriev and S. Ya. Pshezhetskii, *Russ. J. Phys. Chem.*, **34**, 418 (1960).
92. M. T. Dmitriev, *Russ. J. Phys. Chem.*, **35**, 495 (1961).
93. (a) M. T. Dmitriev, *Russ. J. Phys. Chem.*, **40**, 819 (1966); (b) *ibid.*, **40**, 939 (1966).
94. P. Harteck and S. Dondes, *Nucleonics*, **14**, 22 (1956).
95. P. Harteck and S. Dondes, *J. Chem. Phys.*, **27**, 546 (1957).
96. P. Harteck and S. Dondes, *Proc. 2nd Intern. Conf. Peaceful Uses Atomic Energy UN Geneva*, **29**, 415 (1958).
97. P. Harteck and S. Dondes, *J. Phys. Chem.*, **63**, 956 (1959).
98. J. Wright, J. K. Linacre, W. R. Marsh, and T. H. Bates, *Proc. Intern. Conf. Peaceful Uses Atomic Energy, UN Geneva*, **7**, 560 (1955).
99. A. R. Jones, *Radiation Res.*, **10**, 655 (1959).
100. J. K. Linacre, private communication, 1967.
101. P. Harteck and S. Dondes, *Nucleonics*, **14**, 66 (1956).
102. D. A. Flory, *Nucleonics*, **21**, 50 (1963).
103. G. R. A. Johnson, *J. Inorg. Nucl. Chem.*, **24**, 461 (1962).
104. J. A. Hearne and R. W. Hummel, *Radiation Res.*, **15**, 254 (1961).
105. L. M. Dorfman and P. C. Noble, *J. Phys. Chem.*, **63**, 980 (1959).
106. G. A. W. Derwish, A. Galli, A. Giardini-Guidoni, and G. G. Volpi, *J. Chem. Phys.*, **39**, 1599 (1963).
107. S. Gordon, private communication.
108. C. E. Melton, *J. Chem. Phys.*, **45**, 4414 (1966).
109. D. Cleaver, E. Collinson, and F. S. Dainton, *Trans. Faraday Soc.*, **56**, 1640 (1960).
110. F. S. Dainton, T. Skwarski, D. Smithies, and E. Wezranowski, *Trans. Faraday Soc.*, **60**, 1068 (1964).
111. T. Toi, D. B. Peterson, and M. Burton, *Radiation Res.*, **17**, 399 (1962).
112. R. C. Horscroft, *Trans. Faraday Soc.*, **60**, 323 (1964).
113. B. P. Burtt and A. B. Zahlan, *J. Chem. Phys.*, **26**, 846 (1957).
114. A. Luyckx, *Bull Soc. Chim. Belg.*, **43**, 117, 160 (1934).
115. A. R. Anderson and J. A. Winter, "The Effect of Temperature and Pressure on the Vapor Phase γ-Radiolysis of Some Polar Molecules" in *The Chemistry of Ionization and Excitation*, G. R. A. Johnson and G. Scholes, Eds., Taylor and Francis, London, 1967.
116. J. W. Sutherland and H. Kramer, *Abstracts 153rd Am. Chem. Soc. Mtg., April, 1967.*
117. C. H. Cheek and V. J. Linnenbom, *J. Phys. Chem.*, **62**, 1475 (1958).
118. H. D. Smyth, *Rev. Mod. Phys.*, **3**, 347 (1931).
119. J. C. Lorquet, *J. Chim. Phys.*, **57**, 1078 (1960).
120. V. Čermák and Z. Herman, *J. Chim. Phys.* **57**, 717 (1960).
121. J. F. Paulson and R. L. Mosher, *J. Chem. Phys.*, **44**, 3025 (1966).
122. J. F. Paulson, F. Dale, and R. L. Mosher, *Nature*, **204**, 377 (1964).
123. S. Gordon, E. J. Hart, M. S. Matheson, J. Rabani, and J. K. Thomas, *Discussions Faraday Soc.*, **36**, 193 (1963).
124. A. R. Anderson, J. V. F. Best, and D. A. Dominey, *J. Chem. Soc.*, **1962**, 3498.
125. D. L. Baulch, F. S. Dainton, and R. L. S. Willix, *Trans. Faraday Soc.*, **61**, 1146 (1965).
126. P. Harteck, W. Groth, and K. Faltings, *Z. Physik. Chem.*, **41B**, 15 (1938).
127. H. Jucker and E. K. Rideal, *J. Chem. Soc. (London)*, **1957**, 1058.
128. M. H. J. Wijnen, *J. Chem. Phys.*, **24**, 851 (1956).
129. B. H. Mahan, *J. Chem. Phys.* **33**, 959 (1960).

130. (a) P. Warneck, *Discussions Faraday Soc.*, **37**, 57 (1964); (b) *J. Chem. Phys.*, **41**, 3435 (1964).

131. F. S. Feates and R. S. Sach, Harwell Research Report AERE-R4836 (1965).

132. D. L. Baulch and W. H. Breckenridge, *Trans. Faraday Soc.*, **62**, 2768 (1966).

133. T. G. Slanger, *J. Chem. Phys.*, **45**, 4127 (1966).

134. D. R. Stranks, *Proc. 2nd Conf. Peaceful Uses of Atomic Energy*, U.N., Geneva, **7**, 362 (1958).

135. K. Hirota, H. Terasaki, and M. Hatada, *Bull. Chem. Soc., Japan*, **35**, 1762 (1962).

136. J. O. Hirschfelder and H. S. Taylor, *J. Chem. Phys.*, **6**, 783 (1938).

137. P. Harteck and S. Dondes, *J. Chem. Phys.*, **26**, 1734 (1957).

138. J. Sutton, M. Faraggi, and M. Schmidt, *J. Chim. Phys.*, **57**, 643 (1960).

139. S. Weyssenhoff, S. Dondes, and P. Harteck, *J. Am. Chem. Soc.*, **84**, 1526 (1962).

140. F. S. Dainton, *Discussions Faraday Soc.*, **36**, 237 (1963).

141. D. A. Dominey, H. Morley, B. P. K. Sharpe, and R. J. Waite, "The Effect of Methene on the Rate of Exchange of ^{14}C between CO and CO_2 under Radiation" in *The Chemistry of Ionization and Excitation*, G. R. A. Johnson and G. Scholes, Eds., Taylor and Francis, London, 1967.

142. P. F. Knewstubb and T. M. Sugden, *Nature*, **196**, 1312 (1962).

143. W. L. Fite and J. A. Rutherford, *Discussions Faraday Soc.*, **37**, 192 (1964).

144. J. L. Pack and A. V. Phelps, *J. Chem. Phys.*, **45**, 4316 (1966).

145. D. Katakis and H. Taube, *J. Chem. Phys.*, **36**, 416 (1962).

146. O. F. Raper and W. B. De More, *J. Chem. Phys.*, **40**, 1047 (1964).

147. R. A. Young and A. Y.-M. Ung, *J. Chem. Phys.*, **44**, 3038 (1966).

148. K. F. Preston and R. J. Cvetanovic, *J. Chem. Phys.*, **45**, 288 (1966).

149. N. G. Moll, D. R. Clutter, and W. E. Thompson, *J. Chem. Phys.*, **45**, 4469 (1966).

150. R. Lind and J. Wright, *Proc. 3rd Intern. Conf. Peaceful Uses Atomic Energy, UN, Geneva*, **9**, 541 (1965).

151. S. C. Lind and D. C. Bardwell, *J. Am. Chem. Soc.*, **47**, 2675 (1925).

152. S. Dondes, P. Harteck, and S. von Weyssenhoff, *Z. Naturforsch.*, **19a**, 13 (1964).

153. A. C. Stewart and H. J. Bowlden, *J. Phys. Chem.*, **64**, 212 (1960).

154. M. S. B. Munsen, F. H. Field, and J. L. Franklin, *J. Chem. Phys.*, **37**, 1790 (1962).

155. C. E. Melton and G. F. Wells, *J. Chem. Phys.*, **27**, 1132 (1957).

156. R. N. Smith, D. A. Young, E. N. Smith, and C. C. Carter, *J. Inorg. Chem.*, **2**, 829 (1963).

157. A. R. Blake, W. T. Eeles, and P. P. Jennings, *Trans. Faraday Soc.*, **60**, 697 (1964).

158. P. G. Clay, G. R. A. Johnson, and J. M. Warman, *Discussions Faraday Soc.*, **36**, 46 (1963).

Chapter 6

Organic Gases*

G. G. Meisels

Department of Chemistry

University of Houston

* This investigation was supported by the United States Atomic Energy Commission, for whose assistance we are deeply grateful. This is Document ORO-3606-4.

347

I. INTRODUCTION

A. General Comments

It might appear that it is more difficult to elucidate the mechanism of the radiolysis of organic gases than that of inorganic ones because of the larger number of possible intermediates and the more complex distribution of final products. Almost all organic compounds, even the simplest ones such as methane or ethylene, yield dozens of identifiable products even at the lowest conversions. This variety of products can be an advantage since every single product is potentially a diagnostic tool for the unravelling of overall mechanism of radiolysis.

Our presentation shall attempt principally to develop a generalized mechanism for the sequence of events leading from the first steps of energy absorption to the formation of the final, stable products (Section II). This formalism will be followed by an analysis of the information which can be obtained from related fields (Section III), and by a short summary of experimental techniques (Section IV). We shall then proceed to a description of the radiolysis results obtained in some individual pure systems (Section V) and mixtures (Section VI).

B. Historical Development of Radiation Chemistry Interpretations

Early workers (1,2) proposed radiolytic mechanisms based on the formation of ionic clusters, but did not allow for the participation of excited or radical species in the mechanism. A typical sequence proposed for acetylene radiolysis was

$$C_2H_2 \rightarrow C_2H_2^+ + e^- \tag{1}$$

$$C_2H_2^+ + xC_2H_2 \rightarrow (C_2H_2)_{x+1}^+ \tag{2}$$

$$e^- + (C_2H_2)_{x+1}^+ \rightarrow \text{products (polymer)} \tag{3}$$

A major advance was made in the mid-1930's when Eyring, Hirschfelder, and Taylor (3) suggested that the initial formation of ionic species and their reactions might produce free radicals, and that excitation and ionization processes may lead to formation of the same neutral species with approximately equal efficiency. Further chemical changes were then ascribed to the radicals produced by either mechanism. This view was eagerly accepted, since it allowed the interpretation of gas phase radiation chemistry mechanisms in terms of the well known mechanisms of photochemistry and free radical chemistry. Although ionic reactions leading to product formation were considered in detail by these authors (3), this possibility was almost totally neglected for the next twenty years and product formation was interpreted by mechanisms such as

$$C_2H_2 \rightarrow C_2H + H \tag{4}$$

$$C_2H + C_2H_2 \rightarrow C_4H_3 \tag{5}$$

$$H + C_2H_2 \rightarrow C_2H_3 \tag{6}$$

and further addition reactions of the radicals. The rediscovery of ion–molecule reactions by Tal'roze (4), by Stevenson (5), and by Franklin, Field, and Lampe (6) forcefully demanded consideration of ionic reaction steps in chemical product formation.

II. GENERALIZED MECHANISM

A. Initial Event and Initial Species

The deposition of energy in a molecule can lead to immediate ionization or to excitation to electronic levels below and above the ionization limit (see Chapter 1). The species excited to electronic states above the ionization potential ("superexcited" molecules) (7,8) may react, dissociate, or lose an electron ("preionization").

Our chief concern here is a qualitative discussion of the possible events leading to chemical product formation. This can be done conveniently by separating the initial events according to the amount of energy deposited in the molecule. Appropriate dividing points for the discussion are the lowest energy required for fragmentation of the neutral excited molecule, and the ionization potential of the molecule.

1. Energies Below the Lowest Fragmentation Process.

The initial interaction of the degradation spectrum electrons with molecules does not lead to appreciable vibrational and rotational excitation (9). The small extent to which such processes might occur should not lead to deposition of enough energy to cause chemical changes. Therefore, we need concern ourselves only with electronic excitation (Chapter 1). The lowest excited states of organic molecules are normally spin-forbidden triplet states. For molecules of interest here, these energy levels lie between 2 and 9 eV above the ground states. Acetylene and simple derivatives have their lowest excited states at energies of the order of 2–3 eV (10), olefins between 3 and 4 eV and lower if conjugated; simple ketones are of the same order, while alkanes have somewhat higher levels for the lowest excited states. Although molecules excited to such states are not produced abundantly in the initial process, they may play a role in product formation since they may be reached by intersystem crossing, possibly followed by internal conversion or degradation to the lowest state of the system. If the energy level is lower than that of the energetically most favorable fragmentation process, chemical changes may occur, provided the triplet state has a molecular configuration different from that of the ground state. In the case of olefins, for example, the triplet state has a minimum energy configuration with the substituent groups at 90° from each other with respect to the double bond axis (11). Return to the ground state can thus yield either the *cis* or the *trans* isomer (12). If the molecule has insufficient energy to dissociate, it must eventually lose its energy by a low probability radiative process, or by intersystem crossing and eventual vibrational degradation.

2. Energies Above the Lowest Dissociation Process, but Below the Threshold for Ionization. The excitation energy imparted to molecules, by high energy irradiation is usually greater than the lowest fragmentation energy, and the degradation processes mentioned above will, therefore, be in competition with dissociation. This may occur either by the breaking of a simple bond, producing two radical species, or by the elimination of a stable entity such as molecular hydrogen. As mentioned in Chapter 3, such elimination processes then produce either two stable entities, or a stable species and a diradical. For example, the following have been reported (13–16):

$$CH_4^* \rightarrow CH_2 + H_2 \tag{7}$$

$$C_2H_6^* \rightarrow C_2H_4 + H_2 \tag{8}$$

$$CH_3OH^* \rightarrow CH_2O + H_2 \tag{9}$$

The relative probabilities of isomerization–degradation processes and dissociation will depend on the state from which the two processes originate. Excitation to a repulsive state clearly will produce immediate dissociation and no rearrangement. Nondissociative states, where radiationless transitions are required, can be expected to have lifetimes sufficiently long to allow competition between radiationless transitions, radiative return to the ground state, and collisional processes. An excellent summary of the possible photophysical processes is given by Calvert and Pitts (17), and the reader is referred to Chapter 3 of this volume for a detailed description of the modes of decomposition of highly excited molecules.

3. Energies Above the Ionization Threshold. Platzman (18) has repeatedly emphasized an important consequence of the optical approximation: the most abundant radiation interactions are those with the greatest values for $f(E)/E$, the ratio of the oscillator strength to the energy of the transition. Therefore, as noted in Chapter 1, the major portion of excitation processes involves outer shell and bonding electrons. The bulk of the oscillator strength of organic compounds lies very high, between 10 and 30 eV above the ground state, even when states of low energy exist in molecules having double bonds or conjugation. A sizable fraction of the initial species are thought to be superexcited molecules capable of either ionizing or dissociating (7,8). It is now well established that ionization efficiencies of organic molecules can be considerably less than unity, whether energy deposition is by photons (19,20) or by collisions of the second kind (21). It is thought that dissociation and ionization are competitive processes in certain superexcited molecules (18). This expectation can be experimentally tested to some extent, since such a model would result in different ionization efficiencies for isotopically substituted analogues of a molecule (see Chapter 1). Such an effect has been observed in photoionization measurements (22–24), in electron impact studies (25) and in studies of the Jesse effect (21). The latter method suffers from uncertainties in the contribution to ion formation of metastable states and trapped resonance radiation, and simultaneous population of more than one level in the transferring rare gas atoms (26).

Studies of the vacuum ultraviolet photolysis in the ionization region (22,23,27) and of the argon sensitized radiolysis (28) have provided strong evidence that homolytic dissociation of the molecule excited to a superexcited state leads to the formation of the same products as the dissociation of molecules excited to singlet states below the ionization limit.

Ionization of complex organic molecules by collision with 50–100 eV electrons leads in part to excited ions capable of dissociation into fragment ions and neutral products which may dissociate further. This area has been one of considerable interest to mass spectrometrists, and is the general subject of the quasi-equilibrium theory of ionic fragmentation (29,30) (Chapter 2).

The lifetimes of the initial ions with respect to fragmentation are determined by the excess electronic energies with which they are formed.

These are typically rather small (less than 5 eV) (31,32), and the distribution falls off rapidly at high transferred energies. Moreover, the energy deposition distribution is thought to be essentially independent of incident energy at sufficiently high electron energies. This is supported by the classical observation that above bombarding electron energies of 50 eV fragmentation patterns of even relatively complex organic molecules are fairly insensitive to a further increase of several hundred eV (33–36). Therefore, it is usually assumed that energy transfer from 70 eV electrons (the energy used in commercial mass spectrometers) results in an ionic fragmentation pattern typical of that observed in general radiation chemistry. Thus, the ordinary mass spectrometric "cracking patterns" can provide information of great relevance to radiation chemistry. It is important to emphasize, however, that mass spectrometric information obtained in an ordinary analytical instrument is strictly applicable only to ions with lifetimes of the order of 10^{-5} sec, corresponding in a closed system to the collision-free period at a pressure of 0.1 torr ca. As pointed out in Chapter 2, collisional deactivation of the excited ions would be expected to diminish the importance of ionic fragmentation at higher pressures. (At a pressure of ~ 1 atm, the collision free period is about 10^{-10} sec.)

Application of the statistical theory of mass spectra indicates that the fragmentation of the parent ion is rapid, and the importance of most primary decomposition processes should not change drastically between 10^{-5} and 10^{-10} sec (37). Secondary fragmentation requires more than 10^{-10} sec, however, and is therefore more effectively quenched at increased pressures. Superimposed on this quenching will be a general but slow trend towards stabilization of the parent ion at elevated pressures. At high pressures, the excited parent ion may collide with a molecule prior to decomposition and a reaction may occur. (Ion–molecule reactions involving excited parent ions are discussed in Chapter 4.) These trends with pressure have been confirmed by limited results obtained in the irradiation of simple hydrocarbons (38–40).

4. Gaseous Mixtures: Distribution of Absorbed Energy. The division of initial energy absorption between the components of a mixture has often been estimated to be equivalent to the electron fractions of the various components. This approximation arose from the well-known attenuation coefficient for Co^{60} gamma rays, where the chief absorption process, the Compton effect, is proportional to the electron density. The electron fraction of a substance in a mixture gives, therefore, a reasonable approximation (41) to the fraction of absorbed energy taken up by a given species. However, since the division of the energy will to some extent depend on the portion of the oscillator strength residing at lower energies this method will lead to a priori estimates of the energy division which must be considered as rough guidelines only. Experimental evidence suggests that electron fractions are entirely adequate only when comparison is made between compounds of a homologous series (42).

The energy division is actually determined by the degradation spectrum of electrons in the mixture, and the cross sections for individual processes under these specific conditions (43) (see Chapter 1). Neither degradation spectrum nor stopping powers for electrons with such an energy range are known. Approximations have included cross sections calculated for electron–molecule interaction using polarizabilities (43,44) and use of the Bethe stopping power equation for electrons, combined with Bragg's rule (45).

The degradation of electron energy once it is below the lowest energy state of the molecule is slow. The "subexcitation electrons" can interact with even a small amount of impurity or added gas having energy levels lower than the major component, leading to selective excitation and to product formation from the added gas in yields essentially independent of its concentration. It follows that a detailed estimate of the energy division must be made. Although approximate stopping powers may apply to the gross energy distribution between higher excited and ionic states of a binary system, the contribution of the lowest excited state of one system to product formation will differ substantially from that in pure components. For example, some complications may be expected when acetylene is added to methane since there is an energy gap of perhaps 6 eV between the lowest states.

B. Bimolecular Events

1. One Component Systems

a. Primary Products and Original Products. Chemical product formation can occur by a variety of mechanisms. One of the very difficult tasks of radiation chemistry of organic compounds is the extrapolation of the original product and primary species distribution from the final observed product distribution. As has been pointed out by Back (46) and by Ausloos et al. (47), this extrapolation may be exceedingly difficult to achieve, even at conversions of the order of 0.001%. Saturated hydrocarbons are particularly troublesome, because the accumulated products, such as olefins and higher molecular weight hydrocarbons, have a much higher reactivity toward intermediate radical and ionic species than the parent compounds.

b. Chemically Reactive Collisions of Excited Molecules. Atom abstraction reactions and excimer additions in condensed phase photochemistry of organic compounds are well known (17), but no evidence has been presented that such processes are important contributors to gas phase radiation chemistry. It is, however, quite conceivable that such reactions, particularly atom abstractions, may contribute to product formation.

The reactions of highly excited atoms can lead to associative ionization (48), often referred to as Hornbeck-Molnar processes. They are common in the rare gases, but they are much less well established in molecular gases. They have been reported for acetylene excited to 10.2 eV (49).

$$C_2H_2^* + C_2H_2 \rightarrow C_4H_3^+ + H + e^- \tag{10}$$

$$C_2H_2^* + C_2H_2 \rightarrow C_4H_2^+ + H_2 + e^- \tag{11}$$

but the veracity of this report has been questioned seriously since these reactions are endothermic for acetylene with 10.2 eV excitation energy (50). Prerequisites of associative ionization in organic molecules are unsatisfied valences and the existence of high-lying metastable or resonance states (excited states). Since lifetimes of states which may return to the ground state by allowed transitions may be of the order of 10^{-9} sec, such processes might be more important in radiation chemistry than one would anticipate on the basis of mass spectrometric results.

c. Ion–Molecule Reactions. The reactions between ions and neutral molecules frequently occur with collision efficiency. Because of ion-induced dipole forces, rate constants for the collision process are about one order of magnitude larger than those involving only neutral species. These reactions are recognized as being important contributors toward radiation chemical mechanisms, and convincing evidence for their participation in product formation has been presented in a number of systems. Most of our knowledge derives from mass spectrometric evidence, and is summarized in Chapter 4.

Limitations of the applicability of mass spectrometric information result partly from the fact that the reacting ions in a mass spectrometer may have kinetic energies above thermal. This may affect the relative probabilities of alternate reaction channels leading to different products in a given ion–molecule collision.

The application of low pressure mass spectrometric data to radiation chemistry at elevated pressures is also complicated by the change in time scale. The studies of Wexler (51), Kebarle (52,53), and Field and Munson (54–56) indicate clearly that at the higher pressures typical of radiation chemistry gas phase solvation of ions may occur. Moreover, the lifetime of the intermediate complex in condensation reactions appears to be considerably greater than the collision-free period at 1 atm (40,57,58), suggesting the possibility of collisional deactivation of the intermediate ion complex and further reactions of the complex itself.

d. Electron Attachment Reactions. When the energy of the free electron falls below that required to excite the lowest electron level of a molecule, further energy loss can only be by excitation of vibrational and rotational modes, slow processes (59) even when intermediate negative ion states are involved (9).

Electrons in the subexcitation energy range may either be unreactive toward the substrate molecule, attach themselves with the formation of stable negative molecule ions, or undergo a dissociative attachment process (60,61).

e. Neutralization. The charge neutralization process may involve positive ions, and electrons or negative ions. It normally occurs homogeneously in the gas phase; however, at very low dose rates and, therefore, very small steady state concentrations of charged species in the radiation vessel, the dominant neutralization process is diffusion to the walls and neutralization thereon (62). It has been proposed (63) that at very high localized dose rates neutralization on the wall could occur because of convection currents (64). Neutralization rates have been studied extensively for inorganic gases because of the importance in upperatmosphere chemistry (65), but very little has been done on organic gases. Firestone (66) has estimated the rate of neutralization by chemical means. As a rule, rate constants for homogeneous neutralization processes of electrons and positive ions are of the order of 10^{17} mole^{-1} cc sec^{-1} while those for recombination of positive and negative ions are about one order of magnitude smaller (65).

f. Radical Reactions. Radicals can either react with other radicals, with substrate molecules, or with charged species.

The reaction of radicals with each other can lead to combination or to disproportionation. These reactions lead to the destruction of the radical character of the interacting species and occur approximately with collision efficiency ($k_{R+R} \simeq 10^{14}$ mole^{-1} cc sec^{-1}) for most simple hydrocarbon radicals (67,68). Considerably less is known about the reactions of radicals containing heteroatoms, although alkoxy and fluoroalkyl radicals have been investigated to some extent as well (68).

Free radical species may also react with their molecular substrates (67). This may be by an abstraction reaction (metathesis) which changes the nature of the radical by transferring an atom. Activation energies for metathetical reactions vary from very low values of 1 or 2 kcal to as much as 14 kcal. The radical may add to a substrate molecule if it contains double bonds. Activation energies for such processes are very low for the addition of hydrogen atoms to olefins (68–70) while the addition of alkyl radicals is normally competitive with abstraction reactions. Radical addition reactions may lead to polymerization, or to telomerization if the chain length is very short or if chain transfer is important. Such reactions are of interest because they provide a possible means for utilizing radiation for the synthesis of polymers and telomers at relatively high pressures and low dose rates (see Chapter 8).

The dissociation of excited molecules or ions can yield radicals with an epithermal energy distribution, and such species are referred to as "hot" radicals or "hot" atoms. If the dissociation occurred from a highly vibrationally excited ground state appreciable excess kinetic energy is not to be expected in the products on the basis of unimolecular dissociation rate theory (71). However, any fragmentation resulting from a dissociative state may easily lead to retention of excess translational energy in the dissociating partners (see Chapters 2 and 3). Evidence for the participation of "hot" species, particularly hydrogen atoms in product formation, has occasionally been sought by studies of the effect of scavengers. Products resulting from radical reactions in the scavenged radiolysis are compared with those occurring in the same system in a lower energy (photolytic) irradiation where radical energies are close to thermal. For hydrogen atoms this corresponds to the reactions:

$$H + S \rightarrow RS \tag{12}$$

$$H + RH \rightarrow H_2 + R \tag{13}$$

Since reaction (12) with the scavenger S occurs approximately with collision efficiency, an apparent increase in the rate constant ratio k_{13}/k_{12} is ascribed to greater than thermal efficiency of reaction (13) and hence to excess energy in the radical species. However, other mechanisms such as charge transfer and excitation transfer

$$M^+ + S \rightarrow products \tag{14}$$

$$M^+ + M \rightarrow H_2 + products \tag{15}$$

lead to the same formal kinetic relationships and can, therefore, also be held responsible for the observed effects. There is at present little conclusive evidence for the major participation of hot radicals in the radiolysis of organic gases, although as noted in Chapter 3 there is rather conclusive evidence for hot H atom reactions in the vacuum ultraviolet photolysis.

2. Gaseous Mixtures–Energy Transfer

Energy transfer may take any of three forms: excitation transfer, charge transfer, and the formation of a new ion pair. The last requires the existence of an excited state whose level is above the ionization potential of the acceptor molecule. Alternately, an associative ionization process may occur between heteromolecular species.

The occurrence of the three processes mentioned above in mixtures constitutes the principal nontrivial difference between mixtures and pure systems. A simplification of the reaction mechanism may occur if one of the components of the mixture is a scavenger of free radicals, for example, nitric oxide, iodine, and hydrogen iodide, or a charge acceptor, i.e., a species whose ionization potential is lower than that of the reactant ion. Both types of reactions can be used to advantage in attempts to elucidate mechanisms of radiation chemistry. Free radical scavengers may also have low ionization potentials and in such cases they can act both as charge acceptors and as free radical scavengers (72,73).

III. INFORMATION DERIVED FROM RELATED FIELDS

A. Radiation Physics

Physical techniques give us most of our information on the energy absorption and division processes, and the probable nature of the excited species formed. For example, spectroscopic results combined with measurements of the energy spectrum of electrons in an irradiated system ("degradation spectrum") can hopefully lead to an estimate of the nature of the various initial species formed. In addition, the measurement of the average energy expended in the formation of an ion pair (W value) provides a basis for the calculation of total energy absorption in an irradiated sample. In addition, a knowledge of W values permits an estimate of the relative importance of ionization and of excitation to neutral electronic states (74).

B. Spectroscopy

One of the most important functions of spectroscopy is the establishment of the oscillator strength distribution and energy diagrams for the molecule under irradiation. In photon absorption by a molecule, energy is either totally expended in the excitation process or divided between the ionized molecule and the free electron only. Photoelectron spectroscopy (75) permits the direct evaluation of the excitation function of ionized species.

Electron impact spectroscopy is presently the only manageable technique for investigating formation of highly excited species and generalized oscillator strengths at energies above approximately 20 eV (76,77), although cyclotron radiation has been employed successfully (78). At lower energies, electron impact spectroscopy can be used to identify low-lying and optically forbidden levels (79–81). Again, the reader is referred to Chapter 1 for a more complete discussion.

C. Mass Spectrometry

Mass spectrometry chiefly yields information on the formation and the reactions of ionic species as discussed above. In addition, it provides one of the most potent tools for the determination of thermodynamic quantities of uncharged species such as heats of formation of free radicals and bond energies. One of the greatest difficulties in this application is the precise measurement of the energies at which fragment ions are formed ("appearance potentials"). The ionization efficiency curve of organic species is complicated by excitation of vibrational levels. Recent photoionization measurements (82) have clearly shown that ionization efficiency curves are complex, and the reliability of some of the older electron impact data is open to question (83). This problem is well exemplified by the difficulty of measuring even the ionization potential of simple compounds such as methane. This value is now thought to be 12.7 eV, based on the photoionization mass spectrometric measurements of Dibeler and his co-workers (82). For several years prior to that, a value of 13.0 eV was generally accepted. Knowledge of adiabatic ionization potentials is exceedingly important because it permits an assessment of the feasibility of charge transfer and ion–molecule reactions, and is required for theoretical treatments of ion decomposition. An interesting mass spectrometric approach to the radiation chemistry of organic gases is that of Melton and Rudolph (84), which has been described in Chapter 3 and 4.

Although mass spectrometry has been chiefly applied to the study of positive ions, it can yield information on negative ions and electron attachment reactions as well (86). Negative ions are not too common in the simple organic compounds usually studied in gas-phase radiation chemistry. Consequently, there is little known about their formation, and their reactions in ion–molecule collisions. Ion–molecule reactions involving negative ions have been reported in the acetaldehyde, the cyanogen system, in nitromethane, and in mixtures of sulfur dioxide and methyl iodide or nitrobenzene (85,86).

D. Electron Swarms

The analysis of dissociative and non-dissociative electron attachment reactions and equilibrium relationships for the interaction of thermal electrons with molecules has been applied chiefly to complex organic molecules (60,61). Knowledge of the electron attachment coefficient is imperative to an estimate of the concentration of free electrons, and the rate of recombination of oppositely charged ions. The general experimental conditions of the pulse sampling technique (61), and swarm methods (60) are fairly similar to those encountered in radiolytic experiments, and the information derived should be directly applicable to radiation chemistry.

E. Photochemistry

The optical approximation to the estimate of initial species formation suggests an obvious parallel to photochemistry. That portion of the radiation chemistry resulting from neutral excited species should be an average of the photochemistry of the molecules at various wavelengths, weighted according to the excitation spectrum. The consequence of excitation to levels typical of photochemical processes is more easily studied when the complication of ionic processes is avoided. Results from studies of the wavelength dependence of primary processes in photochemistry can be used to assess the nature of the excited species produced by electron impact (see Chapter 3). In recent years photochemistry of organic molecules has been extended to the vacuum ultraviolet (87) as far as the argon resonance lines (11.6 eV) (22,88) and He resonance lines (89), 584 Å (21.2 eV), which are capable of ionizing most molecules commonly of interest to gas-phase radiation chemists. One of the chief advantages of vacuum ultraviolet photochemistry is the monochromatic nature of the radiation. This gives an a priori knowledge of the energy deposited, and permits the direct evaluation of ionization efficiencies. In addition, when the photon energy is judiciously chosen, certain ions (e.g., parent ions) may be uniquely formed in the system, thus facilitating the study of their reactions. Ausloos has employed this technique in gaseous hydrocarbons to considerable advantage (22,23,27,90). For a more complete discussion see Chapter 3.

IV. EXPERIMENTAL TECHNIQUES

A. Irradiation Procedures and Analysis

A variety of experimental techniques have been used in the study of gas phase reactions and are described in Chapter 5. Historically, α particles resulting from radioactive decay of natural isotopes were the first energy source to be employed for systematic radiation chemistry studies (8). Within the last few decades γ-rays and electrons, the latter principally from accelerator sources, have been used for convenience and because of their greater penetration. Only very little work has been carried out with very heavy ionizing particles and with fission recoils.

1. Gamma Rays, X-rays, and Accelerated Electrons. The availability and relatively low expense of Co^{60} gamma radiation sources has been a prime factor in the popularity of this type of radiation, even though it has certain disadvantages. Energy absorption occurs both by direct absorption of the γ-rays in the sample gas and from secondary electrons originating in the cell walls. This can introduce a sizable dosimetry problem. Gas phase dosimetry is discussed in Chapter 5, Section III.

The penetration of γ-rays is considerable, and, therefore, permits a wider variation of pressures for radiation chemical studies. Gamma-ray initiated reactions have, in fact, been studied at pressures up to hundreds of atmospheres (91), which requires container walls of a thickness not penetrable by high energy electrons. When isotope radiation is unavailable, x-rays produced by the impact of high-energy electrons on heavy metal targets can be used to obtain a penetrating form of radiation from accelerators. Such radiation sources are nonhomogeneous, and dose rates will vary among the regions of larger irradiation vessels.

The chief source of electron radiation for chemical studies has been the Van de Graaff electrostatic accelerator, which normally provides a steady beam of monoenergetic electrons up to 3 MeV. Absorption of such radiation provides a steady rate of formation of the intermediate species similar to that induced by absorption of gamma radiation. This has the advantage of permitting a relatively easy assessment of steady state concentrations, provided that the rate constants are known, and that energy absorption is homogeneous throughout the system. Resonant transformers and linear accelerators are also occasionally used as radiation sources. The latter, of course, are widely used for pulse radiolysis studies, which can also be carried out with modified electrostatic accelerators. Pulse techniques provide a direct means of measuring rate constants for reactions of intermediate species, but this technique has been applied only sparingly to gas phase chemistry, largely because of the low stopping power of gases for high energy radiation. Sauer and Dorfman (92) have pioneered in gas phase pulse radiolysis, but found it necessary to do such studies in the presence of high pressures of inert gases to facilitate the formation of initial species.

The dosimetry of electron beams is as difficult as that of Co^{60} γ-rays. An advantage is the absence of the complications arising from cavity ionization, and the necessity to deal only with monoenergetic incident electrons. The applicable stopping powers are easily calculated from the Bethe equation and are considered quite accurate.

Electron irradiation can also be obtained from isotope sources such as the weak beta emitters nickel-63, tritium, and more energetic ones such as cesium-137. The use of the former has been restricted almost entirely to physical measurements of ionization rates, while the latter has been employed only sparingly because of the difficulty associated with dosimetry.

2. Heavy Particles. Occasionally experiments are still carried out with alpha particles from radon or from commercially available solid sources. Such sources seldom have sufficient intensity to permit convenience of use, and are used only rarely. Energy deposition by heavy particles is not homogeneous, and, therefore, diffusion effects must be considered even in the gas phase, particularly at higher pressures.

Dosimetry can be achieved by calculating the total number of decays, and multiplying by the known energy of the disintegration (1). Where heavy particles from accelerators are involved, total ion–beam intensity measurements and stopping powers are normally employed for an estimate of energy absorption.

3. Fission Recoils.   Only few investigators have attempted to use fission recoils as a radiation source, largely because of the complexities of the energy absorption process, and the difficulties of the experimental technique. The chemistry of fission recoils is normally studied by packing radiation vessels with thin glass fibers containing uranium-235 (93). In order to obtain efficient energy deposition in the gas phase, such fibers must have diameters of 5 μ or less. The cell assemblies are then lowered into the neutron flux of a nuclear reactor, and the U^{235} in the fibers undergoes fission. Estimates of energy deposition are exceedingly difficult, and have been a matter of major controversey for long periods (94,95). A major improvement was obtained when thin fuel plates (96,97) were substituted for the fibers. Most recently, sources of californium-252 have become available, a naturally fissioning isotope. To date, only physical measurements have been made with such sources (98–100).

B. Analytical Techniques

Most gas phase radiolyses of organic compounds have been carried out in static systems, and product analyses are made on aliquots of the irradiated samples at periodic intervals or at the end of irradiation. In the last decade the use of gas chromatography for detection and quantitative analysis of products has become common and is often combined with mass spectrometric analysis, particularly when labelled compounds are used to elucidate mechanisms (101). High vacuum techniques such as distillation at low temperatures are also employed occasionally, particularly for hydrogen (102). Flow techniques have been employed chiefly for radiolysis with fission recoils, where sufficient conversion can be achieved per pass so that recirculation is not required (103).

C. Diagnostic Tools

Early investigators could do little but rationalize product distributions in terms of what appeared to be reasonable mechanisms. Within the last decade, a number of techniques have been applied to or developed for gas phase radiation chemistry, and have been material to our present understanding of radiation chemical mechanisms. These methods are usually not unequivocal and must be used in combination with each other for mutual support to suggest a sequence of reaction steps.

1. Radical Scavengers. The discovery that certain simple compounds such as nitric oxide inhibit the formation of products arising from free radical reactions was made decades ago (104). Since then oxygen, nitric oxide, iodine, hydrogen iodide, and hydrogen sulfide have been the most important radical scavengers employed in the gas phase (101).

In using such additives, however, complications may occur. If the ionization energy of the scavenger is lower than that of the molecule under study, these species can act not only as radical scavengers but also as charge acceptors (72,73). The problems of charge exchange can best be avoided by the use of oxygen whose ionization potential (IP $= 12.08$ eV) is higher than that of most organic compounds. Yet another possible complication is the appreciable electron affinity of such reactants as I_2 and HI, which may also attach electrons and thus alter the course of the final neutralization process. A more detailed discussion of gas phase radiolysis scavenging is given in Chapter 3, Section I-B-2.

2. Electron Scavengers. The attachment of electrons to molecules is competitive with neutralization only when the capture cross section for thermal electrons is appreciable and the attachment process is exothermic. Most simple gaseous species such as hydrocarbons and olefins have very small or negligible attachment cross sections, and do not effectively attach electrons. The addition of efficient electron acceptors such as nitrous oxide (N_2O), or sulfur hexafluoride (SF_6), can, therefore, be used to estimate the yield of free electrons and positive ions (105–107), although direct physical measurement is preferred when possible. Electron attachment to nitrous oxide (108) leads to the formation of nitrogen:

$$N_2O + e \rightarrow N_2 + O^- \tag{16}$$

$$O^- + N_2O \rightarrow N_2 + O_2^- \tag{17}$$

$$O^- + N_2O \rightarrow NO + NO^- \tag{18}$$

with $k_{17}/k_{18} = 1.22$ (109). If the sequence (16–18) is followed exclusively, the resultant nitrogen yield is 1.55 times, the yield of free electrons in the system. This was confirmed for unreactive hydrocarbons such as propane. Alcohols or olefins, however, are not inert towards O^-, and interfere with reactions (17) and (18). While this presents an interesting tool for the studies of O^- reactivities, the nature of the reactions is unknown since product distributions have not been determined to date. It is interesting to note that addition of electron scavenger pairs (e.g., DI and N_2O or SF_6 and N_2O) leads to more efficient scavenging in methylcyclohexane vapor than single scavengers (110), demonstrating cooperative action. Electron scavenging can also be of considerable assistance in the analysis of the neutralization process (see Section V-A-1-c).

3. Charge Acceptors. The presence of relatively unreactive intermediate ions can often be recognized by adding a compound of low ionization potential capable of donating an electron to the intermediate, which is then transformed into a stable neutral species which can be analyzed directly (72,73). For example, neutral butenes (ionization potential 9.13–9.58 eV) are produced in the radiolysis of ethylene when charge acceptors (CA) such as dimethylamine (IP 8.24 eV), toluene (IP 8.82 eV) or nitric oxide (IP 9.25 eV) are added:

$$C_2H_4^+ + C_2H_4 \rightarrow C_4H_8^+ \tag{19}$$

$$C_4H_8^+ + CA \rightarrow C_4H_8 + CA^+ \tag{20}$$

This reaction competes with the inefficient step ($k_{20}/k_{21} \geq 100$) (23,57, 58,111).

$$C_4H_8^+ + C_2H_4 \rightarrow C_6H_{12}^+ \tag{21}$$

and the possible dissociation of $C_4H_8^+$, which is quenched at pressures above ca. 100 torr.

Charge transfer can also lead to a reduction in the yield of products resulting from ion–molecule reactions by removing a precursor. This has been suggested by Ausloos and Lias (112) who observed the inhibition of the H_2 transfer reaction

$$\text{cyclo-}C_6H_{12}^+ + C_3D_6 \rightarrow CD_3CDHCD_2H + C_6H_{10}^+ \tag{22}$$

when nitric oxide was added to a cyclohexane–propylene (1 : 0.04) mixture. This was tentatively ascribed to charge transfer from cyclohexane ion (IP, 9.88 eV) to NO. The neutralization of propylene ion by a reaction involving nitric oxide has also been suggested (63).

4. Sensitization by Rare Gases. Mass spectrometric investigations of charge-transfer processes between rare gas ions and organic molecules were pioneered by Lindholm (113) and have suggested the possibility of producing some ionic species preferentially (63,114–116). This approach has been summarized by von Bunau (117). In a closed radiolytic system interpretation of results obtained with an organic additive in a rare gas is complicated by the fact that ionization can often also occur through collisions with rare gas atoms excited to metastable and resonant states and superexcited states (118). Moreover, excited and ionic rare gas dimers may be formed and interfere with the energy transfer to the organic molecules. This means that an organic compound irradiated in the presence of a large excess of rare gas may not receive a well-defined increment of energy in a charge transfer process, as has sometimes been assumed.

5. Application of Electrostatic Fields. The effect of applied electrostatic fields on radiolytic product yields was first investigated by Essex (119) and has been used to a considerable extent recently (62,102,120) as a tool for interpretation of radiolytic systems.

The application of electrostatic fields during radiolysis causes a drift of the ions to the electrodes where all charges are neutralized at sufficiently high voltages. At the same time there is a material shift in the energy distribution of both positive and negative species. While the mean energy of ions is raised slightly by about 0.1 eV, and ionic reactions are, therefore, not greatly affected, electron energies are increased considerably, and increased molecular excitation results. Therefore, an analysis of the effects of the field on radiolytic product yields allows a discrimination between products formed in ionic and in nonionic processes.

Near the onset of secondary ionization, electrostatic fields shift the average electron energies to values of several electron volts (121,122). It is clear that even near the onset of secondary ionization, mean electron energies are only a few electron volts. While this energy is not enough to cause electronic excitation, the distribution of electron energies is probably fairly close to Maxwellian (121). The proportion of electrons with energies in excess of the lowest excitation potential is typically of the order of a few percent at the onset of secondary ionization, allowing the formation of electronically excited species, which may subsequently decompose to yield stable products (see Chapter 3, Fig. 1).

Two complications exist in this respect. The first is the possibility of negative ion formation when electron attachment requires epithermal electrons. Such processes are most common in inorganic gases (123), but have also been considered for organic species (81). The second and more severe complication arises from the inadequacy of the qualitative argument that products arising solely from ionic precursors are not affected or are reduced in yield, while the yields of products arising from excitation processes are enhanced near the onset of secondary ionization. This assumes tacitly that the excitation processes induced by the low energy electrons resulting from acceleration by the applied field are the same as those produced by the primary interaction of radiation. However, while optical selection rules are reasonably well obeyed by high energy radiation as demonstrated by the applicability of the optical approximation (18,124), low energy electrons can readily excite forbidden energy levels as well (59,79). An example of such a discrepancy is readily apparent in ethylene, where a primary process leading to vinyl radical formation is important when applied fields are present, but is virtually absent with high energy radiation alone (120).

Neutralization of ions on the electrode rather than in the body of the gas may change the nature of the neutralization process. For example, the recombination energy may be dissipated on the surface if the neutralization entity can survive at least five to ten vibrational periods (120). The change from bulk to heterogeneous neutralization will be complete as soon as total collection of ions at the electrodes is reached (onset of saturation), long before other effects of the applied field become predominant. It is another matter whether this change manifests itself in the chemical products, that is whether there is a difference in the chemical consequence of neutralization. Ions reaching the electrodes will have undergone a large number of collisions which will have changed their chemical identities by any of the host of ion–molecule reactions possible in organic systems. They will, therefore, typically be carbonium ions or polymeric ions. Polymeric ions will probably not dissociate on homogeneous or heterogeneous neutralization. Simple carbonium ions must yield a radical on neutralization. It should be pointed out that there will be a reduction in the total number of collisions an unreactive ion experiences when a field is present. This will decrease the yield of products which are formed in slow ion–molecule reactions (101). One would thus also expect a reduction of the chain length of ionic polymerization.

The use of applied electrostatic fields can be valuable in spite of the complications and ambiguities of the technique if results are carefully and judiciously interpreted with weight given to auxiliary information obtained from photochemical and other radiation chemical diagnostic

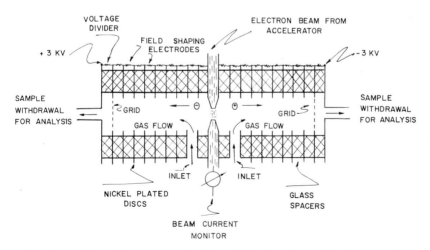

Fig. 1. Experimental arrangement for the analysis of products arising from reactions of positive ions, negative ions, and the effect of electric field strengths (126) on product distribution.

methods. Its use is particularly indicated when other diagnostic techniques such as charge scavenging and isotopic labeling are ambiguous or not applicable.

An interesting and promising development of the applied field method is due to Tal'roze (125,126). The experimental arrangement is shown schematically in Fig. 1. Irradiation with high energy electrons at the center of the vessel leads to formation of ion pairs and neutral reactive intermediates. Charge separation is achieved by applying electrostatic fields across the length of the tube. Since field strengths of the order of one V cm^{-1} torr^{-1} are sufficient for saturation, secondary excitation induced by the field should be of minor importance. A constant flow of gas is maintained to reduce mixing between the reaction zones, and capillaries are inserted in the center and in the electrode regions to allow gas sampling. Continuous mass spectrometric analysis is employed, and the composition at the three points is followed with the field on and off, thus enabling distinction between the products of homogeneous neutralization and discharge at the anode and the cathode.

6. Isotopic Labeling. Perhaps the most powerful technique for the elucidation of ionic reaction mechanisms in the gas phase is the use of deuterium labeling in connection with product analysis, a technique pioneered extensively by Ausloos and his co-workers (101). Its use is particularly advantageous in systems relatively devoid of rapid self-condensation type ion–molecule reactions.

Most experiments using this technique have been carried out with the complete elimination of free radical reactions by the addition of scavengers such as oxygen or nitric oxide. Presuming the effectiveness of such scavengers and absence of interfering reactions, the observed products are assumed to result only from unimolecular dissociation of ions and excited molecules and from bimolecular ionic reactions leading to stable molecules.

For hydrocarbons five important types of group transfer type ion–molecule reactions have been elucidated through the use of this technique:

(*a*) Proton transfer reactions (127), such as

$$CH_5^+ + C_4H_{10} \rightarrow CH_4 + C_4H_{11}^+ \tag{23}$$

followed by

$$C_4H_{11}^+ \rightarrow C_3H_7^+ + CH_4 \quad \text{or} \quad C_2H_5^+ + C_2H_6 \tag{24}$$

and

$$C_2H_5^+ + C_3H_6 \rightarrow C_2H_4 + C_3H_7^+ \tag{25}$$

(*b*) Hydride transfer reactions (128,129) such as

$$C_2H_5^+ + C_4H_{10} \rightarrow C_2H_6 + C_4H_9^+ \tag{26}$$

(c) H_2^- ion transfer reactions (38,130), such as

$$C_2H_4^+ + C_4H_{10} \rightarrow C_2H_6 + C_4H_8^+ \tag{27}$$

(d) Hydrogen molecule transfer (111)

$$C_3H_6 + C_5H_{12}^+ \rightarrow C_3H_8 + C_5H_{10}^+ \tag{28}$$

(e) H atom transfer

$$C_3H_6 + C_5H_{12}^+ \rightarrow C_3H_7 + C_5H_{11}^+ \tag{29}$$

As an example of an application of this approach, consider the radiolysis of propane. When equal mixtures of C_3H_8 and C_3D_8 are irradiated in the presence of a radical scavenger, the product ethane is found to have the isotopic constitution C_2H_6, C_2D_6, C_2HD_5, C_2H_5D, $C_2D_4H_2$, and $C_2H_2D_4$ (128). The formation of the partially deuterated species in the presence of radical scavengers can be reasonably ascribed to the following typical sequence of reactions (128,131):

$$C_3H_8^+ \rightarrow C_2H_5^+ + CH_3 \tag{30a}$$
$$\rightarrow C_2H_4^+ + CH_4 \tag{30b}$$
$$C_2H_5^+ + C_3D_8 \rightarrow C_3D_7^+ + C_2H_5D \tag{31}$$
$$C_2H_4^+ + C_3D_8 \rightarrow C_3D_6^+ + C_2H_4D_2 \tag{32}$$

and the analogous reactions involving the oppositely deuterated and protonated species. By application of simple statistical considerations, it is possible to derive an estimate of the yield of the primary fragment ions $C_2H_5^+$ and $C_2H_4^+$ from the yields of their partially deuterated reaction products.

A further discussion of these reactions as they occur in hydrocarbon systems is given by Ausloos and Lias who have recently reviewed hydrocarbon radiolysis from the point of view of reaction classification (101,129). Since the same ions can often undergo several types of reaction in a given mixture, careful and extensive investigation and judicious selection of mixtures is required to derive the maximum benefit from this approach.

7. Electron Spin Resonance. Electron spin resonance studies can be carried out directly in the gas phase when high concentrations of intermediates are available such as in electrical discharges (132). An indirect approach involves the use of a vapor stream crossed by a primary 40 keV argon beam and condensation of the reactant and product gases on a cold finger immediately after the intersecting point (133). The cold finger and the condensed material are then transferred under vacuum into an ESR cavity. Cyclopentane, cyclohexane, ethanol, and benzene have been investigated in this manner, and the expected radicals were observed. The technique suffers from uncertainty in the relative effectiveness of

electronic interactions and atomic collisions, and time lag between creation
of reactive species and their trapping in the solid.

V. CHEMISTRY OF INDIVIDUAL PURE SYSTEMS

Major emphasis over the last decade has been on the reactions in hydrocarbon systems, and most of our fundamental knowledge is restricted to
this area. In part, this has resulted from the feeling that in such systems
one should find a more limited product distribution since there are only
compounds of C and H possible. This, of course, is a fallacy since even the
simplest hydrocarbons give dozens of products and generally show very
poor material balance between starting material consumed and products
which were quantitatively determined (134,135). A second reason has been
experimental. Most organic compounds have limited volatility at ambient
temperatures, the most convenient condition for radiation chemical
experiments. Only within the last few years have studies at temperatures
above 100°C appeared (136,137).

One of the chief contributors to our knowledge of hydrocarbon
chemistry has been Ausloos, and his view of the subject has recently been
summarized (101). Some of the more detailed discussion of experiments
and arguments which have led to our present understanding are given
there, and will not be repeated here. When pertinent to a complete discussion of a particular compound, we shall briefly review some of the mass
spectrometric and photochemical evidence as well. Since the investigations before the mid-fifties are principally of historical interest, and
listed in Lind's book (1), no reference to the early work will be made.

A. Hydrocarbons
1. Alkanes

a. Methane. The radiation chemistry of the simplest hydrocarbon has
received a great deal of attention. It was the subject of some of the earliest
work (1), and interest has continued until the present day. The recognition
of the problems caused by reaction of radical and ionic intermediates with
accumulated products in the fractional percent conversion range has been
material in furthering our understanding (47,88).

The vacuum ultraviolet photochemistry has been studied by a number
of investigators (13,39,138–141). There is now clear evidence that CH_3,
CH_2, and CH are among the important primary reactive species at photon
energies between 10 and 12.8 eV (139–141). The relative extent of CH
radical formation is about three times larger at the higher photon energy
(141). At low intensities, CH disappears by insertion into methane:

$$CH + CH_4 \rightarrow C_2H_5{}^* \tag{33}$$

The resultant ethyl radical is excited and capable of further dissociation,

$$C_2H_5^* \rightarrow C_2H_4 + H \tag{34}$$

Because of the large excitation energy of the ethyl radical produced in reaction 33, dissociation cannot be quenched at pressures below one atmosphere (141). CH_2 also inserts into methane yielding excited ethane which can dissociate further

$$CH_2 + CH_4 \rightarrow C_2H_6^* \tag{35}$$

$$C_2H_6^* \rightarrow 2CH_3 \tag{36}$$

Quenching of the excited ethane may occur in this instance at lower pressures (39) as deduced from the pressure dependence of products formed by CH_2 insertion in radiolysis studies.

Hydrogen atoms may interact with accumulated products, particularly olefins, and undergo a metathetical reaction with methane

$$H + CH_4 \rightarrow H_2 + CH_3 \tag{37}$$

or disappear by combination with other radicals, either homogeneously or at the walls of the vessel. These processes are competitive with each other, and their relative extent can be estimated for the particular conditions of the experiment. Reaction (37) is relatively inefficient, having a rather large activation energy (68). Methyl radicals are removed almost entirely by radical recombination reactions yielding chiefly ethane. This process also requires a third body, but is essentially second order at pressures above about 50 torr (142).

The mass spectrum of methane is dominated by the formation of CH_4^+ (46%), CH_3^+ (40%), and CH_2^+ (7.5%) which together account for 93% of the total ionization at low pressures and electron energies of 50–70 eV (143). These ions are highly reactive toward the parent gas. The dominant reaction of CH_4^+ is

$$CH_4^+ + CH_4 \rightarrow CH_5^+ + CH_3 \tag{38}$$

As pointed out in Chapter 4 this reaction proceeds mainly through a loose complex so that there is effectively no isotopic scrambling if one of the reaction partners is CD_4 (144,145). Both hydrogen atom and hydrogen ion transfer occur (145,146), the relative extent depending on ion energy (145). There is also a small amount of charge exchange such as

$$CH_4^+ + CD_4 \rightarrow CH_4 + CD_4^+ \tag{39}$$

Reaction (38) has been extensively studied by various techniques including pulsing methods to evaluate rate constants for reactions of thermal CH_4^+ ions, and a value of $k_{38} = 1.2 \pm 0.1 \times 10^{-9}$ cc molecule^{-1} sec^{-1} or

7.4×10^{14} cc mole^{-1} sec^{-1} (147–150) appears established. With ions of higher kinetic energy, alternate reaction paths of CH_4^+ with CH_4 leading to the formation of CH_3^+ may set in (145,151,152), but these are not of primary concern to the radiation chemist.

The reactions of methyl ion with methane yield predominantly ethyl ion

$$CH_3^+ + CH_4 \rightarrow C_2H_5^+ + H_2 \tag{40}$$

one of the earliest reactions reasonably well established in a radiation chemical mechanism (14,135,153). Alternate reactions leading to vinyl ion have been reported (154), but judging from the trend in abundance with ion energy probably require epithermal species (145).

CH_2^+ can undergo several reactions (155), and the relative extent is strongly energy dependent (145), lower energies favoring a more complex distribution. Such processes of fragment ions can be studied conveniently only by tandem mass spectrometers, but here it is not possible to investigate reactions of thermal ions. At a nominal ion energy of 0.3 ± 0.3 eV (uncertain because of contact potentials and other difficulties in handling low energy ions), reaction of CH_2^+ and methane leads to 11% $C_2H_2^+$, 25% $C_2H_3^+$, 37% $C_2H_4^+$, and 27% $C_2H_5^+$ (145).

The mass spectra of methane have also been studied at pressures of a few torr (51,54,155), and confirmed that CH_5^+ and $C_2H_5^+$ do not react with methane to form new products. Their contribution to total ionization is close to that anticipated from the intensities of the primary precursor ions CH_4^+ and CH_3^+, in the low pressure mass spectrum. Secondary features are the disappearance of $C_2H_3^+$ and the appearance of $C_3H_5^+$ in the high pressure mass spectrum. At even higher pressures, up to ca. 200 torr (156), $C_3H_7^+$ becomes an important ion. Since the rate constants for further reactions of the two major secondary ions with methane are probably less than 10^9 cc mole^{-1} sec^{-1} (156), this observation has been ascribed to reaction of the secondary ions with impurities. Ausloos and co-workers (47) suggested that the CH_5^+ entity will react with products of higher carbon number even when they are present in concentrations in the parts per million range.

From the above summaries of photochemical and mass spectrometric evidence it is apparent that more than half a dozen important intermediates and a host of minor ones must contribute to product formation in methane radiolysis. Evidence for their participation, and estimates of yields have now been obtained for several, and the radiolysis of this compound is probably as well understood as that of any other gaseous compound.

The major ionic species CH_5^+ and $C_2H_5^+$ do not react with methane to form different products. Consequently, their reactions with additives can be employed advantageously to estimate the yield of the precursor ions.

The yield of ethyl ion can be assessed in mixtures of methane and C_4D_{10}, where the yield of C_2H_5D can be ascribed to the hydride transfer reaction (47,157)

$$C_2H_5^+ + C_4D_{10} \rightarrow C_2H_5D + C_4D_9^+ \tag{47}$$

In the same system, the yield of C_3D_8 arises from dissociative proton transfer followed by a hydride transfer reaction

$$CH_5^+ + C_4D_{10} \rightarrow CH_4 + CHD_3 + C_3D_7^+ \tag{42}$$

$$C_3D_7^+ + C_4D_{10} \rightarrow C_3D_8 + C_4D_9^+ \tag{43}$$

All these reactions are carried out in the presence of radical scavengers such as oxygen (47), and it has been shown that this is indeed the reaction sequence observed. On the assumption that methanium and ethyl ions arise only from reactions (38) and (40), estimates of $G(CH_4^+) = 1.9$ and $G(CH_3^+) = 1.0$ ion/100 eV at a pressure of 480 torr could be made (47). At the same time, the unreactivity of the ethyl ion toward methane could be demonstrated by noting the minor yield of products arising from the reaction of $C_3H_7^+$ with deuterobutane (14).

Although the CH_5^+ ion is nominally unreactive toward methane, it can transfer protons, which leads to isotopic mixing in the radiolysis of mixtures of CH_4 and CD_4 (158).

$$CH_5^+ + CD_4 \rightarrow CH_4 + CD_4H^+ \tag{44}$$

The transfer is a chain process and is terminated either by reaction with accumulated product or by neutralization. The application of competition kinetics and assumed applicability of rate constants for reaction (44) and its analogs from tandem mass spectrometry has led to the estimate that neutralization by free electrons in this system proceeds with a rate constant of over 4×10^{20} cc mole^{-1} sec^{-1} (66), considerably larger than anticipated by measurements in other systems (65). This could possibly result from heterogeneous neutralization (64). The large rate constant for neutralization suggested for methane requires further experimental support.

While estimates for neutralization rates by electrons have thus been obtained, the fate of the predominant species is uncertain. CH_5^+ may lose either a hydrogen atom or a hydrogen molecule

$$CH_5^+ + e \rightarrow CH_4 + H \tag{45}$$

$$CH_5^+ + e \rightarrow CH_3 + H_2 \tag{46}$$

but no really convincing evidence for either reaction has been presented to date. Neutralization of ethyl ion produced in the radiolysis of deuterium-ethane mixtures has been suggested to yield ethyl radicals directly (66), while the quite similar propyl ions produced in the xenon sensitized radiolysis of propane are thought to lead neither to formation of propylene or propyl radicals (115). Further work is clearly required.

There is little doubt that methyl radicals are important intermediates in methane radiolysis. They are produced by several reactions, including the dissociation of initial excited species into hydrogen atoms and CH_3. Moreover, second order neutral processes such as (36) and, at very low dose rates and conversions (159), reaction (37) lead to methyl radical formation. Ionic processes also contribute through reaction (38), while neutralization may also lead to the formation of CH_3. The total yield of methyl radicals is difficult to determine since every additive one may use to assess its overall yield will also interfere with at least some of the steps by which it is formed. Several attempts at an elucidation of this yield have been made (160–163) by adding olefins, acetylene, and other reactants. A value of 3.3 ± 0.2 CH_3 radical/100 eV has been reported for the ethylene and propylene scavenged radiolysis (160). Since reaction (38) should contribute at least 1.9 methyl radicals/100 eV under these conditions, and since the proton transfer

$$CH_5^+ + C_2H_4 \rightarrow CH_4 + C_2H_5^+ \tag{47}$$

prevents reactions (45) and (46) while all hydrogen atoms add rapidly to the olefin, it may be concluded that direct dissociation of excited methane to produce methyl radical occurs with a yield of 1.4 dissociation/100 eV.

Early estimates of the yield of methylene, CH_2, were based on the ethylene yield in the presence of scavenger on the supposition that the reaction

$$CH_2 + CH_4 \rightarrow C_2H_4 + H_2 \tag{48}$$

occurred quantitatively (39). It is now thought, however, that ethylene arises largely from reactions of CH (141). An estimate of $G(CH_2) = 0.7$ radicals/100 eV has been made by adding propylene (160) and assuming that the products butene-1 and methylcyclopropane can be ascribed solely to the methylene addition reactions to yield excited methylcyclopropane which can subsequently isomerize to butene-1 or be stabilized by collision (164).

The contribution of CH radicals to methane radiolysis product formation has been demonstrated very recently (141). No quantitative estimate of its yield was given because of the simultaneous occurrence of other reactions than (34) but $G(CH)$ is probably of the order of 0.1–0.3 radicals/100 eV.

It is thus seen that all species expected from a priori consideration of photochemical and mass spectrometric information are indeed observed. The total of all initial processes arising from neutral excited states can now be estimated by summing the methyl radical yield arising from unimolecular dissociation, methylene and CH yields, and is approximately $G(\text{Exc.}) \simeq 2.3$ events/100 eV. In combination with $W(CH_4) = 28$ eV/ion pair, corresponding to $G(\text{ions}) = 3.6$ ions/100 eV, a value of 0.64 is obtained for N_{ex}/N_i, the number of chemically recognizable excitation events per ion pair. These values apply at ambient temperatures and pressures of ca. 100 torr to 1 atm.

The effect of pressure on the relative importance of primary processes in methane radiolysis has been investigated by Ausloos (39), using reactions of $C_2H_5^+$ and CH_5^+ with deuterated additives as the diagnostic tool. A gradual decrease of the fragment methyl ion yield is indicated between 14 torr and 103 atm, with $G(CH_3^+) \simeq 0.2$ ions/100 eV estimated at the highest pressure. Although complications from the stabilization of the intermediate ion–molecule reaction complex of reaction (40) $C_2H_7^+$, may exist, this is thought to be unimportant in view of the fact that $C_2H_7^+$ probably reacts like an ethyl ion. An increase in the yield of methanium ion up to a pressure of about 15 atm was noted as evidenced by an increase of the yield of C_3D_8 in CH_4–C_4D_{10} mixtures. A decrease of this product of higher pressures was attributed to the inability of CH_5^+ to react with the additive to form a carbonium ion at higher pressures. In the same study, an increase in the ethane yield between 14 torr and 13 atm was attributed to stabilization of excited ethane produced in reaction (35). The yield of this species was then independent of pressure up to 180 atm, suggesting the absence of a pressure dependence of the neutral dissociation process in which CH_2 is formed.

Increase in pressure also materially affects the product distribution from methane in the absence of scavengers (165), a reduction in hydrogen and ethane yields by a factor of about two being noted when pressure was raised to about 300 atm. Since the extent of reaction with accumulated products cannot readily be assessed in this experiment one cannot expect a quantative interpretation. The change is consistent with the results of Ausloos et al. (39).

Maurin (165) has also investigated the temperature dependence of pure methane radiolysis at 490 torr over the range -78 to $+495°C$. At low dose rates (0.2 Mrad/hr) he observes a doubling of the hydrogen yield over this range, and no material effect on ethane production. This is ascribed to increasing effectiveness of reaction (37). At high dose rates (300–400 Mrad/hr) he observes a decrease with temperature in all yields but that of ethane. However, conversions in high dose rate experiments were excessive (25% or more) so that it is difficult to draw meaningful conclusions. Firestone (66) reports no temperature effect on the chain proton transfer reaction (44) between $-78°C$ and $25°C$.

An interesting side aspect of methane radiolysis pertains to the yield of acetylene. Reports of its formation in small amounts (114,162,163,166) have been questioned (167). In pure methane, $G(C_2H_2)$ is probably less than 0.02 and dose rate dependent if formed at all (166). The radiolysis of methane thus differs sharply from electrical discharges where acetylene is one of the most important condensation products (168).

b. Ethane. The vacuum ultraviolet photochemistry of ethane has been investigated repeatedly (86,87,169,170,171) and as indicated in Chapter 3 the results suggest the following primary processes

$$C_2H_6 \xrightarrow{h\nu} CH_3CH^* + H_2 \tag{49}$$

$$C_2H_6 \xrightarrow{h\nu} C_2H_5^* + H \tag{50}$$

$$C_2H_6 \xrightarrow{h\nu} CH_4 + CH_2 \tag{51}$$

At 1470 Å, process (49) accounts for ca. 85% of the total primary events. Formation of ethylidene is followed by

$$CH_3CH^* \rightarrow C_2H_4^* \tag{52}$$

$$M + C_2H_4^* \rightarrow C_2H_4 + M \tag{53}$$

$$C_2H_4^* \rightarrow C_2H_2 + H_2 \tag{54}$$

while the ethyl radical produced in reaction (50) can either dissociate further into ethylene and a hydrogen atom, or be stabilized by collision.

The mass spectrum of ethane (143) consists of $C_2H_4^+$, (44.8%); $C_2H_3^+$, (14.9%); $C_2H_6^+$ (11.7%); $C_2H_2^+$ (10.3%) and $C_2H_5^+$ (9.7%). The ion-molecule reactions of this compound have been investigated, but not extensively (154,172,173). Some nine reactions of the important ions above have been reported, and it appears that $C_2H_6^+$ and $C_2H_4^+$ may react with ethane with a collision efficiency of only ca. 5% (173).

Ethane radiolysis has been investigated within the last decade by Dorfman (174,175), Yang (176,177), Back (46,62), and Ausloos (169,178). Ethylene and acetylene arise from the neutral dissociation reaction (49) (169). The former is also produced by the hydride transfer reaction

$$C_2H_3^+ + C_2H_6 \rightarrow C_2H_4 + C_2H_5^+ \tag{55}$$

In the radiolysis of ethane–perdeuteroethane–nitric oxide mixtures, reactions (55) and (49) followed by (53) are the only processes leading to ethylene formation. Since there are presumably no ionic processes leading to acetylene formation the yield of step (49) can be estimated as 1.7 events/100 eV. The occurrence of the dissociation step (51) is evidenced by the formation of propane and methane in equivalent amounts after subtraction of ionic contributions to methane formation. A yield of ca.

0.28 methylene radicals/100 eV was estimated in this manner (169). While methyl radicals are observed (177) they almost certainly arise as a complement to methyl ion formation, since the yield of ca. 0.08 radicals/100 eV exactly corresponds to that predicted from the mass spectra (169,177). Summing the yields of neutrals leads to $N_{ex}/N_i \simeq 0.5$.

The effect of pressure on the product distribution suggests that at pressures above 25 atm only parent ions participate in ionic reactions, as evidenced from the absence of reactions ascribed to other ionic intermediates (169). Since the fate of the parent ion is unknown at any pressure, this is to be accepted with reservations, particularly since this conclusion cannot be reached in other similar systems. As in the methane system, the yield of methylene appears to show little dependence on pressure. The yields of ethylene and acetylene decrease sharply with pressure while that of n-butane increases threefold. At even higher pressures (densities to 0.3 g cc^{-1}) the yield of ethylene begins to rise again (179,180). The formation of total hydrogen decreases gradually with increase in pressure, while that in the presence of olefins as free radical scavengers remains essentially constant. This is ascribed to parent ion recapture of electrons (179,180).

While the pressure dependence might suggest that neutral excited ethane dissociation by (50) is increased at the expense of (49), the uncertainties in the nature of the neutralization process and the fate of the ethane ion preclude a definite conclusion. That homogeneous neutralization in the gas phase probably yields hydrogen atoms at low pressures is suggested by the effect of applied electrostatic fields (62), where a reduction in the H atom yield was observed when neutralization occurred at the surface of the radiation vessel. It is quite possible that the reduction in hydrogen yield observed by Peterson et al. (179) can be ascribed to a similar effect of density, suggesting a very short lifetime for the excited species resulting from the neutralization process.

c. Propane. The vacuum ultraviolet photochemistry (86) indicates that excited propane eliminates molecular hydrogen

$$C_3H_8 \xrightarrow{h\nu} C_3H_6{}^* + H_2 \qquad (56)$$

where the $C_3H_6{}^*$ entity may have the structures CH_3CCH_3, $CHCH_2CH_3$, or $CH_2{=}CHCH_3$, the first of these being the most important product of the hydrogen formation reaction. A second dissociation process is

$$C_3H_8 \xrightarrow{h\nu} C_2H_4 + CH_4 \qquad (57)$$

where the ethylene may have the vinylidene structure. Other dissociation modes have also been suggested but are less well established, perhaps because they occur only to a minor extent. In particular

$$C_3H_8 \xrightarrow{hv} C_2H_5^* + CH_3 \qquad (58)$$

has been suggested recently (see Chapter 3).

In the mass spectrum of propane only three ions constituting somewhat over 60% of the total ionization are observed with intensities of 10% or higher (181). These are $C_2H_5^+$, (30.5%); $C_2H_4^+$, (18.1%), and $C_2H_3^+$, (12.3%). Nine different reactions need to be considered before one accounts for 90% of the total ionization, and the parent ion constitutes only 8.9%.

Ion–molecule reactions of propane (62,173,182–187) have also been studied. The most recent study has involved mass and energy resolved ion beams, a necessity in attempts to elucidate the reactions of this host of primary ions (187). $C_3H_8^+$ ion undergoes only charge exchange with propane, while propyl ion is totally unreactive and does not even undergo a thermoneutral hydride transfer process. (Collision efficiency less than 10^{-3}.) The only important reactions of the species CH_3^+, $C_2H_5^+$ and $C_3H_5^+$ are hydride transfers with parent propane, converting these ions to methane, ethane, and propylene, respectively, and forming $C_3H_7^+$ as product ion. $C_2H_3^+$ also chiefly (at least 80%) undergoes hydride transfer yielding ethylene. $C_3H_6^+$ reacts largely by H_2^- transfer and to a small extent by hydride transfer, the former predominating by an order of magnitude:

$$C_3H_6^+ + C_3D_8 \rightarrow C_3H_6D_2 + C_3D_6^+ \qquad (59)$$

Ethylene ion may undergo both H_2^- and H^- transfer as well, the first accounting for about 70% of the reaction of that ion with propane. It is of interest to note that H_2^- reactions were first established by radiation chemical techniques (101), and were only later confirmed by mass spectrometric investigations.

Although one might expect the radiation chemistry of propane to be hopelessly complex because of the variety of primary species, the situation is alleviated by the predominance of hydride and hydrogen anion (H_2^-) transfer reactions which convert most ions to propyl ion and propylene. The high pressure mass spectrum of propane shows that between 0.1 and 1.0 torr ca. 80% of all ions observed are propane and propyl ions (173,184). Fragment ions are almost immediately converted to stable "molecular" products.

Propane radiolysis has received fairly extensive attention (46,63,102, 115,119,128,131,161,181,188–196), at least in part because its ionic fragmentation has been the subject of several detailed applications of the quasi-equilibrium theory of mass spectra (29,37,197,198). The most recent

Table I

Ionic Fragmentation of Propane

	Relative abundance, 70 eV			
	Radiolysis		Mass spectrum	Mass spectrum extrapolated to 1 atm[d]
Ion	760 torr[a]	30 torr[b]	$\sim 10^{-5}$ torr[c]	
$C_3H_8^+$	$(0.25)^e$		0.09	0.098
$C_3H_7^+$	0.10		0.07	0.10_4
$C_3H_5^+$	0.07		0.04	0.08_4
$C_2H_5^+$	0.35	0.41	0.30_5	0.43
$C_2H_4^+$	0.16	0.14	0.18	0.19
$C_2H_3^+$	0.07	0.06	0.12	0.07
Total (1.00)				

[a] Reference 196.
[b] Reference 101.
[c] Reference 143.
[d] Reference 37.
[e] By difference and therefore including all other ions not listed.

investigations suggest that the behavior of propane under high energy irradiation is now fairly well understood (63,198,101). Using the now well-established technique of essentially total product analysis and isotopic substitution (101), a substantially complete ionic breakdown has been estimated and is given in Table I, together with the mass spectrometric fragmention pattern at low pressures, and that estimated from the quasi-equilibrium theory of mass spectra for ca. 1 atm. (37). The agreement between estimated fragmentation and that observed is surprisingly good. There is some discrepancy between experimental evaluations, which may be due in part to the assumption (196) that the reaction branching ratios and yields obtained at 30 torr can be applied at 1 atm without correction.

The contributions of neutral excited molecules to product formation are summarized in Table II. The process distribution is fairly similar to that observed in the vacuum ultraviolet photochemistry, suggesting once more that the decomposition of the superexcited molecule leads to the formation of the same products as that of allowed states below the ionization limit. For the chemically consequential excited states per ion pair a value for N_{ex}/N_i of 0.4 may be derived.

$N_{ex}/N_i \approx 1$ has also been estimated by a direct mass spectrometric technique (199–201). This consists of two ionization chambers. The first is used with a 200 eV electron beam to cause ionization and excitation,

Table II

Neutral Dissociation Processes in Propane Radiolysis

Process	Yield, molecules/100 eV	
	30 torr[a]	760 torr[b]
$C_3H_8 \longrightarrow C_3H_8^+$		
$C_3H_8^* \longrightarrow$		
$\quad C_3H_6 + H_2$	0.47	0.60
$\quad C_2H_6 + CH_2$	$\simeq 0.08$	0.08_5
$\quad C_2H_4 + CH_4$	0.15	0.21
$\quad C_2H_4 + CH_3 + H$	0.56	0.51
$\quad C_2H_3 + ?$	0.16	—
$\quad C_2H_2 + ?$	0.45	—

[a] Reference 102 and 181.
[b] Reference 196.

and all charged products are withdrawn so that they do not enter the second chamber, which can be reached by the neutral products. There a second electron beam is used to analyze the products of dissociation. Because of a large background of neutral parent molecules, synchronous detection is employed. N_{ex}/N_i is estimated by summing neutral fragments after correction for their formation as counterparts of ionic fragmentation, and division into the total analyzed ion current. Semiquantitative confirmation is achieved by following the gas density modulation caused by the change in the first electron beam. However, a large number of assumptions has to be employed and although the method is interesting and gives a qualitative picture of events it is not reliable for estimates of N_{ex}/N_i (101).

The consequences of neutralization are still uncertain. Clearly the neutralization process must involve chiefly propyl and propane ions provided that conversions are sufficiently low that self-scavenging of radicals and ions by accumulated products do not interfere. It was first noted by Woodward and Back (62) that application of electrostatic fields reduced the hydrogen yield from 7.4 to 5.1 molecules/100 eV. At the same time, a reduction of the dose rate also led to a reduction in the yield of hydrogen. The interpretation that neutralization at the walls prevented dissociation normally occurring on neutralization, first suggested by Essex (119), was strongly supported by the work of Johnson and Warman (105–109), who also observed a reduction of the hydrogen yield on addition of the electron scavengers, N_2O, SF_6, and CCl_4, and of the exact same magnitude. It would seem that the negative ion serves the same stabilizing function as the walls of the radiation vessel in the study of Woodward and Back. Apparently the positive ion–negative ion neutralization process leads to less fragmentation than ion–electron recombination or none at all. Since nitrous oxide does not further reduce the yield of hydrogen when propylene is originally added to the system, the electron scavenger therefore reduces the hydrogen atom yield. The reduction in the observed hydrogen formation corresponds to an ion pair yield of ca. 0.6, suggesting that 60% of the ion–electron neutralization processes yield hydrogen atoms. It is tempting to associate this value with the estimated yield of propyl ions in the system. Unfortunately, some doubt on this interpretation has been cast by the suggestion that neutralization occurs heterogeneously at even higher dose rates than those employed by Woodward and Back, and produces hydrogen atoms as well (63). There may not be a real inconsistency since clearly only a portion of the neutralization processes at the wall lead to inhibition of hydrogen atom formation, and it is quite possible that the other events still yield some hydrogen atoms. At the same time, Klots (64) has offered convection currents as an explanation for the observation that neutralization may occur on the walls even at much higher dose rates.

While increase in pressure does not appear to have a major influence on the yield or distribution of products arising from neutral dissociation events, there is a marked effect on ionic fragmentation (38) (Table III). The yield of $C_2H_3{}^+$, which arises by successive fragmentation processes, is monotonically decreased as the pressure is increased from a few torr to 40 atm. The formation of $C_2H_5{}^+$, on the other hand, increases before it is reduced below the level of formation at a few torr, as expected for an ion participating in further fragmentation processes. Ethylene ion is a product of both first and second generation dissociation processes and also decreases with pressure (101).

Table III
Effect of Pressure on Ionic Fragmentation of Propane[a]

	Yield, ions/100 eV	
Pressure	$C_2H_3{}^+$	$C_2H_5{}^+$
0.04 atm.	0.201	1.32
0.3	0.18	1.64
1.7	—	1.77
4.4	—	1.41
16.0	0.07	1.16
43.0	0.01_3	0.49

[a] Reference 38.

d. Other Alkanes. The behavior of the other alkanes parallels that of propane to a large extent. Only relatively limited investigations of the vacuum ultraviolet photochemistry of these compounds have been made (22,27,86,202,203), and the dissociation processes are analogous to those of propane. It is generally observed that at lower energies the molecular hydrogen elimination process to yield the olefin of the same carbon number is enhanced, while other processes become more important at the higher energies more typical of radiation chemistry (202).

The high pressure mass spectrometry of *n*-butane has been investigated by Munson and co-workers (173,205) who observed again that hydride transfer reactions dominate the reaction mechanism. At butane pressures of ca. 0.5 torr and more, butyl ion accounts for more than 75% of the total ionization in *n*-butane and more than 90% in isobutane. The reactivity of these ions toward various hydrocarbons and olefins differs, and it is concluded that the butyl ion in *n*-butane has the secondary structure and that in isobutane it is a tertiary ion (205).

The radiolysis of the butanes has been investigated by Woodward and Back (62), Borkowski and Ausloos (204,206), and by Miyazaki and Shida (207). Use of the deuterium labeling technique has led to an estimate of the distribution of primary events summarized in Table IV. The estimate of primary ion yields from mass spectrometry is clearly very poor for these compounds. There is a very definite effect of pressure on the formation of ionic intermediates in both butanes, (204,207), which can be attributed to quenching of ionic fragmentation. For isobutane, Borkowski and Ausloos (204) have argued that the relative decrease of the ethane and ethylene

Table IV
Primary Ion Yields in Butane Radiolysis

| Primary species | *n*-Butane | | Isobutane | |
	Radiolysis[a], 40 torr	Mass spectrometer[c], Low pressure	Radiolysis[b], 40 torr	Mass spectrometer[c], Low pressure
$C_3H_7^+$	1.4	1.2	2.1	1.55
$C_3H_5^+$	0.22	0.39	0.26	0.67
$C_2H_5^+$	0.46	0.52	0.07	0.09
$C_2H_3^+$	0.14	0.52	0.11	0.45

[a] Reference 206.
[b] Reference 204.
[c] Reference 37.

yields suggests that at higher pressures the parent ion dissociates from a lower level of excitation. This would indicate that the first collision of the excited parent ion only removes a fraction of the excitation energy. Further support for this suggestion was obtained by Ausloos and Lias (22) in photoionization experiments when "inert" gases such as He, Ar, or N_2 were added, species which cannot produce ions at the wavelength of irradiation. Considerable differences in the effectiveness of the various additives in deactivating the excited parent ions were observed, and thought to be explicable only in terms of less than unit collision efficiency for deactivation. Ausloos and Lias also observed a large isotope effect for deactivation of $C_4H_{10}^{+*}$ or $C_4D_{10}^{+*}$ by He in support of their explanation. While the experiments may indicate that gases other than hydrocarbons do not completely remove all vibrational excitation of the ion, it is not established whether encounter with molecules of large polarizability or capable of undergoing an efficient ion–molecule reaction can also lead to partial deexcitation.

A value of 0.3 has been estimated for the number of neutral excited n-butane molecules formed per ion pair and was based on the yields of the molecular products given in Table V of Chapter 3.

Table V

Primary Yields in the Radiolysis of Pentanes

	Molecules/100 eV[a]			
	n-Pentane		Neopentane	
Primary species	Radiolysis[b], 300 torr	Mass spectrum	Radiolysis[c] 30 torr	Mass spectrum
$C_4H_9^+$	0.21	0.15	≥ 1.36	1.75
$C_3H_7^+$	1.31	1.17	0.06	0.03
$C_3H_5^+$	0.127	0.54	≥ 0.25	0.72
$C_2H_5^+$	0.20	0.29	0.29	0.68
$C_2H_3^+$	0.20	0.49	≥ 0.12	0.27
C_5H_{10}	0.32	—	—	—
C_4H_8	0.02	—	0.0	—
C_3H_8	0.025	—	—	—
C_3H_6	0.070	—	≤ 0.09	—
C_2H_6	0.12	—	≤ 0.05	—
C_2H_4	0.133	—	≤ 0.34	—

[a] Based on W (pentane) = 23.5 eV/ion pair.
[b] Reference 202.
[c] Reference 203.

The radiolysis of the pentanes has been studied by Futrell (208,209) and by Ausloos and co-workers (202,203,210) who also investigated the effects of pressure. The most important results are summarized in Table V. Increase in pressure from 5 to 2200 torr n-pentane reduces the yields of allyl, ethyl, and vinyl ions which arise from complex dissociation schemes, while that of butyl ion is substantially increased. Investigations of n-hexane vapor (189,211,212) have not progressed sufficiently to allow a detailed analysis, but hydride transfer reactions of fragment ions again play an important role.

2. Olefins

a. Ethylene. The vacuum ultraviolet photolysis leads essentially to three types of processes (23,86,88):

$$C_2H_4^* \rightarrow C_2H_2 + H_2 \quad (\Delta H = 1.8 \text{ eV}) \tag{60}$$

$$C_2H_4^* \rightarrow C_2H_3 + H \quad (\Delta H = 4.6 \text{ eV}) \tag{61}$$

$$C_2H_4^* \rightarrow C_2H_2 + 2H \quad (\Delta H = 6.3 \text{ eV}) \tag{62}$$

where step (62) may be considered as a very rapid loss of a hydrogen atom from excited vinyl radicals formed in process (61). Processes (60) and (62) dominate the photochemistry. Their relative extent is wavelength dependent, increasing photon energy favoring the most endothermic process (62). Hydrogen atoms produced in (61) and (62) immediately add to ethylene

$$H + C_2H_4 \rightarrow C_2H_5 \tag{63}$$

and this reaction accounts quantitatively for hydrogen atoms under most conditions. At very high light intensities

$$H + C_2H_5 \rightarrow C_2H_6^* \tag{64}$$

may set in and lead to additional ethane and some methyl radical formation if the dissociation

$$C_2H_6^* \rightarrow 2CH_3 \tag{65}$$

is not quenched at elevated pressures. The collision efficiency of (63) is about 0.1 (69), and the ethyl radical produced in this step has excess energy. Under normal conditions this is rapidly removed by collisions, so that the species participating in (64) is in thermal equilibrium with its surroundings. The vinyl radical produced in (61) is most likely to add to ethylene yielding butenyl

$$C_2H_3 + C_2H_4 \rightarrow C_4H_7 \tag{66}$$

and since the yield of this radical is normally minor it will chiefly disappear by reaction with ethyl radicals:

$$C_2H_5 + C_4H_7 \rightarrow C_6H_{12} \tag{67}$$

$$C_2H_5 + C_4H_7 \rightarrow C_2H_4 + C_4H_8 \tag{68}$$

$$C_2H_5 + C_4H_7 \rightarrow C_2H_6 + C_4H_6 \tag{69}$$

The direct combination of vinyl radicals cannot be eliminated totally as a possibility of butadiene formation (213). Ethyl radicals not participating in reactions 67–69 combine or disproportionate

$$2C_2H_5 \rightarrow n\text{-}C_4H_{10} \tag{70}$$

$$2C_2H_5 \rightarrow C_2H_4 + C_2H_6 \tag{71}$$

with $k_{71}/k_{70} \simeq 0.14$ (67,68,214).

Three ions account for ca. 85% of the total ionization of ethylene in the mass spectrometer: $C_2H_4^+$ (38%), $C_2H_3^+$ (23%), and $C_2H_2^+$ (22%). Their ion–molecule reactions have been investigated extensively, not only as far as the reactions of the primary ions are concerned (4–6,154,215), but also with a view toward obtaining an understanding of higher order reactions and ionic polymerization at pressures up to several hundred torr (52,53,58,216–220). Substantial contributions to the understanding of higher order reactions have also been made by radiation chemical techniques (23,40,57,111) and it is therefore most convenient to combine the evidence in the discussion of ionic reactions of ethylene ions.

Reactions of ethylene ion proceed via a relatively long-lived intermediate complex

$$C_2H_4^+ + C_2H_4 \rightarrow [C_4H_8^+] \tag{72}$$

$$[C_4H_8^+] \rightarrow C_3H_5^+ + CH_3 \tag{73}$$

$$[C_4H_8^+] \rightarrow C_4H_7^+ + H \tag{74}$$

with $k_{73}/k_{74} \simeq 10$. The intermediate complex ion or the secondary fragment ion produced in reactions (73) and (74) may react further with ethylene

$$[C_4H_8^+] + C_2H_4 \rightarrow C_4H_8^+ + C_2H_4 \tag{75}$$

$$[C_4H_8^+] + C_2H_4 \rightarrow [C_6H_{12}^+] \tag{76}$$

$$C_3H_5^+ + C_2H_4 \rightarrow [C_5H_9^+] \tag{77}$$

$$C_4H_7^+ + C_2H_4 \rightarrow [C_6H_{11}^+] \tag{78}$$

The importance of the direct process of intermediate hexene ion formation by step (76) is in serious question, and it may not occur at all (216,218,219). The ions produced in (77) and (78) are excited and must probably be stabilized by collision to be observed (216). The rate constant for dissociation of these species is presumably considerably less than 10^7 sec^{-1}, that is, they are much longer lived and therefore more easily deactivated than the intermediate butene ion produced in reaction (72) (38).

The stabilized intermediate butene ion may undergo further ion–molecule reactions but the successive addition steps of higher olefin ions to ethylene are inefficient and butene ion survives more than 100 collisions before forming a hexene ion reaction complex which may dissociate

$$C_4H_8{}^+ + C_2H_4 \rightarrow (C_6H_{12}{}^+) \tag{79}$$

where $(C_6H_{12}{}^+)$ represents a lower energy state of $C_6H_{12}{}^+$ than $[C_6H_{12}{}^+]$ found in reaction (76),

$$(C_6H_{12}{}^+) \rightarrow C_5H_9{}^+ + CH_3 \tag{80}$$

$$(C_6H_{12}{}^+) \rightarrow \text{other products} \tag{81}$$

The rate constant for decomposition of $(C_6H_{12}{}^+)$ has been estimated on the assumption that the competitive deactivation by ethylene occurs with collision efficiency, and is probably of the order of 4×10^8 sec^{-1} (57,58). Further addition reactions of stabilized $C_6H_{12}{}^+$ to ethylene occur with a collision efficiency of well below 1% (58), and the intermediate product octene ions probably dissociate rapidly with a rate constant of 5×10^9 sec^{-1} or greater (57). It is again of interest that this behavior was originally suggested in radiolytic studies and later confirmed by mass spectrometry (58).

Mass spectrometric experiments cannot readily distinguish between the sequences (72,75,79), and (80) and (72,73,77), and collisional stabilization. In both cases the order of reaction and the precursor ions are identical. However, once the rate constant for reactions (73) and (74) are known from independent measurements (23,40,111), it is possible to analyze the events in the ion source of the mass spectrometer by a complete theory of third order reactions. Such considerations have suggested that at low pressures (0.2 torr or less) the second sequence is more representative of events, while at high pressures (greater than a few torr) the first series of reactions is descriptive of the mechanism (221). This has also been supported by recent mass spectrometric results using deuterated ethylenes (222).

Uncertainty of the complete reaction sequence arising from vinyl ion reactions results from ignorance of the lifetime of the intermediate addition complex

$$C_2H_3{}^+ + C_2H_4 \rightarrow [C_4H_7{}^+] \tag{82}$$

$$[C_4H_7{}^+] \rightarrow C_2H_2 + C_2H_5{}^+ \tag{83}$$

For example, it has been suggested that the formation of ethyl ion proceeds in part by a stripping type mechanism which does not involve an intermediate ionic complex whose dissociation may be quenched by collision (58). There is relatively little convincing evidence on the mechanism by which the higher order reactions of vinyl ions occur. One of the difficulties in studying these reactions is the inability to produce vinyl ions exclusively in an atmosphere of ethylene. While an attempt to do so has been made with the charge exchange technique, the results even for the parent ion reactions are at such variance with established mechanisms that little weight can be placed on the findings (220).

Acetylene ion can exchange charge with ethylene

$$C_2H_2^+ + C_2H_4 \rightarrow C_2H_2 + C_2H_4^+ \tag{84}$$

since the ionization potentials are favorable. This reaction has been reported repeatedly (58,218,219) but it is not known whether it involves a long-lived complex. The condensation type ion–molecule reactions involve more than the transfer of a simple species and therefore almost certainly a long-lived complex

$$C_2H_2^+ + C_2H_4 \rightarrow [C_4H_6^+] \tag{85}$$

$$[C_4H_6^+] \rightarrow C_3H_3^+ + CH_3 \tag{86}$$

$$[C_4H_6^+] \rightarrow C_4H_5^+ + H \tag{87}$$

Higher order ion–molecule reactions may involve either the complex ion formed in reaction (85) or the products of its dissociation.

The radiation chemistry of ethylene has been explored in some detail (40,57,72,111,113,120,134,224–229). The yield of parent ions has been estimated as $G(C_2H_4^+) \simeq 1.5$ ions/100 eV (57) by measuring the methyl radical yield and assuming it is formed by the sequence (72,76,79,80) and $k_{80}/k_{81} = 10$. This appeared to be supported by measurements of the yields of stable butenes formed when charge acceptors were added to ethylene radiolysis. Since reactions (76) and (79) are highly inefficient and the lifetime of $[C_4H_8^+]$ is of the order of 0.1 μsec, compounds of low ionization potential can act as charge acceptors (A)

$$C_4H_8^+ + A \rightarrow C_4H_8 + A^+ \tag{88}$$

Stable butene formation can be correlated with the yield of the precursor ethylene ion and leads to $G(C_2H_4^+) \simeq 1.5$ ions/100 eV. Step (88) was examined using over 20 different additives, and good correlation with ionization potential of the additive was observed (40,72). The charge transfer technique can only give minimum values for the yield of butene, and therefore, ethylene ion, since the additive may reduce the extent of butene ion formation by charge transfer from parent ethylene ions. Moreover, the occurrence of other ion–molecule reactions with $C_4H_8^+$ not leading to neutral butenes can lower butene yields. Step (88) was confirmed for nitric oxide as the charge acceptor when $C_4H_8^+$ ions were

created in the vacuum ultraviolet photolysis of cyclobutane at 1048–67 Å
(73). Butenes were reported with an ion pair yield slightly above one when
NO was added, but this ion pair yield high, as the determination was
probably made through measurement of a saturation current at too high
a sample pressure (89). The relative amounts of the various butene isomers
formed was identical to the distribution formed in the radiolysis of ethylene and cyclobutane (229). Mass spectrometric investigations, however,
have seen little charge transfer from butene ion to nitric oxide (58) and
have instead observed chiefly the adduct $C_4H_8NO^+$. It is conceivable that
this ion yields stable butene on neutralization as well, and the two mechanisms (charge transfer and addition–neutralization) would therefore give
the same yield of butene. It is difficult to see how this could apply to all
other additives as well.

The only attempt to measure the yield of vinyl ions has been based on
the formation of molecular methane by nonradical reactions, a product
tentatively ascribed to decomposition of higher molecular weight ions
formed by reaction of $C_2H_3^+$ (57). Methane is clearly produced by ionic
reactions as demonstrated by the applied field technique (120), the effect
of large excesses of scavengers (57), and by analysis of the isotopic
labeling of product methane (40). Reaction of $C_2H_3^+$ leading to methane
formation is the only mode consistent with these observations and is reinforced by results from mass spectrometry (218), and the assignment of
vinyl ion as the precursor appears indicated. On this basis, $G(C_2H_3^+) \simeq 1.0$
ion/100 eV may be estimated at 100 torr, again in good agreement with
expectations from mass spectra (57).

The extent of acetylene ion formation has been estimated using the
molecular hydrogen production from isotopically labeled ethylenes
(40,228). The assignment of $G(C_2H_2^+) \simeq 0.8$ ions/100 eV must be considered as highly tentative at best. Similarly, the value of 1.0 $C_2H_2^+$/100
eV based on material balance of H; H_2; C_2H_3; and C_2H_2 (230) is uncertain.
Acetylene ions almost certainly undergo largely condensation type
ion–molecule reactions, with an upper limit of approximately 50% to be
set on the contribution of charge transfer (reaction 84) (6,58,217–219).
The intermediate addition complex is probably relatively long-lived, and
should therefore be deactivated or react further at the normal pressures of
radiation chemistry. Resultant products would be difficult to observe since
they would be largely highly unsaturated.

The dissociation of neutral excited molecules produced in the radiolysis
proceeds by all processes possible in the photochemistry (reactions
60–62). An assignment of their relative probabilities can be made if the
yields of vinyl ion and acetylene ion are accepted as suggested, and if one
assumes that acetylene ion formation is accompanied by the formation of

Table VI

Dissociation of Neutral Excited Ethylene Molecules

Method of excitation	Energy, eV	Relative occurrence[a], process		
		$C_2H_2 + H_2$ (1.8 eV)	$C_2H_3 + H$ (4.6 eV)	$C_2H_2 + 2H$ (6.3 eV)
Hg^3P_1 photosensitization	4.8 eV	0.92	0.08	—
Hg^1P_1 photosensitization[b]	6.7 eV	0.38	0.48	0.13
Direct photolysis	8.3 eV	0.44	0.11	0.45
Direct photolysis	10.0 eV	≤ 0.26	≥ 0.0	≤ 0.74
Direct photolysis[c]	11.6 eV	~ 0.23	~ 0.0	~ 0.77
Radiolysis	?	≥ 0.10	0.03	≤ 0.87
Slow electrons	4–10 eV	0.38	0.16	0.46

[a] Unless otherwise indicated from the summary in Reference 53.
[b] Reference 213.
[c] Reference 23.

molecular hydrogen. $G(C_2H_2 + H_2) \geq 0.3$, $G(C_2H_3 + H) = 0.1$, and $G(C_2H_2 + 2H) = 2.7$ processes/100 eV are obtained in this manner. A comparison of these values with those obtained by photolytic techniques and by slow electrons (applied fields) is given in Table VI. There is a striking change in the distribution of primary neutral dissociation processes toward the more endothermic reactions as the energy deposited in ethylene is increased. Two aspects are particularly noteworthy: the process distribution in the radiolysis is essentially that observed for dissociation of the superexcited molecule produced at 11.6 eV, while the decomposition induced by the applied field technique is totally atypical of the radiolysis and shows considerably greater resemblance to photolyses in the 7–8 eV range. Applied fields, therefore, appear to induce primarily dissociation from the lower allowed states. Dissociation of the triplet excited molecule would probably be largely quenched at these pressures (120).

The role of neutralization in ethylene radiolysis is quite uncertain. A minor depression of $G(—C_2H_4)$ in the presence of small applied fields (120) which do not affect the yields of the other products suggest that neutralization produces chiefly complex olefinic species which escape detection and are classified as "ionic condensation products" (40). It does not appear that any of the major radicals or neutral species are formed in this manner.

Study of the effect of pressure on the distribution of primary dissociation processes is complicated by the reactivity of ethylene which leads to radical addition, and eventually to polymerization as discussed in Chapter 8.

b. Other Olefins. The vacuum ultraviolet photochemistry of propylene has been investigated (86,231) and the studies have been extended into the region where the parent ion is formed and its reactions are amenable to elucidation (89). Ion–molecule reactions in this compound, investigated by Fuchs (154) and Harrison (232), lead mainly to condensation to C_4^+ and C_5^+ ions. The radiolysis (223,233) has shown that hydrogen, ethylene, and 2,3-dimethylbutane (or 2-methylpentane) are formed with yields in excess of 1.0 molecules/100 eV, with ethane, propane, allene, *n*-hexane, pentene-1, propyne, and several other products formed with G values less than one. The formation of all products was strongly decreased by increase in pressure. Product distributions in the single phase and two-phase regions were compared also. The effect of ca. 2% oxygen suggested that the hydrogen yield could be depressed substantially (ca. 40%), and this portion was therefore ascribed to thermal atom reactions. This does not seem plausible in view of the efficiency of propylene as a hydrogen atom scavenger (191). The substantial yield of *n*-propyl radicals as shown (233) by the formation of *n*-hexane is surprising and was explained by neutralization of $C_3H_7^+$ ions.

The gas phase radiolysis of the butenes has been investigated as an adjunct to liquid phase studies (234) particularly with a view towards establishing ionic condensation steps. Very little dimer was observed, and this is probably in part attributable to the low dose rate employed which would lead to longer free radical and ionic condensation chains than in the liquid phase, where parent-ion recapture is important.

3. Acetylene. The photochemistry of acetylene is interpreted in terms of simultaneous processes involving excited state dimerization (86,235,236) and dissociation to yield either molecular hydrogen and C_2 or C_2H and a hydrogen atom. The relative extents of these processes, which are also thought to occur in the mercury photosensitized photolysis (237–239), are not established. As expected, there is no scission of the triple bond (240) followed by reactions of CH leading to benzene formation, one of the principal products.

Acetylene ion constitutes ca. 75% and C_2H^+ 15% of the ions in the mass spectrum of acetylene. Both ions undergo condensation-type ion–molecule reactions (154,241–246,249), and their behavior at pressures up to several tenths of a torr indicates ionic polymerization reactions similar to those of ethylene (244,246).

Although the radiation chemistry of acetylene has been investigated for decades, there is still no universal agreement on the mechanism by which the major products, cuprene and benzene, are formed. The disappearance of acetylene was for a time used as a gas-phase dosimeter, but the relative extent of formation of the two principal products is pressure, dose rate and temperature dependent, (237,240,247–252). Both $G(—C_2H_2)$ and $G(C_6H_6)$ pass through a maximum at ca. 20–30 torr.

Although the pressure dependence for benzene formation can be explained adequately by a mechanism which allows only for the deactivation of intermediates at the wall (247,248), either an excited state or free radical mechanism may be responsible. An ion–molecule mechanism for benzene formation may be rejected in view of the investigation by Futrell and Sieck (249), who established that the yield of benzene is not increased by the addition of large excesses of those rare gases which induce additional ionization in acetylene by charge exchange. Neon was found to be unique among the rare gases in that it alone enhanced the yield of benzene. This increase was correlated with the production of hydrogen atoms during the ionic fragmentation of acetylene which follows charge exchange with neon ions.

While the excited state mechanism also has attractive features, it cannot readily explain the dose rate dependence, nor can it account for the observation that one third of the benzene produced in the radiolysis of equimolar acetylene–acetylene-d_2 mixtures contains odd numbers of deuterium atoms (237). This almost surely points to the participation of a species such as C_2H_3 or C_2H in the mechanism. An example of such a sequence initiated by radicals which would be second order in intermediates might be:

$$C_2H_2 \longrightarrow C_2H + H \tag{89}$$

$$H + C_2H_2 \longrightarrow C_2H_3 \tag{90}$$

$$C_2H_3 + C_2H_2 \longrightarrow C_4H_7 \tag{91}$$

etc., with termininating steps of radical combination and disproportionation eventually leading to the formation of benzene and other products. A similar sequence which is dose rate independent could be initiated by C_2H (249).

The mechanism for the formation of polymer is still uncertain, and excited state, free radical, and ionic mechanisms have been proposed. Although sensitization of polymer formation is observed for all of the rare gases, krypton cannot produce additional ionization in acetylene by charge exchange. This observation casts some doubt on the role of positive ions in the polymerization scheme, and any mechanism which one can propose at this time must be regarded as highly speculative.

4. Cycloalkanes

a. Cyclopropane. This compound is relatively stable towards attack by hydrogen atoms or other radicals (108,253,254), and has therefore been used as a radiolytic source of hydrogen atoms. In the presence of C_2H_3T one observes the formation of tritiated compounds arising from normal radical reactions of species formed by hydrogen atom addition, and also n-propyl radicals and C_3H_5 radicals from direct radiolysis action on cyclopropane (214).

   b. Cyclobutane. The vacuum ultraviolet photolysis has been investigated at several wavelengths (73). Ethylene is by far the most important product of the dissociation of neutral excited molecules, and results from a molecular detachment process since virtually only C_2H_4 and C_2D_4 are found in the photolysis of cyclobutane and perdeuterocyclobutane. Acetylene is the second most important product and is formed with a yield almost an order of magnitude smaller. It is largely but probably not exclusively produced by a molecular detachment process (16% C_2HD). When the energy of the photolysis is sufficiently high (11.6–11.8 eV) that ionization of the molecule may occur (> 10.3 eV), the formation of $C_4H_8{}^+$ with an ion pair yield close to unity in the presence of nitric oxide strongly suggests that neutralization by charge exchange is the only important reaction in the presence of suitable charge acceptors. The absence of ionic fragmentation at 11.6–11.8 eV is also indicated by the virtual absence of propylene, which would be produced by hydride transfer reaction of $C_3H_5{}^+$, the only energetically allowed fragment ion at these energies. It is clear that cyclobutane ion either opens to the butene structure, or undergoes this transformation on encounter with a suitable charge acceptor.

   The mass spectrum of cyclobutane contains a large number of ions in sizable abundance (Table VII). The radiation chemistry reveals evidence for the participation of several of these ions (229) in product formation. The formation and yield of the parent ion are demonstrated by the enhancement of the butene yield in the presence of charge acceptors, while that of the most important fragment $C_3H_5{}^+$ is clearly evident from the formation of C_3D_5H and C_3H_5D in mixtures of deuterated and protonated cyclobutanes in the presence of radical scavengers, arising from a hydride transfer reaction. The absolute value of the ethylene ion yield is unknown, but its formation is demonstrated in mixtures of cyclobutane-d_8 and cyclopentane, where ethane-d_4 becomes an important product, presumably because of the $H_2{}^-$ transfer from C_5H_{10} to $C_2D_4{}^+$.

$$C_2D_4{}^+ + c-C_5H_{10} \rightarrow C_2D_4H_2 + C_5H_8{}^+ \qquad (92)$$

   The formation of C_2H_5D and C_2D_5H in the radiolysis of such scavenged mixtures indicates the presence of ethyl ion, and can be used for a crude quantitative estimate of its yield. A summary of yield estimates is given in Table VII. Stabilization of cyclobutane ion by collision is strongly suggested by these data.

Table VII

Yield of Primary Ions in Cyclobutane Radiolysis at 20 Torr

Ion	Ions/100 eV	
	Yield, radiolysis[a]	Yield, calculated from mass spectrum and W
$C_4H_8^+$	1.79	0.60
$C_4H_7^+$	—	0.21
$C_3H_5^+$	0.55	0.94
$C_3H_4^+$	0.08	0.06
$C_3H_3^+$	—	0.23
$C_2H_5^+$	0.11	0.09
$C_2H_4^+$	0.06	1.06
$C_2H_3^+$	0.25	0.47
$C_2H_2^+$	—	0.27

[a] Reference 253.

The products of the dissociation of the neutral excited molecule in the radiolysis are difficult to elucidate in this system because there is no reliable evidence that any single product is exclusively formed as a result of the decomposition of a neutral excited cyclobutane molecule. If the products obtained in the radiolysis of cyclobutane-d_8–H_2S mixtures are corrected for the neutral products of parent ion dissociation and ion–molecule reactions, the relative yields of the residual products are in good agreement with those observed in the photolysis at 1236 Å (10 eV), and a value of 0.6–0.7 can be estimated for the ratio of chemically consequential excitation processes to ionization (N_{ex}/N_i) (229).

c. Cyclopentane. Studies of the vacuum ultraviolet photochemistry have indicated that propane and propyl radicals are not products of the dissociation of the neutral excited molecule (255). Only limited studies of the radiolysis have been carried out (112,130). On the basis of the hydrogen transfer reaction

$$c\text{-}C_5H_{10}^+ + C_3D_6 \rightarrow C_3D_6H_2 + C_5H_8^+ \tag{93}$$

an approximate yield of $G(C_5H_{10}^+) \simeq 1.5$ ions/100 eV could be estimated, considerably larger than the 0.54 ions/100 eV estimated from W and mass spectral fragmentation pattern. This is almost certainly a result of the inhibition of ionic fragmentation processes. It is of interest to note that some *cis* and *trans* pentene-2 is observed in the presence of nitric oxide or trimethylamine, evidence for ring opening of the cyclopentane ion. The yield of this product, however, levels off at 0.37 ions/100 eV. Unless the charge acceptors are not efficient in neutralizing all ions for which the reaction is exothermic, this would indicate that only a portion of the ions are capable of opening the ring either before or during the charge transfer process.

Propylene ion undergoes both an H_2^- and an H^- transfer reaction yielding propane and propyl radicals. Since neither product is formed in the photolysis, their total yield can be equated with the formation of the precursor ion and $G(C_3H_6^+) \simeq 1.37$ ions/100 eV results, considerably smaller than the value 1.87 estimated from W and the mass spectrum. This is again ascribed to quenching of fragmentation of parent ion (130). No reliable yields of the other intermediate ions are available as yet.

d. Cyclohexane. The important ions in the mass spectrum of cyclo-hexane are the parent ion (14.6%), $C_4H_8^+$ (20.7%), $C_4H_7^+$ (7.1%), and $C_3H_5^+$ (14.0%). The ion–molecule reactions have been investigated briefly by Henglein (256), by Milhaud and Durup (257,258), and by Abramson and Futrell (259). While the parent ion is unreactive towards cyclohexane, the fragment ions readily undergo H^- and H_2^- reactions. These reactions were, however, established before the system was investi-gated in the mass spectrometer (130,265).

The vapor phase radiolysis of cyclohexane has been investigated re-peatedly (136,137,257,258,260–263), at least in part because of the con-siderable relevance of gas phase studies to an understanding of the mechanism of energy transfer in the popular cyclohexane–benzene system (264). The principal products of cyclohexane vapor radiolysis are hydrogen, ethylene, ethane, cyclohexane, dicyclohexyl, C_4 hydrocarbons, propylene, propane, methane, and acetylene, in decreasing order of importance. Alkylcyclohexanes are probably produced as well, and the formation of propylene and butene has been suggested also (265). The effect of a series of additives on product formation is consistent with the suggestion that ethylene and acetylene arise chiefly from dissociation of the neutral excited molecule, while the alkanes with less than 6 carbon atoms probably have fragment ions and free radicals as precursors. Propylene probably arises in part from a neutralization process. It is of considerable interest that the study by Theard (262) provides evidence for the fate of the parent ion upon neutralization. From the effect of added hydrogen iodide and iodine which are capable of attaching electrons it appears that approximately 1.5 hydrogen molecules per 100 eV arise from neutralization, and it is tempting to correlate this yield with the abundance of the parent ion, for which a yield of this magnitude might be expected by analogy with the other cycloalkanes. Hydrogen is also produced by hydrogen abstraction of H atoms and by molecular detachment not involving neutralization. It is thus clear that no single mechanism will suffice to quantitatively account for the behavior of the hydrogen yield as a function of composition in any mixture.

Milhaud and Durup have radiolyzed mixtures of cyclohexane and cyclohexane-d_{12} in the presence and absence of nitric oxide (257,258). Their observations indicate that about one-half the ethane and nearly 90% of the propane arise from free radical processes, as suggested by Theard. The remainder of the ethane is produced almost exclusively by hydride transfer, while the rest of the propane yield is probably better ascribed to H_2^- transfer. Acetylene consisted of almost 15% C_2HD, ethylene contained about a similar percentage C_2H_3D, and propylene contained 64% $C_3H_3D_3$ in the absence of nitric oxide but less than 6% in its presence. On the basis of these results Milhaud and Durup derive the yields of intermediates listed in Table VIII, and one can estimate $N_{ex}/N_i = 0.3$ (101,265). This is, however, somewhat uncertain since the yield of neutral fragments arising from dissociation of excited parent ions cannot be estimated accurately; moreover, there are still discrepancies in the reported yields.

The effect of pressure on the gamma-radiolysis of cyclohexane has been

Table VIII

Probable Yields of Primary Species in the Radiolysis of
Gaseous Cyclohexane–Cyclohexane-d_{12} Mixtures

	Yield, species/100 eV	
	Radiolysis	Estimated from W and mass spectra
$C_2H_5^{+}$[a]	0.020	0.082
$C_3H_7^{+}$	0.038	0.114
$C_3H_6^{+}$	~ 0.012	0.25
CH_3	0.61	
C_2H_5	0.29	
C_3H_7	0.48	
C_2H_3	0.24	
C_2H	0.6	
C_2H_4	1.0[a]–1.7[b]	
C_2H_2	0.2[a]–0.35[b]	

[a] Reference 258 yields listed include those of the perdeuterated analog.
[b] Reference 262; no cyclohexane-d_{12} present.

investigated at 300°C by Jones (263), covering a density range of 0.008–0.42 g cc^{-1}. This study spanned the critical temperature range as well for mixtures of cyclohexane and benzene, demonstrating a dramatic effect of phase on hydrogen yield. Over the range investigated yields of the C_1–C_3 hydrocarbons decrease by about an order of magnitude while those of hydrogen and cyclohexene are reduced only slightly if at all. The formation of dicyclohexyl, however, is increased. These observations are consistent with a major quenching of dissociation of excited ionic species which cannot undergo fragmentation at the higher pressures.

e. Methylcyclohexane. The radiolysis of methylcyclohexane has recently been investigated by Holtslander and Freeman (110,266,267). From the effect of various scavengers it was concluded that about 50% of the ions yield hydrogen on neutralization. Other important products include methylcyclohexene isomers ($G = 2.0$), whose yield is increased in the presence of ethylene or ammonia, presumably because of protection against radical and ion attack. Ethylene, methane, propylene, and probably cyclohexane constitute the remainder of products formed with yields in excess of 0.5 molecules/100 eV, and their yields can be rationalized in terms of free radical and ionic reactions and molecular detachment processes.

B. Compounds of C, H, and O

1. Alcohols

a. Methanol. As discussed in Chapter 3, the dissociation of excited methanol occurs mainly by way of processes 94 and 95:

$$CH_3OH^* \rightarrow CH_2O + H_2 \tag{94}$$

$$CH_3OH^* \rightarrow CH_3O + H \tag{95}$$

The most important ions in the mass spectrum of methanol are the parent ion (30%), CH_2OH^+ (43%), and CHO^+ (18%). All these ions react rapidly with methanol to form the protonated alcohol, which solvates further at higher pressures in the ion source (55,150,215,268–271). CH_2OH and CH_3O are to be expected as neutral counterparts of the ion–molecule reactions since proton and hydrogen atom transfer in the ion–molecule reaction

$$CH_3OH^+ + CH_3OH \rightarrow CH_3OH_2^+ + (CH_3O \text{ or } CH_2OH) \tag{96}$$

occur with equal probability from the hydroxyl or the methyl group (150,271).

The radiolysis of methanol vapor has been investigated only sparingly, (272,273). Products measured to date include hydrogen ($G \simeq 11$ molecules/

100 eV), carbon monoxide ($G \simeq 0.8$–0.2), and methane ($\simeq 0.3$ molecules/ 100 eV). Ethylene glycol is produced as well, and formaldehyde is probably also one of the products. The addition of electron scavengers such as methyl bromide and carbon tetrachloride reduces the hydrogen yield to ca. 7 molecules/100 eV, possibly because the formation of negative ions by electron capture would inhibit the neutralization reaction:

$$H^+(CH_3OH)_x + e^- \rightarrow H + x\,CH_3OH \tag{97}$$

Recent results by Warman (109) suggest that O^- resulting from electron capture by nitrous oxide leads to further reaction with alcohols, possibly accounting for the inability of N_2O to inhibit hydrogen atom formation (273) by step (97).

The addition of ethylene, propylene or benzene reduces the hydrogen yield to $G(H_2) = 2.1$ molecules/100 eV. The relative efficiencies of the additives indicate that hydrogen atom scavenging is the only mode by which all three scavengers reduce the hydrogen yield. Little can be said as yet about the steps leading to the formation of the other intermediates except what one might speculate from the photochemistry and mass spectra. The results of such attempts at rationalization do give fair agreement with the experimentally observed product distribution (274).

b. Ethanol. The photochemistry of this compound has been investigated only qualitatively (275), and there is not as yet a clear choice of processes. In the mass spectrum, $C_2H_4OH^+$ (16%), $C_2CH_2OH^+$ (41%), $C_2H_3^+$ (9%), $C_2H_3OH^+$ (69%) and CHO^+ (5%) are important ions (271). The ion–molecule reactions have been investigated also (55,215,269–271), and the major reactions are either proton transfer with or without subsequent dissociation, and hydride transfer. At high pressures, solvation is again observed.

Ethanol radiolysis has been investigated by Freeman et al. (276–278) and by Sieck and Johnsen (279). Major product yields and the effect of certain scavengers on them are summarized in Table IX. An appreciable portion of the hydrogen yield is inhibited by olefins and nitric oxide. It seems possible to account for all or at least part of the yields of acetaldehyde, ethylene, and water and some of the carbon monoxide on the basis of proton transfer reactions, while hydrogen atoms probably arise partially from neutralization of the solvated proton. Methane arises principally from the reactions of methyl radicals, and rate constant ratios for abstraction from the hydroxyl and the α-hydrogen could be calculated using specifically labeled ethanol (278). The higher yield of methane in the work of Freeman et al. is consistent with an activation energy for the abstraction reaction of methyl radicals, and a yield of $G(CH_3) = 2.4$

Table IX

Yield of Major Products and Effect of Scavengers on Ethanol, Radiolysis

	Yield, molecules/100 eV absorbed in ethanol					
	Pure compound			1-3% NO added, 25°C	1-3% propylene, 25°C	20% 1,3-pentadiene, 105°C
Product	105°C[a]	25°C[a]	108°C[c]			
H_2	7.5	11.0	7.6	3.5	3.7	1.8
CH_4	2.3	0.9	1.7	0.3_2		0.3
C_2H_2	0.09	0.3	0.03	0.3		0.3_5
C_2H_4	1.2	1.6	0.7	1.6		1.1
C_2H_6	0.2	0.6_5	0.2	0.0_1		0.0_1
CH_3CHO	3.5	4.2	4.5	0.0		1.8
$C_3H_6(OH)_2$ (1,2)	0.9	0.1_6	0.15	0.1_6		0.0
$C_4H_8(OH)_2$ (2,3)	3.1	1.2	1.2	0.9–1.2		0.0
CO	0.6	1.2_4	1.1	0.7_1		0.3_6
CO_2		0.1				
CH_3OH		0.05		0.3_5		
$C_3 + C_4$		0.12		0.0		

[a] Reference 278; Co^{60} gamma rays.
[b] Reference 279; 2 MeV electrons.
[c] Reference 278; Po-α particles.

radicals/100 eV can be estimated from that investigation, while at lower temperatures a corresponding increase in $G(C_2H_6)$ leads to an estimate of 1.7 CH_3/100 eV, in fair agreement with the work at higher temperatures.

2. Ethers

Only diethylether has been investigated to any extent (273,280,281). Important products include hydrogen ($G = 5.9$ molecules/100 eV) 2,3-diethoxybutane ($G = 2.21$), ethylene ($G = 1.8$), acetaldehyde ($G = 1.4$), methane ($G = 1.5$), ethyl-isopropyl ether ($G = 1.1$) as well as a series of other products such as ethane, propane, ethanol, ethyl-*sec*-butyl ether, ethyl-vinyl ether and ethyl-methyl-acetal in yields between 0.2 and 0.8 molecules/100 eV. The addition of pentadiene reduces the formation of hydrogen to 1.64 molecules/100 eV. There is a considerable isotope effect reducing the yield of molecular and atomic hydrogen when the ether is progressively labeled. The molecular hydrogen is chiefly produced by elimination from the methyl group. About 20% of the methane formation appears to occur via methyl radical reactions, presuming the efficiency of pentadiene as a free radical scavenger. The largest portion of the methyl radicals are produced by simple C—C cleavage in the ethyl group. The alpha position is more reactive just as in the case of ethanol. Ethylene, which is almost entirely due to a molecular elimination process, arises from one ethyl group, with α cleavage contributing slightly less than β cleavage.

Product yields show a considerable temperature dependence, particularly above ca. 150°C where a chain reaction leading to ethane, acetaldehyde, n-butane, sec butyl ether and 2,3-diethoxybutane sets in. This can be quantitatively ascribed to a free radical mechanism of the Rice-Herzfeld type (282). The hydrogen yield decreases with temperature while that of ethanol increases by about the same amount, ($G \simeq 1.5$ molecules/100 eV between 30 and 150–200°C). The sum of these yields remains unchanged over the entire temperature range, and this may indicate that a precursor of hydrogen atom formation becomes a precursor of ethanol formation, but any conclusions based on superficial numerology are uncertain at best.

3. Ketones

a. Acetone. While the photochemistry of acetone and other ketones has been investigated fairly extensively (see Chapter 3), little information is available about the radiolysis. It is well known that in the photolysis the dominant dissociation process produces an excited acetyl radical capable of further dissociation:

$$CH_3COCH_3^* \rightarrow CH_3CO^* + CH_3 \tag{97}$$

$$CH_3CO^* \rightarrow CH_3 + CO \tag{98}$$

which can be quenched by collisions at low temperature. A dissociation process of hydrogen atom or molecular hydrogen elimination from one methyl group may begin to contribute at very short wavelengths (283).

The mass spectrum of acetone is dominated by the parent ion (21%) and CH_3CO^+ (56%). Only limited investigations of the ion–molecule reactions have been made (55,56). Again, proton transfer is an important mode of reaction of parent and fragment ions.

Major products in the radiolysis are CO, CH_4, H_2, C_2H_4 and C_2H_6 (284,285). Biacetyl was not observed as a product, and the yield of carbon monoxide was unaffected by temperature up to 153°C and by the presence of scavengers. This indicates that the acetyl radical formed in the first stage of events is excited and therefore dissociates quantitatively by reaction (96). At low temperatures methane and ethane are about 95% removed by free radical scavengers (nitric oxide or ethylene), evidence for their formation by radical processes, which include abstraction from acetone by methyl radicals. In order to account for the reduction in hydrogen yield on addition of scavengers, and the generally high yield of hydrogen, a primary dissociation process leading to hydrogen atom formation was suggested, but it is equally or more plausible that neutralization of protonated acetone is a major source of H atoms (286.)

b. 2 and 3 Pentanone. Scala and Ausloos (286) have investigated both the vacuum UV photolysis and the radiolysis of these compounds, and suggest that the neutral excited pentanone molecules decompose according to the processes given in Chapter 3, Section III-A-2. The observation of 1-butene and allene in significant amounts only in the radiolysis indicates that these are produced by ion–molecule reactions. The conclusion is supported by the observation that the yields of these products are not enhanced by the application of electrostatic fields. Their origin is probably to be found in proton transfer reactions, since

$$C_2H_5^+ + CH_3COCH_3 \rightarrow C_2H_4 + C_3H_7O^+ \tag{99}$$

has been suggested by Sandoval and Ausloos (131). Proton transfer has been shown to be the principal mode of formation of secondary ions in the mass spectrometer (287). Although plausible mechanisms can be offered for the formation of most products, their simultaneous formation from ionic intermediates and neutral excited states precludes a detailed assessment of relative yields of contributing species at this time.

An interesting aspect of this investigation is the analysis of the nature of the products arising from excitation by slow electrons in the presence of applied electrostatic fields. Scala and Ausloos show convincingly that the low energy electrons excite mostly the first upper singlet or triplet state: only the yields of products typical of decomposition of the less excited states are enhanced. Thus the applied field technique yields products more analogous to the long wavelength photolysis than to radiation chemistry or vacuum UV photochemistry (see Fig. 2 in Chapter 3).

c. 2,3 Butanedione (Diacetyl). Radiolysis of this simple diketone appears to lead to product formation chiefly by free radical reactions (288). The major products CO ($G = 7.4$), C_2H_6 ($G = 2.$), acetone ($G = 1.1$) ketene and hydrogen ($G = 0.5$ ea.) are essentially dose, dose rate and pressure independent at 25°C. The rate of hydrogen and ethane formation are essentially temperature independent as well, but the methane yield increases with temperature. The addition of iodine reduces methane by 80%, hydrogen by 20%, and eliminates ethane completely. Only two ion–molecule reactions were reported, one leading to an ion of mass 129, presumably $(CH_3CO)_3^+$, while the other

$$CH_3^+ + (CH_3CO)_2 \rightarrow CH_2CO + (CH_3)_2COH^+ \tag{100}$$

may be held in part responsible for ketone and acetone formation, as well as for a portion of the hydrogen atom yield after neutralization of $(CH_3)_2CO \cdot H^+$. Dissociation of the neutral excited molecule is suggested to yield chiefly two acetyl radicals, one of which may be capable of further dissociation into CH_3 and CO.

4. Esters

a. Methylacetate. The radiolysis of CH_3COOCD_3 has been reported by Ausloos and Trumbore (289). Major volatile products include hydrogen, methane, ethane, and carbon monoxide. Formation of methane and ethane are essentially inhibited by the addition of iodine, while hydrogen becomes all H_2 and D_2 in the presence of the scavengers. Dimethylether is a minor product quenched by iodine. Results were interpreted on the basis of free radical reactions alone, but this is no indication that ionic processes are not important contributors to the mechanism.

C. Compounds Containing Heteroatoms

1. Halocarbons

a. Chlorides and Bromides. The radiation chemistry of such compounds has been investigated by Willard and his associates (290–292), and by Schindler (293).

Alkyl chlorides undergo isomerization which is catalyzed by HCl (292). The following sequence is presumably responsible, using propylchloride as an example.

$$n\text{-}C_3H_7Cl + HCl \rightarrow CH_3CHCH_2Cl + HCl \qquad (101)$$

$$CH_3CHCH_2Cl \rightarrow CH_3CHClCH_2 \qquad (102)$$

$$CH_3CHClCH_2 + HCl \rightarrow i\text{-}C_3H_7Cl + Cl \qquad (103)$$

$$R + C_3H_6Cl \rightarrow RC_3H_6Cl \qquad (104)$$

Since the chain can also be initiated photochemically at 2537 Å, the initial formation of free radicals need not involve ionic precursors, although their participation is certainly possible.

The radiolysis of gaseous CCl_3Br at 108°C (291) leads to the formation of CCl_4 ($G = 7.5$ molecules/100 eV), CCl_2Br_2 ($G = 6.1$ molecules/100 eV), $CClBr_3$ ($G = 0.82$ molecules/100 eV), C_2Cl_6 ($G = 1.2$ molecules/100 eV) and bromine ($G = 1.2$ molecules/100 eV). The formation of C_2 compounds containing bromine can be excluded. Addition of bromine reduces the formation of CCl_4 to 5.1 molecules/100 eV, also decreases that of the other C_1 compounds, and completely eliminates the formation of hexachloroethane. The results indicate that C—Cl bond rupture occurs to a minor extent only, if at all. The interpretation is essentially based only on the participation of free radicals, and important steps include

$$CCl_3 + CCl_3Br \rightarrow CCl_4 + CCl_2Br \qquad (105)$$

$$CCl_3 + CCl_3Br \rightarrow C_2Cl_6 + Br \qquad (106)$$

$$CCl_3 + Br_2 \rightarrow CCl_3Br + Br \qquad (107)$$

and radical recombination reactions. CCl_2Br may participate by another metathetical reaction to account eventually for the formation of chloro-tribromoethane.

b. Iodides. Gevantman and Williams (294) demonstrated that the vapor phase radiolysis of the C_1 to C_4 alkyl iodides leads principally to rupture of the carbon-iodine bond, producing the alkyl radical. For the higher iodides some fragmentation of the C—C bond was observed as well. Only minor amounts of isomerization of the carbon chain were observed in the radical. Later experiments suggested (295,296) that molecular detachment of HI

$$C_2H_5I \rightarrow C_2H_4 + HI \qquad (108)$$

needed to be included in the mechanism. This process has also been suggested to occur in the photochemistry of alkyl iodides.

c. Fluorocarbons. The ion–molecule reactions of tetrafluoroethylene have been investigated in the mass spectrometer up to pressures where higher order reactions participate (297). The dissociation of the intermediate complex, if there is one, bears little resemblance to the fragmentation pattern of the dimer. Reactions of perchloroethylene ion and perchloroacetylene ion lead only to $C_3F_5{}^+$ and $C_3F_3{}^+$, respectively, with normal reaction cross section, while perchlorovinyl ion leads to three different product ions with low efficiency.

Radiolysis of perfluoromethane (298,299) leads to the formation of perfluoromethyl radicals. These can combine to yield perfluoroethane, a reaction which is easily inhibited by the addition of oxygen and, presumably, free fluorine. Fluorine atoms cannot abstract from CF_4, and therefore also lead to a rapid removal of CF_3. In the presence of small amounts of oxygen, the ether CF_3OCF_3 becomes the predominant product.

The radiolysis of hexafluoroethane has been investigated by Kevan and Hamlet (300), who point out that one of the interesting aspects of fluorocarbon radiolysis arises from the entirely different behavior of fluoroalkyl radicals, which, for example, do not disproportionate. The scavengable yields of fluoromethane and the fluoroalkanes are ascribed to simple rupture of C—C and C—F bonds and recombination of the radicals, including reformation of fluoroethane by reaction of C_2F_5 and F. The nonscavengeable yield of fluoromethane is tentatively ascribed to an ionic condensation reaction of fluoroethyl ion or to a fluoride transfer from fluoroethane to CF_3^+. The yield of cyclopropane-F_6, which is actually increased by addition of oxygen, was also ascribed to an ion molecule reaction although alternate mechanisms of formation cannot be excluded.

D. Compounds of C, H, and N

1. Amines. Both the photochemistry (17) and mass spectrometry (56) have been investigated. Few radiation chemical studies have appeared (301,302); methyl, dimethyl-, and trimethylamine (302) yield hydrogen and methane, and as expected the formation of the former decreases while that of the latter increases with methyl substitution. Ammonia is also a product. 81% of the hydrogen formation from CH_3NH_2 and 41% of that from $(CH_3)_3N$ can be inhibited by the addition of ethylene.

2. Azomethane. Ausloos and co-workers (131,303) have investigated the radiolysis of this compound, which is well known as a source of methyl radicals in photochemistry (17). Chief products are nitrogen, ethane, methane, and hydrogen, in decreasing order of importance. The addition of oxygen effectively inhibits the formation of ethane and reduces that of methane by a factor of ca. 8. The sum of the yields of ethane and methane relative to that of nitrogen decreases considerably between 2 and 200 torr azomethane. Increase in temperature decreases ethane but increases methane formation. The main features of the results can be explained by fragmentation of the neutral excited molecule into free radicals:

$$CH_3N_2CH_3^* \rightarrow 2CH_3 + N_2 \tag{109}$$

with subsequent abstraction and combination of methyl radicals to form methane and ethane, respectively. However, ionic or molecular detachment processes of an unspecified nature must be invoked as well to account for the unscavengeable yield of hydrogen.

VI. RADIOLYSIS OF MIXTURES

Addition of specific reactants to a gas under investigation to diagnose the mechanism of product formation is the chief use of mixture radiolysis. This often allows one to estimate the yields of the intermediate species by their conversion into stable products which can be analyzed at the end of irradiation.

Rare gas hydrocarbon mixtures have long received special attention (1,114). An extensive summary of individual rare gas organic systems has recently been given by von Bunau (117). A number of studies have appeared since then concerning the sensitized radiolysis of methane (304,305), ethane (306), propane (63,306), ethylene (307), fluoroethane and fluoropropane (167,308,309). Although reactions are occasionally discussed simply in terms of a general energy transfer efficiency, it is necessary to make allowance for the specific steps which lead to product formation. The radiolysis of pure rare gases (see Chapter 5) leads to the population of resonance levels and forbidden levels, to rare gas excited dimers, to dimeric rare gas ions, and to monoatomic ions in a variety of excited states. At sufficiently high pressures all ions can be converted to the dimers, while no multiple ionized species will survive because of charge exchange. Addition of only trace amounts of organic additives may lead to energy transfer from all these states and reactions may, therefore, be initiated at a series of energies deposited in the molecule. Moreover, complex formation may set in leading to uncharged or charged adducts of the reactant with the rare gas intermediate (309–311). At high concentrations of the organic molecule, its greater polarizability can interfere with the formation of the rare gas dimer, as evidenced by the negative Jesse effect (312). In spite of these complications it is occasionally possible to emphasize the contribution from one particular intermediate ion (63, 114), and this can be helpful in the evaluation of the overall mechanism of radiolysis.

Mixtures consisting of two or more organic compounds have also been investigated extensively. In such systems charge transfer may be anticipated to occur from one parent ion to another if there are no alternative channels of reaction. At the present time there is, however, little or no conclusive evidence substantiating the occurrence of charge transfer from one parent organic ion to a neutral organic compound in the radiolysis. Such charge transfer processes have, however, been clearly indicated in mass spectrometric studies, especially by Lindholm (113). However, in such studies the charge donor has excess kinetic energy and such information may, therefore, not be applicable to a static system where the positive ions will have thermal energy. The major difficulty in obtaining information on charge transfer reactions in the radiolysis of organic systems is the difficulty of associating a particular product with the reacting parent ion in question. Furthermore, fragmentation of organic ions, as noted earlier in this chapter, is usually extensive so that it is difficult to isolate the rather minor charge transfer reactions which may occur from the more extensive chemical effects resulting from reactions of other species. Recently, however, an attempt (313) has been made to establish the occurrence of charge transfer processes in the photoionization of cycloalkanes where fragmentation of the parent ion is negligible. It has in particular been shown that when the difference in ionization potentials of the two reacting species is less than 0.1 eV, charge transfer can occur in either direction, e.g.:

$$C_6D_{12}{}^+ + C_6H_{12} \rightleftarrows C_6D_{12} + C_6H_{12}{}^+ \tag{110}$$

For reacting species with ionization energies which differ in magnitude by more than 0.1 eV only the exothermic charge transfer reaction is shown to be efficient:

$$C_5D_{12}{}^+ + C_6H_{12} \rightarrow C_5D_{12} + C_6H_{12}{}^+ \quad \Delta H \sim {}^-0.6 \text{ eV} \tag{111}$$

Useful information concerning free radicals and ion–molecule kinetics can be derived from radiolysis of mixtures. A compound which is known to produce a given intermediate, such as the hydrogen atom, is irradiated in the presence of additives and the reactions of the intermediate with various other compounds can be studied. This approach has been successful in some instances. For example, hydrogen atom production in the radiolysis of propane has been used to evaluate the rate constants for hydrogen atom addition to olefins (69), and the results were in good agreement with those obtained by other methods where comparison was possible.

Hydrogen atom production has also been achieved in the hydrogen-sensitized radiolysis (230) of simple compounds, and by the rare gas-sensitized hydrogen radiolysis (226,251), which relies on the reaction

$$Ar^+ + H_2 \rightarrow ArH^+ + H \tag{112}$$

with neutralization of the ArH^+ also yielding hydrogen atoms. Complications in both approaches are, of course, the possibility that excited states or the protonated species (ArH^+ or $H_3{}^+$ in pure hydrogen) may react with the additive as well, since the lifetime of the ion is invariably sufficient to allow encounter with the reactant. Transfer of a proton from $H_3{}^+$ or ArH^+ to trace amounts of hydrocarbons has been clearly established (104,230).

Similarly, the relative rates of reaction of a carbonium ion with various compounds have been measured (129) by irradiating a hydrocarbon which forms a given fragment ion in the presence of added foreign compounds.

REFERENCES

1. S. C. Lind, *The Chemical Effects of Alpha Particles and Electrons*, 1st ed., The Chemical Catalog Publishing Co., New York, 1921; *Radiation Chemistry of Gases*, Reinhold, New York, 1961.
2. W. Mund, *Actions Chim. Biol. Radiations*, Paris, **2** (1956).
3. H. Eyring, J. O. Hirschfelder, and H. S. Taylor, *J. Chem. Phys.*, **4**, 479, 570 (1936).

4. V. L. Tal'roze, A. K. Lyubimova, *Dokl. Akad. Nauk SSSR*, **86**, 909 (1952).
5. D. O. Schlissler and D. P. Stevenson, *J. Chem. Phys.*, **24**, 926 (1956).
6. J. L. Franklin, F. H. Field, and F. W. Lampe, *J. Am. Chem. Soc.*, **78**, 5697 (1956).
7. R. L. Platzman, *Radiation Res.*, **17**, 419 (1962).
8. W. P. Jesse and R. L. Platzman, *Nature*, **195**, 790 (1962).
9. J. C. Y. Chen and J. L. Magee, *J. Chem. Phys.*, **36**, 1407 (1962).
10. D. F. Evans, *J. Chem. Soc. (London)*, **1960**, 1735.
11. K. J. Laidler, *The Chemical Kinetics of Excited States*, Clarendon Press, Oxford, 1955, p. 95.
12. R. B. Cundall, in *Progress in Reaction Kinetics*, Vol. II, C. Porter, Ed., Pergamon, New York, 1964.
13. B. H. Mahan and R. Mandal, *J. Chem. Phys.*, **37**, 207 (1962).
14. P. J. Ausloos and S. G. Lias, *J. Chem. Phys.*, **38**, 2207 (1963).
15. H. Okabe and J. R. McNesby, *J. Chem. Phys.*, **34**, 668 (1961).
16. R. P. Porter and W. A. Noyes, Jr., *J. Am. Chem. Soc.*, **81**, 2307 (1959).
17. J. G. Calvert and J. N. Pitts, Jr., *Photochemistry*, Wiley, New York, 1966.
18. R. L. Platzman, *Vortex*, **23**, 8 (1962).
19. G. L. Weissler, in *Handbuch der Physik*, 2nd ed., Vol. 21, Springer, Berlin, 1956, p. 304.
20. R. I. Schoen, *J. Chem. Phys.*, **40**, 1830 (1964).
21. W. P. Jesse, *J. Chem. Phys.*, **41**, 2060 (1964).
22. P. Ausloos and S. G. Lias, *J. Chem. Phys.*, **45**, 524 (1966).
23. R. Gorden, Jr. and P. Ausloos, *J. Chem. Phys.*, **47**, 1799 (1967).
24. J. C. Person, *J. Chem. Phys.*, **43**, 2553 (1965).
25. S. Meyerson, *J. Chem. Phys.*, **39**, 1445 (1963).
26. G. S. Hurst and T. E. Bortner, *J. Chem. Phys.*, **42**, 713 (1965).
27. A. A. Scala and P. Ausloos, *J. Chem. Phys.*, **45**, 847 (1966).
28. G. G. Meisels, *Nature*, **206**, 287 (1965).
29. H. M. Rosenstock, M. B. Wallenstein, A. L. Wahrhaftig, and H. Eyring, *Proc. Natl. Acad. Sci. (U.S.)*, **38**, 667 (1952).
30. H. M. Rosenstock and M. Krauss, in *Mass Spectrometry of Organic Ions*, F. W. McLafferty, Ed., Academic Press, New York, 1963.
31. D. P. Stevenson, *Radiation Res.*, **10**, 610 (1959).
32. W. A. Chupka and M. Kaminsky, *J. Chem. Phys.*, **35**, No. 6, 1991 (1961).
33. J. T. Tate, P. T. Smith, and A. L. Vaughan, *Phys. Rev.*, **48**, 525 (1935).
34. A. Adamczyk, A. J. H. Boerboom, B. L. Schram, and J. Kistemaker, *J. Chem. Phys.*, **44**, 4640 (1966).
35. J.-C. Abbe and J.-P. Adloff, *Phys. Rev. Letters*, **11**, 28 (1964).
36. P. Kebarle and E. W. Godbole, *J. Chem. Phys.*, **36**, 302 (1962).
37. M. Vestal, A. L. Wahrhaftig, and W. H. Johnston, ARL 62-426, Theoretical Studies in Basic Radiation Chemistry. Report to the Office of Aerospace Research, USAF, Sept. 1962.
38. P. Ausloos, S. G. Lias, and I. B. Sandoval, *Discussions Faraday Soc.*, **36**, 66 (1963).
39. P. Ausloos, R. Gorden, Jr., and S. G. Lias, *J. Chem. Phys.*, **40**, 1854 (1964).
40. G. G. Meisels, *Advan. Chem. Ser.*, **58**, 243 (1966).
41. J. Bednar, *Collections Czech. Chem. Commun.*, **30**, 1328 (1965).
42. C. E. Klots, *J. Chem. Phys.*, **44**, 2715 (1966).
43. C. E. Klots, *J. Chem. Phys.*, **39**, 1471 (1963).

44. F. W. Lampe, J. L. Franklin, and F. H. Field, *J. Am. Chem. Soc.*, **79**, 6129 (1957).
45. G. G. Meisels, *J. Chem. Phys.*, **41**, 51 (1964).
46. R. A. Back, *J. Phys. Chem.*, **64**, 124 (1960).
47. P. Ausloos, S. G. Lias, and R. Gorden, Jr., *J. Chem. Phys.*, **39**, 3341 (1963).
48. J. A. Hornbeck and J. P. Molnar, *Phys. Rev.*, **84**, 621 (1951).
49. I. Koyana, I. Tanaka, and I. Omura, *J. Chem. Phys.*, **40**, 2734 (1964).
50. M. S. B. Munson, *J. Chem. Phys.*, **69**, 572 (1965).
51. S. Wexler and N. Jesse, *J. Am. Chem. Soc.*, **84**, 3425 (1962).
52. P. Kebarle and E. W. Godbole, *J. Chem. Phys.*, **39**, 1131 (1963).
53. P. Kebarle and A. M. Hogg, *J. Chem. Phys.*, **42**, 668, 798 (1965).
54. F. H. Field and M. S. B. Munson, *J. Am. Chem. Soc.*, **87**, 3829 (1965).
55. M. S. B. Munson, *J. Am. Chem. Soc.*, **87**, 2332 and 5313 (1965).
56. M. S. B. Munson, *J. Phys. Chem.*, **70**, 2034 (1966).
57. G. G. Meisels, *J. Chem. Phys.*, **42**, 2328 (1965).
58. P. Kebarle, R. M. Haynes, and S. Searles, *Advan. Chem.*, **58**, 210 (1966).
59. H. S. W. Massey and E. H. S. Burhop, *Electronic and Ionic Impact Phenomena*, Oxford Univ. Press, New York, 1952.
60. L. G. Christophorou, R. N. Compton, G. S. Hurst, and P. W. Reinhardt, *J. Chem. Phys.*, **45**, 536 (1966); T. L. Cottrell and I. C. Walker, *Trans. Faraday Soc.*, **63**, 549 (1967).
61. W. E. Wentworth and E. Chen, *J. Gas Chromatog.*, **5**, 170 (1967).
62. T. W. Woodward and R. A. Back, *Can. J. Chem.*, **41**, 1463 (1963).
63. L. I. Bone, L. W. Sieck, and J. H. Futrell, *J. Chem. Phys.*, **44**, 3667 (1966).
64. C. E. Klots and V. E. Anderson, *J. Phys. Chem.*, **71**, 265 (1967).
65. B. H. Mahan and J. C. Person, *J. Chem. Phys.*, **40**, 392 (1964).
66. R. H. Lawrence and R. F. Firestone, *Advan. Chem.*, **58**, 278 (1966).
67. J. A. Kerr and A. F. Trotman-Dickenson, "The Reactions of Alkyl Radicals," in *Progress in Reaction Kinetics*, Vol. I, G. Porter, Ed. Pergamon Press, New York, 1961, p. 107.
68. S. W. Benson and W. B. DeMore, *Ann. Rev. Phys. Chem.*, **16**, 397 (1965).
69. K. Yang, *J. Am. Chem. Soc.*, **84**, 3795 (1962).
70. R. J. Cvetanovic and R. S. Irwin, *J. Chem. Phys.*, **46**, 1694 (1967).
71. C. E. Klots, *J. Chem. Phys.*, **41**, 117 (1964).
72. G. G. Meisels, *J. Chem. Phys.*, **42**, 3237 (1965).
73. R. D. Doepker and P. Ausloos, *J. Chem. Phys.*, **43**, 3814 (1965).
74. R. L. Platzman, *Intern. J. Appl. Radiation Isotopes*, **10**, 116 (1961).
75. D. W. Turner and D. P. May, *J. Chem. Phys.*, **45**, 471 (1966).
76. S. M. Silverman and Edwin N. Lassettre, *J. Chem. Phys.*, **42**, 3420 (1965).
77. J. A. Simpson, S. R. Mielczarek, and J. Cooper, *J. Opt. Soc. Am.*, **54**, 269 (1964).
78. R. P. Madden and K. Codling, *Phys. Rev. Letters*, **10**, 516 (1963).
79. (a) A. Kuppermann and L. M. Raff, *Discussions Faraday Soc.*, **35**, 30 (1963); (b) *ibid.*, *J. Chem. Phys.*, **39**, 1607 (1963).
80. J. A. Simpson and S. R. Mielczarek, *J. Chem. Phys.*, **39**, 1606 (1963).
81. C. R. Bowman and W. D. Miller, *J. Chem. Phys.*, **42**, 681 (1965).
82. V. H. Dibeler, M. Krauss, R. M. Reese, and F. N. Harlee, *J. Chem. Phys.*, **42**, 3791 (1965).
83. B. Steiner, C. F. Giese, and M. G. Inghram, *J. Chem. Phys.*, **34**, 189 (1961).
84. (a) C. E. Melton and P. S. Rudolph, *J. Chem. Phys.*, **47**, 1771 (1967); (b) P. S. Rudolph and C. E. Melton, *J. Phys. Chem.*, **71**, 4572 (1967).

85. C. E. Melton, in *Mass Spectrometry of Organic Ions*, F. W. McLafferty, Ed., Academic Press, New York, 1963, p. 163.
86. A. Henglein and G. A. Muccini, *J. Chem. Phys.*, **31**, 1426 (1959).
87. J. R. McNesby and H. Okabe, in *Advances in Photochemistry*, Vol. 3, W. A. Noyes, G. S. Hammond and J. N. Pitts, Eds., Interscience, (1964), p. 157.
88. A. H. Laufer and J. R. McNesby, *J. Chem. Phys.*, **42**, 3329 (1965).
89. D. C. Walker and R. A. Back, *J. Chem. Phys.*, **38**, 1526 (1963).
90. R. Gorden, Jr., R. D. Doepker, and P. Ausloos, *J. Chem. Phys.*, **44**, 3733 (1966).
91. P. Colombo, L. E. Kukacka, J. Fontana, R. N. Chapman, and M. Steinberg, *J. Polymer Sci.*, **4**, 29 (1966).
92. (a) M. C. Sauer, Jr., and L. M. Dorfman, *J. Am. Chem. Soc.*, **86**, 4218 (1964); (b) *ibid.*, **87**, 3801 (1965).
93. P. Harteck and S. Dondes, *Nucleonics*, **15**, (8), 94 (1957).
94. M. Steinberg, M. Loffelholz, and J. Pruzansky, *Industrial Uses of Large Radiation Sources*, Vol. II, International Atomic Energy Agency, Vienna, 1963.
95. J. W. Sutherland, J. Pruzansky, and M. Steinberg, *Trans. Am. Nuclear Soc.*, **7**, 311 (1964).
96. W. D. Tucker, G. Farber, and M. Steinberg, *Trans. Am. Nuclear Soc.*, **7**, 316 (1964).
97. M. Steinberg, *Chem. Eng. Progr.*, **62**, 105 (1966).
98. R. C. Axtmann and J. T. Sears, *Nucl. Sci. Eng.*, **23**, 299 (1965).
99. R. C. Axtmann and J. T. Sears, *J. Chem. Phys.*, **44**, 3279 (1966).
100. P. M. Mulas and R. C. Axtmann, *Phys. Rev.*, **146**, 296 (1966).
101. P. Ausloos and S. G. Lias, "Gas Phase Radiolysis of Hydrocarbons," in *Actions Chimiques et Biologiques des Radiations*, M. Haissinsky, Ed., Masson et Cie., Paris, 1967, Ch. 5.
102. P. Ausloos and R. Gorden, Jr., *J. Chem. Phys.*, **41**, 1278 (1964).
103. C. W. Farley and W. A. Lloyd, *Trans. Am. Nuclear Soc.*, **5**, 211 (1962).
104. C. N. Hinshelwood, *Kinetics of Chemical Change*, Clarendon Press, Oxford, 1940, Ch. 1–4.
105. G. R. A. Johnson and J. M. Warman, *Trans. Faraday Soc.*, **61**, 512 (1965).
106. J. M. Warman, *J. Phys. Chem.*, **71**, 4066 (1967).
107. J. M. Warman, *J. Phys. Chem.*, **72**, 52 (1968).
108. P. L. Gant and K. Yang, *J. Chem. Phys.*, **32**, 1757 (1960).
109. J. M. Warman, *Nature*, **213**, 381 (1967).
110. G. R. Freeman and W. J. Holtslander, *Chem. Commun.*, in press.
111. G. G. Meisels, unpublished results.
112. P. Ausloos and S. G. Lias, *J. Chem. Phys.*, **43**, 127 (1965).
113. E. Lindholm, *Advan. Chem. Ser.*, **58**, 1 (1966).
114. G. G. Meisels, W. H. Hamill, and R. R. Williams, Jr., *J. Phys. Chem.*, **61**, 1456 (1967).
115. J. H. Futrell and T. O. Tiernan, *J. Chem. Phys.*, **37**, 1694 (1962).
116. L. Kevan, *J. Chem. Phys.*, **44**, 683 (1966).
117. G. v. Bunau, *Fortschr. Chem. Forsch.*, **5**, (2), 347 (1965).
118. V. Čermák and Z. Herman, *Collection Czech. Chem. Commun.*, **30**, 169 (1965).
119. H. Essex, *J. Phys. Chem.*, **58**, 42 (1954).
120. G. G. Meisels and T. J. Sworski, *J. Phys. Chem.*, **69**, 2867 (1965).
121. Healey and Reed, *The Behavior of Slow Electrons in Gases*, Amalgamated Wireless of Australia, Sydney, 1941.
122. T. L. Cottrell and I. C. Walker, *Trans. Faraday Soc.*, **61**, 1585 (1965).

123. (a) G. J. Schulz, *Phys. Rev.*, **125**, 229 (1962); (b) *ibid.*, *J. Chem. Phys.*, **34**, 1778 (1961).

124. R. L. Platzman, "Energy Spectrum of Primary Activations in the Action of Ionizing Radiation," paper presented at the 3rd International Congress of Radiation Research, Cortina d'Ampezzo, Italy, June 27, 1966.

125. I. K. Larin and V. L. Tal'roze, *Russ. J. Phys. Chem.*, **39**, 1105 (1965).

126. V. L. Tal'roze, *Chem. Biol. Action Radiations*, **11** (1967).

127. P. Ausloos and S. G. Lias, *Discussions Faraday Soc.*, **39**, 36 (1965).

128. P. Ausloos and S. Lias, *J. Chem. Phys.*, **36**, 3163 (1962).

129. P. Ausloos, S. G. Lias, and A. A. Scala, "Ion–Molecule Reactions in the Gas Phase," *Advan. Chem. Ser.*, **58**, 264 (1966).

130. R. D. Doepker and P. Ausloos, *J. Chem. Phys.*, **44**, 1951 (1966).

131. I. B. Sandoval and P. Ausloos, *J. Chem. Phys.*, **38**, 2454 (1963).

132. (a) A. Carrington and D. H. Levy, *J. Chem. Phys.*, **44**, 1398 (1966); (b) A. Carrington, D. H. Levy, and T. A. Miller, *J. Chem. Phys.*, **45**, 4093 (1966).

133. D. R. Smith and J. C. Tole, *Can. J. Chem.*, **45**, 779 (1967).

134. G. G. Meisels, *J. Am. Chem. Soc.*, **87**, 950 (1965).

135. F. W. Lampe, *J. Am. Chem. Soc.*, **79**, 1055 (1957).

136. J. M. Ramaradhya and G. R. Freeman, *J. Chem. Phys.*, **34**, 1726 (1961).

137. P. J. Dyne, J. Denhartog, and D. R. Smith, *Discussions Faraday Soc.*, 135 (1963).

138. M. M. Ellington, *J. Chem. Phys.*, **39**, 855 (1963).

139. P. Ausloos, R. E. Rebbert, and S. G. Lias, *J. Chem. Phys.*, **42**, 540 (1965).

140. W. Braun, K. H. Welge, and J. R. McNesby, *J. Chem. Phys.*, **45**, 2650 (1966).

141. R. Gorden, Jr. and P. Ausloos, *J. Chem. Phys.*, **46**, 4823 (1967).

142. J. A. Bell and G. B. Kistiakowsky, *J. Am. Chem. Soc.*, **84**, 3417 (1962).

143. *Catalog of Mass Spectral Data*. American Petroleum Institute Project 44, Texas A & M University, College Station, Texas, 1947–1961.

144. C. D. Wagner, P. A. Wadsworth, and D. P. Stevenson, *J. Chem. Phys.*, **28**, 517 (1958).

145. F. P. Abramson and J. H. Futrell, *J. Chem. Phys.*, **45**, 1925 (1966).

146. T. W. Shannon, F. Meyer, and A. G. Harrison, *Can. J. Chem.*, **43**, 159 (1965).

147. V. L. Tal'roze and E. L. Frankevich, *Zh. Fiz. Khim.*, **34**, 2709 (1960).

148. C. W. Hand and H. von Weyssenhoff, *Can. J. Chem.*, **42**, 195 (1964).

149. J. L. Franklin, Y. Wada, P. Natalis, and P. M. Hierl, *J. Phys. Chem.*, **70**, 2353 (1966).

150. S. K. Gupta, E. G. Jones, A. G. Harrison, and J. J. Myher, *Can. J. Chem.*, **45**, (1967).

151. A. Giardini-Guidoni and L. Friedman, *J. Chem. Phys.*, **45**, 937 (1966).

152. F. P. Abramson and J. H. Futrell, *J. Chem. Phys.*, **46**, 3264 (1967).

153. G. G. Meisels, W. H. Hamill, and R. R. Williams, Jr., *J. Chem. Phys.*, **25**, 790 (1956).

154. V. R. Fuchs, *Z. Naturforsch.* **16a**, 1026 (1961).

155. F. H. Field, J. L. Franklin, and M. S. B. Munson, *J. Am. Chem. Soc.*, **85**, 3575 (1963).

156. R. M. Haynes and P. Kebarle, *J. Chem. Phys.*, **45**, 3899 (1966).

157. F. H. Field and F. W. Lampe, *J. Am. Chem. Soc.*, **80**, 5587 (1958).

158. R. H. Lawrence, Jr. and R. F. Firestone, *J. Am. Chem. Soc.*, **87**, 2288 (1965).

159. L. W. Sieck and R. H. Johnsen, *J. Phys. Chem.*, **67**, 2281 (1963).

160. G. G. Meisels, *Abstracts 150th Meeting of the Am. Chem. Soc., Atlantic City, N. J., Sept. 1965*, p. 17V.

161. K. Yang and P. J. Manno, *J. Am. Chem. Soc.*, **81**, 3507 (1959).
162. R. W. Hummel, *Discussions Faraday Soc.*, **36**, 75 (1963).
163. W. P. Hauser, *J. Phys. Chem.*, **68**, 1576 (1964).
164. J. N. Butler and G. B. Kistakowsky, *J. Am. Chem. Soc.*, **82**, 759 (1960).
165. J. Maurin, *J. Chim. Phys.*, **59**, 15 (1962).
166. R. W. Hummel, *J. Phys. Chem.*, **70**, 2685 (1966).
167. R. H. Johnsen, *J. Phys. Chem.*, **69**, 3218 (1965).
168. H. Wiener and M. Burton, *J. Am. Chem. Soc.*, **75**, 5815 (1953).
169. H. H. Carmichael, R. Gorden, Jr., and P. Ausloos, *J. Chem. Phys.*, **42**, 348 (1965).
170. A. H. Laufer and J. E. Sturm, *J. Chem. Phys.*, **40**, 612 (1964).
171. R. F. Hampson, Jr. and J. R. McNesby, *J. Chem. Phys.*, **42**, 2200 (1965).
172. G. A. W. Derwish, A. Galli, A. Giardini-Guidoni, and G. G. Volpi, *J. Chem. Phys.*, **40**, 5 (1964).
173. M. S. B. Munson, J. L. Franklin, and F. H. Field, *J. Phys. Chem.*, **68**, 3098 (1964).
174. L. M. Dorfman, *J. Phys. Chem.*, **60**, 826 (1956).
175. L. M. Dorfman, *J. Phys. Chem.*, **62**, 29 (1958).
176. K. Yang, *Can. J. Chem.*, **38**, 1234 (1960).
177. K. Yang and P. L. Gant, *J. Phys. Chem.*, **65**, 1861 (1961).
178. L. J. Stief and P. Ausloos, *J. Chem. Phys.*, **36**, 2904 (1962).
179. C. M. Wodetzki, P. A. McCusker, and D. B. Petersen, *J. Phys. Chem.*, **69**, 1045 (1965).
180. C. M. Wodetzki, P. A. McCusker, and D. B. Peterson, *J. Phys. Chem.*, **69**, 1056 (1965).
181. P. Ausloos and S. G. Lias, *J. Chem. Phys.*, **44**, 521 (1966).
182. K. R. Ryan and J. H. Futrell, *J. Chem. Phys.*, **42**, 819 (1965).
183. E. Pettersson and E. Lindholm, *Arkiv Fysik*, **24**, 49 (1963).
184. G. A. W. Derwish, A. Galli, A. Giardini-Guidoni, and G. G. Volpi, *J. Chem. Phys.*, **41**, 2998 (1964).
185. V. Aquilanti and G. G. Volpi, *J. Chem. Phys.*, **44**, 2307 (1966).
186. L. W. Sieck and J. H. Futrell, *J. Chem. Phys.*, **45**, 560 (1966).
187. L. I. Bone and J. H. Futrell, *J. Chem. Phys.*, **46**, 4084 (1967).
188. K. Yang, *J. Chem. Phys.*, **32**, 1892 (1966).
189. R. A. Back and N. Miller, *Trans. Faraday Soc.*, **55**, 911 (1959).
190. B. F. Birdwell and G. W. Crawford, *J. Chem. Phys.*, **33**, 928 (1960).
191. K. Yang, *J. Am. Chem. Soc.*, **84**, 719 (1962).
192. S. G. Lias and P. Ausloos, *J. Chem. Phys.*, **37**, 877 (1962).
193. G. R. A. Johnson and J. M. Warman, *Trans. Faraday Soc.*, **61**, 1709 (1965).
194. L. I. Bone and R. F. Firestone, *J. Phys. Chem.*, **69**, 3652 (1965).
195. L. W. Sieck, N. K. Blocker, and J. H. Futrell, *J. Phys. Chem.*, **69**, 888 (1965).
196. L. I. Bone, L. W. Sieck and J. H. Futrell, "The Chemistry of Ionization and Excitation" G. R. A. Johnson and G. Scholes, Eds., Taylor and Francis Ltd., London, (1967).
197. M. Vestal, A. L. Wahrhaftig, and W. H. Johnston, *J. Chem. Phys.*, **37**, 1276 (1962).
198. M. Vestal, *J. Chem. Phys.*, **43**, 1356 (1965).
199. D. Beck and O. Osberghaus, *Z. Physik.*, **160**, 406 (1960).
200. D. Beck and A. Niehaus, *J. Chem. Phys.*, **37**, 2705 (1962).
201. D. Beck, *Discussions Faraday Soc.*, **36**, 56 (1963).

202. P. Ausloos and S. G. Lias, *J. Chem. Phys.*, **41**, 3962 (1964).
203. S. G. Lias and P. Ausloos, *J. Chem. Phys.*, **43**, 2748 (1965).
204. R. P. Borkowski and P. Ausloos, *J. Chem. Phys.*, **38**, 36 (1963).
205. M. S. B. Munson, *J. Am. Chem. Soc.*, **89**, 1772 (1967).
206. R. P. Borkowski and P. Ausloos, *J. Chem. Phys.*, **39**, 818 (1963).
207. T. Miyazaki and S. Shida, *Bull. Chem. Soc. Japan*, **38**, 716 (1965); *ibid.*, **39**, 2344 (1966).
208. J. H. Futtrell, *J. Phys. Chem.*, **64**, 1634 (1960).
209. J. H. Futrell, *J. Phys. Chem.*, **65**, 565 (1961).
210. A. A. Scala and P. Ausloos, to be published; see ref. 27.
211. J. H. Futrell, *J. Am. Chem. Soc.*, **81**, 5921 (1959).
212. H. A. Dewhurst, *J. Am. Chem. Soc.*, **83**, 1050 (1961).
213. N. L. Ruland, Dissertation, University of Houston, June 1966.
214. H. Umezawa and F. S. Rowland, *J. Am. Chem. Soc.*, **84**, 3077 (1962).
215. V. L. Tal'roze, *Pure Appl. Chem.*, **5**, 455 (1962).
216. S. Wexler, A. Lifschitz, and A. Quattrochi, *Advan. Chem.*, **58**, 193 (1966).
217. (a) P. S. Rudolph and C. E. Melton, *J. Chem. Phys.*, **32**, 586 (1960); (b) C. E. Melton and P. S. Rudolph, *J. Chem. Phys.*, **32**, 1128 (1960).
218. F. H. Field, *J. Am. Chem. Soc.*, **83**, 1523 (1961).
219. S. Wexler, *J. Am. Chem. Soc.*, **86**, 781 (1964).
220. I. Szabo, *Arkiv Fysik*, **33**, 57 (1966).
221. G. G. Meisels and H. F. Tibbals, paper presented at the 15th Annual Meeting on Mass Spectrometry, Denver, Colorado, May, 1967.
222. J. H. Futrell and T. O. Tiernan, paper presented at the 14th Annual Meeting on Mass Spectrometry, Denver, Colorado, May, 1967.
223. R. A. Back, T. W. Woodward, and K. A. McLauchlan, *Can. J. Chem.*, **40**, 1380 (1962).
224. K. Yang and P. J. Manno, *J. Phys. Chem.*, **63**, 752 (1959).
225. B. M. Mikhailov, V. G. Kiselev, and V. S. Bogdanov, *Izv. Akad. Nauk*, **1958**, No. 5, 545–549.
226. F. W. Lampe, *Radiation Res.*, **10**, 691 (1959).
227. M. C. Sauer, Jr. and L. M. Dorfman, *J. Phys. Chem.*, **66**, 322 (1962).
228. P. Ausloos and R. Gorden, Jr., *J. Chem. Phys.*, **36**, 5 (1962).
229. R. D. Doepker and P. Ausloos, *J. Chem. Phys.*, **44**, 1641 (1966).
230. G. G. Meisels and T. J. Sworski, *J. Phys. Chem.*, **69**, 815 (1965).
231. D. A. Becker, H. Okabe, and J. R. McNesby, *J. Phys. Chem.*, **69**, 538 (1965); E. Tschuikow-Roux, *J. Phys. Chem.*, **71**, 2355 (1967).
232. A. G. Harrison, *Can. J. Chem.*, **41**, 236 (1963).
233. M. Trachtman, *J. Phys. Chem.*, **70**, 3382 (1966).
234. P. C. Kaufman, *J. Phys. Chem.*, **67**, 1671 (1963).
235. M. Zelikoff and L. M. Aschenbrand, *J. Chem. Phys.*, **24**, 1034 (1956).
236. L. J. Stief, V. J. DeCarlo, and R. J. Mataloni, *J. Chem. Phys.*, **42**, 3113 (1965).
237. G. J. Mains, H. Niki, and M. H. J. Wijnen, *J. Phys. Chem.*, **67**, 11 (1963).
238. A. G. Sherwood and H. E. Gunning, *J. Phys. Chem.*, **69**, 1732 (1965).
239. M. Tsukuda and S. Shida, *J. Chem. Phys.*, **44**, 3133 (1966).
240. Y. Rousseau and G. J. Mains, *J. Phys. Chem.*, **68**, 3081 (1964).
241. F. H. Field, J. L. Franklin, and F. W. Lampe, *J. Am. Chem. Soc.*, **79**, 2665 (1957).
242. R. Barker, W. H. Hamill, and R. R. Williams, Jr., *J. Phys. Chem.*, **63**, 825 (1959).
243. P. S. Rudolph and C. E. Melton, *J. Phys. Chem.*, **63**, 916 (1959).

244. G. A. W. Derwish, A. Galli, A. Giardini-Guidoni, and G. G. Volpi, *J. Am. Chem. Soc.*, **87**, 1159 (1965).

245. A. Block, *Advan. Mass Spectrometry*, **2**, 48 (1963).

246. M. S. B. Munson, *J. Phys. Chem.*, **69**, 572 (1965).

247. L. M. Dorfman and A. C. Wahl, *Radiation Res.*, **10**, 680 (1959).

248. F. H. Field, *J. Phys. Chem.*, **68**, 1039 (1964).

249. J. H. Futrell and L. W. Sieck, *J. Phys. Chem.*, **69**, 892 (1965).

250. A. Russell Jones, *J. Chem. Phys.*, **32**, 953 (1960).

251. F. W. Lampe, *J. Am. Chem. Soc.*, **82**, 1551 (1960).

252. S. C. Lind and P. S. Rudolph, *J. Chem. Phys.*, **26**, 1768 (1957).

253. H. S. Shiff and E. W. R. Steacie, *Can. J. Chem.*, **29**, 1 (1951).

254. C. F. Smith, B. G. Corman, and F. W. Lampe, *J. Am. Chem. Soc.*, **83**, 3559 (1961).

255. R. D. Doepker and P. Ausloos, quoted in reference 130 (footnote 11).

256. A. Henglein and G. A. Muccini, *Z. Naturforsch.*, **18a**, 753 (1963).

257. J. Milhaud and J. Durup, *Compt. Rend.*, **260**, 6363 (1965).

258. J. Milhaud, Dissertation, University of Paris Center D'Orsay, France, December, 1966.

259. F. P. Abramson and J. H. Futrell, *J. Phys. Chem.*, **71**, 1233 (1967).

260. J. M. Ramaradhya and G. R. Freeman, *Can. J. Chem.*, **39**, 1769 (1961).

261. J. Blachford and P. J. Dyne, *Can. J. Chem.*, **42**, 1165 (1964).

262. L. M. Theard, *J. Phys. Chem.*, **69**, 3292 (1965).

263. K. H. Jones, *J. Phys. Chem.*, **71**, 709 (1967).

264. J. P. Manion and M. Burton, *J. Phys. Chem.*, **56**, 560 (1952).

265. R. D. Doepker and P. Ausloos, *Abstracts, 148th meeting Am. Chem. Soc. Chicago, Ill., 1964*, p. 48V.

266. W. J. Holtslander and C. R. Freeman, *Can. J. Chem.*, **45**, 1649 (1967).

267. W. J. Holtslander and G. R. Freeman, *Can. J. Chem.*, **45**, 1661 (1967).

268. E. Lindholm and P. Wilmenius, *Arkiv Kemi Band*, **20** (22) (1963).

269. K. R. Ryan, L. W. Sieck, and J. H. Futrell, *J. Chem. Phys.*, **41**, 111 (1964).

270. V. Aquilanti, A. Galli, and G. G. Volpi, *Ric. Sci.*, **36**, 267 (1966).

271. L. W. Sieck, F. P. Abrahamson, and J. H. Futrell, *J. Chem. Phys.*, **45**, 2859 (1966).

272. (a) J. H. Baxendale and R. D. Sedgwick, *Trans. Faraday Soc.* **57**, 2157 (1961); (b) J. H. Baxendale and G. P. Gilbert, *Discussions Faraday Soc.*, **36**, 605 (1963).

273. G. M. Meaburn and F. W. Mellows, *Trans. Faraday Soc.*, **61**, 1701 (1965).

274. Z. Prasil, *Collections Czech. Chem. Commun.*, **31**, 3263 (1966).

275. A. J. Harrison and J. S. Lake, *J. Phys. Chem.*, **63**, 1489 (1959).

276. J. M. Ramaradhya and G. R. Freeman, *Can. J. Chem.*, **39**, 1836 (1961).

277. J. M. Ramaradhya and G. R. Freeman, *Can. J. Chem.*, **39**, 1843 (1961).

278. J. J. J. Myron and G. R. Freeman, *Can. J. Chem.*, **43**, 1484 (1965).

279. L. W. Sieck and R. H. Johnsen, *J. Phys. Chem.*, **69**, 1699 (1965).

280. M. K. M. Ng and G. R. Freeman, *J. Am. Chem. Soc.*, **87**, 1639 (1965).

281. K. M. Bansal and G. R. Freeman, *J. Am. Chem. Soc.*, **88**, 4326 (1966).

282. For example, see K. J. Laidler, *Chemical Kinetics*, 2nd ed., McGraw-Hill, New York, 1965, p. 150 ff., or S. W. Benson, *The Foundations of Chemical Kinetics*, McGraw-Hill, New York, 1960, p. 225 ff.

283. A. G. Leiga and H. A. Taylor, *J. Chem. Phys.*, **41**, 1247 (1964).

284. P. Ausloos and J. F. Paulson, *J. Am. Chem. Soc.*, **80**, 5117 (1958).

285. L. J. Stief and P. Ausloos, *J. Phys. Chem.*, **65**, 1560 (1961).

286. A. Scala and P. Ausloos, *J. Phys. Chem.*, **70**, 260 (1966).
287. B. C. de Souza and J. H. Green, *Austral. J. Chem.*, **18**, 1153 (1965).
288. G. J. Mains, A. S. Newton, and A. F. Sciamanna, *J. Phys. Chem.*, **65**, 1286 (1961).
289. P. Ausloos and C. N. Trumbore, *J. Am. Chem. Soc.*, **81**, 3866 (1959).
290. R. F. Firestone and J. E. Willard, *J. Am. Chem. Soc.*, **83**, 3551 (1961).
291. A. H. Young and J. E. Willard, *J. Phys. Chem.*, **66**, 271 (1962).
292. M. Takenhisa, G. Levey, and J. E. Willard, *J. Am. Chem. Soc.*, **88**, 5694 (1966).
293. R. N. Schindler, *Radiochim. Acta*, **2**, 62 (1963).
294. L. H. Gevantman and R. R. Williams, Jr., *J. Phys. Chem.*, **56**, 569 (1952).
295. R. H. Luebbe, Jr. and J. E. Willard, *J. Am. Chem. Soc.*, **81**, 761 (1959).
296. R. Schindler and M. H. J. Wijnen, *Z. Physik. Chem.* **34**, 109 (1962).
297. G. A. W. Derwish, A. Galli, A. Giardini-Guidoni, and G. G. Volpi, *J. Am. Chem. Soc.*, **86**, 4503 (1964).
298. R. W. Fessenden and R. H. Schuler, *J. Chem. Phys.*, **43**, 2704 (1965).
299. J. Fajer, D. R. MacKenzie, and F. W. Bloch, *J. Phys. Chem.*, **70**, 935 (1966).
300. L. Kevan and P. Hamlet, *J. Chem. Phys.*, **42**, 2255 (1965).
301. J. Collin, *Bull. Soc. Chim. Belg.*, **67**, 549 (1958).
302. S. Takumuka and H. Sakurai, *Bull. Chem. Soc. Japan*, **38**, 791 (1965).
303. L. J. Stief and P. Ausloos, *J. Phys. Chem.*, **65**, 877 (1961).
304. R. W. Hummel, *Trans. Faraday Soc.*, **62**, 59 (1966).
305. V. Aquilanti, *J. Phys. Chem.*, **69**, 3434 (1965).
306. G. v. Bunau and R. N. Schindler, *J. Chem. Phys.*, **44**, 420 (1966).
307. A. D. Site and A. Mele, *J. Phys. Chem.*, **69**, 4033 (1965).
308. D. Smith and L. Kevan, *J. Chem. Phys.*, **46**, 1586 (1967).
309. F. H. Field and J. L. Franklin, *J. Am. Chem. Soc.*, **83**, 4509 (1961).
310. P. S. Rudolph, S. C. Lind, and C. E. Melton, *J. Chem. Phys.*, **36**, 1931 (1962).
311. F. H. Field, H. N. Head, and J. L. Franklin, *J. Am. Chem. Soc.*, **84**, 1118 (1962).
312. C. E. Klots, *J. Chem. Phys.*, **46**, 3468 (1967).
313. P. Ausloos and S. G. Lias, "The Chemistry of Ionization and Excitation" G. R. A. Johnson and G. Scholes, Eds., Taylor and Francis Ltd., London (1967), p. 77.

Chapter 7

Organic Liquids*

Richard A. Holroyd

Atomics International
A Division of North American Rockwell Corp.
Canoga Park, California

* Preparation of this review was supported by the Research Division of the United States Atomic Energy Commission.

I. INTRODUCTION

Early studies of the radiolysis of organic liquids were largely concerned simply with establishing the net effect of radiation on a sample, rather than with understanding the nature of the chemical events taking place. Many different organic compounds were irradiated, and the literature of that period lists products and yields in detail (1). The unraveling of the details of the chemical processes occurring in irradiated systems was finally based on various experimental developments which now make possible studies of reactive intermediate species produced by radiation. For example, by the 1950's, mass spectrometers and gas chromatographs were generally available, making possible the detection and analysis of products formed in very small yields. In this period luminescence studies (2) established the presence of electronically excited molecules in irradiated liquids.

The addition of radical scavengers (3,4) demonstrated that radicals were formed in significant yields; subsequently radical reactions were invoked to explain the formation of most products in irradiated systems.

More recently, the irradiation of samples with electron beams from high intensity accelerators has made possible the formation of reactive intermediates in concentrations high enough that their identities and reactions can be directly established through spectroscopy or electron spin resonance (ESR).

The importance of ionic intermediates in organic liquids was established by a series of events which occurred in the early 1960's and confirmed earlier suggestions (5,6). In 1961, information from chemical studies suggested that hydrogen formation in alcohol radiolysis (7,8) was linked to a reaction of electrons. In 1963, the solvated electron was detected in alcohols by pulse radiolysis (9,10) following similar observations of the hydrated electron in water (Chapter 10). While a substantial yield of free ions was found in polar liquids, conductivity studies (11,12) indicated a lower yield for hydrocarbons. These studies were followed by efforts to find other ways to detect ions in liquids, through the use of chemical scavengers (13,14), for instance.

Before discussing the reactions of the intermediate species formed in irradiated organic systems, it is appropriate to consider briefly the probable primary processes by which they are formed. When the energy of an ionizing particle is dissipated in an organic liquid, ions, electrons (which may initially have excess kinetic energy), and excited molecules will be formed. The situation is complicated by the inhomogenous deposition of energy which results in the formation of these species in groups or spurs. In the case of gamma radiolysis, the spur may be visualized as a small, spherical volume which contains several initial species. For more densely ionizing radiation, such as alpha particle radiolysis, these spurs may be close enough together so that a continuous track of ionized and excited species is formed.

Subsequent processes may include decomposition or reaction of the excited molecules and/or ions to form other radical or ionic intermediates. The ejected electrons are rapidly thermalized, and, in a polar medium, the molecules of the liquid become oriented around them, forming so-called "solvated electrons." If an electron is within the field of a positive ion, it will diffuse to the ion and neutralize it, forming an excited species which may decompose. As the intermediates which are present in the spur diffuse, the spur expands in volume, and in-spur combination and disproportionation reactions occur between various intermediates. These reactions during diffusion from the spur occur in times between 10^{-11} and 10^{-8} sec after passage of the ionizing particle. The intermediates which escape the spur to react homogeneously in the body of the liquid are either ions ("free ions"), radicals or long-lived excited molecules. A detailed discussion of these events is given in Chapter 8, Section I-B.

This chapter on fundamental processes in the radiolysis of organic liquids reflects the current state of knowledge regarding intermediates. The principles of the methods used to detect various intermediates as well as their yields and reactions are considered in Sections II through IV. Section V is devoted to a detailed examination of linear energy transfer effects in organic liquids. Section VI is concerned with charge and energy transfer processes in mixtures. Many problem areas and questions which are of current concern to the radiation chemist are noted. Where the subject matter is controversial, the important facts and various alternative explanations are presented.

II. IONIC INTERMEDIATES

A. Detection of Ions

The methods of detecting ionic species in organic liquids during radiolysis are briefly described in this section. Two physical methods which have been used are conductivity and absorption spectroscopy. These methods provide a measure of the yields of those ions which escape the spur and are homogeneously neutralized ("free ions"). On the other hand, the addition of chemical scavengers which react with ions to form products detectable by subsequent chemical analysis can provide a measure of the total yield of free ions and ions which do not escape the spur.

1. Conductivity. Free ions can be detected in liquids of low dielectric constant by measurements of the transient radiation-induced conductivity. If the ions react primarily by bimolecular combination

$$A^+ + B^- \rightarrow \text{neutral species} \tag{1}$$

then the rate of formation of ions can be equated to the rate of combination. Assuming that the concentrations of positive and negative species are equal and that ions are singly charged, one then derives equation (2):

$$G(\text{free ions}) = (100\, k_1 N/\text{dose rate}) \times [C]^2 \tag{2}$$

as an expression of the yield of free ions, where k_1 is the rate constant for reaction (1). The concentration of ions, [C], is in moles/liter; the dose rate is in eV l^{-1} sec^{-1}, N is Avogadro's number, and G is ions formed per 100 eV absorbed. Since the total conductivity, κ, is equal to the concentration of ions times the total ionic mobility, μ, equation (2) can be rewritten as equation (3).

$$G(\text{free ions}) = (100\, k_1 N/\text{dose rate}) \times \kappa^2/\mu^2 \tag{3}$$

According to equation (3) an experimental determination of the yield of free ions in an organic liquid requires a knowledge of the ion recombination rate constant, the conductivity as a function of the dose rate, and the ionic mobility (15). The accuracy of G(free ions) determined this way depends on how well each of these terms has been evaluated.

The conductivity should be proportional to the square root of the dose rate according to equation (3). This has been experimentally verified for dose rates above about 10^{11} eV g^{-1} sec^{-1} (12,15). At lower dose rates the reactions of ions at surfaces invalidate the square root dependence.

The ionic mobility must also be known in order to calculate the free ion yield from conductivity data. A direct experimental determination of the mobility of the ions in irradiated organic liquids obviates any assumption about the nature of the ions. Measurements of mobilities are usually made in cells in which a layer of ions can be induced between a pair of electrodes by means of an x-ray beam (15). The time required for the ions in this layer to reach the electrodes after application of a voltage provides a measure of the mobility. A comparison of measured ionic mobilities reported for *n*-hexane shows quite a range of values (Table I). Takagaki et al. (16) used essentially the same technique as Hummel and Allen but obtained lower values. Secker and Lewis (17) suggest that the high apparent mobilities measured in their study can be explained by the occurrence of liquid motion in the direction of motion of the ions in their cell (18). In view of the results shown in Table I this source of error must have been less important in the cells used by others.

Another method of obtaining the ionic mobility is to estimate it from diffusion coefficients as in Eq. (4). The constant D is the diffusion coeffi-

$$\mu = De/\mathbf{k}T \tag{4}$$

cient of the ion, e is the ionic charge, and $\mathbf{k}$ is the Boltzmann constant. If the ions are of molecular dimensions, the self diffusion coefficient of a

Table I
Ionic Mobilities in *n*-Hexane at 25°

Investigators	μ^+, cm^2/V-sec	μ^-, cm^2/V-sec	Ref.
Le Blanc (1959)	—	1.23×10^{-3}	19
Gzowski and Terlecki (1959)	0.41×10^{-3}	1.3×10^{-3}	20
Takagaki et al. (1964)	0.23×10^{-3}	0.43×10^{-3}	16
Secker and Lewis (1965)	1.0×10^{-3}	2.0×10^{-3}	17
Hummel and Allen (1966)	0.66×10^{-3}	1.27×10^{-3}	15[a]

[a] The reported accuracy of these values is $\pm 5\%$.

molecule of the liquid may be assumed for D. For cyclohexane the individual ionic mobilities are calculated to be 0.54×10^{-3} cm^2 V^{-1} sec^{-1} from equation (4) (12). Since this value is comparable to the measured values for n-hexane, the free ions in hydrocarbons must be approximately of molecular dimensions.

Mobility measurements can also provide information about the nature of the ionic species in organic liquids. In the presence of oxygen the electron is most likely captured by O_2 to form the O_2^- anion. Based on equation (4) and the known value of the diffusion coefficient of oxygen in hydrocarbons (12,21), the negative ion mobility in the presence of oxygen should be about 1.1×10^{-3} cm^2 V^{-1} sec^{-1}. Two different groups have measured ionic mobilities in hexane in the presence and absence of oxygen. Takagaki et al. (16) reported that oxygen doubles the cation mobility but did not report the effect on the anion. Hummel et al. (22) found no effect of oxygen on the negative ion mobility and their measured value is close to that calculated from equation (4). Thus in degassed solutions the negative free ion is either O_2^- or an ion with the same mobility as O_2^-. Although oxygen is without effect, tetrahydrofuran decreases the negative ion mobility by a factor of two (22) indicating electron attachment to this ether. Conceivably other ion–molecule reactions could be detected by mobility measurements (see also Section II-C).

The *recombination rate constant* for liquids of low dielectric constants can be calculated from the theoretical relationship of Debye (23), equation (5), where e is the electronic charge and ϵ is the dielectric constant. For

$$k_1 = 4\pi e\mu/\epsilon \qquad (5)$$

hexane k_1, calculated from this relationship, is 1.1×10^{12} M^{-1} sec^{-1} if the mobility is taken as 1.9×10^{-3} cm^2 V^{-1} sec^{-1}. A value of k_1 determined from measurements of the decay of the transient conductivity in hexane (15) was within experimental error of this theoretical value.

2. Absorption Spectroscopy. Evidence for ionic intermediates in organic liquids has also been obtained in pulse radiolysis experiments. This technique involves the use of a pulse of ionizing radiation combined with suitable equipment to detect the transient intermediates formed in the irradiation cell. The radiation pulse should be intense enough to produce a measurable concentration of intermediates and its duration should be short compared to the lifetime of the intermediates. The detection methods used in pulse radiolysis are generally optical methods such as spectro-photoelectric recording as is employed in flash photolysis (24). Other methods of detection such as conductivity can also be used (25).

a. Solvated Electrons. The existence of the solvated electron was first indicated by studies in 1961 in which it was shown by chemical analysis that there were two kinetically distinguishable precursors of the H_2 formed in the radiolysis of alcohols. The most reactive of these species was shown not to be the hydrogen atom and was termed the solvated electron by Baxendale and Mellows (7) and the polaron by Hayon and Weiss (8).

The solvated electron was identified in the pulse radiolysis of methanol and ethanol (9,10) by its visible absorption spectrum which is very similar to that of the hydrated electron, and by its chemical reactivity. Subsequently, solvated electrons were detected by the pulse radiolysis technique in several other alcohols (26–28), and in other polar liquids including ethylenediamine (29,30), methylamine and ethylamine (31). Visible absorptions which have been observed in liquids of low dielectric constant such as dioxane (32) and diethyl ether (31) have also been attributed to solvated electrons. With micro-second time resolution, the pulse radiolysis technique detects only *free* solvated electrons. That is, in any liquid only a fraction of the electrons which are released by the radiation escape rapid recombination with the positive ions in the spur, and it is only these free ions that are usually detectable.

The *absorption spectrum* of the solvated electron in an alcohol consists of one very broad peak throughout most of the visible region with a maximum between 5270 and 8200 Å (27) (Table II). For the alcohols there is a definite shift in absorption maxima to shorter wavelengths with increasing dielectric constant (33). Anbar and Hart (29) have also considered this shift in absorption maxima for various liquids. They showed that there was a correlation of the position of the absorption maximum for solvated electrons with the maximum in the charge transfer spectrum of iodide ion in that particular solvent. Recent measurements (30) indicate that their proposed linear correlation must be modified.

The lack of structure and great width of the absorption spectrum of the solvated electron are features which are still unexplained. The half-width of the spectrum is about 1 eV. Additional fine structure is not revealed in the spectrum when the temperature is lowered from 25 to $-78°C$ (31). It has been suggested that the breadth of the absorption peak is related to the fact that the solvated electron can be excited to a large number of levels (31) (see Chapter 10).

For mixtures of two alcohols only one absorption peak is observed and this peak maximum is intermediate between those of the two pure alcohols (31). The lack of two peaks indicates that the electron interacts with many molecules as required by the continuum model of Jortner (35), which visualizes the solution as a continuous dielectric medium characterized

Table II

Solvated Electron Absorption Spectra

Solvent	Spectral characteristics of e_{sol}^- [a]		
	λ_{max} Å	E_{max}, kcal	ϵ_{10} at maximum M^{-1} cm^{-1}
Glycerol	5270	54.0	—
Ethylene glycol	5800	49.3	1.4×10^4
Methanol	6300	45.2	1.7×10^4
Ethanol, $-78°$	5800	49.3	—
Ethanol, $25°$	7000	40.8	1.5×10^4
Water	7200	39.6	—
1-Propanol	7400	38.5	1.3×10^4
2-Propanol	8200	34.8	1.4×10^4
Ethylenediamine	> 11,200	< 25	—
Ethylamine	> 11,200	< 25	—
Acetone	(11,000)[b]	(26)[b]	—
Ammonia	14,000	19.7	—

[a] References 29, 30, 33, 34.

[b] Predicted from the correlation in reference 29.

by a macroscopic dielectric constant. Thus, the solvated electron is *not* to be considered as the anion, ROH$^-$; more than one molecule of solvent is necessary to solvate an electron. The solvated electron has been observed in a solution containing only 4% methanol in cyclohexane (36) which demonstrates that aggregates of the appropriate size to solvate the electron exist even in the 4% solution.

The natural *lifetime of a solvated electron* in a pure liquid such as an alcohol is still uncertain. Values that have been reported for lifetimes of solvated electrons in alcohols vary from 1.5 to 5 μsec (10,26,37). These observed lifetimes are likely to be lower limits according to Dorfman (33) because of the presence in the alcohols of trace impurities, ions, and products with which the electron is likely to react. It has been suggested (26,37) that, by analogy with the reactions of the hydrated electron (see Chapter 10), the solvated electron interacts with an alcohol molecule in its solvation sphere and forms a hydrogen atom and a negative ion:

$$e_{sol}^- + ROH \rightarrow H + RO^- \tag{6}$$

b. Aromatic Ions. In pulse radiolysis studies of solutions of aromatic compounds absorption spectra have been observed which demonstrate the formation of solute ions. In the radiolysis of ethanol solutions, negative ions of biphenyl, p-terphenyl, naphthalene, and naphthacene have been identified (28,38). In the radiolysis of cyclohexane solutions, negative ions of biphenyl, anthracene, benzophenone, N,N-dimethylaniline and diamino-durene have been identified (39,40). The formation and reactions of these ions are discussed in Section II-B-1. Although *solute* ions have been reported in pulse radiolysis studies, the molecular ions which must be formed in the radiolysis of pure aromatic liquids, for example $C_6H_6^-$ in benzene, have not been identified by this technique.

B. Reactions of Ions

In the early 1950's it was suggested that ions survive long enough in organic liquids to undergo reactions such as charge transfer (5) and disso-ciative electron capture (6). In recent years chemical reactions of electrons and of positive ions in irradiated liquids have been elucidated by various chemical and physical means.

1. Reactions of Electrons. The electron can be considered to be a free radical; like a radical it undergoes addition and abstraction reactions with solute molecules. These two types of reactions are more conventionally termed electron attachment and dissociative electron attachment (for example, reactions 7 and 8, respectively). A third and very important reaction of the electron is neutralization of positive ions. Neutralization reactions are discussed in Section II-B-2.

$$e^- + C_{12}H_{10} \rightarrow C_{12}H_{10}^- \qquad (7)$$

$$e^- + C_6H_5CH_2Cl \rightarrow C_6H_5CH_2\cdot + Cl^- \qquad (8)$$

a. Reactions of Solvated Electrons. Electron attachment of solvated electrons to aromatic hydrocarbons has been demonstrated in pulse radiolysis studies (28,38). For example the biphenylide ion is formed in the pulse radiolysis of a solution of biphenyl in alcohol. The rate constants for these electron attachment reactions are given in Table III. The rates are very nearly diffusion controlled and increase with the electron affinity of the solute (38).

The anions that are formed by electron attachment to hydrocarbons can either accept a proton from the alcohol or react with other solute molecules by electron transfer (41). The reaction with the alcohol is believed to be proton transfer as in reaction (9), since the corresponding radical is observed to be formed and since the rate of the reaction correlates with the

$$C_{12}H_{10}^- + ROH \rightarrow RO^- + C_{12}H_{11} \qquad (9)$$

acidity of the alcohol (38). The yield of hydrocarbon radicals increases with increasing polarity of the solvent (42). This is reasonable since the yield of solvated electrons also increases with increasing polarity of the

Table III
Reactions of the Solvated Electron

Reaction	Solvent	Rate Constant k $M^{-1} sec^{-1}$ for e_{sol}^-	for e_{aq}^-	Ref. for e_{sol}^-
$e_{sol}^- + m$-Dinitrobenzene	2-Propanol	2.6×10^{10}		43[a]
$e_{sol}^- + ClCH_2CO_2H$	C_2H_5OH	2.0×10^{10}		44[b]
$e_{sol}^- + $ Nitrobenzene	2-Propanol	1.9×10^{10}	3.0×10^{10}	45[a]
$e_{sol}^- + CCl_4$	2-Propanol	1.9×10^{10}	3.1×10^{10}	45[a]
$e_{sol}^- + $ Benzophenone	2-Propanol	1.1×10^{10}		45[a]
$e_{sol}^- + $ Naphthacene	Ethanol	1.0×10^{10}		38
$e_{sol}^- + $ Bromobenzene	2-Propanol	8.7×10^{9}		43[a]
$e_{sol}^- + p$-Terphenyl	Ethanol	7.2×10^{9}		38
$e_{sol}^- + $ Naphthalene	Ethanol	5.4×10^{9}	5.4×10^{9}	38
$e_{sol}^- + $ Benzyl chloride	Ethanol	5.1×10^{9}	5.5×10^{9}	10
$e_{sol}^- + $ Acetone	2-Propanol	5.0×10^{9}	5.9×10^{9}	45[a]
$e_{sol}^- + $ Chlorobenzene	2-Propanol	4.7×10^{9}		43[a]
$e_{sol}^- + $ Biphenyl	Ethanol	4.3×10^{9}		38
$e_{sol}^- + (C_6H_5)_3CH$	Ethanol	2.0×10^{8}		10
$e_{sol}^- + $ Benzene	2-Propanol	1.8×10^{8}	$<7 \times 10^{6}$	45[a]
$e_{sol}^- + $ 2-Propanol	2-Propanol	1.5×10^{6}		45[a]

[a] Rate const. based on relative rates and an assumed value of $k(e_{sol}^- + N_2O \rightarrow)$ = 8.7×10^9 (25).
[b] Relative to $k(e^- + H^+ \rightarrow) = 2 \times 10^{10}$.

solvent (Section II-C). Reaction (9) and others like it are thus of major importance in the formation of radicals in solutions of aromatic hydrocarbons, particularly in polar solvents.

Solvated electrons have been found to react with many other solutes, some of which are listed in Table III. A solute which is very useful in studying the reactions of electrons is nitrous oxide (45–48). The reaction of the solvated electron with nitrous oxide, reaction (10), was first dem-

$$e_{sol}^- + N_2O \rightarrow N_2 + O^- \qquad (10)$$

onstrated to occur in the radiolysis of water (Chapter 10). At millimolar concentrations of N_2O in alcohols the yield of nitrogen is substantial; at higher concentrations $G(N_2)$ rises to values above 3 (Fig. 1). When $G(N_2)$ is plotted vs. the concentration of N_2O on a log–log scale a linear dependence is observed at least up to 0.1 M. It is noted that the yield of N_2 formed at 10^{-4} M nitrous oxide is nearly equal to the yield of "free" solvated electrons for alcohols (see Section II-C). At higher concentrations

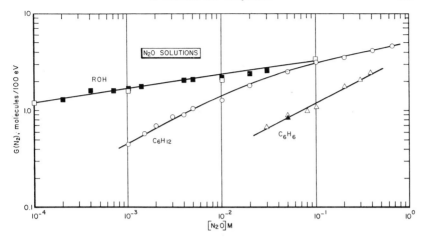

Fig. 1. Yields of nitrogen from the radiolysis of nitrous oxide solutions. Points refer to different solvents: △ benzene (50), ▲ benzene (13), ○ cyclohexane (51), ■ methanol (46,48), □ ethanol (105).

nitrous oxide presumably scavengers those electrons which in the absence of nitrous oxide would have undergone geminate recombination.

Many of the rate constants for reactions of the solvated electrons with solutes given in Table III have been calculated from the rate constant ratios $k(e_{sol}^- + S \rightarrow)/k_{10}$ and an assumed value for k_{10} in propanol of $8.7 \times 10^9 \, M^{-1} \, sec^{-1}$. The ratio of rate constants is obtained from the depression in the nitrogen yield caused by the presence of the solutes during radiolysis. The value assumed for k_{10} is the known rate constant for this reaction in water. This assumption is justified on the basis that rate constants for reactions of the solvated electron are remarkably independent of solvent for those reactions which are diffusion controlled or nearly so (25,47) (compare columns 3 and 4). This generalization does not apply to slower reactions, such as the reaction of the solvated electron with benzene. For example, the rate constant for reaction of the electron with benzene is smaller in methanol than in 2-propanol because of the increased stabilization of the electron in methanol (47). From these studies it is found that the reactivity of aromatic compounds toward the solvated electron is related to their reactivity toward nucleophilic reagents as expressed by the Hammet σ_{para}-function (49). Therefore the solvated electron, like the hydrated electron, is to be considered as a nucleophile which conforms to the general pattern of bimolecular nucleophilic substitution (S_N2).

   b. Reactions of Electrons in Hydrocarbons.   Since nitrous oxide reacts rapidly with solvated electrons in alcohols it is reasonable to expect that an analogous reaction occurs in nonpolar media such as hydrocarbons where the free electrons do not become solvated. Evidence for this was first reported by Scholes and Simic (13) who found that nitrogen was formed and the yield of hydrogen was reduced in the radiolysis of solutions of N_2O in cyclohexane. Their results have been verified by others (50–52) and are illustrated in Fig. 1. The intermediate with which nitrous oxide reacts is most likely the electron since most other intermediates do not react with this solute. For example, the rate of reaction of hydrogen atoms with N_2O is too slow (53) to be important at the concentrations employed in these studies. Charge transfer cannot occur since the gas phase ionization potential of nitrous oxide is greater than the ionization potential of hydrocarbons. Excited molecules were ruled out on the basis that the precursor reacts more readily with solutes of high electron affinity (51).

The yield of nitrogen from cyclohexane solutions of N_2O increases with N_2O concentration to a G value of 4.7 at $0.7M$ nitrous oxide (Fig. 1). $G(N_2)$ actually exceeds the yield of electrons, possibly as a result of a secondary reaction of O^- with N_2O such as reaction (11) (54).

$$O^- + N_2O \rightarrow N_2 + O_2^- \tag{11}$$

The slopes of the lines for hydrocarbons in Fig. 1 are approximately 0.5 at low concentrations; that is $G(N_2)$ is proportional to $[N_2O]^{1/2}$ in this region. This reaction of N_2O with the electron is much more efficient than reactions of solutes with positive ions.

Very similar results have been obtained recently with methyl halides as solutes in the radiolysis of cyclohexane (55), where reaction (12) is operative:

$$e^- + CH_3X \rightarrow CH_3 + X^- \tag{12}$$

The yield of methyl radicals also follows a square root dependence at concentrations of methyl halide less than $0.01M$ and $G(CH_3)$ approaches 4.4 at $1M$ CH_3X. It should be mentioned that hydrogen atoms can also react with methyl iodide to form methyl radicals, but a large share of the methyl radicals apparently are formed by reaction (12).

The electron formed in hydrocarbon radiolysis also reacts with other solutes such as aromatic compounds in reactions analogous to (7) and (8). In the pulse radiolysis studies of solutions of aromatic compounds in cyclohexane, the absorption spectra observed are complex but comparison of the spectra with known absorption spectra of aromatic anions shows that anions can be present (39,40). However since aromatic anions and cations have nearly identical absorption spectra (56), cations may be present as well. The anions are formed by electron attachment as in reaction (7), whereas the cations are formed by charge transfer as in

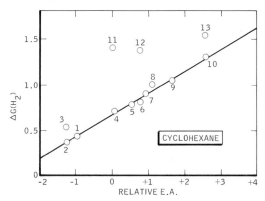

Fig. 2. The change in hydrogen yield from cyclohexane due to the presence of a solute at a concentration of 5 mM. The electron affinity of the solute taken relative to that of chlorobenzene is plotted along the abscissa. Solutes are: *1*, aniline; *2*, benzyl alcohol; *3*, phenanthrene; *4*, *cis*-stilbene; *5*, *trans*-stilbene, *6*, pyrene; *7*, azobenzene; *8*, anthracene; *9*, benzaldehyde; *10*, nitrobenzene; *11*, chlorobenzene; *12*, bromobenzene; *13*, iodobenzene. Reprinted by permission, ref. 57.

reaction (13)

$$RH^+ + C_{12}H_{10} \rightarrow RH + C_{12}H_{10}^+ \tag{13}$$

In the case of biphenyl the yield of these ions is less if N_2O is also present (40). Conversely, it has also been shown that aromatic solutes, if present in solutions of nitrous oxide in cyclohexane, reduce the yield of nitrogen (51). Since N_2O reacts with the electron, this evidence indicates that aromatic solutes react with the electron. This conclusion is further supported by the effect of aromatic solutes in reducing the hydrogen yield in the radiolysis of cyclohexane (51,57–59). The reduction in $G(H_2)$ increases with the electron affinity of the solute as is shown in Fig. 2 (57).

Rate constants for reaction of the electron with solutes in cyclohexane have been measured relative to the rate of reaction with N_2O. Fluorocarbons (except CF_4 and C_2F_6) (60), chlorinated hydrocarbons and aromatic solutes such as stilbene, anthracene, and benzophenone (51,61) are about as reactive as nitrous oxide toward the electron. Benzene is reported to be from 20 (50) to 300 (51) times less reactive than nitrous oxide toward the electron. Thus solvation does not seem to have any effect on the reactivity of the electron, according to the available rate constant data obtained in hydrocarbon and alcohol systems.

2. Reactions of Positive Ions. Direct spectroscopic evidence for the presence of positive ions in organic liquids is scant but there is chemical evidence for their reactions. Positive ions can undergo fragmentation, charge transfer, ion–molecule, and neutralization reactions. With the exception of charge transfer which is considered in Section VI, evidence for the occurrence of these other processes in irradiated liquids is discussed here.

a. Fragmentation. Fragmentation of excited molecular cations occurs in the gas phase during electron bombardment (see Chapter 6). Such fragmentation is expected to be less important in the liquid phase relative to other reactions of the cations since the excited ions can be deactivated by collisions. This effect has been verified by studies concerned with the effect of pressure in the gas phase (62–65). However, there is evidence which indicates that fragmentation of ions does occur to some extent in certain liquids.

In the radiation-induced decomposition of alkanes, if we consider a mode of fragmentation in which the scission of carbon–carbon bonds is asymmetric, it is sometimes found that unequal yields of the two complementary fragment radicals are formed (Section IV). For example, the yield of methyl radicals exceeds the yield of propyl radicals for butane (66) and isobutane (67). Similarly in neopentane $G(CH_3) \gg G(C_4H_9)$ and in isopentane $G(CH_3)$ exceeds $G(C_4H_9)$ and $G(C_2H_5)$ exceeds $G(C_3H_7)$ (68). This nonequivalence of yields suggests that there is a process in which an excited ion or highly excited molecule decomposes to form a radical and another fragment which does not always stabilize but decomposes further. In the case of neopentane it has been suggested (69) that this process is elimination of a methyl radical from the excited molecular ion leaving a *t*-butyl carbonium ion (reaction (14)).

$$neo\text{-}C_5H_{12}{}^+ \rightarrow CH_3 + t\text{-}C_4H_9{}^+ \tag{14}$$

The latter upon neutralization might lose a hydrogen atom and form an olefin, or stabilize to some extent into *t*-butyl or isobutyl radicals, reaction (15). Estimates of the yield of *t*-butyl carbonium ions have been repor-

$$t\text{-}C_4H_9{}^+ + e^- \rightarrow H + i\text{-}C_4H_8,\ t\text{-}C_4H_9,\ \text{or}\ i\text{-}C_4H_9 \tag{15}$$

ted; $G(C_4H_9{}^+)$ is 1.0 in neopentane and 0.4 in 2,2,4-trimethylpentane (70). These yields were based on G(isobutene) from radiolysis of a solution containing both methylamine and SF_6 compared to G(isobutene) from a solution containing only SF_6 in the hydrocarbon. The methylamine converts the $C_4H_9{}^+$ ions to isobutene by proton transfer (see Section b-1 below).

Williams (71) proposed that ionic fragmentation also occurs in the radiolysis of tertiary alcohols. Lower molecular weight ketones are major products from tertiary alcohols whereas aldehydes are formed from *n*-alcohols (72). Ketone formation can be explained by the ionic decomposition step 16 followed by loss of a proton from the fragment ion. Such fragmentation is most likely in tertiary alcohols.

$$
\begin{array}{cc}
\overset{+}{O}H & \overset{+}{O}H \\
| & \| \\
R_1CR_2 \rightarrow R_3\cdot + R_1CR_2 \\
| \\
R_3
\end{array}
\qquad (16)
$$

b. Ion–Molecule Reactions. The role of ion–molecule reactions could more readily be established if it were known what ions were present during radiolysis. Although some fragmentation does occur in liquids as shown above, a large percentage of the excited molecular ions are apparently stabilized by collisions to ground state ions (70). Thus it is important to consider the possible reactions that molecular ions might undergo prior to neutralization.

(1) PROTON TRANSFER REACTIONS. Molecular ions are Brønsted acids in that they can donate a proton to molecules like ammonia and ethanol. This proton transfer reaction is expected to occur readily in liquids in which the molecules have a permanent dipole moment.

In the radiolysis of ethanol, proton transfer reaction (17) followed by neutralization reaction (18) has been proposed (73) to account for the isotopic composition of the hydrogens formed in the radiolysis of C_2H_5OD.

$$C_2H_5OH^+ + C_2H_5OH \rightarrow C_2H_5OH_2^+ + C_2H_5O \qquad (17)$$

$$e_{sol}^- + C_2H_5OH_2^+ \rightarrow H + C_2H_5OH \qquad (18)$$

In organic acids proton transfer is facilitated because of the existence of hydrogen bonded dimers. Proton transfer in acids, reaction (19), results in dissociation of the dimer (74).

$$RCOOH\ldots RCOOH^+ \rightarrow RCOO + RC(OH)_2^+ \qquad (19)$$

Decarboxylation is a major reaction in the radiolysis of acids and is accounted for by the instability of the RCOO radical formed in reaction (19). This is consistent with early studies of the radiolysis of partially deuterated acetic acid which showed that decarboxylation does not occur by a unimolecular reaction such as 20 (75).

$$RCOOH \rightarrow RH + CO_2 \qquad (20)$$

A reaction analogous to (17) has been suggested to occur in the radiolysis of cyclohexane containing ethanol. When the ethanol in such solutions is C_2H_5OD, HD is formed (76) presumably as a result of proton transfer to ethanol, reaction (21).

$$RH^+ + C_2H_5OD \rightarrow C_2H_5ODH^+ + R \qquad (21)$$

Substantial yields of HD are also formed in the radiolysis of solutions of ND_3 in cyclohexane (14,77). This HD is believed to be formed as a consequence of proton transfer to ND_3 from the positive ions which could be either $C_6H_{12}^+$ or $C_6H_{13}^+$. The subsequent neutralization of the ND_3H^+ ion presumably yields a hydrogen atom that can abstract to form

$$RH^+ + ND_3 \rightarrow ND_3H^+ + R \tag{22}$$

molecular hydrogen. Most other processes including charge transfer, electron capture and hydrogen atom abstraction can be ruled out as sources of the observed HD. At an ND_3 concentration of $10^{-4}M$ the yield of HD is 0.076 (Fig. 3). This is precisely the yield to be expected if all free ions ($G = 0.1$) are scavenged and if the ND_3H^+ ion decomposes statistically into either D or H atoms upon neutralization. Apparently only free ions are scavenged at low concentrations. As the concentration of ND_3 increases some of the positive ions that are subject to geminate recombination react according to 22. The yield of HD is proportional to the square root of the ND_3 concentration, as was found to be the case for reactions of the electron with nitrous oxide and methyl iodide. Freeman (78) has calculated theoretical values of $G(HD)$ based on the proton transfer mechanism and obtained a good correlation with the data (see Fig. 3). The fact that a solute at a concentration of 0.01–0.1M can interfere with geminate neutralization has been interpreted to mean that such ions recombine in times less than 10^{-7} sec and that the average ion lifetime is $\sim 10^{-9}$ sec (14).

Proton transfer reactions from a hydrocarbon molecular ion to a

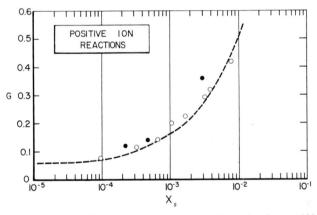

Fig. 3. Reactions of positive ions. Yield of product in molecules per 100 eV vs. X_s, the mole fraction of the solute. ○, yield of HD from solutions of ND_3 in cyclohexane (14,88); ●, yield of $CD_2HCD_2CD_2H$ from solutions of cyclopropane-d_6 in 3-methylpentane (8). 7 Dotted line is theoretical (78).

neutral hydrocarbon molecule have also been suggested (79,80). The available experimental evidence indicates, however, that such proton transfer does not occur readily in most large hydrocarbons. Attempts to observe proton transfer reactions in hydrocarbon vapors with mass spectrometers have been unsuccessful (81,82) except for the now famous examples of methane (83) and ethane (84). Since the rate constant for reaction 23 is 7×10^{11} M^{-1} sec^{-1} (85), CH_5^+ ions are expected to be formed in liquid methane, but this has not as yet been verified experimentally (86).

$$CH_4^+ + CH_4 \rightarrow CH_5^+ + CH_3 \tag{23}$$

Apparently proton transfer reactions are too slow, except in the case of methane and ethane, to compete with geminate neutralization of positive ions in hydrocarbons. However, the *free* positive ion in a hydrocarbon such as cyclohexane could be $C_6H_{13}^+$ since the free ions are sufficiently longlived (at low dose rates) to permit a slow reaction to occur.

(2) H_2 TRANSFER REACTIONS. Another reaction of the molecular cation which has been shown to occur in liquids is the transfer of an H_2 species from the cation to an olefin or to cyclopropane. This reaction is known from gas phase studies (Chapter 4) and evidence for its occurrence in the liquid phase was obtained from studies of the radiolysis of alkanes containing either cyclopropane or olefins (87,88,89).

That H_2 is transferred from the molecular ion to the solute was demonstrated as follows. *The entity transferred is H_2 or D_2* since in the radiolysis of a mixture of C_5H_{12} and C_5D_{12} in the presence of cyclo-C_3D_6 the deuterated propanes that are formed are mainly C_3D_8 and $CD_2HCD_2CD_2H$. *The reactant species is a positive ion* since in the radiolysis the addition of methanol, which reacts with positive ions, reduces the yield of propane. Further in the photolysis of the same mixtures in the gas phase, propane is not formed unless the photon energy is above the photoionization potential. *The reactant is the molecular cation* $C_5H_{12}^+$, not $C_5H_{13}^+$, because the entity transferred in mixtures of C_5H_{12} and C_5D_{12} (1:1) is either H_2 or D_2.

Therefore, in the radiolysis of hydrocarbons containing cyclopropane or olefins, reactions such as (24) and (25) occur.

$$C_5H_{12}^+ + cyclo\text{-}C_3H_6 \rightarrow C_3H_8 + C_5H_{10}^+ \tag{24}$$

$$C_5H_{12}^+ + C_2H_4 \rightarrow C_2H_6 + C_5H_{10}^+ \tag{25}$$

If cyclopropane-d_6 is present in an irradiated perprotonated alkane system, $CD_2HCD_2CD_2H$ is a major product. The dependence of its yield on concentration (Fig. 3) is similar to that for HD formation from ND_3 solutions in that the yield of $CD_2HCD_2CD_2H$ also increases linearly with the square root of the concentration of solute. A similar reaction occurs with low molecular weight olefins and acetylene as solutes. The mechanism of these H_2 transfer reactions is not fully known. It has been noted, however, that H_2 transfer is most efficient when the ionization potential of the additive is comparable to that of the alkane which is being irradiated (87). Thus resonant charge transfer in the reaction complex may facilitate H_2 transfer.

(3) H TRANSFER. The H transfer reaction from molecular ions to alkenes and cyclopropane which was suggested (90) and later demonstrated (91) to occur in the gas phase, has recently been shown to occur in liquid hydrocarbons (89). In the radiolysis of cyclohexane containing cyclopropane-d_6, n-propyl radicals are formed suggesting that reaction (26) occurs. The results indicated that H transfer is a more important reaction of the $C_6H_{12}^+$ ion with cyclopropane than is H_2 transfer.

$$C_6H_{12}^+ + cyclo\text{-}C_3H_6 \rightarrow CD_2HCD_2CD_2\cdot + C_6H_{11}^+ \qquad (26)$$

In any alkane–olefin or alkane–cyclopropane system, H transfer and H_2 transfer are competing reactions of the molecular ion. For example, in the case of cyclopentane, H_2 transfer has been shown to be the principal mode of reaction of the $C_5H_{10}^+$ ions with ethylene (89); therefore, H transfer, reaction (27), must be relatively unimportant in cyclopentane

$$C_5H_{10}^+ + C_2H_4 \rightarrow C_2H_5 + C_5H_9^+ \qquad (27)$$

solutions containing ethylene. The ethyl radicals that are produced in this system (92) are therefore not attributed to reaction (27) but rather to thermal hydrogen atom addition to ethylene (Section IV).

(4) ION–MOLECULE CONDENSATION REACTIONS. In the radiolysis of liquid alkenes a process occurs by which monomer units are coupled directly. The earliest experimental indication of this was the observation that the C_{12} products formed in the radiolysis of 1-hexene were largely monoolefins (93). If hexyl and hexenyl radicals were involved then diolefins and saturated C_{12} products would be formed in addition to the dodecenes.

The molecular condensation reaction was also shown to occur in ethylene from studies of the radiolysis of mixtures of C_2H_4 and C_2D_4 (1:1). The isotopic composition of the products: 1-butene ($G = 0.3$), cyclobutane ($G = 0.1$), and trans-2-hexene ($G = 0.1$) demonstrated that they were formed by molecular dimerization and trimerization reactions (94).

Similarly in the radiolysis of propylene the dimers: 4-methyl-1-pentene ($G = 0.38$) and 1-hexene ($G = 0.14$) are formed principally by molecular dimerization (95).

It has been suggested (96) that in liquid ethylene, the 1-butene, hexenes, octenes, and higher molecular weight olefins could be formed by a reaction which is initiated by $C_2H_4^+$ and terminated by electron combination with $(C_2H_4)_m^+$.

c. Neutralization. In a pure liquid the positive ion is neutralized in the spur by the electron or an anion. Some ions escape this in-spur neutralization and become free ions. However, even in alcohols where the yield of free ions is considerable ($G \sim 1$) about three-fourths of the ions recombine in spurs since the total yield of ions is ~ 4 ion pairs/100 eV. Thus in-spur neutralization is a very important reaction of the positive ions.

In cyclohexane, neutralization apparently leads to the formation of either hydrogen atoms or H_2, since solutes which react with the electron also lower the hydrogen yield. From the considerations discussed above the positive ion which is neutralized may be the molecular ion in all alkanes except methane and ethane. Thus for cyclohexane, for example, we can consider reactions (28) and (29) as likely neutralization reactions (97). Since an intermediate excited molecule is formed in these reactions,

$$e^- + C_6H_{12}^+ \rightarrow C_6H_{12}^* \rightarrow H_2 + C_6H_{10} \qquad (28)$$

$$\rightarrow C_6H_{12}^* \rightarrow H + C_6H_{11} \qquad (29)$$

the products of neutralization can be predicted if it is known how excited molecules of energy equal to the neutralization energy decompose. Optically excited cyclohexane molecules of such energies decompose primarily as in reaction (28) and to some extent as in (29) (see Chapter 3). However, neutralization need not necessarily lead to an excited molecule in an optically allowed state; molecules in other excited states can be formed. For organic molecules in general, neutralization is thought to lead to both singlet and triplet states. For cyclohexane, the formation of a dissociative triplet state could account for the formation of hydrogen atoms and cyclohexyl radicals.

The neutralization of a positive ion by a negative ion is less exothermic than neutralization by an electron, and thus may yield entirely different products from those formed in the charge neutralization by an electron. For example, in a study (98) in which an electron scavenger, CH_3I, was added to *n*-pentane, a drastic reduction was observed in the yields of the alkyl radicals which, in the radiolysis of pure pentane, are probably formed as a result of the decomposition of the n-$C_5H_{12}^*$ species formed in the neutralization of n-$C_5H_{12}^+$. The addition of N_2O to a hydrocarbon system also causes changes in the product distribution (13,50,52,99). However, the interpretation is not always straightforward, since new products may result from a chemical reaction between the negative and positive ions. An increase in G(cyclohexyl) and G(cyclohexene) was noted (50,52,99) in the irradiation of a cyclohexane–N_2O mixture which suggests that reactions of $C_6H_{12}^+$ and O^- (formed in reaction (10)) may lead to the formation of cyclohexyl radicals. These results were explained (50) in terms of the following mechanism:

$$O^- + c\text{-}C_6H_{12}^+ \rightarrow C_6H_{12} + OH \tag{30}$$

$$OH + c\text{-}C_6H_{12} \rightarrow H_2O + C_6H_{11} \tag{31}$$

The occurrence of this mechanism is supported by the observation of H_2O as a product (50,100).

In pure liquids in which the electron reacts rapidly with the substrate the positive ion is expected to be neutralized by an anion. For example, in liquid biphenyl as a consequence of electron capture the neutralization of positive ions probably involves reaction (32).

$$C_{12}H_{10}^+ + C_{12}H_{10}^- \rightarrow C_{12}H_{10}^* + C_{12}H_{10} \tag{32}$$

The radiation stability of such liquids must be in part related to the fact that excited molecules formed in reaction (32) do not decompose (Section III).

In some liquids the molecular cation is expected to undergo proton transfer before neutralization. Thus for alcohols the species which is neutralized is $(ROH)_nH^+$ (see Section II-B-2-b-1). Reactions (33) and (34) have been considered as possible neutralization steps.

$$e_{sol}^- + (ROH)_nH^+ \rightarrow H + nROH \tag{33}$$

$$\rightarrow H_2 + RO + (n-1)ROH \tag{34}$$

Since the yield of D_2 in the radiolysis of C_2H_5OD is very small (73) reaction (33) is more important than (34). Thus the net result of neutralization of the protonated alcohol ion is the formation of a hydrogen atom from reaction (33). It should be pointed out that reaction (33) may proceed in two steps, namely, reaction (6) followed by the reaction of RO^- with the protonated alcohol ion to re-form alcohol, reaction (35).

$$RO^- + (ROH)_nH^+ \rightarrow (n+1)ROH \tag{35}$$

C. Yields of Ionic Species

A large body of data listing yields of free ions is available from conductivity measurements (11,12,15,22,101–103) pulse radiolysis studies (27,33), and determinations of yields with chemical scavengers. It is seen generally that the yield of "free ions" increases as the static dielectric constant of the liquid increases. Data illustrating this point are plotted in Fig. 4. This means that the escape of an electron from the coulombic field of its parent positive ion is facilitated by the polarity of the medium

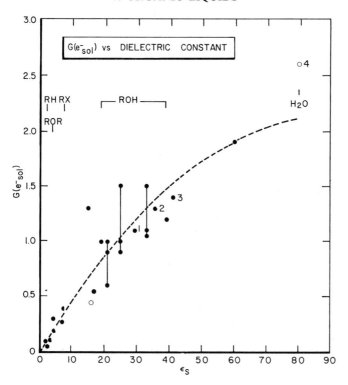

Fig. 4. The yield of solvated electrons in liquids of various static dielectric constant. Open circles are data for water and ammonia. Dotted line is theoretical (78,101). Data was obtained from the following references: alkyl halides and ethers (101,107), ammonia (34), alcohols (7,31,33,44,45,105,108). Points for ethanol–water mixtures are numbered: *1*, 26% water; *2*, 45% water; *3*, 58% water; *4*, 100% water.

in which the ion is formed. In liquids of high dielectric constant, as discussed above, the electron will become solvated.

It is of interest to compare yields of free ions and free electrons determined for a given system by various different methods. For example, conductivity measurements indicate that G(free positive ions) is about 0.1 in cyclohexane. The pulse radiolysis of cyclohexane containing $10^{-3}M$ aromatic solutes (see Section A-2-b above) gives a value for G(free ions) of 0.3–0.4 (39); these yields probably represent the sum of the yields of cations and anions and should be divided by two to obtain the yield of ion pairs. Thus, within the assigned experimental uncertainty of 50%, these yields are considered to be in agreement with the free ion yield determined by conductivity. The yield of electrons in cyclohexane determined by studies with chemical electron scavengers such as N_2O and

CH_3I is about 4 electrons/100 eV (see Section II-B-1). These results do not contradict the results of the conductivity measurements, since these solutes determine not only the so-called "free electrons", but also those electrons which ordinarily would undergo geminate recombinations with positive ions in the spur. Thus, the yield of ions varies with time after passage of an ionizing particle. At very short time intervals (10^{-13} sec) the yield of ions is large, $G \sim 3$ to 4. Geminate recombination of ions occurs rapidly, but high concentrations of solutes can scavenge electrons in this period. At 10^{-9} sec only a fraction of the ions, estimated to be $\frac{1}{12}$ of the total, remain (104). At 10^{-6} sec, pulse radiolysis studies indicate the yield of ions is approximately 0.1 ion pairs/100 eV. Pulse radiolysis studies (109) of ion yields at nanosecond times should provide an experimental test of this description.

In one study the solvated electron yield in ethanol was measured (with scavengers) as a function of temperature (105). This study indicated $G(e_{sol}^-)$ at $-110°C$ is 1.9 compared to a value of 1.5 at $25°C$ and the dielectric constant of ethanol increases with decreasing temperature. There is some uncertainty about this temperature effect at present since a pulse radiolysis study of ethanol (31) indicated the yield of solvated electrons at $-78°C$ is within 5% of the value observed at 25°.

The yield of free ions can be calculated theoretically from the number of ions formed initially, their distribution, and the probability $\phi(r)$ that an ion pair of separation r will escape geminate recombination (101). The electron is considered to be free of the positive ion when it has exceeded the critical distance (r_c) at which the coulombic energy is balanced by the thermal energy of the ions. This distance is given by equation (36a) for singly charged ions. The probability of escape is given by equation (36b).

$$r_c = e^2/\epsilon_s kT \tag{36a}$$

$$\phi(r) = \exp(-r_c/r) = \exp(-e^2/\epsilon_s ykT) \tag{36b}$$

The separation r will have a distribution of values which can be calculated if values of the energy and range of secondary electrons are known. This distribution of separations is not expected to vary from liquid to liquid. Thus, the yield of free ions depends primarily on the value of the dielectric constant. Until the dependence of ion yield on the static value of the dielectric constant shown in Fig. 4 had been demonstrated experimentally, there was uncertainty about what value to use for the dielectric constant in such calculations. In previous theoretical treatments (106) a value of 3 was used for liquids of high static dielectric constant such as water. The experimental results suggest the correct value to use is the static value. Theoretical yields of free ions for various liquids have been calculated in this way and are represented in Fig. 4 as the dotted line. A yield of 3 ion pairs/100 eV gives the best fit to the experimental results (78). The theory satisfactorily represents the yields of free ions for non-polar liquids as determined by the conductivity method, as well as the yields of solvated electrons found in polar liquids (see Section C-2 below).

The time required for geminate recombination of the electrons which are trapped within the field of the positive ions will determine whether or not reactions of scavengers with these ionic species can occur. This theoretical treatment (101) also leads to a relationship (eq. 37) for the time (in seconds) required for geminate recombination. The time required, τ_g, is a function of the dielectric constant, the separation of the ions r (in

$$\tau_g = \epsilon(r^3 - r_0^3)/4.32 \times 10^{-7}\mu \tag{37}$$

cm), and the encounter radius r_0 (in cm) of ion recombination. According to this equation, ion pairs in hydrocarbons separated by no more than 50 Å will recombine in times less than 3×10^{-10} sec. In order for charge scavengers to intercept a major fraction of the ions in such liquids it is necessary that the reaction of the electron with the scavenger S be rapid. Specifically the requirement is that the product of the rate constant for reaction of S with the electron times the concentration of S must be greater than about 10^9 sec^{-1}.

III. EXCITED MOLECULES

In the radiolysis of organic liquids electronically excited molecules are formed. However, the yields of excited molecules and the level of excitation cannot be predicted *a priori*. In radiation chemistry excited molecules of various energies may be formed initially. If a molecule receives energy in excess of its ionization potential it may be excited to a "superexcited" state which does not always autoionize (see Chapter 1). Since optical selection rules do not apply to interactions of slowly moving electrons, excited molecules may be produced in triplet as well as in singlet states. Charge neutralization of positive ions also results in the formation of electronically excited molecules which may be in singlet or triplet states. Determining the yields and roles of excited molecules is one of the challenging tasks of the radiation chemist.

A. Singlet States

In this section the detection, yields, and reactions of excited molecules in singlet states are considered. Much of this information has been obtained from studies of liquid scintillators, that is, of pure liquids or liquids with dissolved fluorescent solutes which emit light upon absorption of ionizing radiation. For a complete discussion of the general topic of scintillation the reader is referred to reviews (110,111).

Organic liquids emit light when subjected to ionizing radiation. The spectral distribution of the light emitted is characteristic of the liquid that is irradiated and includes fluorescence from excited molecules in singlet states (111,112). Only a small fraction of the energy absorbed appears as light; for a pure liquid such as benzene the efficiency of light emission is less than 1% (111). The addition of a solute such as *p*-terphenyl, anthracene, or diphenyloxazole increases the emission yield considerably. This improvement is attributed to energy transfer to the solute which has a higher fluorescence efficiency than the solvent (see Section A-4 below). Because of this amplification effect, scintillation studies are generally carried out with such solutes present.

1. Lifetime Measurements. Upon irradiation of a solution of *p*-terphenyl in benzene (a "scintillator-solution") light which is primarily the fluorescence of *p*-terphenyl is emitted. If the solution is subjected to a pulse of radiation, the lifetimes of the intermediates responsible for the emission can be determined from the decay of the emission. The decay of luminescence intensity is characterized by several lifetimes (113–115). One of these is the lifetime of excited scintillator molecules, another is the lifetime of excited solvent molecules and a third is associated with the phenomenon of delayed fluorescence. The emission lifetime associated with the excited *p*-terphenyl molecules is 2.2 nsec (110,113). This lifetime varies slightly for different solvents and quencher concentrations (116). The second lifetime is a function of the scintillator concentration (111,117). In a benzene solution of terphenyl, excited benzene molecules can either be deactivated, reaction (38), or can transfer energy to the terphenyl

$$B^* \rightarrow B \tag{38}$$

$$B^* + S \rightarrow S^* + B \tag{39}$$

(reaction (39)). The lifetime, τ, of the excited solvent molecule B^* in such a solution is then given by equation (40).

$$1/\tau = k_{38} + k_{39} [S] \tag{40}$$

From measurements of this lifetime at low solute concentrations the rate constants k_{38} and k_{39} can be independently evaluated. The reciprocal of k_{38} is the lifetime of the excited molecule in the pure solvent, k_{39} is the rate constant for energy transfer. Such measurements have been made for benzene and solutions of benzene in cyclohexane (117). The lifetime of the species which transfers energy to the scintillator is 16 nsec for pure benzene and 33 nsec for dilute solutions ($\sim 1\%$) of benzene in cyclohexane. That is, the lifetime of an excited benzene molecule decreases when surrounded by more and more benzene molecules. This self-quenching has been attributed to the formation of excimers (see Section A-3 below). For benzene, it has been established that the excited molecule which transfers energy to the scintillator is in the $^1B_{2u}$ state (111). Photochemical experiments have indicated that the lifetime of this state is ≤ 25 nsec (118), which agrees with the value derived from luminescence decay times.

There is in addition a long-lived component of the fluorescence which can be observed for substances which can be excited to triplet states. This component is termed delayed fluorescence and arises from the interaction of two excited triplet–state molecules. The fluorescence is long-lived since the triplet–state molecules have long lifetimes. Such triplet–triplet annihilation can occur homogeneously. This effect has been reported for solutions of anthracene in benzene (119). The delayed fluorescence of anthracene was observed as a result of the bimolecular annihilation of triplet-state anthracene molecules, reaction (41).

$$^{(3)}A^* + {}^{(3)}A^* \rightarrow {}^{(1)}A^* + A \tag{41}$$

Reactions similar to (41) would also be favored in spurs because of the high local density of excited molecules in such regions. Voltz, Dupont, and King (120) have shown that the time dependence of the fluorescence decay in toluene excited by alpha particles can be accounted for by a diffusion model involving solvent triplet states and a bimolecular reaction similar to (41). The model assumes the existence of both singlet and triplet states in the spur. Fluorescence occurs from excited molecules in singlet states which are formed either directly or by triplet–triplet interactions. The solution obtained to the diffusion equation for times long compared to the lifetime of the excited singlet molecules was found to give a very good fit to the experimental luminescence decay curve in the 20–160 nsec time scale. These results support the concept of spur interactions of excited molecules in aromatic liquids (see Section V). This theory could account for the observation of Sjölin (114) that the decay time of a commercial organic liquid scintillator increased from 25 nsec for electron beams to 47 nsec for proton beams. This increase could result from an increased contribution of delayed fluorescence arising from triplet interactions in the tracks for the proton case. It is known that the relative intensity of the long-lived component increases with LET (110).

2. Rates of Excitation Quenching and Transfer. Studies of the rates of excitation quenching and transfer have revealed some unusual aspects of the energy transfer process in scintillator solutions. The quenching of luminescence occurs by the interaction of a quencher molecule, such as oxygen or carbon tetrachloride, with an excited molecule to remove the excitation (reaction (42)), and may include energy transfer (reaction (39)), where the quencher molecule becomes electronically excited.

$$B^* + Q \rightarrow B + Q \qquad (42)$$

In considering the rates of these reactions, two cases need to be distinguished; the molecule which transfers energy can be an excited *solute* molecule or an excited *solvent* molecule. Rate constants are generally higher in the latter case.

If the excited molecule is a *solute* molecule, such an excited terphenyl molecule in a liquid scintillator solution, rate constants for quenching and transfer can be accounted for by diffusion. In a benzene solution where the scintillator solute is 2,5-diphenyloxazole (PPO), the quenching rate constant $k(PPO^* + O_2 \rightarrow)$ is 3.7×10^{10} M^{-1} sec^{-1} (121). Similarly if the solute is anthracene (A) the quenching rate constant $k(A^* + O_2 \rightarrow)$ is 3.0×10^{10} M^{-1} sec^{-1} (21). It has been shown by Ware (21,122) that the theory of diffusion controlled reactions satisfactorily accounts for this rate of quenching of anthracene fluorescence providing the transient terms as well as the unusually large diffusion coefficient of oxygen are taken into account.

The rate constant of energy transfer from an excited molecule to an acceptor molecule is comparable to the rate of quenching if the excited molecule is a solute molecule. The rate constant for transfer from an excited benzene molecule to a scintillator in a cyclohexane solution containing only 1% benzene has been measured in decay time studies (117). Equation 40 was used and k_{39}, the rate constant, for transfer of excitation from benzene to the scintillator was calculated from the slope of a plot of $1/\tau$ vs. scintillator concentration. The observed value of k_{41} was 1.6×10^{10} M^{-1} sec^{-1} which is comparable to the predicted value based on the rate of diffusional encounters in cyclohexane (118).

When the excited molecule is a solvent molecule, surrounded by unexcited molecules of its own kind, rates of transfer and quenching are generally greater and exceed the diffusion limited rate. For example, the rate constant for quenching of excited benzene molecules by oxygen in benzene, $k(B^* + O_2 \rightarrow)$, is $9 \pm 2 \times 10^{10}$ M^{-1} sec^{-1} (111,121). For quenching of excited benzene by metal perphenyls in benzene the rate constant is 10^{11} M^{-1} sec^{-1} (123). The analogous rate constant for quenching of excited toluene molecules by oxygen is 6×10^{10} (111,124). High quenching rate constants are not limited to aromatic solvents. The rate constant for quenching of excited cyclohexane molecules in cyclohexane by carbon tetrachloride, $k(C_6H_{12}^* + CCl_4 \rightarrow)$, is equal to or greater than 2.5×10^{11} M^{-1} sec^{-1} (125). An excited cyclohexane molecule may not necessarily be involved here however (see Section A-3 below).

Similar high rate constants are observed for energy transfer from excited solvent molecules. An example is transfer from excited benzene molecules to p-terphenyl in benzene solution (reaction (39)). From decay time studies (117) the rate constant was evaluated to be 5×10^{10} M^{-1} sec^{-1}. Thus the rate constant for energy transfer, k_{39} is 3 times greater in pure benzene than in the 1% benzene solution. A comparable high value for k_{39} was derived from steady state luminescence studies of benzene–terphenyl solutions (111). The fact that transfer and quenching rate constants are similar must be taken into account when considering mechanisms (see Section A-3 below).

The decay time measurements (117) show quite convincingly that the rate of energy transfer increases significantly with increasing benzene concentration in the benzene–cyclohexane solution. In other mixtures the variation in rate with concentration is not as great. In a study of the luminescence of a benzene–methanol–PPO solution it was found that the fluorescence intensity was proportional to the benzene concentration from 1 to 100% benzene (126). This indicates the transfer rate is concentration independent in this case. The same conclusion was reached in a study of toluene–cyclohexane mixtures (127). However in a comparable system, p-xylene–hexane, it was found that the rate of transfer increased a factor of 1.7 in going from dilute to concentrated xylene solutions (111).

3. Mechanism of Excitation Transfer. The rate of transfer of energy from an excited *solute* molecule to another solute molecule can be explained by the theory of diffusion controlled reactions. However, observed rate constants for transfer from an excited *solvent* molecule to a solute are higher than diffusion controlled rates. That diffusion plays a role in transfer in the latter case has been shown by studies of viscosity effects. The transfer efficiency as measured in scintillation experiments decreases with increasing viscosity as expected, but the decrease is less than predicted from diffusion theory (128). It appears therefore that rate constant data require that processes other than material diffusion contribute to transfer from excited solvent molecules.

a. Ions. A role for ionic reactions in the quenching of excited solvent molecules has been suggested (111). Differences in quenching rates have been noted in comparing gamma ray with ultraviolet induced luminescence. Quenching is more efficient in the gamma ray case. Such differences can be explained if the quencher interferes not only by transfer of excitation, reaction (39), but also by reaction with ions (equation (43)). If the quencher reacts with the electron, the nature of the neutralization step is changed

$$B^+ + e^- + Q \rightarrow B + Q \qquad (43)$$

and the formation of excited molecules thereby inhibited. Such a mechanism is reasonable since the rate constants for reactions of the electron with typical quenchers such as oxygen and carbon tetrachloride are large (Table III).

The effect of quenchers on product yields is consistent with ionic interactions. In the case of benzene, quenchers reduce the luminescence intensity but do not affect the yield of hydrogen. Thus if the quencher does react with the electron, the electron cannot be precursor of hydrogen. Studies with nitrous oxide as solute have confirmed that the electron is not a precursor of hydrogen (50). When cyclohexane is the solvent, quenchers both reduce the yield of products such as hydrogen (see Section VI) and the luminescence intensity. Thus, in this case if the mechanism of quenching is interaction with the electron, the electron must be a precursor of hydrogen, and the experimental results with nitrous oxide additive indicate that it is (13).

b. Domains. High rate constants for transfer and quenching of excited solvent molecules imply a role of unexcited solvent molecules in the transfer. Lipsky and Burton in 1959 (129) proposed that rapid transfer is a consequence of liquid structure. They pictured in the solvent flickering microcrystalline domains, the existence of which is supported by x-ray (130) and thermodynamic studies (131). An excitation belonging to the entire domain is called an exciton. Excitons can account for high rates of energy transfer from solvent molecules since a collision of a solute molecule with any molecule in a domain is equivalent to a collision with the initially excited molecule.

Domains were also invoked to explain the luminescence phenomena observed for mixtures containing benzene, cyclohexane, terphenyl, and quencher (132,133). The luminescence intensity at certain cyclohexane-benzene ratios exceeds that expected on the basis of the mixture law (see Section VI) but at 1% benzene a minimum in the luminescence is observed. It has been suggested (2) that at 1% benzene the excited benzene molecules lack the efficient mechanism of energy transfer to the terphenyl (involving domains) which exists at higher concentrations and thus are more readily quenched.

c. Excimers. Another mechanism which has been suggested to explain energy transfer in scintillator solutions involves "excimers." Ivanova, et al. (134) showed in 1962 that an excited benzene molecule in liquid benzene associates with an unexcited benzene molecule to form an excited dimer, called an excimer. The existence of excimers is manifested by the concentration dependence of the long wavelength component of benzene fluorescence. In pure benzene the excited monomer B* and the excimer

D* contribute about equally to the fluorescence (117,135). Excimer formation is now well established and is known to occur in all aromatic hydrocarbon solvents typically used in liquid scintillators.

Energy transfer from an excited solvent molecule to a solute can occur from either an excited monomer or the excimer (136). Further, the excitation can be transported via excimer formation (136–139). Birks et al. have proposed a fast excitation migration mechanism based on the formation and dissociation of excimers. As a consequence of reaction (44) the excitation moves from molecule i to molecule j.

$$B_i{}^* + B_j \rightleftarrows D_i{}^* \rightarrow B_{ij} + B_j{}^* \tag{44}$$

Transfer rate constants will vary with benzene concentration since the excitation migration mechanism will prevail at high benzene concentration and the diffusion mechanism will prevail at low concentrations.

Intramolecular excimer formation has been found in diphenyl alkanes and polystyrene (140). In this case an excited phenyl group associates with a neighboring unexcited phenyl group (on the same molecule) to form an excimer. If Birks' migration mechanism applies here, then in the case of polystyrene long-range energy transfer along the chain is possible.

It is also significant to the mechanism that the high values of rate constants for transfer are not unique to excitation by ionizing radiation. Similar effects are observed in luminescence studies in which ultraviolet excitation is used. For example, studies of the luminescence of benzene–terphenyl solutions indicated the rate constant for energy transfer k_{39} is the same for both ultraviolet and gamma radiolysis (129). A similar comparison of the fluorescence of solutions of PPO in cyclohexane and n-hexane has demonstrated that electronic excitation transfer from alkanes to aromatic solutes does occur (141). The PPO solute fluoresced even when the cyclohexane was photolyzed at wavelengths in the vacuum ultraviolet region where the cyclohexane absorbs most of the light. It was concluded from this work that the main interaction of aromatic solutes in cyclohexane both in photolysis and beta radiolysis is a true electronic energy transfer from excited cyclohexane molecules to the solute. For n-hexane, however, the amount of fluorescence observed in photolysis of similar solutions was considerably less than for cyclohexane but not zero. Thus in the case of n-hexane other energy transferring processes than electronic energy transfer must occur to give the observed (142) β-ray induced luminescence. The mechanism of energy transfer in alkanes is unclear; excimers are unknown in this case but domains have been postulated (132,143).

4. Yields of Excited Molecules. For liquid scintillators the light emitted during radiolysis is the characteristic fluorescence of the scintillator solute, that is, a photon is emitted when an excited scintillator molecule returns to the ground state. If the absolute yield of photons is measured, the yield of excited *solute* molecules can be calculated, providing the fluorescence efficiency ϕ_f is known. In cases where the excited solute molecules are formed by excitation transfer from excited solvent molecules the yield of the latter $G(M^*)$ can then be calculated from equation (45);

$$G(M^*) = G(\text{photons})/(\phi_t)(\phi_f) \qquad (45)$$

where ϕ_t is the efficiency of energy transfer in the solution. A measurement of the yield of excited molecules thus requires a knowledge of ϕ_t, ϕ_f and the yield of emitted photons (111).

As shown by equation (45) the yield of light emitted from a liquid during radiolysis is proportional to the fluorescence efficiency. The low light output from a pure liquid such as benzene compared to a solution containing a scintillator like terphenyl is in part due to differences in ϕ_f. The quantum yield of fluorescence is 0.04 for benzene and close to unity for a scintillator like terphenyl (144–146). Thus transfer of the excitation from benzene to terphenyl enhances the probability of light emission by a factor of 25.

The yield of excited molecules in toluene has been calculated from the absolute light emission efficiency. For a solution of hexadecane-1-^{14}C and 2,5-diphenyloxazole in toluene (147) a yield of 793 photons per beta particle was observed. Since the average energy of the beta particles is 49 keV (148), this corresponds to $G(\text{photons}) = 1.6$. In the toluene solution ϕ_t is 0.7 and ϕ_f is 0.95 (123). Thus from equation (49) the yield of excited molecules for beta rays of 49 keV average energy is ~ 2.4 molecules/100 eV. In another study of the same system the yield of excited toluene molecules was found to be 3.2/100 eV (124).

A lower limit for the yield of excited molecules in paraffin oil was obtained in a pulse radiolysis study in which anthracene was the fluorescent solute (149). The yield of excited anthracene molecules in the solution after the pulse was calculated from the fluorescence yield. At 0.87 mM anthracene $G(A^*)$ was 1.1. A yield of excited solvent molecules greater than 1.1 is indicated since ϕ_t is presumably less than unity.

The efficiency of emission of light from liquid scintillators decreases with increasing LET of the ionizing radiation. A direct comparison of the relative fluorescence yields per unit energy loss by alpha and beta particles, which is termed the alpha/beta ratio, has been made for several liquid scintillators (150). The alpha/beta ratio for benzene as solvent was found to be 0.1 and for cyclohexane, 0.06. This decreased efficiency of light emission for alpha particles is attributed to dynamic quenching of excited solvent molecules by other species such as radicals in the particle track (see Section IV-D) rather than to a change in the initial yield of excited molecules with LET. The fact that the alpha/beta ratio is smallest for cyclohexane, for which the radical yield is the largest, supports the role of radicals in dynamic quenching. The light emission efficiency of a scintillator solution has also been measured for fission fragment irradiation and found to be even less than for alpha particles (151).

B. Triplet States

In the radiolysis of solutions it has been established that excited *solute* molecules in triplet states are formed in substantial yields. To the extent that these are formed by triplet–triplet transfer from excited solvent molecules they provide a measure of the yield of *solvent* triplets in the radiolysis of pure liquids. Excitation of triplet states of solute molecules may however occur in other ways than triplet–triplet transfer from solvent triplets. Other mechanisms which may be involved are: charge transfer to solute molecules followed by neutralization, intersystem crossing (152,153) from an excited singlet state of the *solute* molecule, excitation of the solute by low energy electrons, or even triplet transfer from excited product molecules.*

The general methods by which triplets are detected in radiation chemistry are discussed below. Subsequently the mechanisms of formation of solute triplets in various solvents are considered.

1. Detection of Triplet States. *a. Absorption and Emission Spectroscopy.* In pulse radiolysis studies excited molecules can be detected by absorption spectroscopy. Studies to date have been limited to triplet states, the lifetimes of which generally exceed the resolving time of the detection equipment. It is preferable to use a substance as solute for which the triplet absorption spectrum is known in order to facilitate observation and identification of the transient. Fortunately, many such absorption spectra are known. The triplet anthracene absorption spectrum (38), for example, has a strong absorption peak at 4240 Å. Other solutes for which the triplet absorption spectra are known are: naphthalene (4150 Å), 1,2-benzanthracene (4850, 4590, 4280 Å), benzophenone (5300 Å), and biphenyl (3600 Å) (155,156); the numbers in parentheses refer to the wavelengths of absorption maxima. It is also necessary that the absorption coefficient ϵ be sufficiently large that the concentration of excited molecules induced by the radiation will cause an observable change in optical density. Anthracene is particularly well suited for the detection of triplet excited molecules because of the large value of ϵ at 420 mm of $1.2 \times 10^5 \text{ M}^{-1}$

---

* The formation of products in excited triplet states has been observed in the radiolysis of solutions of benzyl chloride dissolved in ethanol (Ref. 154).

cm^{-1} (145). Quantitative application of this method to determine the yield of molecules excited to triplet states depends on the accuracy with which ϵ is known.

This method has been used to detect only *solute* triplet excited molecules, such as triplet naphthalene in a dilute solution of naphthalene in benzene. In principle, *solvent* triplets could be detected by absorption spectroscopy in pulse radiolysis studies but this has not as yet been done. The detection of triplet state benzene in the radiolysis of pure benzene would be of particular interest and its absorption spectrum has recently been reported (159). However, detection of triplet benzene* by the pulse radiolysis method may prove to be difficult if its lifetime is only 10^{-8} sec as is indicated by kinetic studies (158,159).

Triplet states can also be detected by light emission; excited molecules in triplet states emit a characteristic phosphorescence. Advantages of this means of detection are that the triplet excited molecules can be identified from the emission spectrum and the intensity of emission is a measure of the yield of triplets. Triplet biacetyl has been detected in radiolysis by this method; the characteristic green phosphorescence at 520 nm has been observed in the pulse radiolysis of a solution of biacetyl in benzene (162). The emission of delayed fluorescence also provides evidence of the presence of triplet states (119).

b. *Chemical Indicators of Triplet Excited Molecules.* Many solutes are known which react with molecules excited to triplet states. Diphenyl-picrylhydrazyl reacts with photolytically generated triplet states of benzene, toluene, and chlorobenzene and is destroyed in the reaction (163). Ferric chloride has also been reported to scavenge triplets in radiolysis. In a solution of 0.5 mM FeCl$_3$ in benzene $G(-FeCl_3) \sim 3$ (164), which is too great a yield to attribute to free radicals or ions. It has been suggested (164) that FeCl$_3$ is reduced by triplet benzene at approximately millimolar concentrations. These solutes also react with radicals and this reaction must be distinguished in assessing the role of triplets with these solutes.

The *cis-trans* isomerization of 2-butene has been used to measure the yield of triplet excited molecules in the ultraviolet photolysis of benzene in the gas (165) and liquid phases (159). The 2-butenes, as well as other substituted olefins, also undergo *cis-trans* or *trans-cis* isomerization when present as solutes during the radiolysis of hydrocarbons. The yield of isomerization provides a measure of the yield of solute triplet states

* A transient has been observed in the pulse radiolysis of pure benzene which absorbs at wavelengths between 3000 and 3400 Å (160,161) with a maximum at 3200 Å. It has been suggested that this transient is an excited molecule (156) but the absorption spectrum is quite different from that observed by Godfrey and Porter (157) which they assigned to the triplet state of benzene.

Table IV

Yields of Solute Triplets

Solvent	Solute	Technique	G(solute triplets)	Ref.
Alkanes				
Cyclohexane	Naphthalene	Spectroscopy	3.3	170
Cyclohexane	Anthracene	Spectroscopy	0.5[a]	171
Cyclohexane	t-Stilbene	Isomerization	2.3	61
n-Octane	2-Butene	Isomerization	1.2	172
Dodecane	2-Butene	Isomerization	2.1	172
Aromatics				
Benzene	2-Octene	Isomerization	5.0	166,173
Benzene	2-Butene	Isomerization	4.0	159,174
Benzene	Naphthalene	Spectroscopy	2.6	170
Benzene	t-Stilbene	Isomerization	6.0	175, 61, 167
Benzene	$FeCl_3$	Chemical change	1.4	164
Benzene	n-Butyrophenone	Chemical change	> 1.0	169
Benzene	Anthracene	Spectroscopy	0.7[a]	158
Benzene	Diphenyl-cyclopropane	Isomerization	> 1.6	176
Ketones				
Acetone	2-Butene	Isomerization	1.8	172
Acetone	Anthracene	Spectroscopy	0.7[a]–1.0	177
Cyclopentanone	None	Reactivity	0.6[b]	178,179
Cyclohexanone	None	Reactivity	0.6[b]	180
Alcohols				
Ethanol	Anthracene	Spectroscopy	Small	38
Ethers				
Dioxane	Anthracene	Spectroscopy	—	32
Tetrahydro-furan	Anthracene	Spectroscopy		170

[a] The molar extinction coefficient for anthracene triplets is taken as 11.5×10^4 (145).

[b] These values refer to the yield of solvent triplet excited molecules.

(Table IV) in cases where it can be shown that the isomerization is a result of triplet excitation of the olefin. This seems to be the case for the 2-alkenes, since $G(cis \to trans)/G(trans \to cis)$ is approximately the same as the corresponding ratio observed for photochemically produced triplet cis- and trans-alkenes (166). However, cases have been cited where the isomerization of olefin solutes does not involve solute triplet states.

Examples are the isomerization of pure *cis*-stilbene in benzene (167) and of *trans*-stilbene in methanol (168).

   c. Reactions of Triplet Excited Molecules. The yield of triplet state intermediates formed in the radiolysis of a substrate can be assessed from product yield measurements if the triplet decomposes in a characteristic manner. This method of detection which does not require the presence of a solute has been used primarily in the radiolysis of ketones.

   It is known from photochemical studies that the triplet state of cyclopentanone isomerizes to 4-pentenal, reaction 46. It is also known that the triplet state can be quenched by piperylene and the ratio k_{47}/k_{46} is 21 M^{-1} (178).

$$\text{cyclopentanone*} \rightarrow \text{4-pentenal} \qquad (46)$$

$$\text{cyclopentanone*} + \text{piperylene} \rightarrow \text{cyclopentanone} + {}^{(3)}\text{piperylene*} \qquad (47)$$

In the radiolysis of cyclopentanone, 4-pentenal is also formed, piperylene quenches its formation, and k_{47}/k_{46} is 20 M^{-1} (179). Because of this similarity in reactivity, it is concluded that the excited triplet state is involved in 4-pentenal formation in the radiolysis (178). The yield of triplets can then be assessed from G(4-pentenal).

   Another example of this type of study is the radiolysis and photolysis of liquid *n*-butyrophenone. In both cases ethylene is a major product (181). In the photolysis ethylene is formed by photoelimination. Such eliminations may occur from both the singlet and triplet states of excited ketones (182), but in butyrophenone mainly the triplet state is involved (183). Thus, the triplet state butyrophenone molecule accounts for some of the ethylene ($G = 3.5$) formed in the radiolysis.

$$C_6H_5COC_3H_7* \rightarrow C_6H_5COCH_3 + C_2H_4 \qquad (48)$$

   The formation of triplet excited states in the radiolysis of ketones has also been demonstrated in pulse radiolysis experiments (177). In solutions of aromatic molecules, such as anthracene in acetone, the triplet state of the aromatic has been detected. The observed yields of such triplet state molecules are comparable to those determined for cyclopentanone (Table IV). From the rate of formation of triplets following the pulse of radiation the lifetime of the precursor of the aromatic triplet was determined to be 5 μsec. This value is comparable to that indicated by photochemical studies for the lifetime of triplet state acetone in pure liquid acetone (184). The precursor is therefore believed to be triplet state acetone and the mode of excitation of the solute is probably triplet–triplet transfer.

   2. Mechanism of Excitation of Solute Triplet Excited Molecules.
a. Intersystem Crossing. The mechanism of excitation of anthracene and

naphthalene in hydrocarbon solutions has been investigated by McCollum and Nevitt (149). For a solution of anthracene in paraffin oil, it was reported that the ratio of the yield of excited triplet anthracene to the yield of excited singlet anthracene was equal to 0.13, which is the value of this ratio found in photochemical studies (185). Triplet states of anthracene can be formed by the well known photochemical process of intersystem crossing, reaction (49).

$$^{(1)}S^* \rightarrow {}^{(3)}S^* \tag{49}$$

Since the ratio of triplets to singlets is no greater than that observed photochemically, it is not necessary to postulate any other mechanisms to explain triplet state formation in anthracene dissolved in paraffins.

The same conclusion was arrived at from a pulse radiolysis study of solutions of naphthalene in cyclohexane (170). It was observed that the yield of naphthalene triplets went through a maximum of $G = 3.3$ as the concentration of naphthalene increased. The dependence of G(triplets) on concentration is similar to the dependence of the yield of luminescence from liquid scintillators on concentration (186). This similarity suggests a common mechanism of production of solute singlet and triplet states in alkanes and strongly supports the suggestion of McCollum and Nevitt (149) that triplets are formed by intersystem crossing from singlet states.

b. Ionic Reactions. Aromatic solutes in alkanes are known to react with electrons. Therefore, in studies where solutes like anthracene, naphthalene, or stilbene are present, it is to be expected that the neutralization process in alkanes will involve reaction (50), particularly at high solute

$$RH^+ + S^- \rightarrow S^* + RH \tag{50}$$

concentrations. As a result of this reaction excited solute molecules in both singlet and triplet states can be formed. Excited molecules which are initially in singlet states may be converted to the triplet by intersystem crossing, reaction (49).

Recent pulse radiolysis experiments show that large yields of solute triplets are present 3 nanoseconds after the pulse (189); thus a fast excitation process is required. Ion neutralization is sufficiently rapid since most ions recombine in times less than 1 nanosecond in alkanes (see Section II-C).

In a study of the isomerization of t-stilbene in cyclohexane (61) it was observed that the yield of isomerization of t-stilbene approaches a limiting value of $G = 1.4$ above $0.1M$ stilbene. It has been suggested that in this system triplet stilbene molecules are formed as a result of electron capture and neutralization. Ionic precursors are indicated since the concentration dependence of isomerization is similar to that of the yield of nitrogen in nitrous oxide-cyclohexane solutions, and electron scavengers such as CCl_4 reduce the yield of isomerization. The results can be explained if neutralization of the molecular cation RH^+ by the stilbene anion results in the formation of excited stilbene as in reaction (50).

Triplet state anthracene has been detected in the pulse radiolysis of solutions of anthracene in dioxane (32,170). Hydrochloric acid is effective in removing 75% of these triplets indicating ions are likely to be involved. Baxendale, Fielden, and Keene (32) suggested that triplets are formed in dioxane solutions by a reaction analogous to 50.

c. Triplet–Triplet Transfer. Triplet state solute molecules can be formed in radiolysis by transfer of electronic energy from solvent triplet excited molecules to the solute, reaction (51). Such transfer is likely only if the triplet energy level of the solvent molecule is higher than or equal to

$$B^* + S \rightarrow S^* + B \qquad (51)$$

that of the solute (Fig. 5). If this is the case, then energy transfer is efficient. In fact the specific rate constant for energy transfer can be equal to or greater than the diffusional rate of encounters (118). For most of the solutes considered here, the energy level of the triplet state is lower than that of benzene. Thus, in principle, these solutes are excellent acceptors of energy from triplet benzene. 2-Butene is particularly well suited as an acceptor as its triplet energy level is just below that of benzene. Further, the first singlet of 2-butene is above the $^1B_{2u}$ level of benzene (Fig. 5), thus, energy transfer from the $^1B_{2u}$ state of benzene to 2-butene should not occur. This is supported by the fact that 2-butene has no effect on the fluorescence yield of benzene (159). However, transfer from higher singlet states of benzene to 2-butene is energetically possible.

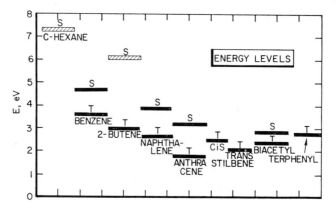

Fig. 5. Singlet (S) and Triplet (T) energy levels of donors and acceptors (187,188). The singlet energy levels of cyclohexane and 2-butene are estimates.

The yield of isomerization of 2-butene in benzene solutions increases with the concentration of 2-butene to a plateau value of approximately 2.0 at $0.1 M$. Radical scavengers such as iodine and electron scavengers such as carbon tetrachloride do not affect the yield of isomerization (172). Charge transfer from benzene to 2-butene is unfavorable energetically. Thus, the triplet–triplet transfer mechanism, reaction (51), seems to be the most likely mechanism of formation of triplet excited butene in benzene solutions. This is supported by the fact that at 24°C the ratio k_{51}/k_{52} is

$$B^* \rightarrow B \tag{52}$$

15 M^{-1} in radiolysis, and 20 M^{-1} in the photolysis of similar solutions (159). Further, this ratio of rate constants has the same temperature dependence in photolysis and radiolysis.

Solute triplets are apparently formed in this case by triplet–triplet transfer and the data indicate a yield of triplet benzene of 4.0.

Cis-trans isomerization is quite general for solutions of 2-alkenes in benzene. The limiting yield of isomerization of 2-pentene, 2-hexene, 2-heptene, and 2-octene is 2.5 molecules/100 eV at alkene concentrations above $0.5 M$ (166,173). This provides additional evidence that the yield of triplet excited molecules in benzene radiolysis is high.

The rate of triplet–triplet transfer (reaction (51)) could be faster than the rate of diffusion since it involves a reaction of an excited solvent molecule. Thus, as was the case for singlet excited molecules (Section III-A) the solvent may also play a role in triplet transfer either through exciton transfer in domains or by triplet excimer formation. There is evidence that triplet benzene forms excimers as do excited singlet molecules (190). Cundall and Griffiths (159) have obtained evidence from a study of the effect of temperature on k_{51}/k_{52} that the solvent participates in triplet transfer. Their results are not consistent with transport of excitation by diffusion but can be accounted for by assuming the triplet excitation is delocalized as, for example, in domains.

Several different groups of investigators have measured the yield of isomerization of stilbene in benzene (61,167,175,191–193). The recent work by Hentz et al. (167) indicates that the yield of isomerization of cis-stilbene cannot be used to quantitatively estimate the yield of triplet molecules. However, there are several facts which show that at low concentrations of stilbene ($\sim 10^{-3} M$) the isomerization is a result of the formation of triplet stilbene. Between 10^{-4} and $10^{-3} M$ the ratio $G(cis \rightarrow trans)/G(trans \rightarrow cis)$ is equal to 1.0 which agrees with the corresponding ratio obtained for photochemically produced *cis*- and *trans*-stilbene (61,175). At these low concentrations the yield of isomerization is reduced by the presence of millimolar quantities of anthracene and enhanced by the presence of

molecules isomerize to stable compounds such as fulvene (198,201) and valence isomers such as benzvalene (202,203). One would, therefore, expect that fulvene would be an initial product of the radiolysis since singlet excited molecules are formed, but it has not been reported. This can be explained since fulvene is not formed in the photolysis of benzene if olefins are present and olefins are products of the radiolysis. Hexafluorobenzene isomerizes to the "Dewar" isomer (hexafluorobicyclo 2.2.0 hexa-2,5-diene) when irradiated (204). Because of the formation of valence isomers, isomerization of substituted benzenes would be expected to occur in the radiolysis. The experimental evidence on this point is inconclusive. Verdin (205) reports a yield of radiation induced isomerization of o-xylene to m-xylene of $G = 0.04$ for massive doses ($\sim 10^{22}$ eV/g). Another report (206) indicates the yield of isomerization is two orders of magnitude greater than this.

Information about the decomposition and reactions of optically excited cyclohexane molecules is available from recent photochemical experiments (207). At wavelengths of 1280 and 1470 Å the main products of the direct photolysis are hydrogen and cyclohexene and thus reaction (54) is the main primary process. The formation of hydrogen atoms and cyclohexyl radicals (reaction (55)) also occurs but to a lesser extent. The decomposition

$$C_6H_{12}{}^* \rightarrow H_2 + C_6H_{10} \ (\sim 86\%) \tag{54}$$

$$C_6H_{12}{}^* \rightarrow H + C_6H_{11} \ (\sim 14\%) \tag{55}$$

of such excited molecules therefore probably accounts for a significant part of the cyclohexene and hydrogen formed in the radiolysis.

IV. RADICALS

Radicals are short-lived neutral molecular fragments which have one or more unpaired electrons. These are formed in the radiolysis of organic liquids by the decomposition of excited molecules and ions and have a significant role in product formation. It is important to distinguish the reactions of radicals in spurs from the reactions of homogeneously distributed radicals. The "free" radicals, which are diffusing in the bulk of the liquid during radiolysis can be detected by the methods that are discussed here. The radicals of much shorter lifetimes which react in spurs are generally not detected by these methods.

A. Methods of Detection and Identification of Radicals

Chemical scavenger techniques such as those described by Schuler (208) are very useful for radical detection. However, an important factor in establishing the existence and identity of radicals in irradiated liquids has been the development of physical methods of detection such as electron spin resonance and absorption spectroscopy. The principles of detection of radicals by these physical and chemical methods are discussed here. Additional experimental details can be found in available reviews of each method (25,209,210). Also included is a discussion of yields and reactions of radicals in several organic liquids.

1. Physical Methods. *a. Electron Spin Resonance.* The detection of radicals by electron spin resonance (ESR) is based on the magnetic properties of radicals. A radical in a magnetic field is oriented and can absorb microwave electromagnetic radiation and reverse its spin. High sensitivity to radicals is possible here since there is no background absorption; that is, molecules which have no unpaired electrons can be thought of as being transparent to the electromagnetic radiation. An isolated electron would show a single absorption line. In a radical, because the electron is affected by the magnetic field of adjacent nucleii, hyperfine structure is observed. It is this coupling to the nucleii that permits radicals to be identified by this technique.

Early applications of electron spin resonance were confined to the detection of radicals in solids (Chapter 9). The development by Fessenden and Schuler (209,211) in 1960 of an ESR method to detect radicals in liquids during radiolysis was a most significant advance. In liquids, in contrast to solids, the ESR absorption lines are narrow (< 1 G). Measurements of the spacing between the narrow well-resolved lines usually permit the investigator to identify the radical intermediates.

Detection of radicals in irradiated liquids by this method requires a high rate of radical production. The necessary rate can be calculated from equation (56) in which the production rate, on the left hand side, is equated to the rate of disappearance of radicals. It is assumed that radicals disappear primarily in bimolecular reactions;

$$10G_R \times (\text{dose rate})/N = 2k_{57}[R]^2 \qquad (56)$$

$$R + R \rightarrow R_2 \qquad (57)$$

the dose rate is in eV ml^{-1} sec^{-1}. For example, if the detection limit for radicals is assumed to be a concentration of $10^{-7}M$ and if k_{57} is taken as 10^9 M^{-1} sec^{-1} (Table X) then the production rate must be at least $10^{-5}M$ sec^{-1}. Thus *in situ* detection is possible with any radiation source which generates radicals at this or a greater rate.

The capability of the ESR technique has been amply demonstrated. Numerous hydrocarbon radicals have now been positively identified in radiolysis experiments. These include alkyl radicals, allylic radicals, and vinyl radicals (212). The reason that this technique has not been used to detect radicals in polar liquids such as alcohols is that high loss of microwave power in these liquids, which are physically inside the ESR cavity during the experiment, precludes the application of the technique.

In addition to its use to identify radicals the ESR technique has supplied other radiation chemical information. In an early study (209), the rate of combination of ethyl radicals was determined from the ethyl radical concentration observed by ESR, the yield of $G(C_2H_5) = 4.4$ (213), and equation (56). Radical yields can alternately be measured by ESR if the lifetime and concentration of radicals as well as the dose rate are substituted into equation (56). The yield of "free" ethyl radicals in liquid ethane at $-175°C$ was determined in this way to be 3.8 radicals/100 eV (214).

Two methods of determining radical lifetimes in ESR experiments have been demonstrated (214). One of these is the analog of the rotating sector technique often used in photochemical experiments (215). The electron beam from a van de Graaff accelerator is electronically pulsed to provide the intermittent irradiation. The radical lifetime is deduced from the change in average concentration of the radicals as the pulsing frequency is varied.

Lifetimes can also be determined by what has been termed the "sampling" technique. This method is so named because the ESR signal is observed only at a given delay time after each irradiation pulse. A series of pulses is required to provide an adequate signal. To obtain the entire radical decay curve the delay time is varied. The lifetime of the ethyl radical in liquid ethane at $-177°C$ was found to be 7.3 msec with the sampling technique (214).

b. Optical Absorption Spectroscopy. Simultaneous to the development of the ESR technique, radicals were also being detected in organic liquids by optical absorption spectroscopy in pulse radiolysis experiments (25). This became possible in about 1960 with the availability of electron accelerator machines capable of producing high current pulses of microsecond duration. High dose rates are necessary in order to produce a sufficiently high concentration of radicals. This is particularly important for detection of radicals by optical absorption spectroscopy since for many radicals the molar extinction coefficient is not very large (Table V).

Several radicals have now been identified in pulse radiolysis studies of organic liquids. However, the association of an optical absorption spectrum with a radical species is not as straightforward as the identification of a radical by ESR. Usually a good deal of other chemical information including the lifetime of the species is necessary in order to assign a spectrum. In some cases the absorption spectrum of the radical may be

Table V

Absorption Spectra of Organic Radicals

Radical	Wavelength of Absorption Maxima in Å	ϵ M^{-1}, cm^{-1}	Ref.
C_6H_{11}	2550 (?)	340	218
$C_6H_{11}O_2$	2750, 2960	$\epsilon_{275} \sim 2000$	219
$C_6H_{10}OH$	3020, 3140, 3330	—	219
CH_3CHOH	2470, 2570, 2660, 2900	$\epsilon_{297} = 230^a$	220
CH_2OH	2900	$\sim 1200^b$	26
Allyl			
	4080	—	221
	3100	—	222
	2250	—	223
Phenyl			
	4500–5300		216
Benzyl			
	3050, 2970, 4530	$\epsilon_{318} \sim 1100^c$	154, 216
Cyclohexadienyl			
C_6H_6OH	3130	3500	150
$C_6H_5C_6H_6$	3230	—	38, 161
	3600	—	160
Ketyl			
$C_6H_5\dot{C}OHC_6H_5$	3300, 5450	$\epsilon_{545} = 5100$	216
		$\epsilon_{330} = 20,000$	

[a] Absorption coefficient for aqueous solution.
[b] $G(CH_2OH)$ is assumed to be 6.
[c] This value is based on product analysis.

known from flash photolysis studies. Such is the case for many aromatic radicals such as benzyl, ketyl, and phenyl radicals (216). Benzyl radicals have been observed in the radiolysis of solutions of benzyl chloride in cyclohexane and alcohols (154,217). Ketyl radicals have been observed in the radiolysis of solutions of benzophenone (28,42,155). The absorption spectrum of the phenyl radical is known (Table V) and phenyl radicals are belived to be intermediates in the radiolysis of benzene but they have not yet been observed in pulse radiolysis studies.

Identification of aliphatic radicals is not as facile as aromatic radicals because little is known about their absorption spectra. In the pulse radiolysis of cyclohexane, a transient which absorbs in the far ultraviolet has been detected. The absorption spectrum (Fig. 6) has been assigned to the

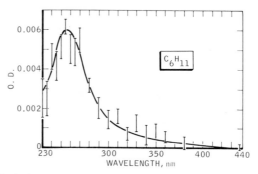

Fig. 6. Optical absorption spectrum of the cyclohexyl radical, optical density vs. wavelength in nm. Maximum at 255 nm may be an experimental artifact, see page 51 of ref. (171). Reproduced by permission, ref. 218.

cyclohexyl radical (218). When oxygen is present the absorption shifts to longer wavelengths (219); the absorbing species here is presumably the cyclohexylperoxy radical. A series of hydroxyalkyl radicals have been detected in the pulse radiolysis of alcohols.

The absorption spectrum of the allyl radical as observed in the flash photolysis of allyl bromide vapor consists of a series of diffuse bands, the strongest being at 4083 Å (221). A stronger absorption at much shorter wavelengths with a maximum at 2250 Å observed in the flash photolysis of various olefins is also attributed to allyl (223). An absorption spectrum with a peak at 3100 Å observed in the pulse radiolysis of allyl bromide in cyclohexane has also been assigned to the allyl radical (222). In all three of these recent studies it is reasonable that allyl radicals were intermediates but it is doubtful that all three absorption systems are due to allyl radicals.

c. Light Emission. Radicals can be determined through their emission spectrum in cases where they are formed in excited states and *lose* the excess energy through light emission. This method of detection is practical only for radicals, such as the benzyl radical, for which the emission spectrum is known (224). Detection of radicals by emission might be observed during steady state irradiation as well as in pulse radiolysis depending on the sensitivity of the detection equipment. In a study of this type the emission from gamma irradiated benzene solutions was monitored as a function of the concentration of various solutes (112) (see Section VI). Light emission was observed even though the normal fluorescence of the benzene was excluded by a filter. It was suggested by the authors that the emitting species may be excited radicals or ions.

2. Chemical Methods of Radical Detection. The chemical methods of radical detection can be classified either as titration or "sampling" techniques. An example of the former is the use of iodine which titrates or scavenges all the "free" radicals. In the sampling technique a radioactively labeled radical such as $^{14}CH_3$ is employed to combine with a small fraction of the "free" radicals. Any one method may be useful either to identify and measure the individual radical yields or to measure the total radical yield. The emphasis here is on methods of the former type which provide more detailed information about fundamental fragmentation processes.

a. Titration Techniques. (1) IODINE. Extensive use has been made of iodine as a radical scavenger in hydrocarbon radiolysis. Iodine is attractive because its reaction (58) with radicals is fast (see Table X) and the yields of

$$R + I_2 \rightarrow RI + I \qquad (58)$$

alkyl iodides formed can in general be determined to provide a quantitative measure of the individual radical yields. Further, the system is kinetically simple; since each radical removes one-half of an iodine molecule, the total radical yield $G(TR)$ is equal to $G(-\frac{1}{2}I_2)$. Schuler (208) has cautioned that the concentration of iodine should be $\leq 10^{-3}M$ since complicating reactions such as electron capture (see Section II and Ref. 209) ensue at higher concentrations. This necessitates analysis of less than micromole quantities of alkyl iodides, which is possible with available detectors for gas chromatography or by the use of iodine-131 and radiochemical techniques.

Detailed information concerning the interaction of iodine has been obtained by pulse radiolysis studies (218). In cyclohexane solutions the iodine atom generated in reaction (58) forms a transient complex, cyclo-$C_6H_{12}\cdot I$, and this transient disappears in a rapid bimolecular reaction which reforms I_2.

One limitation of the iodine technique is that in certain organic liquids the iodide formed in reaction (58) may be unstable; tert-alkyl iodides for example have not been successfully detected with this technique (210). The iodine method has been most useful primarily in those hydrocarbons where such labile iodides would not be expected.

(2) HYDROGEN IODIDE was recognized as a radical scavenger in early studies by Forrestal and Hamill (225) and Schuler (226). The reaction of hydrogen iodide with radicals is also fast; k_{59} is 0.7 k_{58} if R is cyclohexyl (227) and 0.5 k_{58} if R is a radical formed in irradiated *n*-hexane (228). In

$$R + HI \rightarrow RH + I \qquad (59)$$

this method the total yield of radicals is measured by the appearance rather than by the disappearance of iodine. The iodine formation complicates the method in that the iodine competes with HI for radicals. The HI concentration cannot be increased to overcome this difficulty because of complications caused by electron capture and energy transfer phenomena (225,226); however, the corrections necessary to take iodine scavenging into effect can readily be applied (227,228).

When either tritium or deuterium iodide is used then labeled hydrocarbons are formed in reaction (59). The labeled hydrocarbons serve to identify as well as measure the yield of the individual radicals which are present (229). This technique has been used to determine the yields of fragment radicals for the liquid n-alkanes from hexane through hexadecane.

(3) OTHER TITRATION REAGENTS. Many other solutes such as DPPH, anthracene, oxygen, nitric oxide, galvinoxyl, etc., have been used as radical scavengers (208,230,231). In general, these reagents seem to be less reactive than either iodine or HI toward radicals. Anthracene, for example, is reported to be 100 times less reactive than iodine towards radicals (232). There are complications which must be borne in mind here too. For example, it is known that certain of these solutes react with excited molecules (see Section III) and ions (see Section II).

b. Radical Sampling. In the radical sampling technique (233,234) a radioactively labeled radical is generated from solutes such as $^{14}C_2H_4$ or $^{14}CH_3I$ (reactions (60) or (61)) which are present at millimolar concentrations. Low concentrations of the labeled solutes are necessary to avoid

$$e^- + {}^{14}CH_3I \rightarrow {}^{14}CH_3 + I^- \qquad (60)$$

$$H + {}^{14}C_2H_4 \rightarrow {}^{14}C_2H_5 \qquad (61)$$

large perturbations of the system; both solutes react with ions and high concentrations would result in an altered neutralization reaction (Section II-B-2-C). The labeled radical is present during radiolysis at low concentrations ($< 10^{-7}M$) and reacts *in situ* with a small fraction of the "free" radicals which are combining homogeneously. Labeled hydrocarbons, $R_i^{14}CH_3$, which are formed by combination reactions such as (62) characterize the radicals R_i formed from the solvent. It is necessary in

$$^{14}CH_3 + R_i \rightarrow R_i{}^{14}CH_3 \qquad (62)$$

this technique to employ experimental conditions which favor radical–radical reactions. This requires utilizing sufficiently high dose rates that radical–radical reactions predominate and radical addition and abstraction reactions (see Section C below) are relatively unimportant. For saturated hydrocarbons dose rates in excess of 10^{18} eV g^{-1} sec^{-1} must be employed. For olefins, ketones, and other more reactive substrates either much higher dose rates or lower temperatures are required.

It has been shown that the relative yields of these labeled hydrocarbons are proportional to the relative radical yields (234). Thus the ratio of the yields of two radicals R_1 and R_2 is given by expression (64). The factors D/C are the disproportionation to combination ratios for reaction of

$$G(R_1)/G(R_2) = \frac{G(R_1{}^{14}CH_3)}{G(R_2{}^{14}CH_3)} \times \frac{(1 + D_1/C_1)}{(1 + D_2/C_2)} \tag{63}$$

methyl radicals with each radical; i.e., k_{64}/k_{62}.

$$^{14}CH_3 + R_i \rightarrow {}^{14}CH_4 + R_i(-H) \tag{64}$$

Accurate experimental measurements of these disproportionation ratios are not available for the numerous radical reactions encountered in the radiolysis of organic liquids. However disproportionation factors are generally not large and apparently introduce small errors since the results obtained with this "sampling" technique have been shown to be very similar to those obtained by other methods where comparison is possible (210).

Absolute yields of radicals can be determined from the relative yields, as given by equation (63), if any one radical yield is known. A yield that can be measured absolutely is $G(^{14}CH_3)$ as this is the total yield of labeled hydrocarbons. This method has been applied to the determination of radical yields in various hydrocarbons (68,235,236) including alkenes (94,237,238). In the case of cyclohexene it was used to determine the effect of L.E.T. on radical yields. This sampling method which measures radical *yields* should not be confused with the ESR sampling technique (Section A-1-a) which measures radical *lifetimes*.

B. Radical Formation in Various Organic Compounds

1. Alkanes. Information about the yields and identity of the radicals formed in the radiolysis of the *n*-alkanes from methane to hexadecane is now available (210). In gamma radiolysis the total yield is about five radicals/100 eV. In the case of *n*-hexane four-fifths of the radicals are hexyl radicals formed by C—H bond scission, and one-fifth are fragment alkyl radicals, formed by cleavage of the carbon–carbon bonds. The average values for the yields, obtained through three different chemical methods (TI, I_2, and $^{14}C_2H_5$ sampling) are given in Table VI. The percentage of fragment radicals is not a constant for all alkanes but decreases with increasing chain length.

In *n*-alkanes the yield of a fragment radical is sometimes roughly equal to the yield of the complementary radical, as expected for simple carbon–carbon bond scission. There is a definite tendency for carbon–carbon bonds in the middle of the molecule to break down more readily particularly for long chain alkanes. This effect is very predominant in hexadecane (Fig. 7).

In a recent comprehensive survey of the yields of methyl radicals from

Table VI

Radical Yields in the Radiolysis of Some Hydrocarbons

Hydrocarbon	Radicals[a]	G(total radical)	Ref.
Alkanes			
Methane	H(> 2.8), methyl (> 2.3)	—	209,86
Ethane	Methyl (0.6), ethyl (3.8)	—	213,214
n-Hexane	Methyl (0.7), ethyl (0.3), n-propyl (0.3), butyl (0.27), pentyl (0.04), hexyl (4.1)	4.8–5.9	210,228 .
2,2-Dimethylbutane	Methyl (0.99), ethyl (1.2), butyl (0.57), hexyl (1.2)	—	236,241
Cyclopentane	Cyclopentyl (4.0), n-pentyl (0.1), 4-pentenyl (0.1), cyclopentenyl-3 (0.2)	4.6–5.5	210
Cyclohexane	Cyclohexyl (~ 4)	4.8–5.5	210,240
Alkenes			
Ethylene	Ethyl (1.7), vinyl (0.9), 3-butenyl	2.6	94,234
trans-2-Butene	Methyl (0.2), sec-butyl (1.3, trans-1-methylallyl (2.5), cis-1-methylallyl (0.1)	4.0[b]	237
Cyclohexene	Cyclohexyl (1.6), 3-cyclo-hexenyl (3.4)	5.2	238
Cyclohexadiene-1,4	Cyclohexadienyl	—	242,243
Aromatics			
Benzene	Phenylcyclohexadienyl, cyclohexadienyl	0.7–1.0	164,244

[a] Numbers in parenthesis are yields in radicals/100 eV.
[b] The total radical yield is assumed to be 4.

alkanes (239) it was shown that $G(CH_3)$ from n-alkanes is proportional to $(n - 1)^{-2}$ where n is the number of carbon atoms in the molecule (equation (65)).

$$G(CH_3) = 2.04/(n - 1)^2 \qquad (65)$$

The yield of ethyl radicals shows a similar dependence on the length of the alkane chain, as given by expression (66) (see Fig. 8). For n-alkanes $G(C_2H_5) = 3.5G(CH_3)$.

$$G(C_2H_5) = 7.0/(n - 1)^2 \qquad (66)$$

Branched alkanes have been the subject of several extensive studies on

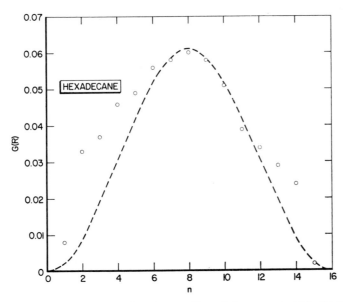

Fig. 7. Radical yields from *n*-hexadecane. Plot of the percent yield of total fragment radicals as a function of *n*, the number of carbon atoms in the radical (212).
Dotted line is a normalized plot of $0.061 \sin^2 n\pi/16$.

radical yields. The yields of methyl radicals were determined for eighteen
branched hydrocarbons by Schuler and Kuntz (239). Individual alkyl
radical yields have been determined for all the pentanes, hexanes, 2,4-
dimethylpentane and 2,2,4-trimethylpentane (68,236). The total radical
yield as in the case of *n*-alkanes is approximately 5 radicals/100 eV.
Branching however enhances carbon–carbon bond scission (241). Fragment alkyl radicals represent an increasingly larger percentage of the total
radical yield as the number of side chains increases. For example the
yield of fragment alkyl radicals is 2.6 for 2,2-dimethylbutane compared to
1.0 for *n*-hexane (Table VI).

This effect of molecular structure on the yields of fragment radicals was
noted by Schuler and Kuntz (239) who related the yield of methyl radicals
to the number of methyl radicals attached at a particular site in a molecule.
They derived an empirical relationship for estimating methyl radical yields
for any alkane. A similar effect of structure was noted to apply to fragment radicals other than methyl in various branched alkanes (236).
Stated simply this effect is that the carbon–carbon bond most likely to
cleave is the one with the most carbon–carbon bonds adjacent to it. With
this fact and the dependence of radical yields on $(n-1)^{-2}$ noted above in
mind, an empirical equation (equation (67)) was derived for estimating the

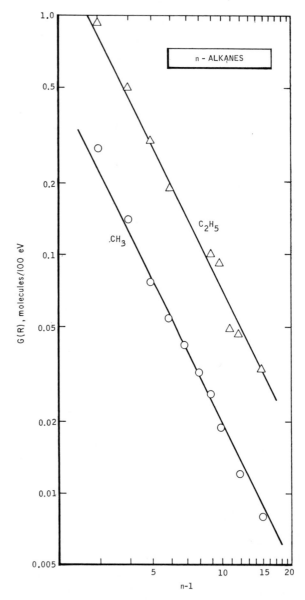

Fig. 8. Methyl and ethyl radical yields in *n*-alkanes and the dependence on $(n-1)^{-2}$ (210,212,239,240). $\bigcirc$, methyl radical yields in *n*-alkanes; $\triangle$, ethyl radical yields in *n*-alkanes. Curves are lines of slope equal to -2.

yield of a fragment radical, R;

$$G(R) = \frac{1}{(n-1)^2} (1.0C_1 + 2.8C_2 + 8.6C_3 + 29C_4) \qquad (67)$$

C_1, C_2, etc. are the number of carbon–carbon bonds in the molecule which have 1,2 etc. adjacent carbon–carbon bonds and which can be dissociated to form the fragment radical R. Equation (67) could equally well be written as (68).

$$G(R) = \frac{1}{(n-1)^2} \sum_{i=1}^{4} 3^{(i-1)}C_i. \qquad (68)$$

Equation (67) has been shown to predict the yield of fragment radicals from alkanes with reasonable accuracy. It has some limitations, it does not apply to the yields of fragments remaining after loss of a methyl or ethyl radical because of unsymmetrical fragmentation. For example, in branched alkanes $G(CH_3) > G$(Parent-minus-methyl). It also predicts that a break at any carbon–carbon bond except the terminal ones in a n-alkane has equal probability. This is true in a particular case such as n-hexane but these radical yields differ somewhat in a long chain n-alkane (Fig. 7).

A theoretical explanation of the observed structural effects on radical yields is not available. It is premature for such considerations since the precursor of fragment radicals has not been specified. However it has been reported (245) that in the photolysis of liquid 2,2,4-trimethylpentane at a wavelength of 1470 Å the ratio of quantum yields of the fragment radicals is very similar to the ratio of yields of fragment radicals observed in the radiolysis of this liquid (236). This similarity suggests that neutral excited molecules may be precursors of fragment radicals in alkanes. It has also been suggested that the probability of dissociation of a particular carbon–carbon bond is related to the distribution of the electrons in the highest occupied molecular orbital of the molecule (236). A comparison of the calculated electron density (246) in this orbital for the various C—C bonds in n-hexadecane with the observed yields of radicals originating from C—C scission at the various sites in the molecule is shown in Fig. 7.

The individual radical yields in the radiolysis of cyclopentane have been measured with both the iodine and radical sampling techniques (68,247). These methods and the ESR studies indicate that the major radical species is cyclopentyl. A minor radical detected by the chemical methods is n-pentyl. It has been shown that this radical is not formed in a secondary free radical reaction; it may be the product of an ionic reaction. The radical sampling method revealed that two additional kinds of C_5 radicals are formed: 3-cyclopentenyl and 4-pentenyl. The formation of 3-cyclopentenyl radicals requires a process involving cleavage of several chemical bonds. This type of process is unique to cycloalkanes; that is, the analogous allylic radical species are not observed in normal and branched alkanes.

In the radiolysis of cyclohexane, cyclohexyl radicals have been detected by electron spin resonance (209), absorption spectroscopy (Fig. 6), and various chemical techniques (210). The yield of cyclohexyl radicals as measured with the iodine technique is ~4 at $10^{-4}M$ iodine (240). By analogy with cyclopentane small yields ($G \sim 0.1$) of n-hexyl, 5-hexenyl, and 3-cyclohexenyl radicals are expected to be formed. Of these radicals only n-hexyl has actually been identified in the radiolysis of cyclohexane (225).

A major unsolved problem in the radiolysis of cyclohexane is the mechanism of formation of cyclohexyl radicals. Cyclohexyl radicals are only a minor product in the photolysis of cyclohexane indicating that the decomposition of neutral excited cyclohexane may not account for the large yields of cyclohexyl radicals observed in the radiolysis.

Thermal Hydrogen Atoms. Detection of thermal hydrogen atoms in alkanes is difficult because they rapidly abstract from the solvent. The lifetime of a hydrogen atom at 25°C in cyclohexane is estimated to be $\sim 10^{-7}$ sec based on the rate constant of reaction (69) (Table IX). The lifetime is expected to be longer at lower temperatures

$$H + C_6H_{12} \rightarrow H_2 + C_6H_{11} \tag{69}$$

However, the only direct evidence for the formation of thermal hydrogen atoms in the radiolysis of liquid alkanes is the observation of the H atom ESR absorption spectrum in liquid methane at −175°C (209).

One of the chemical methods that has been used to measure $G(H)$ involves determining the effect of a solute, which is judged to be a hydrogen atom scavenger, on the yield of H_2. This method gives conflicting values for H atom yields in the radiolysis of cyclohexane (Table VII), and is invalid unless the solute scavenges only thermal hydrogen atoms. Many solutes scavenge precursors of hydrogen other than thermal hydrogen atoms (Sections II and III). The lack of agreement for $G(H)$ obtained with benzene as the solute is attributed in part to an experimental uncertainty; there is disagreement regarding the extent to which benzene reduces the hydrogen yield. Merklin and Lipsky (248) found benzene had little effect at low concentrations and concluded $G(H)$ is small. Most other workers have reported much greater effects of benzene (Section VI).

The total reduction in the hydrogen yield at high concentrations of solute can be taken as an upper limit for $G(H)$; thus the data obtained with iodine or oxygen as scavengers require $G(H) \leq 2.0$. The yield of HI from

Table VII

The Yield of Hydrogen Atoms in the Radiolysis of Cyclohexane

Solute	Range of solute concn. M	Yield measured	$G(H)$, atoms/100 eV	Ref.
Methyl methacrylate	0.01–0.07	$\Delta G(H_2)$	3.1	249
Iodine	0.002–0.04	$\Delta G(H_2)$	~2.0	80
Oxygen	0.005–0.02	$\Delta G(H_2)$	~2.0	80
Benzene	0.05–0.6	$\Delta G(H_2)$	0.2–2.4	248, 250
Iodine	0.02–0.04	$G(HI)$	2.1	251
Hydrogen iodide	0.05–0.7	$G(HD)$	1.7	252
Ethylene	0.01–0.07	$G(C_2H_5)$	1.8	92
Ethylene	0.001–0.3	$G(C_2H_5)$	1.3	253

iodine solutions is in agreement with this value. The depression of the hydrogen yield by millimolar concentrations of either iodine (226) or oxygen (80) appears to be consistent with the rate constant data (Table IX) suggesting this effect is due to hydrogen atom scavenging.

A more direct method of evaluating $G(H)$ consists of measuring the yield of a product HS which is produced when hydrogen atoms react with a solute S. Ethylene has been employed as such a solute and the product formed is the ethyl radical (reaction (61)). The maximum yield of ethyl radicals formed in cyclohexane was found to be 1.8 with the sampling technique (92) and 1.3 with the radioactive iodine technique (253). These values can be criticized since high concentrations of ethylene were necessarily employed and ethylene reacts with positive ions at this concentration (see Section II). However the value of $G(H)$ can also be inferred from the ethyl radical yield at low concentrations of ethylene where ion scavenging is minimal (see Fig. 4). A yield of hydrogen atoms of 1.5 has been estimated from $G(C_2H_5)$ at 10 mM ethylene concentration and the rate constant ratio of $k_{61}/k_{69} = 300$ (253).

A comparable value for $G(H)$ was obtained by Meissner and Henglein (254) in the radiolysis of solutions of D_2S in n-hexane. The yield of HD which could be equated to $G(H)$ at high concentrations of D_2S was 1.5. It has also been shown that the yield of $G(H)$ is about 1.5 for normal and branched alkanes (92). It is concluded that there is an intermediate formed in the radiolysis of alkanes with a yield of 1.5 which reacts with ethylene to form ethyl radicals and with D_2S to form HD.

2. Alkenes. The radical intermediates in irradiated liquid ethylene have been established by ESR and sampling techniques (209,234) to be ethyl, vinyl, and 3-butenyl (Table VI). An interesting temperature effect was observed in the radiolysis of this olefin. Vinyl radicals were readily detected at temperatures below $-160°C$ but not at temperatures near $-130°C$. Instead, 3-butenyl radicals were detected at the higher temperatures. This shows that vinyl radicals add readily to ethylene, reaction (70). From the observed temperature dependence it was found that the energy of activation of reaction (70) is 3.4 kcal/mole which is well below

$$C_2H_3 + C_2H_4 \rightarrow 3\text{-}C_4H_7 \tag{70}$$

that for addition of alkyl radicals to ethylene. The enhanced reactivity of the vinyl radical can be attributed to its unique sp^2 structure (209).

In alkenes other than ethylene, the most important radical intermediate is the allylic species. This fact has been demonstrated by the radical sampling technique (237) and by electron spin resonance studies of liquids (212) and solids (255). For example, in propylene 45% of the radicals are allyl and in the 2-butenes and cyclohexene about two-thirds of the radicals are allylic (Table VI). The 1-methylallyl radicals that are intermediates in the radiolysis of cis- and trans-2-butene are predominately in the same configuration as the parent olefin; for example, the radical formed from trans-2-butene has structure (71). This retention of configuration shows that the precursor of radicals in olefins, whether it is an ion or excited molecule, does not isomerize prior to dissociation or reaction.

$$
\begin{array}{ccc}
CH_3 & & CH \\
\diagdown & & \diagup \\
& CH & CH_2
\end{array}
\tag{71}
$$

The second most important radical that is found in alkenes is the H atom adduct. Thus, cyclohexyl radicals are formed in cyclohexene, butyl radicals in butenes and propyl radicals in propylene (237). These radicals are not attributed to thermal hydrogen atom addition to the olefins because of the large yields of normal radicals that are formed in the radiolysis of 1-alkenes (237). For example in propylene the ratio $G(iso\text{-}C_3H_7)/G(n\text{-}C_3H_7)$ = 2.7 and thermal hydrogen addition to propylene in the liquid phase gives isopropyl and n-propyl radicals in a ratio of 10:1 (233). Further, the yield of thermal hydrogen atoms in the radiolysis of alkenes is low; in cis-2-butene $G(H) \leq 0.4$ atoms/100 eV (237).

Some process other than hydrogen atom addition must be invoked to explain alkyl radical formation in alkenes. Reaction sequence (72–73) has

$$C_4H_8^+ + C_4H_8 \rightarrow C_4H_9^+ + C_4H_7 \tag{72}$$

$$C_4H_9^+ + e^- \rightarrow C_4H_9^* \tag{73}$$

been proposed for isobutene (256), and analogous reactions have been postulated for other alkenes (93,95). The excited butyl radicals formed in reaction (73) may be collisionally deactivated. Some decomposition, as in reaction (74), is to be expected, however.

$$C_4H_9^* \rightarrow C_2H_4 + C_2H_5 \tag{74}$$

In fact for most alkenes the yield of alkyl radicals is less than the yield of allyl radicals.

Methylallyl radicals are also formed in reaction (72) and if this reaction is a proton transfer rather than a H atom transfer reaction then the $C_4H_8^+$ molecular ion must retain the geometric configuration of the olefin from which it was formed to account for retention of configuration in allylic radicals.

3. Aromatic Hydrocarbons. This discussion will be concerned with the radiation chemistry of benzene, since this molecule has been more extensively studied than other aromatic species. Since the total radical yield for benzene is smaller than that observed for alkanes, detection by physical means is more difficult. Attempts to observe radicals in irradiated liquid benzene by the ESR technique have been unsuccessful but in solid benzene cyclohexadienyl radicals and other radicals (209,242) have been detected by ESR. In the pulse radiolysis of benzene as well as chlorobenzene and some methyl substituted benzenes, optical absorption spectra have been observed which were assigned to cyclohexadienyl and substituted cyclohexadienyl radicals (160,161). The substituted cyclohexadienyl radicals, first observed in pulse radiolysis studies (160,161) may have various isomeric structures. For example, the phenyl-cyclohexadienyl radical which is likely to be present in the radiolysis of benzene would be of structure (75a) if formed by phenyl radical addition to benzene. The formation of cyclohexadienyl is an indication that hydrogen atoms are present. The lifetime of a thermal hydrogen atom in benzene should be about 10^{-8} sec if the rate of addition is that given in Table IX. In neither of these studies were phenyl radicals observed. This is not surprising since phenyl radicals add rapidly to benzene.

There is little information available at present from direct physical observation or scavenger techniques as to the identity of individual radicals in other aromatic liquids. Even the total radical yield is uncertain (1). In the pulse radiolysis of pure α-methyl styrene two transients which absorbed at 366 mμ were detected (257). The longer lived transient of 5 msec half-life was assigned to an unidentified radical species. It has been observed that the $C_{12}H_{11}$ radical (presumably structure 75b) is formed in the radiolysis of solutions containing biphenyl, possibly by electron attachment followed by proton transfer to the biphenylide ion (160).

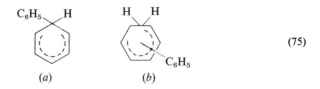

$$(75)$$

(a)   (b)

4. Halogenated Alkanes. It is interesting to compare fluorocarbons with hydrocarbons with respect to radical formation. The relative importance of carbon–carbon bond cleavage in the two types of compounds can be assessed from radical yield measurements. Radical yields in irradiated liquid C_2F_6 have been investigated by the electron spin resonance technique and, in contrast to liquid ethane, the radical spectrum observed was that of CF_3 rather than C_2F_5 (258). Carbon–carbon bond cleavage is thus of major importance in fluorethane.

The structure of CF_3 was deduced from its ESR spectrum to be nonplanar; the electron is in an orbital with considerable s-character, nearly sp^3. Because of its structure CF_3 does not need to rearrange in reacting by H abstraction to form a compound such as CF_3H. This may account for the fact that the activation energy for abstraction of hydrogen by CF_3 is lower than it is for CH_3 radicals (259).

The radiolysis of some other normal and branched perfluoroalkanes have been studied but radical detection techniques have not been employed. The products formed indicate that the carbon–carbon bonds are more susceptible to cleavage than the carbon–fluorine bonds (260). For example, in the radiolysis of perfluorohexane the sum of the yields of intermediate C_7 to C_{11} fluorocarbon products is much greater than the yield of perfluorododecane (261) which would be formed by combination of two perfluorohexyl radicals. If these intermediate and dimer products are formed by combination of radicals, the yield of fragment radicals is greater than that of C_6F_{13} radicals. This contrasts with n-hexane where 80% of the radicals are hexyl radicals. In the radiolysis of perfluorocyclohexane the scission of carbon–fluorine bonds is more important than in other fluorocarbons. The major product is perfluorodicyclohexyl. It has been suggested that this product is formed by combination of cyclohexyl radicals (262,263). The radiation-induced decomposition of perfluorocyclohexane is similar to that of cyclohexane in that there is little carbon–carbon bond scission.

Direct information about the radical intermediates formed in carbon tetrachloride has been obtained by the use of labeled scavengers. The radical CCl_3 was identified as the principal radical intermediate by Schulte (264) who used millimolar concentrations of chlorine-36 as a scavenger. From the yield of exchange he concluded that $G(CCl_3) = 3.5$. A yield of chlorine atoms comparable to this was obtained by Chen et al. (265) who studied the effect of $HCCl_3$ on the Cl_2 yield. Measurements of the total radical yield in carbon tetrachloride with such scavengers as DPPH and anthracene (231,266) are difficult to interpret because of the reaction of these solutes with other intermediates. Ciborowski et al. (266) explained the high observed yield of DPPH consumption in CCl_4 ($G = 23$) mainly as a result of reaction of excited molecules with DPPH and assigned only a small yield to radicals ($G_R = 3$).

In the radiolysis of alkyl halides both carbon–halogen and carbon–hydrogen bond rupture is observed but loss of a halogen atom is the major process (267,268). For example, in the radiolysis of n-propyl iodide $G(n\text{-propyl}) = 4.2$ and fragment radicals account for only 2% of the radicals formed (see Table VIII). Cleavage of carbon–halogen bonds is expected to occur preferentially in the decomposition of excited molecules,

Table VIII
Radical Yields in Halogen and Oxygen Containing Compounds

Compound	Radicals[a]	G(total radicals)	Ref.
Methanol	H(2.5), CH_3(1.0), OH(0.6), CH_2OH	6	7,269,270
Ethanol	H(2.2–2.7), CH_3(0.4), $CH_3\dot{C}HOH$(9), CH_2CH_2OH(0.1)	5–8	37,220, 231,269
n-Propanol	H(1.9), $CH_3CH_2\dot{C}HOH$(4.8)	—	108,271
Acetone	H(<0.6), CH_3(1.3), CH_3CO, $CH_2COCH_2\cdot$(<1)	4	272,273, 274
Ethyl ether	H(1.4), CH_3(0.2), $CH_3\dot{C}HOC_2H_5$	9–12	107,275
Acetic acid	CH_3(2.8)	4	269
Methyl acetate	CH_3(3.4), CH_3O(1.7), CH_3CO(0.5), CH_3CO_2(0.5), $CH_3CO_2CH_2$(0.5)	7	276,277
Carbon tetrachloride	Cl(3), CCl_3(3.5)	3–7	231,264, 265
C_2F_6	CF_3	—	
cyclo-C_6F_{12}	cyclo-C_6F_{11}	—	
Ethyl iodide	C_2H_5(4)	4–6	267,278
$HCCl_3$	Cl(5.4), $CHCl_2$(5.6), CCl_3(0.2)	5	279,266, 274
n-Propyl chloride	CH_3(0.1), C_2H_5(0.1), i-C_3H_7(0.04), n-C_3H_7(4.2), $CH_3CH_2\dot{C}HCl$(2.4), $CH_2CH_2CH_2Cl$(0.9)	—	268

[a] Numbers in parenthesis are yields in radicals/100 eV.

reaction (76), and in electron capture reactions such as (77).

$$C_2H_5I^* \rightarrow C_2H_5 + I \qquad (76)$$

$$C_2H_5I + e^- \rightarrow C_2H_5 + I^- \qquad (77)$$

5. Alcohols. The yields of hydrogen atoms and methyl radicals formed in irradiated alcohols have been inferred from the effects of added H atom and methyl radical scavengers on $G(H_2)$ and $G(CH_4)$. At concentrations less than millimolar, solutes such as benzoquinone reduce the yield of hydrogen formed in reaction 33 by scavenging solvated electrons (Section II). A further reduction in the hydrogen yield in the presence of added benzoquinone at millimolar concentrations in methanol is attributed to the scavenging of hydrogen atoms and corresponds to $G(H) = 2.5$. Other hydrogen atom scavengers such as pentadiene (37), acetone, and chloro-acetic acid (280) give comparable results for the value of $G(H)$ in alcohols. The reduction in the methane yield observed in the presence of added benzoquinone corresponds to $G(CH_3) = 1.0$ for methanol and $G(CH_3) = 0.44$ for ethanol. The methyl radical yield is higher for t-butyl alcohol where $G(CH_3) = 1.7$. Branching enhances the fragment radical yield in alcohols as is the case in alkanes.

Both CH_3O and CH_2OH are formed in irradiated methanol, as was shown in a study in which benzene was used as scavenger for radicals. At 5% benzene both anisole ($G = 0.03$) and α-hydroxy-methyl-cyclo-hexadiene ($G = 0.25$) are formed (281). These are products that result from the addition to benzene of methoxy radicals, reaction (78) and CH_2OH radicals, reaction (79).

$$CH_3O + C_6H_6 \longrightarrow \quad \xrightarrow{\text{Disp.}} \text{anisole} \qquad (78)$$

$$CH_2OH + C_6H_6 \longrightarrow \quad \xrightarrow{\text{Disp.}} \qquad (79)$$

Methoxy radicals should be formed in a proton transfer reaction analogous to reaction (17). The H, OH, CH_3, and CH_3O radicals formed will abstract H atoms from methanol to produce CH_2OH radicals. At room temperature the major radical species in methanol is thus presumed to be CH_2OH; this radical has been detected in pulse radiolysis experiments.

Whether the CH_3O or CH_2OH radicals are also formed as primary decomposition products can be ascertained from the isotopic composition of the hydrogen formed in the radiolysis of partially deuterated alcohols (see also Section II-B-2b). In the radiolysis of CD_3OH, three-fourths of the hydrogen formed is HD (282,283); in the radiolysis of CH_3OD 60% of the hydrogen is HD (287). If it is assumed that hydrogen atoms, once formed, would primarily abstract a hydrogen from the methyl group then this data requires that loss of a hydrogen atom from the hydroxyl group must be a very important primary process.

In ethanol 1-hydroxyethyl radicals have been detected in pulse radiolysis studies. The yield of this species was estimated to be 9 radicals/100 eV by assuming that its molar extinction coefficient at 297 mμ is the same in ethanol as in water (220). Its yield should be approximately equal to the total radical yield in ethanol since other radicals would be expected to abstract hydrogen atoms from ethanol to form 1-hydroxyethyl radicals. It has been estimated from the observed glycol products (37) that $G(2$-hydroxyethyl) $= 0.1$. The formation of ethoxy radicals in the radiolysis of ethanol is supported by studies in which carbon monoxide was used as a radical scavenger (284,285). Ethyl formate is a product and is presumed to arise from scavenging of ethoxy radicals by CO followed by hydrogen atom abstraction, reactions (80) and (81).

$$C_2H_5O + CO \rightarrow COOC_2H_5 \tag{80}$$

$$COOC_2H_5 + C_2H_5OH \rightarrow HCOOC_2H_5 + CH_3\dot{C}HOH \tag{81}$$

The evidence available at present suggests that alkoxy radicals rather than hydroxyalkyl radicals are formed initially in alcohols, but alkoxy radicals readily abstract H atoms (286) to produce hydroxyalkyl radicals.

6. Acetone. An important intermediate in the radiolysis of acetone is the methyl radical. On the basis of scavenger effects on the methane yield (287) and the yields of isotopic methanes formed in the radiolysis of acetone–acetone-d_6 mixtures (288) it was concluded that at least 85% of the methane is formed from reactions of thermal methyl radicals. In an early study $G(CH_4)$ was reported to be 2.5; however later it was shown that traces of water have a profound effect on the yield of methane and $G(CH_4)$ is only 1.7 in dry acetone (272). In the presence of 3 mM iodine $G(CH_4)$ is only 0.4 (272). Based on this reduction by iodine $G(CH_3)$ is 1.3. Since methyl radicals are formed, it is reasonable to expect some acetyl radicals. The formation of acetaldehyde ($G = 0.09$), biacetyl ($G = 0.56$) (272), and acetylacetone (289) indicate acetyl radicals are intermediates.

There is little indication that thermal hydrogen atoms are important intermediates in acetone. If they were formed, then according to the data in Table IX they would mainly abstract from acetone to give H_2. Since $G(H_2)$ is only 0.44 (272), $G(H)$ is estimated to be ≤ 0.6.

Another radical which is likely to be present is acetonyl. The evidence for this is that one of the most important liquid products is acetonylacetone (272,289). This product and methylethylketone are presumably formed by combination of acetonyl with other radicals. The yields of these products suggest that the yield of acetonyl radicals is quite large. Methyl radicals abstract H atoms from acetone to form acetonyl radicals (290). However, acetonyl radicals may be formed to some extent by abstraction of hydrogen from acetone by excited molecules, reaction (82)

$$(CH_3COCH_3)^* + CH_3COCH_3 \rightarrow CH_3\dot{C}OHCH_3 + CH_3COCH_2 \cdot \quad (82)$$

(184,291). This reaction also forms $CH_3\dot{C}OHCH_3$ radicals which must also be present since the dimerization product, pinacol is a product (289).

Values for the total radical yield in acetone as high as 50 have been reported (4). Recent studies have shown that the actual value is much lower. Iodine scavenging results indicate $G(R) = 4.2$ (Table VIII).

7. Radicals in Ethers. Evidence concerning the yields of individual radicals in ethers has not been obtained by any of the chemical techniques described here. The total radical yield measured by the consumption of ferric chloride is $G(R) = 8.9$ to 12.5 (Table VIII). The effect of scavengers such as pentadiene and benzene indicates that $G(H) = 1.4$ in diethyl ether (107).

Considering the products formed in the radiolysis of ethers one can conclude that the major radical intermediate is that formed by loss of a hydrogen atom from the carbon attached to the oxygen atom. For example, in the radiolysis of diethyl ether, 2,3-diethoxybutane $(G = 2.6)$ (107) is a major product, indicating the formation of $C_2H_5O\dot{C}HCH_3$ radicals. In the pulse radiolysis of ethers radicals have also been detected by absorption spectroscopy. In methyltetrahydrofuran an absorption at 270 mμ was attributed (26) to a radical of structure (83a), and in dioxane an absorption at 290 mμ was attributed (32) to a radical of structure (86b).

$$(83)$$

(a)   (b)

C. Secondary Reactions of Radicals

A good deal of information is now available from various areas of chemical kinetics about reactions of radicals and particularly about values of rate constants. Since radical reactions are important in product formation, a knowledge of such reactions is essential for interpretation purposes. For this reason some secondary reactions of radical intermediates that are expected to occur in the radiolysis of liquids are considered briefly here.

1. Combination and Disproportionation. At room temperature and at dose rates typical of ^{60}Co-gamma ray sources ($\sim 10^{16}$ eV g^{-1} sec^{-1}) combination and disproportionation are important reactions of radicals. The rate of combination is diffusion controlled and not too dependent on radical structure. Typical values of rate constants are given in Table X.

The fact that radicals disproportionate as well as combine in the liquid phase has been established in photochemical studies (292–295). In liquid hydrocarbons the ratios of rate constants k(disp.)/k(comb.) are not much different from gas phase values (293). There is evidence however that disproportionation becomes more important in polar liquids and at lower temperatures (293,296,297). Some known values for disproportionation to combination ratios for radicals commonly encountered in the radiolysis of organic liquids are given in a review by Burns and Barker (297). These ratios can also be estimated from empirical relationships (298).

The possible combination and disproportionation reactions of the radical intermediates in benzene radiolysis are considered here as an

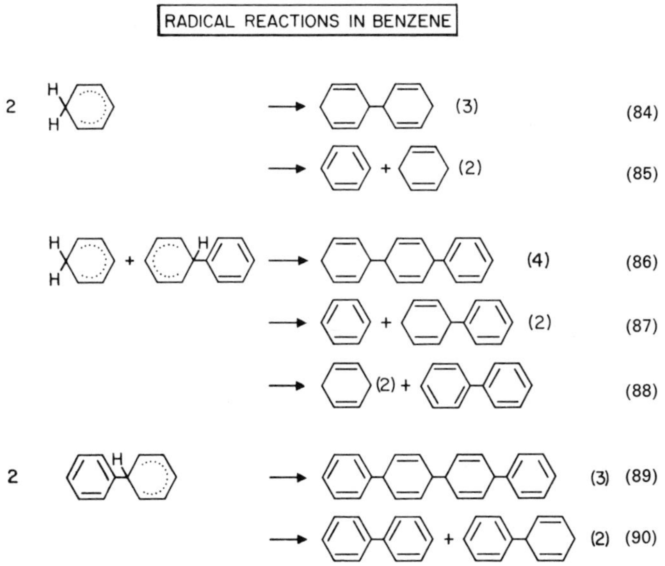

Fig. 9. Radical reactions in benzene.

illustration of product formation. The radicals that are present in benzene are presumed to be cyclohexadienyl and phenylcyclohexadienyl. It is expected that these two radicals will react in various ways as shown in reactions (84–90) (Fig. 9). Although there are only two radicals there are 10 possible combination and 8 possible disproportionation reactions which lead to 15 different products, if all isomers are considered. Many of the products which have been identified in the radiolysis of benzene can be accounted for by this series of reactions. Cyclohexadiene which is the single most important product ($G = 0.12$) (299) is formed in reactions (85) and (88). Biphenyl ($G = 0.06$) (299) is formed in reactions (88) and (90). Phenylcyclohexadiene ($G = 0.07$) is formed in reactions (87) and (90).

2. Abstraction.   Another important reaction of radicals is hydrogen atom abstraction, e.g., reaction (91). Competition of abstraction with other modes of reaction such as combination, reaction (92) will be dose rate dependent. Dose rate effects attributed to this competition have been observed in the radiolysis of hydrocarbons (239,297,300,301), acetone (290), chloroform (302) and other compounds (297).

$$CH_3 + RH \rightarrow CH_4 + R \qquad\qquad (91)$$
$$CH_3 + CH_3 \rightarrow C_2H_6 \qquad\qquad (92)$$

Dose rate effects can best be described by considering an example such as the radiolysis of n-hexane. The radicals present include hydrogen atoms, methyl, ethyl, n-propyl, butyl, pentyl, and hexyl radicals. All of these except hexyl are considered to abstract from hexane. The rate constants for the abstraction reactions differ for the various radicals; in this case $k(H + C_6H_{14}\rightarrow) \gg k(CH_3 + C_6H_{14}\rightarrow) > k(C_2H_5 + C_6H_{14}\rightarrow)$ (see Tables IX and X). Therefore competition of abstraction with combination will become important at different dose rates for each kind of radical. Three different dose rate regions can be defined. At dose rates less than 10^{16} eV g^{-1} sec^{-1}, hydrogen atoms and fragment radicals both abstract hydrogen from alkanes and hexyl radicals combine and disproportionate. Between 10^{17} and 20^{25} eV g^{-1} sec^{-1} alkyl radicals combine but hydrogen atoms still abstract. At very high dose rates, above 10^{25} eV g^{-1} sec^{-1}, it is expected that both hydrogen atoms and alkyl radicals will combine but abstraction will be of minor importance. Experimental verification of this latter point is still lacking however. A more comprehensive discussion of dose rate effects in radiation chemistry is given by Burns and Barker (297).

3. Addition Reactions.   In unsaturated compounds radicals will add to the double bond as well as abstract. Thus in alkenes addition and abstraction reactions will compete with radical combination.

Hydrogen atom addition to alkenes and aromatic hydrocarbons occurs

Table IX
Selected Reactions of Hydrogen Atoms

Reactant	Product	Rate constant, $M^{-1} sec^{-1}$	Ref.
HI	H_2	2×10^{10a}	80
O_2	HO_2	1×10^{10}	303
Ethylene	C_2H_5	$1-2 \times 10^{8b}$	304,305
		3×10^8	253
Propanol-2	H_2	5×10^7	306
Glucose	—	4×10^7	306
Ethanol	H_2	2×10^7	306,307
Benzene	C_6H_7	1×10^{7b}	305
		7×10^{7a}	308
Cyclohexane	H_2	5×10^{6a}	80
Methanol	H_2	2×10^6	307
n-Hexane	H_2	1×10^6	309
		5×10^6	310
Acetone	H_2	9×10^5	311
	CH_3COHCH_3	3×10^5	311
n-Butane	H_2	2×10^{5b}	312
Chloroacetic acid	H_2	2×10^5	303
Acetic acid	H_2	1×10^5	303
Nitrous oxide	N_2	1×10^4	53

[a] Based on relative rates and $k(H + O_2 \rightarrow) = 1.2 \times 10^{10}$ (303).
[b] Gas phase values.

readily, e.g., $k(H + C_2H_4 \rightarrow) \sim 10^8$ (Table IX). In alkyl benzenes the rate of addition of hydrogen atoms to the benzene ring is faster than abstraction from the side chain (317). The rates of other hydrogen atom reactions are given in Table IX.

Vinyl and phenyl radicals approach hydrogen atoms in their reactivity. Vinyl radicals even add to ethylene at $-160°C$ (209). Phenyl radicals add rapidly to benzene. This has been demonstrated in studies of the decomposition of benzoylperoxide in chlorobenzene which showed the only biphenyl formed is monochlorobiphenyl (318). Also in the decomposition of $Pb(C_6H_5)_4$ in benzene-d_6, little undeuterated biphenyl is formed (299). The rapid addition of phenyl radicals to benzene is further supported by the identification of the disproportionation and combination products of phenylcyclohexadienyl radicals in the decomposition of benzoyl peroxide in benzene (319). The reported rate constant for phenyl radical addition to benzene is much greater than that for alkyl radical addition (Table X).

Table X
Selected Radical Reactions in Organic Liquids

Reaction	Solvent	Rate constant, $M^{-1} sec^{-1}$	Temp., °C	Ref.
Recombination + disproportionation[a]				
$C_2H_5 + C_2H_5 \rightarrow$	Ethane	$2k = 3.4 \times 10^8$	-177	214
$C_6H_{11} + C_6H_{11} \rightarrow$	Cyclohexane	$2k = 2.5 \times 10^9$	25	218
$2C_6H_{10}OH \rightarrow$	Cyclohexanol	$2k = 3.4 \times 10^8$	25	313
$2CH_3\dot{C}HOH \rightarrow$	H_2O	$2k = 1.4 \times 10^9$	25	220
$2C_{12}H_{11} \rightarrow$	Benzene	$2k = 2.8 \times 10^9$	25	161
$2C_3H_5 \rightarrow$	Cyclohexane	$2k \sim 2 \times 10^9$	25	222
$2CCl_3 \rightarrow$	CCl_4	$2k = 1 \times 10^8$	30	314
$2CF_3 \rightarrow$	C_2F_6	$2k \cong 10^9$	-95	258
Hydrogen atom abstraction				
$CH_3 + c\text{-}C_6H_{10} \rightarrow$	Cyclohexene	$k = 14$	23	238
$CH_3 + c\text{-}C_6H_{12} \rightarrow$	Cyclohexane	$k = 140$	20	297
$CH_3 + CH_3COCH_3 \rightarrow$	Acetone	$k = 370$	20	297
$C_2H_5 + C_7H_{16} \rightarrow$	Heptane	$k = 4$	20	297
Radical addition				
$CH_3 + c\text{-}C_6H_{10} \rightarrow$	Cyclohexene	$k \cong 20$	23	238
$C_6H_5 + C_6H_6 \rightarrow$	Benzene	$k = 4.8 \times 10^4$	23	161
$C_2H_3 + C_2H_4 \rightarrow$	Ethylene	$k = 5$	-164	209
$CH_3 + C_6H_6 \rightarrow$	Benzene	$k \sim 3^{b}$	20	239,315
Miscellaneous				
$C_6H_{11} + I_2 \rightarrow$	Cyclohexane	$k = 7 \times 10^9$	20	218
$CH_3 + I_2 \rightarrow$	2,2,4-Tri-methylpentane	$k = 3 \times 10^{8b}$	20	239
$R\cdot + HI \rightarrow$	Hexane	$k = 1.5\text{--}2.0 \times 10^8$	20	227,228
$C_6H_{11} + S \rightarrow$	Cyclohexane	$k = 3 \times 10^8$	20	316

[a] Rate of radical disappearance $= 2k(R)^2$.
[b] Based on relative rates.

Alkyl radicals both add to and abstract from alkenes with comparable rates. The rate of addition of methyl radicals to propylene for example is three times the rate of abstraction at room temperature (320). These two rates are even more comparable for methyl radical reaction with cyclohexene (Table X). For some olefins such as 1,4-cyclohexadiene which has very reactive allylic hydrogen atoms, abstraction is expected to predominate. This reaction accounts for the formation of dicyclohexadienyl in the radiolysis of mixtures 1,4-cyclohexadiene in cyclohexene (321). An activation energy of about 7 kcal/mole for addition of alkyl radicals to olefins is typical (322,323). Substitution in the olefin lowers the activation energy. For example, the activation energy for addition of CCl_3 radicals to 2,3-dimethylbutadiene is lower (3.1 kcal/mole) than for addition to butadiene (5.5 kcal/mole) (324).

The addition of radicals to alkenes can be a chain reaction and consequently has several practical applications. In the radiolysis of solutions of alkenes in alcohols, aldehydes, ethers, amines, and hydrocarbons a 1:1 adduct of the alkene with the substrate is usually formed at low alkene concentration. At higher alkene concentrations the products formed are more likely to contain more than one alkene unit (325). In either case there may be a radical chain reaction and the yields of such adducts are large. The radiation induced addition of HBr to ethylene is the basis for the commercial synthesis of ethyl bromide. The reaction is a radical chain involving addition of bromine atoms to ethylene and yields as large as 10^6 molecules/100 eV have been observed (326).

The products of the radiolysis of solutions of tetrachloroethylene in n-pentane are unusual in that the 1:1 adduct is not observed. Instead HCl and $C_7H_{11}Cl_3$ are formed in a chain reaction (327). Apparently the radical formed by addition of pentyl radicals to C_2Cl_4 is unstable and loses a chlorine atom forming the observed olefin product, reaction (93).

$$C_5H_{11} + C_2Cl_4 \rightarrow C_5H_{11}CCl_2CCl_2 \cdot \rightarrow C_5H_{11}CCl{=}CCl_2 + Cl \qquad (93)$$

4. Fragmentation. In general, fragmentation of thermalized radicals is not expected to be an important reaction in the radiolysis of organic liquids. That is, most radicals are stable and decompose only at higher temperatures. For example the energy of activation for the decomposition of an alkyl radical into an olefin and a smaller radical is about 25 kcal/mole (323). For decomposition of acetyl, E_{act} is ~ 15 kcal/mole (323). The acetoxy radical is perhaps the most unstable; its decomposition is generally invoked in the radiation induced decarboxylation of organic acids.

Fragmentation of alkyl radicals is important in the radiation decomposition of liquid hydrocarbons at higher temperatures. This process, which has been termed radiation thermal cracking (328), takes place at about 200°C below the temperature at which pyrolysis occurs. The radiation initiates the process by bond scission. Radiation thermal cracking is not characterized by the high activation energies of pyrolytic reactions (~ 60 kcal/mole) but by lower activation energies characteristic of radical fragmentation reactions. The products of the higher temperature radiolysis of alkanes are methane and low molecular weight hydrocarbons which are largely unsaturated (328–330). The chain reaction is presumably similar to that occurring in the gas phase pyrolysis: that is, smaller radicals abstract a hydrogen atom from the hydrocarbon and the radical formed splits off olefinic units forming smaller radicals, reaction (94).

$$RCH_2\dot{C}HCH_3 \rightarrow R\cdot + C_3H_6 \tag{94}$$

5. Isomerization. Most radicals are not expected to rearrange under ordinary radiation conditions. Simple alkyl radicals, for example, do not isomerize in radiolysis (210,267). At elevated temperatures, larger radicals such as *n*-pentyl do isomerize to *sec*-pentyl but this occurs by an intra-molecular hydrogen atom abstraction, a reaction involving an activation energy of about 11 kcal/mole (331).

Radiation induced rearrangements have been observed to occur, how-ever, in certain halogenated and aromatic compounds and this has been attributed to isomerization of the free radical intermediates. Apparently halogen atoms and phenyl groups can migrate in certain radicals by 1–2 shifts.

In the radiolysis of liquid 2-phenylethyl bromide the phenyl radical has been observed to undergo a 1–2 shift (332). It has been suggested that this

$$C_6H_5CH_2{}^{14}CH_2\cdot \rightarrow C_6H_5{}^{14}CH_2CH_2\cdot \tag{95}$$

is a result of a rearrangement of a phenylethyl radical intermediate, reaction (99). In this case however the yield is small ($G = 0.3$) and the reaction may involve a 1–2 shift in a phenylethyl carbonium ion intermediate.

Radiation induced rearrangements are also observed in the radiolysis of alkyl halides (268,333–335). This reaction is particularly efficient for the chlorides. In the radiolysis of *n*-propyl chloride a chain reaction gives isopropyl chloride in high yield ($G \sim 50$). Originally this was attributed to isomerization of propyl radicals but it has been shown that *n*-propyl and not isopropyl radicals are present during radiolysis (Table VIII). It is now known that the chain isomerization involves a 1–2 shift of a chlorine atom in a chloropropyl radical intermediate, reaction (96), and

$$CH_3\dot{C}HCH_2Cl \rightarrow CH_3CHClCH_2\cdot \tag{96}$$

$$CH_3CHClCH_2\cdot + HCl \rightarrow CH_3CHClCH_3 + Cl \tag{97}$$

the isopropyl chloride is formed as a result of a reaction of HCl with the radical formed in reaction (96).

V. LINEAR ENERGY TRANSFER AND SPUR EFFECTS

In the radiolysis of liquids, product yields are a function of the quality of the ionizing radiation. That is, product yields for radiolysis by heavy charged particles such as protons or alpha particles are usually different from those obtained with fast electrons. These effects are generally referred to as linear energy transfer (LET) effects. The linear energy transfer of a particular type of radiation is the average rate of energy loss by a particle per unit distance traveled. Although the yield of one product may change with LET the overall yield of decomposition of the substrate generally changes little except in the case of aromatic hydrocarbons. These effects are discussed below in terms of the various models which have been proposed to explain the observations. An essential part of several of these models is the assumption that energy is initially deposited inhomogeneously in "spurs" or "tracks." As a consequence of spur formation and the high local concentration of intermediates in such regions, combination reactions of intermediates are favored.

The bimolecular reactions of excited molecules in spurs was discussed in Section III. Evidence for combination of radicals in spurs in low LET radiolysis is considered here in Section V-A.

A. Spur Combination of Radicals in Gamma Radiolysis

Considerable experimental evidence in favor of recombination of radicals in the spur in the radiolysis of organic liquids has now accumulated (278,336,337). The phenomenon is quite general and has been observed in hydrocarbons, ketones, esters, and alkyl halides. The available data indicate only the approximate magnitude of the effect; it is clear that more quantitative data on this subject are needed. By analogy with the terminology of aqueous radiation chemistry the yield of radicals which combine in the spur is referred to here as the "molecular" yield.

In the radiolysis of cyclohexane a major product is dicyclohexyl formed with a yield of 1.8 ± 0.2 molecules/100 eV (297). If we assume that this product is formed by combination of cyclohexyl radicals, reaction (98), then in order to account for the yield of dicyclohexyl, $G(C_6H_{11})$ must be ~ 7.5 since disproportionation, reaction (99), also occurs and $k_{99}/k_{98} = 1.1$

$$C_6H_{11} + C_6H_{11} \rightarrow C_{12}H_{22} \tag{98}$$

$$C_6H_{11} + C_6H_{11} \rightarrow C_6H_{10} + C_6H_{12} \tag{99}$$

(see Section IV-C-1). This yield of cyclohexyl radicals is approximately twice the yield of "free" cyclohexyl radicals. Therefore about one-half of the dicyclohexyl must be formed in some other way, which is suggested to be combination of cyclohexyl radicals in the spurs. If this is the case then G(dicylcohexyl) should be close to 0.9 in the presence of efficient radical scavengers present at low ($\sim 10^{-4}M$) concentrations. Indeed dicyclohexyl is formed in the presence of radical scavengers but the measurements that have been made were at concentrations greater than this. The residual yields observed (see Table XI) should therefore be regarded as minimum values for the extent of spur recombination of radicals.

Table XI
Yields of Products Formed in Spur Reactions

Compound	Product	G	Solute	Solute conc., mM	Radiation type	Ref.
Alkanes						
Cyclohexane	$(C_6H_{11})_2$	0.3	O_2	1.8	Gamma	316
Cyclohexane	$(C_6H_{11})_2$	0.32	I_2	0.9	Gamma	316
Cyclohexane	$(C_6H_{11})_2$	0.43	Benzoquinone	4.0	Gamma	339
Cyclohexane	$(C_6H_{11})_2$	0.90	O_2	1.8	Po alphas	316
Cyclohexane	$(C_6H_{11})_2$	0.80	I_2	4.0	Po alphas	316,340
Neopentane	C_2H_6	0.42	—[a]	—	Gamma	301
Ketones						
Acetone	C_2H_6	0.25	—[a]	—	Fast e^-	290
Acetone	C_2H_6	0.3	I_2	3.6	Gamma	272
Acetone	C_2H_6	0.56	—[a]	—	2 MeV H$^+$	290
Diethyl ketone	C_4H_{10}	0.3	I_2	10	Gamma	287
Esters						
Methyl acetate	C_2H_6	0.36	DPPH	1–8	Gamma	277
Methyl acetate	C_2H_6	0.27	I_2	10	Gamma	341
Alkyl halides						
Methyl iodide	C_2H_6	1.0–1.2	—[b]	—	Gamma	267,337,342
Methyl iodide	C_2H_6	1.2	—[b]	—	32 MeV He^{2+}	343
Methyl iodide	C_2H_6	1.6	—[b]	—	10 MeV He^{2+}	343
Carbon tetrachloride	C_2Cl_6	0.7	$^{36}Cl_2$	0.7	Gamma	264

[a] The dose rate was sufficiently low that the solvent was the radical scavenger.
[b] Iodine is a product and is present in sufficient concentration during radiolysis to scavenge radicals.

Similar information is available for *n*-pentane. The total radical yield as measured by scavengers is near 5, yet the total alkyl radical yield required to account for product yields is 9.7 (338). These data can also be considered as evidence of combination of radicals in spurs.

Evidence for spur recombination of radicals in *n*-hexane was obtained in a different fashion from a study of the effect of dose rate on the yield of C_7–C_{11} products (297). These products are assumed to be formed by combination of fragment radicals with hexyl radicals. At a sufficiently low dose rate the yield of C_7–C_{11} products is expected to decrease because of the increased importance of abstraction by fragment radicals (see Section IV-C-2). The expected falloff is observed at dose rates between 10^{17} and 10^{15} eV ml^{-1} sec^{-1}. However, the sum of the yields of these intermediate products does not go to zero even at dose rates much lower than this. The residual total yield of C_7 to C_{11} products, which is approximately 27% of the high dose rate yield, has been attributed to radical combination reactions in spurs (297).

Several studies have been made of hydrocarbons, acetone, and methyl acetate at dose rates at which it is known that thermal methyl radicals only abstract from the substrate to form methane (297). However, at these dose rates ethane as well as other combination products of methyl radicals are observed to be formed in significant yields (Table XI). In neopentane, for example, ethane, isopentane, and 2,2-dimethylbutane are formed even at low dose rates (68). The yields of these products correspond to a methyl radical yield of 1.2 which can be termed the molecular methyl radical yield. Since the yield of "free" methyl radicals is 2.3 it is concluded that one-third of the methyl radicals combine with radicals in the spur. A similar result was obtained for 2,2,4-trimethylpentane (236). The molecular methyl radical yield was found to be 0.47 methyl radicals/100 eV and the yield of "free" methyl radicals is 0.69, indicating that $\sim 40\%$ of the methyl radicals combine with other radicals in the spur.

The situation in ketones is analogous to that in hydrocarbons in that the ketone itself acts as a methyl radical scavenger and the homogeneous combination of methyl radicals cannot compete with abstraction, reaction (100), at the dose rates that have been employed (272,297). Nonetheless,

$$CH_3 \cdot + CH_3COCH_3 \rightarrow CH_4 + CH_3COCH_2 \cdot \qquad (100)$$

ethane is a product ($G = 0.3$) of the radiolysis of acetone (Table XI) and its yield cannot be reduced by scavengers. Again this has been attributed to radical combination in the spurs (287). Actually the ethane could be formed by a molecular elimination or by a cage recombination of radicals in spurs. The relative importance of these various mechanisms was examined in studies of the radiolysis of mixtures of acetone and acetone-d_6 (344).

Some of the ethane was CH_3CD_3, which can arise from combination of radicals in a spur where originally several molecules are dissociated, but cannot be formed by the other mechanisms. This intermolecular ethane formation is a process which is unique to radiolysis and does not occur in the photolysis of liquids (345). In the mixed acetone experiment, C_2H_6 and C_2D_6 were also formed either by combination of methyl radicals in spurs or by an immediate cage combination. When the temperature was reduced to 77°K, where diffusion is severely limited the intermolecular product CH_3CD_3 was all but eliminated, whereas the cage recombination still occurred (344).

In the radiolysis of alkyl iodides (RI) the product R_2 is usually formed, which indicates that combination of alkyl radicals occurs. For example, in the radiolysis of ethyl iodide, butane is formed (337,342). The origin of this butane cannot be homogeneous combination of ethyl radicals since iodine, a radical scavenger, is also a product of the radiolysis. Excited molecule reactions are ruled out since butane is not formed in the photolysis of ethyl iodide. Ionic reactions have been eliminated on the basis that butane is not formed in the radiolysis of the vapor (337). The butane can therefore only be accounted for by spur combination of ethyl radicals.

B. Radical Diffusion Model

1. Theory. The diffusion model is well known in radiation chemistry because of its use in describing LET effects in aqueous systems. In brief, the model considers that initially an inhomogeneous distribution of intermediates exists in the liquid. For radiation of low LET the intermediates are generated mainly in isolated spurs along the particle track. For radiation of high LET a string of spurs which overlap into a track is formed. The intermediates in the spurs can react with one another, or with a component of the solution or they can diffuse away. Because there is spur overlap at high LET, there are enhanced in-spur combination reactions.

The general models and mathematical calculations necessary to apply the diffusion model have been described by Kupperman (346) and are discussed in Chapter 8, Section I-B. Application of these models to organic systems has been limited by the lack of experimental data, particularly rate constant data. The theoretical calculations (297,316,347,348) that have been made usually have employed the simplest one-radical model in which it is assumed that only one radical R is formed. The distribution of the radicals in the spur is assumed to be Gaussian initially as well as during track expansion. Solution of the diffusion equation permits one to calculate the fraction of the radicals R which form R_2 in the spur.

2. Radical and Molecular Yields. Theoretical calculations based on the radical diffusion model indicate that about 50% of the alkyl radicals react together in spurs in the gamma radiolysis of alkanes (297,348). The experimental data on spur combination of radicals (Section V-A), which indicates from 30 to 50% of the radicals react in spurs is thus in reasonable agreement with the model.

The molecular yield of radicals should increase and the free radical yield should decrease with increasing LET according to this model. The data in Table XI indicate that molecular yields of radicals increase with LET. For example, in the radiolysis of acetone the yield of ethane is greater for 2 MeV protons than for gamma rays at comparable dose rates. This is a true spur effect since there is no reason for cage recombination to become more important at high LET. Similarly in the radiolysis of methyl iodide the yield of ethane, as well as the yields of other products, increases with LET. In the radiolysis of cyclohexane the yield of dicyclohexyl is approximately three times higher for alpha particle irradiation than for gamma rays at comparable concentrations of radical scavengers. This increase in the unscavengeable dicyclohexyl radical yield has been shown to be consistent with the one-radical diffusion model (316).

The variation in the molecular yield is expected to be complemented by a decrease in the yield of "free" radicals with increasing LET. Data on this point are limited at present. In the radiolysis of n-hexane, 2,2,4-trimethylpentane and cyclohexane, the free radical yield as measured by the uptake of iodine is less for irradiation by 10 MeV He^{2+} ions than for gamma rays (349). For cyclohexane G(cyclohexyl) is 4 for gamma rays and a recent report indicates G(cyclohexyl) $\sim$ 2 for alpha particle irradiation (340). The alpha particle value was measured at an iodine concentration of $\sim$ 10 mM. Radical yields have been measured in the radiolysis of cyclohexene with the radical sampling technique (238). The apparent yields of cyclohexyl and 3-cyclohexenyl radicals were found to decrease considerably with increasing LET.

The effects of LET on the products of the radiolysis of cyclohexane are considered here in light of the radical diffusion model. Until recently the radiolysis of cyclohexane was thought to be independent of LET. This was based on early studies which showed that the major products, $G(H_2)$, $G(C_6H_{10})$, and $G(C_{12}H_{22})$, were independent of LET (350,351). In these studies relatively high total doses ($\sim 10^{21}$ eV/ml) were employed. It was later shown that the yields of these products decrease with dose (339,352). Therefore the high conversions may have masked any LET effects in these early studies.

Reinvestigations of the cyclohexane radiolysis at low conversions (297,353,354) have shown that dicyclohexyl and cyclohexene are both formed in reduced yields at high LET (Fig. 10). The hydrogen yield is nearly constant to an LET of about 20 eV/Å, as was indicated in the early

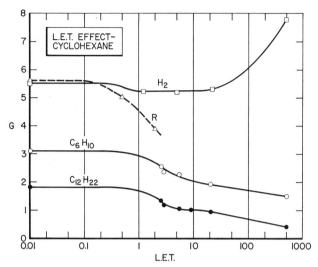

Fig. 10. Cyclohexane: product and radical yields in molecules/100 eV as a function of LET in eV/Å (297,339,340,350,354–356). ●, $G(C_{12}H_{22})$; ○, $G(C_6H_{10})$; □, $G(H_2)$ at a total dose of $\sim 10^{20}$ eV/g; △, $G(C_6H_{11}I)$.

work. In the radiolysis with fission fragments (354), for which the LET is ~ 500 eV/Å, the yields of hydrocarbon products decrease even further while $G(H_2)$ increases to 7.7.

In terms of a radical mechanism the dicyclohexyl and a part of the cyclohexene are considered to be formed in bimolecular reactions of cyclohexyl radicals, reactions (98) and (99). If only cyclohexyl radicals were present there would be no effect of LET on the *total* yield of dicyclohexyl or cyclohexene in the absence of scavengers. Only the fraction of the cyclohexyl radicals which react in the spur would be expected to increase with LET. However, the observed decrease in the total yields of dicyclohexyl and cyclohexene can be explained in terms of the radical diffusion model if a significant yield of hydrogen atoms is present initially. If the hydrogen atoms react with each other and with cyclohexyl radicals (357,358) then the net yield of cyclohexyl radicals will be reduced at high LET due to competition of reactions (102) and (103) with (98,99, and 101).

$$H + C_6H_{12} \rightarrow H_2 + C_6H_{11} \qquad (101)$$

$$H + H \rightarrow H_2 \qquad (102)$$

$$H + C_6H_{11} \rightarrow C_6H_{12} \qquad (103)$$

The decrease in the cyclohexene yield is equal to 1.1 times the decrease in the yield of dicyclohexyl. This is consistent with the radical mechanism since the ratio of these two decreases should be a constant equal to

k_{99}/k_{98}. Thus this ratio of rate constants is equal to 1.1. A similar value of 1.1 ± 0.3 for this disproportionation to combination ratio was obtained from the effect of oxygen and benzoquinone on these product yields (339). These values are in agreement with a recent photochemically determined value of this ratio of 1.1 ± 0.1 (359).

Some important facts remain unaccounted for by the radical model and suggest that other LET dependent processes must be occurring. The competition of reactions (102) and (103) with (101) would result in a decrease in $G(H_2)$ at high LET, but an increase is observed. Also, unidentified hydrogen-deficient products are formed at high LET. The yields of these unknown products are even greater in fission fragment irradiation where the yield of hydrogen unaccounted for is 5.8! This contrasts with the gamma ray radiolysis of cyclohexane where a material balance is obtained (352). Data on the hydrogen yields from the radiolysis of $C_6H_{12}-C_6D_{12}$ mixtures show that most of the hydrogen at high LET is formed in bimolecular processes (354). There are still several unanswered questions concerning the mechanism of radiolysis of cyclohexane. A more detailed examination of the products of the radiolysis of cyclohexane at high LET would be useful. A better understanding of the processes occurring will also require considerations of possible interreactions of the ions and excited molecules which can exist in the track of an ionizing particle (see Section V-F).

C. Excited Molecule Diffusion Model and Aromatics

Attempts have been made to account for LET effects in aromatics in terms of a diffusion model where instead of radicals the interaction of other intermediates such as excited molecules are considered (297,360,361). The radical diffusion model is inapplicable in aromatics primarily because radical yields are low and because it does not predict an increase in total decomposition with increasing LET. Aromatic liquids are unique in this respect in that the total disappearance rate increases with LET (Figs. 11 and 12). For other liquids including alkanes (Fig. 10), alcohols (Fig. 14) and water the disappearance rate is nearly independent of LET.

In the radiolysis of benzene the principal product is a mixture of high molecular weight components, which is commonly called "polymer." For low LET radiation this polymer formation consumes 0.8 molecules of benzene per 100 eV (164,362). Other products and their yields at low LET include hydrogen ($G = 0.039$), acetylene ($G = 0.02$), biphenyl ($G = 0.07$), cyclohexadiene ($G = 0.12$), and phenylcyclohexadiene (164). Cyclohexadiene is formed in higher yield than any other single compound at low LET (299). Studies of the radiolysis of 1:1 mixtures of benzene and benzene-d_6 have established that most of the hydrogen is formed in a

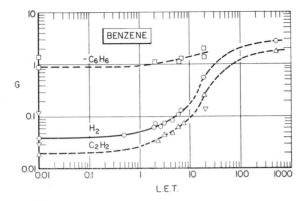

Fig. 11. Benzene: product yields in molecules/100 eV as a function of LET in eV/Å. Solid line is theoretical $G(H_2)$ from diffusion model calculations (360). Dashed lines are representative of data (297,299,367–369). $\bigcirc$, $G(H_2)$; $\triangle$, $G(C_2H_2)$; $\square$, $G(-C_6H_6)$; $\triangledown$, $G(cyclohexadiene)$.

bimolecular process and that acetylene is formed in a unimolecular process (299,363–365).

The yields of both hydrogen and acetylene increase by two orders of magnitude as the LET increases from 0.01 to 500 eV/Å, yet the ratio $G(H_2)/G(C_2H_2)$ is always 2 over the entire range. At high LET the yields of these two gases increase slightly with temperature (366), whereas there is no effect of temperature for gamma radiolysis. The increase in $G(-benzene)$ with increasing LET is comparable to the increase in $G(H_2)$ (see Fig. 11). The magnitude of the LET effect for other products has not been studied in detail. $G(biphenyl)$ is roughly independent of LET (367). Some preliminary experiments have indicated that $G(C_6H_8)$ increases slightly with LET (368). Very similar LET effects have been observed in the radiolysis of toluene (370) and terphenyl (371,372).

As an explanation of the observed effects in aromatic molecules it has been suggested (297,361) that excited molecules can participate in bimolecular as well as unimolecular reactions in the spurs. Such a mechanism is supported by the evidence for bimolecular reactions of excited triplet state molecules in spurs (Section III-A). The general mechanism which has been proposed by Burns and co-workers involves a bimolecular reaction which leads directly to products or to radicals (reaction (104)). This competes with the first order deactivation step (105). Calculations

$$C_6H_6{}^* + C_6H_6{}^* \rightarrow \text{products or radicals} \tag{104}$$

$$C_6H_6{}^* + C_6H_6 \rightarrow 2C_6H_6 \tag{105}$$

based on the excited molecule diffusion model have been made by Burns and Jones (360). With assumed values for the rate constants and other parameters, the model is consistent with the experimental dependence of $G(H_2)$ on LET up to 6 eV/Å (solid line Fig. 11). A small spur size of about 10 Å radius is required to fit the experimental data.

The increase in the yield of polymer with LET is more difficult to explain with this model. If radicals are involved in the formation of high molecular weight products then a possible explanation is that reaction (104) also leads to a phenyl and a cyclohexadienyl radical. This implies that the total radical yield should increase with LET but experimental data are not available to verify this. Excited molecule condensation reactions (Section III-C) may also become more important at high LET since in-spur reaction of excited molecules with unsaturated products will become more likely.

A modification in the excited molecule diffusion model is necessary at very high LET as a result of solvent depletion in the track. That is, a large fraction of the benzene molecules in the track of an ionizing particle become excited or ionized. Consequently the efficiency of the deactivation step, reaction (105) will decrease. The magnitude of this effect may be estimated as follows. For a small track radius of 10 Å and if $G(C_6H_6^*) = 6$, 50% of the molecules in the particle track are excited at an LET of 18 eV/Å. If the track radius is taken as 30 Å then 50% of the molecules in the particle track are excited initially at an LET of 160 eV/Å. Higher conversions will occur at even higher LET. Such high rates of energy loss can also lead to the phenomenon of multiple ionization and excitation of a single molecule (363). Boyd and Connor (371,372) have attributed LET effects in aromatics such as terphenyl to the effect of solvent depletion on the reactions of excited intermediates. In the solvent depletion model of Boyd an excited molecule or ion is deactivated by collision, reaction (106),

$$T^+ + T \rightarrow 2T \qquad (106)$$

$$T^+ \rightarrow I \qquad (107)$$

at low LET but unimolecularly converts to a reactive intermediate at high LET, reaction (107). This intermediate is then responsible for product formation. This model is in some ways similar to that proposed by Burns and co-workers but differs in principle in that the LET effect is attributed here to competition between two reactions which are both first order in the concentration of excited molecules.

The LET dependence of product formation in o-terphenyl shows some interesting high temperature effects. Recent work of Boyd et al. has shown that the decomposition is LET *independent* above 350°C at dose rates of $\sim 5 \times 10^{17}$ eV ml^{-1} sec^{-1} (371,373). This is largely a result of the increased effectiveness of gamma rays at high temperatures (Fig. 12). G(-terphenyl)

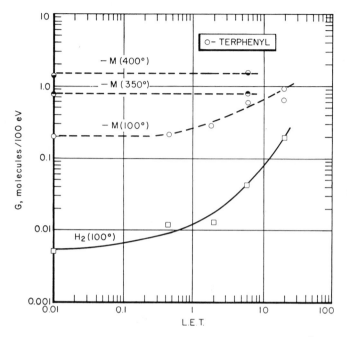

Fig. 12. *o*-Terphenyl: product yields as a function of LET in eV/Å (360,369,371, 374). □, $G(H_2)$, circles are $G(\text{-}o\text{-terphenyl})$ at various temperatures; ○, 100–150°C; ◐, 350°C; ◑, 400°C.

is 0.2 at 100°C but increases to 1.5 at 400°C. Thus the yield of decomposition of *o*-terphenyl at high temperature is of the same order of magnitude as product yields in the radiolysis of saturated hydrocarbons at low temperatures. This increase in the radiolytic decomposition of terphenyl at high temperature has been attributed to the decomposition of a temperature sensitive intermediate by Boyd et al. (371) and to thermal spikes (see Section V-E below) by Scarborough et al. (374).

The excited molecule model accounts for several features of the LET dependence of the radiolysis but does not account for effects at very high LET. As in the radical diffusion model one should perhaps not limit theoretical considerations to the interactions of one type of intermediate only.

D. End of Track Effects

Various other models have been proposed to explain observed LET effects in aromatic liquids. A point made recently by Schuler (363) is that such effects are not necessarily due to differences in the average LET of the ionizing particle. Instead increased product yields in the irradiation of aromatic molecules with heavy charged particles is attributed to the phenomena occurring at the end of the track of such particles.

An explanation of this end of track effect requires an understanding of the slowing down process for such ionizing particles. For protons and alpha particles the average energy loss per length of track increases with decreasing energy of the particle, that is, the LET increases with decreasing energy. A point is reached, however, where the LET reaches a maximum (the Bragg maximum) which for alpha particles is at 0.5 MeV and for protons is at a lower energy. At energies less than the Bragg maximum the LET of the particle decreases due to the occurrence of reversible charge exchange (e.g., reactions (108) and (109)), which reduces the effective charge on the particle.

$$He^{2+} + e^- \rightleftarrows He^+ \qquad (108)$$

$$He^+ + e^- \rightleftarrows He \qquad (109)$$

Evidence for an end of track effect has been obtained from measurements of the differential yield of hydrogen $G_i(H_2)$ for toluene (370) and benzene (363). The differential hydrogen yield may be thought of as the yield of hydrogen for irradiation of an extremely thin layer of liquid such that the charged particle loses only a small amount of energy in passing through it. Experimentally, the differential yield is determined in a larger cell from the integral hydrogen yield measured at various initial beam energies.

The differential yields for heavy particle irradiation *at high energies* are not significantly different from the low LET yields. For toluene $G_i(H_2)$ is 0.16 (370) for irradiation by protons of energy greater than 0.5 MeV and $G(H_2)$ is 0.13 (375) for gamma rays. For benzene $G_i(H_2)$ is 0.048 (363) for irradiation by helium ions of energy greater than 20 MeV and $G(H_2)$ is 0.039 for gamma rays. The integral yields of hydrogen for heavy particle irradiation are higher than these values; for example in benzene $G(H_2) = 0.1$ for a 20 MeV helium ion (363). Therefore in order to account for these integral yields much of the damage must occur near the end of the track. Specifically, considerable decomposition is caused by protons of energy less than 0.5 MeV and by alpha particles of energies less than 10 MeV.

Various models have been proposed to explain the end of track effects. One model suggested by Schuler (363) is based on the reversible charge exchange of slowly moving particles (reactions (108) and (109)). Such exchange occurs readily and results in a transport of electrons along the path of the ionizing particle. This leads to charge separation. The average number of these charge cycles which a particle goes through is comparable to the excess number of hydrogen molecules generated at the end of the track. The excess hydrogen yield could then be explained if one assumed that these separated ion pairs efficiently generated hydrogen. However, the average charge separation is only 20 Å and thus capture of the electron by the positive ion would be expected to occur with a high probability in benzene.

Another suggestion is based on the energy lost by a charged particle in elastic collisions with nuclei. At low energies the mechanism of energy loss by heavy charged particles ceases to be ionization and excitation and a substantial amount of energy (~ 20 eV for protons) is lost in elastic collisions with hydrogen and carbon atoms. Chemical bonds are very likely to be broken as atoms recoil from such collisions. If this process were efficient, a 20 keV proton could conceivably generate ~ 1000 hydrogen atoms. If these were to all form H_2 this process could account for the excess hydrogen yield at the end of the track. Since these "knocked-out" hydrogen atoms can add to benzene we expect this process also contributes to the increased loss of benzene at high LET.

Finally it has been suggested (361) that since the LET is greatest near the end of the track the high yields of H_2 are due to an LET effect. Measurements of the differential yield of hydrogen, $G_i(H_2)$, for benzene at very low helium ion energies (between 0.1 and 1.5 MeV) have recently been made (361). $G_i(H_2)$ does not rise to the very high values anticipated by the charge exchange or elastic collision models but reaches a maximum value of 0.5 near 0.5 MeV (417), which is the energy for which the LET is a maximum (Bragg maximum). These data indicate that $G_i(H_2)$ is largest at the points along the track of the particle where it is losing energy most rapidly by ionization. These data are quite convincing and suggest that the end of track effect is a true LET effect and that elastic collision and charge exchange processes are of minor importance in H_2 production. Whatever the explanation of end of track effects, the increase in other products such as acetylene and the increase in total decomposition also need to be accounted for. This discussion has focused primarily on the hydrogen yields for which quantitative data are available at present.

E. Thermal Spikes

Yang, Strong, and Burr (369) have suggested that the dependence of product yields on LET is a consequence of thermal spikes. This concept of a thermal spike arises from a consideration of the relatively large amounts of energy lost by the ionizing particle per unit distance traveled. The thermal spike models that have been considered require that this energy be degraded to thermal energy within a small volume close to the particle track. Thus, high rates of energy loss lead to the formation of momentarily "hot" regions called thermal spikes. Chemical effects in these spikes are attributed either to pyrolysis of the substrate or to acceleration of the reactions of intermediates. Application of the thermal spike model requires a knowledge of the activation energy of the reaction which is thermally accelerated. The extent of this reaction must then be integrated over the duration and volume of the thermal spike to determine the enhanced yield (361,376).

Because of the large value of the coefficient for thermal diffusion $(D = 10^{-3} \text{ cm}^2 \text{ sec}^{-1})$ the high temperatures do not persist for long. In gamma radiolysis the calculated thermal spike temperature rise is only $35°$ for spherical spurs of 50 eV energy and initial radius $\gamma_0 = 30$ Å. As a result of thermal diffusion the temperature decreases rapidly and is only $13°$ above ambient at 2×10^{-11} sec after passage of the ionizing particle (361).

A thermal spike model of low LET radiolysis has been postulated in the radiolysis of toluene to account for the increase in the bimolecular yield of H_2 with temperature (377). The same model has been used to account for the increase in $G(\text{-terphenyl})$ with temperature in the electron radiolysis of o-terphenyl (374). In this low LET thermal spike model it is considered that a small fraction of the gamma ray energy is associated with delta rays or "short tracks" (378) which are the high LET component of gamma radiolysis. It is assumed that a 1 keV delta ray loses all of its energy in a volume of 10 Å radius. The validity of the low LET model assumptions and calculations have been questioned and the temperature effects observed may have other explanations (367,379).

For high LET radiation much higher initial temperature rises are calculated from the thermal spike model. Again, cooling off of the spikes is rapid (361). The quantitative calculations (369,376) based on this model that have been made to explain the LET effects in benzene cannot be compared with experiment because of the unrealistic value chosen for the thermal diffusion coefficient. Thus a judgment of this model must await further calculations.

Several *a priori* objections to the thermal spike model can be raised. The most important has to do with the assumption that all energy *lost* by the ionizing particle is deposited as *heat* along the track of the particle. This energy is of course initially in electronic excitation; that is, ions and excited molecules are formed. These intermediates retain significant amounts of energy for times long compared to the lifetime of thermal spikes. For benzene, if 6 excited molecules are formed per 100 eV, then about 30% of the energy deposited is retained in excited molecules which have lifetimes of $\sim 10^{-8}$ sec (Section III). Another 30% of the energy is dissipated in the formation of ions which may have average lifetimes of 10^{-11} sec. The times involved, at least in the case of excited molecules, are sufficient for diffusion and dispersion of the energy. Thus, much of the energy lost by the ionizing particle is not localized in time and space along the particle track as required by thermal spike model calculations.

F. General Diffusion Model

If radicals and excited molecules recombine or react in spurs then other intermediates such as ions should also be considered to undergo diffusion recombination in spurs (380). In alcohols, ions are sufficiently long lived that spur reactions of ions are expected to occur.

In the radiolysis of methanol the yields of hydrogen and formaldehyde increase at high LET but the yield of ethylene glycol decreases. Carbon monoxide becomes a very important product at high LET (Fig. 13). If the LET effect is to be understood in terms of a general diffusion model then with increasing LET, bimolecular reactions of intermediates present in the spur will be favored. The ions most likely to be present are e_{sol}^- and $(CH_3OH)_nH^+$. The radicals most likely to be present are H and CH_3O (Section IV). Thus in-spur reactions such as (110) to (113) and (33) would be expected to occur. Reactions (110) and (111) would explain the increase in the yield of formaldehyde with LET; all four reactions would cause a decrease in the yield of glycol at high LET. Little is known about

$$CH_3O + CH_3O \rightarrow CH_2O + CH_3OH \tag{110}$$

$$CH_3O + H \rightarrow H_2 + CH_2O \tag{111}$$

$$CH_3O + e_{sol}^- \rightarrow ? \tag{112}$$

$$H + e_{sol}^- \rightarrow ? \tag{113}$$

$$e_{sol}^- + (CH_3OH)_nH^+ \rightarrow H + nCH_3OH \tag{114}$$

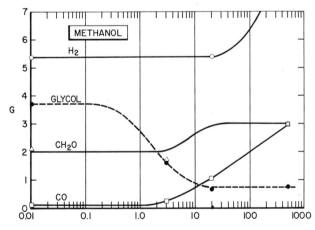

Fig. 13. Methanol: product yields in molecules/100 eV as a function of LET in eV/Å (7,72,381–383): ○, $G(H_2)$; ●, $G(CH_2OH)_2$; △, $G(CH_2O)$; □, $G(CO)$.

reactions of radicals with solvated electrons; such reactions should occur readily since radicals are electrophilic in nature. Reaction (113) has been observed in water and occurs at the diffusion controlled rate (Chapter 10). In-spur neutralization (reaction (33)) will also be enhanced at high LET. Increased neutralization of ions in tracks would be most evident as lowered yields of *free* solvated ions at high LET. In hydrocarbons the free ion yield decreases with increasing LET (22).

In a general diffusion model not only ion–ion and excited molecule-excited molecule reactions but also interspecies reactions such as ion–radical and ion-excited molecule should be considered. Reactions of radicals with excited molecules have been suggested to account for the low yield of luminescence of scintillators in aliphatic hydrocarbons at high LET. Thus in a general diffusion controlled spur model of hydrocarbon radiolysis it is necessary to consider all possible bimolecular reactions of radicals, excited molecules, and ions. Theoretical considerations must necessarily involve more sophisticated "many-radical" models (380).

VI. CHARGE AND ENERGY TRANSFER IN MIXTURES

In the radiolysis of binary mixtures of organic liquids it is often found that there is either enhanced or reduced decomposition of one component due to the presence of the other. These observed decreases or increases are in excess of those expected from the mixture law (see below) and are often interpreted as an indication that either charge or energy transfer occurs. The basis for such conclusions are examined here and some examples given of cases in which it has been concluded that charge or energy transfer occur.

It must be borne in mind that many interactions may occur in mixtures. In addition to energy transfer, charge exchange, and electron capture, radical and ion–molecule reactions such as those discussed in Sections II-B and IV-C can also occur. These effects need to be taken into account in arriving at any conclusions about energy or charge transfer. The necessary condition for charge exchange to occur is that the ionization potential of the acceptor must be lower than that of the donor. For energy transfer to occur the acceptor molecule must have a low-lying energy level as was discussed in Section III. For electron capture to occur the specific rate constant for reaction of the electron with the solute must be large (see Section II-B).

A. The Mixture Law

To date a great deal of the evidence for charge and energy transfer is based on the mixture law. The mixture law states that energy is initially absorbed in a mixture in proportion to the electron fraction of the components. Thus the predicted yield of a product or intermediate is given by the yield of that product or intermediate in the radiolysis of the pure component times the electron fraction of that component in the mixture. The yield of a product P in a mixture of A and B should be given by equation (115).

$$G(P) = G_A(P)\epsilon_A + G_B(P)\epsilon_B \tag{115}$$

where ϵ_A and ϵ_B are the respective electron fractions of A and B. Deviations from this mixture law may indicate that transfer of either charge or energy has occurred. As pointed out in Chapter 1 the mixture law is an assumption. There are alternate explanations such as the optical approximation (384, 385) in which effects observed in mixtures are attributed to preferential initial absorption of the energy in a particular component.

1. Mixture Law Behavior.   In the radiolysis of certain mixtures of benzene with other aromatics the mixture law is followed. For example, in mixtures of benzene and toluene the yields of hydrogen (Fig. 14), methane, biphenyl, bibenzyl, bitolyl, and phenyltoluene all follow the mixture law (5,386). This behavior is observed even though charge transfer is possible since the ionization potentials of the components differ by ~ 0.4 V. In the radiolysis of benzene–pyridine mixtures the principal product is polymer. The yield of polymer from mixtures is a linear function of the electron fraction (Fig. 15). In this case the ionization potentials of the two components are comparable. In mixtures of benzene with benzene-

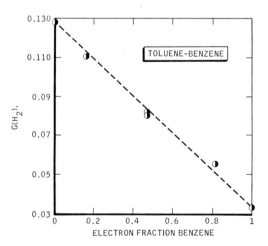

Fig. 14. Benzene–toluene mixtures. Hydrogen yields in molecules/100 eV.
(Reproduced by permission, reference 5.)

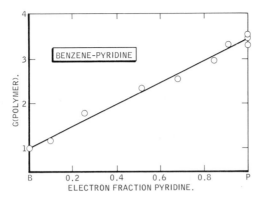

Fig. 15. Benzene–pyridine mixtures. G(polymer) in molecules of monomer converted to polymer per 100 eV as a function of the electron fraction of pyridine. (Reproduced by permission, reference 387.)

d_6 the yields of hydrogen (364,365) biphenyl and phenylcyclohexadiene (299,363) are approximately a linear function of the electron fraction.

Mixture law behavior has been demonstrated in the radiolysis of an alkane mixture by measurements of the radical yields with the iodine scavenging technique. In cyclopentane–cyclohexane solutions the yields of cyclohexyl and cyclopentyl radicals follow the mixture law (355). Thus in this case there is no apparent transfer even though charge transfer to cyclohexane is exothermic by 0.6 eV.

2. Deviations from the Mixture Law. In contrast to the examples cited above there are numerous cases in which deviations from the mixture law have been observed.

Detection of the intermediates formed from a particular component provides a direct approach to the study of effects in mixtures. For example, positive deviations from the mixture law have been detected in the ethane–ethylene system at $-160°C$ by *in situ* measurements of the vinyl and ethyl radical concentrations with the electron spin resonance technique (Section IV-A). The ethylene used was deuterated and $G(C_2D_3)$ at 10% ethylene-d_4 was six times larger than the value predicted from the mixture law (388). Thus some interaction occurs which results in enhanced ethylene decomposition in such mixtures. Charge transfer has been shown to occur in alkane–alkene mixtures in the solid phase (389).

Other techniques for the detection of intermediates such as pulse radiolysis can be applied to energy transfer effects in the study of mixtures (Section III-B and 390).

A new technique reported by Horikiri and Saigusa involves the measurement of the gamma-induced luminescence emitted *from each component*

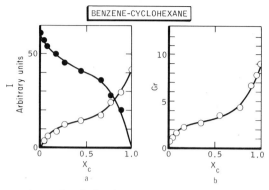

Fig. 16. Comparison of radiation induced light emission with radical yields from benzene–cyclohexane mixtures. (a) Emission intensity (fluorescence excluded) vs. mole fraction of cyclohexane; ●, benzene emission, ○, cyclohexane emission. (b) G(—DPPH) vs. mole fraction of cyclohexane. (Reproduced by permission, reference 112.)

of a mixture where the fluorescence is excluded (112). With this technique they compared the luminescence of each component of the mixture with the total radical yield as measured by DPPH disappearance. Mixtures of benzene with cyclohexane (Fig. 16), ethanol, methyl acetate, and acetone have been investigated with this method. A strikingly close parallel was observed between the radical yields and the emission intensity of the solutions of cyclohexane, ethanol, and methyl acetate throughout the entire concentration range (Fig. 16). The authors make the point that a correlation is observed only if the normal fluorescence of the components of the mixture is excluded. Because of this parallel Horikiri and Saigusa suggested that the species which emit the observed light are either excited radicals or ions. They conclude along with others (123,391) that "energy transfer processes exist between highly excited states of two components of mixtures."

B. Excitation Transfer

In the radiolysis of mixtures it is difficult to distinguish charge transfer from excitation transfer. Excited molecules can be formed upon ion neutralization and the interaction of a solute with ions can have the same effect as interaction with excited molecules. In the examples considered below there is evidence to suggest that excitation transfer plays a chemically significant role.

1. Cyclohexane–Benzene. Some of the earliest work in the radiolysis of organic liquids dealt with the cyclohexane–benzene system (5,392).

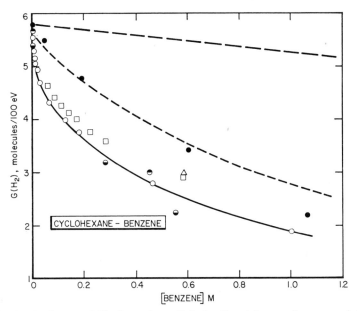

Fig. 17. Hydrogen yields from the radiolysis of cyclohexane–benzene mixtures. Dashed line is $G(H_2)$ based on the mixture law. Dotted line is $G(H_2)$ calculated from the expression: $5.6\epsilon_{C_6H_{12}}/[G(H_2) - 0.04] = 1 + 0.84[C_6H_6]$. Data from indicated references: ●, (248); △, (396); ◐, (250); ○, (356); ◓, (5); □, (80).

Although there is still some experimental uncertainty, it is clear that the yield of hydrogen exhibits negative deviations from the mixture law at all concentrations (Fig. 17). It was pointed out by Manion and Burton (5) that this reduction of $G(H_2)$ could be attributed to either charge or energy transfer to the benzene. Charge exchange is exothermic by 0.7 eV and the energy levels are favorable for energy transfer (Fig. 5); thus both can occur. The reality of excitation transfer from excited cyclohexane molecules to benzene has been demonstrated in photochemical studies (207).

Other products of the radiolysis of cyclohexane such as cyclohexene and dicyclohexyl are also formed in yields much less than predicted by the mixture law. For example, 1% benzene reduces $G(C_{12}H_{22})$ by 60% (250,393). A recent report indicates the reduction is only 17% at this concentration (394) but in any case there is a negative deviation. Certain products such as phenylcyclohexane (200,356,395) and dicyclohexadienyl (250) are formed in mixtures which are not formed in either pure component. These facts suggest that hydrogen atom and radical addition to benzene must be taken into account in evaluating the contribution of energy transfer to the observed effects.

In benzene rich mixtures it is found that G(biphenyl) is well below the

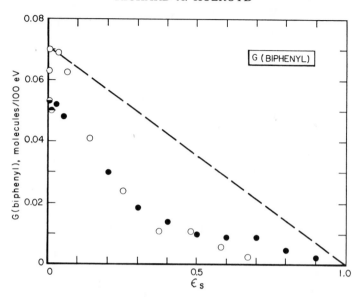

Fig. 18. Yields of biphenyl from the radiolysis of benzene as a function of the electron fraction of solute. Dashed line represents the mixture law. Solutes are: ○, Ni(CO)$_4$ (397); ●, cyclohexane (200); ◐, p-terphenyl (398); ◑, 2-butene (398).

mixture law values (Fig. 18). One might conclude from this data alone that there is *energy transfer from benzene to cyclohexane* except for the data on the yields of H$_2$, C$_6$H$_{10}$, and C$_{12}$H$_{22}$. This example emphasizes that conclusions about energy transfer cannot be made from the variation in the yield of a single product without consideration of other information. The explanation for the negative deviation for G(biphenyl) is not clear but it cannot be charge or energy transfer. It is suggested in Section IV that biphenyl results from phenyl radical addition to benzene. Thus, cyclohexane may reduce the biphenyl yield by scavenging phenyl radicals. Certain other solutes such as Ni(CO)$_4$ exhibit very similar effects when mixed with benzene (397,398) and a common interaction could be involved.

More definitive information about the benzene–cyclohexane system has been obtained from the use of deuterated cyclohexane (399,400). If cyclohexane containing a few percent of C$_6$D$_{12}$ is irradiated, the hydrogen formed contains D$_2$ and HD. The D$_2$ can be considered a unimolecular product since in dilute solutions both hydrogen atoms must come from the same molecule. The HD is formed in a bimolecular process in which the hydrogen atoms come from different molecules. The presence of 6% benzene in such a solution reduces the hydrogen yield by 50% but does not affect the percentage of D_2 and HD in the total hydrogen formed (399). That is, the bimolecular and unimolecular processes are quenched by benzene to comparable extents. On the basis that it is unlikely that benzene would quench two different precursors of hydrogen with comparable efficiency, Dyne concluded that benzene quenches a common precursor of HD and D_2. The precursor was later suggested (97) to be ions: $C_6H_{12}^+$ and e^-.

The same effect on $G(H_2)$ would be achieved if benzene quenched excited cyclohexane molecules, reaction (116). This reaction would compete with decomposition of the cyclohexane, represented by reaction (54). For such a competition the yield of hydrogen is given by the yield

$$C_6H_{12}^* + C_6H_6 \rightarrow C_6H_{12} + C_6H_6^* \qquad (116)$$

$$C_6H_{12}^* \rightarrow H_2 + C_6H_{10} \qquad (54)$$

$$G(H_2) = G(C_6H_{12}^*)\, \epsilon_{C_6H_{12}}/(1 + k_{116}[C_6H_6]/k_{54}) \qquad (117)$$

of the precursor divided by: $1 + k_{116}[C_6H_6]/k_{54}$ (401). If the yield of hydrogen formed from benzene is neglected one obtains equation (117). The dotted line in Fig. 17 is a plot of this equation with $G(C_6H_{12}^*)$ equal to 5.6 and $k_{116}/k_{54} = 0.84$ M^{-1}. This equation accounts satisfactorily for the data of Merklin and Lipsky (248) but is in disagreement with data reported by several others. The reason for this experimental uncertainty regarding the extent of reduction of $G(H_2)$ by benzene at low concentrations is not clear. Several other workers have reported similar values for k_{116}/k_{54} of 0.78 M^{-1} (250) and 0.65 M^{-1} (402). From these studies the lifetime of the cyclohexane species which is quenched by benzene can be estimated to be $\sim 0.8 k_{116}^{-1}$ sec.

2. n-Alkane–Benzene. The cyclohexane–benzene system may not be a typical example of an alkane–benzene mixture but very little work has been done on the effect of benzene on the radiolysis of other alkanes for comparison. n-Hexane has been studied to some extent and benzene is much more effective in reducing the yield of hydrogen at low concentrations in this case. A benzene concentration of 5 mM is reported to reduce $G(H_2)$ from hexane by 20% (328) which corresponds to a $\Delta G(H_2)$ of 1.0. This effect could be hydrogen atom scavenging but the rate data in Table IX indicate that benzene is not that efficient a hydrogen atom scavenger. Benzene depresses the yields of all products including C_2's to C_4's, hexenes, C_8's to C_{12}'s and H_2 to the same extent in the radiolysis of n-hexane (328,403). The similarity of the effect of benzene on all of the products of the radiolysis of n-hexane suggests a single interaction of some sort. This conclusion is supported by the work of Dyne et al. (400) who studied the effect of benzene on the radiolysis of solutions of C_6D_{12}

in *n*-pentane. Additional data on other alkanes would be of interest to clarify the role of benzene in mixtures with alkanes.

3. Argon–Ethylene.   Another example where excitation transfer probably occurs is in solutions of hydrocarbons in liquid argon or xenon. In liquid argon solutions, ethylene is efficiently decomposed; $G(-C_2H_4) \cong 11$ at an ethylene electron fraction of only 0.0043 (404). The product distribution formed in such mixtures in which the ethylene concentration is in the 1–100% region differs markedly from that observed for mixtures containing below 1% ethylene (Fig. 19), which shows that different processes occur in the two concentration regions. At concentrations of ethylene between 0.05 and 1.0% the yields of the products are constant. Klassen (404) has suggested that in view of the short lifetimes of most ions ($\sim 10^{-10}$ sec) this is the region in which energy is transferred to ethylene only from neutral excited species such as Ar^* and Ar_2^* and that charge transfer reactions can only be important for solute concentrations greater than 1%.

4. Cyclohexene–1,3-Cyclohexadiene.   Some interesting results have been obtained in studies of mixtures of cyclohexene and 1,3-cyclohexadiene (321). The products of the radiolysis of cyclohexene such as hydrogen,

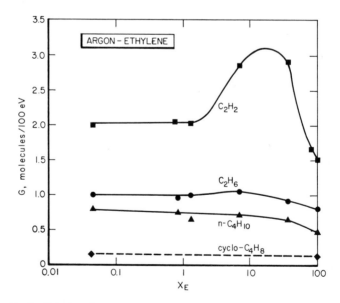

Fig. 19. Radiolysis of liquid argon–ethylene solutions at $-161°C$. Yields of products in molecules per 100 eV absorbed in the solution vs. electron percent ethylene. (Reproduced by permission, reference 404.)

dicyclohexenyl, etc. exhibit large negative deviations from the mixture law and the decomposition of the cyclohexadiene is in excess of mixture law expectations. Large yields ($G \cong 6$) of a true dimer of cyclohexadiene are formed at low concentrations of cyclohexadiene. This dimer which has been identified as tricyclo[4.4.2^{7,10}.0] dodeca-2,8-diene is apparently formed by cycloaddition of one (excited) cyclohexadiene molecule to another. Such cycloadditions are common in photochemical reactions of related conjugated molecules (405). Its formation suggests that energy transfer from cyclohexene to 1,3-cyclohexadiene occurs. However, excited cyclohexadiene molecules could also be formed by neutralization following charge exchange.

C. Charge Transfer

There is evidence that charge exchange occurs in certain mixtures. For alkanes the evidence is based primarily on the yields of hydrogen in mixtures containing deuterated components and on a correlation with ionization potential differences.

The occurrence of either charge or energy transfer, if it is based on measurements of the hydrogen yield, can be detected more readily if one of the components is deuterated. In one study of this type mixtures containing an alkane and a few percent of C_6D_{12} were irradiated (400,406). It was found that the value of $G(D_2)$ from such mixtures depended on the solvent alkane. More specifically $G_{C_6D_{12}}(D_2)$, the number of molecules of D_2 formed per 100 eV absorbed by the C_6D_{12}, varied by a factor of 30 depending on the alkane solvent. $G_{C_6D_{12}}(D_2)$ was 0.33 for C_6H_{12} (IP = 9.9 eV) as solvent. For certain alkanes such as n-pentane (IP = 10.3 eV), cyclopentane (IP = 10.5 eV), and 2,2-dimethylbutane (IP = 10.0 eV), $G_{C_6D_{12}}(D_2)$ was found to be greater than 0.33. This was interpreted as evidence for charge or energy transfer to C_6D_{12}. For other alkane solvents such as 2,2,4-trimethylpentane (IP = 9.8), n-decane and decalin, $G_{C_6D_{12}}(D_2)$ was less than 0.33 indicating transfer from C_6D_{12} to the solvent molecules. The yield of HD behaved similarly, for example, if $G_{C_6D_{12}}(D_2)$ was large $G_{C_6D_{12}}(HD)$ was also large.

In later work (407) it was pointed out that alkanes for which $G_{C_6D_{12}}(D_2)$ was greater than 0.33 all had gas phase ionization potentials greater than cyclohexane and conversely alkane solvents for which $G_{C_6D_{12}}(D_2)$ was less than 0.33 all had ionization potentials less than that of cyclohexane. Transfer effects were observed even where the ionization potential difference was only a tenth of an electron volt. On the basis of this correlation with ionization potentials Hardwick (407) identified this process as charge transfer. It has been noted that this transfer process is not very efficient. It is much less efficient than the interaction of solutes such as

N_2O with ions. For alkane solutes present at a concentration of 5 mole %
in n-hexane only about one-tenth of the ions undergo charge transfer
reactions (97,407).

Similar effects have been noted (408) in mixtures containing deuterated
alkanes other than cyclohexane. Liquid mixtures of either ethane (IP =
11.6 eV), propane (IP = 11.1 eV) or butane (IP = 10.6 eV) with one of the
corresponding fully deuterated compounds were irradiated in order to
explore the ionization potential correlation further. As in the use of
cyclohexane the yields of D_2 and HD served as a measure of charge
transfer either to or from the deuterated component. Enhanced yields of
D_2 and HD were observed where the ionization potential of the non-
deuterated alkane was much greater than the ionization potential of the
deuterated alkane, diminished yields of D_2 and HD were observed where
the ionization potential was much lower. For cases where the ionization
potentials were nearly equal, little if any effect was observable. This study
indicated that the efficiency of transfer is proportional to the magnitude
of the difference in the ionization potentials.

More recently, Kudo and Shida (409) demonstrated charge transfer
from n-hexane (IP = 10.17 eV) and from cyclopentane (IP = 10.51 eV)
to 2,2,4-trimethylpentane (IP = 9.84 eV) through an analysis of changes in
the yields of hydrocarbon products.

The ionic nature of this transfer process is supported by the fact that
the transfer is inhibited by solutes which are positive ion scavengers.
Ammonia, for example, decreases the enhanced yields of D_2 and HD
observed in mixtures of deuterated hydrocarbons in solvents of higher
ionization potential (408). Similarly in the presence of 0.6% cyclopropane,
a positive ion scavenger (see Section II), little or no transfer from isopentane (IP = 10.3) to isohexane (IP = 10.1) is observed (87). Thus this
evidence indicates that charge transfer occurs in alkane mixtures, but that
it is inefficient particularly when ionization potentials are comparable.

There are many other examples of mixtures where the component of
higher ionization potential transfers energy to the component of lower
ionization potential. In the ethane–ethylene system discussed above charge
transfer to ethylene is exothermic by 1.1 eV. In solutions of hydrocarbons
in argon charge transfer to the hydrocarbon is exothermic. Similarly
ionization potentials are favorable for charge transfer to benzene in the
benzene–cyclohexane system. However it has not been shown in all these
cases that the effects occurring include charge exchange.

There are also many examples where energy is transferred to the component of higher ionization potential. In mixtures of cyclohexane
(IP = 9.9) with CCl_4(IP = 11.4) the decomposition of the CCl_4 is
enhanced (410). Other systems which behave similarly are mixtures of

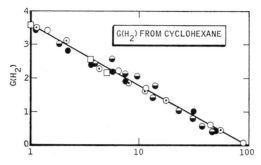

Fig. 20. Hydrogen yields in the radiolysis of cyclohexane containing various solutes. Solutes are: ●, methyl iodide; ○, chloroform; ◉, benzyl acetate; ◖, benzene; □, carbon tetrachloride (225).

benzene (IP = 9.2) with CCl_4 (411), cyclohexane with methyl chloride (IP = 10.7) (412), and toluene (IP = 9.2) with carbon disulfide (IP = 10.4) (413). Since in these cases charge transfer is endothermic it is unlikely to occur, but electron capture and excitation transfer can occur. The electron reacts readily with solutes such as CCl_4 and CS_2 (303). Since electron capture by a solute changes the neutralization reaction (Section II-B) it can result in decreased decomposition of the hydrocarbon. New products which are not characteristic of the radiolysis of either component are formed in these mixtures. Interpretation of the results obtained in these systems is difficult at present because changes in the neutralization reaction, radical scavenging reactions and the role of excitation transfer must all be taken into account. Figure 20 illustrates that solutes such as carbon tetrachloride reduce $G(H_2)$ from cyclohexane to the same extent as benzene does. Yet the ionization potential of carbon tetrachloride is greater than that of cyclohexane and the ionization potential of benzene is less. The curve in Fig. 20 is not a universal relationship but does represent the H_2 yield for the solutes listed.

D. Intramolecular Energy Transfer

A topic related to energy transfer in mixtures of alkanes and aromatics is the radiolysis of pure compounds in which an aromatic group is chemically bonded to an alkane. Phenylcyclohexane may be thought of as an intimate mixture of benzene and cyclohexane where the electron fraction of benzene is 0.47. At this concentration, $G(H_2)$ is 0.57 for the physical mixture and 0.26 for the phenylcyclohexane (248). Thus the benzene ring is more effective in reducing the hydrogen yield where it is bonded to the alkyl group.

Since chemical bonding results in more effective energy transfer than is obtained in mixtures it is interesting to consider over how many chemical bonds this effect can be transmitted. One approach is to measure $G(H_2)$ for compounds with varying numbers of methylene groups between the phenyl and cyclohexyl group (structure (118a)). Another approach is to study the radiolysis of various alkyl benzenes and measure $G(H_2)$ as a function of the length of the side chain (structure (118b)). As n increases in

(118a)

(118b)

these compounds $G(H_2)$ increases but the fraction of energy initially absorbed in the alkyl groups also increases, according to the mixture law. Thus one should consider $G_{alkyl}(H_2)$, the yield of hydrogen per 100 eV absorbed in the alkyl group (414). The value of $G_{alkyl}(H_2)$ is approximately 0.5 for all compounds of type (118a) that have been studied and also for compounds of type (118b) if $n < 4$. If n increases beyond 5, $G_{alkyl}(H_2)$ begins to increase. Thus for long chains the effect of benzene on $G(H_2)$ decreases. The effect of chain length is also illustrated by data on $G(H_2)$ from isomeric phenylnonanes. For example $G(H_2)$ is 0.59 for 1-phenyl-nonane and 0.42 for 5-phenylnonane (415). Thus the more stable isomer is the one which has the phenyl group located in a central position. Similar effects have been observed in the radiolysis of alcohols and ketones (416).

ACKNOWLEDGMENTS

Preparation of this review was supported by the U.S. Atomic Energy Commission. The author would like to express his appreciation to the numerous people whose comments and letters were helpful in the preparation of this chapter. Special thanks go to R. B. Ingalls who read and criticized the manuscript and to J. Kropp who read and criticized Section III.

REFERENCES

1. A. J. Swallow, *Radiation Chemistry of Organic Compounds*, Pergamon Press, Oxford, 1960.
2. M. Burton, P. J. Berry, and S. Lipsky, *J. Chim. Phys.*, **52**, 657 (1955).
3. E. N. Weber, P. F. Forsyth, and R. H. Schuler, *Radiation Res.*, **3**, 68 (1955).
4. A. Prevost-Bernas, A. Chapiro, C. Cousin, Y. Landler, and M. Magat, *Discussions Faraday Soc.*, **12**, 98 (1952).
5. J. P. Manion and M. Burton, *J. Phys. Chem.*, **56**, 560 (1952).
6. R. R. Williams, Jr. and W. H. Hamill, *Radiation Res.*, **1**, 158 (1954).
7. J. H. Baxendale and F. W. Mellows, *J. Am. Chem. Soc.*, **83**, 4720 (1961).

8. E. Hayon and J. J. Weiss, *J. Chem. Soc.*, **1961**, 3962.
9. I. A. Taub, M. C. Sauer, Jr., and L. M. Dorfman, *Discussions Faraday Soc.*, **36**, 206 (1963).
10. I. A. Taub, D. A. Harter, M. C. Sauer, Jr., and L. M. Dorfman, *J. Chem. Phys.*, **41**, 979 (1964).
11. A. O. Allen and A. Hummel, *Discussions Faraday Soc.*, **36**, 95 (1963).
12. G. R. Freeman, *J. Chem. Phys.*, **39**, 988 (1963).
13. G. Scholes and M. Simic, *Nature*, **202**, 895 (1964).
14. F. Williams, *J. Am. Chem. Soc.*, **86**, 3954 (1964).
15. A. Hummel and A. O. Allen, *J. Chem. Phys.*, **44**, 3426 (1966).
16. T. Takagaki, K. Hayashi, K. Hayashi, and S. Okamura, *Nippon Hoshasen Kobunshi Kenkyu Kyokai Nenpo*, **6**, 13 (1964–65).
17. P. E. Secker and T. J. Lewis, *Brit. J. Appl. Phys.*, **16**, 1649 (1965).
18. E. Gray and T. J. Lewis, *Brit. J. Appl. Phys.*, **16**, 1049 (1965).
19. O. H. LeBlanc, *J. Chem. Phys.*, **30**, 1443 (1959).
20. O. Gzowski and J. Terlecki, *Acta Phys. Polon.*, **18**, 191 (1959).
21. W. R. Ware, *J. Phys. Chem.*, **66**, 455 (1962).
22. A. Hummel, A. O. Allen, and F. H. Watson, Jr., *J. Chem. Phys.*, **44**, 3431 (1966).
23. P. Debye, *Trans. Electrochem. Soc.*, **82**, 265 (1942).
24. G. Porter, in *Technique of Organic Chemistry*, Vol. VIII, part 2, 2nd ed., Interscience, New York, 1963, p. 1055.
25. L. M. Dorfman and M. S. Matheson, in *Progress in Reaction Kinetics*, Vol. III, Pergamon Press, London, 1965, Ch. 6.
26. F. S. Dainton, J. P. Keene, T. J. Kemp, G. A. Salmon, and J. Teply, *Proc. Chem. Soc.*, **1964**, 265.
27. M. C. Sauer, Jr., S. Arai, and L. M. Dorfman, *J. Chem. Phys.*, **42**, 708 (1965).
28. G. E. Adams, J. H. Baxendale, and J. W. Boag, *Proc. Royal Soc. (London)*, **277**, 549 (1964).
29. M. Anbar and E. J. Hart, *J. Phys. Chem.*, **69**, 1244 (1965).
30. L. R. Dalton, J. L. Dye, F. M. Fielden, and E. J. Hart, *J. Phys. Chem.*, **70**, 3358 (1966).
31. S. Arai and M. C. Sauer, Jr., *J. Chem. Phys.*, **44**, 2297 (1966).
32. J. H. Baxendale, E. M. Fielden, and J. P. Keene, *Science*, **148**, 637 (1965).
33. L. M. Dorfman, in *Solvated Electron*, Am. Chem. Soc. Publications,, Washington, D.C., 1965, Ch. 4.
34. D. M. J. Compton, J. F. Bryant, R. A. Cesena, and B. L. Gehman, in *Pulse Radiolysis*, Academic Press, New York, 1965, p. 43.
35. J. Jortner, *Radiation Res.*, *Suppl. No. 4*, 24 (1964).
36. T. J. Kemp, G. A. Salmon, and P. Wardman, in *Pulse Radiolysis*, Academic Press, London, 1965, p. 247.
37. J. J. J. Myron and G. R. Freeman, *Can. J. Chem.*, **43**, 381 (1965).
38. S. Arai and L. M. Dorfman, *J. Chem. Phys.*, **41**, 2190 (1964).
39. J. P. Keene, E. J. Land, and A. J. Swallow, *J. Am. Chem. Soc.*, **87**, 5284 (1965).
40. G. Scholes, M. Simic, G. E. Adams, and J. W. Boag, *Nature*, **204**, 1187 (1964).
41. S. Arai, D. A. Grev, and L. M. Dorfman, *J. Chem. Phys.*, **46**, 2572 (1967).
42. J. P. Keene, T. J. Kemp, and G. A. Salmon, *Proc. Royal Soc. (London)*, **A287**, 494 (1965).
43. W. V. Sherman, *J. Phys. Chem.*, **70**, 2872 (1966).
44. E. Hayon and M. Moreau, *J. Phys. Chem.*, **69**, 4053 (1965).
45. W. V. Sherman, *J. Phys. Chem.*, **70**, 667 (1966).

46. H. Seki and M. Imamura, *J. Phys. Chem.*, **71**, 870 (1967).
47. W. V. Sherman, *J. Am. Chem. Soc.*, **88**, 1567 (1966).
48. H. Seki and M. Imamura, *Bull. Chem. Soc. Japan*, **38**, 1229 (1965).
49. D. H. McDaniel and H. C. Brown, *J. Org. Chem.*, **23**, 420 (1958).
50. S. Sato, R. Yugeta, K. Shinsaka, and T. Terao, *Bull. Chem. Soc. Japan*, **39**, 156 (1966).
51. W. V. Sherman, *J. Chem. Soc. (A)*, **1966**, 599.
52. N. H. Sagert and A. S. Blair, *Can. J. Chem.*, **45**, 1351 (1967).
53. F. S. Dainton and S. A. Sills, *Proc. Chem. Soc. (London)*, **1962**, 223.
54. J. M. Warman, *Nature*, **213**, 381 (1967).
55. J. M. Warman, R. H. Schuler, and K.-D. Asmus, in *Advan. Chem.*, in press.
56. T. Shida and W. H. Hamill, *J. Chem. Phys.*, **44**, 2375 (1966).
57. W. V. Sherman, *Nature*, **210**, 1285 (1966).
58. J. G. Burr, F. G. Goodspeed, and C. W. Warren, *J. Am. Chem. Soc.*, **89**, in press.
59. F. C. Goodspeed and J. G. Burr, *J. Am. Chem. Soc.*, **87**, 1643 (1965).
60. L. A. Rajbenbach, *J. Am. Chem. Soc.*, **88**, 4275 (1966).
61. R. R. Hentz, D. B. Peterson, S. B. Srivastava, H. F. Barzynski, and M. Burton, *J. Phys. Chem.*, **70**, 2362 (1966).
62. P. Ausloos, S. G. Lias, and I. B. Sandoval, *Discussions Faraday Soc.*, **36**, 66 (1963).
63. D. P. Stevenson, *Radiation Res.*, **10**, 610 (1959).
64. T. Miyazaki, *J. Phys. Chem.*, **71**, 4282 (1967).
65. T. Miyazaki and S. Shida, *Bull. Chem. Soc. Japan*, **38**, 716 (1965).
66. C. E. McCauley and R. H. Schuler, *J. Am. Chem. Soc.*, **79**, 4008 (1957).
67. B. Y. Yamamoto, A. F. Sciamanna, and A. S. Newton, Report UCRL-9924 (1961).
68. R. A. Holroyd and G. W. Klein, *J. Am. Chem. Soc.*, **87**, 4983 (1965).
69. T. F. Williams, *Trans. Faraday Soc.*, **57**, 755 (1961).
70. K. Tanno, T. Miyazaki, K. Shinsaka, and S. Shida, *J. Phys. Chem.*, **71**, 4290 (1967).
71. F. Williams, *Nature*, **194**, 348 (1962).
72. W. R. McDonell and A. S. Newton, *J. Am. Chem. Soc.*, **76**, 4651 (1954).
73. J. J. J. Myron and G. R. Freeman, *Can. J. Chem.*, **43**, 1484 (1965).
74. D. R. Howton and G. S. Wu, *J. Am. Chem. Soc.*, **89**, 516 (1967).
75. J. G. Burr, *J. Phys. Chem.*, **61**, 1481 (1957).
76. J. W. Buchanan and F. Williams, *J. Chem. Phys.*, **44**, 4377 (1966).
77. W. R. Busler, D. H. Martin, and F. Williams, *Discussions Faraday Soc.*, **36**, 102 (1963).
78. G. R. Freeman, *J. Chem. Phys.*, **43**, 93 (1965).
79. L. Kevan and W. F. Libby, in *Advances in Photochemistry*, Vol. II, Interscience, New York, 1964, p. 183 ff.
80. C. E. Klots, Y. Raef, and R. H. Johnsen, *J. Phys. Chem.*, **68**, 2040 (1964).
81. F. P. Abramson and J. H. Futrell, *Abstracts 152nd Meeting of the American Chemical Society, New York, Sept. 1966*, V. 122.
82. J. H. Green and D. M. Pinkerton, *J. Phys. Chem.*, **68**, 1107 (1964).
83. D. P. Stevenson and D. O. Schissler, *J. Chem. Phys.*, **23**, 1353 (1955).
84. A. Henglein and G. A. Muccini, *Z. Naturforsch.*, **17a**, 452 (1962).
85. See Chapter 4, Table 4.
86. H. A. Gillis, *J. Phys. Chem.*, **71**, 1089 (1967).
87. A. A. Scala, S. G. Lias, and P. Ausloos, *J. Am. Chem. Soc.*, **88**, 5701 (1966).
88. P. Ausloos, A. A. Scala, and S. G. Lias, *J. Am. Chem. Soc.*, **88**, 1583 (1966).

89. P. Ausloos, A. A. Scala, and S. G. Lias, *J. Am. Chem. Soc.*, **89**, 3677 (1967).
90. R. D. Doepker and P. Ausloos, *J. Chem. Phys.*, **42**, 3746 (1965).
91. F. P. Abramson and J. H. Futrell, *J. Phys. Chem.*, **71**, 1233 (1967).
92. R. A. Holroyd, *J. Phys. Chem.*, **70**, 1341 (1966).
93. P. C. Chang, N. C. Yang, and C. D. Wagner, *J. Am. Chem. Soc.*, **81**, 2060 (1959).
94. R. A. Holroyd and R. W. Fessenden, *J. Phys. Chem.*, **67**, 2743 (1963).
95. C. D. Wagner, *Tetrahedron*, **14**, 164 (1961).
96. C. D. Wagner, *J. Phys. Chem.*, **66**, 1158 (1962).
97. P. J. Dyne, *Can. J. Chem.*, **43**, 1080 (1965).
98. P. R. Geissler and J. E. Willard, *J. Am. Chem. Soc.*, **84**, 4619 (1962).
99. R. Blackburn and A. Charlesby, *Nature*, **210**, 1036 (1966).
100. Y. Okada, *J. Phys. Chem.*, **68**, 2120 (1964).
101. G. R. Freeman and J. M. Fayadh, *J. Chem. Phys.*, **43**, 86 (1965).
102. G. R. Freeman, *J. Chem. Phys.*, **39**, 1580 (1963).
103. A. Jahns and W. Jacobi, *Z. Naturforsch.*, **21a**, 1400 (1966).
104. G. R. Freeman, *J. Chem. Phys.*, **46**, 2822 (1967).
105. J. C. Russell and G. R. Freeman, *J. Phys. Chem.*, **71**, 755 (1967).
106. A. H. Samuel and J. L. Magee, *J. Chem. Phys.*, **21**, 1080 (1953).
107. M. K. M. Ng and G. R. Freeman, *J. Am. Chem. Soc.*, **87**, 1635 (1965).
108. R. A. Basson and H. J. van der Linde, (a) *Nature*, **210**, 943 (1966); (b) *J. Chem. Soc., A*, **1967**, 28.
109. J. K. Thomas and R. V. Bensasson, *J. Chem. Phys.*, **46**, 4147 (1967).
110. J. B. Birks, *Theory and Practice of Scintillation Counting*, 1st ed., MacMillan, New York, 1964.
111. S. Lipsky, in *Physical Processes in Radiation Biology*, Academic Press, New York, 1964, p. 215 ff.
112. S. Horikiri and T. Saigusa, *Bull. Inst. Chem. Res., Kyoto Univ.*, **43**, 45 (1965).
113. H. Dreeskamp and M. Burton, *Z. Elektrochem.*, **64**, 165 (1960).
114. P. G. Sjölin, *Nucl. Instrum. Methods*, **37**, 45 (1965).
115. M. Burton, P. K. Ludwig, M. S. Kennard, and R. J. Povinelli, *J. Chem. Phys.*, **41**, 2563 (1964).
116. M. Burton, A. K. Ghosh, and J. Yguerabide, *Radiation Res., Suppl. No. 2*, 462 (1960).
117. M. A. Dillon and M. Burton, in *Pulse Radiolysis*, Academic Press, London, 1965, p. 259 ff.
118. N. Turro, *Molecular Photochemistry*, W. A. Benjamin, New York, 1965.
119. J. Nosworthy and J. P. Keene, *Proc. Chem. Soc.*, **1964**, 114.
120. R. Voltz, H. Dupont, and T. A. King, *Nature*, **211**, 405 (1966).
121. C. L. Braga, M. D. Lumb, and J. B. Birks, *Trans. Faraday Soc.*, **62**, 1830 (1966).
122. W. R. Ware and J. S. Novros, *J. Phys. Chem.*, **70**, 3246 (1966).
123. D. B. Peterson, T. Arakawa, D. Walmsley, and M. Burton, *J. Phys. Chem.*, **69**, 2880 (1965).
124. H. R. Lukens, Jr., *Nucl. Applications*, **1**, 597 (1965).
125. C. R. Mullin, M. A. Dillon, and M. Burton, *J. Chem. Phys.*, **40**, 3053 (1964).
126. G. Hoefer, *Z. Physik.*, **181**, 44 (1964).
127. V. Bar and A. Weinreb, *J. Chem. Phys.*, **29**, 1412 (1958).
128. A. Weinreb, *J. Chem. Phys.*, **35**, 91 (1961).
129. S. Lipsky and M. Burton, *J. Chem. Phys.*, **31**, 1221 (1959).
130. P. H. Bell and W. P. Davey, *J. Chem. Phys.*, **9**, 441 (1941).
131. A. Münster, *Trans. Faraday Soc.*, **46**, 165 (1950).

132. J. Nosworthy, J. L. Magee, and M. Burton, *J. Chem. Phys.*, **34**, 83 (1961).
133. M. Burton, M. A. Dillon, and C. R. Mullin, *J. Chem. Phys.*, **41**, 2236 (1964).
134. T. V. Ivanova, G. A. Mokeeva, and B. Ya. Sveshnikov, *Opt. Spectry.*, **12**, 325 (1962).
135. J. B. Birks, C. L. Braga, and M. D. Lumb, *Proc. Royal Soc. (London)*, **283**, 83 (1965).
136. J. G. Carter and L. G. Christophorou, *J. Chem. Phys.*, **46**, 1883 (1967).
137. J. B. Birks, J. M. de C. Conte, and G. Walker, *IEEE Trans. Nucl. Sci.*, **3**, 148 (1966).
138. J. B. Birks and J. M. de C. Conte, *Proc. Royal Soc. A (London)*, **303**, 85 (1968).
139. L. G. Christophorou and J. G. Carter, *Nature*, **209**, 678 (1966).
140. F. Hirayama, Report ANL-6938 (1965) pp. 27–35.
141. U. Laor and A. Weinreb, *J. Chem. Phys.*, **43**, 1565 (1965).
142. S. F. Kilin and I. M. Rozman, *Opt. Spectry.*, **17**, 380 (1964).
143. M. Burton, *Strahlentherapie*, **51**, Suppl., 1 (1962).
144. C. A. Parker and C. G. Hatchard, *Analyst*, **87**, 664 (1962).
145. M. Windsor and W. R. Dawson, *Mol. Cryst.*, **3**, 165 (1967).
146. E. J. Bowen, in *Advances in Photochemistry*, Vol. I, Interscience, New York, 1963, p. 32.
147. J. W. Hastings and G. Weber, *J. Optical Soc. Am.*, **53**, 1410 (1963).
148. G. J. Hine and G. L. Brownell, *Radiation Dosimetry*, Academic Press, New York, 1956.
149. J. D. McCollum and T. D. Nevitt, Report ASD-TDR-63-616 (1963).
150. I. B. Berlman, R. Grismore, and B. G. Oltman, *Trans. Faraday Soc.*, **59**, 2010 (1963).
151. E. Matovich, *Trans. Am. Nucl. Soc.*, **8**, 64 (1965).
152. M. Kasha, *Radiation Res.*, *Suppl. No. 2*, 243 (1960).
153. A. A. Lamola and G. S. Hammond, *J. Chem. Phys.*, **43**, 2129 (1965).
154. R. L. McCarthy and A. MacLachlan, *Trans. Faraday Soc.*, **56**, 1187 (1960).
155. F. S. Dainton, T. J. Kemp, G. A. Salmon, and J. P. Keene, *Nature*, **203**, 1050 (1964).
156. J. H. Baxendale et al., *Nature*, **201**, 468 (1964).
157. T. S. Godfrey and G. Porter, *Trans. Faraday Soc.*, **62**, 7 (1966).
158. J. Nosworthy, *Trans. Faraday Soc.*, **61**, 1138 (1965).
159. R. B. Cundall and P. A. Griffiths, *Trans. Faraday Soc.*, **61**, 1968 (1965).
160. L. M. Dorfman, I. A. Taub, and R. E. Buehler, *J. Chem. Phys.*, **36**, 3051 (1962).
161. A. MacLachlan and R. L. McCarthy, *J. Am. Chem. Soc.*, **84**, 2519 (1962).
162. R. B. Cundall and P. A. Griffiths, *Chem. Commun.*, **1966**, 194.
163. L. R. Griffith, Report-UCRL-3422 (1956).
164. E. A. Cherniak, E. Collinson, and F. S. Dainton, *Trans. Faraday Soc.*, **60**, 1408 (1964).
165. R. B. Cundall, F. J. Fletcher, and D. G. Milne, *Trans. Faraday Soc.*, **60**, 1146 (1964).
166. M. A. Golub and C. L. Stephens, *J. Phys. Chem.*, **70**, 3576 (1966).
167. R. R. Hentz, K. Shima, and M. Burton, *J. Phys. Chem.*, **71**, 461 (1967).
168. A. Torikai, K. Fueki, and Z. Kuri, *Nippon Kagaku Zasshi*, **87**, 391 (1966).
169. W. G. Brown, *Chem. Commun.*, **1966**, 195.
170. T. J. Kemp, G. A. Salmon, and F. Wilkinson, *Chem. Commun.*, **1966**, 73.
171. E. J. Land, in *Pulse Radiolysis*, Academic Press, New York, 1965, p. 293 ff.
172. R. B. Cundall and P. A. Griffiths, *Discussions Faraday Soc.*, **36**, 111 (1963).

173. N. A. Golub, C. L. Stephens, and J. L. Brash, *J. Chem. Phys.*, **45**, 1503 (1966).
174. R. B. Cundall and P. A. Griffiths, *J. Phys. Chem.*, **69**, 1866 (1965).
175. E. Fischer, H. P. Lehmann, and G. Stein, *J. Chem. Phys.*, **45**, 3905 (1966).
176. W. G. Brown, *J. Phys. Chem.*, **69**, 4401 (1965).
177. S. Arai and L. M. Dorfman, *J. Phys. Chem.*, **69**, 2239 (1965).
178. P. Dunion and C. N. Trumbore, *J. Am. Chem. Soc.*, **87**, 4211 (1965).
179. D. L. Dugle and G. R. Freeman, *Trans. Faraday Soc.*, **61**, 1174 (1965).
180. A. Singh and G. R. Freeman, *J. Phys. Chem.*, **69**, 666 (1965).
181. P. J. Coyle, *J. Phys. Chem.*, **67**, 1800 (1963).
182. P. J. Wagner and G. S. Hammond, *J. Am. Chem. Soc.*, **87**, 4009 (1965).
183. E. J. Baum, J. K. S. Wan, and J. N. Pitts, Jr., *Abstracts of the 149th Meeting American Chemical Society*, Detroit, April, 1965, S77.
184. R. F. Borkman and D. R. Kearns, *J. Am. Chem. Soc.*, **88**, 3467 (1966).
185. G. Porter, *Proc. Chem. Soc.*, **1959**, 291.
186. M. Furst and H. Kallman, *Phys. Rev.*, **85**, 816 (1952).
187. G. S. Hammond, et al., *J. Am. Chem. Soc.*, **86**, 3197 (1964).
188. J. S. Brinen, J. G. Koren, and W. G. Hodgson, *J. Chem. Phys.*, **44**, 3095 (1966).
189. J. W. Hunt and J. K. Thomas, *J. Chem. Phys.*, **46**, 2954 (1967).
190. J. W. vanLobenSels and J. T. Dubois, *J. Chem. Phys.*, **45**, 1522 (1966).
191. H. P. Lehmann, G. Stein, and E. Fischer, *Chem. Commun.*, **1965**, 583.
192. R. A. Caldwell, D. G. Whitten, and G. S. Hammond, *J. Am. Chem. Soc.*, **88**, 2659 (1966).
193. L. I. Samokhvalova and V. A. Krongauz, *Dokl. Akad. Nauk SSSR*, **168**, 154 (1966), from NSA, **20**, 40961 (1966).
194. R. Chang and C. S. Johnson, Jr., *J. Chem. Phys.*, **46**, 2314 (1967).
195. W. A. Noyes, Jr. and D. A. Harter, *J. Chem. Phys.*, **46**, 674 (1967).
196. K. E. Wilzbach and L. Kaplan, *J. Am. Chem. Soc.*, **88**, 2066 (1966).
197. D. Bryce-Smith and A. Gilbert, *Chem. Commun.*, **1966**, 643.
198. D. Bryce-Smith and H. C. Longuet-Higgins, *Chem. Commun.*, **1966**, 593.
199. S. Gordon, A. R. Van Dyken, and T. F. Doumani, *J. Phys. Chem.*, **62**, 20 (1958).
200. T. Gäumann, *Helv. Chim. Acta*, **44**, 1337 (1961).
201. H. J. F. Angus, J. M. Blair, and D. Bryce-Smith, *J. Chem. Soc.*, **1960**, 2003.
202. K. E. Wilzbach and L. Kaplan, *J. Am. Chem. Soc.*, **87**, 4004 (1965).
203. K. E. Wilzbach, J. S. Ritscher, and L. Kaplan, *J. Am. Chem. Soc.*, **89**, 1031 (1967).
204. J. Fajer and D. R. MacKenzie, *J. Phys. Chem.*, **71**, 784 (1967).
205. D. Verdin, *J. Phys. Chem.*, **67**, 1263 (1963).
206. A. I. Gebshtein, M. I. Temkin, G. G. Shoheglova, T. V. Llyukhina, and M. A. Proskurnin, in *All-Union Conference on Radiation Chemistry, Moscow, 1957*, from AEC-TR-2925.
207. R. A. Holroyd, J. Y. Yang, and F. M. Servedio, *J. Chem. Phys.*, **46**, 4540 (1967).
208. R. H. Schuler, *J. Phys. Chem.*, **62**, 37 (1958).
209. R. W. Fessenden and R. H. Schuler, *J. Chem. Phys.*, **39**, 2147 (1963).
210. R. A. Holroyd, in *Aspects of Hydrocarbon Radiolysis*, Academic Press, London, in press.
211. R. W. Fessenden and R. H. Schuler, *J. Chem. Phys.*, **33**, 935 (1960).
212. R. H. Schuler and R. W. Fessenden, *Proceedings of the Third International Congress on Radiation Research*, Cortina, Italy, 1966, from RRL-2310-180.
213. H. A. Gillis, *J. Phys. Chem.*, **67**, 1399 (1963).
214. R. W. Fessenden, *J. Phys. Chem.*, **68**, 1508 (1964).

215. R. G. Dickenson, in *Photochemistry of Gases*, Reinhold, New York, 1941, p. 202f.

216. E. J. Land, in *Progress in Reaction Kinetics*, Vol. III, Pergamon Press, Oxford, 1965, p. 386ff.

217. M. S. Matheson and L. M. Dorfman, *J. Chem. Phys.*, **32**, 1870 (1960).

218. M. Ebert, J. P. Keene, E. J. Land, and A. J. Swallow, *Proc. Royal Soc. (London)*, **A287**, 1 (1965).

219. R. L. McCarthy and A. MacLachlan, *J. Chem. Phys.*, **35**, 1625 (1961).

220. I. A. Taub and L. M. Dorfman, *J. Am. Chem. Soc.*, **84**, 4053 (1962).

221. C. L. Currie and D. A. Ramsay, *J. Chem. Phys.*, **45**, 488 (1966).

222. E. J. Burrell, Jr. and P. K. Bhatlacharyya, *J. Phys. Chem.*, **71**, 774 (1967).

223. A. B. Callear and H. K. Lee, *Nature*, **213**, 693 (1967).

224. A. T. Watts and S. Walker, *J. Chem. Soc.*, **1962**, 4323.

225. L. J. Forrestal and W. H. Hamill, *J. Am. Chem. Soc.*, **83**, 1535 (1961).

226. R. H. Schuler, *J. Phys. Chem.*, **61**, 1472 (1957).

227. I. Mani and R. J. Hanrahan, *J. Phys. Chem.*, **70**, 2233 (1966).

228. D. Perner and R. H. Schuler, *J. Phys. Chem.*, **70**, 2224 (1966).

229. R. H. Schuler and D. Perner, *Abstracts of the 148th Meeting of the American Chemical Society, Chicago, 1964*, V121.

230. R. H. Schuler, *J. Phys. Chem.*, **68**, 3873 (1964).

231. J. Laplane-Masanet and N. Ivanoff, *Proceedings of the 1962 Tihany Symposium-Radiation Chemistry*, Hungarian Academy Sciences, Budapest, 1964, p. 75.

232. A. Charlesby and D. G. Lloyd, *Proc. Royal Soc. (London)*, **A249**, 51 (1958).

233. R. A. Holroyd and G. W. Klein, *Intern. J. Appl. Radiation Isotopes*, **13**, 493 (1962).

234. R. A. Holroyd and G. W. Klein, *Intern. J. Appl. Radiation Isotopes*, **15**, 633 (1964).

235. R. A. Holroyd and G. W. Klein, *J. Am. Chem. Soc.*, **84**, 4000 (1962).

236. R. A. Holroyd, *J. Am. Chem. Soc.*, **88**, 5381 (1966).

237. R. A. Holroyd and G. W. Klein, *J. Phys. Chem.*, **69**, 194 (1965).

238. W. G. Burns, R. A. Holroyd, and G. W. Klein, *J. Phys. Chem.*, **70**, 910 (1966).

239. R. H. Schuler and R. R. Kuntz, *J. Phys. Chem.*, **67**, 1004 (1963).

240. R. H. Schuler, private communication.

241. H. A. Dewhurst, *J. Am. Chem. Soc.*, **80**, 5607 (1958).

242. R. W. Fessenden and R. H. Schuler, *J. Chem. Phys.*, **38**, 773 (1963).

243. M. K. Eberhardt, G. W. Klein and T. G. Krivak, *J. Am. Chem. Soc.*, **87**, 696 (1965).

244. F. Antoine, *Compt. Rend.*, **258**, 4742 (1964).

245. R. A. Holroyd, *Abstracts 153rd Meeting American Chemical Society, Miami Beach, 1967*, R55.

246. N. D. Coggeshall, *J. Chem. Phys.*, **30**, 595 (1959).

247. J. Dauphin, *Proceedings Conference on Radioisotopes in the Physical Sciences and Industry*, Vol. III, IAEA-UNESCO, Copenhagen, 1960, p. 471.

248. J. F. Merklin and S. Lipsky, *J. Phys. Chem.*, **68**, 3297 (1964).

249. T. J. Hardwick, *J. Phys. Chem.*, **66**, 1611 (1962).

250. G. R. Freeman, *J. Chem. Phys.*, **33**, 71 (1960).

251. G. Meshitsuka and M. Burton, *Radiation Res.*, **10**, 499 (1959).

252. J. R. Nash and W. H. Hamill, *J. Phys. Chem.*, **66**, 1097 (1962).

253. J. L. McCrumb and R. H. Schuler, *J. Phys. Chem.*, **71**, 1953 (1967).

254. G. Meissner and A. Henglein, *Ber. Bunsenges Physik. Chem.*, **69**, 264 (1965).

255. D. R. Smith and J. J. Pieroni, *J. Phys. Chem.*, **70**, 2379 (1966).
256. F. W. Lampe, *J. Phys. Chem.*, **63**, 1986 (1959).
257. M. Katayama and M. Hatada, *Nippon Kagaku Zasshi*, **87**, 37 (1966), from NSA, **20**, 4994 (1966).
258. R. W. Fessenden and R. H. Schuler, *J. Chem. Phys.*, **43**, 2704 (1965).
259. J. R. Majer and J. P. Simons, in *Advances in Photochemistry*, Vol. II, Interscience, New York, 1964, p. 165.
260. R. E. Florin, L. A. Wall, and D. W. Brown, *J. Res. Natl. Bur. Std.*, **64A**, 269 (1960).
261. T. M. Reed, J. C. Mailen, and W. C. Askew, Report TID-22421 (1965).
262. D. R. Mackenzie, B. W. Bloch, and R. H. Wiswall, Jr., *J. Phys. Chem.*, **69**, 2526 (1965).
263. M. B. Fallgatter and R. J. Hanrahan, *J. Phys. Chem.*, **69**, 2059 (1965).
264. J. W. Schulte, *J. Am. Chem. Soc.*, **79**, 4643 (1957).
265. T. H. Chen, K. Y. Wong, and F. J. Johnston, *J. Phys. Chem.*, **64**, 1023 (1960).
266. S. Ciborowski, N. Colebourne, E. Collinson, and F. S. Dainton, *Trans. Faraday Soc.*, **57**, 1123 (1961).
267. R. H. Schuler and R. C. Petry, *J. Am. Chem. Soc.*, **78**, 3954 (1956).
268. H. L. Benson, Jr. and J. E. Willard, *J. Am. Chem. Soc.*, **88**, 5689 (1966).
269. G. E. Adams, J. H. Baxendale, and R. D. Sedgwick, *J. Phys. Chem.*, **63**, 854 (1959).
270. I. Zwiebel and R. H. Bretton, *J. Am. Inst. Chem. Engrs.*, **10**, 339 (1964).
271. R. A. Basson and H. J. van der Linde, *Chem. Commun.*, **1967**, 91.
272. R. Barker, *Trans. Faraday Soc.*, **59**, 375 (1963).
273. M. Roder, K. Go, N. A. Bakh, and L. T. Bugaenko, *Kinetika i Kataliz*, **5**, 976 (1964), from NSA **19**, 13464 (1965).
274. E. A. Cherniak, E. Collinson, F. S. Dainton, and G. M. Meaburn, *Proc. Chem. Soc.*, **1958**, 54.
275. E. Collinson, J. J. Conlay, and F. S. Dainton, *Discussions Faraday Soc.*, **36**, 153 (1963).
276. M. Haissinsky and M. Magat, *Selected Constants—Radiolytic Yields*, Pergamon Press, Oxford, 1963.
277. R. W. Hummel, *Trans. Faraday Soc.*, **56**, 234 (1960).
278. E. O. Hornig and J. E. Willard, *J. Am. Chem. Soc.*, **79**, 2429 (1957).
279. H. R. Werner and R. F. Firestone, *J. Phys. Chem.*, **69**, 840 (1965).
280. G. E. Adams and R. D. Sedgewick, *Trans. Faraday Soc.*, **60**, 865 (1964).
281. A. Ekstrom and J. L. Garnett, *J. Phys. Chem.*, **70**, 324 (1966).
282. J. G. Burr, C. K. Dalton, and R. A. Meyer, *Abstracts of the 137th Meeting of the American Chemical Society, Cleveland, Ohio, 1960*, 42R.
283. L. M. Theard and M. Burton, *J. Phys. Chem.*, **67**, 59 (1963).
284. B. A. Basson, *Nature*, **211**, 629 (1966).
285. B. A. Basson, A. G. Maddock, and F. Marta, in press.
286. M. H. J. Wijnen, *J. Chem. Phys.*, **27**, 710 (1957).
287. P. J. Ausloos and J. Paulson, *J. Am. Chem. Soc.*, **80**, 5117 (1958).
288. J. D. Strong and J. G. Burr, *J. Am. Chem. Soc.*, **81**, 775 (1959).
289. J. Kucera, *Collections Czech. Chem. Commun.*, **30**, 3080 (1965).
290. R. Barker, *Nature*, **192**, 62 (1961).
291. W. A. Noyes, Jr., *Radiation Res., Suppl. No. 1*, 164 (1959).
292. R. R. Kuntz and G. J. Mains, *J. Am. Chem. Soc.*, **85**, 2219 (1963).
293. P. S. Dixon, A. P. Stefani, and M. Szwarc, *J. Am. Chem. Soc.*, **85**, 3344 (1963).

294. C. E. Klots and R. H. Johnsen, *Can. J. Chem.*, **41**, 2702 (1963).
295. J. W. Falconer and M. Burton, *J. Phys. Chem.*, **67**, 1743 (1963).
296. P. S. Dixon, A. P. Stefani, and M. Szwarc, *J. Am. Chem. Soc.*, **85**, 2551 (1963).
297. W. G. Burns and R. Barker, in *Progress in Reaction Kinetics*, Vol. III, Pergamon Press, Oxford, 1965, Ch. 7.
298. R. A. Holroyd and G. W. Klein, *J. Phys. Chem.*, **67**, 2273 (1963).
299. T. Gäumann, *Helv. Chim. Acta*, **46**, 2873 (1963).
300. R. H. Schuler and G. A. Muccini, *J. Am. Chem. Soc.*, **81**, 4115 (1959).
301. R. A. Holroyd, *J. Phys. Chem.*, **65**, 1352 (1961).
302. N. E. Bibler, *Abstracts of the 153rd Meeting American Chemical Society, Miami Beach, April, 1967*, R51.
303. M. Anbar and P. Neta, *Intern. J. Appl. Radiation Isotopes*, **16**, 227 (1965).
304. J. M. Brown, P. B. Coates, and B. A. Thrush, *Chem. Commun.*, **1966**, 843.
305. K. Yang, *J. Am. Chem. Soc.*, **84**, 3795 (1962).
306. A. Appleby, G. Scholes, and M. Simic, *J. Am. Chem. Soc.*, **85**, 3891 (1963).
307. J. P. Sweet and J. K. Thomas, *J. Phys. Chem.*, **68**, 1363 (1964).
308. P. V. Phung and M. Burton, *Radiation Res.*, **7**, 199 (1957).
309. D. Perner and R. H. Schuler, *J. Phys. Chem.*, **70**, 317 (1966).
310. T. J. Hardwick, *J. Phys. Chem.*, **65**, 101 (1961).
311. S. Nehari and J. Rabani, *J. Phys. Chem.*, **67**, 1609 (1963).
312. K. Yang, *J. Phys. Chem.*, **67**, 562 (1963).
313. R. L. McCarthy and A. MacLachlan, *Trans. Faraday Soc.*, **57**, 1107 (1961).
314. H. W. Melville, J. C. Robb, and R. C. Tutton, *Discussions Faraday Soc.*, **14**, 150 (1953).
315. W. A. Pryor, *Free Radicals*, McGraw-Hill, New York, 1966, p. 224.
316. R. Blackburn, A. Charlesby, and J. F. Read, *Radiation Res.*, **28**, 793 (1966).
317. T. Tiedeman and R. B. Ingalls, *J. Phys. Chem.*, **71**, 3092 (1967).
318. C. Walling, *Free Radicals in Solution*, Wiley, New York, 1957.
319. D. F. DeTar and R. A. Long, *J. Am. Chem. Soc.*, **80**, 4742 (1958).
320. M. Miyoshi and R. K. Brinton, *J. Chem. Phys.*, **36**, 3019 (1962).
321. B. R. Wakeford and G. R. Freeman, *J. Phys. Chem.*, **68**, 2992 (1964).
322. R. A. Holroyd, *J. Phys. Chem.*, **66**, 730 (1962).
323. J. A. Kerr, *Chem. Rev.*, **66**, 465 (1966).
324. C. S. H. Chen and R. F. Stamm, *J. Org. Chem.*, **28**, 1580 (1963).
325. L. H. Gale and C. D. Wagner, *J. Am. Chem. Soc.*, **86**, 4531 (1964).
326. D. Armstrong and J. W. T. Spinks, *Can. J. Chem.*, **37**, 1210 (1959).
327. L. A. Rajbenbach and A. Horowitz, *Chem. Commun.*, **1966**, 769.
328. A. V. Topchiev, *Radiolysis of Hydrocarbons*, English ed., Elsevier, New York, 1964.
329. E. O. Guernsey, H. Shaw, and W. E. Smith, *Am. Inst. Chem. Engr. J.*, **9**, 744 (1963).
330. P. J. Lucchesi, B. L. Tarmy, R. B. Long, D. L. Baeder, and J. P. Longwell, *Ind. and Engr. Chem.*, **50**, 879 (1958).
331. L. Endrenyi and D. J. LeRoy, *J. Phys. Chem.*, **70**, 4081 (1966).
332. D. P. Thornhill and C. C. Lee, *Can. J. Chem.*, **41**, 2482 (1963).
333. M. Takehisa, G. Levey, and J. E. Willard, *J. Am. Chem. Soc.*, **88**, 5694 (1966).
334. W. S. Wilcox, *Radiation Res.*, **10**, 112 (1959).
335. J. W. T. Spinks and R. J. Woods, *Introduction to Radiation Chemistry*, Wiley, New York, 1964, p. 331.
336. A. Chapiro, *Radiation Res.*, **6**, 11 (1957).

337. H. A. Gillis, R. R. Williams, Jr., and W. H. Hamill, *J. Am. Chem. Soc.*, **83**, 17 (1961).
338. A. E. deVries and A. O. Allen, *J. Phys. Chem.*, **63**, 879 (1959).
339. S. K. Ho and G. R. Freeman, *J. Phys. Chem.*, **68**, 2189 (1964).
340. R. Blackburn and A. Charlesby, *Proc. Royal Soc. (London)*, **293**, 51 (1966).
341. P. Ausloos and C. N. Trumbore, *J. Am. Chem. Soc.*, **81**, 3866 (1959).
342. R. N. Schindler, *Radiochim. Acta*, **2**, 73 (1963).
343. J. Sturm and H. A. Schwarz, *Radiation Res.*, **17**, 531 (1962).
344. P. J. Ausloos, *J. Am. Chem. Soc.*, **83**, 1056 (1961).
345. R. E. Rebbert and P. Ausloos, *J. Phys. Chem.*, **66**, 2253 (1962).
346. A. Kupperman, in *Chemical and Biological Action of Radiation*, Vol. V, Academic Press, London, 1961.
347. P. Claes and S. Rzad, *Bull. Soc. Chim. Belges*, **74**, 220 (1965).
348. P. Claes and S. Rzad, *J. Phys. Chem.*, **69**, 1780 (1965).
349. R. H. Schuler, *J. Phys. Chem.*, **63**, 925 (1959).
350. R. H. Schuler and A. O. Allen, *J. Am. Chem. Soc.*, **77**, 507 (1955).
351. H. A. Dewhurst and R. H. Schuler, *J. Am. Chem. Soc.*, **81**, 3210 (1959).
352. P. J. Dyne and J. A. Stone, *Can. J. Chem.*, **39**, 2381 (1961).
353. W. G. Burns and J. R. Parry, *Nature*, **201**, 814 (1964).
354. A. W. Boyd and H. W. J. Connor, *Can. J. Chem.*, **42**, 1418 (1964).
355. G. A. Muccini and R. H. Schuler, *J. Phys. Chem.*, **64**, 1436 (1960).
356. J. A. Stone and P. J. Dyne, *Radiation Res.*, **17**, 353 (1962).
357. R. R. Hentz, J. Y. Chang, and M. Burton, *J. Phys. Chem.*, **69**, 2027 (1965).
358. W. A. Cramer, *J. Phys. Chem.*, **71**, 1112 (1967).
359. W. A. Cramer, *J. Phys. Chem.*, **71**, 1171 (1967).
360. W. G. Burns and J. D. Jones, *Trans. Faraday Soc.*, **60**, 2022 (1964).
361. W. G. Burns and R. Barker, in *Aspects of Hydrocarbon Radiolysis*, Academic Press, London, in press.
362. W. N. Patrick and M. Burton, *J. Am. Chem. Soc.*, **76**, 2626 (1954).
363. R. H. Schuler, *Trans. Faraday Soc.*, **61**, 100 (1965).
364. S. Gordon and M. Burton, *Discussions Faraday Soc.*, **12**, 88 (1952).
365. P. J. Dyne and W. M. Jenkinson, *Can. J. Chem.*, **40**, 1746 (1962).
366. W. G. Burns and C. R. V. Reed, *Trans. Faraday Soc.*, **59**, 101 (1963).
367. R. H. Schuler and T. Gäumann, *J. Phys. Chem.*, **65**, 703 (1961).
368. J. R. Hite, *Dissertation Abstracts*, **B27** (2), 457 (1966).
369. J. Y. Yang, J. D. Strong, and J. G. Burr, *J. Phys. Chem.*, **69**, 1157 (1965).
370. J. Hoigne, W. G. Burns, W. R. Marsh, and T. Gäumann, *Helv. Chim. Acta*, **47**, 247 (1964).
371. A. W. Boyd, H. W. J. Connor, and O. A. Miller, Report AECL-2589 (1966).
372. A. W. Boyd and H. W. J. Connor, *Trans. Am. Nucl. Soc.*, **8**, 421 (1965).
373. A. W. Boyd and H. W. J. Connor, Report AECL-2258 (1965).
374. J. M. Scarborough and R. B. Ingalls, *J. Phys. Chem.*, **71**, 486 (1967).
375. R. B. Ingalls, *J. Phys. Chem.*, **65**, 1605 (1961).
376. T. Wolfram and J. A. Brinkman, Report NAA-SCPP-64-32 (1964).
377. R. B. Ingalls, P. Spiegler, and A. Norman, *J. Chem. Phys.*, **41**, 837 (1963).
378. A. Mozumder and J. L. Magee, *J. Chem. Phys.*, **45**, 3332 (1966).
379. J. Y. Yang and J. G. Burr, *J. Chem. Phys.*, **44**, 1307 (1966).
380. R. Schiller, *J. Chem. Phys.*, **43**, 2760 (1965).
381. D. A. Landsman and J. E. Butterfield, Report AERE-R-3625 (1961).
382. G. Meshitsuka and M. Burton, *Radiation Res.*, **8**, 285, (1958).

383. M. Imamura, S. U. Choi, and N. N. Lichtin, *J. Am. Chem. Soc.*, **85**, 3565 (1963).
384. J. Lamborn and A. J. Swallow, *J. Phys. Chem.*, **65**, 920 (1961).
385. A. J. Swallow, in *Energy Transfer in Radiation Processes*, Elsevier, Amsterdam, 1966, p. 121.
386. J. Hoigne and T. Gäumann, *Helv. Chim. Acta*, **47**, 590 (1964).
387. C. E. Klots and R. H. Johnsen, *J. Phys. Chem.*, **67**, 1615 (1963).
388. R. W. Fessenden and R. H. Schuler, *Discussions Faraday Soc.*, **36**, 147 (1963).
389. J. P. Guarino and W. H. Hamill, *J. Am. Chem. Soc.*, **86**, 777 (1964).
390. R. E. Buehler, T. Gäumann, and M. Ebert, in *Pulse Radiolysis*, Academic Press, London, 1965, p. 279.
391. E. A. Rojo and R. R. Hentz, *J. Phys. Chem.*, **69**, 3024 (1965).
392. C. S. Schoepfle and C. H. Fellows, *Ind. Eng. Chem.*, **23**, 1396 (1931).
393. H. A. Dewhurst, *J. Phys. Chem.*, **63**, 813 (1959).
394. R. Blackburn and A. Kabi, *Chem. Commun.*, **1966**, 862.
395. J. G. Burr and F. C. Goodspeed, *J. Chem. Phys.*, **40**, 1433 (1964).
396. M. Burton, J. Y. Chang, S. Lipsky, and M. P. Reddy, *Radiation Res.*, **8**, 203 (1958).
397. H. F. Barzynski, R. R. Hentz, and M. Burton, *J. Phys. Chem.*, **69**, 2034 (1965).
398. S. B. Srivastava, *Radiochim. Acta*, **5**, 50 (1966).
399. P. J. Dyne and W. M. Jenkinson, *Can. J. Chem.*, **39**, 2163 (1961).
400. P. J. Dyne, J. Denhartog, and D. R. Smith, *Discussions Faraday Soc.*, **36**, 135 (1963).
401. A. O. Allen, *Radiation Chemistry of Water and Aqueous Solutions*, D. Van Nostrand, Princeton, 1961, p. 29.
402. G. Meshitsuka, F. Takemura, T. Sakai, and K. Hirota, *Tokyo Metropolitan Isotope Centre Ann. Rept.*, **1**, 61 (1962).
403. V. I. Makarov, L. S. Polak, N. Y. Chernyak, and A. S. Shcherbakwa, *Neftekhimiya*, **6**, 58 (1966), from *Chem. Abstr.*, **64**, 16878e (1966).
404. N. V. Klassen, *J. Phys. Chem.*, **71**, 2409 (1967).
405. A. Mustafa, *Chem. Rev.*, **51**, 1 (1952).
406. P. J. Dyne and J. Denhartog, *Can. J. Chem.*, **40**, 1616 (1962).
407. T. J. Hardwick, *J. Phys. Chem.*, **66**, 2132 (1962).
408. J. A. Stone, A. R. Quirt, and O. A. Miller, *Can. J. Chem.*, **44**, 1175 (1966).
409. T. Kudo and S. Shida, *J. Phys. Chem.*, **71**, 1971 (1967).
410. J. A. Stone and P. J. Dyne, *Can. J. Chem.*, **42**, 669 (1964).
411. G. K. Oster and H. P. Kallmann, *Nature*, **194**, 1033 (1962).
412. W. V. Sherman, *Chem. Commun.*, **1966**, 489.
413. W. Ando, K. Sugimoto, and S. Oae, *Bull. Chem. Soc. Japan*, **37**, 353 (1964).
414. K. H. Jones, W. Van Dusen, Jr., and L. M. Theard, *Radiation Res.*, **23**, 128 (1964).
415. A. Zeman and H. Heusinger, *J. Phys. Chem.*, **70**, 3375 (1966).
416. T. Isoya, K. Hirota, and M. Hatada, *Bull. Chem. Soc. Japan*, **39**, 1878 (1966).
417. W. G. Burns, private communication.

Chapter 8

Principles of Radiation-Induced Polymerization

Ffrancon Williams

Department of Chemistry, University of Tennessee
Knoxville, Tennessee

I. INTRODUCTION

The subject of radiation-induced polymerization can be approached from two contrasting viewpoints. First, radiation may be regarded primarily as a useful technique for initiating polymerization, with the emphasis placed on the nature of the polymeric materials that are formed rather than on the underlying chemistry. This empirical approach is well

515

represented by the majority of the investigations reported in the literature, and while such studies contribute useful data, they afford little guidance to a basic understanding of the subject in terms of elementary reactions. On the other hand, it has been realized by several workers that the intensive study of radiation-induced polymerization in selected monomer systems can provide powerful insight into some of the basic processes of radiation chemistry, and that this information in turn can enrich our knowledge of polymerization mechanisms in general. It is the author's hope that this chapter generally reflects this second viewpoint.

A. Historical

Only brief mention will be made here of the history of radiation-induced polymerization since this aspect has been treated exhaustively by Chapiro in his well-known monograph (1). Following some early experiments of a phenomenological character, the first detailed kinetic work in the field was carried out almost simultaneously by two groups. Collinson, Dainton, and their co-workers made a thorough study of the radiation-induced polymerization of water-soluble vinyl monomers in aqueous solution (2), while Magat and Chapiro investigated the polymerization of monomers in bulk and in organic solvents (3,4). Both groups exploited polymerization as a suitable chemical amplifier for the sensitive detection of free radical intermediates. Thus Dainton's group obtained information on the yields of H and OH radicals in water, and on the presumed spatial distribution and reactions of these radicals in the spur. Magat and Chapiro compiled yield values for radical initiation in organic systems, and compared the results with those obtained independently by the scavenger technique using the stable free radical, 1,1-diphenyl-2-picryl hydrazyl (DPPH). To a large extent, these methods have been supplanted by other more direct techniques of observing and characterizing free radicals and other intermediates. Also, some of the quantitative conclusions reached in this early work no longer apply with particular force. However, these contributions played an important part in the development of radiation chemistry up to 1955, and the significance of the rate-constant ratios derived for radical polymerization remains unimpaired.

Another significant development in the field was the successful initiation of solid-state polymerization. An early contribution to this subject was made by Mesrobian et al. (5) who made a particularly detailed study of crystalline monomers, including acrylamide and the alkali metal salts of acrylic acid. In due course more work was reported on other crystalline and amorphous monomers, and on the method of graft copolymerization (6) whereby the polymerization of a monomer can be initiated by radicals produced by previous or concomitant irradiation of a polymer. By now there are many practical variants to this procedure leading to the production of a new polymer chain grafted on to the backbone of the original polymer. The properties of graft copolymers continue to excite commercial interest and this area remains promising for radiation applications in the polymer industry.

Several authors (7) have commented upon the late recognition of ionic intermediates in radiation-induced polymerization. Here it will suffice to mention that before 1955, there seemed to be compelling arguments from both theoretical and experimental sources against the idea of long-lived transient ions in media of low dielectric constant, and this climate discouraged exploration of the field. Even after radiation-induced ionic polymerization had been demonstrated at low temperatures (195°K), it was argued by many that ionic mechanisms could be of little importance at room temperature or above. Further developments have shown conclusively that no such restriction applies to the ionic polymerization of several monomers in *dry* systems (8).

Since polymerization is a chain reaction, it is axiomatic that the reaction rate can be depressed by the presence of inhibitors or retarders in relatively low concentrations. If compounds of this nature are generated by irradiation, then the effect should be revealed by proper kinetic analysis. On the other hand, because it often is extremely difficult to remove the last vestiges of certain impurities from the monomer or solvent before irradiation, the kinetic results may not be reproducible. A considerable effort has to be devoted to rigorous techniques of sample preparation in order to obtain reliable and systematic data. Unfortunately, comparatively few studies have emphasized this aspect, and it is not unusual to find reports on absolute rates of polymerization differing by large factors. Although this situation is particularly prevalent in ionic polymerization, it is also found in cases of radical polymerization. Even the achievement of reproducibility does not of itself guarantee that the effects of adventitious impurities are absent. From these general remarks it should be evident that experimental results on radiation-induced polymerization should be subjected to the most careful scrutiny before they can be used to draw valid theoretical conclusions. It is regrettable that an alarming number of misinterpretations and errors have been allowed to accumulate in the literature because of experimental laxity and spurious phenomena.

B. Intermediates and Recombination Kinetics

It is a truism that the mechanism of most chemical reactions cannot be inferred from product analysis alone, and the study of radiation chemistry provides no exception. Of recent years, there has been an increased emphasis on the part played by specific intermediates in radiation-induced reactions and at least in some systems (9), the subject has developed to the point where more is known about the reactions of some intermediates than about the nature of the final reaction products. This situation indicates the pow.er of modern techniques in providing information about the chemistry of transient species.

A satisfactory theory of radiation chemistry must deal with the time scale of primary and secondary processes which follow the passage of an energetic charged particle. Excited molecules, ions, and electrons are usually considered to represent the first generation of reactive species; the subsequent reactions of these intermediates give rise to neutral stable molecules and free radicals, the latter undergoing combination and other processes to yield stable products. At least in the condensed phase, the initial distribution of intermediates formed by high energy radiation is exceedingly inhomogeneous. For typical secondary electrons produced by γ radiation, a multiple number of ionizations and/or excitations are produced in well-separated groups lying along the track, and these localized regions of high concentration are referred to as spurs. The average spur contains about three such primary events.

Magee (10) has summarized the chronological order of processes in a spur. A modified version of Magee's description is shown in Fig. 1. As the time increases, the diffusion of the neutral radicals leads to an increase in spur size until ultimately a homogeneous distribution is set up in the steady state. Thus the reactions between neutral radicals must be treated by diffusion theory so as to take account of the time-dependent growth of the spur (11,12). In particular, the competition between radical–radical and radical–solute (scavenging) reactions depends on the local radical concentration during the expansion of the spur, and it is clear that for any given solute concentration, radical scavenging assumes increasing importance in the approach to the steady state where homogeneous kinetics apply. In other words, very high values of the product $k_s[S]$, where k_s is the rate constant and $[S]$ is the scavenger concentration, are needed to scavenge those radicals which otherwise would undergo combination in the early history of the spur, whereas the scavenging of radicals undergoing recombination in a medium of homogeneous distribution is effective at $k_s[S]$ values which are lower by several orders of magnitude (13). The preceding remarks are qualitative generalizations based upon the detailed theoretical work of Magee and his co-workers (11) on the prescribed diffusion model of the radiation chemistry of water. Similar considerations hold for the simpler problem regarding the probability of geminate recombination following thermal or photochemical generation of radicals in pairs (14); this was described in the earlier literature as the "cage effect." Many experimental studies have been carried out on the radiolysis of water to check the predictions of the prescribed diffusion model (12), and

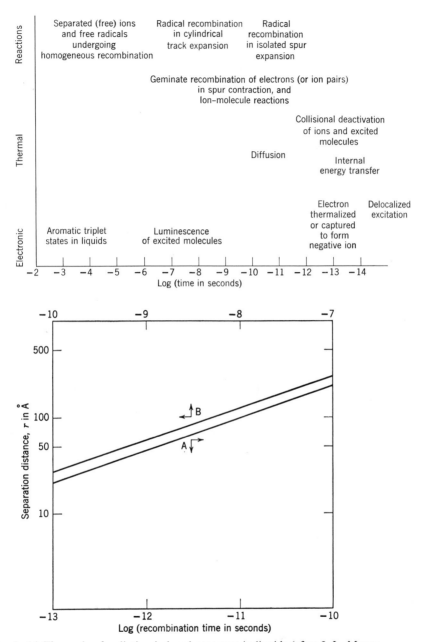

Fig. 1. (*a*) Timescale of radiation-induced processes in liquids (after J. L. Magee, *Ann. Rev. Phys. Chem.*, **12**, 389 (1961); *Discussions Faraday Soc.*, **36**, 232 (1963)). (*b*) Calculated times for recombination of ions and electrons in the nonpolar liquids as a function of initial separation distances, Plot A (11); Plot B (15).

satisfactory agreement is generally obtained within the adjustable range of the *a priori* assumptions regarding initial spur size and the rate constants for radical reactions. For a more detailed discussion of the limitations of this model, see Chapter 10, Sections IV–VI.

Because of coulombic interactions between charged species, the diffusion model for neutral radicals cannot be applied to the problem of ion–electron recombination without considerable modification. In the absence of electron capture either by solutes or by the solvent itself, the solution to the theoretical problem hinges on the nature of the electron in the condensed phase. The earliest treatment of Samuel and Magee (11) regarded the electron as having transport properties similar to those of electrons in gases. According to this description, the time for recombination after thermalization of the electron within 15 Å of the positive ion was deduced to be only ca. 10^{-13} sec.

Persistent evidence for ionic reactions in irradiated hydrocarbons has led to a reconsideration of the question. Studies based on the use of ion scavengers indicate that the lifetimes of many positive ions are greater than 10^{-10} sec, even in liquids of low dielectric constant (15). To explain these findings, it has been suggested (15) that the electron may be trapped in the liquid so that its motion corresponds more closely to that of a molecular ion than to a "free" electron in a gas. On this basis, the time τ_g for geminate recombination from an initial separation distance r between positive ion and trapped electron is given approximately by the relation (15),

$$\tau_g = r^3/3Dr_c \tag{1}$$

where D is the sum, $D_+ + D_-$, of the diffusion coefficients for the ionic species, and r_c is the distance at which the electrostatic and thermal energies are equal, as defined by the expression

$$r_c = e^2/\epsilon kT \tag{2}$$

where e is the charge on the electron, ϵ is the dielectric constant of the medium, k is the Boltzmann constant and T is the temperature. The experimental results on ion (proton) scavenging by ammonia $-d_3$ (15) and ethanol $-d$ (16) in cyclohexane suggest that ions undergo geminate recombination in times between 10^{-7} and 3×10^{-11} sec. By means of relation (1) and taking $D = 2.5 \times 10^{-5}$ cm^2 sec^{-1}, a value typical of diffusion coefficients in solvents of centipoise viscosity, the experimental estimates for the lifetime distribution can be converted to a corresponding distribution of r values between 280 and 19 Å. These limits are reasonable because they conform to simple theoretical expectations. Thus, the upper limit of 280 Å is almost exactly equal to r_c in cyclohexane, and this "critical" distance can be considered very roughly as the maximum separation from which the probability of geminate recombination (17), $[1 - \exp(-r_c/r)]$, still remains significant. The lower value of 19 Å corresponds to a limiting distance (15) at which the motion of the trapped electron may no longer be governed by the Nernst-Einstein expression for the proportionality between force and drift velocity, so that equation (1) would not apply at shorter distances. At such short range (≤ 20 Å), the electron can gain an increment of energy greater than kT as a result of the strong coulombic attraction during its relaxation under Brownian motion. If this excess energy is not dissipated to the surrounding molecules before the next relaxation period, the trapped electron will undergo a strong acceleration towards the positive ion, and under these circumstances its probable lifetime will be less than 10^{-11} sec. To summarize, the experimental results appear to be consistent with a simple model for a trapped electron in liquid hydrocarbons. This interpretation derives support from direct observations of trapped electrons in γ-irradiated 3-methylpentane glasses by optical (18) and electron spin resonance (19) spectroscopy (see Chapter 9).

Although the great majority of ions and electrons undergo geminate recombination in hydrocarbons, a small fraction of these species (free ions) would be expected to escape their mutual coulombic attraction and set up a steady state. Of course, the proportion of ions that escape should be much smaller than for neutral radicals in irradiated hydrocarbons. Onsager (17) has shown that the escape probability for a single pair of ions undergoing Brownian motion and separated by a distance r is $\exp(-r_c/r)$, so that the "free" or separated ions are not derived merely from those few ions and electrons which have achieved rather large initial charge separations, as in the case of delta rays. Again, the theoretical justification for the application of diffusion theory to the problem of ion–electron separation in radiolysis would seem to depend on the model of a trapped electron for the incipient stage of the process. As discussed in Chapter 7, the most direct evidence for "free" ions comes from measurements of electrical conductivity induced in hydrocarbons during irradiation (20,21); the results of these measurements indicate that the yield of separated ions is about 0.1 ions/100 eV ($G = 0.1$), and, therefore, considerably less than the total yield (ca. 2.6) of ion pairs as determined by a chemical method at high scavenger concentrations (16). Strong chemical evidence for free ions comes from the results of ionic polymerization (22) which show that kinetic chain lengths can be as high as 10^6 in well-purified systems; such large values are plainly inconsistent with the short lifetimes attributed to ions undergoing geminate recombination.

The preceding discussion has concentrated on the general theoretical problem of radical and ionic reactions in the condensed phase, and has drawn particular attention to the kinetic complications arising from the inhomogeneity of radical and ion production in spurs. We may outline the position as follows. In solvents possessing large (static) dielectric constants, ϵ, the probability of ionic separation is enhanced. In solvents of low dielectric constant, most ions and electrons ($G \sim 3$) recombine geminately in the parent spurs within a timescale of 10^{-7} to 10^{-11} sec, allowing only a small fraction ($G \sim 0.1$) of charged species to escape the coulomb field and set up a steady-state population of free ions whose mean lifetimes are greater than 10^{-3} sec at conventional dose rates (Table II). See Chapter 7 for further discussion.

Since the majority of ions appear to undergo geminate recombination in hydrocarbons, the disappearance of the ions will give rise to excited molecules which should decompose to give a distribution of free radicals generally similar to that envisaged in the Samuel-Magee treatment. Of course, it must be borne in mind that free radicals are not the only products which may result from neutralization processes, and that molecular products can arise directly by such reactions. Another point that needs to be emphasized is that neutralization by a negative ion rather than an electron can modify the resulting reactions. An effect of this type has been shown very clearly by Geissler and Willard (23) by the addition of alkyl halides to hydrocarbons. They observed a pronounced decrease in the free radical yield from the radiolysis of n-pentane in the presence of ethyl iodide, and this has been attributed to a less disruptive neutralization by the iodide ion in comparison with the high energy of electronic recombination. Despite the problems regarding the nature of the initial species in the spur, it is reasonable to consider that diffusion theory can be applied to the problem of hydrocarbon radical recombination in the expanding spur. On this basis, the Ganguly-Magee model (11) predicts that half the total number of radicals would undergo radical recombination in the spur when $k_s[S]$ is below 10^6 sec^{-1}, and scavenger competition can reduce this "geminate" recombination as $k_s[S]$ increases from 10^8–10^{11} sec^{-1}. This is equivalent to saying that most of the inhomogeneous radical recombination occurs during the first 10^{-8} sec when the spur diameter increases from 5 to 50 Å, and that the spur reaction is virtually over by the time that a diameter of 500 Å is reached.

From this discussion on spur reactions, it is interesting to note that scavengers compete with the ionic and radical recombination reactions in hydrocarbons in the same range of their respective $k_s[S]$ values (10^8–10^{10} sec^{-1}). This correspondence in the experimental data has been pointed out previously by Dyne (24). However, it should be clear from the preceding considerations that this result comes about because of the similarity in the times for inner spur contraction (ions) and expansion (radicals) rather than from the possibility of a unified kinetic description applicable to both ions and neutral radicals. Nevertheless, the approximate correlation in the times for spur contraction ($r^3/3Dr_c$) and expansion ($\pi r^2/16D$) follows as a consequence of the closely related diffusion coefficients in any particular solvent, and of the close identity in the limiting values (ca. 50 Å) for the "effective" spur radii (r) within which most of the recombination takes place.

It is useful to summarize the extent of spur reactions in graphical form. In Fig. 2, an experimental curve is shown for the yield, G_s, of scavengeable *ions* in hydrocarbons (16) as a function of τ_s ($= 1/k_s[S]$), the mean time for scavenging. It may be surmised that for ion scavenging in liquids of higher dielectric constants, the corresponding curve would be shifted horizontally to a longer time scale with some additional broadening in the overall distribution, but a quantitative description is lacking. For comparison with Fig. 2, the calculations of Ganguly and Magee (11) are given in Fig. 3 for the yield of scavengeable *radicals* derived from the expansion of spurs in the tracks of incident 0.5 MeV β particles. It should be mentioned that scavenging curves such as Figs. 2 and 3 do not provide exact measures of the time-resolved distributions of species undergoing spur recombination in the absence of scavenger. The use of a perturbing scavenger implies competition kinetics with the result that $G_s(\tau_s)$ does not coincide exactly

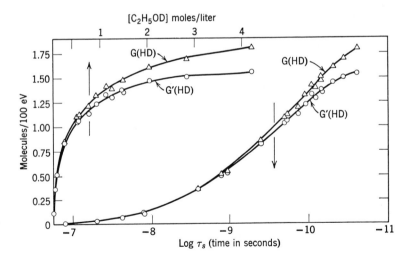

Fig. 2. Radiolysis yields of HD plotted as a function of C_2H_5OD concentration in cyclohexane (16). The lower scale of the abscissae for log τ_s is calculated on the assumption of $k_s \simeq 10^{10}\ M^{-1}\ \mathrm{sec}^{-1}$ for a diffusion-controlled reaction. $G(HD)$ refers to experimental yield and $G'(HD)$ to the corrected yield.

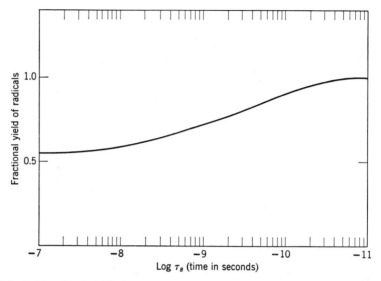

Fig. 3. Fractional yield of scavengeable neutral radicals as a function of τ_s (curve based on the calculations of A. K. Ganguly and J. L. Magee, *J. Chem. Phys.*, **25**, 129 (1956) and Fig. 2 of this reference).

with $G(t)$, the yield of species with lifetimes greater than t in the absence of scavenger, where the two functions are plotted on the same time scale. This is illustrated by the curves shown in Fig. 4, where curve A represents a calculated G_s for the simplest case of scavenger competition with a first-order recombination process characterized by a single rate constant $(1/\tau_g)$ of 10^9 sec^{-1}. It is easy to show that in this case, G_s assumes the form

$$G_s = G_0(\tau_g)/(\tau_g + \tau_s) \qquad (3)$$

where G_0 is the total yield of species undergoing recombination. On the other hand, $G(t)$ represented by curve B is given by the exponential relation,

$$G(t) = G_0 \exp(-t/\tau_g) \qquad (4)$$

It is seen that the two curves begin to coincide closely only as τ_s and t each become of the order of $\tau_g/10$, but at longer times there is a significant difference. Therefore, the method of "time sampling" by means of a typical scavenging curve $G_s(\tau_s)$ does not constitute an exact reflection of the distribution $G(t)$, particularly for $\tau_s > \tau_g$. This conclusion is even more appropriate for the general problem of a distribution characterized by several time constants (τ_g) of recombination.

Turning now to the question of homogeneous recombination, a stationary-state situation will hold for ions and neutral radicals quite independently of each other, and the equations are essentially of the same type.

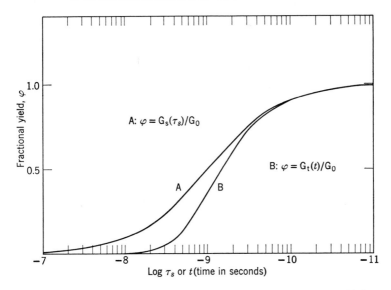

Fig. 4. Fractional yield curves. Refer to text and equations (3) and (4) for definitions and notation.

For ions we may write

$$R_i = k_t[M^+][M^-] \tag{5}$$

where R_i is the rate of initiation and k_t is the termination rate constant for bimolecular reaction between two ions. Assuming charge conservation, the stationary state concentration is given by

$$[M^+] = (R_i/k_t)^{1/2} \tag{6}$$

and the mean lifetime τ_{ss} in the stationary state is defined by

$$\tau_{ss} = [M^+]/R_i = (1/k_t R_i)^{1/2} \tag{7}$$

Expressions corresponding to (6) and (7) may also be used for radicals if we define the rate constant k_t for radical disappearance according to the equation, (25)

$$-\frac{d}{dt}[R\cdot] = 2\left[\frac{k_t}{2}\right][R\cdot]^2 = k_t[R\cdot]^2 \tag{8}$$

By this definition, k_t can be equated to the theoretical value from the diffusion theory of Smoluchowski (26),

$$k_t(\text{radicals}) = 4\pi\sigma D \tag{9}$$

where σ is the effective reaction distance and D is the total diffusion coefficient for the reacting species. For univalent ions in solvents of low dielectric constant where $r_c \gg \sigma$, the general Debye equation (27) for the termination (recombination) rate constant between oppositely charged univalent ions,

$$k_t = 4\pi r_c D/[1 - \exp(-r_c/\sigma)] \qquad (10a)$$

reduces to the simple form

$$k_t \text{ (ions)} = 4\pi r_c D \qquad (10b)$$

Calculations of stationary-state recombination times for ions and radicals are summarized in Tables I and II, respectively. The numerical values of the various parameters are given as footnotes to the tables.

Table I
Recombination Times for Stationary State of Free Ions in the Radiolysis of Hydrocarbon Liquids at 300°K[a]

Mrad/hr	Dose rate I, eV ml^{-1} sec^{-1}	R_i, M sec^{-1}	τ_{ss}, sec	$[M^+]$, M
100	1.4×10^{18}	2.3×10^{-6}	0.9×10^{-3}	2×10^{-9}
20	2.8×10^{17}	4.7×10^{-7}	2.0×10^{-3}	9×10^{-10}
1	1.4×10^{16}	2.3×10^{-8}	0.9×10^{-2}	2×10^{-10}
0.2	2.8×10^{15}	4.7×10^{-9}	2.0×10^{-2}	9×10^{-11}
0.01	1.4×10^{14}	2.3×10^{-10}	0.9×10^{-1}	2×10^{-11}

[a] $\tau_{ss} = (1/k_t R_i)^{1/2}$; $R_i = G_i I/100N$; $G_i = 0.1$; $k_t = 4\pi r_c(D_+ + D_-) = 5.3 \times 10^{11} M^{-1}$ sec^{-1}; $r_c = e^2/\epsilon kT = 280$ Å ($\epsilon = 2$ and $T = 300°$K); $D_+ + D_- = 2.5 \times 10^{-5}$ cm^2 sec^{-1}.

Table II
Recombination Times for Stationary State of Free Radicals of Low Molecular Weight in the Radiolysis of Simple Hydrocarbon Liquids at 300°K[a]

Mrad/hr	Dose rate I, eV ml^{-1} sec^{-1}	R_i, M sec^{-1}	τ_{ss}, sec	$[M\cdot]$, M
100	1.4×10^{18}	1.2×10^{-4}	1.3×10^{-3}	1.6×10^{-7}
20	2.8×10^{17}	2.4×10^{-5}	3.0×10^{-3}	7×10^{-8}
1	1.4×10^{16}	1.2×10^{-6}	1.3×10^{-2}	1.6×10^{-8}
0.2	2.8×10^{15}	2.4×10^{-7}	3.0×10^{-2}	7×10^{-9}
0.01	1.4×10^{14}	1.2×10^{-8}	1.3×10^{-1}	1.6×10^{-9}

[a] $\tau_{ss} = (1/k_t R_i)^{1/2}$; $R_i = G_i I/100N$; $G_i = 5.0$; $k_t = 4\pi\sigma(D_A + D_B) = 4.7 \times 10^9 M^{-1}$ sec^{-1}; $\sigma = 2.5$ Å; $D_A + D_B = 2.5 \times 10^{-5}$ cm^2 sec^{-1}.

Reference to Tables I and II shows that at the same dose rate, the stationary state concentration of ions is about two orders of magnitude below that due to radicals, while the mean lifetimes, τ_{ss} of ions and radicals are comparable. These results are caused by a lower G_i and a higher k_t for ions relative to radicals.

A point of additional interest is that when the local concentrations of transient ions and radicals participating in spur recombination processes are averaged over time and space, these "average" values are negligibly small in comparison with the appropriate values taken from Tables I and II for homogeneous recombination. This "average" concentration $[M_g]$ arising from inhomogeneous processes is essentially equal to the summation

$$[M_g] = \sum_g R_g \tau_g \qquad (11)$$

where R_g is the specific rate of production (conc/time) of species with a recombination time τ_g, averaged over the entire (bulk) volume. Thus, the distribution of very short lifetimes (τ_g from 10^{-7} to 10^{-10} sec) for spur recombination in liquids means that $\sum_g R_g \tau_g$ will only begin to approach $R_i \tau_{ss}$ (Tables I and II) when the dose rates become exceedingly large. Hence, in the usual practical range of dose rates, the true stationary state concentration of ions will greatly predominate over the "average" contribution of inhomogeneous processes despite the fact that in hydrocarbons, $\sum_g R_g$ is about thirty times greater than R_i for ions. For radicals in hydrocarbons this disparity between the homogeneous and inhomogeneous contributions to the "average concentration" is even larger on account of the similarity between R_i and $\sum_g R_g$ in this case.

Before taking up the discussion on the various classes of transient intermediates in radiation-induced polymerization, it is appropriate to mention that the major experimental techniques and approaches which have been used in this field are the same as those used in liquid and solid phase radiation chemistry, namely chemical scavenging, optical spectroscopy, electron spin resonance spectroscopy, luminescence determinations, mass spectrometry, and the measurement of radiation-induced conductivities. These techniques are discussed in detail in Chapters 7 and 9.

C. Terminology

Before taking up the detailed development of the subject, it is useful to review some of the terminology and basic definitions which are widely used in the field of radiation-induced polymerization. We shall be concerned almost exclusively with addition and ring-opening polymerizations, and especially the addition polymerization of vinyl, vinylidene, and diene monomers. Of course, the singular characteristic of addition polymerization is that the polymer is formed by a chain reaction in which the active propagating species is regenerated at each addition process. However, it does not necessarily follow that the molecular weights of polymer molecules are directly related to the length of the *kinetic* chain, for there are many cases where the molecular weight distributions are governed by chain transfer processes.

Some terms and symbols familiar to radiation and polymer chemistry are grouped together in Table III for convenience. Under stationary-state conditions where the reaction *rates* show little or no dependence on time, we have the simple relations (12) and (13), where N is Avogadro's number.

$$G(-m) = 100R_pN/I \tag{12}$$

$$G_i = 100R_iN/I \tag{13}$$

The relation between R_p and R_i is generally a function of the mechanism

Table III
Glossary of Terms and Symbols

Symbol	Definition	Usual units
R_p	Rate of polymerization $\dfrac{-d[m]}{dt}$	M sec^{-1} (% conversion min^{-1})
I	Dose rate	eV g^{-1} sec^{-1} (rad hr^{-1})
R_i	Rate of initiation	M sec^{-1}
$G(-m)$	Yield of monomer molecules converted to polymer per 100 eV of energy input	Molecules/100 eV
G_i	Yield of initiating species	Molecules/100 eV
ν	Kinetic chain length	Molecules reacted per initiator
k_p	Propagation rate constant	M^{-1} sec^{-1}
k_{tr} or k_f	Transfer rate constant	M^{-1} sec^{-1}
k_t	Termination rate constant for bimolecular reaction between two ions or radicals	M^{-1} sec^{-1}
k_{tx}	Termination rate constant for reaction between one active species and molecular terminator	M^{-1} sec^{-1}
$\langle M_n \rangle$	Number-average molecular weight	
$\langle M_w \rangle$	Weight-average molecular weight	
$\overline{DP}_n$	$\dfrac{\langle M_n \rangle}{M_1}$ where M_1 is molecular weight of monomer	

of the chain reaction, and will be discussed later in specific cases. With the aid of the above definitions and equations, it follows that v can be expressed in terms of G values or rates, according to the composite relation (14). Since G_i can never exceed the order of magnitude of the G values (in

$$v = G(-m)/G_i = R_p/R_i \qquad (14)$$

the range up to 10) encountered in systems where chain reactions are absent, it follows immediately that $G(-m)$ values are a convenient measure of the kinetic chain length of the reaction. As a useful corollary to this, a polymerization reaction with a $G(-m)$ value much less than 100 cannot be strictly interpreted in terms of rate expressions applicable only to the generation of *long* kinetic chains.

Under conditions where chain transfer processes are *absent*, a knowledge of the number-average molecular weight $\langle M_n \rangle$ can be usefully employed to relate $G(-m)$ and G_i, according to (15). Apart from the above restriction, the limitation on the accuracy to which G_i may be obtained from (15)

$$G(-m) = G_i \overline{DP}_n = G_i \frac{\langle M_n \rangle}{M_1} \qquad (15)$$

is set by the experimental determination of $\langle M_n \rangle$. Unfortunately, the quantitative reliability of some methods used to determine $\langle M_n \rangle$ leaves much to be desired, and this is especially true of determinations from intrinsic viscosity measurements by means of empirical equations obtained for particular polymer fractions whose molecular weight *distributions* may be substantially different from those of the polymer to be determined. Because the importance of molecular weight distribution is paramount in polymer chemistry, the statistical relations for $\langle M_n \rangle$ and $\langle M_w \rangle$ are given in (16) and (17). Here, n_i refers to the number of molecules of species i

$$\langle M_n \rangle = \sum_i n_i M_i \Big/ \sum_i n_i \qquad (16)$$

$$\langle M_w \rangle = \sum_i n_i M_i^2 \Big/ \sum_i n_i M_i \qquad (17)$$

with molecular weight M_i. Since the weight w_i of species i is given by $n_i M_i$, relation (17) is easily derived from the definition of $\langle M_w \rangle = \sum_i \omega_i M_i / \sum_i \omega_i$. It is clear that the ratio of $\langle M_w \rangle$ to $\langle M_n \rangle$ is greater than 1 for distributions other than *monodisperse*, and the numerical magnitude of this ratio is taken as an index of the type of distribution encountered. For the frequently assumed Poisson or "most probable" distribution, $\langle M_w \rangle / \langle M_n \rangle = 2$.

II. RADIATION-INDUCED POLYMERIZATION

In the broad perspective of polymerization research, the method of radiation-induced polymerization occupies a position which cannot easily be classified alongside the other more conventional methods of initiation. With few exceptions, the other techniques of addition polymerization are characterized by the generation of a *specific* kind of intermediate, usually free radicals or ions. Free radicals can be produced by a variety of methods including direct photolysis of some monomers (e.g., methyl methacrylate), the thermal decomposition of labile compounds such as benzoyl peroxide and azobisisobutyronitrile, or by redox reactions between peroxides and transition metal ions. The reader is referred to excellent summaries (25,28) of free radical polymerization for detailed discussions of theory and experiment. The subject of cationic polymerization has also been adequately reviewed (29,30), and although in this case there is certainly a great variation in the nature and reactivity of the cationic "catalysts" which are commonly used (boron trifluoride, perchloric acid, iodine, etc.), these reagents have common features which makes the general classification useful. A similar situation holds for anionic polymerization (31) where typical initiators include butyl lithium, dispersions of alkali metal, and the sodium naphthalenide reagent. As expected, ionic polymerization by chemical catalysts shows a much greater dependence on solvent conditions than free radical polymerization.

By contrast with chemical, thermal, or photolytic initiation, it must be admitted that high-energy radiation is considerably less specific in its primary action. Since a wide variety of reactive intermediates are known to be formed in this latter case (Section A), it might be concluded that there is a complete lack of specificity in polymerization, giving rise to complicated overall kinetics from the simultaneous operation of more than one mechanism. Quite contrary to this expectation (which indeed appears to be assumed all too readily by those unfamiliar with radiation chemistry), it turns out that such complications are a rarity, and even where two mechanisms are operating together as for bulk styrene, one mechanism predominates under well-defined conditions. Indeed, it would be exceedingly difficult to arrange the converse to apply. That this situation holds is not so surprising if it is recalled that the rates of chain reactions are strong functions of the rate constants for the respective propagation and termination steps, and that the absolute values of k_p and k_t are specified by the nature of the intermediate species. We shall see later that it is just for this type of situation in styrene monomer that the irradiation method provides an original contribution to the important question of the absolute reactivities of free radicals and ions.

The polymerization of certain cyclic compounds (epoxides, ethers, imines, sulfides) by ring opening differs in several respects from that of addition polymerization. Perhaps the most marked difference is that this type of polymerization is rarely, if ever, initiated by a simple free radical catalyst, whereas a wide range of cationic and anionic catalysts are effective. Thus, ring-opening polymerization is generally diagnostic of ionic polymerization, and early radiation studies (32) on the polymerization of the cyclic trimer hexamethylcyclotrisiloxane were of particular importance in calling attention to the general possibilities of radiation-induced ionic polymerization.

The unique possibilities of high energy radiation for the initiation of solid state polymerization have been thoroughly explored in a number of systems. As might be expected, the results are strongly influenced by the molecular arrangement in the solid and the term *topotactic* polymerization is used to describe cases where the periodicity in the lattice structure controls the formation of a stereoregular polymer (33).

A. Transient Species in Radiation-Induced Polymerization

Free radicals and ions are the usual active species responsible for simple addition polymerization. We have already seen in Chapter 7 that there is abundant evidence for free radical intermediates in the radiation chemistry of many compounds, including liquid hydrocarbons of low molecular weight for which the results are particularly conclusive (34). Of course, the detailed interpretation of radiation-induced polymerization in terms of free radical processes has been common practice for twenty years, originating from experiments such as those of Dainton (35) on the polymerization of acrylonitrile in aqueous solution. It is of historical interest to recall that the initiation of polymerization in certain monomers has been used as a diagnostic test for the presence of both free radicals and ions in radiation chemistry. In such cases, the interpretation rests rather heavily on a knowledge of the comparative polymerizability of the various monomers to conventional free radical and ionic initiators, and corroborative evidence is required to fortify conclusions reached in this manner (*vide infra* Section II-B-2).

It must be admitted that little direct evidence exists on the molecular structure of intermediates responsible for radiation-induced polymerization. Even in catalytic polymerization, it is only recently that free radical (36) and anionic (37) propagating species have been detected unambiguously, the former by the ESR technique and the latter by spectrophotometry. There is virtually no doubt that a corresponding demonstration of free radical propagation also could be achieved by ESR in a liquid system under irradiation, but such experiments do not appear to have been undertaken. On the other hand, many ESR studies have been reported on the free radicals present in the radiation-induced polymerization of solid monomers, but the hyperfine structure in the ESR spectra obtained from many polycrystalline solids consist of broad lines too poorly resolved to permit a unique interpretation of the free radical structure. Undoubtedly, the best possibilities for the structural identification of free radicals in solids lie in the study of single crystals (38), and this approach has been used by Morawetz and his colleagues (39) in an investigation of the radiation-induced polymerization of barium methacrylate dihydrate. Important information on the direction of propagation in the monomer crystal can be deduced from careful ESR studies of this kind, and the results bear directly on the strong relationship between crystal geometry and ease of polymerizability.

Important advances have been made in the application of mass spectrometry to the study of ionic reactions in the gas phase at pressures up to several hundred torr, and this work has contributed directly to an understanding of the early reactions in an ionic polymerization chain. Ethylene has been the subject of detailed studies (40–42) and the general conclusion is that the primary ion–molecule reaction gives a dimeric species $C_4H_8^{+} \cdot$. Kebarle (42) has shown that at pressures of only a few torr of ethylene, $C_4H_8^{+\cdot}$ reacts further with ethylene and builds up ions with as many as 14 carbon atoms.

$$(C_2H_4)_m^{+\cdot} + C_2H_4 \rightarrow (C_2H_4)_{m+1}^{+\cdot} \qquad (18)$$

It is found that the rate constants of successive addition steps decrease continuously. At very low pressures, $C_4H_8^{+\cdot}$ dissociates to give $C_3H_5^{+}$ and $CH_3 \cdot$. Wexler (40) considers that at very low pressures, most of the larger polymeric ions (up to $C_7H_{13}^{+}$) can be accounted for by stepwise addition polymerization from the $C_3H_5^{+}$ carbonium ion which in this case functions as the actual initiator. While these results have an important bearing on ionic processes for olefins and vinyl monomers, it should be emphasized that the mere detection of positive ions with relatively high masses from a given monomer provides no guarantee that cationic polymerization predominates in the overall radiation chemistry. Indeed, it seems fairly well established that the polymerization of ethylene in the gas phase is largely a free radical process (43). Considerable mass-spectrometric work at elevated pressures has also been reported on acetylene (44), and here again the dimeric ion $C_4H_4^{+\cdot}$ appears to give fragment ions, only this time by loss of H or H_2. However, the role of these ions in the well-known radiation-induced polymerization of acetylene to cuprene (45) is uncertain. Isobutene is known to undergo cationic polymerization readily. The pioneering researches of Tal'roze (46) established the formation of $C_4H_9^{+}$ from the parent isobutene ion,

$$C_4H_8^{+} \cdot + C_4H_8 \rightarrow C_4H_9^{+} + C_4H_7 \cdot \qquad (19)$$

Recently it has been shown (47) that the $C_4H_9^+$ carbonium ion undergoes an addition reaction with C_4H_8:

$$C_4H_9^+ + C_4H_8 \rightarrow C_8H_{17}^+ \qquad (20)$$

Apart from revealing some of the initial reactions in olefin polymerization, mass-spectrometric studies have also provided direct information about proton transfer reactions from hydrocarbon ions to molecules such as water and ammonia (48). Such reactions have been postulated to explain the great sensitivity of radiation-induced cationic polymerizations to traces of water (8) and ammonia (49). The results from mass spectrometry (48) show that low concentrations (1 mole %) of water or ammonia in methane can exert a profound effect on the ionic distribution pattern. In the case of the water–methane mixture at pressures approaching 2 torr, the majority of the ions were shown to be composed of H_3O^+ and its hydrates (Fig. 5), whereas for ammonia, the ammonium ion NH_4^+ and the corresponding ammonia solvates were formed in great abundance. Thus proton scavenging by water and ammonia from hydrocarbon ions may occur when such reactions are exothermic.

Pulse radiolysis has been used to search for the intermediates in the radiation-induced polymerization of isobutylene, styrene, and α-methylstyrene. Burrell (50) reported that an absorption spectrum with a maximum at 297 mμ could be detected in isobutylene after pulse radiolysis. This spectrum was attributed to the *t*-butyl cation (trimethylcarbonium ion) on the basis of a previous assignment by Rosenbaum and Symons (51). However, more recently Olah and his collaborators (52) have definitely shown that the *t*-butyl cation does not absorb in this region, and that it is probably the polymeric products of isobutylene which give rise to absorption at about 290 mμ. It follows that the interpretation (50) of the 297 mμ absorption in terms of the *t*-butyl cation must be discarded, and this clears up the problem of accounting for the long decay time (0.1 sec) which had been attributed to the disappearance of the ion by a first-order process.

Several pulse radiolysis studies have been reported on styrene and α-methylstyrene, and the results on these two monomers are so similar that they can be discussed together. Katayama and his co-workers (53) observed absorption bands at about 350 mμ and 420 mμ in very dry samples of α-methylstyrene. Further work by the same author (54) revealed two absorption bands at 366 mμ and 546 mμ which were extremely sensitive to the initial water content of the sample, while another band at shorter wavelength was unaffected by water. These observations are paralleled to some extent by the work of Metz, Potter, and Thomas on bulk styrene (55), showing the existence of two transient species with

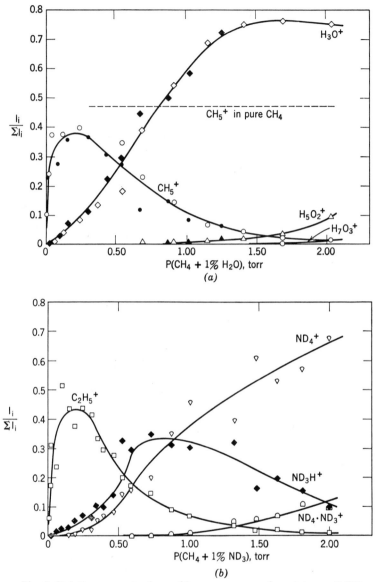

Fig. 5. Relative concentrations of ions vs. pressure for mixtures of CH_4
with (a) 1% H_2O and (b) 1% ND_3 (48).

absorption maxima at 325 mμ and 370 mμ, the former remaining unaffected while the latter can be removed by water addition. Further
evidence that the absorptions are due to different species was obtained
from a study of their decay characteristics (55), the 370 mμ absorption

decaying with a half-life of about 4 μsec whereas the shorter wavelength band disappeared more slowly ($t_{1/2} \simeq 0.5$ msec). It seems fairly clear from this and other work by Schneider and Swallow (56) that the absorption in the vicinity of 320 mμ is due to neutral radicals, since the band position correlates very well with the spectra of α-substituted benzyl radicals (57,58) which have a structure similar to that usually described for the propagating radical in styrene (25). The assignment of the water-sensitive band at about 360–370 mμ is less certain although the authors (54,55) have attributed this to the radical anion formed by electron capture. Hamill and his co-workers (59) have found that the radical anion of styrene in 3-methyl-pentane glass absorbs at 400 mμ, while work in this laboratory (60) has shown that the radical anion of α-methylstyrene in 3-methylpentane and in methyltetrahydrofuran glasses exhibits an absorption maximum of 411 mμ. Moreover, Keene et al. (61) observed a short-lived absorption at 400 mμ for styrene in cyclohexane, and similar results (56) have been obtained from the pulse radiolysis of styrene and α-methylstyrene solutions in saturated hydrocarbons. A possible reason why the apparent positions of λ_{max} for the rapidly decaying species in the bulk monomers do not correspond to those obtained for the radical anion spectra in glasses may be of experimental origin. The pulse radiolysis data are subjected to a careful time-resolution analysis, but this procedure suffers from obvious limitations when overlapping absorptions are involved. Some spectrophotometric evidence for the presence of positive ions has been obtained in γ-irradiated rigid glasses. Solid solutions of styrene (59) and α-methylstyrene (60) in 3-methylpentane give rise to absorption bands at 460 and 475 mμ, respectively on irradiation; these bands are also produced if an electron scavenger such as n-propyl chloride is added in excess so that the corresponding radical anions are not formed. Further, the 475 mμ band is not produced by the irradiation of α-methylstyrene in methyltetrahydrofuran glass (60), although the radical anion is formed in this matrix. On the basis of Hamill's rules (59), these results suggest that the long wavelength bands are due to positive ions. Presumably these are the cation radicals formed by ionization of the monomer, but this assignment cannot be verified because the spectra of these species has not been obtained in an independent manner. It is reasonable to expect that the ionic species in the rigid 3-methylpentane glass should consist of the radical anion and radical cation, since the solute molecules are isolated from each other and bimolecular reactions between these ions and monomer molecules are prevented. On the other hand, it is somewhat surprising that no evidence has been obtained from the pulse radiolysis work for carbanion and carbonium ion species. The former is well known from anionic polymerization studies on styrene and α-methylstyrene (62), with a strong absorption band in the

340 mμ region, the precise position depending upon the alkali metal counterion and the solvent. Also the dimethyl phenyl carbonium ion derived by proton addition to α-methylstyrene is now well established (52) as having absorption bands at 390 mμ (ϵ = 1,400) and 326 mμ (ϵ = 11,000). In contrast, the methyl phenyl carbonium ion corresponding to the conjugate acid of styrene could not be observed by the NMR method even under optimum conditions of extreme acidity (52), and consequently there is no foundation for the previous belief (29) that this carbonium ion absorbs at 420 mμ. In all probability, this latter absorption, which has been widely associated with the polymerization of styrene by conventional cationic catalysts, is actually caused by the more stable and nonpropagating polymeric ions in the acid system (63). Thus, the absorption spectrum for the propagating cation in styrene is unknown at the present time.

Returning to further consideration of the pulse radiolysis results on styrene and α-methylstyrene in the light of this information about carbanions and carbonium ions, it would appear that the present experimental data are insufficiently decisive to allow firm conclusions to be drawn about the exact nature of the intermediates present. Assuming that the monomer radical anion is indeed responsible for the fast decaying component in the absorption, then the expected product of this fast reaction is surely a dimer with separated carbanion and radical centers as represented by $\cdot$MM$^-$. Species of this type with alkali metal cations have been described by Szwarc (64), and they are believed to function as intermediates in the formation of "living" α-methylstyrene tetramer (65). The important point for the pulse radiolysis study is that the absorption spectrum of the structurally independent carbanion should be manifested as the monomer radical anion disappears by reaction, yet this demonstration has not been achieved. In other words, it is desirable to identify both the initiating and the propagating species in the reaction sequence. Finally on this topic, a few remarks are needed on the precise implication of pulse radiolysis experiments for the kinetics of radiation-induced polymerization. It has been suggested (53) that the detection of a specific type of intermediate automatically confirms the mechanism of polymerization. Nothing could be further from the truth; where there are *a priori* reasons for supposing that more than one type of radiation-induced intermediate can cause polymerization as in the case of styrene, the most significant contribution to the rate will generally come from the species having the highest product of stationary state concentration and propagation rate constant. Although the detection and concentration determination of each intermediate species is highly desirable from a kinetic viewpoint, it must be emphasized that the pulse radiolysis technique does not extend to measuring the separate rates of polymerization for the individual active species. Therefore, the results of pulse radiolysis must be supplemented by more routine kinetic measurements of polymerization in order to allow discrimination between the contributions of the various intermediates to the overall polymerization. The most significant contribution of pulse radiolysis to this field is likely to come in unraveling the initiation and the propagation steps, so as to enable the elementary reactions to be described with greater certainty.

In Chapter 7 the methods used to obtain free-ion yields from conductivity and ion mobility measurements were discussed. The information allows the steady state concentration of ions n to be calculated from the general relation,

$$\kappa = ne(\mu_+ + \mu_-) \tag{21}$$

where κ is the radiation-induced conductivity, μ_+ and μ_- are the mobilities of the positive and negative ions, respectively, and e is the electronic charge, it being assumed that the ions are singly charged. The rate of production of ions R_i (equal to $G_iI/100$) is then readily calculated from equation (6) if k_t is known; this rate constant may be obtained either directly from decay studies or calculated from the theoretical value given in equation (10b) which is also equivalent to the expression due to Langevin,

$$k_t = (4\pi e/\epsilon)(\mu_+ + \mu_-) \tag{22}$$

This method then leads to an estimate of G_i, the yield of free or separated ions undergoing homogeneous recombination. A satisfactory method for the determination of ion mobilities during radiation-induced ionic polymerization would be useful in providing direct information about the size of the propagating ions, and such measurements also could serve to discriminate between the relative importance of cationic and anionic propagation, subject to the consideration of *chain transfer*. However, the experimental measurement of ionic mobilities in liquids (66) is difficult even for a uniform population of small ions, and extension to deal with a distribution of polymeric ions might require new techniques. No developments along these lines have been reported, but at least in principle, the approach would seem to offer an opportunity to study ion–molecule reactions in the liquid phase.

Even without mobility data, the results of electrical conductivity can be used to obtain the mean lifetime, τ_{ss}, of ions in the stationary state. Combining relations (21) and (22), the specific conductivity κ can be expressed as

$$\kappa = (\epsilon/4\pi)nk_t \tag{23}$$

and recalling from (6) and (7) that

$$\tau_{ss} = 1/nk_t \tag{24}$$

where $n = \sum [M^+] = \sum [M^-]$, we obtain

$$\tau_{ss} = \epsilon/(3.6 \times 10^{12}\pi\kappa) \tag{25}$$

where a numerical factor of 9×10^{11} is included to convert κ in ohm^{-1} cm^{-1} to units of sec^{-1}. A rigorous derivation of (25) has been given (67) for an ionic distribution where k_t is not treated as a constant but depends on the particular pair of ions undergoing recombination; this general relation for τ_{ss} in terms of ϵ and κ is usually referred to as the Maxwell timelag for the decay of charges in an electrolytic conductor (68). The value of the rate of initiation R_i can be expressed in terms of τ_{ss} and k_t according to equation (26),

$$R_i = \tau_{ss}^{-2}k_t^{-1} \tag{26}$$

but k_t must now be considered as a suitable average over all possible recombination pairs. Therefore, while τ_{ss} is obtainable directly from conductivity, a calculation of R_i also depends on a knowledge of k_t which must be obtained experimentally, either through the ionic mobilities (equation (22)) or conceivably from pulse radiolysis. The use of conductivity data in connection with the kinetics of ionic polymerization is illustrated by recent work (67,69) on styrene, α-methylstyrene, and isobutyl vinyl ether which is discussed later in this chapter.

B. Kinetics and Mechanism

1. General Kinetic Relations. We have seen that the intermediate stationary-state concentrations [Z] for free ions and for free radicals undergoing bimolecular recombination separately can be expressed in each case by the same general equation,

$$[Z] = (R_i/k_t)^{1/2} \tag{27}$$

although the parameters are unrelated in the two instances, of course (cf. Tables I and II). This equation will apply also to the case where the intermediate participates in addition polymerization, provided that the value of the recombination rate constant k_t is independent of R_i. Since the rate of polymerization R_p for long kinetic chains is given by

$$R_p = k_p[Z][M] \tag{28}$$

where k_t is the appropriate propagation rate constant and [M] is the monomer concentration, combination of (27) and (28) leads to the well-known standard expression,

$$R_p = k_p\left(\frac{R_i}{k_t}\right)^{1/2}[M] = R_i k_p[M]\tau_{ss} \tag{29}$$

This equation holds for many examples of thermal, photo-, and radiation-induced *free radical* polymerizations in solution and in the vapor phase, as evidenced by the half-power dependence of R_p on the initiation rate and hence on the dose rate for a radiation-induced polymerization. The same equation (29) does not apply to ionic polymerization unless free ions are responsible for the major part of the polymerization, which is rarely the case in catalytic studies. It was believed until very recently that propagating ions underwent a process of unimolecular termination, even in radiation-induced ionic polymerization, and that R_p should always be proportional to the first power of dose rate in this case. Recent work (69) has shown that this generalization is unfounded, and that the dependence of R_p on dose rate may approach the theoretical value predicted by (29) for free ions when the monomer is extremely dry.

The assumptions on which (29) rests must now be examined more fully. They are as follows:

1. k_p is independent of chain length for any given type of active species,
2. k_t is independent of R_i, and
3. all active species disappear by bimolecular recombination with their own kind (radicals) or conjugate species (ions).

Assumptions (1) and (2) are commonly made in general discussions of free radical polymerization (25) to avoid intractable expressions. Nevertheless, it is necessary to consider that the reactivity of the initial radical or ion formed by radiation may differ significantly from that of the propagating species to which it gives rise. In other words, the rate constant of the first addition, or initiation, reaction, k_i, could be significantly different from k_p for the succeeding steps. The case $k_i > k_p$ would not affect (29) but the converse situation would reduce the propagating lifetime from τ_{ss} to $\tau_{ss} - (1/k_i[M])$ and the kinetic chain length v from

$$v = k_p[M]\tau_{ss} \tag{30}$$

to

$$v' = k_p[M]\tau_{ss} - (k_p/k_i) + 1 \tag{31}$$

A corresponding change in R_p would give the relation,

$$R_p' = \left(\frac{k_p}{k_t^{1/2}}\right)[M]R_i^{1/2} - \left(\frac{k_p - k_i}{k_i}\right)R_i \tag{32}$$

From equation (31) it can be seen that this effect of slow initiation needs to be taken into account only when the rate constant ratio k_p/k_i becomes a nontrivial correction to the first term. For many polymerizations at relatively low dose rates where v is of the order of 10^3 or greater, k_p/k_i would have to exceed 10 to account for an appreciable deviation from the simple kinetic scheme. The above expressions for v' and R_p' are valid only for the case,

$$\tau_{ss} > (k_i[M])^{-1} \tag{33}$$

for otherwise no chain reaction ensues. That is, the limiting value of R_p' is zero when the initiation is so slow that no polymerization occurs. Since R_p' is of the form $aR_i^{1/2} - bR_i$, where a and b are constants, the magnitude of the second term increases relative to the first with increase in dose rate. Thus deviations due to $k_i < k_p$ will be reflected more strongly at high dose rates. A further consequence of equation (32) is that the exponent x for the dose rate dependence of R_p' according to

$$R_p' \propto R_i^x \qquad (34)$$

will drop below 0.5 as the dose rate increases to the point where the second term in equation (32) becomes significant.

The second assumption (2), that k_t remains independent of R_i, may also break down at high dose rates. Benson and North (70) have argued convincingly that the reason this assumption works out so well in practice for many free radical polymerizations is that k_t is largely independent of radical size beyond a degree of propagation of about 100. However, at very high dose rates the chain length is lowered and consequently the assumption becomes untenable for free radical polymerizations. In this case, k_t increases with R_i and the dependence of R_p on R_i through equation (29) should give rise to negative deviations ($x < 0.5$) from the square root relation even in circumstances where assumption (1) holds. Analytical expressions have been derived by Benson and North (70) to cover this situation. Assumption (2) might be expected to hold reasonably well for a free ion polymerization in which only the cation or the anion is the main propagating species, since the value of k_t would then be largely determined by the greater mobility of the nonpropagating species.

Chapiro (1) has considered the kinetics of free radical polymerization at high dose rates from a slightly different point of view, by applying the stationary state approximation separately to the reactions of the primary (initiating) and propagating radicals. But under conditions where a significant fraction of primary radicals undergo termination by homogeneous radical recombination, the time for initiation becomes comparable to the overall lifetime τ_{ss}; for nonzero ν, this implies $k_i < k_p$ so the situation reduces to the case discussed previously (equation 32) and the same general conclusion is reached (1) regarding the dependence of R_p' on dose rate. The treatment given by Chapiro (1) takes account of a variable k_t by using three separate constants to describe the different modes of recombination of primary and propagating radicals. This approach arbitrarily restricts the dependence of k_t on R_i to the limiting case of slow initiation, and therefore does not cover the possible breakdown of assumption (2) under conditions where (1) may remain in force ($k_i \geq k_p$).

To sum up, for free radical polymerization the assumption (2) is expected to fail at low v corresponding to high dose rates. If assumption (1) does not apply ($k_p > k_i$), an additional deviation from the half-power dependence on R_i will also occur through equation (32) as the dose rate is increased. In ionic polymerization by free ions where propagation is due mainly to one charged species, assumption (2) may hold reasonably well but the consequences of $k_p > k_i$ should be similar to those discussed above.

Turning to assumption (3), there are two important cases where competing termination processes have to be considered. First, the active propagating species may undergo unimolecular termination. In radical polymerization this process is referred to often as degradative chain transfer and involves the formation of a more stable radical. If this mechanism of chain termination predominates over recombination, R_p depends on the first power of R_i according to the expression,

$$R_p = R_i \frac{k_p[M]}{k_u} \tag{35}$$

since the lifetime of the propagating species will be given by $1/k_u$, where k_u is the internal or unimolecular termination rate constant. At high dose rates it is possible for the lifetime of the active species to be reduced below $1/k_u$ by the usual bimolecular termination mechanism, so the R_p dependence on increasing dose rate could then conceivably change from first to half power for a case of this kind. However, there are practical limitations to the high dose rates which can be used conveniently for studies of radiation-induced polymerization.

A second alternative mechanism of termination involves a transfer of the active species to an impurity or scavenger such that the reaction product is nonpropagating. This case is particularly important for radiation-induced ionic polymerization where low concentrations of ionic scavengers may be present in the system. A detailed kinetic analysis of this case (69) leads to the equation,

$$R_p = \frac{R_i k_p[M]}{(R_i k_t)^{1/2} + k_{tx}[X]} \tag{36}$$

where k_{tx} is the bimolecular rate constant for transfer or reaction with the ionic scavenger or impurity, X. It may be noted that equation (36) reduces to equation (29) at high dose rates provided that $k_{tx}[X]$ is not too large. Therefore, the variation of R_p with dose rate will depend on the relative magnitude of the two terms in the denominator of equation (36). A large scavenger concentration will result in $R_p \propto R_i^{1.0}$, while R_p should depend on $R_i^{0.5}$ in the limit of zero scavenger concentration. This latter condition is much more difficult to achieve in free ion than in free radical polymerization. A more detailed discussion is deferred to specific examples of this case.

The preceding kinetic relations based on the use of the stationary state approximation ignore any contribution to the chain reaction from active species undergoing geminate recombination in liquids. In Section I-B it was pointed out that the "space-time average" of the intermediate concentration arising from inhomogeneous processes was generally very small in comparison to the steady state value. Only in one special case is it perhaps necessary to consider the possibility that spur processes contribute significantly to the chain reaction. This could occur for an ionic polymerization where there is a large concentration [X] of retarder which depresses the "stationary state R_p" according to equation (36). Remembering that the number (G yield) of ions undergoing geminate recombination G_g in liquids of low dielectric constant greatly exceeds the G_i value for free ions, and that the k_p for ionic propagation may be very large ($> 10^6 \, M^{-1} \, \text{sec}^{-1}$), it is feasible that $G(-m)$ represented by

$$G(-m) = \sum_j G_j \tau_j k_p [M] \tag{37}$$

could include an appreciable yield from the short-lived geminate ions, and become equal to or even exceed the corresponding yield from the free ions. This latter condition will hold when the product sum $\sum_g G_g \tau_g$ taken over all geminate ions equals $G_i \tau_s$, where $\tau_s = 1/k_{tx}[X]$. Since $\tau_s (> \tau_g)$ may be reduced below 10^{-7} sec at high concentrations ($> 10^{-3} M$) of an efficient ionic scavenger, there is no difficulty in satisfying the above requirement for $\sum_g G_g > G_i$. A probable example of this behavior is furnished by the cationic polymerization of cyclopentadiene at high ammonia concentration (124). Of course, the rate of polymerization is strongly retarded and depends on the first power of dose rate in these circumstances. The reason for the lack of any parallel case of this kind in radical polymerization is due to the lower k_p values (10^1–$10^3 \, M^{-1} \, \text{sec}^{-1}$) which apply in this instance, although the radical lifetimes (10^{-6} to 10^{-10} sec) in spur recombination are roughly comparable to those in the ionic case. Thus the spur contribution to $G(-m)$ through equation (37) for radical processes can be safely neglected.

Under certain physical conditions, the kinetics of polymerization reactions become considerably more complicated than those cases already considered. Even in liquid systems, simple treatments based on stationary state kinetics cannot be applied uniformly to cases where the polymer precipitates from solution. Similar limitations apply to solid state polymerization, either in the glassy state or in crystals. Special problems also arise in vapor phase studies due to precipitation of the polymer onto the walls of the vessel. In general, kinetic studies are not very revealing in these instances, although they may be undertaken for practical reasons. Conclusions based on the kinetic form of the results may turn out to be false unless the system is examined carefully for possible deviations from assumptions inherent in the kinetic treatment.

2. Criteria of Mechanisms. Information about the mechanism of a radiation-induced polymerization may be obtained in three general ways:

1. By comparing the characteristics of the polymerization with the known behavior of the monomer in catalytic studies.

2. From the influence of external parameters (temperature, dose rate) on the polymerization rate.

3. By evaluating the effects due to additives, such as free radical or ionic scavengers in low concentrations.

Usually it is necessary to consider that all modes of initiation may occur simultaneously under irradiation, although the rate of polymerization by one particular mechanism will generally predominate for any given set of experimental conditions. The problem then is to determine the mechanism of the propagation reaction which contributes most to the overall rate, and a solution ultimately depends on kinetic measurements.

The known polymerizability of a monomer with different types of initiators (71) provides a useful guide to the interpretation of radiation-induced polymerization, at least for some monomer systems. Table IV lists the classification given by Schildknecht (71) for the reactivity of monomers with cationic, free radical, and anionic reagents in homopolymerization. While some monomers such as styrene show no specificity, experience has shown that isobutylene, β-pinene, cyclopentadiene, and the vinyl ethers display a marked preference for polymerization by cationic catalysts. Some monomers, e.g., α-methylstyrene, polymerize readily by anionic and cationic catalysts, whereas others, e.g., methyl methacrylate,

Table IV

Response of Vinyl-Type Monomers in Homopolymerization
to Different Initiating Systems

Cationic	Free radical	Anionic
Isobutylene	Vinyl chloride	Nitroethylene
Cyclopentadiene	Vinyl acetate	Vinylidene cyanide
Alkyl vinyl ethers	Acrylonitrile	Acrylonitrile
β-Pinene	Methyl methacrylate	Methyl methacrylate
α-Methylstyrene	Ethylene	α-Methylstyrene
Styrene	Styrene	Styrene
Butadiene	Butadiene	Butadiene

are especially reactive to anionic and free radical reagents. Thus the nature and degree of specificity varies from monomer to monomer. This classification, although qualitative and empirical, has provided some good hints regarding the suitability and choice of monomers for mechanistic studies in radiation-induced polymerization.

At first sight, it might be expected that the determination of reactivity ratios (72,73) in radiation-induced copolymerization would offer an attractive method for investigating mechanism. Unfortunately, these ratios are not unique for a particular mechanism, for they fluctuate according to solvent and catalyst conditions, especially in ionic polymerization. On this point, it is important to draw attention to the likelihood that the conditions which obtain in radiation-induced *ionic* polymerization may differ substantially from those encountered in catalytic studies, so analogies should not be pursued too far even if the mechanisms appear to be similar. That reactivity ratios in copolymerization are affected by the type of initiator has been known for a long time (72,73), but the use of these ratios as a criterion of mechanism in radiation-induced polymerization is not always straightforward. This may be illustrated by reference to the radiation-induced copolymerization of styrene and methyl methacrylate which produced a copolymer composition exactly characteristic of a free radical mechanism (74). While this result implies the absence of ionic initiation in this system, it was taken also as fairly conclusive evidence against the role of ionic mechanisms in radiation-induced polymerization. However, later work (75,76) has established that pure, dry styrene undergoes radiation-induced ionic polymerization in bulk at a rate considerably greater than the free radical polymerization which occurs in wet samples of monomer. Thus in radiation studies, it is dangerous to extrapolate from one system to another on the basis of the type of phenomenological behavior which is encountered in catalytic polymerization. In the preceding case (74), the radiation-induced cationic polymerization of styrene may have been inhibited by methyl methacrylate or by traces of water. This great sensitivity to various foreign substances sets radiation-induced cationic polymerization apart from the corresponding catalytic studies. To anticipate further discussion, it can be remarked that although the chemical nature of the species may be identical in the two cases, the variations in physical nature and distribution of the ions in the form of free ions, ion pairs, etc. can lead to very different kinetic behavior. However, the difficulty of making generalizations in this field is indicated by the finding (75) that the reactivity ratios for the radiation-induced copolymerization of the styrene–α-methylstyrene and styrene–isobutyl vinyl ether systems *are* in good agreement with those obtained by cationic catalysis. It may be that a good correlation exists only for cases such as these where both monomers are known to undergo homopolymerization by the same mechanism. This would also explain the apparent predominance of the free radical mechanism in the styrene–methyl methacrylate irradiation (cf. Table IV).

The determination of the number-average molecular weight, or $\bar{D}P_n$, of the polymer in homopolymerization can provide additional information to that obtained from kinetic measurements. This comes about because the magnitude of $\bar{D}P_n$ is affected by nondegradative chain transfer processes as well as by chain termination. General relations for $\bar{D}P_n$ and ν are given in (38) and (39), respectively,

$$\bar{D}P_n = \frac{k_p[Z][M]}{k_f[Z][M] + k_s[Z][S] + k_u[Z] + k_{tx}[Z][X] + k_t[Z]^2} \quad (38)$$

$$\nu = \frac{k_p[Z][M]}{k_u[Z] + k_{tx}[Z][X] + k_t[Z]^2} \quad (39)$$

In the denominators, we have included the sum of all the processes which may contribute to chain breaking (in equation (38)) and to chain termination (in equation (39)). The latter were discussed in the previous section. The first two terms in the denominator of (38) represent nondegradative chain transfer to monomer and to solvent with the respective rate constants k_f and k_s. Combining equations (38) and (39), we obtain

$$\frac{1}{\bar{D}P_n} = \frac{1}{\nu} + \frac{k_f}{k_p} + \frac{k_s}{k_p}\frac{[S]}{[M]} \quad (40)$$

In the absence of significant chain transfer, the last two terms of equation (40) drop out and the equation reduces to the simple form consistent with equations (14) and (15). Therefore G_i, which is given by the quotient $G(-m)/\nu$, can be calculated in this simple case from the experimental quantity $G(-m)/\bar{D}P_n$.

Let us consider first the problem of chain transfer in bulk monomer systems. In this case, $G(-m)$ and $\bar{D}P_n$ are related through equation (41),

$$\frac{1}{\bar{D}P_n} = \frac{G_i}{G(-m)} + \frac{k_f}{k_p} \quad (41)$$

provided of course that these quantities refer to the same experimental conditions of total dose, dose rate, and concentration [X] of a terminating species which may be present as an adventitious impurity. From a practical point of view, it is desirable therefore to determine $G(-m)$ and $\bar{D}P_n$ for the same irradiated sample. The importance of the transfer constant for transfer to monomer, k_f/k_p, in equation (41) can only be gauged from the absolute magnitude of $G(-m)/\bar{D}P_n$. Where this quantity exceeds 10 for a radical reaction or 1 for an ionic process, it is fairly safe to assume that chain transfer is important in determining the molecular weight. For example, in the radiation-induced cationic polymerization of bulk α-methylstyrene (8), the value of $G(-m)/\bar{D}P_n$ exceeds 100, so in this case there is no doubt that $\bar{D}P_n$ is governed solely by the transfer constant in equation (41). In fact, the $\bar{D}P_n$ in this particular case (8) turns out to be independent of $G(-m)$ even when the polymerization is very strongly retarded by either water (8) or ammonia (77). We can easily state the necessary condition,

$$G(-m) < G_i(\bar{D}P_n)_0 \tag{42}$$

in order that the value of $\bar{D}P_n$ should be reduced from the limiting value $(\bar{D}P_n)_0$ as determined by k_p/k_f. Since G_i is of the order of 0.1 in ionic polymerization, it follows that for α-methylstyrene where the limiting $(\bar{D}P_n)_0$ is no greater than about 50, $G(-m)$ must be reduced to less than 5 to observe an effect of this type. At such low $G(-m)$ values, other nonionic contributing reactions must be considered and this prevents a meaningful analysis in terms of equation (41).

The observation that $\bar{D}P_n$ is controlled by chain transfer to monomer for isobutylene, α-methylstyrene, styrene, β-pinene, and isobutyl vinyl ether provides corroborating evidence for the view that the radiation-induced polymerization of these monomers is largely, if not exclusively, cationic in character, since proton transfer processes are known to be strongly characteristic of such cationic polymerizations. Other evidence about the nature of chain transfer can be obtained from infrared and NMR spectroscopic studies on the molecular structure of the end groups in the polymer. For example, the absence of terminal vinylidene $\diagup\!\!\!\!^{\diagdown}\!C\!\!=\!\!CH_2$ groups in the low molecular weight homopolymers from α-methylstyrene and β-pinene (78) indicates the absence of a transfer reaction involving allylic radicals.

We have seen that in bulk monomer systems, G_i cannot be determined from equation (41) when chain transfer predominates. Turning to the problem of solution polymerization, v is now a function of the monomer concentration $k_p[M]\tau$ according to (39) where the lifetime τ of the kinetic chain is defined by

$$\tau^{-1} = k_f + k_{tx}[X] + k_t[Z] \tag{43}$$

Neglecting chain transfer to solvent, equation (40) reduces to the form

$$\frac{1}{\bar{D}P_n} = \frac{k_f}{k_p} + \frac{1}{k_p[M]\tau} \tag{44}$$

and $k_p\tau$ may be evaluated from a plot of $1/\bar{D}P_n$ against $1/[M]$, assuming that τ remains independent of monomer composition and that other variables such as [X] and dose rate are held constant. Hence G_i may be calculated from the expression

$$G_i = \frac{1}{k_p \tau} \frac{G(-m)}{[M]} \tag{45}$$

This approach has been used to obtain G_i in a study (79) of the radiation-induced cationic polymerization of styrene in dichloromethane at $-78°C$.

Effects arising from the variation of temperature and dose rate have been commonly employed as mechanistic criteria in radiation-induced polymerization (80). It is true that the rates of most free radical polymerizations depend on the half power of R_i and are subject to an overall activation energy in the region of 5 kcal/mole. Departures from this trend are suggestive but certainly do not prove that alternative mechanisms apply. Further, it has been frequently asserted (81) that rates (R_p) for ionic mechanisms are characterized by dose rate dependencies close to unity and by low or even negative temperature coefficients. Indeed, if this pattern were always reliable, it would be possible to distinguish between ionic and free radical mechanisms without much difficulty. A related generalization assumed that in radiation studies, free radical mechanisms predominated at room temperature and that ionic mechanisms could be important *only* at low temperatures. Recent work summarized in Ref. 69 has shown that there are important exceptions to this statement. The main flaw in the old arguments was the rigid belief that ionic polymerizations terminated unimolecularly. This view was held despite the lack of real evidence in support. Such an interpretation could satisfactorily explain the observed dependence of R_p on the first power of dose rate. However the same result is predicted by theory (cf. equation (36)) when the chain termination step is controlled by reaction with an impurity whose concentration remains reasonably constant during the experiment. The recognition that radiation-induced ionic polymerizations are extremely sensitive to water (8), coupled with the ubiquitous character of the latter substance, has clarified the subject. Under extremely anhydrous conditions, it has been shown that the rates of ionic polymerizations are considerably enhanced and depend on a power of dose rate close to 0.5, as expected for a bimolecular termination mechanism between free ions corresponding to the limiting case of equation (36) where $k_{tx}[X]$ can be neglected.

Although a radical reaction is generally associated with a dependence of R_p on the half power of I, a first power relation would apply if termination were to occur mainly by degradative chain transfer (equation (35)). Therefore the R_p dependence on I may take on any value between 1.0 and 0.5 depending upon the kinetic form of the termination reaction, irrespective of whether the mechanism is radical or ionic. It must be concluded that the dose rate dependence is in itself an unreliable index to mechanism.

Perhaps the only generalization which can be made about temperature effects is that radiation-induced polymerizations occurring at low temperatures (-80 to $0°C$) with small activation energies (less than about 2 kcal/mole) are much more likely to be ionic than free radical in character. Examples in this category include bulk isobutylene and cyclopentadiene, and styrene in dichloromethane (79,82). Some bulk monomers can only be studied in the liquid phase at relatively high temperatures owing to the value of their crystalline melting points, but it would be incorrect to assume that radical mechanisms always dominate in these circumstances, and that the overall activation energies of ionic polymerizations *must* be negligible.

To summarize, dose rate and temperature effects *per se* do not afford a clear demarcation between ionic and free radical mechanisms, and it is only in conjunction with other evidence that such studies provide valuable insight into the detailed polymerization mechanism.

Efforts to interpret the effects of additives, or scavengers, in radiation chemistry have given rise to much controversy in the past (83). Within the last few years there has grown a better understanding of the nature and role of intermediates with the result that scavengers can be employed in a more selective manner than before, and conclusions derived from their proper use now carry considerable weight. In polymerization studies, certain additives may cause the rate to be greatly diminished and such substances are loosely classified as inhibitors or retarders depending on the degree of this effect.

The well-known free radical scavengers include oxygen, nitric oxide, galvinoxyl (84), DPPH, ferric chloride, and *p*-benzoquinone. Low concentrations ($10^{-3}M$) of these reagents are sufficient to intercept most of the free radicals in the steady state (14). For stable organic radicals such as galvinoxyl and DPPH, the rate constant for radical scavenging approaches that for the recombination of transient radicals, and since the stationary state concentration of transient radicals in the absence of scavenger is unlikely to exceed $10^{-6}M$ in practice (see Table II), it is obvious that the scavenger competes very efficiently at millimolar concentration.

Similar kinetic considerations apply to the scavenging of free ions by ionic scavengers which may be subdivided into positive ion and electron scavengers. The former group includes substances which act by causing positive charge exchange or proton transfer from the "primary" ion to the scavenger. Examples of proton scavengers are water, ammonia, alcohols, and amines, while substances with low ionization potentials such as nitric oxide and hydrogen sulfide (85) are suitable electron donors in charge exchange processes. Electron scavengers include a wide range of aromatic hydrocarbons and their derivatives, alkyl halides, nitrous oxide, sulfur hexafluoride, and carbon dioxide.

The ability to discriminate between radical and ionic polymerization processes by selective scavenging has proved extremely useful, but considerable care is needed in the application of the method and the interpretation of the results. Several free radical scavengers, including oxygen, p-benzoquinone, and DPPH, appear to be capable of interfering with ionic reactions in some systems, so these reagents are unsuitable for the purpose. As a general rule, free radical scavengers are endowed with molecular properties (paramagnetism, conjugated structures, etc.) such as to confer on these substances a high degree of reactivity to other intermediates beside neutral free radicals. Molecular iodine is probably the least complicated radical scavenger but it shows some reactivity for electrons, and it is also a mild polymerization catalyst on its own account in some monomers such as isobutylene where it forms charge-transfer complexes (86). Ideally the scavenging reagent should be highly specific in its action, it should not induce separate reactions, and it should not give products which further interfere. Such a combination is hard to find for a free radical scavenger in radiation-induced polymerization. Perhaps the only area of radiation chemistry where some radical scavengers in low concentration can be employed without complications is in the determination of primary radical yields for systems of low dielectric constant not undergoing chain reactions. The yield of free ions is small in this case and by keeping the scavenger concentration below $10^{-3}M$, possible interference with the geminate ionic recombination process can be avoided.

In contrast to the behavior of many free radical scavengers, the action of ion scavengers such as water and ammonia is much less ambiguous since the relative simplicity of these latter molecules limits the number of likely reaction paths. On account of their large bond dissociation energies, water and ammonia are inert to organic free radicals. We have already referred (see Fig. 5) to the mass-spectrometric evidence (48) that ammonia and water participate as bases in proton transfer reactions from some hydrocarbon ions. This work supports the interpretation (8,49) that the strong retardation of radiation-induced polymerization by water and ammonia (Fig. 6) is due to chain termination by proton transfer or Lewis acid–base reactions. Thus, the sensitivity of a radiation-induced polymerization to water concentrations of $10^{-3}M$ or lower definitely indicates an ionic reaction. As already mentioned, clear-cut support for ionic processes comes from the findings that the rates of the radiation-induced polymerization of α-methylstyrene, β-pinene, styrene, isobutyl vinyl ethers, and other monomers at room temperature are affected by the drying technique and strongly retarded by added water. Further, since retardation by water is

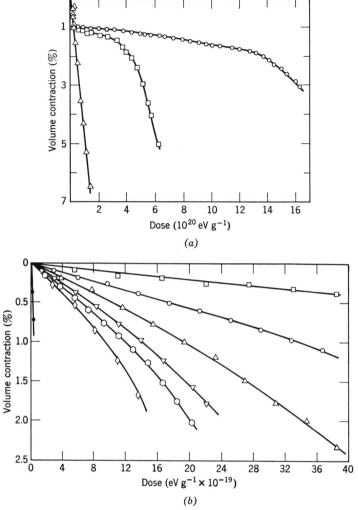

Fig. 6. Effect of water and ammonia on radiation-induced ionic polymerization: (*a*) addition of water to β-pinene polymerization (78); $\triangle$, water absent; $\square$, 5.4 × $10^{-3}M$; $\bigcirc$, 1.2 × $10^{-2}M$. (*b*) addition of ammonia to cyclopentadiene polymerization (88); $\bullet$, ammonia absent; $\diamondsuit$, 1.6 × $10^{-4}M$; $\odot$, 2.0 × $10^{-4}M$; $\triangledown$, 2.5 × $10^{-4}M$; $\triangle$, 7.3 × $10^{-4}M$; $\oslash$, 2.3 × $10^{-3}M$; $\boxdot$, 2.0 × $10^{-2}M$.

consistent with both cationic and anionic propagation, it is possible to resolve this question through the use of ammonia and amines as specific scavengers for positive ions. A further advantage of this approach is that it allows the determination of propagation rate constants by kinetic analysis.

We shall now consider the dependence of R_p and $\bar{D}P_n$ on the concentration [X] of a retarder which causes chain termination. From the general relation for ν in equation (39) and since $R_p = R_i\nu$, we obtain

$$R_p = R_i \frac{k_p[M]}{k_u + k_{tx}[X] + k_t[Z]} \tag{46}$$

The previous expressions (35) and (36) are reduced forms of equation (46) for restricted cases of termination. After inversion, equation (46) may be rewritten,

$$\frac{1}{R_p} = \frac{1}{R_i} \frac{k_{tx}[X]}{k_p[M]} + f(R_i) \tag{47}$$

where $f(R_i)$ includes the dose-rate dependent term associated with the bimolecular recombination of the intermediate Z. For the variation of R_p with [X] at a constant dose rate, the expression becomes

$$\frac{1}{R_p} = \frac{1}{(R_p)_0} + \frac{1}{R_i} \frac{k_{tx}[X]}{k_p[M]} \tag{48}$$

where $(R_p)_0$ is the unretarded rate of polymerization for $[X] = 0$. Alternatively this equation may be rewritten in terms of G values as

$$\frac{1}{G(-m)} = \frac{1}{(G(-m))_0} + \frac{1}{G_i} \frac{k_{tx}[X]}{k_p[M]} \tag{49}$$

where $(G(-m))_0$ is the unretarded yield at the same dose rate.

Starting from equation (38) and by similar steps, we obtain the corresponding relation for $\bar{D}P_n$ as a function of [X] in the form of the well known Mayo equation (87),

$$\frac{1}{\bar{D}P_n} = \frac{1}{(\bar{D}P_n)_0} + \frac{k_{tx}[X]}{k_p[M]} \tag{50}$$

where $(\bar{D}P_n)_0$ is the degree of polymerization in the absence of retarder. We have already seen that $\bar{D}P_n$ will not be very sensitive to the retarder concentration where $(\bar{D}P_n)_0$ assumes a low value governed by regenerative chain transfer to monomer. Therefore the Mayo plot will not give a very accurate result for k_{tx}/k_p in these circumstances. An extreme case of this behavior is encountered with α-methylstyrene. In this case (77), equation (50) cannot be applied (see condition (42)) to obtain meaningful data, and a kinetic analysis can be made only in terms of equation (49). On the other hand, the lesser importance of monomer transfer processes for cyclopentadiene (49,88,124) allowed the ratios k_{tx}/G_ik_p and k_{tx}/k_p to be obtained from the functional dependence of $G(-m)$ and $\bar{D}P_n$, respectively, on the concentration of ammonia retarder.

Finally, it is necessary to mention one limitation to the validity of equations (49) and (50). If these expressions are used to describe results obtained for finite irradiation doses, then a correction term for the possible depletion of retarder during the experiment should be introduced. In practice this is rarely possible, but the above objection may be circumvented by extrapolating to initial $G(-m)$ yields before substitution in equation (49). The determination of initial $\bar{DP}_n$ values for use in equation (50) would be extremely tedious experimentally, so a compromise has to be made by working with the lowest total doses required by practical considerations.

III. POLYMERIZATION IN THE VAPOR PHASE

Although the vapor phase polymerization of a few monomers, and especially ethylene, is of great industrial importance, the experimental results are frequently too diverse and complicated to allow a conventional kinetic analysis. In large measure, this nonideal kinetic behavior can be attributed to the heterogeneous character of these systems. Since high polymers inevitably precipitate out from the vapor, the growing intermediate may undergo drastic changes in physical environment during propagation, and almost all the conditions required by the steady state treatment are vitiated in these circumstances. These comments apply equally well to the catalytic and radiation-induced polymerization of gaseous monomers.

In view of the detailed studies which have been made by mass-spectrometric techniques of the ionic processes in ethylene and acetylene (89), it is surprising and disappointing that so little is known about the extent to which ionic mechanisms contribute to the overall polymerization of these monomers under macroscopic conditions. The remarks made earlier in connection with pulse radiolysis are also pertinent here, for it has been implied all too often in the literature that the demonstration of a few ion–molecule addition reactions in the mass spectrometer substantiates a mechanism of ionic polymerization in radiolysis. Admittedly it is difficult to bridge the gap between the starting processes and the overall reaction in the case of vapor phase polymerization, but remarkably little use has been made of selective scavenging techniques in the macroscopic studies. Perhaps the most revealing comment on the unsatisfactory state of the subject is provided by quoting from conclusions reached in two recent papers on acetylene. Field (90), who has also made extensive mass-spectrometric studies in this area (91), investigated the radiation chemistry over a range of pressure and temperature and found that $G(-C_2H_2)$ and $G(C_6H_6)$ varied together. There is general unanimity that benzene is not formed by an ion–molecule mechanism, so from the preceding observation and others, Field (90) questions the usual interpretation that the polymer (cuprene) is formed by an ionic mechanism. On the other hand, Volpi and his co-workers (89) infer from a very detailed kinetic study of ion–molecule reactions up to the formation of $C_{14}H_{11}^+$ in the mass spectrometer that the high reactivity of all ions in acetylene constitutes strong support for an ionic polymerization mechanism. Thus, although facts are not in dispute, diametrically opposed conclusions are reached from different directions. The situation calls for more crucial experiments to be performed by conventional techniques such as ion scavenging.

A brief review is given below of some aspects of the vapor phase polymerizations of ethylene and isobutylene by radiation.

A. Ethylene

Chapiro has reviewed the earlier literature (92) so here we shall concentrate on some recent observations. The considerable progress made in elucidating the mechanisms which apply to the radiolysis of ethylene at low pressure (93,94) has been reviewed in Chapter 6. The alkane products appear to arise from reactions of free radicals, including addition to ethylene and bimolecular recombination or disproportionation between radicals. Evidence for this well-known reaction sequence has been obtained (93) from radical scavenging experiments with oxygen, nitric oxide, and iodine. This interpretation is further confirmed by a kinetic analysis of pressure and dose rate effects, showing that the rate constant for termination must be of the order of 10^7 times greater than that for addition to ethylene, in agreement with the known values for the rate constants of these radical processes, but at variance with estimates for possible corresponding ionic processes. This argument has been extended by Meisels (95) to cover the radiation-induced polymerization of ethylene at high pressure and very low dose rates. By taking the rate constants for radical propagation and termination to be 2×10^3 M^{-1} sec^{-1} and 1×10^{11} M^{-1} sec^{-1}, respectively, $G(-m)$ can be expressed by the equation,

$$G(-C_2H_4) = 20 + 0.9 \times 10^3 \sqrt{[C_2H_4]/I} \tag{51}$$

where $[C_2H_4]$ refers to the ethylene concentration in moles cc^{-1} and I is the dose rate in Mrad/hr. The first term represents the substantial yield of ethylene disappearance in the absence of any propagation by radical addition to ethylene. This equation agrees extremely well with the polymerization data of Hayward and Bretton (92) which satisfy the relation

$$G(-C_2H_4) = 16 + 1 \times 10^3 \sqrt{[C_2H_4]/I} \tag{52}$$

Thus it appears that although ionic- and excited-molecule processes play an important role in determining the distribution of primary radicals in the radiolysis of ethylene, the stepwise formation of higher alkanes at low pressure and of polymer at high pressures (typically 400 kg/cm^2) are both adequately described by kinetic parameters characteristic of free radical reactions in this system. This interpretation does not dispose of the eventual fates of the higher molecular weight ions seen in the mass spectrometric studies (40–42,91). However, as pointed out by Kebarle (42), the rate constant for the production of $C_{2n}H_{4n}^+$ ions decreases with increasing n. Therefore, high molecular weight polymers may not be produced by an ionic mechanism. Such a view would not be qualitatively inconsistent with the low molecular weight, monoolefinic, branched polymers obtained by Wagner (96) from the irradiation of ethylene in the solid state at 77°K. These products surely do not arise from radical processes since mixtures of C_2H_4 and C_2D_4 gave polymer molecules up to decene containing multiples of d_4 units. In accordance with Wagner's interpretation (96), it appears that the reactions leading to these low polymers at 77°K are related to the ionic reactions observed in the mass spectrometer (42).

Finally on this topic, some recent papers by Machi and his colleagues (97) indicate that under certain conditions at least, the high pressure polymerization of ethylene at 70–400 kg/cm² cannot be described by a simple stationary-state treatment. At 40° or below, the polymer molecular weight is observed to increase with irradiation dose, and the kinetics suggest the formation of propagating radicals with long lifetimes. Neither of these effects were observed at temperatures exceeding 100°C.

The radiation-induced copolymerization of ethylene with carbon monoxide at high pressures has been described (98). The copolymer composition remains between 40 and 52% CO as the initial composition of the gas changes from 5 to 95% CO. These results are consistent with the reactivity ratios r_1 (ethylene) = 0.04 and r_2 (carbon monoxide) = 0, the latter value being assumed from the observation that the homopolymerization of gaseous CO is considerably slower than for ethylene under the same conditions. At 20°C, 680 atm pressure, and a dose rate of 2×10^5 rads/hr, the G value for monomer disappearance in a 39% CO mixture was 3.6×10^4. Thus copolymerization occurs with an appreciable kinetic chain length under high pressures. The dependence of polymerization rate on the 0.63 power of dose rate is consistent with the interpretation in terms of a free radical mechanism. From the decrease in overall rate and $1/r_1$ with temperature, it has been suggested (98) that the addition of CO to a terminal ethylene propagating species involves an equilibrium complex.

B. Isobutene

Mund and his collaborators (99) studied the radiolysis products of isobutene vapor, and the formation of a low molecular weight polymer was attributed to free radical processes. However, a very recent investigation by Okamoto, Fueki, and Kuri (100) has shown that ionic polymerization can occur when the monomer is pure and dry. $G(—C_4H_8)$ values exceeding 1000 were obtained for irradiations carried out at a pressure of 600 torr and a dose rate of 1.8×10^5 rad/hr. The rate of polymerization increased with dose rate according to a first power relation. The polymer was a viscous, colorless liquid with a viscosity molecular-weight M_v ($= 1.85 M_n$) of 6000, and its infrared spectrum corresponded exactly to that of polyisobutene prepared by chemical catalysis. From the $G(-m)$ values and the molecular weight data, it is apparent that chain transfer controls the molecular weight of the polymer. Evidence that the polymerization involves ionic intermediates comes from the effect of additives on the reaction. Ammonia exerts a strong retarding effect, a concentration of $10^{-3} M$ causing $G(-m)$ to be reduced to less than 100. This result, which can probably be attributed to a proton transfer reaction, confirms that the polymerization mechanism is cationic. The polymerization is also retarded by nitric oxide and nitrous oxide, although the retarding effect of these additives is less than for ammonia at comparable concentrations. The action of nitrous oxide may be due to electron capture, since this process would prevent the electron from reaching the walls of the vessel. It is plausible that the propagation lifetime may be reduced in these circumstances owing to a high efficiency for homogeneous charge neutralization between O^- and the propagating cation. The polymerization was found to be sensitized by rare gases such as xenon, and this has been attributed to increased ionization of isobutene by charge and excitation transfer processes. Although Thompson and Schaeffer (101) found that the radiation-induced exchange of hydrogen and deuterium by ionic intermediates (H_3^+, H_2D^+, etc.) was inhibited by xenon because of the proton transfer reaction,

$$H_3^+ + Xe \rightarrow H_2 + XeH^+ \tag{53}$$

the proton affinity of xenon (102) is certainly less than that of isobutylene so a similar chain-breaking step is ruled out for this polymerization, and additives of much higher proton affinity (e.g., ammonia) are necessary to remove the proton from the propagating species in isobutylene.

It is striking that the $G(-m)$ value ($\geq 10^3$) for isobutylene vapor at $\simeq 1$ atm. pressure is at least one order of magnitude greater than the disappearance yield at the same pressure for other simple unsaturated hydrocarbons (e.g., ethylene, acetylene) (103) which are believed to undergo ionic polymerization under these conditions. This suggests that the ionic propagation rate constant for isobutylene is correspondingly larger. An estimate of k_p for isobutylene vapor polymerization may be obtained by applying equation (49) to the data obtained by Okamoto, Fueki, and Kuri (100) for the retardation by ammonia. $G(-m)$ was reduced to 650 and 100 in the presence of 10^{-4} and 10^{-3} mole % ammonia, respectively.

Taking the G yield for ion pairs to be 3, the ratio k_p/k_{tx} is calculated to be about 5×10^{-4}. In the gas phase, the rate constant k_{tx} for exothermic proton transfer to ammonia is probably in the range from 10^{10} to 10^{12} $M^{-1} \sec^{-1}$, so k_p is of the order of 10^7 $M^{-1} \sec^{-1}$ or higher. This estimate is consistent with the values reported for radiation-induced ionic polymerization in the liquid phase (69). Even a qualitative analysis of the ammonia retardation data for ionic polymerization is sufficient to show that while the propagation rate constant is large on the absolute scale, it is still significantly smaller than the rate constant for proton transfer to ammonia. Thus it is erroneous to suppose that ion–molecule polymerization occurs on every collision under macroscopic conditions, as often suggested on the basis of reaction studies in the mass spectrometer. It is worth emphasizing that a carbonium-ion mechanism invariably applies to monomers which are known to undergo radiation-induced cationic polymerization with *long* kinetic chains. On the other hand, there is considerable uncertainty about the exact mechanism of the relatively *short* ionic polymerization chains observed in the mass spectrometer. To clarify this subject, mass spectrometric studies are needed on monomers (isobutylene, butadiene, cyclopentadiene, etc.) that polymerize efficiently by the carbonium ion mechanism.

A novel technique (104,105) has been described for injecting isobutylene ions into the liquid monomer. Following photoionization in the vapor, the positive ions are caused to drift into the liquid by the application of an electric field across the cell. At a temperature of $-135°C$, a light intensity of 3×10^{16} quanta/sec, and a potential difference of 350 V, the quantum yield for monomer disappearance to polymer is calculated from the authors' data (105) to be about 10. The molecular weight of the polyisobutylene produced in these experiments was 3×10^6, so the low overall quantum yield must be due to a low efficiency of initiation. The yield of polymer was proportional to the time and intensity of irradiation, but independent of the applied voltage. All the observations are consistent with the conclusion that the polymerization occurs in the liquid phase. The mechanism of termination is of particular interest since presumably there is a net positive charge in the liquid monomer, and under these conditions, neutralization reactions between positive ions and electrons in the liquid phase are minimized. One could envisage the possibility of "living" cationic polymerization in this system, but there is no evidence in support. The molecular weight appears to be controlled by proton transfer to monomer, or by precipitation in some circumstances. The latter situation applies when the polymerization is carried out in ethyl chloride which is known to be a rather poor solvent for polyisobutylene, with the result that the molecular weight is lowered to 6×10^4. A recent report (106) describes some additional experiments with isobutylene by means of this technique. Examination of the liquid monomer by vapor phase chromatography after irradiation revealed the presence of C_8 and C_{12} hydrocarbons. From this result, the authors (106) discuss the reactions of the dimer ion by proton transfer and hydride–ion transfer, which they incorrectly describe as neutralization processes. There is no mention of polymer formation in this study (106), in contrast to the work of Schlag and Sparapany (104,105), who found that the molecular weight of the polymer was extremely low only in the initial stage of irradiation, and attributed this result to termination by adventitious impurities. Of course, this type of impurity effect which may cause retardation and molecular weight reduction is quite analogous to that observed in studies of radiation-induced cationic polymerization. In this light, it appears probable that the formation of dimer and trimer products from isobutylene is caused by ionic termination with a relatively large concentration of impurities, thereby preventing appreciable propagation. Hence the significance of these particular results (106) is highly suspect.

IV. POLYMERIZATION IN SOLUTION

There is an extensive literature on the subject of radiation-induced polymerization in solution, and a comprehensive survey will not be attempted. Since the potential range of systems is so great, an introductory classification is given in terms of solvent characteristics. Two representative systems are then singled out for a detailed discussion.

In Section II-B-2, we discussed the polymerizability of monomers by the different mechanisms, and it was noted that some monomers polymerize almost exclusively through one particular mechanism under certain conditions. Apart from the intrinsic reactivity of the monomer to the different types of intermediates produced by irradiation, the propagation of polymerization in solution is affected primarily by the capacity of the solvent to sustain or quench a particular mechanism. Thus, at least in principle, a greater degree of selectivity may be obtained in solution polymerization when compared to the irradiation of monomers in bulk. However, the nature of the intermediates and products from the irradiation of the solvent must also be considered, and in some cases, severe complications can set in due to a "reactivity transfer" between solvent and monomer which affects the primary processes leading to the formation of the transient species. In other words, it cannot be generally assumed that the starting processes originate from the separate action of radiation on monomer and solvent. In the older literature, such interaction effects were described under the all-embracing term of "energy transfer" (107). Beset by these possible complications, the kinetics of solution polymerization often presents formidable problems of interpretation, and relatively few systems have yielded satisfactory conclusions. As in many other branches of chemistry, a careful choice of reaction system is necessary for clarity.

Although limited to a small number of water-soluble monomers, aqueous systems offer some advantages for the study of radiation-induced polymerization. First, much is known about the radiation chemistry of water (108) in terms of radical yields. Secondly, water is inert to almost all organic free radicals, and chain transfer processes do not have to be considered. Finally, radiation-induced ionic polymerization processes are entirely suppressed in the presence of substantial water concentrations. On the other hand, one severe drawback is the tendency for many polymers to precipitate in hydroxylic solvents, for this effect can give rise to kinetic abnormalities.

Organic solvents may be chosen so as to favor either cationic or anionic propagation. Okamura and his co-workers (109) were among the first to use chlorinated hydrocarbons for cationic polymerizations, and amines or ethers for anionic processes. To a large extent, these particular choices are consistent with the general experience gained in catalytic studies, but it should be added that the suitability of these solvents is indicated also by radiation chemical studies. Thus, many organic halides undergo dissociative electron capture to form negative halide ions which would not be expected to initiate anionic propagation. Similarly, cationic polymerization is prevented in ethers or amines on account of these solvents having a large proton affinity. By and large, most organic solvents will support free radical polymerization although several solvents participate in chain transfer reactions resulting in a decrease of the polymer molecular weight (cf. equation (40)).

Where the rate constants are known from other work, and the radiation-induced polymerization follows simple kinetics as represented by equation (29), then the results may be used to calculate R_i. For solution polymerization, R_i is a function of both monomer and solvent concentrations, and the precise form of this dependence has to be evaluated by experiment since a simple additive relation cannot be assumed from first principles. Considerable work (110) has been done on the radiation-induced free radical polymerization of styrene, methylmethacrylate, and vinyl acetate in solution, monomers for which the rate constants are well established. Studies on the bulk monomers allow the determination of the radical yield from monomer, G_R^M. This value may then be applied to the results of solution polymerization in order to calculate G_R^S, the radical yield from solvent, subject to the limitation that a simple additive relation holds for R_i as a function of monomer and solvent concentration. In the event that this latter condition does not hold and some type of reactivity transfer applies, the nature of the interaction may be deduced from the dependence of R_i on the composition of the solution. This approach represents an attempt to obtain information about radical yields by using polymerization as a chemical sensor. Unfortunately, the method suffers from inherent shortcomings which affect its quantitative reliability. Included among these is the dubious assumption that the rate constants are unchanged by the monomer to solvent ratio. The comment might be made that chain reactions are generally unsuitable for measuring primary yields because the experimental yield $G(-m)$ is also proportional to the kinetic chain length ν (equation (14)) which can be very sensitive to the reaction conditions. Clearly, the use of a scavenger which does not undergo a chain reaction is preferable for the purpose. The more important contribution which can be obtained from the study of chain reactions is the kinetic chain length itself, and hence the relation of this quantity to the lifetime of the transient intermediate. It is probably a fair summary of the present position to say that the radical yields derived from polymerization studies are of limited accuracy, and the method has become obsolete.

In contrast to these studies (110) which have been mainly concerned with the determination of radical initiating yields from the overall polymerization rates, some investigations have broken new ground by providing determinations of rate constants in free radical polymerization. As an excellent illustrative example, we shall discuss the polymerization of acrylamide in aqueous solution.

A. Acrylamide in Aqueous Solution

The radiation-induced and photosensitized polymerization of this monomer has been described by Collinson et al. (111) and by Dainton and Tordoff (112). The interpretation of earlier work by Collinson and Dainton (2) on the radiation-induced polymerization of acrylonitrile in aqueous solution had been criticized for failing to consider the kinetic effects due to heterogeneity. Acrylamide polymerization is free from this objection because the polymer is soluble in water, and the kinetics were found to conform with the standard equation (29) for homogeneous polymerization (111). Actually, this result throws further doubt on the conclusions reached in the acrylonitrile investigation about the alleged nonuniformity in the distribution of initiating species. On this question we must be careful to point out that while the *initial* distribution of radicals is in fact nonuniform, this distribution is most unlikely to be reflected in the polymerization kinetics because of the short lifetime for the radical undergoing recombination in the spur. As already mentioned, the study of polymerization kinetics is a blunt tool for learning about the yield and distribution of initiating species in radiation chemistry. Inspection of equation

(37) shows that it is the long-lived radical species which dominate the contribution to the polymerization chain.

In the acrylamide work, the rate of initiation was determined independently by measuring the reduction of ferric ion in acidified solutions of ferric perchlorate and monomer. As the concentration of Fe(III) increased to greater than $10^{-3}M$, the polymerization rate decreased and the exponent of the dose rate dependence changed gradually from 0.5 to 1.0. This behavior conforms to equation (36) for the case of a radical terminator X. At high Fe(III) concentrations (10^{-3} to $10^{-2}M$) such that all the polymerization chains undergo exclusive termination with ferric ions, the yield of ferrous ions is constant and corresponds directly to the yield, G_i, of initiating radicals. By this means, G_i was determined to be 8.2 radicals/100 eV, and it is reasonable to assume that this value applies also to the unretarded polymerization in the absence of ferric ion. Hence R_i could be calculated and the values combined with the corresponding results for the unretarded R_p according to equation (29) to obtain the ratio $k_p/k_t^{1/2} = 4.7$ $M^{-1/2} \sec^{-1/2}$ at 25°C. The termination rate constant k_t was obtained from the hydrogen peroxide photosensitized reaction (112) by using the rotating sector method for the determination of the chain lifetime. At 25°C, $k_t = 1.45 \times 10^7 \; M^{-1} \sec^{-1}$ by this latter method, so k_p was finally evaluated as $1.8 \times 10^4 \; M^{-1} \sec^{-1}$. Thus the k_p for acrylamide has been obtained by combining the results of three separate but related experimental determinations. It is of considerable interest that this k_p is one of the highest values obtained for the free radical polymerization of vinyl monomers at 25°C. From the dependence of acrylamide polymerization rate on ferric ion concentration, it was possible to establish that both the species Fe^{3+} and $Fe^{3+}OH^-$ acted as terminators with respective rate constants of $2.0 \times 10^3 \; M^{-1} \sec^{-1}$ and $1.1 \times 10^4 \; M^{-1} \sec^{-1}$ for their reactions with polyacrylamide radicals at 25°C. The use of ferric ion as an ideal inhibitor or retarder of radical polymerization has been extended by Bamford et al. (113) to cover the catalyzed polymerization of styrene, acrylonitrile, methyl methacrylate, and other monomers by azobisisobutyronitrile in nonaqueous solutions containing ferric chloride. It is interesting that the rate constants for the reactions of propagating radicals with Fe(III) do not follow the same trend as for the reactivity of these radicals in chain transfer with toluene, so the electron-donating ability of the radical is probably the significant requirement for reaction with Fe(III). These termination rate constants are comparable in magnitude to those for radical propagation, and considerably lower than for diffusion-controlled reactions. Consequently, the rate of polymerization only becomes sensitive to ferric ion concentrations greater than $10^{-5}M$. At this concentration level, few practical problems are encountered in quantitative work, and there is the further advantage that it is possible to avoid serious depletion of retarder during irradiation.

B. Styrene in Dichloromethane at $-78°C$

This system has been the subject of several studies in the last decade (114–121). Interest originated from the idea (114) that the cationic polymerization of styrene would predominate over the free radical mechanism at a sufficiently low temperature in a solvent of high electron affinity. This expectation has been realized but the results should not be interpreted as meaning that these experimental conditions are *essential* to support cationic polymerization.

Although there is general agreement among the different investigators that the polymerization is cationic, there are differences in the detailed nature and interpretation of the results. Charlesby and Morris (119) have reported that the polymerization rate is affected by the drying technique, a general effect for radiation-induced ionic polymerization which had been previously described for the bulk polymerization of α-methylstyrene (8). The results of Charlesby and Morris (120) on samples dried by sodium are in fair agreement with those of Ueno et al. (121), although the latter group holds that the polymerization is not as sensitive to water as for the polymerization of bulk monomers at room temperature, and that routine drying procedures are adequate to obtain reproducible results at $-78°C$. At a dose rate of 40 krad/hr, the R_p for a $2.2M$ solution of styrene in dichloromethane is given as 6×10^{-3} and $5.7 \times 10^{-3} M \, hr^{-1}$ in Refs. 120 and 121, respectively. Charlesby and Morris (120) report that at 40 krad/hr, R_p depends on the 1.5 power of monomer concentration, but at a dose rate of 3.2 Mrad/hr, R_p is said to be proportional to the first power of monomer concentration. These relationships are applicable only to the results below $1.5M$ monomer concentration. On the other hand, Ueno et al. (121) found R_p to depend on the first power of monomer up to $2.2M$ at a dose rate slightly below 40 krad/hr. This apparent discrepancy in the dependence on monomer concentration also detracts from the good agreement in the absolute rates under the specific conditions mentioned above. However, a closer analysis reveals that the published data of Charlesby and Morris (120) are too limited to support their conclusion about the 1.5 order in monomer concentration.

At dose rates less than 0.5 Mrads/hr, R_p depends on the first power of dose rate according to most of these investigations (115–121). An exception is the polymerization in the presence of Aerosil (120), a silica gel of high surface area, for which the dose rate exponent is said to be 0.5. This case also corresponds to higher rates of polymerization (by a factor of between 4 and 10) in comparison to the results obtained in the absence of Aerosil. From the available data (120), R_p in the presence of Aerosil changes approximately from a half power to a first power dependence on dose rate above 0.5 Mrads/hr, whereas the opposite trend is evident from the results obtained in the absence of solid additive. The dependence on monomer concentration in the presence of Aerosil appears to change from half to first order with increasing dose rate. The complexity of these results rules out a simple interpretation based on the kinetics alone.

The most significant conclusions come from an analysis (120) of the molecular weights as a function of monomer concentration by means of Mayo plots based on equation (44),

$$\frac{1}{\overline{DP}_n} = \frac{k_f}{k_p} + \frac{J}{k_p[M]} \tag{54}$$

where J now refers to the reciprocal of the kinetic chain lifetime τ. The ratio k_f/k_p derived from the intercept of a plot of $1/\overline{DP}_n$ against $1/[M]$ is almost the same for the results with and without Aerosil, the value being 2×10^{-3} in the former case and between 2×10^{-3} and 5×10^{-3}, depending on dose rate, in the latter. This corresponds to a limiting value of $\overline{DP}_n \simeq 500$ at high monomer concentration when chain transfer to monomer determines the molecular weight. The slopes of the Mayo plots J/k_p in the two instances vary slightly with dose rate (proportional to $I^{0.2}$) and fall within the range 3.4×10^{-3} to $5.1 \times 10^{-3} M$ without Aerosil, and between 0.6×10^{-3} and $1.5 \times 10^{-3} M$ with Aerosil. From these results it is likely that the effect of Aerosil is to decrease J with k_p remaining constant, and hence to increase the kinetic chain lifetime. Corroborative evidence is provided by the increased rate of polymerization in the presence of Aerosil, and it is concluded that Aerosil decreases the termination rate. Moreover, the observation that the R_p dependence on dose rate below 0.5 Mrad/hr changes from a first power to half power relation when Aerosil is present is also consistent with a lengthening of the kinetic chain arising from a decrease in the magnitude of $k_{tx}[X]$ as discussed in connection with equation (36). The calculation of G_i was carried through on the basis of equation (45), which now may be rewritten,

$$G_i = \frac{JG(-m)}{k_p[M]} \tag{55}$$

In this expression, J/k_p is known from the slope of the appropriate Mayo plot of molecular weight data, and $G(-m)$ is known as a function of monomer concentration [M] from the kinetic data at each dose rate. It is interesting that the values of G_i obtained from the results with and without Aerosil at 40 krads/hr are about 0.1 and 0.2, respectively, at low monomer concentrations, and the values apparently decrease in each case above $2M$. The authors (120) allege that the G_i values calculated at this low dose rate are typical of those over the whole dose rate range studied, but this cannot be strictly correct since they find that J/k_p varies as $I^{0.2}$ with and without Aerosil, and that R_p depends on $I^{0.5}$ with Aerosil, and on $I^{1.0}$ without Aerosil. In view of the equation,

$$R_p = k_p[M]G_iI/J \tag{56}$$

where I is expressed in appropriate units, the sum of the dose rate exponents for corresponding values of R_p and J/k_p should be unity if G_i is taken to be independent of dose rate. Aside from this inconsistency which must presumably arise from uncertainties in the intensity exponents because of experimental scatter, the results point to a simple free ion mechanism of polymerization in the presence of Aerosil. The only role of Aerosil which has been clearly established is that of reducing chain termination (120), an effect which can be readily attributed to the removal of water or other adventitious impurities from the solution. This latter interpretation differs from that of the authors (120), who suggest that positive ions are stabilized on the surface of the additive. This particular conclusion was reached in the earlier paper (119) to account for an apparent increase of G_i in the presence of Aerosil, on the assumption that chain transfer to monomer is negligible. Despite the fact that the more detailed later results (120) show this assumption to have been incorrect, the authors (120) persist with their conclusion, solely on the basis of the kinetic differences observed in the presence of Aerosil. However, in this latter case, the dependence on dose rate conforms to the free ion theory, and the order in monomer concentration cannot be regarded as a crucial test of mechanism. From the value (120) of $J/k_p = 0.6 \times 10^{-3}M$ with Aerosil at 40 krads/hr, and taking J as about 20 sec^{-1} corresponding to a mean lifetime τ_{ss} of 5×10^{-2} sec for free ions at this dose rate (see Table I), we arrive at a crude estimate of $k_p \simeq 3 \times 10^4 \ M^{-1}$ sec^{-1} from the present data. This value is about two orders of magnitude lower than k_p for the radiation-induced ionic polymerization of bulk styrene at room temperature (69). It is possible that a solvent effect could give rise to such a large difference in k_p for ionic polymerization by free ions, but in view of the tentative estimate provided by the above calculation, further speculation is unwarranted. At least it can be said that the magnitude of this derived value for k_p does not contradict our suggestion that a free ion mechanism is adequate to explain the experimental data in the presence of Aerosil (120).

If we accept the above thesis that Aerosil reduces the concentration of terminating impurities in the solution, then the corresponding lower absolute rates and their dependence on the first power of dose rate in the absence of Aerosil are fully explained by a predominant termination with impurity. The suggestion which has been frequently made (120) that an intensity exponent of unity denotes termination by the "original gegenion" is theoretically unsound, for we have seen in Section II-B-1 that spur processes (geminate recombination) can make little contribution to ionic polymerizations under conditions of mild retardation. The only circumstance under which unimolecular termination by the original gegenion could be important comes from the possibility of ion–pair propagation, in which case the oppositely charged ions must preserve sufficient stability to coexist during the period of propagation growth, a situation that is believed to occur in many ionic polymerizations induced by chemical catalysts. Some early ad hoc suggestions (8) were made along these lines for radiation-induced ionic polymerization but these ideas have been discarded (122) in the light of recent progress, and the concept of ion pairs with more than a fleeting existence is of doubtful validity for the vast majority of simple systems in radiation chemistry. It is hardly necessary to add that there is no firm experimental evidence to support ion–pair propagation in radiation polymerization.

Perhaps the most difficult question connected with radiation-induced polymerization in solution concerns the precise mechanism of initiation. In radical polymerization, the initiating species generally include the free radicals derived from both solvent and monomer, but the complication of reactivity transfer affects the absolute and relative yields of these species. The problem is even more acute for ionic polymerization, but some aspects have been clarified in the recent work of Ueno et al. (121). These authors found that the radiation-induced ionic polymerization of styrene occurs in several organic solvents at $-78°C$, including carbon disulfide, ethyl bromide, monofluorotrichloromethane, as well as dichloromethane. These results dispose of the previous suggestion by Sheinker and Abkin (123) that the mechanism of initiation necessarily involves a proton transfer to monomer from the ionized solvent molecule. The possibility of polymerization occurring through the radiation-induced formation of hydrogen chloride was also excluded for the dichloromethane system (121). The G_i values calculated on the basis of complete utilization of the energy absorbed in solution are in the range between 0.1 and 0.2 for styrene concentrations up to $2M$ in several solvents. These results (121) parallel those of Charlesby and Morris (120), and also coincide with the determination (124) of G_i for the free ion polymerization of cyclopentadiene at $-78°C$. From these results, it appears that the chemical nature of the ion is of secondary importance *in these particular systems*, and that physical separation to give free ions is the required characteristic. However, one cannot rule out the possibility that monomer ions are formed preferentially by reactivity transfer from solvent to monomer. Chain transfer from propagating styrene cations to dichloromethane has been shown to be negligible (120). Finally, there appears to be no simple correlation of R_p for the radiation-induced polymerization of styrene with the dielectric constant of the solvent (121). An effect of dielectric constant is commonly observed in catalytic cationic polymerization (125), but it is hardly surprising that this result finds no parallel in radiation studies because the mechanism is entirely different in the two cases. At the same time, it would be wrong to dismiss the effect of dielectric constant altogether in radiation polymerization since this property should affect the probability for physical separation of the ions (Chapter 7). The failure to achieve an appreciable rate of styrene polymerization in nitroethane (121), a solvent of high dielectric constant ($\epsilon = 28$), is probably caused either by ineffective initiation, or by retardation from radiolysis products. Such complexities can all be traced to the fact that the solvent is not a simple diluent in radiation studies, and strong variations in solvent behavior should be expected on this account.

V. POLYMERIZATION OF BULK MONOMERS

This subject divides logically into a classification of monomers according to the particular mechanism which makes the major contribution to the overall polymerization rate.

A free radical mechanism applies to the radiation-induced polymerization of methyl methacrylate and vinyl acetate at or above room temperature. These monomers have been the subject of detailed studies, and the work has been thoroughly reviewed by Chapiro (1). Since the individual rate constants are known for these monomers from other work, it has been possible to derive the radiation chemical yields for the production of free radicals. These radical polymerizations exhibit the kinetic features of equation (29), provided that the rate measurements are confined to low conversion so that the change in the viscosity of the system is small. Styrene undergoes a radical polymerization when the monomer is not dried rigorously before irradiation. Acrylonitrile polymerizes in the liquid state by a free radical mechanism but the polymer is insoluble and simple kinetics do not apply. In this case the G_R value has been determined from molecular weight data according to equation (15).

The main impact of this work has been to demonstrate that the techniques of polymerization can be used to provide information about the radiation chemical yield of radicals derived from pure monomers. The kinetics of vinyl polymerization by radical mechanisms had been developed in exhaustive detail before the advent of systematic radiation studies, so it was natural for the main emphasis in the latter work to become concentrated on the determination of radical yields.

In Table V the radical yields derived by the polymerization method are compared with those obtained independently by the disappearance of

Table V

Radical Yields from Irradiated Monomers

Monomer	G_R	$G(-DPPH)$
Vinyl acetate	12	6–9
Methyl methacrylate	11.5	5.5–6.7
Acrylonitrile	1.2–5.6	5.0
Styrene	0.69	0.66

DPPH in bulk monomer systems; the results are taken from the compilation by Chapiro (1). It is evident that fair agreement exists between the results obtained by the two methods. The G_R value for styrene is consistent with the low G yields which are often associated with the breakdown of aromatic molecules by high energy radiation. For vinyl acetate and methyl methacrylate, the G_R values are considerably higher than for styrene, and are similar to the $G(-m)$ yields for the radiolysis of *nonpolymerizable* compounds having a similar molecular structure.

Since R_p is a function of $k_p/k_t^{1/2}$ and of $R_i^{1/2}$ (proportional to $G_R^{1/2}$) according to equation (29), the absolute values of k_p and k_t are particularly relevant in comparisons of free radical polymerization rates for different monomers. In Table VI we have listed these rate constants for the monomers of interest.

The larger values of $k_p/k_t^{1/2}$ and G_R for both vinyl acetate and methyl methacrylate relative to styrene mean, of course, that the former monomers are more readily polymerized by radiation through the free radical mechanism. Various other monomers such as methyl acrylate and the *n*-alkyl methacrylates are also known to have k_p values (25) in the same range (720–360 M^{-1} sec^{-1} at 30°C) so it follows that these monomers should also

Table VI

Rate Constants for Free Radical Polymerization

Monomer	k_p M^{-1} sec^{-1}	k_t M^{-1} sec^{-1}	Temp., °C	Ref.
Vinyl acetate	1012	5.9×10^7	25	167
Methyl methacrylate	286	2.4×10^7	30	167
Acrylonitrile	1960		60	25
Styrene	44	4.75×10^7	25	167

respond to radiation-induced free-radical polymerization, and this is found to be the case (1). On the other hand, some monomers have distinctly low k_p values in comparison with those we have considered. For example, the corresponding rate constants for p-methoxystyrene and 4-vinyl pyridine are less (25) than k_p for styrene. Unfortunately, even for free radical polymerization, there is no information available on the absolute rate constants for many monomers, and the literature provides only qualitative estimates of polymerizability.

This brief discussion of free radical polymerization is intended to point out the common ground which exists between the radiation work and other studies. It is very satisfying that no single case is known of a vinyl monomer susceptible to conventional free radical polymerization which does not polymerize by high energy radiation under the appropriate conditions of temperature and dose rate. Further, in the few cases for the well known monomers where a quantitative correlation has been sought, the radiation results, as exemplified by the reasonableness of the G_R values, imply at least a semiquantitative agreement with the knowledge available on absolute rate constants. This summary of the position for free radical polymerization provides an introduction into the subject of ionic polymerization by radiation. Unquestionably, it is this latter area of radiation polymerization which has shown the most striking advances in the last decade, and it is timely to review this subject in some detail.

Ionic mechanisms of catalytic polymerization are not so well defined as for the free radical case, and considerable variations are found in rates of polymerization depending on the nature of the catalyst and solvent. Thus, it has been quite impossible to approach radiation-induced ionic polymerization in a manner similar to that of the free radical studies mentioned above. The only relevant information which can be gleaned from the catalytic work is qualitative in character, excepting the very recent determinations of k_p for the anionic polymerization of styrene (126,127) by free ions and ion pairs. Against this background, the radiation method has proved to be uniquely suited for pioneering investigations into the more quantitative aspects of ionic polymerization. This opportune development has been aided by two main factors. First, high energy radiation provides the only means of generating free ions in a medium of low dielectric constant, such as a hydrocarbon monomer; these ions lead a transient existence and their kinetic behavior is described by the simple theory of homogeneous recombination as discussed in Chapter 7. Secondly, the separate contributions of free radical and ionic mechanisms to radiation-induced polymerization are identifiable through the methods of Section II-B-2. The problem is further simplified by the tendency for some monomers to undergo polymerization almost exclusively by one type of mechanism (cf. Table IV). Moreover, it turns out that for quite a few monomers, the absolute rates of radiation-induced cationic polymerization are extremely large, sometimes surpassing the high rates obtained for the free radical polymerization of reactive monomers (e.g., vinyl acetate) under the same conditions of dose rate. Little progress could have been achieved if the ionic rates had turned out generally to be considerably lower than in the free radical studies under comparable conditions. It is remarkable that nature provides this fortunate combination of circumstances to facilitate the study of ionic polymerization by the radiation technique.

At the outset of this Chapter, attention was drawn to the report (32) in 1956 concerning the radiation-induced polymerization of hexamethyl-cyclotrisiloxane in the solid state; this was probably the first real indication that ionic polymerization could be initiated by high energy radiation. The authors of this paper drew attention to the likelihood of an ionic mechanism for this ring-opening polymerization, but the suggestion rested entirely on the well-known fact that this monomer is polymerizable by ions but not by free radicals. Since the polymerization appeared to be confined to the solid state, little supporting evidence could be adduced from detailed kinetic studies. However, there are more recent unpublished reports (128) that the polymerization also occurs in the liquid state under sufficiently anhydrous conditions; such findings lend strong support to the suggested ionic mechanism. While on the subject of ring-opening polymerization, mention should also be made of the important studies on trioxane by Okamura, Hayashi, and their co-workers (129). The behavior of trioxane parallels that of hexamethylcyclotrisiloxane in that polymerization occurs readily and reproducibly just below the melting point of the monomer, but the reaction is erratic in the liquid state even when stringent drying procedures are followed (130).

Following quickly after the study on hexamethylcyclotrisiloxane came a series of papers on the polymerization of isobutene; this work provided the first example of a radiation-induced ionic reaction in the liquid phase. Although studies on isobutene were underway in several different laboratories at about the same time, priority for the announcement belongs to Davison, Pinner, and Worrall (131). These early investigations of iso-butylene polymerization have been reviewed in detail by Pinner (7), and since this work is now largely of historical interest, we shall concentrate on the present position.

A. Isobutene

A major effort has been devoted to studying the effects of solid additives on the polymerization. Interest in this aspect of the work arose from the observation (132) that the yields could be markedly enhanced by in-corporating a fine suspension of inorganic oxides, most notably zinc oxide, in the monomer during irradiation (132). This effect was interpreted as being due to the capture of electrons by the solid, thereby affording the positive center a greater lifetime to undergo propagation. This hetero-geneous mechanism certainly appeared attractive at the time, and the idea was warmly embraced by a number of authors. Collinson, Dainton, and Gillis (133) proposed that in the polymerization of isobutene without additive, submicroscopic flakes of glass or the surface of the vessel could function to provide centers for electron capture. Similarly, Davison et al. (131) attributed the lack of reproducibility in the absence of additive to a surface effect. This reasoning was made all the more plausible by the prevailing status of theory (11) regarding the short recombination time of electrons in liquids. So the notion became firmly entrenched in many quarters that radiation-induced ionic polymerization constitutes an unusual reaction involving a heterogeneous component. While the practical basis for this view has always been insecure, it is remarkable how many ingenious explanations have been devised to preserve the hypothesis in the face of every confrontation with the experimental facts.

The original and most convincing argument for electron trapping by solids in isobutene was the observation (132) that G_i increased from about 0.2 to values ranging up to 2.9 when additive was introduced into the monomer. This calculation of G_i rested solely on equation (15) after correction of the molecular weight for radiation-induced degradation of the polymer, and ignored the possible consequences of chain transfer. In these experiments (132), $G(-m)$ values close to 8×10^2 were reported for the monomer alone, and up to 9.6×10^3 in the presence of 41 wt % zinc oxide. The molecular weight of polyisobutene did not change systematically with additive, and the viscosity-average value remained constant at about 5×10^5 both with and without zinc oxide. Dalton et al. (134) repeated this work using various preparations of zinc oxide which had been preheated to temperatures between 500 and 1200°C, and they found considerably higher $G(-m)$ values up to 3×10^6 at a dose rate of 410 rads/hr, although the molecular weights still remained of the order of 10^6. From these results (134), the use of equation (15) gave a wide spread of G_i values depending on irradiation conditions; the apparent initiation yields exceeded two hundred in some cases, and the authors drew the obvious conclusion that the molecular weight is governed by chain transfer under these conditions. They also concluded that termination processes are suppressed by zinc oxide, since $G(-m)$ in its absence was only 2.1×10^3, a value which roughly agreed with those obtained earlier (131). This observation of efficient chain transfer to monomer in isobutene paralleled the findings previously reported for the radiation-induced polymerization of rigorously dried α-methylstyrene (8) and β-pinene (135) in the absence of additive. Since it is well known that proton transfer processes are characteristic of carbonium ions in polymerization, such results provided further evidence for the ionic nature of these polymerizations. Now this effect of chain transfer for isobutene in the presence of zinc oxide rules out the previous argument (132) that the additive causes increased initiation. Notwithstanding this fact, Dalton et al. (134) retained the concept of electron trapping but modified the idea to include the possibility of initiation through the ionization of isobutene "absorbed" on the surface of the zinc oxide. This suggestion was further discussed in two more recent papers (136,137) which present even higher $G(-m)$ values (up to 10^7 at 390 rads/hr) and molecular weights (up to 6.5×10^6) at $-78°C$. A strong inverse dependence of molecular weight on temperature has been noted (136) corresponding to a *negative* activation energy of 5.7 kcal/mol for the plot of log M_v against $1/T$; this is presumably the difference in activation energy between the addition and proton transfer steps in the reaction of the propagating ion with monomer. Another interesting feature of these later results (136) is the exponent of 0.66 for the dose rate dependence of R_p. Both these latter results are expected, the dose rate exponent implying a strong contribution from homogeneous bimolecular termination between charged species. To explain a linear dependence of R_p and of the molecular weight on the concentration of isobutylene in *n*-hexane containing 3.5 mole/liter of zinc oxide (0.17 m² surface area/g) at $-78°C$, two distinct kinds of propagation reaction have been suggested (137) but the theoretical arguments are unconvincing to say the least, since they only extend the previous set of hypotheses (134,136) surrounding the mechanism in the presence of zinc oxide.

Now the salient phenomenological features of the isobutene polymerization with zinc oxide are very similar to the effects observed by Charlesby and Morris (120) for the polymerization of styrene in the presence of Aerosil. We reviewed this work in Section IV-B, and it is pertinent to recall that there again, the original explanation of electron trapping (132) was modified to that of "positive-ion stabilization" (119), the explanation being in this case that glass powder did not enhance the polymerization rate although it would be expected (by the authors (119)) to capture electrons. There is little to gain in trying to follow the intricacies of these ad hoc explanations concerning the role of solid additives in radiation-induced ionic polymerization, so we shall confine the remainder of our discussion on this question to the established facts and their strict interpretation in accordance with the usual standard of scientific logic.

The features common to the effects of zinc oxide on isobutene and of Aerosil (silica gel) on styrene in dichloromethane can be summarized as follows: (*1*) R_p is enhanced by the additive and the dose rate exponent approaches 0.5; (*2*) the molecular weights are limited by chain transfer. Both observations are accounted for by decreased termination in the presence of additive, and on this one conclusion there is complete agreement. A very simple explanation was proposed by the writer and his colleagues in 1960 to cover this point. As the polymerizations of α-methylstyrene and of β-pinene had been shown to be extremely sensitive to the drying technique and very strongly retarded by water, it was suggested (8) that the effect of the solid additives on isobutylene might be attributable to dehydration. The most effective solids such as zinc oxide, silica gel, and alumina (138) are known to be excellent drying agents.

More recent work on the polymerization of isobutene with and without solid additive appears to support the above suggestion. Stannett et al. (139) noted that despite a wide variation in rates for the polymerization without solid, higher $G(-m)$ values were generally obtained for samples which had been rigorously dried. The average $G(-m)$ of 3.6×10^3 recorded in this work (139) for the well-dried samples is slightly greater than the value of 2.1×10^3 due to Dalton et al. (134), and above the range between 1.54 and 0.8×10^3 mentioned by David, Provoost, and Verduyn (138) for runs without solid additive. It is significant that Stannett et al. (139) obtained one value for $G(-m)$ amounting to 1.66×10^4, and in the same study, the viscosity molecular weights of the polymer formed at $-78°$ were found to lie between 3–10 million, depending on dose. Extrapolation to zero dose so as to correct for radiation-induced degradation gave a value of the order of 100 million. This range of experimental molecular weights corresponds rather closely to the results obtained in the presence of zinc oxide (136) and alumina (138).

In discussing the ionic polymerization of monomers without solid additive, it is well to remember that the term "rigorously dried"means different things to different investigators, and that few objective comparisons are available on the efficacy of drying agents. This general problem has been discussed by Plesch (140). An additional problem in the present case is the necessity for the removal of adsorbed water from the surface of the glass vessel, because even the best drying procedures are nullified unless this precaution is taken before the dry monomer enters the vessel. Bakeout methods at 500°C are strongly recommended for this purpose. In view of these experimental difficulties, it is hardly surprising that there is considerable variation in the published rates of radiation-induced ionic polymerization.

We now turn to the crucial investigation of Kristal'nyi and Medvedev (141) on the mechanism of zinc oxide and alumina effects on isobutylene polymerization. These authors carried out three series of experiments at $-78°$ in which the purified monomer was either (1) irradiated alone, (2) *pretreated* with the solid additive (ZnO or Al_2O_3) and irradiated alone, or (3) irradiated with the solid. Series (2) gave essentially the same polymerization rates as series (3), with $G(-m)$ values from 3.2 to 4.1×10^5 at a dose rate of 1.6×10^4 rads/hr. The absolute magnitude of these results compares very favorably with the data of Dalton (136) and of David et al. (138) for polymerization with additive, especially when normalized for the difference in dose rate according to the dependence of R_p observed by Dalton (136). In addition, series (1) gave $G(-m)$ in the range 0.71–1.20 × 10⁵, values which are considerably higher than those obtained in the previous studies without additive. In a further set of revealing experiments (141), water concentrations of 1.6 × 10⁻³ mole % and above were added to monomer, and the procedure repeated as under conditions (1), (2), and (3) above. In this case, series (1) gave $G(-m)$ yields between 600 and 1100, results which are typical of the conversion yields obtained by the other workers (134,138) in the absence of additive. On the other hand, when the wet monomer was pretreated or irradiated with solid additive as under (2) and (3), the conversion yields increased to $G(-m) \simeq 2 \times 10^5$ for each series, this value being quite similar to the data obtained under the same conditions (2) and (3) for monomer to which water had not been added in the first place. These results enable the following clear-cut conclusions to be drawn. It is established that the addition of water sharply diminishes the yield. Since this conversion yield of $\simeq 1000$ molecules/100 eV does not depend on the *added* water concentration between 1.6 × 10⁻³ and 8 mole %, it is likely that the polymerization rate is retarded only by water *dissolved* in the monomer. The fact that the polymerization rate was the same in series (2) and (3) proves that the accelerating action of the two oxides is due to an interaction (with the monomer) which may take place *in the absence of radiation*. Finally, because contact with ZnO and Al_2O_3 restored the yield of water-saturated monomer to the high value obtained in the first set of experiments without water, it is reasonable to infer that the solids adsorb inhibiting impurities, including water, from the monomer. The evidence from this elegant set of logical experiments amounts to a complete rebuttal of the case for the "heterogeneous" mechanism of isobutylene polymerization in the presence of additives.

A striking feature of the results obtained by Kristal'nyi and Medvedev (141) is the high conversion yield for isobutylene even before pre-treatment with the solid oxides. The authors claim to have used a careful low temperature distillation method to prepare highly purified isobutylene. In a companion paper (142), the reported yields in the presence of various inorganic oxides are actually somewhat lower ($G(-m) = 10^3$–10^4) and these data must refer to another batch of monomer because it is noted (142) that more thoroughly purified isobutylene polymerizes with higher rates (both in the presence of ZnO and without additives) as mentioned above (141). These observations suggest that some inhibiting impurities, other than water, were *not* efficiently removed by the inorganic solids in these particular experiments (142). It is clear that isobutylene must be "very pure" and "very dry" to obtain high polymerization rates.

The general interpretation given above is strongly reinforced by independent work on the polymerization of dry isobutylene without solid additive. Ueno et al. (143) have obtained $G(-m)$ values in the range 0.66–2.5×10^5 at $-78°C$ after drying the monomer over barium oxide preheated to $350°C$ under high vacuum. Again very high molecular weights were obtained ($\bar{D}P_n \simeq 10^5$) at $-78°C$. At $0°C$, the yield ($G(-m) = 1.3 \times 10^5$) was comparable to that at $-78°C$ but the molecular weight was considerably lower ($\bar{D}P_n = 2 \times 10^3$). These temperature effects are identical to those observed in the polymerization with zinc oxide (136), and contradict the results obtained earlier (131) without additive. Finally, recent work (144) in this laboratory has shown that $G(-m)$ values of the order of 10^6–10^8 can be obtained at $-78°$ and $0°C$ when the vapor of sodium–potassium alloy under reflux is used as the drying agent for isobutylene according to the method of Plesch (140). In this case, the monomer was never in contact with any added solids, and this eliminates the very remote possibility that the high yields may be caused by a "carry-over" of fine solid particles into the reaction vessel during distillation. Further work is in progress on this system at the present time to obtain a fuller description of the polymerization kinetics in the absence of additive.

To summarize, it seems altogether clear that the enhancement effect of solid additives on the polymerization rate is largely, if not entirely, attributable to their action in the removal of retarding impurities, including water, from the isobutene. We can do no better than quote from an earlier communication (145): "Since the homogeneous polymerization is very sensitive to impurities, the mere enhancement of the rate by the addition of a solid does not constitute a *special effect* unless it can be clearly shown that the rate of polymerization for the monomer alone is lower despite every attempt at rigorous drying and purification." On the basis of this criterion, we must conclude that no acceptable scientific case has been made out for such an effect; consequently, Charlesby's hypothesis (132) of electron trapping on solids (and its many variants) must be regarded as of dubious significance for radiation-induced ionic polymerization.

B. Cyclopentadiene

Some important general conclusions about radiation-induced ionic polymerization have been reached from a study of this monomer at $-78°C$. Cyclopentadiene is polymerizable by a variety of cationic catalysts but is relatively unreactive to free radicals. Since the radiation-induced polymerization occurs at low temperatures, an ionic mechanism is indicated. The strong retardation of the polymerization by ammonia and amines constitutes compelling evidence for cationic propagation (49,88), and from a quantitative study of this effect, the propagation rate constant has been estimated (124).

Considering that the mechanism is ionic, the polymerization rate for this monomer is remarkably reproducible when standardized techniques of drying and manipulation are employed. Thus Bonin et al. (124) have reported $G(-m) = 2.12 \pm 0.3 \times 10^4$, while Bates (146) obtained $G(-m)$ = $1.78 \pm 0.15 \times 10^4$ and a later value of $2.11 \pm 0.28 \times 10^4$, the pretreatment methods being very similar in each case and including the drying of the monomer by passage through a column of silica gel. An attempt at additional drying on sodium mirrors did not appreciably affect the conversion yield (124), but an interpretation of this finding must be withheld because there is some doubt about the absolute drying efficiency by the sodium mirror technique. The conversion yield was independent of temperature between -78 and $-29°C$ but decreased markedly at $0°C$ and above. Further, samples of cyclopentadiene left standing at room temperature and subsequently irradiated at $-78°C$ gave anomalously low yields. From these results it appears probable that the dimer which is produced by a thermal reaction at the higher temperatures may be acting as a polymerization retarder. Diisobutylene is known to be an effective retarder in isobutylene polymerization (131). The possibility cannot be ruled out that the termination reaction for cyclopentadiene polymerization at $-78°C$ may also come from the reaction of the propagating ion with dimer rather than with water or some other impurity. However, every effort was made to keep the temperature of the cyclopentadiene as low as possible ($-40°C$) during the distillations on the vacuum line, so as to minimize the extent of this spontaneous dimerization. The rate of polymerization depended on the first power (1.0 ± 0.1) of dose rate in this work (124), and this has been interpreted in terms of termination by impurities. Linear plots of conversion versus dose were obtained at each dose rate up to 2% conversion. Even in this range, the entire sample gelled at $-78°C$, and the isolated polymer always contained an insoluble fraction. Consequently, it was impossible to obtain a true molecular weight for the polymer, but measurements on the sol fraction gave a value of 5.8×10^5. The conversion yield per unit dose did not vary according to the size and shape of the irradiation vessel, so it has been inferred that heterogeneous effects are unimportant in this system (124).

The high value of $G(-m)$ and the attainment of good reproducibility by means of a fairly simple experimental technique are features which make this system especially suitable for studying the effects of scavengers. The potency of ammonia as a retarder of cationic polymerization was first demonstrated in this system (49), and this evidence helped to establish the general idea that hydrocarbon ions generated by irradiation of aprotic systems may be recognized by their acidic properties (88). Ammonia is an ideal positive-ion scavenger in organic radiation chemistry not only because of its high proton affinity, but also on account of its lack of reactivity with electrons and free radicals. The utility of ammonia also extends to quantitative work, for it is reasonable to assume that the acid–base reaction between a hydrocarbon ion and ammonia proceeds at a diffusion-controlled rate which can be estimated from simple transport theory. These ideas form the necessary background to the quantitative study of retardation by ammonia. Before taking up this aspect of the work, it is interesting to note that ammonia at $10^{-3}M$ concentration depressed $G(-m)$ to 250, whereas a stable free radical (DPPH) concentration of $5 \times 10^{-3}M$ in cyclopentadiene gave $G(-m) \simeq 8500$ which represents only about a twofold reduction in the yield from "pure" monomer. Collinson et al. (133) reported that DPPH did not affect $G(-m) \simeq 300$ for isobutylene polymerization, but since this value is now known to be much too low for the pure monomer, this observation does not exclude the possibility of some retardation by DPPH at lower concentrations of adventitious impurities. However, the persistence of a substantial yield in the presence of a large DPPH concentration is strong evidence for an ionic polymerization. The ionic character of the cyclopentadiene reaction is also indicated by the less pronounced retarding effects of $2 \times 10^{-3}M$ oxygen and $10^{-3}M$ anthracene in comparison with $10^{-3}M$ ammonia. A slight retardation of a radiation-induced cationic polymerization by added DPPH, oxygen, or anthracene is not entirely unexpected. Specifically, oxygen may readily undergo radiation-induced reactions to give organic oxygen compounds with high proton affinities (48), while DPPH and anthracene can be considered weak bases.

The quantitative interpretation of the kinetic and molecular weight data as a function of ammonia concentration rests on the application of equations (15), (49), and (50). Since the polymer formed in the absence of ammonia is of very high molecular weight ($M_v > 10^6$), chain transfer to cyclopentadiene monomer is infrequent so that equation (15) holds to a good approximation. For ammonia concentrations between 2.3×10^{-4} and $2.5 \times 10^{-3}M$, the quotient $G(-m)/\bar{D}P_n$ leads to $G_i = 0.22 \pm 0.07$, although the value is subject to an overall uncertainty of about a factor of two arising from the molecular weight determinations by viscometry. This yield can be compared to the average $G(\text{RNH}_2)$ of 0.30 ± 0.10 derived by end-group analysis using high concentrations of tritium-labeled ammonia. Despite the lack of precision, these yields can be considered to be in fair agreement with the yield of free ions deduced from the conductivity experiments (20,21) on saturated hydrocarbons. At least it can be inferred that the initiation yield is considerably less than 3, which implies that most ions undergo rapid geminate recombination without forming any appreciable polymer. It will be recalled that somewhat similar G_i values (0.19) were obtained in some of the earliest studies (131) on isobutylene before chain transfer had been recognized in that system. In retrospect, it seems probable that this low G_i value for isobutylene is correct because the calculations refer to $G(-m)$ values (not exceeding 10^3) which imply very efficient termination by adventitious impurities. This has the effect of reducing or eliminating the importance of chain transfer in controlling the molecular weight of the polyisobutylene, and under these conditions, equation (15) is applicable. Hence the approximate agreement in the G_i values for cyclopentadiene and isobutene polymerizations under strongly retarded conditions is expected.

A more detailed analysis of the cyclopentadiene polymerization has been given (88,124) in terms of equations (49) and (50). For strong retardation by ammonia, equation (49) reduces to

$$G(-m) = G_i(k_p[M]/k_r[NH_3]) \tag{57}$$

However, the plot of $G(-m)$ against $[M]/[NH_3]$ does not pass through the origin (88), and this deviation has been explained (124) as a possible consequence of the contribution of some geminate ions to the polymerization yield. Such a contribution is insensitive to low ammonia concentrations (cf. Section II-B-1 and discussion of equation (37)) and comes from those ions which have a sufficient lifetime (ca. 10^{-7} sec) to undergo short chain propagation when k_p is large. The value of the ratio $G_i k_p/k_r$ from the slope of the above plot is 3.1×10^{-3} molecules/100 eV and combining this result with the entirely independent determination of $k_r/k_p = 16.8$ from the Mayo plot (equation (50)), G_i becomes 0.05 ions/100 eV. This latter value (G_x in Ref. 124) could be regarded as the true yield of free ions undergoing homogeneous polymerization whereas the previous results for G_i based on equation (15) and on endgroup analysis do not distinguish between the contribution of free and geminate ions to the overall polymerization.

An estimate of k_r has been made along the lines of the Smoluchowski-Debye theory of diffusion-controlled reactions (26,27) for the ion–dipole reaction between the propagating cation and ammonia. This approach (124) gives $k_r = 9.7 \times 10^9 M^{-1}$ sec^{-1} and from the previous value of the rate constant ratio k_r/k_p it follows that $k_p = 6 \times 10^8\ M^{-1}$ sec^{-1}. Considering all the approximations involved, it is unlikely that the accuracy of this estimate is better than a factor of ten but even so, the large value can be considered to be highly significant for the understanding of cationic polymerization by free ions (124).

C. α-Methylstyrene and Styrene

It is now evident that the early emphasis on low temperatures and the presence of finely divided solids for the radiation-induced ionic polymerization of bulk monomers can be traced in large part to the strictly practical requirement of a low water concentration in the monomer. Thus, at low temperatures such as $-78°C$, ionic polymerization is somewhat easier to detect because of the limited solubility of water in the monomer, whereas at higher temperatures, the contribution of ionic polymerization may be entirely suppressed unless the monomer is rigorously dried.

The first indication of the importance of ionic polymerization in dry monomer systems irradiated at room temperature came from studies on β-pinene (78,147) and α-methylstyrene (8). For both monomers, the rate of polymerization was found to depend critically on the drying technique, and the highest rates were obtained by passage of the monomer through silica gel or alumina. Other common drying agents such as calcium hydride and magnesium perchlorate gave inferior conversion yields. Confirmation that the polymerization yields are affected by the reagent efficiency in drying was provided by the strong retarding effect of deliberately added water. The $G(-m)$ values for water-saturated β-pinene and α-methylstyrene samples were reduced by a factor of 10^3 from the corresponding values obtained with well-dried samples. These clear-cut findings led to the suggestion (8) that latent ionic processes might also be realized in other systems under conditions of rigorous drying, and this proposal has been amply fulfilled in later work on isobutyl vinyl ether (148), styrene (75,76), and cyclohexene oxide (149) at room temperature. Thus it is now well established that radiation-induced ionic polymerization can proceed at room temperature, and that such reactions are particularly sensitive to the presence of water.

There are two aspects of this water effect that merit some general discussion. First, the strong retardation of radiation-induced ionic polymerization stands in marked contrast to the effect of water in catalytic polymerization. In this latter case, traces of water often play an essential part in the formation of the initiating complex, and there are examples (150) where polymerization does not proceed in the complete absence of water. However, there is evidence that even in catalytic polymerization, relatively high concentrations of water reduce the polymerization rate from the optimum value attained at lower concentrations (151), and it is important to emphasize that this particular difference in behavior between the catalytic and radiation polymerization is phenomenological rather than fundamental since there is abundant evidence from many sources for the reactivity of carbonium ions with water (152). This point has not always been clearly understood, and the incongruous suggestion has sometimes been made that an optimum concentration of water might also be necessary in radiation-induced ionic polymerization. However, there is no experimental or theoretical foundation for this analogy, and since it is quite clear that hydrocarbon ions can function as proton donors, the idea of a cocatalyst is superfluous in the radiation work.

A second point that has given some difficulty in the past is the question of how very low concentrations ($< 10^{-3}M$) of water can persist to affect radiation-induced ionic polymerization. Some authors (119) have argued that for a low initial water concentration, the rate of polymerization should increase after *all* the residual water has reacted. Such an argument implies that the reaction products from the termination and charge neutralization processes are themselves inactive in inducing further termination. For proton scavengers such as water and ammonia, this situation is most unlikely to occur because a neutralization reaction of the form,

$$H_3O^+ + Y^- \rightarrow H_2O + HY \tag{58}$$

leads to the regeneration of the original scavenger. In the above reaction, H_3O^+ may be solvated by additional water molecules. By this simple mechanism, a proton scavenger mediates in transferring protons from positive hydrocarbon ions to the negative ions in the system without itself being used up in the process. There is evidence in some cases (77,78) that the efficiency of the proton scavenger is reduced as the irradiation proceeds, but experience suggests that it is wrong to expect that the scavenger can be completely destroyed, and this is probably the main reason for the persistent effect of water. Therefore the effects of adventitious ion scavengers in radiation-induced polymerization are entirely different in scope from the action of free radical inhibitors (153) as the latter are generally consumed during irradiation.

The mechanism of the radiation-induced polymerization of α-methylstyrene has been a controversial subject, and we shall review the various studies in chronological order. In 1960, Hirota et al. (154) reported a number of observations which led them to postulate a radical mechanism. The main evidence was drawn from the effects of solvents and additives. The authors found that the polymerization was sensitized by carbon tetrachloride, retarded by naphthalene, and inhibited by oxygen, nitrobenzene, and p-benzoquinone. To explain the first power dependence of R_p on dose rate, the authors suggested that termination occurred by degradative chain transfer (cf. equation (35)). The conclusion in favor of the radical mechanism was made more palatable by qualitative reports in the literature of α-methylstyrene polymerization by redox initiators (155) and by ultraviolet light, although this monomer is generally considered to be difficult to polymerize by radical initiators. The interpretation given by Hirota et al. (154) was criticized by Bates et al. in the light of the first evidence (8) concerning the effect of water on the radiation-induced polymerization. Bates et al. (8) pointed out that the water effect could not possibly be reconciled with a radical mechanism. In a detailed paper (8), Best, Bates, and Williams also argued that Hirota's assumption of degradation chain transfer would require exceptionally high radical initiation yields to account for the high rates of polymerization observed for the well-dried monomer. On the other hand, the low $\bar{D}P_n$ of the polymer and the high rates of polymerization are satisfied by a cationic mechanism involving regenerative chain transfer through proton transfer reactions, an interpretation (8) which is also consistent with the evidence of the water effect. Brownstein, Bywater, and Worsfold (156) confirmed the great sensitivity of the radiation-induced polymerization to traces of water, but suggested that the mechanism might involve anionic propagation because the polymer molecular weights ($\bar{M}_n$ between 1800 and 5100 depending on conversion) (8) are somewhat higher than those usually attained in conventional cationic polymerization. In 1962, Hirota and Takemura (157) reported $G(-m)$ values between 200 and 600 for runs in which the monomer had been dried over sodium mirrors, as compared to an average value of 8000 obtained by Bates et al. (8) using the silica gel method. This comparison was said (157) to be inconsistent with a water effect, on the dubious argument that sodium should be the better drying agent. The other $G(-m)$ values obtained in this work (157) were even lower (50–80), including the results of some experiments with zinc oxide, and in view of these low yields, the surprising conclusion (157) that a radical mechanism *predominates* is obviously untenable. Baumann and Metz (158) noted that higher amounts of a methanol-soluble polymer fraction ($\bar{D}P_n < 6$) are produced from monomer samples previously equilibrated with atmospheric water, but in this case the rate of conversion is only of the order of 1% per 10^8 rads ($G(-m) \simeq 10$) in agreement with the results of Bates et al. (8). For well-dried samples, it had previously been shown (8) that over 90% of the initial monomer could be recovered as methanol-insoluble polymer from a run carried to near complete conversion. Baumann and Metz (158) considered some alternatives to the "simple water effect" in order to explain the higher efficiency of silica gel relative to other drying agents in enhancing the subsequent polymerization rate, and suggested that silica gel and alumina may remove some unidentified inhibitor from the monomer. Of course, the observation of retardation by water (8) does not exclude the possibility of retardation by other impurities in the monomer. As far as the pretreatment is concerned, the question is whether the conversion yields are normally limited by residual water in the monomer as suggested by Bates et al. (8), or by the presence of other impurities. Further work tends to support the former proposition. Thus Hirota and Katayama (159) obtained $G(-m)$ values between 3.3×10^3 and 3.7×10^4 in the course of fourteen runs using sodium films, the average value of 1.6×10^4 being slightly higher than in the work of Bates et al. (15). These later results (159) by the sodium drying technique are considerably higher than those which were reported previously from the same laboratory (157); further, enhanced yields up to $G(-m) = 7.2 \times 10^4$ at a dose rate of 1.2×10^5 rads/hr were obtained (159) with 2 wt % zinc oxide present, a finding which is also at variance with the previous work (157). More recently, Metz (160) has developed a rigorous technique of sample preparation which leads to extremely high rates of polymerization for both styrene and α-methyl-styrene monomers. This technique employs silica gel drying coupled with a thorough bakeout of all the glassware under extremely high vacuum (10^{-7} torr) before the monomer is admitted into the system through a breakseal. This refinement in procedure has resulted in even higher $G(-m)$ yields with values of 7.8×10^4 and 5.1×10^5 at dose rates of 2.4×10^5 and 6.4×10^3 rads/hr, respectively, this variation of $G(-m)$ with dose rate following a negative half power relation as predicted by the dependence of R_p on $R_i^{1/2}$ in the limiting case of negligible impurity termination (equation (36) where [X] = 0). It is perhaps significant that the highest value of $G(-m)$ obtained with zinc oxide (159) is in reasonable agreement with these latter results (160), and this suggests again (cf. isobutene) that no special role needs to be postulated for zinc oxide beyond one of efficient dehydration. From all these results, it is clear that large $G(-m)$ values can be attained with several different but efficient drying agents (sodium, sodium–potassium alloy, silica gel, and zinc oxide) provided that stringent precautions are taken in the purification of the monomer and in the removal of adsorbed water from the surface of glass vessels.

The preceding results point clearly to an ionic mechanism for the radiation-induced polymerization of dry α-methylstyrene. Since the polymerization of this monomer is known from catalytic work to take place by both anionic and cationic mechanisms, further experiments are needed to discriminate between these propagation mechanisms in the radiation case. There are three independent observations which strongly suggest that the cationic mechanism predominates. First, the strong retardation by low concentrations of ammonia and amines (77) provides good evidence for cationic propagation, whereas these bases should have little or no effect on the anionic process. Secondly, the high efficiency of chain transfer to monomer is unusual for anionic polymerization, but finds a ready explanation in terms of the well-known proton transfer reactions in cationic polymerization. Lastly, the copolymerization experiments of Ueno et al. (75) on α-methylstyrene–styrene mixtures have shown that the reactivity ratios are characteristic of a cationic mechanism in dry systems where high rates are achieved, but the ratios become typical of a radical mechanism in wet systems. It should be mentioned that the results of other copolymerization experiments with methyl methacrylate (157) and with isobutene (159) have been interpreted in terms of radical and anionic mechanisms, but this work is questionable because reactivity ratios were not determined and the only evidence consists of some qualitative comparisons of infrared spectra of polymers prepared under different conditions. The observations of strong retardation by DPPH, nitrobenzene, and naphthalene have also given rise to various suggestions (156,159) of anionic and radical mechanisms but experience shows that this type of evidence is unreliable. For example, it has been found (139) that naphthalene and nitro compounds also retard the radiation-induced cationic polymerizations of isobutene (in bulk) and styrene (in chlorinated solvents) at $-78°C$.

A detailed interpretation of the water effect in α-methylstyrene polymerization was first given in terms of two elementary reactions. It was recognized immediately (8) that termination of the propagating carbonium ion could account for the effect, and that the molecular weight of the polymer need not be drastically affected by a reduction in rate when termination is less frequent than chain transfer (cf. equation (42)). A quantitative study revealed certain apparent irregularities in the simple kinetic mechanism, and to account for this deviation, it was also proposed (8) that water interfered with the initiation step. At the time (1960) it was difficult to understand the real reason for the prolonged lifetime of ions undergoing ionic polymerization, and the writer proposed an ad hoc explanation in terms of ion pairs which could undergo a slow geminate recombination and thus allow propagation to proceed. In other words, it was thought that recombination could be governed by the rate of a relatively slow *chemical* reaction between the ions rather than by the time for diffusive approach under the influence of the coulombic interaction (equation (1)). Further work (15) already described in Sections I and II has confirmed the adequacy of the latter model. Moreover, the *raison d'être* for the first idea has been made redundant by the recognition (122) of the part played by free ions in radiation-induced polymerization. In the light of these developments and also the subsequent realization (122) that the complicated kinetics of the water effect may reflect the association of water in the monomer, the postulate (8) concerning the interaction of water with the initiating species must be abandoned, and the role of water can be ascribed entirely to its great efficiency as a terminating agent.

To elucidate the kinetics of radiation-induced cationic polymerization at room temperature, ammonia is a much more satisfactory reagent than water. The advantage of ammonia lies not only in its specificity for reaction with hydrocarbon cations, but also in its tendency to remain unassociated in organic monomers. Similar arguments apply to the use of amines, and for trimethylamine, linear conversion–dose plots were obtained down to very low amine concentrations (10^{-5} to $10^{-6}M$) from which it is possible to obtain a reliable value for the initial rate (or $G(-m)$) in the presence of added retarder. The effect of trimethylamine retardation is interpretable by simple kinetics, and k_p has been obtained (77) from a plot of $1/G(-m)$ against the amine concentration (equation (49)). The slope of this linear plot corresponds to the quotient $k_r/G_i k_p[M]$ and has the numerical value of $3.7 \times 10^3\ M^{-1}$ (molecules/100 eV)$^{-1}$ for α-methylstyrene. As before (124), k_r is estimated from diffusion theory ($8.4 \times 10^9\ M^{-1}\ \text{sec}^{-1}$), but in this case for α-methylstyrene, an assumed value of G_i (0.1) must be used because the frequency of chain transfer prevents a direct determination from the polymerization data. The value of $k_p = 3 \times 10^6\ M^{-1}\ \text{sec}^{-1}$ derived by this approach (77) is in excellent agreement with an independent value of $4 \times 10^6\ M^{-1}\ \text{sec}^{-1}$ calculated (69) from measurements on absolute rates (R_p) and ionic lifetimes. The principle of this latter method will be described for the polymerization of styrene.

As befits its general importance, styrene is probably the most intensively studied monomer in the field of radiation-induced polymerization. We have already referred to the ionic polymerization in chlorinated solvents at $-78°C$ (Section IV-B) and more briefly to the free radical polymerization in bulk at room temperature. All the early results on styrene polymerization at or near room temperature could be explained quantitatively by the free radical mechanism (1), and the only indications of an ionic mechanism came from work at low temperatures in various solvents, and much more dubiously from observations on solid state polymerization in the presence of additives. This was the accepted position as late as 1961–1962, and in fact it was generally believed that for several bulk monomers, including styrene, ionic polymerization was virtually confined to the solid state by means of a mechanism involving a strong dependence on the geometry of the crystal lattice. In support, results were presented (161) showing a precipitous decrease in the polymerization rate as the monomer passed through the transition to the liquid state. At least for styrene, it is now known (75) that this effect is an artifact due to the presence of water.

Following the discovery of the water effect in α-methylstyrene (8), several investigators attempted to obtain enhanced rates of polymerization for well-dried samples of styrene. Worsfold (162) found that the rates of polymerization of pure styrene after drying attained values at least a factor of ten greater than those quoted by Chapiro (1) as characteristic of the free radical mechanism, and that the rates increased still further when the irradiations were carried out in brass or stainless steel vessels. In 1963, Metz and Johnson (163) reported some preliminary results which also revealed a rate enhancement, after a silica gel treatment, by a factor between 4 and 20. Although they did not comment on the significance of their results, Hirota and Takemura (157) also recorded polymerization rates for styrene greater than the earlier data (164) in the literature under comparable irradiation conditions. Through the use of sodium–potassium alloy, Ueno et al. (75) obtained a further increase in the rates $(G(-m) \simeq$

1.4×10^4) over the liquid range between -11 and $37°C$, and observed no maximum in the rate for dry monomer at the melting point ($-30.6°C$) as had been claimed by Chen (161). By contrast, Ueno et al. (75) found that the rate of polymerization in the crystalline state is unaffected by the presence of water, and considerably lower than for the dry liquid monomer. A high conversion yield ($G(-m) = 4.4 \times 10^4$) was also obtained (75) for monomer irradiated in the presence of zinc oxide. More recent work (75,76) has shown that by devoting careful attention to *all aspects* of the preparative procedure, extremely high rates can be achieved, and under these conditions, the dependence of R_p on dose rate approaches the half power dependence predicted by the free ion mechanism. It is apparent from this historical background that the polymerization behavior of pure styrene on exhaustive drying parallels that of α-methylstyrene in every respect, and the only difference in the kinetics is that styrene undergoes a higher "residual" rate of free radical polymerization in wet systems. It is again clear that barium oxide, sodium–potassium alloy, and silica gel can all be extremely effective drying agents under optimized conditions of sample preparation.

The propagation mechanism for styrene polymerization in a dry system has been discussed in extenso by Ueno et al. (75). Of course, it is obvious that a radical mechanism does not suffice to explain the greatly increased rate on drying, and this effect is logically interpreted as the contribution of an ionic mechanism. This ionic contribution is quenched either by the addition of water to dry monomer or by the presence of adventitious water in a wet system ($[H_2O] \sim 10^{-3}M$) exposed only to normal drying techniques. Further evidence for this ionic polymerization (which exceeds the concomitant radical polymerization by a factor of greater than 100) comes from the relatively small temperature dependence of R_p, but an accurate estimate of the activation energy for propagation is hindered by apparent changes in the effective concentration of residual retarder (probably water) as a function of temperature, even for rate measurements on the same sample. The detailed evidence and accompanying arguments in support of the conclusion that a cationic mechanism predominates for styrene have been presented in considerable detail (75), and this follows along similar lines to the previous discussion on α-methylstyrene. The molecular weight of the polystyrene formed by the ionic mechanism is also limited by chain transfer with $\overline{DP}_n \simeq 500$ compared to a corresponding value below $\simeq 50$ for α-methylstyrene. This difference also accords with the interpretation by the proton transfer mechanism because of the greater stability of the carbonium ion derived from α-methylstyrene relative to that from styrene.

The determination of absolute rate constants for cationic polymerization by catalysts is beset with many difficulties because of the variable character of the ionic intermediates in such cases. It appears that the nature of the ionic species is controlled by the catalyst and solvent, and no clear demarcation between the role of free ions and ion pairs has been achieved in conventional cationic polymerization. Indeed, it has been recently suggested (165) that a kind of "pseudocationic" polymerization can proceed through neutral esters in perchloric acid-monomer systems. In anionic polymerization by the "living polymer" technique, the relative contributions of free ions and ion pairs have been measured for styrene polymerization in tetrahydrofuran (126,127). It is interesting that k_p for free anions in this latter system is of the order of $10^5 \ M^{-1} \sec^{-1}$. The feasibility of using the technique of radiation-induced polymerization to gain information about the reactivity of free ions was first demonstrated for cyclopentadiene and α-methylstyrene by the competitive kinetic method based on the retarding effect of ammonia and amines. Recently, independent determinations of k_p have been made from rate and conductivity measurements during the stationary state set up on continuous irradiation (67,69), and this work will now be summarized.

A general treatment (69) of the kinetics of radiation-induced ionic polymerization leads to expression (36) for the rate of polymerization in the presence of an impurity of concentration [X]. This equation is based on the simplifying assumption that a single rate constant k_t can be used to describe the termination of polymeric ions by charge neutralization. It is convenient to rewrite equation (36) in the form

$$R_p = R_i \tau' k_p [\text{M}] \tag{59}$$

where τ' is defined by

$$\frac{1}{\tau'} = \frac{1}{\tau_{ss}} + \frac{1}{\tau_s} \tag{60}$$

and we have

$$\tau_{ss} = (1/R_i k_t)^{1/2} \tag{61}$$

and

$$\tau_s = 1/k_{tx}[\text{X}] \tag{62}$$

The physical significance of τ' is that it corresponds to the mean lifetime of the propagating ion when the latter is subject to termination by the independent processes of charge recombination and scavenging by impurity. It is evident that the above equation for R_p reduces to equation (29) for no impurity termination ($\tau' = \tau_{ss}$) and to the previous expression,

$$R_p = R_i k_p [\text{M}]/k_{tx}[\text{X}] \tag{63}$$

for strong retardation (cf. equation (48) when $(R_p)_0 \gg R_p$).

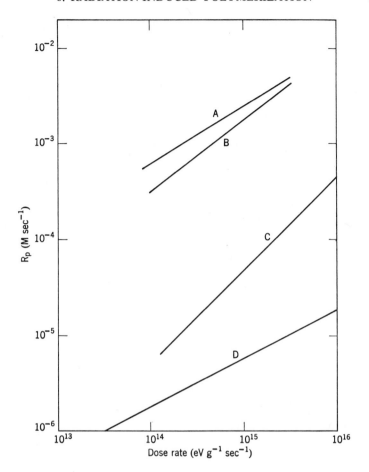

Fig. 7. Radiation-induced polymerization of styrene. R_p as a function of dose rate under different conditions. (A) and (B) refer to different preparations of dry styrene and the dose rate exponents are 0.62 and 0.77, respectively; (C) refers to dry styrene containing $\simeq 10^{-5} M$ $(CH_3)_3N$ for which the dose rate exponent is 0.97; temperature 15°C. (D) refers to wet styrene and the dose rate exponent is 0.5; temperature 20°C. (A), (B), and (C) are taken from K. Ueno, F. Williams, K. Hayashi, and S. Okamura, *Trans. Faraday Soc.*, **63**, 1478 (1967); (D) from A. Chapiro, *Radiation Chemistry of Polymeric Systems*, Interscience, New York, 1962, p. 162.

In the limit of no termination by retarder, the determination of τ_{ss}, R_p, and R_i enables k_p to be calculated from equation (29). By definition, τ_{ss} is the mean lifetime of ions undergoing charge neutralization, and this quantity can be derived from conductivity data according to relation (25). In the present case for styrene (and also for α-methylstyrene), R_i is calculated using $G_i = 0.1$ since this is the most reliable value for the yield of free ions in hydrocarbons. We are assuming here that all cations which achieve separation are capable of initiating polymerization without significant delay $(k_i > k_p)$, as discussed in Section II-B-1. The experimental results (75) for R_p of styrene as a function of dose rate are shown in Fig. 7. In the upper curve corresponding to the highest rates obtained under conditions of rigorous drying, the exponent for the dose rate dependence is 0.62, and therefore higher than the limiting value of 0.5 which would be expected in the complete absence of retarding impurities. The effect of using equation (29) to calculate k_p under conditions of slight retarder termination $(\tau_{ss} > \tau')$ results in a value which is lower than the true value for the rate constant. However, from the above relation for τ_{ss} and τ_s, it is clear that for any given level of retarder concentration [X], the value of τ' approaches τ_{ss} most closely at the highest available dose rate. In other words, k_p will show an apparent increase with dose rate if termination by retarder is at all significant. Accordingly, a limiting value of k_p should be reached at a sufficiently high dose rate. In practice this limit is often difficult to attain because very high dose rates would be required even at low retarder concentrations (*vide infra*), so in this case, the calculated value at the highest observable dose rate necessarily represents the closest approximation. A typical set of calculations for styrene (69) is given in Table VII, from which it can be deduced that the value of k_p is at least 3.5×10^6 M^{-1} sec^{-1}.

Table VII
Radiation-Induced Polymerization of Styrene[a]

I	eV cm^{-3} sec^{-1}	1.0×10^{14}	1.0×10^{15}
R_i	M sec^{-1}	1.66×10^{-10}	1.66×10^{-9}
κ[b]	ohm^{-1} cm^{-1}	1.22×10^{-12}	4.15×10^{-12}
$\tau_{ss} = \epsilon/3.6 \times 10^{12}\pi\kappa$	sec	0.175	0.0514
R_p[c]	M sec^{-1}	6.4×10^{-4}	2.6×10^{-3}
$k_p = R_p/[M]\tau_{ss}R_i$	M^{-1} sec^{-1}	2.5×10^6	3.5×10^6
$k_t = \tau_{ss}^{-2}R_i^{-1}$	M^{-1} sec^{-1}	2.0×10^{11}	2.3×10^{11}

[a] Temperature 15°C; [M] $= 8.75M$; $\epsilon = 2.41$
[b] Ref. 67.
[c] Ref. 75.

Since k_t can be estimated from equation (26), it becomes possible to evaluate the effect of impurity termination on R_p through the calculation of τ' for given values of $k_{tx}[X]$. This procedure is equivalent to the use of equation (36). These calculated curves of R_p as a function of dose rate for various concentrations of [X] are shown in Fig. 8, where we have used the

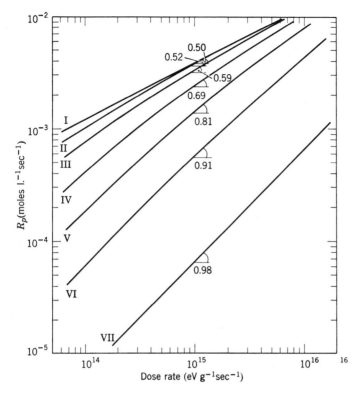

Fig. 8. Calculated dose rate dependence of R_p for styrene as a function of impurity concentration [X] (69). I[X] = 0; II[X] = $10^{-10}M$; III[X] = $3 \times 10^{-10}M$; IV[X] = $10^{-9}M$; V[X] = $3 \times 10^{-9}M$; VI[X] = $10^{-8}M$; VII[X] = $10^{-7}M$. $G_i = 0.1$; $k_t = 2 \times 10^{11}\, M^{-1}\, sec^{-1}$; $k_{tx} = 10^{10}\, M^{-1}\, sec^{-1}$; $k_p = 5 \times 10^6\, M^{-1}\, sec^{-1}$; $R_p \propto I^n$; $R_i = G_i I/100$; $R_p = k_p[M] \dfrac{R_i}{k_{tx}[X] + (R_i k_t)^{1/2}}$.

diffusion-controlled value of $10^{10}\, M^{-1}\, sec^{-1}$ for k_{tx}, and k_p is set at $5 \times 10^6\, M^{-1}\, sec^{-1}$ on the basis of the results in Table VII. The value of k_t ($2 \times 10^{11}\, M^{-1}\, sec^{-1}$) is much higher than for the corresponding recombination of polymeric free radicals (Table VI). This difference is greater than can be accounted for by the r_c and σ terms in the theoretical equations (10b) and (9), respectively, so it must be inferred that one of the ionic species has a diffusion coefficient characteristic of a low molecular weight species, and that presumably it does not propagate efficiently. In terms of information derived from the polymerization study, this description fits the anion. The curves of Fig. 8 show that the dose-rate exponent of R_p increases from 0.5 to 1.0 as the impurity concentration increases from below $10^{-10}M$ to $10^{-7}M$; at the same time the absolute value of R_p

decreases at any given dose rate. At least qualitatively, this is the behavior observed for styrene in Fig. 7, and it can be seen that even under conditions where ionic termination occurs exclusively by reaction with impurity, the rate of ionic polymerization greatly exceeds that of the free radical reaction.

Some comment is called for on the apparent failure to observe a significant contribution due to anionic polymerization of styrene. If the value of k_p for free anions ($\simeq 10^5\ M^{-1}\ \text{sec}^{-1}$) in tetrahydrofuran were relevant to the bulk polymerization, then the comparison of k_p values for cationic and anionic propagation would lead us to expect a contribution of not more than 3% to the overall rate, and this possibility cannot be excluded on the available evidence.

Similar calculations to those in Table VII have been presented (69) for α-methylstyrene and isobutyl vinyl ether. The limiting factor to the accuracy of the derived k_p values lies in the dependence of R_p on dose rate. Ideally the exponent should be 0.5, and some recent data (160) for α-methylstyrene conform to this expectation. For isobutyl vinyl ether (148), the "best" plot (from the highest absolute values of R_p) gives an exponent of 0.62, so the k_p value calculated in this case again must be regarded as a lower limit. A compilation of propagation rate constants is given in Table VIII, from which it is clear that k_p for free ions is much larger than for ion pairs in catalytic polymerization. Another interesting fact is that the measured k_p values for styrene are in the order, free cation > free anion > free radical, which implies that the activation energy for cationic propagation is extremely low, especially since the rate constant for the free cation is almost of the same order as the pre-exponential factor for bimolecular addition processes. In connection with the quantitative aspects of this subject, it is appropriate to remark upon the success of the radiation technique in providing new information hitherto unavailable through the use of more conventional chemical techniques.

D. Other Monomers

The preceding monomers covered in this Section by no means exhaust the possibilities for radiation-induced cationic polymerization. Fairly detailed studies have been made on butadiene (172–174), β-pinene (78), and 1,2-cyclohexene oxide (175). The evidence for the radiation-induced ionic polymerization of butadiene is not entirely satisfactory since it is based (172) on the distribution of unsaturated groups (largely *trans*-1,4 and vinyl) in the polymer. Indeed, it was claimed (172) that since this distribution apparently differs from that obtained by other methods of polymerization, ionic polymerization by irradiation must be so fast (ca. 10^{-13} sec) that thermodynamic equilibrium is not attained. This explanation was made at

Table VIII
Propagation Rate Constants in Addition Polymerization

Monomer	Catalyst	Solvent	Temp., °C	k_p, M^{-1} sec^{-1}	Activation energy, kcal/mole	Ref.
Cyclopentadiene	Radiation Free cation	Bulk	−78°	6×10^8	<2	124
α-Methylstyrene	Radiation Free cation	Bulk	0°	4×10^6		69
	Free cation	Bulk	30°	3×10^6	$\simeq 0$	77
	Anionic ion pair (Na$^+$ counterion)	Tetrahydrofuran (THF)	25°	2.5		166
Styrene	Radiation Free cation	Bulk	15°	3.5×10^6	$\simeq 0$	69
	Free anion	THF	25°	$0.65\text{–}1.3 \times 10^5$		126,127
	Free radical	Bulk	20°	35	7.8	167
	Anionic ion pair (Na$^+$ counterion) Cationic	THF	25°	80		126
	(H$_2$SO$_4$, HClO$_4$)	C$_2$H$_4$Cl$_2$	25°	7.7–17	8.3	168,169
	(SnCl$_4$)	C$_2$H$_4$Cl$_2$	30°	0.41		170
Isobutyl Vinyl Ether	Radiation Free cation	Bulk	30°	3×10^5		69,148
	Cationic (I$_2$)	C$_2$H$_4$Cl$_2$		6.5		171

a time when these processes were not understood, and is untenable in the light of recent progress. The published conversion yields for butadiene (172–174) are rather low (0.21–0.74% per Mrad or $G(-m) = 37$–132), and this result together with the first-order dependence of R_p on dose rate suggests the strong likelihood of impurity termination. Chlorinated solvents are reported (173) to enhance the yield by about a factor of two whereas n-heptane had no effect and naphthalene acted as an inhibitor. The G_i values calculated from equation (15) are from 0.18 to 0.25. It is particularly difficult to reach firm conclusions when the conversion yields are so low ($G(-m)$ less than 100), and similar reservations (7) apply to some early suggestions of ionic polymerization in isoprene and other dienes.

The polymerizations of β-pinene (78) and 1,2-cyclohexene oxide (175) are both strongly affected by water, so there is no doubt that ionic intermediates are involved. A special feature of the β-pinene polymerization (78) is that the addition to the *exo*-cyclic double bond is followed by a ring-opening isomerization which generates the propagating carbonium ion in a position remote from the original addition site. The original paper (78) should be consulted for descriptions of these skeletal rearrangements. Another interesting feature of this polymerization is that a crystalline polymer of high melting point precipitates out during irradiation. Few epoxide polymerizations have been induced by irradiation, so the high conversion yields obtained for 1,2-cyclohexene oxide are particularly notable (175). Attempts to polymerize ethylene oxide and propylene oxide by irradiation have been unsuccessful, while other radiation-induced polymerizations by ring opening (oxetane derivatives, trioxane) appear to occur more readily in the solid state (176). However, these results may be conditioned by the difficulty of removing traces of water and other impurities from the liquid monomers.

It is necessary to end this discussion of cationic polymerization on a cautionary note. It appears that some monomers which are polymerizable by cationic catalysts do not polymerize efficiently by radiation (177). Examples include 3-methylbutene-1 (178) and methyl vinyl ether. This behavior may be due either to inherent causes, as would be the case if k_p were low, or to termination by impurities.

E. Anionic Polymerization

In comparison with the cationic studies described above, much less information is available on the radiation-induced anionic polymerization of bulk monomers. It has already been mentioned that there is no solid evidence for the anionic polymerization of bulk styrene. In only two cases, acrylonitrile (at $-78°C$) and nitroethylene, does there seem to be some experimental support for anionic propagation.

The polymerization of acrylonitrile at $-78°C$ may be regarded as the anionic counterpart to the case of isobutene, but the information is much less detailed and consequently the interpretation leaves much to be desired. Acrylonitrile polymerizes very readily by the radical mechanism at room temperature so the search for an ionic mechanism has been carried out at low temperature (179) following the suggestions of Magat (114). As with isobutene, it has been found (180,181) that certain metal oxides (ZnO, α-Al$_2$O$_3$) increase the rate of polymerization in the liquid phase at $-78°$. However, the highest conversion yields ($G(-m) \sim 2.6 \times 10^3$, or 3.26% per hr at a dose rate of 2.28×10^5 rads/hr) obtained in the presence of additive (180) are considerably lower than for isobutene. In the absence of additive the conversion yields are lower by a factor of 30. The action of the additive has been traced to the requirement of a large surface area and a preliminary degassing temperature of $400°$. Alumina and silica are the most effective solids in promoting the yield under these conditions (180). Abkin and his co-workers (181) claim that there is a correlation between the semi-conductor properties of various additives and their efficiency in enhancing the rates of radiation-induced ionic polymerization. Different mechanisms have been postulated to explain the action of n- and p-type metal oxides, and it is held that polymerization occurs in the chemisorbed layer on the oxide surface (181). As in the case of isobutene, much of this work with additives is open to the serious criticism that little attention has been paid to whether the solid additive is really necessary under conditions of rigorous drying and careful sample preparation. The point should also be made that chain reactions are quite unsuitable for physicochemical studies on the fundamental properties of solids under irradiation.

Nitroethylene is known to be one of the most reactive monomers in anionic catalysis. The bulk polymerization of this monomer (182) proceeds at room temperature with a high conversion yield ($G(-m) \simeq 4.6 \times 10^4$), and the reaction is strongly retarded by low concentrations ($10^{-2}M$ or less) of hydrogen bromide. This effect of acid retardation confirms anionic propagation in this system, and the quantitative behavior fits a kinetic description similar to that given for the cationic polymerization of cyclopentadiene (124). The calculated G_i value of 2.3 (average) for nitroethylene is considerably higher than that obtained for cyclopentadiene and other monomers in systems of lower dielectric constant, so this result is consistent with the other observations concerning the increase in yield of free ions with dielectric constant (183). Assuming diffusion-controlled termination by HBr, an estimate of 4×10^7 M^{-1} sec^{-1} has been obtained (182) for the propagation rate constant. This value is at least two orders of magnitude greater than that quoted for the anionic polymerization of styrene by free ions (Table VIII). It has been shown (184) that nitroethylene captures electrons in 2-methyltetrahydrofuran glasses during gamma irradiation, but the nature of the initiating ion is uncertain. A mass-spectrometric investigation of negative ion–molecule reactions in this monomer has also been reported (184). As already remarked earlier, more information of this type is needed on systems where ionic polymerization has been shown to occur under macroscopic conditions.

VI. OTHER ASPECTS OF RADIATION-INDUCED POLYMERIZATION

This concluding section lists some significant omissions from the scope of our treatment, which has been mainly concerned with the kinetics of radiation-induced polymerization.

A. Copolymerization

Some interesting work is represented by the contributions of Tabata, Oshima, and their colleagues (185) on the copolymerization of tetra-fluoroethylene with ethylene, propylene, etc.

B. Polymerization of Crystalline Monomers

The importance of this field has been stressed in the preceding discussion but a unified treatment is beyond the scope of this article. Essentially the most significant experiments are those based on determinations of the molecular structure of the resulting polymer, and the establishment of the relation between lattice geometry in the monomer and polymer crystals. Some recent papers on the polymerization of trithiane (186,187) exemplify this approach, and attention is directed to the extremely elegant x-ray work of Carazzolo and his colleagues (187). Earlier work on trioxane (129) and the acrylates (5) has been reviewed in detail (33). As Morawetz (see discussion of second paper given in Ref. 129) has noted, our knowledge of the kinetics and mechanism of solid state polymerizations is in a very primitive stage because the usual methods and techniques (Section II-B) used in the liquid state are inapplicable to reactions in the solid state. Another related development (188) is the polymerization of oriented monomers in urea canal complexes.

C. Polymerization in the Glassy State

This subject has been developed extensively by Chapiro and the French school (189). Under certain conditions, a very favorable balance of prop-agation to termination can be maintained in a rigid solvent, resulting in efficient polymerization under *non*-steady state conditions. The kinetics aspects are complicated, but interesting.

D. Oligomerization

We have excluded addition reactions without chain character, but the conclusions derived from the study of ionic condensation reactions in 1-olefins (190) are of particular significance for an understanding of ionic initiation (75).

The writer would like to acknowledge the support of his research by the Atomic Energy Commission under Contract No. AT-(40-1)-2968 at The University of Tennessee, and by the National Science Foundation under Grant GF-216 as a Visiting Scientist to Kyoto University, 1965–1966.

REFERENCES

1. A. Chapiro, *Radiation Chemistry of Polymeric Systems*, Interscience, New York, 1962.
2. E. Collinson and F. S. Dainton, *Discussions Faraday Soc.*, **12**, 212 (1952).
3. A. Chapiro, M. Magat, A. Prévot-Bernas, and J. Sebban, *J. Chim. Phys.*, **52**, 689 (1955).
4. A. Chapiro and M. Magat, *Actions Chimiques et Biologiques des Radiations*, Vol. 3, Masson et Cie., Paris, 1958, p. 63.
5. R. B. Mesrobian, P. Ander, D. S. Ballantine, and G. J. Dienes, *J. Chem. Phys.*, **22**, 565 (1954); A. J. Restaino, R. B. Mesrobian, H. Morawetz, D. S. Ballantine, and D. J. Metz, *J. Am. Chem. Soc.*, **78**, 2939 (1956).
6. H. A. Krässig and V. Stannett, *Fortschr. Hochpolymer-Forsch.* (*Advan. Polymer Sci.*), **4**, 111 (1965).
7. S. H. Pinner, *The Chemistry of Cationic Polymerization*, P. H. Plesch, Ed., Pergamon Press, Oxford, 1963, p. 613.
8. T. H. Bates, J. V. F. Best, and F. Williams, *Nature*, **188**, 469 (1960); *Trans. Faraday Soc.*, **58**, 192 (1962).
9. E. J. Hart, *Ann. Rev. Nucl. Sci.*, **15**, 125 (1965).
10. J. L. Magee, *Ann. Rev. Phys. Chem.*, **12**, 389 (1961); *Discussions Faraday Soc.*, **36**, 232 (1963).
11. A. H. Samuel and J. L. Magee, *J. Chem. Phys.*, **21**, 1080 (1953); A. K. Ganguly and J. L. Magee, *J. Chem. Phys.*, **25**, 129 (1956).
12. A. Kupperman, *The Chemical and Biological Actions of Radiations*, Vol. V, Chap. 3, Academic Press, New York, 1961, p. 85.
13. H. A. Schwarz, *J. Am. Chem. Soc.*, **77**, 4960 (1955).
14. R. M. Noyes, *J. Am. Chem. Soc.*, **77**, 2042 (1955).
15. F. Williams, *J. Am. Chem. Soc.*, **86**, 3954 (1964).
16. (a) J. W. Buchanan and F. Williams, *J. Chem. Phys.*, **44**, 4371 (1966); (b) G. R Freeman, *J. Chem. Phys.*, **46**, 2822 (1967).
17. L. Onsager, *Phys. Rev.*, **54**, 554 (1938).
18. J. B. Gallivan and W. H. Hamill, *J. Chem. Phys.*, **44**, 1279, 2378 (1966).
19. K. Tsuji, H. Yoshida, and K. Hayashi, *J. Chem. Phys.*, **46**, 810 (1967); M. Shirom, R. F. C. Claridge, and J. E. Willard, *J. Chem. Phys.*, **47**, 286 (1967); K. Tsuji and F. Williams, *J. Am. Chem. Soc.*, **89**, 1526 (1967); D. R. Smith and J. J. Pieroni, *Can. J. Chem.*, **45**, 2723 (1967); J. Lin, K. Tsuji, and F. Williams, *J. Am. Chem. Soc.*, **90**, 2766 (1968).

20. A. O. Allen and A. Hummel, *Discussions Faraday Soc.*, **36**, 95 (1963); *J. Chem. Phys.*, **44**, 3426 (1966).
21. G. R. Freeman, *J. Chem. Phys.*, **39**, 988 (1963).
22. F. Williams, *Discussions Faraday Soc.*, **36**, 254 (1963).
23. P. R. Geissler and J. E. Willard, *J. Am. Chem. Soc.*, **84**, 4627 (1962).
24. P. J. Dyne, *Can. J. Chem.*, **43**, 1080 (1965).
25. C. H. Bamford, W. G. Barb, A. D. Jenkins, and P. F. Onyon, *The Kinetics of Vinyl Polymerization by Radical Mechanisms*, Butterworths, London, 1958.
26. M. von Smoluchowski, *Z. Physik. Chem.*, **92**, 129 (1917).
27. P. Debye, *Trans. Electrochem. Soc.*, **82**, 265 (1942).
28. J. C. Bevington, *Radical Polymerization*, Academic Press, New York, 1961.
29. P. H. Plesch, editor, *The Chemistry of Cationic Polymerization*, Pergamon Press, Oxford, 1963.
30. D. C. Pepper, *Quart. Rev.*, **8**, 88 (1954).
31. S. Bywater, *Pure Appl. Chem.*, **4**, 319 (1962).
32. E. J. Lawton, W. T. Grubb, and J. S. Balwit, *J. Polymer Sci.*, **19**, 355 (1956).
33. H. Morawetz, *J. Polymer Sci. C*, **1**, 65 (1963); M. Magat, *Polymer*, **3**, 449 (1962).
34. R. W. Fessenden and R. H. Schuler, *J. Chem. Phys.*, **39**, 2147 (1963).
35. F. S. Dainton, *J. Phys. Coll. Chem.*, **52**, 490 (1948).
36. H. Fischer, *Z. Naturforsch.*, **19a**, 866 (1964).
37. M. Szwarc and J. Smid, *Progress in Reaction Kinetics*, **2**, 219 (1964).
38. J. R. Morton, *Chem. Rev.*, **64**, 453 (1964).
39. J. H. O'Donnell, B. McGarvey, and H. Morawetz, *J. Am. Chem. Soc.*, **86**, 2322 (1964).
40. S. Wexler and R. Marshall, *J. Am. Chem. Soc.*, **86**, 781 (1964).
41. P. Kebarle and A. M. Hogg, *J. Chem. Phys.*, **42**, 668 (1965).
42. P. Kebarle, R. M. Haynes, and S. Searles, *Advan. Chem. Series*, **58**, 210 (1966).
43. J. Dauphin, J. Grosmangin, and J. C. Petit, *Intern. J. Appl. Radiation Isotopes*, **18**, 285, 297 (1967).
44. C. E. Melton and P. S. Rudolph, *J. Chem. Phys.*, **30**, 847 (1959); *ibid.*, **32**, 586 (1960); S. Wexler, A. Lifshitz, and A. Quattrochi, "Ion–Molecule Reactions in the Gas Phase," *Advan. Chem. Series*, P. Ausloos, Ed., **58**, 193 (1966).
45. S. C. Lind, *Radiation Chemistry of Gases*, Reinhold, New York, 1961.
46. V. L. Tal'roze, *Pure Appl. Chem.*, **5**, 455 (1962).
47. V. Aquilanti, A. Galli, A. Giardini-Guidoni, and G. G. Volpi, *Trans. Faraday Soc.*, **63**, 926 (1967).
48. M. S. B. Munson and F. H. Field, *J. Am. Chem. Soc.*, **87**, 4242 (1965).
49. M. A. Bonin, W. R. Busler, and F. Williams, *J. Am. Chem. Soc.*, **84**, 2895 (1962).
50. E. J. Burrell, Jr., *J. Phys. Chem.*, **68**, 3885 (1964).
51. J. Rosenbaum and M. C. R. Symons, *J. Chem. Soc.*, **1961**, 1.
52. G. A. Olah, C. U. Pittman, Jr., R. Waack, and M. Doran, *J. Am. Chem. Soc.*, **88**, 1488 (1966).
53. M. Katayama, M. Hatada, K. Hirota, H. Yamazaki, and Y. Ozawa, *Bull. Chem. Soc. Japan*, **38**, 851 (1965).
54. M. Katayama, *Bull. Chem. Soc. Japan*, **38**, 2208 (1965); *J. Chem. Soc. Japan, Pure Chem. Section* (Jan., 1966).
55. D. J. Metz, R. C. Potter, and J. K. Thomas, *J. Polymer Sci. A*, **5**, 877 (1967).
56. Ch. Schneider and A. J. Swallow, *Proc. 2nd Tihany Symp. Radiation Chemistry*, p. 471 (1966); *Polymer Letters*, **4**, 277 (1966).

57. G. Porter and E. Strachan, *Trans. Faraday Soc.*, **54**, 1595 (1958).
58. T. Shida and W. H. Hamill, *J. Am. Chem. Soc.*, **88**, 3689 (1966).
59. J. P. Guarino and W. H. Hamill, *J. Am. Chem. Soc.*, **86**, 777 (1964).
60. J. Lin, K. Tsuji, and F. Williams, *Trans. Faraday Soc.*, in press.
61. J. P. Keene, E. J. Land, and A. J. Swallow, *J. Am. Chem. Soc.*, **87**, 5284 (1965).
62. C. Geacintov, J. Smid, and M. Szwarc, *J. Am. Chem. Soc.*, **84**, 2508 (1962); see also Ref. 166.
63. S. Bywater and D. J. Worsfold, *Can. J. Chem.*, **44**, 1671 (1966).
64. M. Szwarc, *J. Polymer Sci. C*, **1**, 339 (1963).
65. C. L. Lee, J. Smid, and M. Szwarc, *J. Phys. Chem.*, **66**, 904 (1962).
66. O. Gzowski, *Z. Physik. Chem.*, **1962**, 221, 228.
67. Ka. Hayashi, Y. Yamazawa, T. Takagaki, F. Williams, K. Hayashi, and S. Okamura, *Trans. Faraday Soc.*, **63**, 1489 (1967).
68. L. Onsager, *J. Chem. Phys.*, **2**, 599 (1934).
69. F. Williams, Ka. Hayashi, K. Ueno, K. Hayashi, and S. Okamura, *Trans. Faraday Soc.*, **63**, 1501 (1967).
70. S. W. Benson and A. M. North, *J. Am. Chem. Soc.*, **84**, 935 (1962).
71. C. E. Schildknecht, *Polymer Processes*, Interscience, New York, 1956, p. 203.
72. F. R. Mayo and C. Walling, *Chem. Rev.*, **46**, 191 (1950); T. Alfrey, J. J. Bohrer, and H. Mark, *Copolymerization*, Interscience, New York, 1952.
73. R. B. Cundall in Ref. 29, Chap. 15, p. 549.
74. W. H. Seitzer, R. H. Goeckermann, and A. V. Tobolsky, *J. Am. Chem. Soc.*, **75**, 755 (1953).
75. K. Ueno, K. Hayashi, and S. Okamura, *Polymer Letters*, **3**, 363 (1965); *Polymer*, **7**, 431 (1966); K. Ueno, F. Williams, K. Hayashi, and S. Okamura, *Trans. Faraday Soc.*, **63**, 1478 (1967).
76. R. C. Potter, C. L. Johnson, D. J. Metz, and R. H. Bretton, *J. Polymer Sci.*, **4**, 419, 2295 (1966).
77. E. Hubmann, R. B. Taylor, and F. Williams, *Trans. Faraday Soc.*, **62**, 88 (1966).
78. T. H. Bates, J. V. F. Best, and F. Williams, *J. Chem. Soc.*, **1962**, 1531.
79. A. Charlesby and J. Morris, *J. Polymer Sci. C*, **4**, 1127 (1963).
80. C. S. H. Chen and R. F. Stamm, *J. Polymer Sci.*, **58**, 369 (1962).
81. See, e.g., Ref. 7, p. 647.
82. K. Ueno, K. Hayashi, and S. Okamura, *Polymer*, **7**, 451 (1966).
83. P. J. Dyne, J. Denhartog, and D. R. Smith, *Discussions Faraday Soc.*, **36**, 135 (1963), and related discussion.
84. P. D. Bartlett and T. Funahashi, *J. Am. Chem. Soc.*, **84**, 2596 (1962).
85. G. Meissner and A. Henglein, *Ber. Bunsenges. Physik. Chem.*, **69**, 3 (1965).
86. See S. Freed and K. M. Saucier, *J. Am. Chem. Soc.*, **74**, 1273 (1952); see also Ref. 133.
87. F. R. Mayo, *J. Am. Chem. Soc.*, **65**, 2324 (1943).
88. W. R. Busler, D. H. Martin, and F. Williams, *Discussions Faraday Soc.*, **36**, 102 (1963).
89. G. A. W. Derwish, A. Galli, A. Giardini-Guidoni, and G. G. Volpi, *J. Am. Chem. Soc.*, **87**, 1159 (1965).
90. F. H. Field, *J. Phys. Chem.*, **68**, 1039 (1964).
91. F. H. Field, *J. Am. Chem. Soc.*, **83**, 1523 (1961).
92. See p. 230 of Ref. 1; also J. C. Hayward and R. H. Bretton, *Chem. Eng. Progr.*, **50**, 73 (1954).
93. G. G. Meisels, *J. Chem. Phys.*, **42**, 2328 (1965).

94. G. G. Meisels and T. J. Sworski, *J. Phys. Chem.*, **69**, 2867 (1965).
95. G. G. Meisels, *Amer. Chem. Soc. Div. Polymer Chem.*, *Polymer Preprints*, **5**, No. 2, p. 896 (1964).
96. C. D. Wagner, *J. Phys. Chem.*, **65**, 2276 (1961); *ibid.*, **66**, 1158 (1962).
97. S. Machi, M. Hagiwara, M. Gotoda, and T. Kagiya, *J. Polymer Sci. A*, **3**, 2931 (1965).
98. P. Colombo, L. E. Kukacka, J. Fontana, R. N. Chapman, and M. Steinberg, *Am. Chem. Soc. Div. Polymer Chem.*, *Polymer Preprints*, **5**, No. 2, 914 (1964).
99. W. Mund, C. Guidee, and J. Vanderauwere, *Bull. Acad. Belg. Cl. Sci.*, **41**, 805 (1955); W. Mund and P. Huyskens, *ibid.*, **36**, 610 (1960).
100. H. Okamoto, K. Fueki, and Z. Kuri, *J. Phys. Chem.*, **71**, 3222 (1967).
101. S. O. Thompson and O. A. Schaeffer, *J. Am. Chem. Soc.*, **80**, 553 (1958); *Radiation Res.*, **10**, 671 (1959).
102. F. W. Lampe, J. L. Franklin, and F. H. Field, *Progr. Reaction Kinetics*, **1**, 67 (1961).
103. See Ref. 45, Chap. 9, p. 163.
104. E. W. Schlag and J. J. Sparapany, *J. Am. Chem. Soc.*, **86**, 1875 (1964).
105. J. J. Sparapany, *J. Am. Chem. Soc.*, **88**, 1357 (1966).
106. L. Kevan and N. S. Viswanathan, *J. Am. Chem. Soc.*, **89**, 2482 (1967).
107. See Ref. 1, Chap 7, p. 283.
108. A. O. Allen, *Radiation Chemistry of Water and Aqueous Solutions*, Van Nostrand, New York, 1961.
109. S. Okamura, T. Higashimura, and S. Futami, *Isotopes Radiation (Japan)*, **1**, 216 (1958).
110. Summarized in Ref. 1, p. 250.
111. E. Collinson, F. S. Dainton, and G. S. McNaughton, *Trans. Faraday Soc.*, **53**, 476, 489 (1957).
112. F. S. Dainton and M. Tordoff, *Trans. Faraday Soc.*, **53**, 499 (1957).
113. C. H. Bamford, A. D. Jenkins, and R. Johnston, *Proc. Roy. Soc. (London)*, **A239**, 214 (1957).
114. M. Magat, *Makromol. Chem.*, **35**, 159 (1960); *J. Polymer Sci.*, **48**, 379 (1960).
115. A. Chapiro and Y. Tsuda, *J. Chim. Phys.*, **60**, 59 (1963).
116. A. P. Sheinker, M. A. Yakovleva, E. V. Kristal'nyi, and A. D. Abkin, *Dokl. Akad. Nauk. SSSR*, **124**, 632 (1959).
117. A. Chapiro and V. Stannett, *J. Chim. Phys.*, **56**, 830 (1959); *ibid.*, **57**, 35 (1960).
118. S. Okamura and S. Futami, *Intern. J. Appl. Radiation Isotopes*, **8**, 46 (1960).
119. A. Charlesby and J. Morris, *Proc. Roy. Soc. (London)*, **A273**, 387 (1963).
120. A. Charlesby and J. Morris, *Proc. Roy. Soc. (London)*, **A281**, 392 (1964).
121. K. Ueno, H. Yamaoka, K. Hayashi, and S. Okamura, *Intern. J. Appl. Radiation Isotopes*, **17**, 513 (1966).
122. F. Williams, *Discussions Faraday Soc.*, **36**, 257 (1963).
123. A. P. Sheinker and A. D. Abkin, *Vysokomolekul. Soedin.*, **3**, 716 (1961); A. D. Abkin, A. P. Sheinker, M. K. Yakovleva, and L. P. Mezhirova, *J. Polymer Sci.*, **53**, 39 (1961).
124. M. A. Bonin, W. R. Busler, and F. Williams, *J. Am. Chem. Soc.*, **87**, 199 (1965).
125. D. C. Pepper, *Trans. Faraday Soc.*, **45**, 397 (1949).
126. D. N. Bhattacharyya, C. L. Lee, J. Smid, and M. Szwarc, *Polymer*, **5**, 54 (1964); *J. Phys. Chem.*, **69**, 612 (1965).
127. H. Hostalka and G. V. Schulz, *Polymer Letters*, **3**, 175, 1043 (1965).
128. Private communication from J. A. Ghormley.

129. K. Hayashi and S. Okamura, *Makromol. Chem.*, **47**, 230 (1961); S. Okamura, K. Hayashi, and Y. Kitanishi, *J. Polymer Sci.*, **58**, 925 (1962).
130. K. Ueno, H. Tsukamoto, K. Hayashi, and S. Okamura, *Polymer Letters*, **5**, 395 (1967); private communication from K. Hayashi.
131. W. H. T. Davison, S. H. Pinner, and R. Worrall, *Proc. Roy. Soc. (London)*, **A252**, 187 (1959).
132. R. Worrall and A. Charlesby, *Intern. J. Appl. Radiation Isotopes*, **6**, 8 (1958); R. Worrall and S. H. Pinner, *J. Polymer Sci.*, **34**, 229 (1959); A. Charlesby, S. H. Pinner, and R. Worrall, *Proc. Roy. Soc. (London)*, **A259**, 386 (1960).
133. E. Collinson, F. S. Dainton, and H. A. Gillis, *J. Phys. Chem.*, **63**, 909 (1959).
134. F. L. Dalton, G. Glawitsch, and R. Roberts, *Polymer*, **2**, 419 (1961).
135. T. H. Bates and F. Williams, *Nature*, **187**, 665 (1960).
136. F. L. Dalton, *Polymer*, **6**, 1 (1965).
137. J. A. Bartlett and F. L. Dalton, *Polymer*, **7**, 107 (1966).
138. C. David, F. Provoost, and G. Verduyn, *J. Polymer Sci. C*, **1**, 1135 (1963); *Polymer*, **4**, 391 (1963).
139. V. Stannett, F. C. Bahstetter, J. A. Meyer, and M. Szwarc, *Intern. J. Appl. Radiation Isotopes*, **15**, 747 (1964).
140. P. H. Plesch in Ref. 29, Chap. 18, p. 675.
141. E. V. Kristal'nyi and S. S. Medvedev, *Vysokomolekul. Soedin*, **7**, 1373 (1965); *Polymer Sci. USSR*, **7**, 1523 (1966).
142. E. V. Kristal'nyi and S. S. Medvedev, *Vysokomolekul. Soedin.*, **7**, 1377 (1965); *Polymer Sci. USSR*, **7**, 1527 (1966).
143. K. Ueno, A. Shinkawa, K. Hayashi, and S. Okamura, *Ann. Rept. Japan Assoc. Radiation Res. Polymers*, (1965–1966); *Bull. Chem. Soc. Japan*, **40**, 421 (1967).
144. R. B. Taylor and F. Williams, *J. Am. Chem. Soc.*, **89**, 6359 (1967).
145. F. Williams, *Am. Chem. Soc. Div. Polymer Chem.*, *Polymer Preprints*, **7**, No. 1, 47 (1966).
146. T. H. Bates, *Nature*, **197**, 1101 (1963).
147. Summarized by W. G. Burns, *Proc. 1st Tihany Symp. Radiation Chemistry, 1962*, Akademiai Kiado, Budapest, p. 31 (1964).
148. M. A. Bonin, M. L. Calvert, W. L. Miller, and F. Williams, *Polymer Letters*, **2**, 143 (1964).
149. D. Cordischi, M. Lenzi, and A. Mele, *J. Polymer Sci. A*, **3**, 3421 (1965).
150. A. G. Evans and G. W. Meadows, *Trans. Faraday Soc.*, **46A**, 327 (1950).
151. J. B. Rose, *J. Chem. Soc.*, **1956**, 546.
152. C. K. Ingold, *Structure and Mechanism in Organic Chemistry*, Bell, London, 1953.
153. J. C. Bevington, N. A. Ghanem, and H. W. Melville, *J. Chem. Soc.*, **1955**, 2822.
154. K. Hirota, K. Makino, K. Kuwata, and G. Meshitsuka, *Bull. Chem. Soc. Japan*, **33**, 251 (1960).
155. G. C. Lowry, *J. Polymer Sci.*, **31**, 187 (1958).
156. S. Brownstein, S. Bywater, and D. J. Worsfold, *Makromol. Chem.*, **48**, 127 (1961).
157. K. Hirota and F. Takemura, *Bull. Chem. Soc. Japan*, **35**, 1037 (1962).
158. C. G. Baumann and D. J. Metz, *J. Polymer Sci.*, **62**, S141 (1962).
159. K. Hirota and M. Katayama, *Ann. Rept. Japan Assoc. Radiation Res. Polymers*, **5**, 205 (1963–1964).
160. D. J. Metz, *Advan. Chem. Series*, **66**, 170 (1967).

161. C. S. H. Chen, *J. Polymer Sci.*, **58**, 389 (1962); see also discussion following this paper.
162. Private communication from D. J. Worsfold, February 20, 1962.
163. D. J. Metz and C. L. Johnson, *Am. Chem. Soc. Div. Polymer Chem.*, *Polymer Preprints*, **4** (2) 440 (1963).
164. D. S. Ballantine, A. Glines, D. J. Metz, J. Behr, R. B. Mesrobian and A. J. Restaino, *J. Polymer Sci.*, **19**, 219 (1956).
165. A. Gandini and P. H. Plesch, *Proc. Chem. Soc.*, **1964**, 240; *J. Chem. Soc.*, **1965**, 4765, 4826; *Polymer Letters*, **1965**, 127; see also Ref. 63.
166. D. J. Worsfold and S. Bywater, *Can. J. Chem.*, **36**, 1141 (1958).
167. M. S. Matheson, E. E. Auer, E. B. Bevilacqua, and E. J. Hart, *J. Am. Chem. Soc.*, **73**, 1700 (1951).
168. M. J. Hayes and D. C. Pepper, *Proc. Roy. Soc. (London)*, A **263**, 63 (1961).
169. D. C. Pepper and P. J. Reilly, *J. Polymer Sci.*, **58**, 639 (1962).
170. N. Kanoh, T. Higashimura, and S. Okamura, *Makromol. Chem.*, **56**, 65 (1962).
171. S. Okamura, N. Kanoh, and T. Higashimura, *Makromol. Chem.*, **47**, 35 (1961).
172. W. S. Anderson, *J. Phys. Chem.*, **63**, 765 (1959).
173. R. L. Webb, *Proc. Intern. Symp. Radiation-Induced Polymerization*, Battelle Memorial Institute, A.E.C. Report TID-7643, p. 109 (1962).
174. Y. Tabata, H. Sobue, and E. Oda, *J. Phys. Chem.*, **65**, 1645 (1961).
175. D. Cordischi, A. Mele, and A. Somogyi, *Proc. 2nd Tihany Symp. Radiation Chemistry*, p. 483 (1966); see also Refs. 130 and 149.
176. K. Hayashi and S. Okamura, *Proc. Intern. Symp. Radiation-Induced Polymerization*, A.E.C. Report TID-7643, p. 150 (1962).
177. F. Williams, *Proc. 7th Conf. Radioisotopes, Japan*, Paper C/RC-4, p. 642 (1966).
178. J. P. Kennedy, K. Ueno, K. Hayashi, and S. Okamura, *J. Macromol. Chem.*, **1**, 243 (1966).
179. H. Sobue and Y. Tabata, *J. Polymer Sci.*, **43**, 459 (1960).
180. F. Provoost, *Polymer*, **6**, 515 (1965).
181. L. P. Mezhirova, A. D. Abkin, A. I. Popova, L. P. Tolstoukhova, and A. P. Sheinker, paper B21 presented at 20th Intern. Congr. Pure and Applied Chemistry, Moscow, summarized in A.E.C. Report TID-22360 (1965).
182. H. Yamaoka, F. Williams, and K. Hayashi, *Trans. Faraday Soc.*, **63**, 376 (1967).
183. G. R. Freeman and J. M. Fayadh, *J. Chem. Phys.*, **43**, 86 (1965).
184. K. Tsuji, H. Yamaoka, K. Hayashi, H. Kamiyama, and H. Yoshida, *Polymer Letters*, **4**, 629 (1966); H. Yamaoka et al., *ibid.* **5**, 329 (1967).
185. Y. Tabata, K. Ishigure, K. Oshima, and H. Sobue, *J. Polymer Sci. A*, **2**, 2235 (1964); Y. Tabata, K. Ishigure, and H. Sobue, *ibid.*, **2**, 2243 (1964).
186. J. B. Lando and V. Stannett, *Polymer Letters*, **2**, 375 (1964).
187. G. Carazzolo and M. Mammi, *Polymer Letters*, **2**, 1057 (1964); M. Mammi, G. Carazzolo, and G. Valle, *ibid.*, **3**, 807 (1965).
188. J. F. Brown and D. M. White, *J. Am. Chem. Soc.*, **82**, 5671 (1960).
189. See Y. Amagi and A. Chapiro, *J. Chim. Phys.*, **59**, 537 (1962).
190. P. C. Chang, N. C. Yang, and C. D. Wagner, *J. Am. Chem. Soc.*, **81**, 2050 (1959); C. D. Wagner, *Trans. Faraday Soc.*, **64**, 163 (1968).

Chapter 9

Organic Compounds in the Solid State*

John E. Willard

University of Wisconsin
Madison, Wisconsin

* The preparation of this chapter was supported in part by the United States Atomic Energy Commission under Contract AT(11-1)-32 and by the W. F. Vilas Trust of the University of Wisconsin.

I. INTRODUCTION

A. Phase-Dependent Phenomena in the Radiation Chemistry of Organic Compounds

If samples of the vapor, liquid, and solid states of a compound are exposed for equal time to the same flux of γ radiation, the energy deposited per gram is essentially the same for each. Differences caused by the small shifts in ionization potential and energy levels resulting from stronger intermolecular forces in the condensed phases are negligible. The energy of gamma ray photons is converted to kinetic energy of electrons by the Compton and photoelectric processes, and this energy is dissipated by ionization and excitation of molecules within 10^{-15} sec or less.

In contrast to these primary processes, the subsequent chemical events often depend on the phase. The density of the medium determines the molar concentrations and spatial distribution of the reaction intermediates, thus controlling the relative probabilities of competing chemical, luminescent, and deactivation events. In addition, solid substances may have unique ability to transfer energy or charge over many molecular diameters, and to affect the fate of reaction intermediates by trapping them at points of imperfection in the solid. Temperature differences, often required to allow observations in different phases, may affect both the rate constants for reactions of intermediate species and their equilibrium constants for solvation and complexing. Some examples of phenomena which illustrate these generalizations will be cited here as an introduction to the field.

If a molecule in a gas is ionized by a high energy electron, the positive ion formed has negligible probability of recapturing the electron it loses, because the mean free path of the latter is too long. Consequently, the positive ion may experience many collisions with neutral molecules before neutralization and thus have the opportunity to decompose or undergo an ion–molecule reaction. An identical ionization event in the liquid phase has a high probability of terminating by prompt return of the electron to its "own" positive ion, because the kinetic energy of the electron is reduced to thermal energies at a distance such that the coulomb attraction remains greater than thermal energies (see also Chapters 7 and 9). For this reason the G value for escape of electrons from recombination with the parent cation in liquid hydrocarbons (as determined by electrical conductivity during irradiation) is only 0.1–0.2 (1,2), whereas it is about 3–4 in the gas phase. Although in a solid hydrocarbon at 77°K the electrons must be thermalized at an even shorter distance from the parent positive ions than in the liquid, more electrons escape prompt recombination. G(trapped e^-), observed by infrared absorption in γ irradiated 3-methylpentane glass at 77°K, is 0.8 (3). Structural features of the rigid amorphous medium, which serve as potential wells for trapping the electrons, or induced traps made by solvation of the electron by induced dipoles, must be responsible. These traps are of sufficient depth to compete with the coulomb forces which return nearly all the electrons to their parent positive ions when the irradiated sample is a liquid at 300°K. The polar organic solid, 2-methyltetrahydrofuran, stabilizes electrons even more efficiently, G(trapped e^-) being nearly 3, as observed by both ESR and infrared absorption (4,5).

An effect of the density of condensed phases as compared to a gas is that the ionized and excited molecules and molecular fragments are produced at high local concentrations in tracks and spurs (see Chapter 7), thus increasing the probability of their reaction with each other rather than with the solvent or with scavengers. Increased density of the medium also increases the probability that the complementary partners formed by rupture of a given molecule will be trapped within the same solvent cage and so recombine with each other (6–9) or undergo geminate disproportionation (10,11).

The steady state molar concentration of a reaction intermediate which is homogeneously distributed, and whose concentration is controlled by second order recombination, is proportional to the square root of the density of the medium, assuming constant dose rate and a rate constant which is independent of density. The average lifetime of such intermediates is, therefore, inversely proportional to the square root of the density. Hence the probability that they will react with the solvent or scavengers rather than with each other decreases with increasing density. Such effects may be dramatic when the probability of a reaction in a condensed phase is compared with that of the same reaction in the gas phase (12).

B. Value of Organic Solid State Studies

Radicals, ions, and electrons produced by the irradiation of liquids typically react with each other or with the solvent or solutes in times of microseconds or less. Consequently, they may only be observed, if at all, by very fast measurement using pulse radiolysis techniques (13,14) (see Chapter 7). In the solid state at sufficiently low temperatures, the removal of reaction intermediates can be slowed to rates such that they can be observed over periods of minutes, or even years. Their spectra and reactions may then be studied by common spectrophotometric, ESR, electrical conductivity, and luminescence measurements. At 77°K and below, where many of these investigations are conducted, reaction of thermal radicals with most solvents is precluded by activation energy limitations. A methyl radical in thermal equilibrium with a matrix at 77°K will achieve an energy of 6 kcal mole^{-1} (which is a minimum for abstraction of H) only about once a week.

The characteristics of the formation and decay of the reaction intermediates are revealing as to the properties of the organic solid state matrices used. They indicate, among other features, the presence of traps of different energies and the possibility of trapping electrons close to positive ions. They show major differences between radiation induced chemical processes in the glassy and polycrystalline states of the same compound. They suggest that long-range energy transfer catalyzes radical recombination events.

C. Scope of This Chapter

This chapter reviews the significance, techniques, results, and potentialities of the study of the radiation chemistry of organic solids, exclusive of polymers. Representative references are given, which in turn contain relevant references to work not mentioned because of space limitations.

Discussions covering the literature prior to 1959 on the formation and trapping of free radicals, including radicals formed in organic solids by radiation, are available in a book containing chapters by seventeen authoritative contributors (15). Other relevant reviews include a chapter on *Ionic Processes in γ-Irradiated Solids at −196°* by Hamill (16), *The Radiation Chemistry of Polymeric Systems* by Chapiro (17), the *Physics and Chemistry of the Organic Solid State* edited by Weissberger (18), several papers in *Proceedings of the Second Tihany Symposium on Radiation Chemistry* (19) and a brief early summary of the effects of phase on the products of irradiation of organic systems (20).

The experiments discussed in subsequent sections of this chapter will include examples of production of trapped intermediates and stable products by ionizing radiation, by direct photolysis of solutes, by the reaction of hot hydrogen atoms from the photolysis of HI, and by the reaction of electrons produced by the photoionization of tetramethyl-*p*-phenylenediamine. Methods of observing the intermediates include ESR and optical spectrophotometry, both of which are nondestructive methods; and electrical conductivity and thermoluminescence, which reveal charged species when the condition of the matrix is such that they can migrate and undergo neutralization.

II. METHODOLOGY

A. Experimental Problems Encountered in Solid Phase Radiation Chemistry

Most investigations of the radiation chemistry of organic solids, other than polymers, have been made with low molecular weight compounds. This favors production of relatively few and predominantly simple reaction intermediates and products, increasing the possibility of mechanistic interpretations of general significance. While simplifying the products, the use of low molecular weight compounds complicates the experiments by requiring operation at low temperatures, to hold the samples sufficiently below their melting points to prevent excessive migration of trapped radicals, ions, and electrons. Liquid nitrogen, hydrogen, and helium form convenient coolants for maintaining 77, 20, and 4°K, respectively. Continuously variable temperature control from 77 to 300°K for ESR (21) or optical measurements (22) can be provided by a variable flow rate of nitrogen gas cooled by passing through a coil in liquid nitrogen.

A cryostat in which the sample temperature is controlled by electrical heating in competition with cooling by liquid nitrogen has proved very useful for spectrophotometric measurements between 80 and 300°K (23). For operation at temperatures intermediate between 4, 20, and 77°K commercial equipment is available (24) which provides a cooling capacity of 4 W, produced by gas expansion. A vacuum shield with windows may be made (25) for this unit, such that a sample may be γ-irradiated and then examined on a recording spectrophotometer, while holding the temperature at any desired value in the range. An ESR cavity has also been designed for use with the device (26a).

Temperatures in the range just below 77°K can be obtained by rapid pumping on liquid nitrogen (64°K), or by bubbling helium through liquid nitrogen (26).

At 77°K and below, most cements are too brittle to hold optical windows to a glass or metal cell body. Silicone rubber cements, such as General Electric RTV-112 adhesive sealant (27) have furnished at least a partial solution to this problem.

A detailed discussion of low temperature equipment and techniques for trapped radical investigations has been given by Mauer (28).

An experimental problem peculiar to radiation chemistry is that species produced by irradiation of the walls of the sample container give ESR or optical absorption spectra superimposed on that of the sample. Liquid samples may be irradiated in one end of a tube, and poured to the other for measurement, but this cannot be done for solids which are frozen in position, as is usually the case for the work discussed in this chapter. The problem may be minimized by using sample tubes made from special high purity fused silica (29). For ESR measurements, the problem has some-times been avoided by using small solid pellets of sample formed by allowing drops of the liquid to fall into liquid nitrogen, or by freezing and irradiating a thin cylinder of degassed liquid in a very thin-walled glass tube which is broken off before introduction to a liquid nitrogen dewar in the ESR cavity. Smith has cancelled out the ESR signal due to the irradi-ated sample tube by an identical, but empty, irradiated tube, using a dual cavity technique (30). Another method is to store the signal from an empty irradiated tube in an electronic memory unit and electronically subtract it from that of the tube containing the sample. For optical spectra, the compensation for cell darkening, if necessary, is often accomplished by using neutral density filters or an empty irradiated cell in the reference beam. Radiation effects on the walls of the sample container may be essentially eliminated by incorporating H^3 (tritium) in the sample to serve as a radiation source, thus avoiding the need for external radiation (31,32). The range of the beta rays of tritium is about 1 μ in a condensed medium, making the energy expended in the walls negligible compared to that in the sample. The optical and ESR absorptions induced in fused silica by γ-irradiation have been investigated (33).

B. Types of Solids Investigated

Both pure compounds and solvents containing dilute solutes have been used in studying the radiation chemistry of organic solids. The solute may be chosen to have an ionization energy lower than that of the solvent so that it will immobilize positive charge, or to have a high electron capture cross section. If the extinction coefficient of the anion or cation formed is known the optical absorption spectrum can be used for quantitative evaluation of the amount of charge captured. Solutes known to capture electrons (or positive charge) can be used to determine whether the charge on unidentified reaction intermediates is positive or negative. The solute decreases the yield of intermediates which have the same charge as that acquired by the solute. Such techniques have been used with great effectiveness by Hamill and co-workers (16).

The majority of investigations on frozen organic compounds to date have been made with compounds which form glassy rather than crystalline solids when cooled. Because they are transparent, glasses are better suited to spectrophotometric analysis than polycrystalline samples, which are, at best, translucent. Single crystals of at least some of the compounds of interest are very difficult to grow (e.g., C_2H_5I) and, when prepared, tend to crack when cooled to 77°K (34).

Glasses may have an additional advantage over the crystalline form in that dilute solutes may be less apt to agglomerate during rapid solidification in liquid nitrogen. There is, however, little information on this subject. An illustration of the agglomeration problem is the fact that when well-purified 3-methylpentane containing 10^{-3} mole % iodine is rapidly cooled in liquid nitrogen, a transparent glass is produced. This has an optical absorption peaking at 5200 Å, similar to that in the liquid, but if the sample is then warmed to 100°K, and returned to 77°K within a few minutes, a more intense and complex series of bands appears at lower wavelengths, suggesting agglomeration of the iodine into colloidal scattering centers (35). Other studies have shown that bromine agglomerates in some matrices and not in others (36).

There are known to be dramatic differences between the glassy and polycrystalline forms of several γ-irradiated alkyl halides with respect to both the stable products (37–39) and the trapped intermediates observed by ESR (40). Consequently it is encouraging that it has now proved possible to obtain meaningful optical absorption spectra of γ-irradiated polycrystalline compounds by using thin samples, high intensity light sources and neutral density filters in the reference beam to compensate for the scattering in the analyzing beam (25,41,42).

Glasses which have been used as matrices for low temperature photochemistry and radiation chemistry include (43,44): 3-methylpentane, isopentane, alcohols, 2-methyltetrahydrofuran (MTHF) and various mixtures such as ether–isopentane–alcohol (EPA). Straight chain hydrocarbons crystallize rather than form glasses. A number of alkyl halide glasses have been used in the pure state or with dissolved halogens (37,38,40,45).

EPA, though a popular matrix in early investigations, suffers from the fact (46) that it is difficult to free it of oxidizing impurities which have led to spurious results. Most commonly used at present are 3-methylpentane, as a nonpolar matrix, and 2-methyltetrahydrofuran as a somewhat polar matrix. It is important that these compounds be rigorously purified, especially of CO_2 (47) before use.

The viscosities of many organic glasses potentially useful as trapping matrices have recently been measured as a function of temperature (44). The viscosities of two binary hydrocarbon mixtures have been studied as a function of composition (44b,48). These data are being used to determine whether decay rates of trapped intermediates correlate with viscosity.

III. TRAPPED ELECTRONS

A. Methods of Observation and Broad Conclusions

A most significant aspect of the radiation chemistry of solid organic compounds is the evidence of their ability to trap electrons and positive charges. An ESR spectrum attributed to physically trapped electrons has been reported for each of several γ-irradiated alcohols at 77°K (49–52), methylcellosolve (53), 2-methyltetrahydrofuran (4,22,30), 2-methylpentene-1 (54), other straight chain and branched chain terminal olefins (55), 3-methylpentane (56–59), and for 3-methylpentane in which tetramethyl-p-phenylenediamine has been photoionized (60). In some matrices the trapped electrons have been studied with the aid of their visible-infrared absorption spectra (3,5,22,61), electrical conductivity (53,62), and lumi-

Table I

Yields of Trapped Electrons from γ Irradiation of
Organic Glasses at 77°K

Matrix	Type of measurement		Approx. time for 50% decay
	ESR	Infrared	
3-Methylpentane	~0.5[a]	~0.8[d]	~10 min[f]
MTHF	2.6[b]	3.0[e]	50 hr
2-Methylpentane	0.7[c]		5 hr

[a] Reference 59.
[b] Reference 4.
[c] Reference 54.
[d] Reference 3.
[e] Reference 5.
[f] Very sensitive to small changes in temperature near 77°K.

nescence when neutralized (3,53,63–65). Of these techniques, ESR and infrared have furnished quantitative information on yields (Table I). All four methods, coupled with scavenger and photobleaching techniques, have contributed evidence on trap depths, distance of diffusion before trapping, and proximity of positive and negative charges to each other. The results of these investigations provide considerable evidence that significant fractions of the electrons ejected by ionizing radiation in the solid organic glasses studied escape from their parent positive ion and diffuse hundreds or thousands of angstroms before neutralization or trapping. When trapped, 50% or more are sufficiently close to a positive ion so that the probability of combining with that particular ion is unity, independent, over a wide range, of the number of other ions present in the system.

The fact that the G values of trapped electrons in organic glasses (Table I) are higher than $G(e^-)$ for electrons observable by electrical conductivity in liquid hydrocarbons (as discussed in Section I-A) indicates that one or all of the following factors must operate in the glasses to lower the probability of prompt geminate recombination: (*1*) electrons at energies below the ionization potential of the solvent must have a lower cross section for energy loss per collision in the solid matrix than in the liquid because there are fewer vibrational and rotational states to which energy can be transferred; (*2*) the depth of traps, resulting either from self-trapping by induced dipoles, or from imperfections in the matrix structure, must be great enough to compete with the coulomb force of the parent ion; (*3*) as the concentration of trapped charges increases with increasing dose the presence of competing coulomb fields weakens the directional effect on trapped electrons.

In the absence of effects such as noted in the last paragraph, most electrons ejected during irradiation must recombine with the parent ion after being slowed to thermal energy. For typical organic glasses at 77°K, with a high frequency dielectric constant of about 2, the separation between positive and negative charge required to reduce the energy of coulomb attraction to the thermal energy, kT, is about 1000 Å, while the distance required to slow an electron from a typical initial energy of 10 eV to energy kT by inelastic collisions may be estimated to be about 20 Å (66). The question of geminate recombination of electrons vs. escape has been considered by a number of investigators dealing with a number of media (e.g., 1,2,61,66–71).

It is significant to note that in a sample which has received a typical dose of 10^{19} eV g^{-1}, the average separation of trapped positive charges, assuming $G(+) = 1$, is about 40 molecules (200 Å). Thus an electron in the medium will always be within 100 Å or less of a positive ion.

B. Spatial Distribution of Trapped Electrons from γ Irradiation

1. Evidence from Decay Rates and Photobleaching. Extensive evidence indicates that 50% or more of the electrons trapped in γ-irradiated 3-methylpentane or 2-methyltetrahydrofuran (MTHF) glass at 77°K are in sufficiently close proximity to a positive ion that when they are released from their traps either thermally or by absorption of light they are un-equivocally committed to combining with that particular partner. Convincing proof is that the initial decay rates (as shown by infrared absorption for 3-methylpentane (61) and MTHF (70), and by ESR for MTHF (72)) are directly proportional to the dose and not to the square of the dose. When normalized for the dose, the decay curves are superimposable for the first 50% of decay or more (Figs. 1 and 2). Thus the decay of each electron is by a process which is independent of the number of positive charges in the system. The curves of Figs. 1 and 2 are not exponential as required for a single type of first order decay, indicating that there must be many configurations of ion pairs (space-wise or solvation-wise) with varying probabilities per unit time of recombination. The same situation seems to prevail in γ-irradiated 3-methylpentane glass containing 2 mole % 2-methylpentene-1, since the e^- decay curves for samples varying by 10-fold in initial e^- concentration are superimposable (73b), although it was initially thought that this decay was second order (16,61).

The initial quantum yields (electrons removed per photon absorbed) for photobleaching of the infrared spectrum of trapped electrons in both 3-methylpentane glass (3) and 2-methyltetrahydrofuran glass (70) are independent of γ dose over the ranges tested. They decrease as the fraction bleached increases, the photobleaching curves being superimposable for

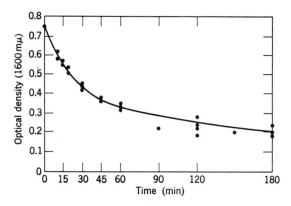

Fig. 1. Spontaneous (dark) decay of trapped electrons in γ-irradiated 3-methyl pentane at −196°C. The data have been normalized by dividing the observed optical densities (OD) by the irradiation time in minutes (4×10^{16} eV ml^{-1} sec^{-1}) (61).

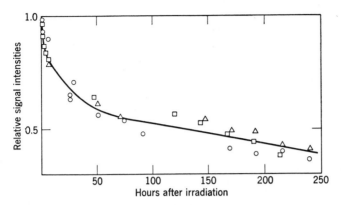

Fig. 2. Thermal decay of trapped electrons in methyltetrahydrofuran following irradiation at 77°K, as measured by the ESR signal. The signal intensities for three γ-doses have been normalized to unity for the initial measurement. $\bigcirc$, 9.8×10^{18} eV ml^{-1}; $\triangle$, 1.7×10^{19} eV ml^{-1}; $\square$, 3.4×10^{19} eV ml^{-1} (72).

50% or more of the bleaching, after normalizing for dose. The fact that the quantum yields are independent of the initial concentration of positive ions seems to require, as in the case of thermal decay, that each electron is predestined to combine with a certain positive ion, rather than randomly. If the fate of an electron following detrapping by a photon were determined by competition between traps and randomly distributed positive ions, the quantum yield for neutralization would increase with positive ion concentration.

The decrease in quantum yield with fraction bleached, indicates that there is a range of probabilities that photon absorption will cause neutralization of an ion pair.

It is reported (16) that 0.02 mole % of an electron scavenger such as biphenyl in γ-irradiated 3-methylpentane glass raises the quantum yield of bleaching to unity and maintains it higher than 0.75 until all the electrons are removed. This would be readily explained if the electrons travel thousands of angstroms through the matrix after photodetrapping before capture or neutralization. This explanation is inconsistent with the evidence that each electron is trapped close to a positive ion with which it is destined to combine. The inconsistency would be resolved if biphenyl can immobilize positive charge thus preventing positive charge from migrating to the vicinity of trapped electrons. The concentration of biphenyl is, however, surprisingly low for such an effect. The question warrants further investigation.

2. Evidence from ESR. Analysis of the situation characteristics of the ESR spectrum of the trapped electron indicates that the average distance of such electrons from neighboring ions or radicals containing unpaired electrons is about 50 Å (72). Reasoning from the photobleaching experiments, Dyne and Miller (70) suggest that a considerable fraction of the electrons may be closer than this, with another fraction substantially further away.

3. Evidence from Electrical Conductivity. A quite different type of evidence which also suggests that a major fraction of the trapped electrons in γ-irradiated 3-methylpentane glass at 77°K are in close proximity to positive ions, comes from observation of the electrical conductivity during warmup (62). The charge movement revealed by peaks of current flow, as electrons are thermally detrapped during warmup, and the effects of low concentrations of electron scavengers on these peaks, suggest that about 10^{-4} of the electrons travel of the order of 10^4 Å in the direction of the field, while most of the remainder undergo neutralization by combination with a nearby positive ion without significant movement in the direction of the applied field.

4. Evidence that Combining Ion Pairs Are Not Geminate Pairs. The evidence that a trapped electron present after gamma irradiation of 3-methylpentane or MTHF glass is destined to combine with a particular positive ion rather than undergoing random recombination does not require that it combine with its geminate partner or even with an ion from the parent spur. Indeed, such combination appears to be an improbable event, since the presence of as little as 0.1 mole % biphenyl during irradiation in 3-methylpentane (61) or 2-methyltetrahydrofuran (5,70) is sufficient to capture all of the electrons responsible for the infrared spectrum of the electron.

5. A Working Model. These observations suggest the following model as a working hypothesis: In the absence of charge scavengers an electron ejected in 3-methylpentane travels, on the average, thousands of molecular diameters, until it eventually becomes trapped in a cage of self-induced dipoles probably in a void in the matrix. At the same time the positive charge migrates either by electron transfer or proton transfer, until it is trapped. Migration of the positive charge away from its point of origin may be assisted by the coulombic repulsion of other positive ions in close proximity in the same spur. Because of the favorable electrostatic field, the probability is high that the positive ion in its cage of induced dipoles will be localized adjacent to a trapped electron, though usually not its "own." The ion pairs so situated will undergo self-"annihilation" on standing, independent of the concentration of other ions in the system. The rate of annihilation may be increased by warming, or by absorption of light, because activation energy is required for reorganization of the induced solvent cages. The decay curves for trapped electrons do not give a linear log $[e^-]$ vs. t first order plot because the thickness and molecular orientation of the polarized solvent layer separating the charges is different for different ion pairs. These differences result in a series of different first order rate constants.

Not *all* of the electrons which are trapped in γ-irradiated 3-methylpentane containing biphenyl are constrained to combine with a predestined positive ion. This is demonstrated by an experiment with 10^{-2} mole % biphenyl, which is insufficient to capture all of the electrons during irradiation (61). Following irradiation, the concentration of trapped electrons (as indicated by the optical density at 16,000 Å) decayed and that of biphenylide ion (observed at 4080 Å) increased. The data indicate that about 25% of the electrons which were trapped in the matrix following irradiation decayed by combination with 10^{-2} mole % biphenyl rather than with positive ions. This represents about 16% of the sum of the electrons captured by biphenyl during irradiation plus those present as trapped e^- following irradiation. Gallivan and Hamill (63) consider that the scatter of the points representing long decay times in Fig. 1 indicates a dose dependence other than first order, which is evidence for a portion of the electrons decaying by random encounters.

C. Spatial Distribution of Electrons from the Photoionization of Tetramethyl-*p*-Phenylenediamine (TMPD)

In the study of electrons produced by photoionization of tetramethyl-*p*-phenylenediamine in 3-methylpentane matrices, experiments based on the photoselectivity of polarized light (74) have produced evidence that the electrons recombine with their parent positive TMPD ions when detrapped by infrared illumination, whereas the effects of low concentrations of electron scavengers lead to the conclusion that the electrons travel thousands of angstroms (75).

1. Polarized Light Experiments. McClain and Albrecht (74) produced a population of oriented TMPD cations by a 10 min illumination of 0.02 mole % TMPD in 3-methylpentane glass with vertically polarized light at 3180 Å. Bleaching of the trapped electrons with infrared radiation produced stimulated recombination emission which was *polarized*. When this emission was complete, the matrix still retained about 35% of the cation population, the remaining electrons being too strongly trapped (75, 76) to be released by infrared radiation. The sample was then given a 1 sec irradiation with horizontally polarized light, producing a concentration of the "horizontal" population of TMPD$^+$ about 0.5% of that of the remaining "vertical" population. Bleaching with infrared then produced luminescence due to the detrapping of the new electrons. It was reasoned that if these combined randomly with the cations some 99.5% would combine with the "vertically polarized" ions, and vertically polarized luminescence would be observed. However, less than 5% (possibly none) of the luminescence resulted from reaction of the detrapped electrons with the "vertical" cations. The authors interpret this result to mean either that each electron is predestined to undergo recombination with its own parent ion, or that the positive ions remaining in the matrix after exhaustive infrared bleaching do not form attractive sites for neutralization of mobile electrons in the matrix. The latter might be the case if they are closely paired with the remaining electrons. It is known that tetramethyl-p-phenylenediamine is an electron scavenger and it may be that the electrons equivalent to the 35% of the parent ions which are not removed by infrared bleaching undergo attachment to tetramethyl-p-phenylenediamine molecules. If this is the case the vertically oriented cations should have been available to the electrons produced by the horizontally polarized light, if these underwent random migration rather than geminate recombination when detrapped by infrared illumination.

2. Scavenger Experiments. The fact that 2×10^{-2} mole % biphenyl completely prevents recombination of e^- with parent cations (either spontaneously at 77°K or by photobleaching) after photoionization of 2×10^{-3} mole % TMPD in 3-methylpentane (75) indicates that the electrons migrate thousands of molecular diameters before combination with a positive ion. This seems to indicate that the polarized photoselection experiments cannot be taken as establishing geminate recombination, and that the 35% of the parent ions which are not bleached by infrared must be so situated, that they have a very low cross section for capture of "new" electrons.

The electrical conductivity of samples of photoionized TMPD in 3-methylpentane glass (62) measured during warm-up does not show either of the peaks (I and II of Fig. 3) attributed to electron migration in γ-

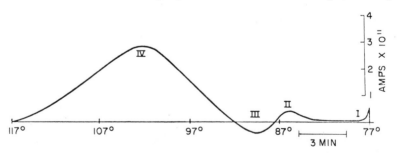

Fig. 3. Current flow as a function of temperature during warmup of 3-methylpentane with applied potential following γ irradiation at 77°K (62).

irradiated pure 3-methylpentane. This indicates that none of the electrons produced by photoionization are as free to move under the influence of an electric field as those responsible for peaks I and II from γ-irradiated samples. However, the electrons released from TMPD give an infrared absorption spectrum, spontaneous decay rate, decay of delayed luminescence (61), and saturation and photobleaching of their ESR signal (60) indistinguishable from electrons formed by γ irradiation of 3-methylpentane.

D. Evidence for Different Types of Trapping Sites

1. General. Curves of the type shown in Figs. 1 and 2 suggest the existence of a variety of configurations for trapping of electrons and positive ions, each with a different first order rate constant for intrapair combination. These configurations may be assumed to differ with respect to the distance of separation of the charges and the organization of the intercharge solvation layers. The infrared spectrum for electrons in the different types of sites does not, however, distinguish between them. There is no dramatic difference in the shape of the spectrum before and after extensive decay or bleaching in either 3-methylpentane (3) or 2-MTHF (70).

As noted below, however, electrical conductivity and luminescence studies on γ-irradiated 3-methylpentane glass indicate the presence of three or more distinctly different populations of trapped electrons.

2. Electrical Conductivity. When the electrical conductivity of γ-irradiated 3-methylpentane glass is monitored as the glass is allowed to stand at 77°K and then warmed, current flow is observed at 77, ~85°K, ~89, and ~105°K (Fig. 3) (62). The current flow at 89°K is in the opposite direction from the currents observed at 77, 85, and 105°K. These currents are attributed to detrapping of two populations of physically trapped electrons (77 and 85°K), thermal randomization at ~89°K of ion pairs oriented when the field is applied at 77°K, and movement of molecular or radical ions (105°K).

Current measured at 77°K immediately after γ irradiation of 3-methylpentane glass (corrected for the polarization current observed in unirradiated samples) indicates the presence of a species which decays almost completely on standing for fifteen minutes following γ irradiation. It is reduced by about 50% by the presence of as little as 10^{-2} mole % of the electron scavenger CCl_4, and completely by 1 mole %, indicating that the charge carriers must have average trajectories of 10^4 molecular diameters, or more. 2-Methylpentene-1 in 3-methylpentane glass at a concentration of 2 mole % is known to increase $G(e^-)$ as observed by infrared absorption, but to slow down the decay of electrons by immobilizing positive charge. This additive does not alter the yield and decay rate of the electron population indicated by peak I of Fig. 3, suggesting that the current may be due to the movement of both positive charge and electrons. According to this interpretation the absence of effect from 2-methylpentene-1 results then from the compensating effects of enhancing the electron population and immobilizing the positive charge.

In discussing the electrical conductivity results, it is important to note that the relative amount of charge collected is not a measure of the relative number of electrons in the different traps. The measured currents are proportional to the number of charges which move under the influence of the field times the distance which they move. Up to 8×10^4 V cm^{-1}, the highest potentials used, the areas of the peaks are proportional to the voltage.

Peak II of Fig. 3, representing the current induced during warm-up at 85°K shown on an enlarged scale in Fig. 4, is eliminated by low concentrations of electron scavengers, is doubled in area by the presence of 2 mole % of 2-methylpentene-1, and is eliminated by infrared illumination at 77°K prior to warmup. For these reasons peak II is attributed to trapped electrons. Infrared irradiation of γ-irradiated 3-methylpentane glass does not produce current when 0.2 mole % biphenyl is present to capture electrons but current flows when electrons are released from the biphenyl by 4100 Å radiation. These electrons (in contrast to those released by the

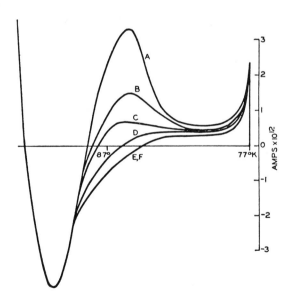

Fig. 4. Effect of CH_3I on the "85°K" peak of the warmup current from γ-irradiated 3-methylpentane. Mole % CH_3I: (A) 0; (B), 0.02; (C), 0.05; (D), 0.2; (E), 1.0; (F), 1.3. Dose, 6.2×10^{18} eV g^{-1}; field, 4×10^3 V cm^{-1} (62).

γ radiation) are at least partially trapped in the matrix without being captured by biphenyl, as indicated by the current peak induced by infrared irradiation subsequent to the 4100 Å radiation.

The current flow attributable to molecular ions, peak IV of Fig. 3, is unaffected by the presence of methyl halides or CCl_4 as scavengers, but shows additional peaks on the high temperature side when biphenyl or TMPD are present, suggesting that biphenylide and $TMPD^-$ do not achieve significant mobility as the matrix warms until it reaches a higher temperature than that required for the smaller ions.

The fact that 3-methylpentane glass containing photoionized TMPD does not produce current flow either at 77 or 85°K when warmed suggests that the trapped electrons are so close to the parent ion that as the matrix relaxes they undergo neutralization without migration under the influence of the field. However, current spikes are induced when the electrons are detrapped by infrared radiation (77) suggesting that the latter gives sufficient energy for escape from the parent partner, while thermal detrapping does not. At present such a conclusion is speculative, since it is not known what fraction of the electrons contribute to the photoconductivity spikes. If this is small, it is possible that the warmup conductivity peak produced by the same electron would be too broad to detect.

A dramatic difference between the large positive peak at about 105°K (IV of Fig. 3) observed following photoionization of TMPD in 3-methylpentane glass, and the corresponding peak from γ-irradiated pure 3-methylpentane, is that the former is almost completely removed by the presence of 1 mole % CH_3I or CCl_4 whereas the latter is unaffected. This has been tentatively attributed to complexing of the TMPD by the alkyl halide, with resultant trapping of the photoionized electrons (by capture by the alkyl halide) (62) in such close proximity to the positive ion that very little movement of the ions can occur in the direction of the applied field before they neutralize each other.

Attempts to obtain information on charge trapping in γ-irradiated MTHF by electrical conductivity, similar to that obtained in 3-methylpentane, are complicated by the relatively high conductivity of the unirradiated compound and the nature of the polarization and polarization relaxation peaks observed on warmup (53,78).

3. Luminescence. Three or more types of electron traps in pure γ-irradiated 3-methylpentane glass at 77°K are indicated by luminescence (64).

(*1*) Spontaneous emission of light immediately after a 5 min irradiation falls to 50% of its initial intensity in about 2 min, 25% in 6 min, and 12% in 20 min suggesting a population of traps of similar type but with a spectrum of rate constants for neutralization of the electrons. These electrons are rapidly bleached by exposure to infrared radiation. (2) If the sample is exposed to infrared illumination at 77°K, 50 min after γ irradiation, by which time the "isothermal" luminescence has essentially all decayed, a sharp increase in luminescence occurs which decays very rapidly at first and then more slowly during continuing illumination for 50 min. (3) On warming a sample which has been bleached with infrared radiation at 77°K, luminescence is observed, with a major peak at about 84°K, and a shoulder at about 93°K. All of the luminescence peaks are absent when a low concentration of CCl_4 is present in the 3 MP matrix during irradiation. The total luminescence spectrum extends from 3500 Å to above 4700 Å with the maximum at about 4250 Å. It was initially considered (63) that the luminescence from γ-irradiated 3-methylpentane glass must be due to neutralization of fragment ions or ions of impurities. It was presumed that addition of an electron to the 3-methylpentane ion would produce bond rupture rather than photon emission. In a more recent paper (79), it is proposed that the state from which luminescence occurs in the glass consists of cation-electron pairs, stabilized by interaction with the matrix molecules to give a state of the system lower in energy than the first excited state of a 3-methylpentane molecule.

It has also been shown (79) that γ-irradiated 2-methylpentane glass has nearly the same luminescence spectrum, peaking at about 4250 Å, as 3-methylpentane. It is a point of interest that whereas the spectrum of isopentane glass peaks at about 4000 Å, that of polycrystalline isopentane peaks at 4900 Å. It has also been found (53,64,79) that the luminescence of γ-irradiated glasses is changed by annealing the glass at 77°K prior to irradiation. Preirradiation annealing does not affect the yields and decay rates of trapped electrons as observed by infrared (61) or electrical conductivity (62).

Luminescence of γ-irradiated 3-methylpentane glass occurs as the result of prompt neutralization processes *during* γ irradiation as well as during the decay of trapped electrons following irradiation (80). The total luminescence during a short irradiation is of the same order of magnitude as that following the irradiation, in plausible agreement with the fact that $G(e^-$ trapped) is about 1 while G(ionization) is presumably about 3. The experiments on luminescence during irradiation were made with x-rays (80), of energy below the threshold energy for producing electrons capable of emitting Cerenkov radiation in 3-methylpentane glass.

The trapped electrons produced by photoionization of TMPD in 3-methylpentane glass, like those produced by γ irradiation of 3-methylpentane, give evidence for at least three types of traps, one empties spontaneously at 77°K, and one requires ultraviolet radiation (76). The kinetics of trap emptying have been followed by observing the recombination luminescence radiation. This is characteristic of the known singlet and triplet states of TMPD. These must be formed in the neutralization step with a certain branching ratio. The rate of disappearance of the parent cation correlates with the decay of stimulated emission. The implications of these results are considered in terms of a model requiring correlated charge pairs (81). An earlier discussion of the thermoluminescence of γ-irradiated alcohol and hydrocarbon glasses, with a literature summary has been given by Magat and co-workers (82a).

4. Correlation Among Experimental Observations. The electrical conductivity, infrared absorption, and isothermal luminescence of γ-irradiated 3-methylpentane glass are all reduced by the presence of electron scavengers. Infrared illumination at 77°K produces a burst of current and of luminescence, following which both the infrared absorption and 85°K electrical conductivity peak are absent. The conductivity peak at 85°K and the infrared absorption are increased by the presence of 2-methylpentene-1, which serves to immobilize positive charge. When biphenyl is present in the 3-methylpentane matrix during γ irradiation, the 85°K conductivity peak and the infrared absorption are eliminated, but both appear when the sample is photolyzed with 4100 Å light which is capable of removing the electron from the biphenylide ion. A further similarity between the conductivity and luminescence results is that both effects tend toward saturation with increasing dose in the range between 10^{18} and 10^{19} eV g^{-1}.

The spontaneous decay of the infrared absorption attributed to trapped electrons in 3-methylpentane at 77°K is interpreted to be a composite of relatively rapid first order processes which approach completion in one hour and a slower second order process (16) or processes which, for typical irradiations, are slow enough so that 25% or more of the optical density of the sample at 1660 mμ which is present immediately after irradiation is still present after several hours. The electrical conductivity results do not correlate well with the infrared or luminescence decay time studies after the first rapid decay. Peak I of Fig. 3 falls to a negligible value in less than 15 min. Peak II is not appreciably altered in area by 5 hr standing at 77°K before the start of warmup. It may be that the apparent lack of correlation results from a fortuitous combination of decay rates and electrical conductivity contributions by different populations of trapped electrons, such that the electrons which decay most slowly as seen by infrared have the migration characteristics to produce peak II. In comparing the infrared absorption and luminescence results with the electrical conductivity peaks it must be reemphasized that the areas under the latter do not indicate the relative numbers of charges represented by the peaks, because the charge collected depends both on the number of charges which move and the distance which they move. Recent investigations (82b) indicate that the rate of thermal decay of trapped electrons in γ-irradiated 3-methylpentane glass at 77°K varies with the size of the sample. This presumably results from a variation in the nature of the trapping sites with rate of cooling when the sample is immersed in liquid nitrogen. Thus, use of identical samples may serve to resolve some of the apparent inconsistencies which have been observed between different methods of measurement of electron decay rates.

There is a striking contrast in the effect of electron scavengers on the electrical conductivity and luminescence from γ-irradiated 3-methylpentane. A CCl_4 concentration of 6.6 mole % does not affect the large conductivity peak above 90°K (IV of Fig. 3, but as little as $4 \times 10^{-3}M$ (5×10^{-4} mf) eliminates the luminescence which appears in the same warmup region. It appears that neutralization of the ionic species responsible for peak IV of Fig. 3 may also be responsible for the luminescence observed in the same temperature region. If this is the case, scavenging of the electrons and positive charges by CCl_4 substitutes ions of essentially the same mobility, but without the property of luminescing on neutralization. The luminescence peak observed on warmup cannot be identified with the conductivity peak at 85°K since the former is not removed by infrared illumination.

E. Stability of Traps

It is too early in the development of this field to be able to make any systematic catalogue of either the absolute or relative stabilities of electrons in different types of electron trapping sites, but some observations are of interest. The evidence available comes from a comparison of isothermal decay rates, estimates of activation energy from the shape of warmup detrapping curves, and the threshold for photobleaching.

From luminescence curves, activation energies of ca. 0.05 eV for detrapping in γ-irradiated 3-methylpentane at 77°K have been estimated (63,64). The energy threshold for photobleaching in 3-methylpentane (61) comes at 0.7 eV or greater, and that in methyltetrahydrofuran at 1.4 eV (70). It appears that there is a lower energy pathway for spontaneous self-neutralization of ion pairs than for photoactivated neutralization. If the thermally activated process required 0.7 eV or more, spontaneous decay at 77°K would require years rather than minutes. Fueki (83a) finds that calculated electron trapping energies in methyltetrahydrofuran, as a function of cavity radius are in plausible agreement with the threshold energies necessary to detrap the electrons by light (70). The number of trapping sites available to electrons produced by γ irradiation of hydrocarbon glasses appears to be limited, i.e., the electron apparently cannot "dig its own hole" at any point in the medium by simply surrounding itself by a cage of induced dipoles, but requires a void or special type of irregularity in the matrix. The evidence for this is that the electron concentration rises to a maximum with increasing radiation dose *and then falls* rather than remaining at a steady state (83b). This indicates that: (1) the number of trapping sites is limited; (2) electrons trapped in these sites are vulnerable to reaction with some product of the irradiation; (3) when such reaction occurs the site is either removed or remains occupied in such a way that it cannot trap another electron. It has been suggested that the electrons are removed by the reaction $H + e^- \rightarrow H^-$ (83b).

The effect of polarity of the medium on the stability of electron traps appears to be reflected by the fact that the initial 50% of the electrons in γ-irradiated 3-methylpentane decay in 10–20 min at 77°K as compared to 5 hr for the same fraction in methyltetrahydrofuran while free radicals decay faster in methyltetrahydrofuran than in 3-methylpentane. The decay rates of trapped electrons in hydrocarbon matrices and the values of $G(e^-)$ are both increased by the presence of polar additives (83c).

F. Chemical Trapping. Electron Attachment and Dissociative Electron Capture

When molecules which react with electrons to form stable negative ions by simple electron attachment or by dissociative electron capture are present in an organic matrix during irradiation, they scavenge the electrons in competition with the physical trapping and neutralization processes. For example, when biphenyl is present in 3-methylpentane or methyltetrahydrofuran glass during irradiation, an optical absorption appears which is identical with that of biphenylide ion produced by the reaction of biphenyl and an alkali metal in an organic solvent (16). From the known extinction coefficient of biphenylide ion the extinction coefficient of trapped electrons in their infrared absorption band can be determined. This can be accomplished by comparing the optical density of the infrared absorption band with that of the biphenylide band when electrons in the physically trapped state are converted to biphenylide by bleaching with infrared radiation, and then converted back to the physically trapped state by illumination with 4100 Å light, which photoionizes the biphenylide. This shift back and forth can be carried out many times on the same sample (61,70).

Capture of electrons by CO_2 in 3-methylpentane glass is indicated by an ESR singlet observable following γ irradiation (59) or photoionization of tetramethyl-p-phenylenediamine (47). The CO_2^- singlet has also been observed in γ-irradiated CH_3OH (84).

Dissociative capture yields of the alkyl halides $(RX + e^- \rightarrow R + X^-)$ are indicated by the free radicals observable by ESR following γ irradiation (9,85) or photoionization of tetramethyl-p-phenylenediamine in 3-methylpentane glass containing alkyl halides. Accompanying these are yields of RH, thought to be produced by hot reaction of R plus solvent molecules at the time of the capture process (73,86,87).

As little as 0.1 mole % biphenyl captures all the electrons available for capture during the γ irradiation of 3-methylpentane or methyltetrahydrofuran. G(biphenylide) in solutions of biphenyl in 3-methylpentane is ca. 1.6 (16) as compared to $G(e^-)$ of about 0.8 in pure 3-methylpentane determined by infrared in the absence of biphenyl (3). G(biphenylide) in methyltetrahydrofuran as determined from optical measurements is ca. 3.0, which is similar to the values of 2.6 for $G(e^-)$ from pure methyltetrahydrofuran and $G(e^-) + G$(biphenylide) from biphenyl solutions in methyltetrahydrofuran, determined by ESR (4). About 0.8 mole % alkyl halide is required to capture all electrons produced during γ irradiation of 3-methylpentane, and somewhat higher to capture all electrons in methyltetrahydrofuran (73). In 3-methylpentane all alkyl halides above methyl show G(trapped radicals)$_{max}$ from dissociative capture = ca. 1.1 (73); the corresponding yield in methyltetrahydrofuran is 2.1 ± 0.3. The yields of capture are greater than these by the yields of RH formed by hot radicals resulting from the dissociative capture, plus the yield, if any, of nondissociative capture. The yield of CH_4 from the γ irradiation of 3-methylpentane glass containing 1 mole % CH_3I is $G = 0.56$, while that from pure 3-methylpentane is 0.06, indicating that $G(CH_4)$ from the dissociative capture process is 0.5 (86). The yield of C_2H_6 from C_2H_5I or C_2H_5Cl in γ-irradiated 3-methylpentane glass is 0.3 (73), of which 0.15 is the contribution from pure 3-methylpentane. The ratios of RH production to trapped radical production for CH_3I and C_2H_5I in γ-irradiated 3-methylpentane glass are about 0.6 and 0.15, respectively. For radicals produced by dissociative capture by alkyl halides in 3-methylpentane at 77°K, using electrons produced by the photoionization of TMPD, the corresponding ratios are 2.8, 1.7, 0.28, for CH_3Cl, C_2H_5Cl, and C_3H_7Cl, respectively (87), suggesting that the energy of dissociation is used less effectively for H abstraction by the more complex radicals. The reason for the difference in the ratio of alkane to free radicals from the γ radiolysis and the TMPD experiments is, as yet, not clear. It has been suggested (88) that the reactivity of methyl radicals formed by the photochemical dissociation of CH_3I in matrices at 4°K may be ascribed to the vibrational energy associated with the fact that it is born in the tetrahedral form, whereas the planar form is the stable form of the free radical. Since the exothermicity of dissociative electron capture by the alkyl halides is probably not more than a few kcal/mole, a similar configurational energy contribution seems to be required to explain the observed alkane formation.

In studies of reactions of photochemically produced hot methyl and ethyl radicals (89,90) in the gas and liquid phases it has been found that the methyl radicals are more efficient than ethyl in abstracting H, as is the case when the radicals are produced by dissociative e^- capture in the solid state.

IV. FATE OF POSITIVE CHARGE

For every electron which is physically or chemically trapped in an organic solid there must also be a trapped positive charge. For the reasons advanced in discussing electron trapping (Section III-B), a large fraction of the positive charges in pure 3-methylpentane and pure methyltetrahydrofuran must be trapped in ion pair configurations which make them predestined to decay by combination with a particular electron. It is not known whether the predominant cation in pure 3-methylpentane is $C_6H_{14}{}^+$, $C_6H_{13}{}^+$, or $C_6H_{15}{}^+$, but the first of these is favored as best able to account for observed charge transfer processes. Whichever species exists, it must be stabilized as a solvated entity in rather close proximity to a solvated electron.

Because of its apparent inability to undergo charge transfer, the positive species in methyltetrahydrofuran and alcohols is considered to be an ion formed by proton transfer from the original ion to a neutral molecule.

There is evidence that positive charge can be transferred many molecular diameters in 3-methylpentane glass by resonance charge transfer, but becomes trapped when it encounters a molecule with lower ionization potential such as 2-methylpentene-1 (16). By decreasing the mobility of the positive charge such additives increase the yield of physically trapped electrons, or, if an appropriate additive such as biphenyl is present, the yield of the anion. Increasing concentrations of 2-methylpentene-1 in 3-methylpentane glass at 77°K, up to 2 mole %, increase the trapped electron yield, but the yield decreases at higher concentrations (91). This has been interpreted to mean that charge can be transferred from a 2-methylpentene-1 molecule to another 2-methylpentene-1 separated by several molecular diameters.

Some compounds (e.g., biphenyl, alkyl iodides, tetramethyl-p-phenylenediamine) can capture either electrons or positive charge. Thus, the spectrum of a γ-irradiated solution of biphenyl in 3-methylpentane may be a composite of the similar and overlapping spectra of the anion and cation (75).

Solutions of CCl_4 in 3-methylpentane glass appear to reveal the existence of differing and highly specific types of positive charge transfer within the same matrix. A peak at 4800 Å following γ irradiation at 77°K appears to be due to a cation formed from the CCl_4 (91), yet it is not affected by increasing toluene concentration, which causes growth in the toluene cation peak (92). Since CCl_4 is presumed to have a higher ionization potential than 3-methylpentane, the authors have been forced to conclude that an excited positive state can be transmitted through the matrix. The 4800 Å CCl_4 cation peak is absent following γ irradiation at 20°K (25), but absorption ascribed to biphenyl and ethyl iodide cations is unaffected by the change in temperature, suggesting that the mechanism for *excited* charge transfer is selectively sensitive to the temperature or density of the matrix.

Shida and Hamill (41,93–97) have reported extensive systematic investigations of positive charge transfer and solute positive ion formation for a number of γ-irradiated solid systems at 77°K, including: aromatic amines in polycrystalline CCl_4 (41); aromatic hydrocarbons in glassy butyl chloride, methyltetrahydrofuran and 3-methylpentane (94); aliphatic and aromatic ketones in the pure polycrystalline state and in glassy solvent matrices (95); butadiene and its homologs in *sec*-butyl chloride and methyltetrahydrofuran glasses (96); vinyl, vinylene, and vinylidene olefins in alkyl chloride and alkane glasses (97).

$G(H_2)$ from the radiolysis of methane present at 0.75 mole % in solid argon at 100°K is 1.3 based on total energy absorbed in the matrix (98). This is interpreted to be the result of resonant charge transfer or excitation transfer by the argon matrix. Because in the radiolysis of a CH_4–CD_4–Ar mixture at 20°K the hydrogen consisted mainly of H_2 and D_2, the formation of hydrogen cannot be ascribed to disproportionation and can only be explained by the occurrence of the process $CH_4^* \rightarrow CH_2 + H_2$. The formation of CH_2 is corroborated by the isotopic analysis of the ethane formed. Resonant charge transfer (terminated by the process $C_6H_{12}^+ + e^- \rightarrow C_6H_{10} + H_2$) has also been postulated to account for the value of 2.4 for $G(H_2)$ from the irradiation of a solid krypton matrix containing 0.3 mole % cyclohexane at 77°K (99). Similar results have been observed for other hydrocarbons.

Scala et al. (100) find that the radiolysis of i-C_5H_{12}, or 3-methylpentane, containing 5 mole % of deuterated cyclopropane, $(CD_2)_3$, at 77°K produces $CD_2HCD_2CD_2H$ in yields which are sharply increased by the presence of the electron scavenger CCl_4 and decreased by the presence of CH_3OH. Their results afford convincing evidence for H_2 transfer reactions of the type $C_5H_{12}^+ + (CD_2)_3 \rightarrow C_5H_{10}^+ + C_3D_6H_2$, in these systems, and in others containing olefins in place of cyclopropane. They indicate that such reactions may contribute to positive charge trapping in matrices such as 3-methylpentane containing an olefin.

There is evidence that positive ion chain reactions may occur in the solid phase. Davis and co-workers (101) ascribe hydrocarbons of average formula $C_{20}H_{40}$, produced in the radiolysis of methane at $77°K$, to such a process. The yield is 0.32 methane molecules converted per 100 eV absorbed.

V. TRAPPED FREE RADICALS

A. Mechanism of Formation

The types of reaction by which free radicals may be produced by ionizing radiation include:

1. Decomposition of excited neutral molecules
2. Neutralization of ions accompanied by decomposition
3. Dissociative electron capture processes
4. Proton transfer reactions, such as $RH^+ + RH \rightarrow R + RH_2^+$
5. Hot atom or hot radical abstraction reactions, such as $\underline{H} + RH \rightarrow H_2 + R$ and $\underline{CH_3} + RH \rightarrow CH_4 + R$
6. H atom addition to unsaturated compounds

Each of these can occur in the gas, liquid, or solid state but the relative yields in the solid state may be different than for the same compound in the liquid or gaseous state because: (*1*) deexcitation of excited molecules competes more effectively with decomposition; (*2*) certain crystal lattices favor transfer of energy or charge from the point of deposition to molecules of a dilute solute, thus favoring decomposition of the solute; (*3*) physical trapping of electrons in competition with prompt geminate recombination may prolong the life of the positive ion sufficiently to allow proton transfer and other ion molecule reactions to occur; (*4*) hot atom and hot radical reactions in a matrix of oriented molecules may result in preferential attack on a specific bond in the target molecule; (*5*) caging effects may favor geminate recombination of fragments; (*6*) trapping of charges may reduce the energy available from neutralization processes.

It is often difficult to evaluate which of the five radical production processes are significant in a system, and which of the several phenomena just enumerated are responsible for observed differences in yield between the liquid and solid state.

There is definitive evidence from ESR spectra for the production of radicals by dissociative capture when dilute solutions of alkyl halides in 3-methylpentane glass at $77°K$ are exposed to electrons produced by photoionization of tetramethyl-*p*-phenylenediamine (9). The formation of RH in these systems is strong evidence that some of the radicals can react as hot radicals to abstract hydrogen from the solvent (73,86,87). Formation of radicals by hot hydrogen atoms produced photochemically from HI in 3-methylpentane glass is conclusively demonstrated by their ESR spectra (102), and there is evidence for radical formation by the addition of thermal H atoms to olefins in this system.

The data available indicate that trapped radicals are not formed as a result of the neutralization processes following release of trapped electrons in γ-irradiated 3-methylpentane or MTHF glass, since the area of the ESR free radical spectra remain constant during photobleaching of the electrons (73,103). Data showing that the free radical spectrum from MTHF *decreases* when the trapped electron signal is bleached (4) are in contrast to the findings of other laboratories (73,103).

Studies of the products of the vacuum-ultraviolet photolysis of C_3H_8 at 20 and 77°K (10,11) show that primary fragmentation following the absorptions of a photon can yield H, H_2, CH_2, CH_3, or CH_4, as in the gas phase, indicating that, even in the solid state, decomposition of highly excited molecules can compete effectively with deactivation. The products of the solid state photolyses furnish evidence for the occurrence of geminate disproportionation reactions such as $CH_3 + C_2H_5 \rightarrow CH_4 + C_2H_6$. Such processes presumably occur in times too short for the radicals to be observed by ESR, and, as in the case of abstraction reactions by hot radicals, can only be detected by analysis of stable products.

B. Detection

Most radicals of interest in the radiation chemistry of simple organic solids do not have optical spectra in readily accessible wavelength regions, although a few, including benzyl (104), have been observed in this way.

By far the most useful means of studying radicals in organic solids is electron spin resonance spectrometry. Such investigations were pioneered by Gordy and co-workers, who investigated a variety of simple organic compounds following x irradiation in the solid state at 77°K (105a,105b) and also many complex systems of biological interest (105c). Smaller and Matheson (106) identified the radicals formed by the γ radiolysis of a series of alkanes, olefins, alcohols, ethers, and alkyl halides at 77°K and considered the implications for radiation chemistry. Alger, Anderson, and Webb (107) explored not only the ESR spectra but the optical spectra, photoconductivity, photo and thermal bleaching, and luminescence of some 50 alcohols, ketones, ethers, esters, acids, paraffins, and alkyl halides following irradiation with electrons or x-rays at 77°K. Following these early investigations many papers have reported ESR spectra of organic radicals produced in organic solids by ionizing radiation. Of these, some have been concerned primarily with establishing correlations between spectral characteristics, radical structure, and ESR theory. Others have used ESR for establishing the presence of easily identified radicals as radiolytic intermediates, and for studying their production and reactions. We shall discuss some examples of such studies which throw new light on radiation-induced processes in the solid state. A few useful review articles and books dealing with the ESR technique are listed in references 108–113.

C. Kinetics of Radical Reactions

1. General. In this section we consider the fates of radicals produced in an organic matrix by reactions of the types 1–5 of Section V-A. The processes by which such radicals may be consumed include: (*1*) prompt geminate recombination within the parent cage (i.e., too rapidly to be observed); (*2*) slow geminate recombination within the parent cage (the rate being determined by the time required for appropriate geometrical orientation of the fragments); (*3*) combination within the parent spur, but not with a fragment within the parent cage other than the sibling partner (as in the preceding cases, the rate of decay of the radicals is dependent on the first power of the initial radical concentration, but in this case the apparent rate constant may decrease with time); (*4*) diffusion until reaction occurs with a fragment from another track (in which case the decay rate is second order); (*5*) reaction with a matrix molecule to abstract hydrogen (precluded, on normal time scales, at 77°K and below for any processes with activation energy of more than ~ 5 kcal mole^{-1}). At the time of writing, published investigations of the growth and decay of radicals, such as are needed to select from these alternatives, are almost nonexistent. A few growth curves for the increase of radical concentration during irradiation of solid CH_3OH and ethylene glycol have been determined (107,114,115). Continuing investigations of growth and decay characteristics of radicals produced by ionizing radiation are in progress in at least a few laboratories (116,117), and closely related work on the decay of CH_3 radicals produced by photoionization of TMPD has appeared (9).

2. Radiation-Catalyzed Removal of Radicals. Of particular interest in the studies which have been completed is evidence for *radiation-catalyzed* removal of radicals, and for first order decay in the absence of radiation.

Gamma irradiation of glassy ethyl iodide at 77°K produces a well-defined, six-line ESR spectrum attributable to C_2H_5 radicals (40,118,119). With continuing irradiation, this spectrum grows at a decreasing rate eventually leveling off to a steady-state plateau. At a dose rate of 3×10^{16} eV g^{-1} sec^{-1} this is reached, after about 90 min of irradiation, the radical concentration being about 8×10^{-2} mole %. When irradiation is stopped, the initial rate of decay of ethyl radicals is less by a factor of 2 or 3 than the initial rate of growth at the start of irradiation. This indicates that radiation-catalyzed removal of radicals occurs, since the rate of decay during irradiation in the plateau region must equal the rate of growth.

The concentration of radicals at the plateau is approximately proportional to the square root of the dose rate [as measured over a range from about 10^{12} eV g^{-1} sec^{-1} to 10^{15} eV g^{-1} sec^{-1} (10 C/ml) using samples of tritiated C_2H_5I glass undergoing self-irradiation (Fig. 5) (32)]. This concentration would be proportional to the first power of the dose rate if no radiation-catalyzed decay occurred, since the thermal decay curves of γ-irradiated samples are superimposable if normalized by the first power of the dose received.

Ermolaev and Voevodsky (114) have interpreted the kinetics of growth of radicals during the γ irradiation of ethylene glycol at 77°K to indicate preferential localization of free radicals near defect sites where other

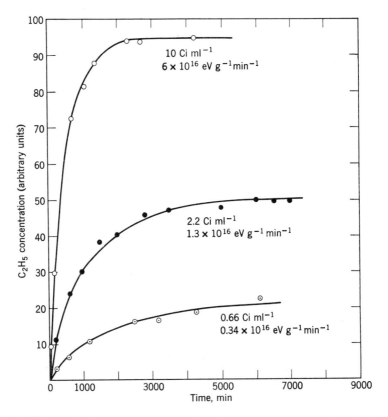

Fig. 5. Ethyl radical concentration as a function of dose rate and time in glassy tritiated ethyl iodide undergoing self-irradiation at 77°K (33).

radicals have already been formed, with each cluster disappearing by recombination when it reaches a critical size as irradiation continues. These workers have applied spin echo ESR techniques to demonstrate the presence of radical clusters in the γ-irradiated matrix (116).

No model of the radiation catalyzed recombination yet proposed is very satisfying. Extensive radical diffusion is not expected at 77°K in the matrices studied, and it is ruled out by the fact that the decay of radicals following irradiation is not second order. Therefore the Voevodsky cluster model requires that energy or charge migrate from the place of deposition to a preferred locale for radical formation. In terms of this model, the preferred location may be either the spots where the first radicals are formed or sites of irregularities in the matrix. This model was developed in an effort to explain the fact that growth plots of radical concentration vs. dose show abrupt changes in slope from an initial nearly linear portion to a second nearly linear portion of much lower slope. This phenomenon has been observed both in ethylene glycol (114) glasses and crystals and in ethyl iodide (117) glasses. Qualitatively, the cluster model explains the breaks in the growth curve either on the assumption that the break comes when favorable sites for radical formation have become saturated and removed by the radical combination processes, or on the assumption that when a critical population has built up within clusters the rate of removal of radicals by combination suddenly increases.

A simpler model assumes that the radicals are formed in conventionally distributed tracks and spurs, and that radiation-catalyzed removal of radicals occurs because energy or charge deposited in the matrix during continuing irradiation migrates to the spurs and hastens radical combination reactions within them. There are, however, arguments against either exciton transfer or electron transfer in a glassy ethyl iodide matrix. Using the simpler model, it may be considered that breaks in the radical growth curves are associated with saturation of different radical decay processes. Except for methyl radicals, radical decay in solid matrices seems to involve a spectrum of recombination probabilities with different first order rate constants, the first portion of decay being much faster than later portions.

The rate of production of Br_2 by the radiolysis of C_2Br_6, at 25°C and a dose rate of 10^{16} eV g^{-1} sec^{-1}, decreases by a factor of 10 in the dose range between 0.2×10^{19} eV g^{-1} and 5×10^{19} eV g^{-1} (as observed by spectrophotometric determinations on the crystals) (42). Since thermal decay of Br_2 is slow at this temperature, the decrease in G value can only be attributed to an increased rate of radiation-catalyzed removal of Br_2 as the concentration of Br_2 and radicals increases.

When C_2Br_6 is subjected to repeated γ doses of 8.4×10^{19} eV g^{-1}, alternated with thermal annealing for 10 min at 100°C, the growth and removal of Br_2 shown in Fig. 6 is observed. Ten minutes at 100°C removes all of the Br_2 annealable at 100°C. These data indicate two populations of reactive species in close proximity to Br_2 in the irradiated crystals. The population of reactive species which is not removed by annealing at 100°K does not increase significantly with successive irradiations, and so must be at a plateau which is maintained by radiation-catalyzed removal of Br_2. It is suggested that the reactive species is C_2Br_4, which achieves a steady state of production and removal during irradiation, but that the activation energy for the $C_2Br_4 + Br_2 \rightarrow C_2Br_6$ reaction is too high for it to be significant as a thermal reaction at 100°K. The thermally annealable Br_2 may be that in close proximity to C_2Br_5 radicals, with which it can react with relatively low activation energy by the reaction $C_2Br_5 + Br_2 \rightarrow C_2Br_6 + Br$. The observed rapid drop (42) in $G(Br_2)$ with dose indicates that this thermally annealable CBr_5–Br_2 population, like the C_2Br_4–Br_2 population, undergoes radiation-catalyzed reaction.

Following neutron irradiation of C_2Br_6, some 30% of the [82]Br atoms

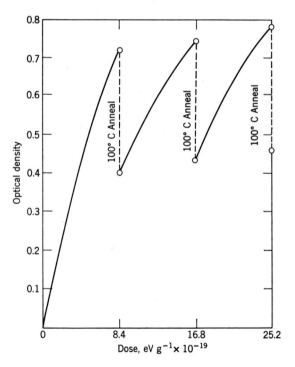

Fig. 6. Effects of repeated γ irradiation and thermal annealing on the Br_2 concentration in a C_2Br_6 pellet (184 mg cm^{-2}). Irradiations and optical measurements at 25°C. Thermal annealing time 10 min at 100°C (42).

which have broken their parent C—Br bonds (as a result of recoil from the γ ray or emission of conversion and Auger electrons) may be extracted in the inorganic phase if the C_2Br_6 is dissolved in CS_2 and washed with aqueous reducing agent (120). If, however, the C_2Br_6 is exposed to 10^{20} eV g^{-1} of γ irradiation before or after neutron irradiation, and annealed for an hour at 100°C before dissolving, the yield may be reduced to as little as 4%, the extent of reduction depending on the extent of γ irradiation. This indicates that, as γ-produced imperfections undergo reorganization at 100°K, energy or charge is transmitted to the sites of chemically metastable ^{82}Br atoms in a way to assist their reentry into organic combination.

a. Factors Which Determine Decay Rates.   The decay rates of trapped radicals in organic matrices depend on the species of the radical, the nature of the matrix molecules, the phase of the matrix (glassy or polycrystalline), and the temperature. In addition the rates and kinetics of decay depend on the mechanism of radical production, because this determines whether the process of radical removal is by reaction with: (a) geminate partners in the parent cage (first order kinetics); (b) geminate or spur partners with varying probabilities of fruitful encounter ("composite first order" kinetics); (c) radicals from elsewhere in the solution (second order); (d) the solvent, or a scavenger solute (first order).

b. Continuum of First Order Decay Processes Shown by Most Radicals. With the exception of CH_3 radicals, all radicals studied show a fractional rate of decay which decreases as the fraction decayed increases. It is sometimes possible, and tempting, to analyze the decay curve in terms of a second order decay or a composite of first order and second order decays. Often this is not warranted, as shown by the fact that an identical curve can be synthesized from other assumed combinations of first and second order decays. More importantly, it has been shown in a number of instances that the decay curves for samples which have received greatly different doses, and hence have greatly different concentrations of radicals are superimposable if plotted on a scale of fraction of radicals decayed as a function of time. This is true, for example, for the ethyl radicals in glassy ethyl iodide samples which have been irradiated at 3×10^{16} eV g^{-1} sec^{-1} for 1 min and 60 min and observed carefully over a period of 10 days, during which about 98% of the C_2H_5 signal decayed in each case. Less precise measurement over a period of one year (122) suggests that the fractional rates may continue constant indefinitely for all samples observed.

The fact that decay curves starting with greatly different concentrations can be superimposed when normalized indicates that the fate of every radical is determined by reaction with other radicals from the same track (reaction with matrix molecules being excluded by activation energy considerations). If intertrack reactions were significant they would be second order, and the fractional decay rate would depend on the intial concentration [as is the case for C_2H_5 radicals observed by ESR during the radiolysis of liquid ethane (123)]. The fact that the apparent half-life for radical decay increases with time of decay (except for the CH_3 cases studied) indicates that different members of the radical population in a track have different probabilities per unit time of encountering a reactive partner or of achieving the configuration necessary for reaction, i.e., potentially reactive pairs are not all trapped with the same proximity or orientation.

   c. *Pure First Order Decay of Methyl Radicals from Methyl Halides.* Methyl radicals produced by dissociative electron capture in 3-methylpentane glass at 77°K, using electrons produced either by photoionization of TMPD (9) or by γ radiolysis (73), decay by a pure first order process showing the same 16 min half life throughout the decay of 90% or more of the radicals (9). The half-life is independent of whether the radicals are produced from CH_3Cl, CH_3Br, CH_3I, or CD_3Br, and is the same if they are produced by photodissociation of CH_3I by 2537 Å light or by dissociative electron capture. It is also independent of the initial concentration of radicals produced over a wide range tested.

   The first order decay precludes the possibility that the radicals decay by random combination with other radicals after diffusion. The facts cannot be explained by assuming that the radicals disappear in a pseudo first order reaction with the solvent, because the activation energy for hydrogen abstraction is too high for such reactions to be significant at 77°K. The absence of such reaction is experimentally confirmed by the fact that the concentration of 3-methylpentyl radicals observable by ESR does not grow as the methyl radical spectrum decays. It therefore appears necessary to conclude that in every dissociation event, the methyl radical is confined within the parent solvent cage together with the halogen ion or atom which was dissociated from it and that each pair is predestined to undergo recombination.

   If the model just described is correct, it implies that the rate of recombination of the caged pairs is governed by the probability of the CH_3 radicals acquiring a configuration suitable for reaction, against the constraining forces of the matrix. As would be expected, the rate of decay is decreased at temperatures lower than 77°K and increased at temperatures higher than 77°K. It is also increased when the viscosity of the matrix is decreased by diluting the 3-methylpentane with isopentane, as shown in Fig. 7 (124), the temperature being held constant at 77°K. The plateau of Fig. 7 is yet to be explained.

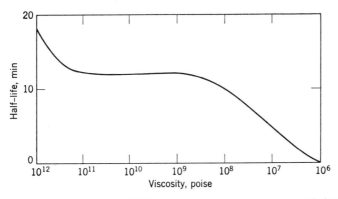

Fig. 7. Effect of viscosity on half-life of methyl radical decay in 3-methylpentane–isopentane mixtures at 77°K. The methyl radicals were formed by dissociative electron capture by methyl halides, using electrons from the photoionization of TMPD (124).

The forces which impede recombination of caged pairs are not the same as those responsible for the macroscopic viscosity of the matrix. This is illustrated by two observations: (1) the rates of decay of both methyl and ethyl radicals formed by dissociative electron capture are faster in MTHF glass at 77°K than in 3-methylpentane glass, although the viscosity (44b) of the MTHF is some 10^8 times greater than that of 3-methylpentane; (2) the decay of methyl radicals is much slower in perdeutero 3-methyl-pentane (C_6D_{14}) glass at 77°K than in the protiated matrix (C_6H_{14}) (87), although the viscosities (44b) of the two are identical within the error of measurement. Thus the constraints upon a radical which lower the frequency with which it achieves the configuration necessary for combination with an adjacent reactive fragment must depend upon the rigidity of individual matrix molecules with respect, for example, to the rotation of alkyl groups.

d. Second Order Decay of 3-Methylpentyl Radicals. 3-Methylpentyl radicals produced by the γ radiolysis of 3-methylpentane decay with second order kinetics (87). This indicates that they combine at random rather than with radicals from the parent spur. The radicals are assumed to be formed by the $C_6H_{14} \leadsto C_6H_{13} + H$ or $C_6H_{14}^+ + C_6H_{14} \rightarrow C_6H_{13} + C_6H_{15}^+$ process. If the first reaction occurs, the H atoms readily diffuse away from the site of formation and combine with other H atoms or radicals (no trapped H atoms are observed by ESR). If the second reaction occurs, it again leaves the radical without a geminate partner capable of recombination. The second order rate constants for decay of C_6H_{13} in protiated 3-methylpentane glass and C_6D_{13} radicals in deuterated 3-methylpentane glass are the same, within the accuracy of the measurements (87), indicating that the diffusion processes necessary for second order recombination are not controlled by the molecular rigidity which controls the first order processes discussed in preceding sections.

D. Free Radical Yields

1. Measurement. Use of electron spin resonance spectra to determine the absolute yields of trapped free radicals requires accurate measurement of the area under the ESR absorption curve and comparison with the area from a sample containing a known concentration of stable free radicals measured under identical conditions. To obtain the absorption curve, the first derivative signal normally displayed by ESR equipment may be integrated either graphically or electronically. Diphenylpicrylhydrazyl (DPPH) and galvinoxyl are convenient stable free radicals to use as standards. Known numbers of radicals may be measured either gravimetrically or by spectrophotometry. The area under the ESR absorption curve of a solution of known concentration of the standard, measured using the same sample geometry used for the unknowns, serves to calibrate the ESR equipment in terms of concentration of spins per area of signal. It is not satisfactory to use such solutions at 77°K after freezing because fractional crystallization with resulting inhomogeneous distribution within the ESR sample tube, often occurs. This problem may be avoided by using a solution to standardize a solid reference at room temperature for subsequent use at 77°K. Reference materials which have been used include galvinoxyl ground with sucrose or KCl (73). Such mixtures are normally not perfectly homogeneous and therefore must be reproducibly positioned in the dewar and the ESR cavity to make valid comparisons at different temperatures. If there is no significant change in the Q of the cavity, a factor of 4 greater sensitivity at 77°K than 300°K is observed, in agreement with the calculated difference in population of spin states at the two temperatures. To avoid deterioration of DPPH or galvinoxyl samples in solution they should be stored at low temperature in the absence of air and light.

When the radical concentrations in different ESR tubes are to be compared, care must be taken to select tubes of identical diameter or to calibrate the tubes against one another, using samples of the same solution of radicals. Before making yield determinations, it is mandatory to obtain spectra at a sufficient number of microwave power levels to insure that measurements are made at levels below the onset of saturation. For some species this is very low, as for trapped electron in 3-methylpentane and biphenylide ions in 3-methylpentane where, in each case, it is less than 1×10^{-6} W. Careful choice of spectrometer gain, modulation, scan rate and chart speed is important to obtain optimum accuracy in measuring low intensity peaks on the wings of spectra and to minimize error in determining the areas of relatively sharp peaks.

For comparison of the relative yields of the *same* radical in different samples, relative peak heights may be used. When two radicals are present in the same system, it may be possible to determine their independent yields if one gives a low broad spectrum and the other a narrow, sharper line spectrum, as in the mixture of methyl and 3-methylpentyl radicals produced by the γ radiolysis of a methyl halide in 3-methylpentane glass. In this case, the area following decay of the methyl radical may be subtracted from the sum of the areas before decay, it being ascertained that the intensity of the 3-methylpentyl signal intensity does not change. This is done by observing the wings of its spectrum which is beyond the methyl radical signal. In other special cases it is possible to evaluate the contribution of one radical to the total in a mixture even though selective decay does not occur, by choosing regions of the spectrum where it is known that the contributions of the two species are vastly different.

2. Yield of Secondary 3-Methylpentyl Radicals from 3-Methylpentane. The yield of free radicals from the γ radiolysis of pure liquid n-C_6H_{14}, as measured by the yield of $RI(I^{131})$ using $I_2(I^{131})$ scavenger, is 8 radicals per 100 eV absorbed (125). By contrast the yield of radicals from the radiolysis of 3-methylpentane glass at 77°K, as measured by ESR, is $G = 1.6$ (73). Radical yields at 77°K may be reduced compared to those at 300°K by the greater probability that radicals will be removed by combination with thermal H atoms, since the latter are precluded from abstracting hydrogen at 77°K. Geminate recombination of radicals will also be more prevalent at 77°K than at 300°K. The mechanism of radical formation in the solid is not known with certainty but it may be proton transfer, i.e., $C_5H_{14}^+ + C_6H_{14} \rightarrow C_6H_{13} + C_6H_{15}^+$. Within the limits of discrimination, the ESR spectrum represents a single radical, and not a mixture.

ESR evidence as to the identity of the radical is ambiguous. The observed spectrum could be accounted for by removal of the 3-methyl group (126), of a secondary hydrogen (102), or of the tertiary hydrogen (59). The uncertainty now seems to have been resolved (121) by comparing the ESR spectra obtained when each of the four possible 3-methylpentyl iodides and chlorides is irradiated in the pure state and in 3-methylpentane glass. The secondary halides are the only compounds which produce spectra which duplicate that from the γ radiolysis of pure 3-methylpentane glass. This indicates that the radical is $CH_3CHCH(CH_3)CH_2CH_3$, formed by loss of a secondary H from 3-methylpentane. Selective removal of a secondary H atom seems to occur in γ irradiation of several other singly branched alkane glasses (121) and also when the H is removed by abstraction by hot H (102), CH_3 (121), or Cl (102b). These observations indicate an effect of the matrix on the vulnerability of bonds, since in the gas phase bond rupture is more random, with some preference for the tertiary position.

3. Yields of Alkyl Radicals from Alkyl Halides. As with the hydrocarbons studied, alkyl halides, for which comparisons are available, yield lower G values for free radical production in the solid state at $77°K$ than in the liquid at $300°K$. Thus G(thermal free radicals) for liquid C_2H_5I at $300°K$ is 4.2 (127) while G(trapped radicals) in the glass at $77°K$ is 2.0 (32,119). Trapped radical yields for several alkyl chlorides bromides, and iodides have been reported by Ayscough and Thompson (118). There are major differences (40,119) between the types of radicals produced in the glassy state as compared to the polycrystalline state of some alkyl halides.

4. Radical Yields from Dissociative Electron Capture by Alkyl Halides in 3-Methylpentane and MTHF Glasses. The yield of trapped alkyl radicals (R) produced by the $RX + e^- \rightarrow R + X^-$ reaction in γ irradiated organic matrices must equal the total number of electrons minus the following (*1*) the number of electrons which undergo prompt recombination with the parent compound; (*2*) the number of radicals which form RH by a hot abstraction process; and (*3*) the number of nondissociative electron attachment events. It is a striking, and somewhat surprising, fact that of 10 alkyl halides radiolyzed in 3-methylpentane glass, all show the same yield of dissociative attachment when present at concentrations above 0.8 mole %, although the yields of RH formation by hot abstraction decrease with increasing complexity of the radical (as discussed in Section III-F). The *relative* yields of trapped radicals are reproducible to $\pm 10\%$, the absolute yield being 1.1 ± 0.3 molecules per 100 eV.

Radiolysis of MTHF glass at $77°K$ produces radicals from the matrix with $G = 2.1 \pm 0.3$ (4,73). $G(C_2H_5)$ produced with 2 mole % C_2H_5I in the matrix is 2 ± 0.3. The concentration of alkyl halide required to achieve maximum e^- capture is about 2 mole %, which is higher than for 3-methylpentane (73).

The relative cross sections for dissociative capture by C_2H_5I, CH_2Cl_2, $CHCl_3$, and CCl_4 in γ-irradiated 3-methylpentane at $77°K$, as shown by competitive capture experiments using mixtures of two alkyl halides in the same matrix, are 1, 1.3, 2.0, and 2.6 (73).

Whereas the G values for alkyl radical production from CH_3Cl, CH_3Br, and CH_3I by the γ radiolysis of their solutions in 3-methylpentane glass at concentrations greater than 0.8 mole % are equal, the yields for radicals formed from the same molecules as a result of photoionization of TMPD in the same matrix fall in the order CH_3Cl, CH_3Br, and CH_3I (9). This may result from different degrees of complexing of the CH_3X molecules with TMPD (9).

5. Comparison of Yields from Aliphatic and Aromatic Compounds.
Comparison of the free radical yields from the solid state radiolysis of
eleven saturated hydrocarbons ($G = 1.6$–5.6), four aliphatic alcohols
($G = 2.8$–12), and four compounds with non-conjugated double bonds
($G = 3$–3.8) with the yields from 16 aromatic compounds ($G = 0.03$–0.3)
shows that the yields from the latter are always at least an order of magnitude lower than those from the nonaromatic compounds (Voevodsky
and Molin, 128). For the aromatic compounds the radical yield decreases
with increasing conjugation in the molecule. The authors have interpreted
their results in terms of radical formation predominantly by C—H bond
rupture, with the latter being possible only when the energy of the first
singlet state of the molecule is in excess of the C—H bond energy. For the
aromatic compounds this energy requirement is not met, or barely met.
For the nonconjugated molecules the first singlet state has energy 1 to 6 eV
in excess of the C—H bond energy, thus being favorable both to radical
formation and to the production of highly reactive hot hydrogen atoms.

E. Chemical Reactions of Radicals in the Solid State

1. General. In addition to the reactions in which free radicals disappear
by combination with other reaction fragments in a solid matrix, discussed
in Section V-C-3, they may undergo photochemical decomposition
(e.g., $CH_2OH \xrightarrow{h\nu} CH_2O + H$; $CHO \xrightarrow{h\nu} CO + H$ (129)); isomerization
($(CH_3)_2CHCH_2 \rightarrow (CH_3)_3C$ (130,131)); oxidation ($R + O_2 \rightarrow RO_2$ (132));
and simple relaxation from a strained geometry to another geometry with
a different ESR spectrum. If the matrix is stable at sufficiently high
temperatures, the radicals may also abstract hydrogen. Changes in the
nature of free radical spectra in matrices held at 77°K or warmed have
been reported for a variety of systems (55,107,119,133–135).

Magat and co-workers (132) have shown that reactions of trapped
radicals are favored as a matrix passes through solid phase transition
temperatures during warmup.

The equilibrium between triphenylmethyl radicals and oxygen
($Ph_3COO \rightleftharpoons Ph_3C + O_2$) in γ-irradiated triphenylmethyl halides and
triphenylmethylacetic acid has been studied as a function of temperature
by electron spin resonance (136).

2. Reactions with Scavengers. Yields of free radicals from the radiolysis
of liquid hydrocarbons can be determined by using iodine as a scavenger
$[R + I_2 \rightarrow RI + I_2]$, and measuring the amount of RI formed (see
Chapter 7, Section IV-B). It might be expected that free radical yields in
the solid state could be determined similarly either by analysis of a γ
irradiated solution of a halogen in a solid hydrocarbon, or of the solution
obtained by dissolving a γ-irradiated solid hydrocarbon at a temperature
below its melting point in a solvent containing a halogen. However, solid

solutions of 0.1 mole % $Br_2(Br^{82})$ in polycrystalline hexane show no formation of organic bromides in excess of the amount ascribable to reaction of Br_2 with the olefinic products (137) which occurs on melting. Since the activation energy for reaction of free radicals with Br_2 is probably less than 1 kcal/mole, the result indicates that the radicals are predestined to combine with other radicals within their own spur, or that the Br_2 is in $(Br_2)_n$ aggregates sufficiently large to lower the effective concentration significantly (36).

The $G[I_2]$ value for solid isopropyl iodide varies with prior irradiation in the liquid phase in a way to suggest that I_2 and HI compete for free radicals in the solid during irradiation or during warming to the melting point (20). HI and I_2 are both complexed by alkyl iodides. Consequently no aggregation of these solutes would be expected in isopropyl iodide.

F. Fate of Hydrogen Atoms

Brown, Florin, and Wall (138) have observed H atoms in γ-irradiated CH_4 at 4°K by their ESR signal. The atoms are produced with an initial G value of 0.9 and reach saturation at 0.2 mole %. They are stable for days at 4°K but decay within hours at 20°K. The yields of hydrogen atoms and methyl radicals from the γ radiolysis of Ar and Kr matrices at 4°K containing low concentrations of CH_4 or CD_4 indicate rather efficient decomposition resulting from energy or charge transfer from the matrix to the methane (139). A series of papers has dealt with the trapping sites for H atoms in rare gas matrices as studied at 4°K by ESR (140).

Although ESR evidence for the presence of H atoms in γ-irradiated organic solids at 77°K and 4°K has been sought none has been found (58, 141), except in the case of CH_4, despite the fact that the H doublet signal is easily observed from empty Pyrex or quartz tubes exposed to γ or mercury arc irradiation at 77°K (142). This means either that H atoms are not produced in organic solids at 77°K or that they are rapidly removed (by abstraction, combination, or addition reactions).

$G(H)$ from the radiolysis of 3-methylpentane at 77°K must be at least 0.8 if the observed G(3-methylpentyl radicals), which is 1.6 (73), results only from radiolytic rupture of C—H bonds and by abstraction of an H atom from C—H bonds by hot H atoms formed by the radiolytic rupture. Increased effectiveness of energy removal from excited molecules and increased caging efficiency in the solid state, would be expected to reduce H atom yields below the value of ($G = 4$) found in the liquid state (125, 143). It may be that no H atoms are formed in the glass. In this case the observed 3-methylpentyl radicals must all be formed by proton transfer $(C_6H_{14}{}^+ + C_6H_{14} \rightarrow C_6H_{13} + C_6H_{15}{}^+)$.

In a pure hydrocarbon at 77°K any hydrogen atoms which are formed during radiolysis must either abstract hydrogen from a matrix molecule by a hot reaction or combine with another H atom or a radical. Reactions of thermalized H atoms with matrix molecules at 77°K are precluded by the activation energies of 8 kcal mole^{-1} (144) or more, commonly required for abstraction of hydrogen from alkanes by hydrogen atoms. An atom at 77°K acquires an energy of even 6 kcal mole^{-1} only about once a week. It has been demonstrated that hot H atoms produced by the photolysis of HI in 3-methylpentane glass at 77°K with 2537 Å radiation are able to abstract H from 3-methylpentane with a quantum yield of about 0.2 (102,145) and that thermalized H atoms in the same system add to alkenes (102).

Several laboratories (84,146–148) have postulated H atoms as reaction intermediates in the radiolysis of solid alcohols. Johnsen and co-workers (147) found that thermal H atoms generated from H_2 at a heated tungsten ribbon (149) produce no appreciable reaction with solid C_2H_5OH at 77°K. However, when solid ethanol which had been γ irradiated was exposed to H atoms, the normal value of G(glycol) = 0.94 was reduced to 0.02 and G(aldehyde) was raised from 2.64 to 4.70. This is interpreted as due to the disproportionation reaction $CH_3CHOH + H \rightarrow H_2 + CH_3CHO$, replacing the dimerization reaction $2CH_3CHOH \rightarrow (CH_3CHOH)_2$. The authors conclude that thermal H atoms can diffuse freely through the matrix but cannot react with matrix molecules.

In a similar experiment, Shida and Hamill (84) allowed H atoms to fall on unirradiated solid CH_3OH at 77°K and found formaldehyde, but not ethylene glycol. They do not indicate whether precautions were used to prevent hot H atoms from the filament from striking the CH_3OH surface.

When CH_3OD containing aromatic hydrocarbons (RC_6H_5) is γ irradiated at 77°K RC_6H_5D radicals were detected. This was first interpreted (146) to indicate rupture of the OD bond, followed by addition of the D to the ring. It was subsequently shown (84) that aromatic compounds can capture electrons to form anions which undergo a deuteron transfer reaction of the type $C_6H_6^- + CH_3OD \rightarrow C_6H_6D + CH_3O^-$. Photobleaching of solvent-trapped electrons in glassy CH_3OH produces H atoms ($e^- + CH_3OH + h\nu \rightarrow CH_3O^- + H$) which can add to solute olefins, producing the corresponding radicals, observable by ESR (84).

VI. COMPARISON OF γ RADIOLYSIS IN GLASSY AND POLYCRYSTALLINE MATRICES

A. The Alkyl Halides

Pure dry ethyl iodide can be frozen to a clear glass by rapid cooling at 77°K. If it is then allowed to warm to about 110°K, it spontaneously changes to an opaque polycrystalline mass. When two samples of ethyl iodide which are identical, except that one is in the glassy form and the other in the polycrystalline form, are given identical γ-irradiations at 77°K, the yields of I_2 (37,39) and HI (39) (Table II), and the ESR spectra (Fig. 8) (40,119) are all very different in the glassy than in the crystalline state.

Table II

Yields of HI and I_2 from the Liquid, Glassy, and Crystalline States
of Alkyl Iodides (39)

| Compound | G values | | | $\dfrac{G_{glass}}{G_{liq}}$ | $\dfrac{G_{cryst}}{G_{liq}}$ | $\dfrac{G_{glass}}{G_{cryst}}$ |
	Liquid	Glassy	Crystalline			
		$G(HI)$				
C_2H_5I	0.41	1.9	1.4	4.7	3.4	1.4
n-C_3H_7I	0.11	2.6	2.3	24	21	1.1
n-C_4H_9I	0.15	1.3	1.4	8.7	9.3	0.93
		$G(I_2)$				
C_2H_5I	2.1	1.2	0.48	0.57	0.23	2.6
n-C_3H_7I	1.5	0.50	0.50	0.33	0.33	1.0
n-C_4H_9I	1.6	1.1	1.2	0.69	0.76	0.91
		$G(HI)/G(I_2)$				
C_2H_5I	0.21	1.6	2.9			
n-C_3H_7I	0.07	5.3	4.6			
n-C_4H_9I	0.09	1.1	1.1			
		$G(HI) + 2G(I_2)$				
C_2H_5I	4.6	4.3	2.4	0.93	0.52	1.8
n-C_3H_7I	3.1	3.6	3.3	1.2	1.1	1.1
n-C_4H_9I	3.4	3.5	3.8	1.0	1.1	0.92

Similar differences in ESR spectra of the glassy and crystalline states have been observed for certain other alkyl iodides (Table III). It is notable that they occur only for those straight chain iodides with even numbers of C atoms. This alternating character from odd to even C number in the chain has an analog in the alternation in degree of resolution of the ESR spectra of γ-irradiated normal paraffin hydrocarbons in the C_{11} to C_{16} range (150), though the latter is less dramatic.

On warming, the trapped radicals in the glassy state of the alkyl iodides disappear at a much lower temperature than those in a polycrystalline matrix. The latter are sometimes observable even after the crystals are partially melted.

Table III

Comparison of ESR Spectra of γ-Irradiated Glassy and
Polycrystalline Alkyl Iodides (119)

Compound	State	No. of lines	Total spread, gauss	Approx. temp. of rapid anneal, $^{\circ}K$	Mp, $^{\circ}K$
C_2H_5I	Glass	6	160	100	165
	Cryst.	30	1000	158	165
$i\text{-}C_3H_7I$[a]	Cryst.	19	500	145	182
$n\text{-}C_3H_7I$	Glass	6	160	108	172
	Cryst.	7	160	172	172
$n\text{-}C_4H_9I$	Glass	6	160	106	170
	Cryst.	24	700	170	170
$n\text{-}C_5H_{11}I$	Glass	6	160	118	187
	Cryst.	6	160	187	187
$n\text{-}C_6H_{13}I$	Glass	6	160	123	205
	Cryst.	15	350	205	205
$n\text{-}C_7H_{15}I$[a]	Cryst.	6	160	213	225
$n\text{-}C_8H_{17}I$[a]	Cryst.	15	350	227	227

[a] All attempts to obtain the glass were unsuccessful.

The G value for production of trapped C_2H_5 radicals in C_2H_5I glass (Fig. 8A) and the G value for production of the radical giving the complex spectrum found in polycrystalline samples (Fig. 8B) are both about 2, based on measurements of the areas of the ESR absorption curves (151) although earlier estimates (7,119) suggested that the yield in the glass was lower.

It has been shown that the complex spectra of the type of Fig. 8B require a radical containing both hydrogen atoms and an iodine atom (40). Species to be considered include CH_3CHI, $C_2H_5I^+$, and $C_2H_5I^-$. For several reasons (40) it appears probable that $C_2H_5I^+$ and/or $C_2H_5I^-$ must be responsible for the spectrum. One reason is that these species seem best able to explain a striking change in spectrum which occurs when the samples are momentarily warmed to 145°K and then returned to 77°K. As seen in Fig. 9, this treatment results in the appearance of a six-line spectrum, typical of the C_2H_5 radical as seen in a glassy sample. This is accompanied by a decrease in the area of the wide complex spectrum. It has been suggested (40) that the reaction responsible for the spectral change may be $C_2H_5I^+ + I^-$ (or $C_2H_5I^-$) $\rightarrow C_2H_5$ (or $2C_2H_5$) $+ I_2$.

It is significant that spectral changes on momentary warming, such as described in the preceding paragraph, occur only in samples which have

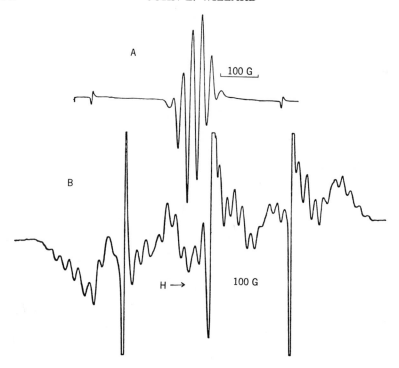

Fig. 8. ESR signals from γ-irradiated ethyl iodide at 77°K. *A*, Irradiated and examined in the glassy state, signal level 1; *B*, irradiated and examined in the polycrystalline state, signal level 160. The three highest lines of *B* are signals from the irradiated suprasil tube; the outer two being the hydrogen doublet (40,119).

been supercooled in the liquid or to the glassy state, followed by rapid spontaneous crystallization. Polycrystalline samples frozen slowly at the melting point in the presence of a seed crystal do not show the effect (Fig. 9). This indicates that release of energy stored as strains in the crystals contributes to the movement of the radicals required to cause the spectral change. A "momentary anneal" phenomena analogous to that of Fig. 9 for ethyl iodide has also been observed for *i*-propyl iodide (40).

Magat and co-workers (132) have discussed the effects of solid state transitions on the reactions of trapped radicals.

Alkyl bromides (38,119,151,152) and chlorides (151a) give complex spectra on the wings of the simple alkyl radical spectra in a number of cases. An extensive survey (151b) using wherever possible both the glassy and polycrystalline forms of the same compound, does not show the pronounced variations with C atom number and physical form observed for the iodides.

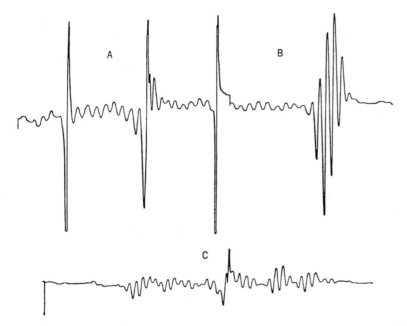

Fig. 9. "Momentary anneal" effects on two samples of polycrystalline ethyl iodide frozen by different procedures. *A*, Spectrum of both samples following γ irradiation at 77°K before annealing, signal level 40; *B* (sample rapidly crystallized by warming a glass), and *C* (sample slowly crystallized at the melting point), after warming to 147°K for 1 min and returning to 77°K. Signal levels 40 and 160 for *B* and *C*, respectively (40).

This work has revealed (151) unique ESR spectra of a number of γ-irradiated alkyl halides, extending to magnetic fields far below and above those expected for alkyl radicals. The species most likely to be responsible appear to be X, X_2^-, and X_2^+.

All of the glassy and polycrystalline alkyl halides examined after γ irradiation, except for the polycrystalline iodides with even number of C atoms, give a central pattern attributable to the free radical produced by loss of a halogen atom, as observed by Ayscough and Thompson (118), in addition to the spectra now observed on the wings of the central pattern.

B. Cause of Glass–Crystal Differences

Differences between the products of irradiation in the glassy and poly-crystalline states must result from differences in density or molecular alignment. These may affect caging efficiency, efficiency of energy transfer or charge transfer, and the relative probability of attack on different parts of a molecule by hot fragments from an adjacent molecule.

Collinson, Conlay, and Dainton (153) have investigated the reduction of $FeCl_3$, I_2, and DPPH when dilute solutions in a series of organic solvents are γ irradiated in the liquid, glass, and polycrystalline states. For all systems in the liquid or glassy states and for some in the polycrystalline state, G(-solute) is nearly independent of the concentration of solute and of the temperature, over a wide range. By contrast, G(-solute) for some poly-crystalline systems at 77°K (including $FeCl_3$ in benzene, diphenyl ether, diphenyl methane, and phenetole; and I_2 in diphenyl ether) increases by a factor of ten or more over the concentration range from $2 \times 10^{-4}M$ to $7 \times 10^{-4}M$ and increases with *decreasing* temperature below the melting point. The results require that charge or energy can be transferred over distances of 50–100 Å (but not much farther) in the crystalline matrix and that such transfer be more efficient the lower the temperature of the crystalline matrix. They also indicate that the phenomenon is not general but requires some type of specific matching of the matrix-solute combina-tion. Reasons are given for believing that excitons corresponding to the molecular state $^1B_{2u}$ (for benzene) are responsible. Interaction of these with solutes whose absorption spectrum overlaps the exciton spectrum causes the chemical changes of solute observed. Several systems show the unique behavior when in the polycrystalline state but not in the supercooled liquid or glassy state at the same temperature.

Other examples of effective transfer of chemically useful energy over many molecular diameters in a γ-irradiated matrices are seen in the magnitude of the deuterium atom ESR signal in γ-irradiated H_2O con-taining 0.1 mole % HDO at 4°K (154) and in the high yields of H and CH_3 produced by the γ irradiation of low concentrations of CH_4 in Ar or Kr at 4°K (138).

VII. SOLID STATE REACTIONS OF IONS AND COMPLEXES FORMED FROM ALKYL HALIDES

Whereas the ESR spectra of γ-irradiated alkyl iodide glasses or solutions of alkyl iodides in 3-methylpentane at 77°K, taken at low gain (Fig. 8), show predominantly the signal of a single alkyl radical, optical spectra re-veal four additional species in the pure alkyl halides (45) and some six (86, 155,156) in 3-methylpentane matrices. These may be distinguished from each other and from the radical observed by ESR by their absorption maxi-ma, by differences in their stability with respect to aging and photobleach-ing, and by the effects which charge scavengers have on their yields. One indication of the independence of the C_2H_5 radical observed by ESR in C_2H_5I from the species responsible for the 7500 Å peak is that the intense blue color of γ-irradiated C_2H_5I fades very little during a year of standing at 77°K while the ESR signal decays to 10^{-3} of its initial value (40).

The presence of a number of different metastable intermediates is not surprising since alkyl iodides may be expected to trap electrons as I^- and RI^- reactions and to trap positive charge as RI^+. In addition, the γ-irradiated matrix may contain I_2, HI, and I, and products of intraspur reactions of these species and the charged species with each other and with the molecules of the matrix.

All of the alkyl iodides tested give a peak at 4050 Å in the pure glass and one at 3950 Å when dissolved in 3-methylpentane glass. These grow after irradiation and appear to be attributable to I_2^- (45,86) or a charge transfer complex of the type $RI \cdot I$ (155). The exact position of the other peaks depends on the nature of the hydrocarbon portion of the iodide molecule. The organic iodides all produce two peaks which are consistent with the presence of an RI^+ species which converts to an $(RI)_2^+$ or R_2I^+ species (45,86,155,156).

In contrast to the alkyl iodides, the γ-irradiated alkyl bromide solids at 80°K contain only two species having absorption-peaks in the 3000–8000 Å region, and the chlorides none (Table IV). The two maxima for the

Table IV
Absorption Maxima in γ-Irradiated Alkyl Halide
Glasses at 80°K (45)

	Absorption		Maxima, Å	
Ethyl iodide	7500	5250	4050	< 3500
n-Butyl iodide	7500	5200	4050	< 3500
Isobutyl iodide	7600	5175	4050	< 3500
n-Pentyl iodide	7600	5200	4050	< 3500
n-Butyl bromide		6000	3750	
n-Butyl chloride		None		

bromides are at different wavelengths than any of those for the iodides, further confirming the conclusion that the absorbing species are all halogen-containing entities.

It is noteworthy that γ-irradiated solutions of alkyl iodides in MTHF glass at 80°K are transparent at all wavelengths in the range where six absorption peaks are seen at 3-methylpentane glass. This difference cannot be attributed to failure of electrons to migrate far enough to be captured in MTHF since $G(R)$ from the $RI + e^- \rightarrow R + I^-$ reaction is 2. It must, then, be either the result of failure of the positive charge to migrate to the RI molecules in MTHF or of differences in the degree of clumping of the RI molecules in the two media, or both. Failure of charge migration would lessen RI^+ formation and the absence of clumps would minimize the possibility of forming species requiring two or more RI molecules. If clumping occurs in the nonpolar 3-methylpentane it might be expected to be less significant in the polar MTHF.

VIII. SOME ADDITIONAL EXAMPLES OF PHASE EFFECTS

A. Choline Chloride Chain Decomposition

A surprising radiation sensitivity of crystalline choline chloride $[(CH_3)_3NCH_2CH_2OH]^+Cl^-$ was first recognized as a result of its unusually rapid decomposition to trimethylamine hydrochloride and acetaldehyde when labeled with C^{14}. G values as high as 55,000 have been found for the crystalline material, whereas in solution the compound exhibits normal radiation stability. The nature of the free radicals produced in the crystalline compound and the kinetics of their decay have been studied by ESR and a mechanism has been proposed to account for the dramatic chain reaction (157).

B. CCl_3Br and C_2H_5I Phase and Temperature Effects

$G(Br_2)$ from the radiolysis of CCl_3Br and $G(I_2)$ from the radiolysis of C_2H_5I show contrasting temperature and phase effects. They illustrate well the fact that the influence of phase on yields is dependent on the species investigated.

In the temperature range from -190 to $-121°C$ $G(Br_2)$ from solid CCl_3Br is 0.12, independent of temperature (158). From $-78°$ to $98°C$ $G(Br_2)$ increases to 3.5 with a linear Arrhenius plot giving an apparent activation energy of 2 kcal mole^{-1}. Although this activation energy suggests a diffusion-controlled process, there is no change in $G(Br_2)$ or its temperature dependence on crossing solid phase transitions at -35.5 and $-13.5°C$, or in going from the solid to the liquid state at $-5.6°C$.

$G(I_2)$ from crystalline C_2H_5I changes from 1 to 2 in the temperature range from -190 to $-120°C$ (37). In the liquid at $-78°$ it is 4 and remains at this value, essentially independent of temperature up to $108°C$.

REFERENCES

1. (a) A. Hummel and A. O. Allen, *J. Chem. Phys.*, **44**, 3426 (1966); *J. Chem. Phys.*, **46**, 1602 (1967); (b) W. F. Schmidt and A. O. Allen, *Science*, **160**, 30 (1968).
2. F. R. Freeman and J. M. Fayadh, *J. Chem. Phys.*, **43**, 86 (1965).
3. D. W. Skelly and W. H. Hamill, *J. Chem. Phys.*, **44**, 2891 (1966).
4. D. R. Smith and J. J. Pieroni, *Can. J. Chem.*, **43**, 2141 (1965).
5. M. R. Ronayne, J. P. Guarino, and W. H. Hamill, *J. Am. Chem. Soc.*, **84**, 4230 (1962).
6. F. W. Lampe and R. M. Noyes, *J. Am. Chem. Soc.*, **76**, 2140 (1954).

7. R. H. Luebbe, Jr. and J. E. Willard, *J. Am. Chem. Soc.*, **81**, 761 (1959).
8. R. L. Strong and J. E. Willard, *J. Am. Chem. Soc.*, **79**, 2098 (1957).
9. R. F. C. Claridge and J. E. Willard, *J. Am. Chem. Soc.*, **87**, 4992 (1965).
10. A. A. Scala and P. Ausloos, *J. Chem. Phys.*, **47**, 5129 (1967).
11. R. E. Rebbert and P. Ausloos, *J. Chem. Phys.*, **46**, 4333 (1967).
12. M. Takehisa, G. Levey, and J. Willard, *J. Am. Chem. Soc.*, **88**, 5694 (1966).
13. L. M. Dorfman and M. S. Matheson, "Pulse Radiolysis," in *Progress in Reaction Kinetics*, Vol. III, G. Porter, Ed., Pergamon Press, New York, 1965.
14. M. Ebert, J. P. Keene, A. J. Swallow, and J. H. Baxendale, Eds., *Pulse Radiolysis*, Proceedings of the International Symposium held at Manchester, April, 1965, Academic Press, New York, 1965.
15. A. M. Bass and H. P. Broida, Eds., *Formation and Trapping of Free Radicals*, Academic Press, New York, 1960.
16. W. H. Hamill, "Ionic Processes in γ-Irradiated Organic Solids at $-196°$," in *Radical Ions*, L. Kevan, Ed., Interscience, New York, 1967.
17. A. Chapiro, *Radiation Chemistry of Polymeric Systems*, Interscience, New York, 1962.
18. A. Weissberger, Ed., *Physical Chemistry of the Organic Solid State*, Interscience, New York, Vol. I, 1963; Vol. II, 1965.
19. J. Dobo and P. Hedvig, Ed., *Proc. 2nd. Tihany Symp. Radiation Chemistry*, Akademiai Kiado, Budapest, 1967.
20. T. O. Jones, R. H. Luebbe, Jr., J. R. Wilson, and J. E. Willard, *J. Phys. Chem.*, **62**, 9 (1958).
21. As provided in the Varian variable temperature device: Varian Associates, 611 Hansen Way, Palo Alto, California.
22. See for example: F. S. Dainton and G. R. Salmon, *Proc. Roy. Soc. (London)*, **A285**, 319 (1965).
23. T. O. Jones and J. E. Willard, *Rev. Sci. Inst.*, **27**, 1037 (1956).
24. "Cryotip Refrigerators," Air Products and Chemicals, Inc., Advanced Products Department, Allentown, Pa.
25. R. F. C. Claridge, R. M. Iyer, and J. E. Willard, *J. Phys. Chem.* **71**, 3527 (1967).
26. (a) J. A. Weil and P. Schindler, *Rev. Sci. Inst.*, **38**, 659 (1967); (b) F. W. Lytle and J. T. Stover, *Science*, **148**, 1721 (1965).
27. Silicone Products Division, General Electric Company, Waterford, New York.
28. F. A. Mauer, "Low Temperature Equipment and Techniques," in *Formation and Trapping of Free Radicals*, A. M. Bass and H. P. Broida, Eds., Academic Press, New York, 1960, pp. 117–168.
29. Such silica is supplied under the trade name Suprasil by the Amersil Quartz Division of Engelhard Industries, Inc., 685 Ramsay Ave., Hillside, N.J.
30. D. R. Smith and J. J. Pieroni, *Can. J. Chem.*, **42**, 2209 (1964).
31. J. Kroh, B. C. Green, and J. W. T. Spinks, *Can. J. Chem.*, **40**, 413 (1962).
32. P. Ogren, Ph.D. thesis, University of Wisconsin, 1967.
33. C. M. Nelson and R. A. Weeks, *J. Am. Ceramic Soc.*, **43**, 396 (1960) and **43**, 399 (1960).
34. P. K. Wong and J. E. Willard, unpublished.
35. T. Lantz and J. E. Willard, unpublished.
36. R. M. Iyer and J. E. Willard, *J. Am. Chem. Soc.*, **88**, 4561 (1966).
37. E. O. Hornig and J. E. Willard, *J. Am. Chem. Soc.*, **79**, 2429 (1957).
38. R. M. A. Hahne and J. E. Willard, *J. Phys. Chem.*, **68**, 2582 (1964).
39. H. J. Arnikar and J. E. Willard, *Radiation Res.*, **30**, 204 (1967).

40. H. W. Fenrick and J. E. Willard, *J. Am. Chem. Soc.*, **88**, 412 (1966).
41. T. Shida and W. H. Hamill, *J. Chem. Phys.*, **44**, 2369 (1966).
42. R. M. Iyer and J. E. Willard, *J. Chem. Phys.*, **46**, 3501 (1967).
43. I. Norman and G. Porter, *Proc. Roy Soc. (London)*, **A230**, 399 (1955).
44. (a) H. Greenspan and E. Fischer, *J. Phys. Chem.*, **69**, 2466 (1965); (b) A. C. Ling and J. E. Willard, *J. Phys. Chem.*, in press.
45. R. F. C. Claridge and J. E. Willard, *J. Am. Chem. Soc.*, **88**, 2404 (1966).
46. S. V. Filseth and J. E. Willard, *J. Am. Chem. Soc.*, **84**, 3806 (1962).
47. P. M. Johnson and A. C. Albrecht, *J. Chem. Phys.*, **44**, 1845 (1966).
48. J. R. Lombardi, J. W. Raymonda, and A. C. Albrecht, *J. Chem. Phys.*, **40**, 1148 (1964).
49. C. Chachaty and E. Hayon, *Nature*, **200**, 59 (1963).
50. C. Chachaty and E. Hayon, *J. Chim. Phys.*, **1964**, 1115.
51. F. S. Dainton, G. A. Salmon, and J. Teply, *Proc. Roy. Soc. (London)*, **A286**, 27 (1965).
52. F. S. Dainton, S. A. Salmon, P. Wardman, and U. Zucker, "Radiolysis of Glassy Methanol and *n*-Propanol at 77°K," in *Proc. 2nd Tihany Symp. Radiation Chemistry*, J. Dobo and P. Hedvig, Eds., Akademiai Kiado, Budapest, 1967, p. 247.
53. I. Kósa-Somogyi, M. Gécs, and M. Vizesy, "Radical Reactions and Radio-thermoluminescence Phenomena in Organic Glasses Irradiated at −196°C," in *Proc. 2nd Tihany Symp. Radiation Chemistry*, J. Dobo and P. Hedvid, Eds., Akademiai Kiado, Budapest, 1967, p. 275.
54. D. R. Smith and J. J. Pieroni, *J. Phys. Chem.*, **70**, 2379 (1966).
55. D. R. Smith, F. Okenka, and J. J. Pieroni, *Can. J. Chem.*, **45**, 833 (1967).
56. K. Tsuji, H. Yoshida, and K. Hayashi, *J. Chem. Phys.*, **46**, 810 (1967).
57. K. Tsuji and F. Williams, *J. Am. Chem. Soc.*, **89**, 1526 (1967).
58. D. R. Smith and J. J. Pieroni, *Can. J. Chem.*, **45**, 2723 (1967).
59. M. Shirom, R. F. C. Claridge, and J. E. Willard, *J. Chem. Phys.*, **47**, 286 (1967).
60. J. Lin, K. Tsuji, and F. Williams, *J. Chem. Phys.*, **46**, 4982 (1967).
61. J. B. Gallivan and W. H. Hamill, *J. Chem. Phys.*, **44**, 1279 (1966).
62. B. Wiseall and J. E. Willard, *J. Chem. Phys.*, **46**, 4387 (1967).
63. M. Burton, M. Dillon, and R. Rein, *J. Chem. Phys.*, **41**, 2229 (1964).
64. K. Funabashi, P. J. Herley, and M. Burton, *J. Chem. Phys.*, **43**, 3939 (1965).
65. D. W. Skelly and W. H. Hamill, *J. Chem. Phys.*, **43**, 3497 (1965).
66. A. H. Samuel and J. L. Magee, *J. Chem. Phys.*, **21**, 1080 (1953).
67. F. Williams, *J. Am. Chem. Soc.*, **86**, 3954 (1964).
68. J. L. Magee, *Discussions Faraday Soc.*, **36**, 247 (1963).
69. M. Magat, *Discussions Faraday Soc.*, **36**, 256 (1963).
70. P. J. Dyne and O. A. Miller, *Can. J. Chem.*, **43**, 2696 (1965).
71. R. L. Platzman, *Natl. Res. Council Publ.*, **305**, 22 (1953).
72. D. R. Smith and J. J. Pieroni, *Can. J. Chem.*, **43**, 876 (1965).
73. (a) M. Shirom and J. E. Willard, *J. Phys. Chem.* **72**, 1702 (1968); (b) M. Shirom and J. E. Willard, unpublished.
74. W. M. McClain and A. C. Albrecht, *J. Chem. Phys.*, **44**, 1594 (1966).
75. J. B. Gallivan and W. H. Hamill, *J. Chem. Phys.*, **44**, 2378 (1966).
76. W. M. McClain and A. C. Albrecht, *J. Chem. Phys.*, **43**, 465 (1965).
77. G. E. Johnson and A. C. Albrecht, *J. Chem. Phys.*, **44**, 3162 (1966); *ibid.*, **44**, 3179 (1966).
78. A. C. Ling and J. E. Willard, unpublished.

79. O. Janssen and K. Funabashi, *J. Chem. Phys.*, **46**, 101 (1967).
80. D. D. Wilkey, C. N. Oster, and J. E. Willard, unpublished.
81. A. C. Albrecht, P. M. Johnson, and W. M. McClain, in *Proc. Intern. Conf. Luminescence*, Budapest, 1966.
82. (a) A. Deroulede, F. Kieffer, and M. Magat, *Israel J. Chem.*, **1**, 509 (1963); (b) K. Tsuji and F. Williams, *J. Am. Chem. Soc.*, **90**, 2766 (1968).
83. (a) K. Fueki, *J. Chem. Phys.*, **44**, 3140 (1966); (b) M. Shirom and J. E. Willard, *J. Am. Chem. Soc.*, **90**, 2184 (1968); (c) M. A. Bonin, J. Lin, K. Tsuji, and F. Williams, *Proc. Intern. Conf. Radiation Chemistry*, Argonne Natl. Lab., Aug., 1968, Advan. Chem. Ser. Vol. 81.
84. T. Shida and W. H. Hamill, *J. Am. Chem. Soc.*, **88**, 3689 (1966).
85. D. W. Skelly, R. G. Hayes, and W. H. Hamill, *J. Chem. Phys.*, **43**, 2795 (1965).
86. R. F. C. Claridge and J. E. Willard, *J. Am. Chem. Soc.*, **89**, 510 (1967).
87. W. G. French and J. E. Willard, unpublished.
88. C. D. Bass and G. C. Pimentel, *J. Am. Chem. Soc.*, **83**, 3754 (1961).
89. R. D. Doepker and P. Ausloos, *J. Chem. Phys.*, **41**, 1865 (1964) and references given therein.
90. R. E. Rebbert and P. Ausloos, *J. Chem. Phys.*, **46**, 4333 (1967).
91. J. P. Guarino and W. H. Hamill, *J. Am. Chem. Soc.*, **86**, 777 (1964).
92. D. W. Skelly and W. H. Hamill, *J. Phys. Chem.*, **70**, 1630 (1966).
93. T. Shida and W. H. Hamill, *J. Chem. Phys.*, **44**, 2375 (1966).
94. T. Shida and W. H. Hamill, *J. Chem. Phys.*, **44**, 2369 (1966).
95. T. Shida and W. H. Hamill, *J. Am. Chem. Soc.*, **88**, 3683 (1966).
96. T. Shida and W. H. Hamill, *J. Am. Chem. Soc.*, **88**, 5371 (1966).
97. T. Shida and W. H. Hamill, *J. Am. Chem. Soc.*, **88**, 5376 (1966).
98. P. Ausloos, R. E. Rebbert, and S. G. Lias, *J. Chem. Phys.*, **42**, 540 (1965).
99. J. A. Stone, *Can. J. Chem.*, **43**, 809 (1965).
100. A. A. Scala, S. G. Lias, and P. Ausloos, *J. Am. Chem. Soc.*, **88**, 5701 (1966).
101. D. R. Davis, W. F. Libby, and W. G. Meinschein, *J. Chem. Phys*, **45**, 4481 (1966).
102. (a) S. Aditya and J. E. Willard, *J. Am. Chem. Soc.*, **88**, 229 (1966); (b) R. Arce and J. E. Willard, unpublished.
103. A. Salmon, *Discussions Faraday Soc.*, **36**, 284 (1963).
104. W. H. Hamill, J. P. Guarino, M. R. Ronayne, and J. A. Ward, *Discussions Faraday Soc.*, **36**, 169 (1963).
105. (a) C. F. Luck and W. Gordy, *J. Am. Chem. Soc.*, **78**, 3240 (1956); (b) W. Gordy and C. G. McCormick, *J. Am. Chem. Soc.*, **78**, 3243 (1956); (c) W. Gordy, *Radiation Res.*, *Suppl. 1*, **1959**, 491.
106. B. Smaller and M. S. Matheson, *J. Chem. Phys.*, **28**, 1169 (1958).
107. R. S. Alger, T. H. Anderson, and L. A. Webb, *J. Chem. Phys.*, **30**, 695 (1959).
108. M. C. R. Symons, "The Identification of Free Radicals," in *Advances in Physical Organic Chemistry*, Vol. I, V. Gold, Ed., Academic Press, London, 1963, p. 284.
109. D. E. O'Reilly and J. H. Anderson, "Magnetic Properties," in *The Chemistry and Physics of the Solid State*, Vol. II, D. Fox, M. M. Labes, and A. Weissburger, Eds., Interscience, New York, 1965, p. 121.
110. D. J. E. Ingram, *Free Radicals as Studied by Electron Spin Resonance*, Butterworths, London, 1958.
111. M. Bersohn and J. C. Baird, *An Introduction to Electron Spin Resonance*, W. A. Benjamin, New York, 1966.
112. C. P. Poole, Jr., *Electron Spin Resonance*, Interscience, New York, 1967.

113. (a) J. E. Wertz and J. R. Bolton, *Electron Spin Resonance*, ACS Short Course Lectures, American Chemical Society, 1967; (b) P. B. Ayscough, *Electron Spin Resonance in Chemistry*, A. Methuen, London, 1967.

114. V. K. Ermolaev and V. V. Voevodsky, "Effect of the Phase State on the Radiolysis of Organic Solids," in *Proc. 2nd Tihany Symp. Radiation Chemistry*, Akademiai Kiado, Budapest, 1967.

115. V. K. Ermolaev, Yu. N. Molin, and N. Ya. Buben, *Kinetika i Kataliz*, 3, 315 (1962).

116. V. V. Voevodsky, Rajzimring, Yu. D. Tzvetkov, Chmelinsky, Semonov, and Yhidomirov, private communication.

117. P. Ogren, N. Nazhat, and J. E. Willard, unpublished.

118. P. B. Ayscough and C. Thompson, *Trans. Faraday Soc.*, 58, 1477 (1962).

119. H. W. Fenrick, S. V. Filseth, A. L. Hanson, and J. E. Willard, *J. Am. Chem. Soc.*, 85, 3731 (1963).

120. (a) K. E. Collins and J. E. Willard, *J. Chem. Phys.*, 37, 1908 (1962); (b) K. E. Collins and G. Harbottle, *Radiochim. Acta*, 3, 21 (1964); (c) K. E. Collins, in *Chemical Effects of Nuclear Transformations*, Vol. I, International Atomic Energy Agency, Vienna, 1965.

121. D. J. Henderson, Ph.D. thesis, University of Wisconsin, 1968.

122. H. W. Fenrick, Ph.D. thesis, University of Wisconsin, 1966.

123. (a) R. W. Fessenden and R. H. Schuler, *J. Chem. Phys.*, 39, 2147 (1963); (b) R. W. Fessenden, *J. Phys. Chem.*, 68, 1508 (1964).

124. R. F. C. Claridge and J. E. Willard, unpublished.

125. P. R. Geissler and J. E. Willard, *J. Am. Chem. Soc.*, 84, 4627 (1962).

126. K. Fueki and Z. Kuri, *J. Am. Chem. Soc.*, 87, 923 (1965).

127. R. J. Hanrahan and J. E. Willard, *J. Am. Chem. Soc.*, 79, 2434 (1957).

128. V. V. Voevodsky and Yu. N. Molin, *Radiation Res.*, 17, 366 (1962).

129. S. B. Millikan and R. H. Johnsen, *J. Phys. Chem.*, 71, 2116 (1967).

130. P. B. Ayscough and H. E. Evans, *J. Phys. Chem.*, 68, 3066 (1964).

131. M. Iwasaki and K. Toriyama, *J. Chem. Phys.*, 46, 2852 (1967).

132. R. Bensasson, M. Durup, A. Dworkin, M. Magat, R. Marx, and H. Szwarc, *Discussions Faraday Soc.*, 36, 177 (1963).

133. V. V. Voevodsky, *Abstracts, 4th Intern. Symp. Free Radical Stabilization*, Washington, D.C., 1959.

134. T. Ohmae, S. Ohnishi, H. Sakurai, and I. Nitta, *J. Chem. Phys.*, 42, 4053 (1965).

135. P. B. Ayscough, R. G. Collins, and T. J. Kemp, *J. Phys. Chem.*, 70, 2220 (1966).

136. C. L. Ayers, E. G. Jansen, and F. J. Johnston, *J. Am. Chem. Soc.*, 88, 2610 (1966).

137. R. M. Iyer and J. E. Willard, *J. Phys. Chem.*, 71, 3070 (1967).

138. D. W. Brown, R. E. Florin, and L. A. Wall, *J. Phys. Chem.*, 66, 2602 (1962).

139. W. V. Bouldin, R. A. Patten, and W. Gordy, *Phys. Rev. Letters*, 9, 98 (1962).

140. See, for example, and further references: S. N. Foner, E. L. Cochran, V. A. Bowers, and C. K. Jen, *J. Chem. Phys.*, 32, 963 (1960).

141. See for example, V. V. Voevodsky, in *5th Intern. Symp. Free Radicals, Uppsala, Sweden*, Almquist and Wiksell, Stockholm, 1961; paper A-1.

142. See, for example: (a) R. Livingston and J. Weinberger, *J. Chem. Phys.*, 33, 499 (1960); (b) L. Kevan and C. Fine, *J. Am. Chem. Soc.*, 88, 869 (1966).

143. T. J. Hardwick, *J. Phys. Chem.*, 65, 101 (1961).

144. B. A. Thrush, "Reaction of Hydrogen Atoms in the Gas Phase," in *Progress in Reaction Kinetics*, Vol. III, G. Porter, Ed., Pergamon Press, Oxford, 1965, p. 63.

145. J. R. Nash, R. R. Williams, Jr., and W. H. Hamill, *J. Am. Chem. Soc.*, **82**, 5974 (1960).

146. J. A. Leone and W. S. Koski, *J. Am. Chem. Soc.*, **88**, 656 (1966).

147. R. H. Johnsen, A. K. E. Hagopian, and H. B. Yun, *J. Phys. Chem.*, **70**, 2420 (1966).

148. F. S. Dainton, G. A. Salmon, and J. Tepley, *Proc. Roy. Soc.*, **286A**, 27 (1965).

149. (a) R. Klein and M. D. Scheer, *J. Am. Chem. Soc.*, **80**, 1007 (1958); (b) M. D. Scheer and R. Klein, *J. Phys. Chem.*, **65**, 375 (1961); (c) R. Klein, M. D. Scheer, and R. Kelley, *J. Phys. Chem.*, **68**, 598 (1964).

150. N. Ya. Chernyak, N. N. Bubnov, L. S. Polyck, Yu. D. Tsvetkov, and V. V. Voevodsky, *Zh. Optika i Spektroskopiya*, **6**, 564 (1959) (A.E.C.-tr-3825).

151. (a) R. Egland and J. E. Willard, unpublished; (b) R. Egland and J. E. Willard, *J. Phys. Chem.*, **71**, 4158 (1967).

152. F. W. Mitchell, B. C. Green, and J. W. T. Spinks, *J. Chem. Phys.*, **36**, 1095 (1962).

153. E. Collinson, J. J. Conlay, and F. S. Dainton, *Discussions Faraday Soc.*, **36**, 153 (1963).

154. H. A. Judeikis, J. M. Flournoy, and S. Seigel, *J. Chem. Phys.*, **37**, 2272 (1962).

155. J. P. Mittal and W. H. Hamill, *J. Am. Chem. Soc.*, **89**, 5749 (1967).

156. E. P. Bertin and W. H. Hamill, *J. Am. Chem. Soc.*, **86**, 1301 (1964).

157. For references see: M. A. Smith and R. M. Lemmon, *J. Phys. Chem.*, **69**, 3370 (1965).

158. R. F. Firestone and J. E. Willard, *J. Am. Chem. Soc.*, **83**, 3551 (1961).

Chapter 10

Water and Aqueous Solutions

M. Anbar

The Weizmann Institute of Science
Rehovoth, Israel

I. INTRODUCTION

Water is undoubtedly the most important constituent of the biosphere of our planet. Thus it is not surprising that the understanding of its physical and chemical properties became one of the primary objectives of modern science. Every conceptional or technological development in science found water as one of its first objects. Once a new form of energy was discovered, namely x-ray and nuclear energy, the investigation of its action on water followed soon after. The first report on the action of alpha particles on water dates back to 1901, only five years after the discovery of x-rays and natural radioactivity (1). The radiolytic decomposition of water by beta and gamma rays was observed and examined before the end of the first decade of the century (2). Over sixty years have passed since, and over 3000 papers have been published on different phases of the radiolysis of water and aqueous solutions. Nevertheless, the number of scientists in this field is increasing and, as we shall see from this review, the subject is still at its early stages of development. In spite of the extensive research carried out, it may be stated that there is as yet no conclusive answer to any of the salient problems in this field.

From the standpoint of the physical chemist, water is an ideal medium with which to study the radiolytic behavior of condensed systems. It is a liquid consisting of triatomic molecules, two of which are hydrogen atoms and therefore the number of possible primary and secondary products is limited to less than 10. Moreover, only three stable radiolytic products may be formed, namely, molecular hydrogen, hydrogen peroxide, and molecular oxygen. There are very few chemical species which have been studied as thoroughly as water, so that most of the physicochemical information necessary for the interpretation of its radiolytic behavior is available. Water is also the only condensible reagent that can be purified in large quantities to a level of submicromolar concentration of impurities (i.e., to the level of parts per billion). In order to obtain meaningful results on the primary processes of radiolysis the radiolytic chemical change should be kept down to a minimum. This minimum is determined by the sensitivity of the analytical methods and even more so by the level of impurities present in the radiolyzed system. In short, if water were not as easily available to the radiation chemist as it is, he would have had to synthesize it.

Water is the most abundant constituent of living matter and a substantial fraction of the absorbed energy of ionizing radiation in living systems is converted into chemical energy in water molecules. Thus, the understanding of the elementary processes which take place in irradiated water as well as the chemical behavior of the primary and secondary products of its radiolysis is a prerequisite for the interpretation of extra and intracellular radiobiological behavior at the molecular level.

The history of radiation chemistry of water up to 1960 has been previously told in sufficient detail (3–6) and there is little point to repeat it here. The field of radiation chemistry of water in this decade is so lively and full of excitement that the time has not yet come for a detailed and objective historical description of the development of recent ideas in this field. We cannot help but present a concise and to a great extent subjective description of the developments in this field in recent years.

Two major developments have taken place during this decade in the field of radiation chemistry of water. Water has been one of the first substances to be examined by pulse radiolysis (7) and by far the material most extensively studied by this novel technique. Pulse radiolysis, which facilitates the direct and quantitative evaluation of the behavior of radiolyzed systems, led to one of the most important landmarks in radiation chemistry of water, and perhaps in chemistry in general—the discovery of the hydrated electron (8).

Since the two excellent reviews of Hart and Platzman (5) and of Allen (6) have been published in 1961, there have appeared a number of review articles on the radiation chemistry of water (9–11) including the recent comprehensive review of Haissinsky (12). Further developments in the field in the last five years found their expression in a number of conferences, two of which were exclusively devoted to the radiation chemistry of water (13,14) whereas in others (15–22), special sessions were designated to this rapidly developing subject.

In this chapter we shall try to look into the radiolytic behavior of water from the phenomenological standpoint of its chemistry. We shall also discuss in brief the chemical behavior of the free radicals produced in the radiolysis of water. We shall try to describe the experimental facts as clearly and as critically as possible, and only after doing that shall we try to present plausible interpretations of these findings. Emphasis will be put on experimental information accumulated since 1962 which has hitherto not been thoroughly evaluated.

II. THE IDENTITY OF THE PRODUCTS OF RADIOLYSIS OF WATER

The very first investigations of the radiolysis of water by alpha particles have shown that it decomposes to form hydrogen, oxygen, and hydrogen peroxide. The same products were also found in water radiolyzed by beta and gamma radiation (i.e., at low linear energy transfer-LET). Subsequently, it has been found that *pure*, oxygen free water is virtually stable to beta and gamma radiation and no permanent chemical change is detectable under these radiations. On the other hand, under alpha radiation one finds net decomposition of water even in the absence of any solute. Thus, there apparently exists a qualitative difference between the radiolysis of water by alpha and beta radiation.

In the presence of relatively low concentrations of certain inorganic solutes (10^{-5} to $10^{-3}M$), e.g., bromide or nitrite ions, one finds also at low LET that the radiolytic yields of molecular hydrogen and hydrogen peroxide increase linearly with the absorbed dose (4–6). It must be concluded, therefore, that hydrogen and hydrogen peroxide are produced also under low LET conditions but that these products are eliminated by some other reactive species which may possibly be removed by added solutes.

These reactive species could be identified by their chemical properties. It was found that when small concentrations of aliphatic solutes (10^{-4} to $10^{-2}M$) are present in the radiolyzed water at neutral pH, the yield of molecular hydrogen increases by more than 100% (4–6). When the hydrogen atoms of the aliphatic solute were labeled with deuterium the excess hydrogen produced was composed of HD molecules (23). From kinetic experiments at high ionic strength it was concluded that the precursor of the HD is uncharged. It was further found that the yield of HD or of the additional H_2 is independent of solute concentration. The only explanation for this series of observations is the formation of *hydrogen atoms* from radiolyzed water. The identity of H atoms has been corroborated in the radiolysis of ice where these species have been identified by electron spin resonance (ESR) (24). Furthermore they have been shown to be identical in their kinetic behavior with H atoms produced by dissociation of gaseous hydrogen molecules (25).

In addition to H atoms another reducing species is produced in radiolyzed water. This species reduces carbonyl compounds to the corresponding alcohols, carbon dioxide to CO_2^-, H_3O^+ to hydrogen atoms and nitrous oxide to nitrogen. On the other hand there is no reaction between this reducing species and aliphatic alcohols, ethers or amines (26), all of which undergo hydrogen abstraction by H atoms. Kinetic investigations have shown that the rates of reaction of the second reducing species, formed in the radiolysis of water, with different reactants differ by orders of magnitude from those of H atoms, even in cases where the same products are formed. For instance the second reducing species reacts with hydrogen peroxide about 200 times faster than H atoms (6,9) and with nitrate ions and with acetone 1000 (27,28) and 10,000 (28,29) times faster, respectively. On the other hand, H atoms react with phenol or aniline 1000 times faster than the second reducing species (30,32). From the salt effect on its rate of reaction with various charged substrates it was found that it carries a single negative charge (33). This has been confirmed by studying the conductivity of water under pulse radiolysis (34). The chemical and physicochemical properties of the second reducing agent formed in radiolyzed water fit those predicted for a *hydrated electron* e_{aq}^-, which has been expected to be formed under these conditions (35). The characteristic absorption spectrum of this species which has been demonstrated under pulse radiolytic (8,36) and flash photolytic (37,38) conditions, is comparable with the predicted spectrum of e_{aq}^- deduced from the absorption spectrum of the electron solvated in ammonia (35).

The reactions of hydrated electrons cited above may be summarized as follows:

$$e_{aq}^- + R_1R_2CO \rightarrow R_1R_2CO^- \rightarrow R_1R_2COH \tag{1}$$

$$e_{aq}^- + CO_2 \rightarrow CO_2^- \tag{2}$$

$$e_{aq}^- + N_2O \rightarrow N_2O^- \rightarrow N_2 + O^- \rightleftarrows N_2 + OH + OH^- \tag{3}$$

$$e_{aq}^- + H_3O^+ \rightarrow H_3O \rightarrow H + H_2O \tag{4}$$

$$e_{aq}^- + H_2O_2 \rightarrow H_2O_2^- \rightarrow OH + OH^- \tag{5}$$

$$e_{aq}^- + NO_3^- \rightarrow NO_3^{2-} \rightarrow (NO_2)_{aq} \tag{6}$$

$$e_{aq}^- + RPh \rightarrow RPh^- \rightarrow RPhH + OH^- \tag{7}$$

The presence of an oxidizing species formed in the radiolysis of water could be inferred from the formation of oxidized products of added solutes (under conditions where the reduction of the oxidized products was inhibited by additives which remove hydrogen atoms and hydrated electrons). Thus in neutral radiolyzed solutions, in the presence of dissolved oxygen or nitrous oxide, one can oxidize low concentrations (10^{-4} to $10^{-2}M$) of iodide and bromide to iodine (6,39) and bromine (6,40), of chloride in acid solution to chlorine (6,41) or of formate to oxalate and carbon dioxide (42). In hydrogen-containing radiolyzed solution the oxidation of H_2 to H atoms takes place (6,42–45). Carbon monoxide was claimed to be oxidized to carbon dioxide in radiolyzed solution accompanied by the formation of H atoms (46,47). It was also shown that the oxidizing species produces hydroxylated derivatives of aromatic compounds and an intermediate PhHOH has been identified under pulse radiolysis (48,49). Further it was shown by kinetic analysis of salt effect that the oxidizing species in radiolyzed water, which is identical with the product of photolysis of H_2O_2 (43), is an uncharged species in the pH range 0–10 (43,50), but it undergoes dissociation to a basic form with a pK = 11.7 (51). The *OH radical* which has been also identified by ESR in irradiated ice (24,52), is evidently the oxidizing species which is liable for all the reactions described above (see equations 8–15).

$$OH + I^- \longrightarrow OH^- + I \tag{8}$$

$$H^+ + OH + Cl^- \longrightarrow H_2O + Cl \tag{9}$$

$$OH + HCOO^- \longrightarrow H_2O + COO^- \tag{10}$$

$$OH + H_2 \longrightarrow H_2O + H \tag{11}$$

$$OH + CO \longrightarrow H + CO_2 \tag{12}$$

$$OH + PhH \longrightarrow PhHOH \tag{13}$$

$$H_2O_2 \xrightarrow{hv} 2OH \tag{14}$$

$$OH + OH^- \rightleftharpoons H_2O + O^- \tag{15}$$

It may be concluded that the radiolysis of water produces, in addition to the H_2 and H_2O_2, also H atoms, OH radicals and hydrated electrons. In dilute solutions (up to $10^{-2}M$) these five species account practically for all the chemical change which takes place in water following the absorption of beta or gamma radiation.

In addition to OH radicals another oxygen containing intermediate has been detected in the radiolysis of water, especially under high LET conditions (53). This radical is the precursor of molecular oxygen in the presence of oxidizing additives, such as Cu^{2+}, under conditions where H_2O_2 is being quantitatively reduced (54). The chemical behavior of this radiolytic product is identical with that of the *perhydroxyl radical*, HO_2, which may be produced in radiolyzed systems by one of the following reactions

$$H + O_2 \rightarrow HO_2 \tag{16}$$

$$e_{aq}^- + O_2 \rightarrow O_2^-; \quad O_2^- \rightleftarrows HO_2 \tag{17}$$

$$OH + H_2O_2 \rightarrow H_2O + HO_2 \tag{18}$$

HO_2 radicals thus produced have been directly identified in frozen solutions (55) and their absorption spectrum determined by pulse radiolysis (56).

Another species which has been considered as a possible intermediate in the radiolysis of water is an *excited water molecule*, H_2O^*, which has a lifetime long enough to react with solutes in the moderate range of concentrations (10^{-2} to $1.0M$). The existence of this species has been suggested to explain the kinetic behavior of a number of systems (57–62). Although the existence of a long-lived triplet excited H_2O^* molecule cannot be excluded on physical grounds (63), and there was even a reported fluorescence of irradiated water attributed to a long lived H_2O^* (64), the evidence is still too scarce to allow a conclusive answer to the question of the participation of excited water molecules at the "chemical stage" ($> 10^{-8}$ sec) in radiolyzed water. There is little doubt, on the other hand, that short lived excited water molecules participate in the "physical stage" of radiolysis, namely in processes which take place within $< 10^{-13}$ sec and result in the formation of the "primary" products of radiolysis described above.

Another species which may play a role in the radiolysis of moderately concentrated solutions ($> 10^{-2}M$) are *subexcitation electrons* (65), namely, electrons with energies ranging between thermal energies and the minimum energy necessary to excite water to its first electronic excitation level. As the rate of thermalization of electrons is relatively slow (of the order of 10^{-13} sec) and in view of their fast motion through the medium, they have a chance to react with solutes before undergoing complete thermalization. There are very few systems the chemical behavior of which suggested the participation of subexcitation electrons. The formation of carbon monoxide from concentrated formic acid (66) and of H_2 from concentrated iodide solutions (67) (an observation confirmed by Meyerstein) (68)

$$2H_2O + e_{se}^- + I^- \rightarrow I + H_2 + 2OH^- \qquad (19)$$

have been attributed to these species.

III. DETERMINATION OF THE RADIOLYTIC YIELDS OF THE "MOLECULAR" AND RADICAL PRODUCTS

The yield of H_2 is determined directly by gasometric or gas chromatographic techniques (6) and that of H_2O_2 colorimetrically by one of the conventional analytical procedures, such as the I_3 method (69). Owing to the simultaneous formation of e_{aq}^-, H and OH radicals, the molecular products cannot be determined in pure water, as they are removed by the following reactions (20–28).

$$e_{aq}^- + H_2O_2 \rightarrow OH^- + OH \qquad k_1 = 1.2 \times 10^{10} \text{ M}^{-1} \text{ sec}^{-1} \quad \text{(ref. 70)} \quad (20)$$

$$H + H_2O_2 \rightarrow H_2O + OH \qquad k_2 = 9 \times 10^7 \text{ M}^{-1} \text{ sec}^{-1} \quad \text{(ref. 71)} \quad (21)$$

$$OH + H_2O_2 \rightarrow H_2O + HO_2 \qquad k_3 = 2.3 \times 10^7 \text{ M}^{-1} \text{ sec}^{-1} \quad \text{(ref. 72)} \quad (22)$$

$$OH + H_2 \rightarrow H_2O + H \qquad k_4 = 6 \times 10^7 \text{ M}^{-1} \text{ sec}^{-1} \quad \text{(ref. 45)} \quad (23)$$

In acid medium one has to add:

$$e_{aq}^- + H_3O^+ \rightarrow H + H_2O \qquad k_5 = 2.4 \times 10^{10} \text{ M}^{-1} \text{ sec}^{-1} \quad \text{(ref. 73,74)} \quad (24)$$

and in alkaline solutions:

$$H + OH^- \rightarrow e_{aq}^- \qquad k_6 = 1.8 \times 10^7 \text{ M}^{-1} \text{ sec}^{-1} \quad \text{(ref. 75)} \quad (25)$$

$$O^- + H_2 \rightarrow OH^- + H \qquad k_7 = 1.6 \times 10^8 \text{ M}^{-1} \text{ sec}^{-1} \quad \text{(ref. 51)} \quad (26)$$

$$e_{aq}^- + HO_2^- \rightarrow OH^- + O^- \qquad k_8 = 3.5 \times 10^9 \text{ M}^{-1} \text{ sec}^{-1} \quad \text{(ref. 76)} \quad (27)$$

$$O^- + HO_2^- \rightarrow OH^- + O_2^- \qquad k_9 = 9 \times 10^8 \text{ M}^{-1} \text{ sec}^{-1} \quad \text{(ref. 72)} \quad (28)$$

The radicals may be eliminated by the addition of various reactants but it is required that also the products of these reactions should not compete for the "molecular" products. The reactivity of the additive should be comparable to or higher than that of H_2 or H_2O_2; thus, it might compete effectively for the radicals, as long as the molecular products did not build up to a concentration comparable with that of the protective additives. Bromide ions fulfil these requirements and enable the quantitative determination of the "molecular" hydrogen.

$$OH + Br^- \rightarrow OH^- + Br \qquad k_{10} = 1.2 \times 10^9 \quad \text{(ref. 77)} \quad (29)$$

$$Br + Br^- \rightarrow Br_2^- \qquad k_{11} \sim 10^{10} \quad \text{(ref. 77)} \quad (30)$$

$$Br_2^- + H_2 \rightarrow HBr + Br^- \qquad k_{12} < 10^5 \quad \text{(refs. 77, 78)} \quad (31)$$

$$Br_2^- + H \rightarrow HBr + Br^- \qquad k_{13} \sim 10^{10} \quad \text{(ref. 79)} \quad (32)$$

$$Br_2^- + e_{aq}^- \rightarrow 2Br^- \qquad k_{14} = 1.3 \times 10^{10} \quad \text{(ref. 77)} \quad (33)$$

Owing to the fact that $k_{12} \ll k_4$, k_7, H_2 is not attacked to any appreciable extent, whereas Br_2^- is eventually reduced by e_{aq}^- and H atoms. Oxygen, on the other hand reacts with e_{aq}^- or H to give O_2^- or HO_2 and the latter disproportionates to give H_2O_2 and O_2

$$e_{aq}^- + O_2 \longrightarrow O_2^- \qquad\qquad k = 1.9 \times 10^{10}\ M^{-1}\,sec^{-1} \quad \text{(ref. 73)} \quad (34)$$

$$H + O_2 \longrightarrow HO_2 \qquad\qquad k = 2 \times 10^{10}\ M^{-1}\,sec^{-1} \quad \text{(ref. 71)} \quad (35)$$

$$O_2^- + O_2 \xrightarrow{H_2O} HO_2^- + O_2 + OH^- \quad k = 1.5 \times 10^7\ M^{-1}\,sec^{-1} \quad \text{(ref. 56)} \quad (36)$$

Under these conditions $G(H_2O_2) = G_{H_2O_2} + \frac{1}{2}(G_H + G_{e^-} - G_{OH})$.

There are a number of indirect methods to determine the radiolytic yield of H_2O_2 using oxygen saturated solutions (6,9). Alternatively one may use a reagent such as tetranitromethane in the presence of oxygen which removes *all* free radicals quantitatively and leaves the "molecular" H_2O_2 intact (80). A relatively straightforward estimate of the yield of primary yield of H_2O_2 may also be obtained from the yield of oxygen generated in the presence of ceric sulfate acid solution (81). The only direct procedure to determine the radiolytic yield of H_2O_2 is to measure the formation of $H_2O_2^{18,18}$ from H_2O^{18} in the presence of $H_2O_2^{16,16}$ as protective reagent, and extrapolating to zero $H_2O_2^{16,16}$ concentration (82).

The radiolytic yield of H atoms can be determined directly by measuring the yield of HD formed from a deuterium-carrying substrate such as D_2 (83), CD_3OH (84,85), or $DCOO^-$ (86,87) or indirectly by conversion of H atoms to hydrated electrons (reaction 25) and measuring the yield of the latter by pulse radiolysis (88). Further, one may derive the yield of H atoms from the increase in the total yield of hydrogen generated in water on addition of aliphatic solutes at low concentrations (89).

The yield of hydrated electrons from the radiolysis of water may be determined directly spectrophotometrically by pulse radiolysis, provided its molar extinction coefficient is accurately known (88). The rate of production of e_{aq}^- may also be determined from the yield of reduced products such as N_2 from N_2O (57), isopropanol from acetone, or of nitroform ions from tetranitromethane (89,90). The formation of OH radicals by the reaction

$$e_{aq}^- + N_2O \longrightarrow N_2O^- \xrightarrow{H_2O} N_2 + OH + OH^-$$

may also be used as a measure for the yield of e_{aq}^-.

There is no direct way to determine the radiolytic yield of OH radicals. The reactions of OH with H_2, H_2O_2, I^-, Br^-, and $Fe(CN)_6^{-4}$ are the best systems to measure the primary yield of OH radicals indirectly.

$$
\begin{aligned}
OH + H_2 &\rightarrow H_2 + H \\
H + OH^- &\rightarrow e_{aq}^- \quad \text{(ref. 88)}
\end{aligned}
\qquad (37)
$$

$$OH + H_2O_2 \rightarrow H_2O + HO_2$$
$$HO_2 \rightarrow H^+ + O_2^- \qquad (38)$$
$$O_2^- + C(NO_2)_4 \rightarrow C(NO_2)_3^- + NO_2 + O_2 \quad \text{(ref. 80)}$$

$$OH + I^- \rightarrow I + OH^-$$
$$I + I^- \rightarrow I_2^- \quad \text{(ref. 72)} \qquad (39)$$

$$OH + Br^- \rightarrow Br + OH^- \quad \text{(ref. 80)}$$
$$Br + Br^- \rightarrow Br_2^- \qquad (40)$$

$$OH + Fe(CN)_6^{4-} \rightarrow OH^- + Fe(CN)_6^{3-} \quad \text{(refs. 75, 91)} \qquad (41)$$

All these reactions have been studied by pulse radiolysis. The "classical" procedures (4,6,9) are based on overall yields of oxidized products or on balance of material:

$$2G_{H_2O_2} + G_{OH} = 2G_{H_2} + G_{e_{aq}^-} + G_H \qquad (42)$$
are less reliable

In Table I we have summarized the most reliable values of the yields of the different radiolytic products cited above as obtained under different conditions. The implications of the change of yield with LET described in this table will be discussed below in Section V.

IV. THE "CLASSICAL" DIFFUSION MODEL OF THE RADIOLYSIS OF WATER

Radiation chemistry has two main objectives. One is to investigate the chemical behavior of matter under radiation and the second is to explain the *mechanism* of the radiolytic processes. Many of the investigations of the effects of different factors on radiolytic yields have been carried out in order to prove one or another of the postulated mechanisms of radiolysis. Because of this one finds in many cases conflicting statements on the chemical behavior of radiolyzed systems. It is necessary therefore, at this stage, to describe in short the proposed mechanisms of the radiolysis of water which have developed up to date.

There are in principle two approaches to our problem. One is that of the physicists who are trying to extrapolate information on the structure of matter and on the interaction of radiation with gases to a condensed polar medium like water (5,35,65,97). The other approach is to derive information from the identity of the radiolytic products, their yields, their specific rates of reaction and from the effect on their yields of added reactants to the radiolyzed system. From this information chemists try to extract a mechanism consistent with all the experimental facts (4,6,9). Unfortunately

Table I

The Radiolytic Yields of H_2O_2, H_2, OH, H, and e_{aq}^- under Different Conditions[a]

Radiation	pH	$G_{H_2O_2}$	G_{H_2}	G_{OH}	G_H	$G_{e_{aq}^-}$	$G_H + G_{e_{aq}^-}$	G_{HO_2}	$G(-H_2O)$[b]	$G(-H_2O)_{corr}$[c]	Ref.
Co^{60} gamma or	0–2	0.8	0.45	2.95	0.6	3.05	3.65	0.008[f]	4.55	4.55	92
fast electrons	4–9	0.75[d]	0.45	2.8	0.6	2.8	3.4		4.3	4.3	88, 80
(>1 MeV)	12	0.75	0.40	2.9	0.55	3.05	3.6		4.4	4.4	88
H^3 beta	0.5	1.0	0.6	2.1			2.9		4.1	4.3	93
18 MeV D$^+$	0.5	1.03	0.7	1.75			2.4		3.85	4.1	94
8 MeV D$^+$	0.5	1.2	1.05	1.45			1.7		3.85	4.3	94
32 MeV He^{2+}	0.5	1.25	1.15	1.05			1.3		3.55	4.0	94
5.5 MeV He^{2+}	0.5	1.34[e]	1.57	0.5			0.6	0.11	3.5	4.2	95
B$^{10}(n, \alpha)$ Li^7 recoils	0.5	1.55	1.05	0.45			0.25		3.55	4.3	96

[a] The radiolytic yields are expressed in G units, i.e., the number of molecules or radicals produced per 100 eV absorbed energy.

[b] $G(-H_2O) = 2G(H_2O_2) + G_{OH} + 3G(HO_2)$.

[c] $G(-H_2O)_{corr} = G(-H_2O) + G(H_2O_2) - 0.8$.

[d] Reference 52.

[e] Reference 95.

[f] Reference 105.

no mechanism derived from chemical data accounts for all observations, nor does it agree in full with the predictions based on physical grounds.

There is one mechanism which has been in accord with most experimental observations gathered until recently and it is still generally accepted as the best approximation of the actual mechanism of the radiolysis of water (98). We are referring here to the "diffusion model" based on the radical theory of Weiss (99) and the expanding spur kinetics (3), which was formulated by Samuel and Magee (101) developed in the computations of Ganguly and Magee (102), Flanders and Fricke (103) and those of Dye and Kennedy (104) and which culminates in the computations of Kuppermann (105). We shall here present this model in its revised form (106) which takes into consideration the existence of the hydrated electron and consequently the uneven distribution of radicals within the spur (107,108).

According to this mechanism fast electrons dissipate their energy by ionization and excitation of water molecules. The excited water molecules undergo deexcitation without producing any chemical products. The ionized water molecules and the electrons extracted from them interact with water to give OH radicals and hydrated electrons.

$$H_2O + e_{fast}^- \rightarrow H_2O^+ + e^- + e_{fast}^- \tag{43}$$

$$H_2O^+ + H_2O \rightarrow H_3O^+ + OH \tag{44}$$

$$e^- + \text{water} \rightarrow e_{aq}^- \tag{45}$$

The estimated half-lives of these processes are $t_{1/2}$ (43) $\simeq 10^{-16}$ sec (109), $t_{1/2}$ (44) $\simeq 10^{-14}$ sec (110), and $t_{1/2}$ (45) $\simeq 10^{-10}$ sec (111), respectively. These reactions are followed by the following secondary reactions:

$$e_{aq}^- + H_3O^+ \rightarrow H + H_2O \qquad k = 2.4\ 10^{10}\ M^{-1}\ sec^{-1} \quad \text{(ref. 73)} \tag{24}$$

$$e_{aq}^- + OH \rightarrow OH^- \qquad k = 3.0\ 10^{10}\ M^{-1}\ sec^{-1} \quad \text{(ref. 74)} \tag{46}$$

$$e_{aq}^- + e_{aq}^- \rightarrow H_2 + 2OH^- \qquad k = 4.5\ 10^9\ M^{-1}\ sec^{-1} \quad \text{(ref. 74)} \tag{47}$$

$$OH + OH \rightarrow H_2O_2 \qquad k = 5 \times 10^9\ M^{-1}\ sec^{-1} \quad \text{(ref. 45)} \tag{48}$$

These precede the ternary reactions

$$e_{aq}^- + H \rightarrow H_2 \qquad k = 2.5 \times 10^{10}\ M^{-1}\ sec^{-1} \quad \text{(ref. 74)} \tag{49}$$

$$OH + H \rightarrow H_2O \qquad k = 7 \times 10^9\ M^{-k}\ sec^{-1} \quad \text{(ref. 45)} \tag{50}$$

as well as the reactions (20–23) involving H_2O_2 and H_2 described in Section III.

Now it was assumed that the energy of the fast electrons is dissipated within the medium in localized portions which produce an average of 6 radicals (OH and e_{aq}^-) in a limited volume (101,102), called "spur," with an estimated radius of about 30 Å (106). The initial (up to 10^{-10} sec) distribution of OH radicals, H_3O^+ ions, and hydrated electrons within

From the H/D isotope effect on the formation of H_2 and H atoms from the radiolysis of water by low energy alphas and heavier nuclei it was concluded (114) that this H_2 cannot be formed primarily from the $e_{aq}^- + e_{aq}^-$ reaction and that the H atoms do not originate from the $e_{aq}^- + H_3O^+$ reaction as previously assumed (6). These results suggest that the diffusion model which seemed to offer a quantitative explanation of the dependence of the radiolytic yields on LET is still open to serious criticism.

B. The Effect of Added Solutes

It has been stated above that H_2 and H_2O_2 cannot accumulate in irradiated water unless the free radicals, which are produced simultaneously, are removed by some reactant added to the solution. Now, once the concentration of such an additive is considerably increased, it is found that the yield of either H_2 or H_2O_2 decreases, in spite of the fact that there is no reaction between the additive and the molecular products. The dependence of the decrease in the yield of the molecular products on the concentration of the additive is not linear and it may be roughly expressed by the function $G = G^0 - \alpha[\text{additive}]^{1/n}$ where α is a function of the additive, and $n = 3$ for most systems studied at low LET (6,9,98,106,115). At high LET, n has been claimed to increase by a factor of 2 (116). It should be noted, however, that the dependence of n on LET is a consequence of the calculation of Ganguly and Magee (102) and the results of Burton and Kurien (116) for 24 MeV electrons which were found to depend on $n = 2.6$ can easily be fitted on a cube plot ($n = 3$). According to the diffusion model, on the other hand, the cube root dependence is fortuitous and coincidental to 1 MeV electrons only (117). In any case for a given LET, within a limited range of concentrations, generally $10^{-4}-10^{-1}M$, α is constant and depends only on the nature of the solute. At concentrations higher than $0.5M$ the value of α diminishes (or that of n increases); the deviations however are much smaller than predicted by the diffusion model calculations (106), especially if corrected for the changes in ionic activity (118). There is, however, little support for the suggestion (118,119) that a given solute has *two* different α values at low and at high solute concentrations (120).

A semiquantitative correlation between the values of α for a limited number of additives and their rates of reaction with H atoms and OH radicals has been suggested as evidence that the H atoms and OH radicals are the precursors of the "molecular" H_2 and H_2O_2, respectively (6,121). The discovery of the hydrated electron resulted in a modification of the interpretation of the mechanism of formation of H_2, adding reactions (47) and (49) to the scheme, the former reaction becoming now the main source for H_2 (106,122,123). A reevaluation of the α factors of different additives compared with their rates of reaction with e_{aq}^- and H atoms, has revealed serious inconsistencies with the assumptions of the diffusion model. Further disagreement was manifested when a more critical study of the effect of additives on the yield of molecular H_2O_2 was undertaken.

It was shown that the effect of NO_3^- on G_{H_2} is pH independent up to $0.8N$ H_2SO_4 (124,125), which is completely unexpected if e_{aq}^- in the spur is converted (by reaction (24)) to an H atom which is far less reactive toward NO_3^- than e_{aq}^- $(k_{H+NO_3^-} = 1 \times 10^7$ compared with $k_{e_{aq}^- +NO_3^-} = 1.1 \times 10^{10}\,M^{-1}\,sec^{-1})$ (126,127). In N_2O saturated solutions the decrease in G_{H_2} with N_2O concentration is again independent of pH (125a,b) in spite of the very large difference between $k_{e_{aq}^- +N_2O} = 8.7 \times 10^9$ and $k_{H+N_2O} = 2.2 \times 10^5\,M^{-1}\,sec^{-1}$ (73,127a). The evidence against reaction (47) as the main source for the "molecular" hydrogen is corroborated by the fact that H/D isotope effect on the formation of the H_2 is also pH independent down to pH $= 0$ (128), and its value is lower by about a factor of 2 from that expected according to the diffusion model (129). These facts which are hard to reconcile with the diffusion model (130), are supported by the fact that $Co(II)_{aq}$ at pH $= 4$ has a relatively slight effect on G_{H_2} (inducing a decrease from $G_{H_2} = 0.45$–0.40 at $0.1M$ (68), compared with a decrease to about $G_{H_2} = 0.26$ (68,132) at the same concentration of NO_3^-) Tl^+ was also found to have no effect, within 5%, on G_{H_2} both in neutral and acid solution (133). On the other hand, $Co(NH_3)_6^{3+}$ and $Fe(CN)_6^{3-}$ exhibit an effect on G_{H_2} comparable with that of NO_3^- (68). All the ions cited have comparable reactivities toward e_{aq}^-, and their rates of reaction with hydrated electrons are either diffusion controlled or very close to the diffusion controlled limit $(k_{e_{aq}^- +X} = 1.1, 1.2, 2.9, 3.0, 9.0, 0.3 \times 10^{10}\,M^{-1}\,sec^{-1}$ for NO_3^-, Co_{aq}^{II}, Ni_{aq}^{II}, Tl^+, $Co(NH_3)_6^{3+}$, and $Fe(CN)_6^{3-}$, respectively) (29,73,134,135). Were H_2 produced from e_{aq}^- in the spurs it is expected that all these reactants would have a comparable effect on G_{H_2}. It should be noted that $Co(NH_3)^{3+}$ which is very effective in reducing G_{H_2}, is practically nonreactive toward H atoms ($k = 1.6 \times 10^6$ $M^{-1}\,sec^{-1}$) (136), whereas Tl^+ which is far more reactive toward H atoms ($k = 1.0 \times 10^8\,M^{-1}\,sec^{-1}$) (68) has no effect on G_{H_2} even in acid solution (133). These findings exclude also H atoms as the main precursors of molecular H_2. Interestingly enough when it was tried to compute G_{H_2} by the diffusion model using recent rate constants and diffusion coefficients, it was not possible to account for more than 60% of the experimental yield and it was necessary to assume that 40% of the "molecular" H_2 is produced by a pathway different from reactions (47), (49), and (52) (98).

Considering the effect of additives on $G_{H_2O_2}$ one encounters a similar situation. Tl^+ ions are more effective in reducing $G_{H_2O_2}$ than $Fe(CN)_6^{4-}$

and I^- (123,133,137). The specific rates of reaction of these three reactants are comparable and are probably diffusion controlled ($k_{OH+X} = 0.8$, 1.7, 1.0×10^{10} M^{-1} sec^{-1} for Tl^+, $Fe(CN)_6^{4-}$, and I^-, respectively) (72,138). On the other hand, iodide and bromide ions which differ in their reactivity with OH radicals by a factor of 5, show an identical effect on $G_{H_2O_2}$ (98). Considering organic additives it was shown that the acrylamide and benzamide are as effective as iodide in diminishing $G_{H_2O_2}$ (133). On the other hand 2-propanol (123), methanol (138), and ethanol (50,139) do not affect $G_{H_2O_2}$ even at 0.5–1.0M concentrations. Thus the effect of these reactants on $G_{H_2O_2}$ cannot be correlated with their rates of reaction with OH radicals ($k_{X+OH} = 1.9$, 2.1, 0.5, 0.7, and 1.7×10^9 M^{-1} sec^{-1} for acrylamide, benzamide, methanol, ethanol, and isopropanol, respectively) (72,140).

There exists another series of most interesting observations on the effect of added H_2O_2 on $G_{H_2O_2}$ (82,141). It has been found that the yield of $H_2O_2^{18,18}$ from H_2O^{18} diminishes with increasing concentration of $H_2O_2^{16,16}$ (141). $G^0(H_2O)_2^{18,18} - G(H_2O_2^{18,18}) = \Delta G(H_2O_2^{18,18})$ depends on $[H_2O_2^{16,16}]^{1/3}$ (83). At the same time $H_2O_2^{16,18}$ is being produced so that $G(H_2O_2^{16,18}) + 2G(H_2O_2^{18,18}) = G^0(H_2O_2^{18,18}) = G_{H_2O_2}$ (82). A quantitative analysis of the results shows that $H_2O_2^{16,18}$ cannot be formed from an interaction between HO^{18} radicals and $H_2O_2^{16,16}$, as practically all HO^{18} radicals are scavenged by $H_2O_2^{16,16}$ before any $H_2O_2^{16,18}$ is being formed (82). Thus the precursor of $H_2O_2^{18,18}$ which is being scavenged by $H_2O_2^{16,16}$ produces at the same time $H_2O_2^{16,18}$ and this precursor *cannot* be an OH radical.

In analogy with Tl^+, which diminishes $G_{H_2O_2}$ without affecting G_{H_2}, it was found that NO_3^- ions which diminish G_{H_2} most effectively, have no effect whatsoever on $G_{H_2O_2}$ (123,142). G_{H_2} and $G_{H_2O_2}$ are thus affected independently by different reactants which coincidentally may also be reactive toward e_{aq}^-, H atoms, or OH radicals.

It may be concluded that there is ample evidence that the "molecular" H_2 is *not* produced primarily by the $e_{aq}^- + e_{aq}^-$, $e_{aq}^- + H$, and $H + H$ reactions (47, 49, and 52) and that the OH + OH reaction is *not* the main pathway for the formation of "molecular" H_2O_2 under low LET conditions. Further, it does not seem plausible that H_2 and H_2O_2 are formed from a single precursor, such as an excited water molecule, by a bimolecular process,

$$H_2O^* + H_2O^* \rightarrow H_2 + H_2O_2 \qquad (53)$$

as has been previously suggested (122,143).

We shall end this description of the effect of solutes on the radiolytic yields with a short discussion of the behavior of G_H. The conversion of e_{aq}^- to H atoms by reaction (24) is according to the diffusion model a secondary or tertiary reaction, which may be suppressed by the addition of e_{aq}^- scavengers according to kinetics competition. A certain yield of H atoms is produced in neutral solution which cannot be accounted for by reaction (24) in the bulk of solution. These H atoms which were shown to be chemically identical with those produced by reaction (24) (10,144,145), are formed with a different H/D isotope effect (114,128,129). This implies that they are not formed from reaction (24) occurring in spurs as is predicted by the diffusion model. There are contradictory reports on the effect of solutes on G_H, the primary yield of H atoms. On one hand it was claimed that G_H is unaffected by e_{aq}^- scavengers like NO_3^- (146) and by (147,147a) OH^- which should eliminate H_3O^+ in the spur (147,147a,147b). On the other hand it was claimed that NO_3^- at high concentration affects G_H in a manner parallel to G_{H_2} (148) and that G_H diminishes at high pH (98). It may be concluded that the effect of solutes on G_H is still an open question, as is also the mechanism of formation of these H atoms.

C. The Effect of Other Parameters on the Radiolytic Yields

The effect of dose rate at low LET on the molecular yields has been investigated and the results (149) are in disagreement with the predictions of the older computations of the diffusion model (105). Further, pulse radiolysis data at high dose rates show deviations from the behavior expected from the diffusion model (150). These results may be interpreted by assuming the existence of H_3O radicals as short-lived precursors of H atoms. The latter species will yield H_2 on dimerization (150).

The effect of pressure on the yields of the different radiolytic products has been recently investigated (151). The results, which show no change in these yields up to about 7 kilobars, are surprising if one assumes that according to the diffusion model reaction (24) which is not diffusion controlled and which has a considerable $\Delta V^\ddagger$ has to compete in the spur with radical–radical diffusion controlled processes, e.g., reactions (46,47,49).

Temperature has also little effect on the radiolytic yields (152,153). As G_{H_2} was found temperature independent in the range 8–96°C (152), this may indicate again that H_2 does not originate from reactions (47), (49), and (52) competing with the recombination reactions (46) and (50) and with reaction (24), as required by the diffusion model, because H atom reactions have been shown to have about half the activation energy of e_{aq}^- reactions (154) (probably because H atoms may diffuse through water without requiring the breaking and making of hydrogen bonds).

Radiolytic yields in D_2O compared with those in H_2O are of considerable interest. The most reliable parameter for a quantitative evaluation of the effect of deuterium substitution is the ratio $R_x = G_x^{H_2O}/G_x^{D_2O}$. From the extensive data in literature one may derive $R_{H_2} = 1.21 \pm 0.09$, $R_{H_2O_2} =$

1.16 ± 0.09, $R_{OH} = 1.00 \pm 0.08$, $R_H = 0.94 \pm 0.07$, and $R_{-H_2O} = 1.01 \pm 0.07$ (154a). If the difference between G_{H_2} and G_{D_2} was due to a broader distribution of the electrons in the spur in D_2O as has been tentatively assumed (104) resulting in less $e_{aq}^- + e_{aq}^-$ and $e_{aq}^- + H$ recombination, one would expect $R_{H_2} \gg R_{H_2O_2} < 1.0$ and $R_{-H_2O} < 1.0$, because less $e_{aq}^- + OH$ recombination should take place in the extended spur. Moreover, if H_2O_2 was formed by the $OH + OH$ recombination the significant difference between $G_{H_2O_2}$ and $G_{D_2O_2}$ should be accompanied by a significant increase of G_{OD} over G_{OH} which has not been observed. It must be concluded therefore that the observed radiolytic yields in D_2O are inconsistent with the predictions of the diffusion model.

VI. MECHANISMS OF RADIOLYSIS OF WATER IN VIEW OF THE MORE RECENT EXPERIMENTAL FINDINGS

In Section IV we have described the classical free radical mechanism and the diffusion model. Although no alternative or revised comprehensive mechanism has been formulated up to date it seems that time has come for a thorough reconsideration of the whole subject. Such an ambitious task is obviously beyond the scope of this review. It may take another five to ten years before a mechanism is elaborated which will account for all the experimental findings, many of which have to be rechecked for methodical and other errors. We shall now mention in brief some of the modifications proposed for the revision of the existing free radical diffusion theory. Each such revision suggests a series of elementary steps leading to primary and occasionally also to secondary products different from H, OH, and e_{aq}^-.

We shall start with the mechanism suggested by Magee in 1963 (155)

$$e^-(\text{fast}) + H_2O \rightarrow H_2O^+ + e^- + e_{aq}^- \tag{54}$$

$$H_2O^+ + H_2O \rightarrow H_3O^+ + OH \tag{55}$$

$$e^- + H_3O^+ \rightarrow H_3O \tag{56}$$

$$H_3O \rightarrow H_2O + H \tag{57}$$

$$\approx H_3O_{aq}^+ + e_{aq}^- \tag{58}$$

all these reactions take place inside the spur, the track, or blob (112) within $< 10^{-10}$ sec. The main difference between this mechanism and the "classical" one is the existence of H_3O as an intermediate and the mode of formation of e_{aq}^-. There is little doubt that H_3O may exist as a short-lived intermediate in water; its presence has been demonstrated in the gas phase (155a), postulated in pulse radiolysis (154) and its formation in reaction (56) has been implied from the H/D isotope effect (128). On the other hand there is no independent evidence that H_3O is a precursor of

e_{aq}^-. In any case it is unlikely that this is the main pathway for the formation of e_{aq}^-. Moreover, this scheme does not resolve the main objections to the diffusion model.

Sworski has extended the "H_3O theory" to explain the formation of the "molecular" hydrogen (125).

$$H_3O + H_3O \rightarrow H_2 + 2H_2O \tag{59}$$

As this reaction has been shown to be inconsistent with the experimental findings, another species was added to this mechanism, namely H_2O^* which is considered to be identical with the radical pair $H_3O \cdot OH$ (125b).

$$H_3O^+ + e_{aq}^- + OH \rightarrow H_3O \cdot OH \rightarrow 2H_2O \tag{60}$$

$$H_3O \cdot OH + H_3O \rightarrow H_2 + OH + 2H_2O \tag{61}$$

Sworski's scheme has a few drawbacks. It does not account for the formation of H_2O_2, it requires part of the H_2 to be formed from recombination of H_3O in the bulk of the solution ("interspur reactions") (125), it does not try to explain the chemical reasons for the preferential reactivity of certain scavengers with the precursor of H_2. A further refinement of Sworski's ideas was the suggestion that H_2O^* decomposes by a unimolecular process to give H_2 (156a). From the standpoint of energetics such H_2O^* could not be formed from $H_3O^+ + e_{aq}^-$ as suggested. Further, this hypothesis does not seem to remove the other objections cited above.

A completely different scheme of elementary reactions has been proposed by Voevodski (157), based on the formation of excited or hot H atoms and OH radicals, in analogy with species obtained as the results of nuclear transformations.

$$e_{fast}^- + H_2O \rightarrow H^* + OH^* + e_{fast}^- \tag{62}$$

$$H^* + H_2O \rightarrow H_2 + OH \tag{63}$$

$$H^* + H_2O \rightarrow H + H_2O \tag{64}$$

$$OH^* + H_2O \rightarrow H_2O_2 + H \tag{65}$$

$$OH^* + H_2O \rightarrow OH + H_2O \tag{66}$$

This scheme is rather unlikely as it fails to explain two most important facts in the behavior of radiolyzed water: First, it does not account for the formation of hydrated electrons; second, it does not explain the effect of additives on the molecular yields, because H^* or HO^* are expected to undergo deactivation or react with water on first collision, thus only *very* high concentrations of additives could compete with water for these hot radicals.

The effect of solutes on the "molecular" yields has been interpreted by Bednar (158,159) as the result of a kind of collective excitation or exciton migration within the aqueous system resulting in the formation of excited solvated solute molecules. These may break down to give either fragments of the solute or excited water molecules. This mechanism would mean a complete revision of one of the main arguments of the diffusion theory, namely, the interpretation of the effect of solutes on the yields of H_2 and H_2O_2. There is, however, little physical evidence for such a mechanism. The use of the Uncertainty Principle as an argument for the existence of collective excitation is unjustified and results from a misunderstanding of the physical meaning of this principle (160). Furthermore, the claims for the reaction $[nH_2ONO_3]^* \rightarrow NO_2 + O$, which might be used as evidence for Bednar's proposed mechanism, could not be experimentally verified (142).

A reaction which has not been considered thus far in this review has been proposed by Platzman (161) as a pathway for the formation of "molecular" H_2

$$e^-_{subexit} + H_2O \rightarrow H^- + OH \tag{67}$$

$$H^- + H_2O \rightarrow H_2 + OH^- \tag{68}$$

This is a rather plausible mechanism as it predicts the low H/D isotope effect in the formation of molecular H_2 (128,129), and as far as the reactivity of H^- is concerned it is expected to react faster with double electron acceptors, like NO_3^-, NO_2^-, or H_2O_2, than with single electron acceptors like Ni^{2+}, Co^{2+}, or Tl^+ just as has been found for the precursor of H_2. The reactivity of H^- toward the different species can be inferred from the redox potentials of the corresponding oxidants; $E^0(NO_3^- - NO_2^-) = -0.01$ V, $E^0(H_2O_2 - H_2O) = -1.77$ V and $E^0(HNO_2 - H_2N_2O_2) = -0.86$ V (162). The reduction of Ni^{2+} or Co^{2+} to their zero oxidation state by H^- is thermodynamically forbidden $(E^0(Ni^{2+} - Ni^0) = +2.50$ V, $E^0(Co^{2+} - Co^0) = +2.77$ V) (162) and evidently $Tl^+ \rightarrow Tl^-$ is not feasible in aqueous medium. On the other hand, $E^0(Cu^{2+} - Cu^0) = -0.337$ (162) allows the reduction by H^-, and actually Cu^{2+} is quite effective in diminishing G_{H_2} (121). The main drawback of this mechanism is the assumption that reaction (33) is considerably slower than diffusion controlled, otherwise no solute could ever compete with water for H^-. The small isotope effect observed, which has been used as an argument for the very fast rate of this reaction (129), could stem from an addition of H^- to H_2O to give $HHOH^-$ as a short lived intermediate, which decomposes to $H_2 + OH$ without any appreciable H/D isotope effect.

The formation of the "molecular" hydrogen via reactions (32) and (33) implies that reaction (33) is considerably slower than the diffusion controlled limit, otherwise no solute could ever compete with water for H^-. Reaction (33) was assumed to proceed by a diffusion controlled rate in view of its H/D isotope effect which is close to unity (129). It may be conceived however, that the rate determining step in reaction (33) involves primarily the addition of H^- to H_2O rather than the H—OH bond cleavage. Alternatively one has to assume that the scavengers react with subexcitation electrons, the reactivity of which can hardly be predicted at present. The lower yield of "molecular" hydrogen in D_2O (see Section V-C) may be due to a smaller cross section of D_2O to undergo dissociative electron capture.

Reactions (32) and (33) have been suggested as the source of the "molecular" hydrogen formed from water in the gas phase (163), the existence of which has been confirmed in a number of studies (164–166). In the gas phase evidently no nonhomogeneous kinetics in "spurs" are feasible, thus the hydrogen produced in the presence of e^- and H scavengers cannot originate from a radical–radical reaction. Although other mechanisms for the formation of the "molecular" H_2 in the gas phase have been proposed (167), reactions (67) and (68) are still considered as plausible (168). It is most interesting to note that the observed G_{H_2} in water vapor (~ 0.5) is identical, within the experimental error, with G_{H_2} in liquid water, an agreement which can hardly be considered as fortuitous. Recent experiments (168a) which have been used as evidence for the prevalence of a monomolecular mechanism in the formation of the "molecular" hydrogen in the radiolysis of water vapor, are utterly nonconclusive because the observed identity between the isotopic composition of the water and the "molecular" hydrogen implies only that the latter is being formed with an isotope effect of unity, as expected for the hydride mechanism in the gas phase, and no conclusions can be drawn concerning the number of water molecules involved in the formation of H_2.

If about half of the molecular H_2 in liquid water were formed by reaction (68) which has an H/D isotope effect close to unity (129), and the other half by radical–radical reactions, one would obtain the experimental value (128,129) for the H/D isotope effect on H_2 formation (169). This suggestion is in agreement with $G_{H^-} = 0.2 \pm 0.1$ estimated by Platzman (161). However, the observed *decrease* in the H/D isotope effect at high LET (114), where the radical–radical pathway should gain weight, presents a serious objection to this hypothesis. It may, therefore, just as well be assumed that reaction (67) involves an appreciable H/D isotope effect and that a much larger fraction of the "molecular" H_2 is produced via reactions (67) and (68). It has been further shown that the H/D isotope effect on the formation of the "molecular" hydrogen at low LET is independent of the ionic strength (68,154a). If there were contributions of both reactions (47) and (49) to the formation of H_2 in the spur, their relative rates would change at high ionic strength, resulting in an increase in the H/D isotope effect. Such an increase was not observed however.

An alternative theory has been recently suggested by W. H. Hamill (private communication) to explain the formation of the molecular H_2. According to this theory there is a third type of electrons involved in the radiolysis of water, namely thermalized nonhydrated electrons, which move very rapidly throughout the medium and readily react with H_2O^+ before the latter has a chance for proton transfer. The result of this reaction would be highly excited H_2O, part of which produce "hot" H atoms. The latter form H_2 by reaction (63). The effect of scavengers in this case would be to eliminate the thermalized nonhydrated electrons before they have a chance to react with H_2O^+. One drawback of this theory, which is the presumed short half-life of H_2O^+ in water, may be overcome by assuming a reaction between thermalized nonhydrated electrons and H_3O^+ resulting again in "hot" H atoms. The peculiar behavior of scavengers would then be due to the different kinetic behavior of the non-hydrated electrons compared with their hydrated analogs. This theory, which is evidently also consistent with the measured H/D isotope effects, requires a lifetime of $> 10^{10}$ sec for the thermalized electron before it undergoes hydration—an assumption which is not in serious conflict with any experimental data.

Whereas a number of alternative pathways have been proposed to explain the formation of the "molecular" H_2, little consideration has been devoted to alternative pathways for the production of the "molecular" H_2O_2. Hydrogen peroxide may be formed, by a nonradical–radical mechanism via reactions (69) and (71).

$$e_{\text{fast}}^- + H_2O \rightarrow H_2O^{**} + e^- \qquad (H_2O^{**} = \text{a superexcited } H_2O^{170}) \qquad (69)$$

$$H_2O^{**} \rightarrow OH^+ + H + e^- \qquad (70)$$

$$OH^+ + H_2O \rightarrow H_3O^+ + O \qquad (71)$$

$$O + H_2O \rightarrow H_2O_2 \qquad (71a)$$

The formation of OH^+ ions has been demonstrated in silent discharge (171) and in the mass spectrometer, where the appearance energy of the OH^+ ion is 18.7 eV (172–174), they are formed with a relatively large cross section, about 25% of that of H_2O^+ for 100 eV electrons (173). Reaction (70) is in fact the dissociation of a superexcited molecule as predicted by Platzman (156,170). It is plausible that the ratio of yields OH^+/H_2O^+ in liquid water will be comparable with that in the gas phase, as H_2O^{**} has little chance to undergo deactivation before dissociation. As H_2O^+ is considered as the main precursor of OH radicals in liquid water, the comparability of OH^+/H_2O^+ in the gas phase with $G_{H_2O_2}/G_{OH}$ in liquid water may be of significance.

The hypothesis that OH^+ is the precursor of H_2O_2 is consistent with the effect of different additives on $G_{H_2O_2}$. It is expected that aliphatic compounds which are excellent OH scavengers should be relatively non-reactive toward O atoms. On the other hand, olefins and aromatic compounds should be much more reactive toward this species. The metal ions like $Fe(CN)_6{}^{-3}$ which are capable of undergoing a single electron transfer, are thus expected to be less effective than Tl^+ which undergoes preferentially a double electron transfer to Tl^{3+}. Further, the isotopic experiments (82), described above (Section V-B), can be understood if OH^+ is the precursor of H_2O_2.

$$O^{18} + H_2O_2^{16/16} \rightarrow HO^{16}O^{18}H + O^{16} \tag{72}$$

$$O^{16} + H_2O \rightarrow H_2O_2^{16/18} \tag{73}$$

Reaction (72) is an electrophilic attack of O on H_2O_2 which results in isotopic exchange, the analogous attack of O on H_2O results in the formation of H_2O_2 (reaction (73)).

This proposed mechanism, unlike the diffusion theory, is consistent with the finding that the yield of "molecular" hydrogen peroxide is considerably lower in D_2O than in H_2O. Reaction (35) which is a dissociative ionization will have a lower probability in D_2O compared with the nondissociative ionization (23) (cf. 156).

It may be inferred that O atoms at their ground state $O(^3P)$ react much faster with molecular oxygen to give ozone than with water. This results in a decrease in quantum yield of photolysis of ozone in water at 3130 Å (175), where 60% of the O atoms produced are $O(^3P)$ (176). It is assumed, however, that those oxygen atoms which escape reaction with O_2, form H_2O_2 quantitatively. The lower yield of H_2O_2 observed in the photolysis of ozone at longer wave lengths (175), may be due to a secondary decomposition of H_2O_2 either by photolysis or by reaction with excited O_2 molecules produced under these conditions. The occurrence of reaction (37) in the gas phase may be inferred from the fact (141) that water molecules act as inhibitors of the radiolytic isotopic exchange chain reaction $O_2^{16,16} + O_2^{18,18} \rightarrow 2O_2^{16,18}$. It is rather unlikely, in any case, that H^- and OH^+ may be formed in a single event:

$$e_{fast}^- + H_2O \rightarrow H^- + OH^+ + e^- \tag{74}$$

in view of the fact that OH^+ is most probably the product of a super-excited molecule, whereas H^- is formed by a dissociative electron capture of a subexcitation electron. This means that according to our scheme H_2 and H_2O_2 are produced by two independent processes.

The newly suggested pathways for the formation of "molecular" H_2 and H_2O_2 in radiolyzed water differ in principle from the "classical" approach. We have been considering, in all cases, "unimolecular" processes (involving only solvent molecules as the second reagent) in contrast to the bimolecular reactions implied in the diffusion theory. If bimolecular reactions occur in our new scheme such as

$$OH^+ + H^- \rightleftharpoons H_2O^* \rightarrow H + OH \tag{75}$$

these would *diminish* the molecular yield rather than increase it.

We have now to reconcile the chemical evidence which favors H^- or e^- and OH^+ as precursors of H_2 and H_2O_2, respectively, with the kinetic behavior of radiolyzed aqueous solutions. It has been pointed out once and again (125a) that the cube root relation is an empirical description of the effect of solutes on the yields of the molecular products. It has been further pointed out (125,176) that the yield of the molecular products at low LET in the concentration range 10^{-2} to $5M$ may be satisfactorily described by the general function,

$$\frac{1}{G_M} = \frac{1}{G_M{}^0} + \alpha[\text{solute}] \tag{76}$$

where $G_M = G_{H_2}$ or $G_{H_2O_2}$ and $G_M{}^0$ is the molecular yield extrapolated to [solute] = 0. This function may be derived from the following mechanism, exemplified for the molecular hydrogen

$$H^- + H_2O \xrightarrow{k_1} H_2 + OH^- \quad \text{(a first order process)} \tag{68}$$

$$H^- + S \xrightarrow{k_2} H^+ + S^{2-} \quad \text{(S = solute)} \tag{77}$$

leading to

$$\frac{1}{G_{H_2}} = \frac{1}{G_{H_2}{}^0} + \frac{k_2}{k_1}\frac{1}{G_{H_2}{}^0}[\text{S}] \tag{78}$$

Thus there is no contradiction between the experimental behavior of moderately concentrated aqueous solutions and the H^- or $e^- - OH^+$ pathways for the formation of H_2 and H_2O_2.

It has been pointed out, however, that $G_{H_2}{}^0$ obtained by extrapolation is smaller than the observed G_{H_2} (125), which suggests that part of the hydrogen is produced by another mechanism. Interestingly enough the diffusion model treatment leads to the same conclusion, namely, that the "molecular" hydrogen is formed by two parallel pathways (98). It seems that the effect of reactive solutes at low concentrations (10^{-4} to $10^{-2}M$) may be better explained by the diffusion model treatment, whereas at higher solute concentrations the system follows the homogeneous first order kinetics. One cannot, therefore, exclude altogether radical–radical recombination reactions as a source for H_2 and H_2O_2, especially at high LET. At low LET it is expected that the effect of diffusion kinetics will be manifested at relatively low solute concentrations ($< 10^{-2}M$), whereas at high LET this effect will become pronounced at higher concentrations of solutes. This explanation of the effect of additives at low solute concentrations is in accord with the observed effect of solutes on photochemical processes (178) where a pair of radicals exists in a "cage" (179). Recent experiments have actually demonstrated the existence of radical pairs of the type $e_{aq}^- + H_3O^+$ in the radiolysis of water (180).

There exists another discrepancy between the photolysis and radiolysis of water. The quantum yield of photolysis to give H + OH is about 0.5 at 1850 Å (181). This is a rather high quantum yield for photolysis so close to the threshold of absorption of water in the UV (182). On the other hand it may be concluded that the dissociation of excited water molecules to produce H + OH in liquid water is very small, in contrast to the predominance of this reaction in the radiolysis of water in the gas phase. It is suggested that owing to the polymeric structure of liquid water there is an efficient dissipation of the excitation energy of H_2O^* formed by radiolysis within the water aggregates. The photolytic excitation takes place predominantly in monomeric H_2O molecules which exist in a small concentration in liquid water, in equilibrium with the aggregates. The monomeric H_2O^* has a much smaller probability to undergo deexcitation before dissociation. This mechanism is conceivable if the extinction coefficient of monomeric water is considerably higher than that of water in aggregates. The latter condition is plausible in view of the fact that at 185 nm water vapor has a molar extinction coefficient over 3 orders of magnitude higher than that of liquid water at 25°C (183). Furthermore, the extraordinarily high temperature coefficient of the extinction of water in the UV (184) supports the suggested monomer–polymer equilibrium in liquid water.

If the dissociative electron capture of subexcitation electrons in water, reaction (32), takes place primarily by monomeric water, in view of the analogy between the optical extinction coefficients and the cross sections of electron capture (170), one could attribute considerable lifetimes to e_{se}^- and consequently the action of scavengers on G_{H_2} could be explained by their reactions with subexcitation electrons.

Two additional major objections to the diffusion model should be mentioned here, which are in no conflict with our new interpretation. One is the fact that at low LET no $e_{aq}^- + e_{aq}^-$ or $e_{aq}^- + H$ reactions, taking place by second order reaction kinetics at high local concentrations, could be detected even within 1 nsec (185) or less (186) from the instance of a pulse of radiation. This observation excludes the generation of a major part of the hydrated electrons in spurs. Secondly, there is the discrepancy between the average amount of energy dissipated in water by electrons required by the diffusion theory (sufficient to produce an average of 6 radicals per spur) (105) and the significantly lower amounts of energy experimentally found to be dissipated in single interactions in low molecular weight materials similar to water (187,188).

Our interpretation of the behavior of radiolyzed water does not exclude the presence of "spurs" or "blobs" but it is believed that their contribution to the overall radiolytic yields at low LET is rather small, while most of the radiolytic process takes place in separate events homogeneously distributed throughout the medium. These events include (in the order of decrease in the energy involved):

1. The formation of superexcited water molecules, H_2O^{**}, which dissociate to form OH^+, H, and e^-, which, in turn, react to form H_2O_2 and e_{aq}^-. Some H_2O^{**} may also dissociate to $H^* + OH$ or $H + OH^*$, which may in turn produce some H_2 and H_2O_2.

2. The ionization of water to give $H_2O^+ + e^-$; followed by the formation of H_3O^+, OH, and e_{aq}^-.

3. The excitation of water to give H_2O^*. This may undergo dissociation to an $H + OH$ pair in a cage which, in most cases, in the absence of additives, will recombine.

4. The dissociative electron capture of subexcitation electrons by H_2O to yield $H^- + OH$; followed by the reaction of H^- with water to form H_2 and OH^-.

5. The dissociative electron capture of thermalized nonhydrated electrons by H_3O^+, resulting in the formation of "hot" H atoms. These may in turn produce H_2 on reaction with water. It is difficult at present to assess the relative contributions of steps *4* and *5*.

At high LET the local density of radicals and excited water molecules is high enough to make recombination reactions predominant, thus the kinetic behavior of the system will follow the pattern predicted by the diffusion model. As has been stated in Section V-A. it seems likely that also at high LET the distribution of the radicals *within* the tracks is homogeneous. This is in accord with our hypothesis that the majority of the primary radiolytic products are formed by "monomolecular" processes and that at high LET the overall local concentration of such separate events is increased.

It has been stated that the diffusion model explains a lot of data quite successfully. "There are some difficulties, but a model that explains this much cannot be all wrong" (189). My feeling is that our suggested mechanism of radiolysis removes many of those difficulties and although it is also not completely free from objections, it seems to be a better approximation of the mechanism of radiolytic processes in aqueous media.

VII. THE CHEMICAL BEHAVIOR OF THE
RADIOLYTIC PRODUCTS

Previous reviews of the radiation chemistry of aqueous solutions have described the radiolytic behavior of a large number of inorganic and organic solutes (4–6,9). These studies were primarily concerned with the identity of products and their yields. Not many additional systems have been investigated thoroughly, in recent years, for their radiolytic chemical changes. On the other hand, the last five years have been prosperous in kinetic studies on the rates of reaction of hydrated electrons, hydroxyl radicals and hydrogen atoms with hundreds of inorganic and organic compounds (135). These developments allow us a different and more fundamental approach to the radiation chemistry of aqueous solutions: Instead of trying to catalog the chemical behavior of hundreds of radiolyzed aqueous systems, only the chemical behavior of e_{aq}^-, H, and OH radicals will be described. The radiolytic fate of any given solute at low concentration may then be predicted semiquantitatively even in the absence of any direct chemical information. According to this approach radiation chemistry of aqueous solutions should be considered as the sum of the chemical behavior of the "primary" radicals produced in radiolyzed water. The chemistry of radiolyzed aqueous solutions, which takes place at time intervals longer than 10^{-8} sec, thus becomes an integral part of general chemistry.

A. Hydrated Electrons

The chemical behavior of the hydrated electron has been reviewed not too long ago (26) thus we shall limit ourselves to a brief summary only. The general pattern of the reactions of e_{aq}^- may be expressed by the equations

$$A + B^- \tag{79}$$

$$e_{aq}^- + AB \longrightarrow AB^- \xrightarrow{H_2O} ABH + OH^- \tag{80}$$

$$\xrightarrow{X} AB + X^- \tag{81}$$

$$\xrightarrow{AB^-} AB + AB^{2-} \tag{82}$$

In all cases the electron is accommodated in a vacant orbital of the substrate molecule. The product thus formed may be a stable species or it may be a short-lived intermediate, which splits up into two fragments one of them being an anion. Alternatively AB^- formed will undergo neutralization in water or it may transfer an electron to an acceptor. When the acceptor happens to be another molecule of AB^- the reaction is defined as a disproportionation reaction. The following will exemplify these different types of reaction.

$$e_{aq}^- + Cu^{2+} \longrightarrow Cu^+ \tag{83}$$

$$e_{aq}^- + CH_3Cl \longrightarrow CH_3Cl^- \longrightarrow CH_3 + Cl^- \tag{84}$$

$$e_{aq}^- + (CH_3)_2CO \longrightarrow (CH_3)_2CO^- \longrightarrow (CH_3)_2CHO + OH^- \tag{85}$$

$$e_{aq}^- + CO_2 \longrightarrow CO_2^- \xrightarrow{CH_3Cl} CO_2 + CH_3 + Cl^- \tag{86}$$

$$e_{aq}^- + O_2 \longrightarrow O_2^- \xrightarrow{O_2^- + H^+} O_2 + H_2O_2 \tag{87}$$

The rates of e_{aq}^- reaction are dependent on the availability of a vacant orbital on the substrate molecule (26). This has been demonstrated for a number of organic (30,190,191) and inorganic (29,192) homologous systems. It has been shown that most reactions of hydrated electrons ranging in specific rates from 10^{10} to 10^5 M^{-1} sec^{-1} have a comparable energy of activation in the temperature range 0–100°C, equal to that of diffusion through water (193,194). This implies that the rates of the slow electron transfer reactions of e_{aq}^- are limited either by their entropies of activation or by the transmission coefficient. Most of the reactions of hydrated electrons may possibly be considered as extramolecular electron transfer processes, analogous to intramolecular electron transfer, which may be forbidden to a certain degree. The probability of electron transfer to a given molecule, which is evidently a function of the availability of a vacant orbital, is thus not expected to be temperature dependent in the given temperature range. It has been suggested (194a) that hydrated electron reactions involve electron tunnelling and that the slower reactions are limited by transmission coefficients smaller than unity.

B. Hydrogen Atoms

The chemical behavior of hydrogen atoms may be summarized in the following examples:

$$H + Br_2 \rightarrow HBr + Br \quad \text{(abstraction)} \tag{88}$$

$$H + CH_3CH_2OH \rightarrow H_2 + CH_3CHOH \quad \text{(abstraction)} \tag{89}$$

$$H + CH_2{=}CH{-}CH_2CH_3 \rightarrow CH_3CHCH_2CH_3 \quad \text{(addition)} \tag{90}$$

$$H + C_6H_6 \rightarrow C_6H_7 \quad \text{(addition)} \tag{91}$$

$$H + H \rightarrow H_2 \quad \text{(addition–radical recombination)} \tag{92}$$

$$H + O_2 \rightarrow HO_2 \quad \text{(addition–radical recombination)} \tag{93}$$

$$H + Cu^{2+} \rightarrow Cu^+ + H^+ \quad \text{(charge transfer)} \tag{94}$$

$$H + H_2O_2 \rightarrow H_2O + OH \tag{95}$$

$$H + OH^- \rightleftharpoons e_{aq}^- + H_2O \tag{96}$$

These reactions include first of all abstraction reactions, some of which—like hydrogen abstraction—may be defined as oxidation processes. Then come additions to double bonds and to free radicals or biradicals, all of which may be considered as reducing processes. The third category of H atom reactions are reduction processes which proceed probably via electron transfer mechanisms. Finally there are the $H + OH^-$ and $H + F^-$ reactions (85,194b) which yield hydrated electrons and which have no simple analog in chemistry.

The rates of reaction of hydrogen abstraction reactions by hydrogen atoms from aliphatic compounds have been studied extensively (78,195) and were found to depend on a combination of the C—H bond strength and the stability of the free radical produced. The abstraction of halogen atoms from haloaliphatic compounds was investigated kinetically (196) and the parameters which determine these rates are analogous to those which determine hydrogen abstraction. The kinetics of addition of H atoms to aromatic compounds has been investigated (32) and the H atom, unlike the hydrated electron, was found to behave as an electrophilic reagent. It should be noted that in general the rates of addition reactions are faster by one to two orders of magnitude than the hydrogen abstraction reactions.

The kinetics of reaction of H atoms with substituted cobaltic complexes have shown these reactions to proceed by an atom transfer rather than an electron transfer mechanism (197). The mechanisms of the $H + Cu^{2+}$, $H + Fe^{3+}$, and $H + Fe^{2+}$ reactions are still open for discussion. The mechanisms of the $H + OH^-$ and $H + F^-$ reactions have been discussed elsewhere (85) and it was concluded that they involve the formation of the H—OH and H—F bonds in the rate determining step.

C. Hydroxyl Radicals

Hydroxyl radicals like H atoms may react via atom abstraction, radical addition or electron transfer, as is exemplified in the following processes:

$$OH + CH_3CH_2OH \rightarrow H_2O + CH_3CHOH \quad \text{(abstraction)}$$
$$OH + H_2 \rightarrow H_2O + H \quad \text{(abstraction)}$$
$$OH + CH_2{=}CHCOOH \rightarrow CH_2OH{-}CHCOOH \quad \text{(addition)}$$
$$OH + C_6H_6 \rightarrow C_6H_6OH \quad \text{(addition)}$$
$$OH + H \rightarrow H_2O \quad \text{(addition–radical recombination)}$$
$$OH + OH \rightarrow H_2O_2 \quad \text{(addition–radical recombination)}$$
$$OH + I^- \rightarrow OH^- + I \quad \text{(electron transfer?)}$$
$$OH + Fe(CN)_6{}^{4-} \rightarrow OH^- + Fe(CN)_6{}^{3-} \quad \text{(electron transfer)}$$
$$OH + OH^- \rightleftharpoons H_2O + O^- \quad \text{(acid–base equilibrium)}$$

The hydrogen abstraction reactions by OH radicals from aliphatic compounds has been investigated systematically (195) and the transition state of these reactions seem to have the structure $HO\cdots H\cdots R$. OH radicals evidently behave as typical electrophiles (198) in addition reactions, and as in the case of H atoms, the addition reactions are faster than the abstraction reactions. The OH adduct to aromatic compounds has been demonstrated by pulse radiolysis (48,49,199). It should be noted that hydrogen abstraction from a given compound by OH radicals proceeds considerably faster than abstraction by H atoms, the reason being, most probably, the difference between the bond strength of the HO—H and H—H bonds. The OH + $Fe(CN)_6^{4-}$ reaction (91) may be considered as a typical electron transfer reaction, whereas the OH + halide reactions involve most probably the formation of an intermediate of the type HOX^- analogous to Cl_2^- or I_2^- (41). The OH + $OH^- \rightleftharpoons O^- + H_2O$ equilibrium has been investigated kinetically using the OH + $Fe(CN)_6^{4-}$ reaction as monitor and a pK of 11.7 has been estimated for the OH radical (91).

D. Concluding Remarks

The description of the chemical behavior of e_{aq}^-, H, and OH radicals in this chapter has been far from extensive. The reason is that the chemistry of each of these species is today a subject in its own right. These species may be produced by a variety of pathways, radiation chemistry being just one of them. Although they determine the fate of solutes in radiolyzed aqueous solutions, their chemistry should be treated comprehensively as a part of general chemistry independently of the radiation chemistry of water.

REFERENCES

1. P. Curie and A. Debriene, *Compt. Rend.*, **132** 770 (1901); F. Geisel, *Ber.*, **35**, 3608 (1902); **36**, 342 (1903).
2. M. Kernbaum, *Compt. Rend.*, **148**, 705 (1909); **149**, 116 (1909).
3. D. E. Lea, *Actions of Radiation on Living Cells*, Cambridge University Press, Cambridge, 1946.
4. I. V. Vereshihinskii and A. K. Pikaev, *Introduction to Radiation Chemistry*, Moscow Akad. Sciences, 1962, Ch. 4; English transl., Israel Program for Scientific Translations, Jerusalem 1964.
5. E. J. Hart and R. L. Platzman, in *Mechanisms in Radiology*, Vol. 1, M. Errera and A. Forssberg, Eds., Academic Press, New York, 1961, Ch. 2.
6. A. O. Allen, *The Radiation Chemistry of Water and Aqueous Solutions*, Van Nostrand, Princeton, N.J., 1961.
7. M. S. Matheson and L. M. Dorfman, in *Progress in Reaction Kinetics*, Vol. III, G. Porter, Ed., Pergamon Press, New York, 1965, p. 239.
8. E. J. Hart and J. W. Boag, *J. Am. Chem. Soc.*, **84**, 4090 (1962); E. J. Hart, *Science*, **146**, 19 (1964).

9. J. W. Spinks and R. J. Woods, *An Introduction to Radiation Chemistry*, Wiley, New York, 1964, Ch. 8.
10. H. A. Schwarz, *Advan. Radiation Biol.*, **1**, 1 (1964); *Ann. Rev. Phys. Chem.*, **16**, 347 (1965).
11. E. J. Hart, *Ann. Rev. Nucl. Sci.*, **15**, 125 (1965).
12. M. Haissinsky, in *Actions Chim. Biol. Radiations*, **11** (1967).
13. NAS-NRC Conference on Basic Mechanisms in the Radiation Chemistry of Aqueous Media, Gatlinburg, May 1963; *Radiation Res.*, Suppl. 4 (1964).
14. 5th Informal Conference on the Radiation Chemistry of Water–Notre Dame Oct. 1966, COO-38-519 (1967).
15. 2nd International Congress of Radiation Research, Harrogate, July 1962, *Abstracts* (1962).
16. Tihany Symposium on Radiation Chemistry, Sept. 1962; *Proc. Hung. Acad. Sci.*, Budapest (1964).
17. Fundamental Processes in Radiation Chemistry, Notre Dame, Sept. 1963, *Discussions Faraday Soc.*, **36**, (1963).
18. *Proc. Pulse Radiolysis Symp., Manchester, April 1965.*
19. *150th Meeting of the American Chemical Society—Symposium on Solvated Electron, Atlantic City, Sept. 1965*, Am. Chem. Soc., *Advan. Chem. Ser.*, **50** (1965).
20. *Proc. 2nd Tihany Symp. Radiation Chem. May 1966*, Hungarian Acad. of Sciences, Budapest, 1967.
21. 3rd International Congress of Radiation Research Cortina D'ampezzo June 1966, Proceedings of symposia. *Radiation Research*, G. Silini, Ed., North Holland Publ. Co., 1967.
22. Symposium of Radiation Research Newcastle Sept. 1966. Proceedings—*The Chemistry of Ionization and Excitation*, G. R. A. Johnson and G. Scholes, Eds., Taylor and Francis Ltd., London, 1967.
23. C. Lifshitz and G. Stein, *J. Chem. Soc.*, **1962**, 3706.
24. J. Kroh, B. C. Green, and J. W. T. Spinks, *Can. J. Chem.*, **40**, 413 (1962).
25. J. Jortner and J. Rabani, *J. Am. Chem. Soc.*, **83**, 4868 (1961); *J. Phys. Chem.*, **66**, 2081 (1962).
26. M. Anbar, ref. 19, p. 55.
27. J. K. Thomas, S. Gordon, and E. J. Hart, *J. Phys. Chem.*, **68**, 1524 (1964).
28. S. Nehari and J. Rabani, *J. Phys. Chem.*, **67**, 1609 (1963).
29. M. Anbar and E. J. Hart, *J. Phys. Chem.*, **69**, 973 (1965).
30. M. Anbar and E. J. Hart, *J. Am. Chem. Soc.*, **86**, 5633 (1964).
31. E. J. Hart, S. Gordon, and J. K. Thomas, *J. Phys. Chem.*, **68**, 1271 (1964).
32. M. Anbar, D. Meyerstein, and P. Neta, *Nature*, **209**, 1348 (1966).
33. G. Czapski and H. A. Schwarz, *J. Phys. Chem.*, **66**, 471 (1962).
34. K. H. Schmidt and W. L. Buck, *Science*, **151**, 70 (1966).
35. R. L. Platzman, in *Basic Mechanism in Radiobiology*, NCR Publication 305, 1953.
36. J. P. Keene, *Radiation Res.*, **22**, 1 (1964).
37. M. S. Matheson, W. A. Mulac, and J. Rabani, *J. Phys. Chem.*, **67**, 2613 (1963).
38. L. I. Grosswiner, G. W. Swenson, and E. F. Zwicker, *Science*, **141**, 1042, 1180 (1963).
39. M. Anbar, D. Meyerstein, and P. Neta, *J. Phys. Chem.*, **68**, 2967 (1964).
40. C. Cheek and V. Linnebom, *J. Phys. Chem.*, **67**, 1856 (1963).
41. M. Anbar and J. K. Thomas, *J. Phys. Chem.*, **68**, 3829 (1964).

42. E. J. Hart, *J. Am. Chem. Soc.*, **76**, 4198, 4312 (1954); *ibid.*, **83**, 567 (1961).
43. C. J. Hochanadel, *Radiation Res.*, **17**, 286 (1962).
44. D. Katakis and A. O. Allen, *J. Phys. Chem.*, **68**, 3107 (1964).
45. J. K. Thomas, J. Rabani, M. S. Matheson, and E. J. Hart, *J. Phys. Chem.*, **70**, 2409 (1966).
46. F. S. Dainton and T. J. Hardwick, *Trans. Faraday Soc.*, **53**, 333 (1957).
47. G. Buxton and W. K. Wilmarth, *J. Phys. Chem.*, **67**, 2835 (1963).
48. L. M. Dorfman, I. A. Taub, and D. A. Harter, *J. Chem. Phys.*, **41**, 2954 (1964).
49. D. F. Sangster, *J. Phys. Chem.*, **70**, 1712 (1966).
50. A. Hummel and A. O. Allen, *Radiation Res.*, **17**, 302 (1962).
51. J. Rabani and M. S. Matheson, *J. Phys. Chem.*, **70**, 761 (1966).
52. P. N. Moorthy and J. J. Weiss, ref. 19, p. 180.
53. D. Donaldson and N. Miller, *Trans. Faraday Soc.*, **52**, 562 (1956).
54. E. J. Hart and P. Walsh, *Radiation Res.*, **1**, 342 (1954).
55. J. Kroh, *J. Am. Chem. Soc.*, **83**, 2201 (1961).
56. G. Czapski and L. M. Dorfman, *J. Phys. Chem.*, **68**, 1169 (1964).
57. F. S. Dainton and D. B. Peterson, *Proc. Roy. Soc. (London)*, **A267**, 443 (1962).
58. F. S. Dainton and W. S. Watts, *Proc. Roy. Soc. (London)*, **A275**, 447 (1963).
59. M. Anbar and D. Meyerstein, *J. Phys. Chem.*, **68**, 1713 (1964).
60. E. Hayon, *Trans. Faraday Soc.*, **60**, 1059 (1964).
61. F. S. Dainton and S. R. Logan, *Trans. Faraday Soc.*, **61**, 715 (1965).
62. F. S. Dainton, A. R. Gibbs, and D. Smithies, *Trans. Faraday Soc.*, **62**, 3170 (1966).
63. A. Kuppermann, ref. 13, p. 15.
64. D. N. Sitharamarao and J. F. Duncan, *J. Phys. Chem.*, **67**, 2126 (1963).
65. R. L. Platzmann, *Radiation Res.*, **2**, 1 (1955).
66. D. Smithies and E. J. Hart, *J. Am. Chem. Soc.*, **82**, 4775 (1960).
67. A. R. Anderson and B. Knight, ref. 15, presented by H. C. Sutton in *Radiation Effects in Physics Chemistry and Biology*, M. Ebert and A. Howard, Eds., North Holland Publ. Co., 1963, p. 56.
68. D. Meyerstein, Ph. D. thesis, Jerusalem, 1965.
69. J. A. Ghormley, *Radiation Res.*, **5**, 247 (1956).
70. E. J. Hart and E. M. Fielden, ref. 19, p. 253.
71. J. P. Sweet and J. K. Thomas, *J. Phys. Chem.*, **68**, 1363 (1964).
72. J. K. Thomas, *Trans. Faraday Soc.*, **61**, 702 (1965).
73. S. Gordon, E. J. Hart, M. S. Matheson, J. Rabani, and J. K. Thomas, *J. Am. Chem. Soc.*, **85**, 1375 (1963).
74. J. P. Keene, ref. 17, p. 304.
75. M. S. Matheson and J. Rabani, *J. Phys. Chem.*, **69**, 1324 (1965).
76. W. D. Felix, B. H. Gall, and L. M. Dorfman, *J. Phys. Chem.*, **71**, 384 (1967).
77. M. S. Matheson, W. A. Mulac, J. L. Weeks, and J. Rabani, *J. Phys. Chem.*, **70**, 2092 (1966).
78. M. Anbar, D. Meyerstein, and P. Neta, *J. Chem. Soc. (London)*, **1966**, [A] 572.
79. H. C. Sutton, G. E. Adams, J. W. Boag, and B. D. Mochael, ref. 18, p. 61.
80. A. O. Allen and B. Bielski, ref. 14, p. 9.
81. T. J. Sworski, *Radiation Res.*, **4**, 483 (1956).
82. M. Anbar, I. Pecht, and G. Stein, *J. Chem. Phys.*, **44**, 3645 (1966).
83. P. Riesz and E. J. Hart, *J. Phys. Chem.*, **63**, 858 (1959).
84. G. Navon and G. Stein, *J. Phys. Chem.*, **69**, 1390 (1965).
85. M. Anbar and P. Neta, *Trans. Faraday Soc.*, **63**, 141 (1967).

86. A. Appelby, G. Scholes, and M. Simic, *J. Am. Chem. Soc.*, **85**, 3891 (1963).
87. G. Scholes and M. Simic, *J. Phys. Chem.*, **68**, 1781, 1738 (1964).
88. E. M. Fielden and E. J. Hart, *Radiation Res.*, **32**, 564 (1967).
89. G. Scholes and M. Simic, *Nature*, **199**, 276 (1963).
90. J. Rabani, W. A. Mulac, and M. S. Matheson, *J. Phys. Chem.*, **69**, 53 (1965).
91. J. Rabani and M. S. Matheson, *J. Am. Chem. Soc.*, **86**, 3175 (1964).
92. Ref. 14, p. 27.
93. E. Collinson, F. S. Dainton, and J. Kroh, *Nature*, **187**, 475 (1960).
94. N. F. Barr and R. H. Schuler, *Radiation Res.*, **7**, 302 (1957).
95. (a) M. Lefort and X. Tarrago, *J. Phys. Chem.*, **63**, 833 (1959); (b) M. Carmo Anta, M. H. Mariano, and M. Limaesantos, *J. Chim. Phys.*, **61**, 577 (1964).
96. M. Lefort, *Ann. Rev. Phys. Chem.*, **9**, 123 (1958).
97. R. L. Platzman, ref. 21, p. 20.
98. H. A. Schwarz, ref. 14, p. 51.
99. J. Weiss, *Nature*, **153**, 748 (1944).
100. G. L. Kochanny, A. Timnik, C. J. Hochanadel, and C. D. Goodman, *Radiation Res.*, **19**, 462 (1963).
101. A. H. Samuel and J. L. Magee, *J. Chem. Phys.*, **21**, 1080 (1953).
102. A. K. Ganguly and J. L. Magee, *J. Chem. Phys.*, **25**, 129 (1956).
103. D. A. Flanders and H. Fricke, *J. Chem. Phys.*, **28**, 1126 (1958).
104. P. J. Dyne and J. M. Kennedy, *Can. J. Chem.*, **36**, 1518 (1958).
105. A. Kuppermann, in *Actions Chimiques et Biologiques des Radiations*, Vol. 5, M. Haissinsky, Ed., Masson, Paris, 1961, p. 85; ref. 21, p. 212.
106. H. A. Schwarz, ref. 13, p. 89.
107. L. H. Gray, *J. Chim. Phys.*, **48**, 172 (1951).
108. H. Frohlich and R. L. Platzman, *Phys. Rev.*, **92**, 1152 (1953).
109. H. A. Dewhurst, A. H. Samuel, and J. L. Magee, *Radiation Res.*, **1**, 62 (1954).
110. F. W. Lampe, F. H. Field, and J. L. Franklin, *J. Am. Chem. Soc.*, **79**, 6132 (1957).
111. P. J. Coyle, F. S. Dainton, and S. R. Logan, *Proc. Chem. Soc.* (*London*), **1964**, 219.
112. A. Muzumder and J. L. Magee, *Radiation Res.*, **28**, 203 (1966).
113. J. P. Sweet and J. K. Thomas, *J. Phys. Chem.*, **68**, 1363 (1964).
114. M. Anbar and D. Meyerstein, *Trans. Faraday Soc.*, **61**, 263 (1965).
115. T. J. Sworski, *J. Am. Chem. Soc.*, **76**, 4687 (1954).
116. M. Bunton and K. C. Kurien, *J. Phys. Chem.*, **63**, 899 (1959).
117. Farhataziz, *J. Phys. Chem.*, **70**, 2696 (1966).
118. H. A. Mahlman, *J. Chem. Phys.*, **32**, 601 (1960).
119. R. Schiller, ref. 20, p. 105.
120. R. G. Sowden, *J. Am. Chem. Soc.*, **79**, 1263 (1957).
121. H. A. Schwarz, *J. Am. Chem. Soc.*, **77**, 4960 (1955).
122. E. Hayon, *Nature*, **194**, 737 (1962).
123. E. Hayon, *Trans. Faraday Soc.*, **61**, 723 (1965).
124. H. A. Mahlman, ORNL-3488 (1963).
125. T. J. Sworski, ref. 19, p. 263; (a) ref. 14, p. 28. (b) H. A. Mahlman and T. J. Sworski, ref. 22, p. 258.
126. G. Navon and G. Stein, *J. Phys. Chem.*, **69**, 1384 (1965).
127. (a) J. K. Thomas, S. Gordon, and E. J. Hart, *J. Phys. Chem.*, **68**, 1524 (1964); (b) G. Czapski and J. Jortner, *Nature*, **188**, 50 (1960).
128. M. Anbar and D. Meyerstein, *J. Phys. Chem.*, **69**, 698 (1965).

129. M. Anbar and D. Meyerstein, *Trans. Faraday Soc.*, **62**, 2121 (1966).
130. V. M. Byakov, B. V. Ershler, and L. T. Bugaenko, ref. 20, p. 87.
131. G. Scholes, ref. 17, p. 311.
132. H. A. Mahlman and J. W. Boyle, *J. Chem. Phys.*, **27**, 1434 (1957).
133. M. Faraggi, D. Zehavi, and M. Anbar, to be published.
134. J. H. Baxendale, E. M. Fielden, and J. P. Keene, *Proc. Roy. Soc. (London)*, **A286**, 320 (1965).
135. M. Anbar and P. Neta, *Intern. J. Appl. Radiation Isotopes*, **18**, 493 (1967).
136. M. Anbar and D. Meyerstein, *Nature*, **205**, 989 (1965).
137. G. Hughes and C. Willis, ref. 17, p. 223.
138. B. Cercek, M. Ebert, and A. J. Swallow, *J. Chem. Soc.*, [A] **1966**, 612.
139. M. Anbar, unpublished result.
140. K. Chambers, E. Collinson, F. S. Dainton, and W. Seddon, *Chem. Commun.*, **1966**, 498.
141. M. Anbar, S. Guttmann, and G. Stein, *J. Chem. Phys.*, **34**, 703 (1961).
142. D. Zehavi, M. Faraggi, and M. Anbar, to be published.
143. Ref. 4, p. 64.
144. M. Anbar and D. Meyerstein, *J. Phys. Chem.*, **68**, 3184 (1964).
145. J. Sutton and M. Moreau, ref. 20, p. 95.
146. E. Hayon, *J. Phys. Chem.*, **68**, 1242 (1964).
147. J. Sutton, ref. 22, p. 312.
147. (a) G. Scholes, ref. 22, p. 313; (b) E. M. Fielden, ref. 22, p. 314 (1967); E. J. Hart and E. M. Fielden, *J. Phys. Chem.* (in press).
148. H. Couraqui and J. Sutton, *Trans. Faraday Soc.*, **62**, 2111 (1966).
149. Ref. 4, p. 106.
150. J. K. Thomas, ref. 13, p. 111.
151. R. R. Hentz, Farhataziz, D. J. Milner, and M. Burton, *J. Chem. Phys.*, **46**, 2995 (1967); *ibid.*, **47**, 374 (1967).
152. S. Gordon and E. J. Hart, *Proc. 2nd Intern. Conf. Peaceful Uses Atomic Energy, Geneva, Sept., 1958*, **29**, 13 (1958).
153. Ref. 6, p. 48.
154. J. K. Thomas, ref. 14, p. 56.
154a. M. Anbar and D. Meyerstein, 19th Farkas Memorial Symposium, Jerusalem, Dec., 1963, to be published.
155. J. L. Magee, ref. 13, p. 20.
155a. C. E. Melton and H. W. Joy, *J. Chem. Phys.*, **46**, 4275 (1967).
156. R. L. Platzman, *Radiation Res.*, **17**, 419 (1962).
156a. T. J. Sworski, ref. 22, p. 315.
157. V. V. Voevodski, ref. 16, p. 392.
158. J. Bednar, ref. 16, p. 325.
159. J. Bednar and S. Lukac, *Collections Czech. Chem. Commun.*, **29**, 341 (1964).
160. R. L. Platzman at the discussions of the Cortina Conference, 1966.
161. R. L. Platzman, ref. 15, p. 128.
162. W. M. Latimer, *Oxidation Potentials*, Prentice-Hall, Englewood Cliffs, N.J., 1962.
163. J. H. Baxendale and G. P. Gilbert, *J. Am. Chem. Soc.*, **86**, 516 (1964).
164. A. R. Anderson, B. Knight, and J. A. Winter, *Trans. Faraday Soc.*, **62**, 359 (1966); *Nature*, **209**, 199 (1966).
165. J. Y. Young and I. Marcus, *J. Am. Chem. Soc.*, **88**, 1625 (1966).
166. G. R. A. Johnson and M. Simic, *J. Phys. Chem.*, **71**, 1118 (1967).

167. I. Santar and J. Bednar, ref. 20, p. 17.
168. (a) I. Santar and J. Bednar, *Collections Czech. Chem. Commun.*, **32**, 953 (1967); (b) G. R. A. Johnson and M. Simic, ref. 22, p. 211.
169. D. Meyerstein, ref. 14, p. 67.
170. R. L. Platzman, ref. 21, p. 41.
171. F. M. Loomis and W. H. Brand, *Phys. Rev.*, **49**, 55 (1936).
172. E. Cottin, *J. Chim. Phys.*, **56**, 1024 (1959).
173. R. I. Reed, *Ion Production by Electron Impact*, Academic Press, New York, 1962, p. 78.
174. F. Fiquet-Fayard and P. M. Guyon, *J. Chim. Phys.*, **60**, 1069 (1963).
175. H. Taube, *Trans. Faraday Soc.*, **53**, 656 (1957).
176. W. B. De More and O. F. Raper, *J. Chem. Phys.*, **44**, 1780 (1966).
177. T. J. Sworski, *J. Am. Chem. Soc.*, **86**, 5034 (1964).
178. J. Jortner, M. Ottolonghi, and G. Stein, *J. Phys. Chem.*, **65**, 1232 (1961); *ibid.*, **66**, 2029, 2037, 2042 (1962); *ibid.*, **67**, 1271 (1963); *ibid.*, **68**, 247 (1964).
179. R. M. Noyes, *J. Am. Chem. Soc.*, **78**, 5486 (1956).
180. J. K. Thomas and R. V. Bensasson, *J. Chem. Phys.*, **46**, 4147 (1967).
181. U. Sokolov and G. Stein, *J. Chem. Phys.*, **44**, 3456 (1966).
182. G. R. A. Johnson and M. Simic, ref. 22, p. 21.
183. W. Wanatabe and M. Zelikoff, *J. Opt. Soc. Ann.*, **43**, 753 (1953); B. A. Thompson, P. Harteck, and R. R. Reeves, *J. Geophys. Res.*, **68**, 6431 (1965).
184. M. Halmann and I. Platzner, *J. Phys. Chem.*, **70**, 580 (1966).
185. J. K. Thomas, ref. 14, p. 77.
186. J. W. Hunt and J. K. Thomas, *Radiation Res.*, **32**, 149 (1957).
187. O. Klemperer in *Physical Processes in Radiation Biology*, L. Augenstein, Ed., Academic Press, New York, 1964, p. 127.
188. R. D. Binkhoff, Ref. 187, p. 145.
189. H. A. Schwarz, ref. 14, p. 66.
190. M. Anbar and E. J. Hart, *J. Phys. Chem.*, **69**, 271 (1965).
191. E. J. Hart, E. M. Fielden, and M. Anbar, *J. Phys. Chem.*, **71**, 993 (1967).
192. M. Anbar and E. J. Hart, *Am. Chem. Soc. Adv. Chem. Ser.*, in press.
193. M. Anbar, Z. Alfassi, and C. Reisler, *J. Am. Chem. Soc.*, **89**, 1263 (1967).
194. M. Anbar and E. J. Hart, *J. Phys. Chem.*, **71**, 3700 (1967).
194. (a) M. Anbar, *Quart. Rev.*, 1968, in press; (b) M. Anbar and E. J. Hart, *J. Phys. Chem.*, **71**, 4163 (1967).
195. M. Anbar, D. Meyerstein, and P. Neta, *J. Chem. Soc. (London)*, (A) **1966**, 742.
196. M. Anbar and P. Neta, *J. Chem. Soc. (London)*, (A) **1967**, 834.
197. M. Anbar and D. Meyerstein, *Nature*, **206**, 818 (1965).
198. M. Anbar, D. Meyerstein, and P. Neta, *J. Phys. Chem.*, **70**, 2660 (1966).
199. I. A. Taub and L. M. Dorfman, *J. Am. Chem. Soc.*, **84**, 4053 (1962).

Author Index

Numbers in parentheses are reference numbers and show that an author's work is referred to although his name is not mentioned in the text. Numbers in *italics* indicate the pages on which the full references appear.

A

Abbe, J.-C., 352(35), *404*

Abkin, A. D., 561(116), 564, 591, *596, 598*

Abramson, F. P., 112(142), 114(142), *118*, 183(47), 193(84), 195(84,88), 205, 206, 211, 212, 214, 215(47,163), 216–218, 219(166), 220(47,166), 227, 228(143), 261(143), *272–274, 280*, 369(145), 370 (145,152), 392, 394(271), 395(271), *407, 410*, 429(81), 430(91), *506, 507*

Adamczyk, A., 352(34), *404*

Adams, G. E., 419(28), 421(28,40), 424 (40), 425(40), 455(28), 469 (269), 470(280), *505, 511*, 657(79), *682*

Aditya, S., 624(102), 633(102), 637(102), *647*

Adloff, J.-P., 352(35), *404*

Air Products and Chemicals, Inc., 603 (24), *645*

Akimoto, H., 140(57), *169*

Akopyan, M. E., 84(100), *117*

Albrecht, A. C., 606(47,48), 611, 615(77), 616(76), 617(81), 619(47), *646, 647*

Alder, P., 40, 42(108), *57*

Alfassi, Z., 678(193), *685*

Alfrey, T., 544(72), *595*

Alger, R. S., 624, 635(107), *647*

Al-Joboury, M. I., 79(77–81), 83, 85(77–81), 102(80), *116*

Allen, A. O., 415(11), 417, 418(15,22), 432(11,15,22), 481(338), 483(350), 484 (350), 493(22), 499(401), *505, 513, 514*, 521(20), 558(108), 575(20), *594, 596*, 601(1), 607(1), *644*, 652(6), 653, 654(6), 655(6,44,50), 657(6), 658(6,80), 659(6, 80), 660(80), 661(44), 662–664(6), 666 (50), 667(153), 677(6), *680, 682, 684*

Allison, R., 38(99), *57*

Altshuller, A. P., 251(294), *278*

Amagi, Y., 592(189), *598*

Amenu-Kpodo, F. K., 257(313), 258(313), *279*

American Chemical Society, 653(20), *681*

American Petroleum Institute, 77(60), 102(60), 107(60), *116*, 369(143), 374 (143), 377(143), *407*

Anbar, M., 291(37,44), 292, 310, 328(44), 331, 333, 334, 342, *343*, 419, 420(29), 475(303), 503(303), *505, 512*, 654(26,29, 30,32), 655(39,41), 656(59), 657(78), 658 (82,85), 664(114), 665(29,128,129,133, 135,136), 666(82,133,139,141,142), 667 (114,128,129,144), 668(128,154a), 670 (128,129,142), 671(114,128,129,154a), 673(82,141), 677(26,135), 678(26,29,30, 190–194a), 679(32,78,85,194b–197), 680 (41,195,198), *681–685*

Ander, P., 516(5), 592(5), *593*

Anders, L. R., 203(136), *274*

Anderson, A. R., 285(21,22), 290(29), 291, 301, 302, 303(31,47,75), 304(31,75, 81), 305, 306, 321(21), 322(115), 324, 327(22,124), 328(22,124), 329, 330, 332 (22), 335(29), 337(21), 338, 339, 340 (21), 341(21), *341–344*, 657(67), 671 (164), *682, 684*

Anderson, J. H., 625(109), *647*

Anderson, T. H., 624, 635(107), *647*

Anderson, V. E., 355(64), 371(64), 379 (64), *405*

Anderson, W. S., 588(172), 590(172), *598*

Ando, W., 503(413), *514*

Angus, H. J. F., 452(201), *509*

Antoine, F., 460(244), *510*

Appell, J., 87(114), *117*

Appleby, A., 475(306), *512*, 658(86), *683*

Appleyard, R. K., 40

Aquilanti, V., 183(101), 195(100,101), 225, 231(101,205–208), 255(100,101,205, 206), 256(101), 257–2 59(206), 260(100,

Subject Index

A

α, polarizability, 176

Absorption, of light, 2–7

Absorption cross section, definition, 2
   threshold behavior, 13

Absorption spectra of aklyl halides, 642, 643
   of aromatic negative ions, 420
   of biphenyl ions, 621
   character of above ionization threshold, 6
   character of in far ultraviolet, 10
   of cyclohexyl radical, 456
   detection of radiolytic intermediates by, 357, 414, 418–420, 602
   of isobutene polymer, 533
   quantitative determination of charge captured in solid by, 605, 606
   of radicals, 454–456, 624
   of solvated electron, 419, 420, 422 (table)
   of trapped electron, 608–610, 613, 617
   of triplet excited molecules, 443, 444
   use of data in radiation chemistry, 357

Acetaldehyde, ion, reactions, 258(table)
   W-value, 40

Acetic acid, radiolysis, liquid phase, 427, 469(table)
   reaction with H-atom, rate constant, 475(table)

Acetone,–benzene mixture, liquid radiolysis, 495–496
   H–atom scavenger, in alcohols, 470
   ion, fragmentation, 75, 96, 397
   photolysis, gas phase, 155, 397
   –propanol mixture, negative ion formation, 422(table)
   radiolysis, gas phase, 397
   liquid phase, dose rate effect, 474
      LET effects, 483
      product yields, 480(table)
      radical yields, reactions, 469, 471–472, 481–483
      scintillator solutions, 445–446

reaction, with H-atom, rate constant, 475(table)
   with hydrated electron, 654, 658, 678
   solvated electron in, absorption spectrum, 420
   triplet state, lifetime, 446

Acetylene,–argon mixtures, ion–molecule reactions in, 234
   dosimeter, 290, 388, 389
   "effective atomic number," 53
   excited states of, 349, 350
   ion, fragmentation, 96, 102, 388
      reactions, 257(table), 262(table), 263 (table), 265(table)
   –methane mixtures, energy distribution in, 353
   photolysis, 153–154, 388
   radiolysis, gas phase, cluster theory interpretation of, 348
      ion–molecule reactions in, 388, 532, 552, 553
      polymerization, 290, 532, 552, 553
      pressure effect, 388, 389, 532
      rare gas-sensitized radiolysis, 389
   W-value, 40

Acid, organic, radiolysis, liquid, 427
   solid, 624

Acrylamide, polymerization, 516, 559, 560
   water solution, radiolysis, 666

Acrylates, polymerization, 593

Acrylic acid, reaction with OH radical, 679

Acrylonitrile, polymerization, 531, 543 (table), 559, 560, 565, 566(table), 590, 591

Actinometry, carbon dioxide, 130, 161
   photoionization region, 128,129
   sodium salicylate, 38

Activated complex, configurations for ionic decomposition, 62, 71–73
   configurations for ionic isomerization, 70

Activation energy, hydrated electron reactions, 678